Advances in Intelligent and Soft Computing
151

Editor-in-Chief

Prof. Janusz Kacprzyk
Systems Research Institute
Polish Academy of Sciences
ul. Newelska 6
01-447 Warsaw
Poland
E-mail: kacprzyk@ibspan.waw.pl

T0074069

For further volumes:
http://www.springer.com/series/4240

Sigeru Omatu, Juan F. De Paz Santana,
Sara Rodríguez González, Jose M. Molina,
Ana M. Bernardos,
and Juan M. Corchado Rodríguez (Eds.)

Distributed Computing and Artificial Intelligence

9th International Conference

 Springer

Editors
Sigeru Omatu
Graduate School of Engineering
Osaka Prefecture University
Sakai, Osaka
Japan

Jose M. Molina
Applied Artificial Intelligence Group
Universidad Carlos III de Madrid
Colmenarejo, Madrid
Spain

Juan F. De Paz Santana
Department of Computer Science
Faculty of Science
University of Salamanca
Salamanca
Spain

Ana M. Bernardos
Data Processing and Simulation Group
ETSI. Telecomunicación
Universidad Politécnica de Madrid
Madrid
Spain

Sara Rodríguez González
Department of Computer Science
Faculty of Science
University of Salamanca
Salamanca
Spain

Juan M. Corchado Rodríguez
Department of Computer Science
Faculty of Science
University of Salamanca
Salamanca
Spain

ISSN 1867-5662
ISBN 978-3-642-28764-0
DOI 10.1007/978-3-642-28765-7
Springer Heidelberg New York Dordrecht London

e-ISSN 1867-5670
e-ISBN 978-3-642-28765-7

Library of Congress Control Number: 2012933101

Printed on acid-free paper

Springer is part of Springer Science+Business Media (www.springer.com)

Preface

The International Symposium on Distributed Computing and Artificial Intelligence 2012 (DCAI 2012) is a stimulating and productive forum where the scientific community can work towards future cooperation in Distributed Computing and Artificial Intelligence areas.

This conference is a forum to present applications of innovative techniques for solving complex problems. Artificial intelligence is changing our society. Its application in distributed environments, such as the internet, electronic commerce, environment monitoring, mobile communications, wireless devices, distributed computing, to mention only a few, is continuously increasing, becoming an element of high added value with social and economic potential, in industry, quality of life and research. These technologies are changing constantly as a result of the large research and technical effort being undertaken in both universities and businesses. The exchange of ideas between scientists and technicians from both the academic and industry sector is essential to facilitate the development of systems that can meet the ever-increasing demands of today's society.

This symposium is continuing to grow and prosper in its role as one of the premier conferences devoted to the quickly changing landscape of distributed computing, artificial intelligence and the application of AI to distributed systems. This year's technical program will present both high quality and diversity, with contributions in well-established and evolving areas of research.

This year, 178 papers were submitted from over 22 different countries (Portugal, Japan, Spain, South Korea, Australia, United Kingdom, Malaysia, Canada, Algeria, Germany, Russia, China, Finland, Brazil, France, Iran, Switzerland, Mexico, Tunisia, Senegal, Poland, Italy), representing a truly "wide area network" of research activity. The DCAI'12 technical program has selected 92 papers (89 long papers, 3 short papers). For this particular occasion, four special issues published by the Journal of Artificial Intelligence (IJAI), the International Journal of Imaging and Robotics (IJIR), the International Journal of Interactive Multimedia and Artificial Intelligence (IJIMAI) and the International Journal of Management and Production Engineering Review (MPER), will cover extended versions of the most highly regarded works.

The present edition brings together past experience, current work and promising future trends associated with distributed computing, artificial intelligence and their application in order to provide efficient solutions to real problems. This symposium is organized by the Bioinformatics, Intelligent System and Educational Technology Research Group (http://bisite.usal.es/) of the University of Salamanca. The present edition was be held in Salamanca, Spain, from 28th to 30th March 2012.

We thank the sponsors (IEEE Systems Man and Cybernetics Society Spain, AEPIA *Asociación Española para la Inteligencia Artificial*, APPIA *Associação Portuguesa Para a Inteligência Artificial*, CNRS *Centre national de la recherche scientifique*), the Local Organization members and the Program Committee members for their hard work, which was essential for the success of DCAI'2.

March 2012   Sigeru Omatu
Salamanca   Juan F. De Paz Santana
                                                                                  Sara Rodríguez González
                                                                                            Jose M. Molina
                                                                                          Ana M. Bernardos
                                                                            Juan M. Corchado Rodríguez

Organization

General Chairs

José M. Molina	Universidad Carlos III de Madrid, Spain
James Llinas	State University of New York, USA
Andre Ponce de Leon F. de Carvalho	University of Sao Paulo, Brazil
Ajith Abraham	Norwegian University of Science and Technology

Scientific Chair

Sigeru Omatu	Osaka Institute of Technology, Japan

Organizing Committee

Juan F. De Paz (Chairman)	University of Salamanca, Spain
Sara Rodríguez (Co-Chairman)	University of Salamanca, Spain
Juan M. Corchado (Co-Chairman)	University of Salamanca , Spain
Javier Bajo	Pontifical University of Salamanca, Spain
Dante I. Tapia	University of Salamanca, Spain
Fernando de la Prieta Pintado	University of Salamanca, Spain
Davinia Carolina Zato Domínguez	University of Salamanca, Spain
Cristian I. Pinzón	University of Salamanca, Spain
Rosa Cano	University of Salamanca, Spain
Emilio S. Corchado	University of Salamanca, Spain
Eugenio Aguirre	University of Granada, Spain
Manuel P. Rubio	University of Salamanca, Spain
Belén Pérez Lancho	University of Salamanca, Spain
Angélica González Arrieta	University of Salamanca, Spain
Vivian F. López	University of Salamanca, Spain
Ana de Luís	University of Salamanca, Spain
Ana B. Gil	University of Salamanca, Spain
Mª Dolores Muñoz Vicente	University of Salamanca, Spain
Jesús García Herrero	University Carlos III of Madrid, Spain
José Ramón Casar Corredera	Polytechnic University of Madrid, Spain
Manuel Felipe Cátedra Pérez	University of Alcalá, Spain
Javier Ortega-García	Autonomous University of Madrid, Spain

Scientific Committee

Adriana Giret	Politechnich University of Valencia, Spain
Alberto Fernández	University Rey Juan Carlos, Spain
Álvaro Herrero	University of Burgos, Spain
Álvaro Luis Bustamante	University Carlos III of Madrid, Spain
Ana Bernardos	Polytechnic University of Madrid, Spain
Ana Carolina Lorena	Federal University of ABC, Brazil
Ana Cristina Bicharra	University Federal Fluminense, Brazil
Ângelo Costa	University of Minho, Portugal
Antonio Berlanga	University Carlos III of Madrid, Spain
Antonio Moreno	University Rovira y Virgili, Spain
Antonio Ortega	University of Southern California, USA
Antonio Pereira	Instituto Politécnico de Leiria, Portugal
Araceli Sanchís	University Carlos III of Madrid, Spain
Ayako Hiramatsu	Osaka Sangyo University, Japan
B. Cristina Pelayo García-Bustelo	University of Oviedo, Spain
Bianca Innocenti	University of Girona, Spain
Bogdan Gabrys	Bournemouth University, UK
Bruno Baruque	University of Burgos, Spain
Carina González	University of La Laguna, Spain
Carlos Carrascosa	Politechnich University of Valencia, Spain
Carlos Delgado	University of Alcalá, Spain
Carmen Benavides	University of Leon, Spain
Changjiu Zhou	Singapore Polytechnic, Singapore
Daniel Gayo Avello	University of Oviedo, Spain
Daniel Glez-Peña	University of Vigo, Spain
Daniel Ramos Castro	Autonomous University of Madrid, Spain
Dante I. Tapia	University of Salamanca, Spain
David Griol Barres	University Carlos III of Madrid, Spain
Davide Carneiro	University of Minho, Portugal
Davinia Carolina Zato Domínguez	University of Salamanca, Spain
Dídac Busquets	University of Girona, Spain
Dongshik Kang	Ryukyu University, Japan
Doroteo Torre Toledano	Autonomous University of Madrid, Spain
Eladio Sanz	University of Salamanca, Spain
Eleni Mangina	University College Dublin, Ireland
Eliseo García	University of Alcalá, Spain
Emilio Corchado	University of Burgos, Spain
Enrique Martí Muñoz	University Carlos III of Madrid, Spain
Eugenio Aguirre	University of Granada, Spain
Eugénio Oliveira	University of Porto, Portugal
Evelio J. González	University of La Laguna, Spain
Faraón Llorens Largo	University of Alicante, Spain
Fernando de la Prieta Pintado	University of Salamanca, Spain
Fernando Díaz	Univesity of Valladolid, Spain

Fidel Aznar Gregori	University of Alicante, Spain
Florentino Fdez-Riverola	University of de Vigo, Spain
Francisco Pujol López	Polytechnic University of Alicante, Spain
George Cybenko	Dartmouth College, USA
Germán Gutiérrez	University Carlos III, Spain
Gonzalo Blazquez Gil	University Carlos III of Madrid, Spain
Gonzalo de Miguel	Polytechnic University of Madrid, Spain
Gregori Vázquez	Polytechnic University of Cataluña, Spain
Grzegorz Bocewicz	Koszalin University of technology, Poland
Helder Coelho	University of Lisbon, Portugal
Ichiro Satoh	Thammasat University, Japan
Ivan López Arévalo	Lab. of Information Technology Cinvestav, Mexico
Jamal Dargham	University of Malaysia, Saba, Malaysia
James Llinas	State University of N.Y. at Buffalo, USA
Javier Bajo	Pontifical University of Salamanca, Spain
Javier Carbó	University Carlos III of Madrid, Spain
Javier Galbally Herrero	Autonomous University of Madrid, Spain
Javier Martínez Elicegui	Telefónica I+D, Spain
Javier Portillo	Polytechnic University of Madrid, Spain
Jesús García Herrero	University Carlos III of Madrid, Spain
Joao Gama	University of Porto, Portugal
Joaquín González Rodríguez	Autonomous University of Madrid, Spain
Jørgen Bach Andersen	Aalborg University, Denmark
José Luis Guerrero	University Carlos III of Madrid, Spain
José M. Molina	University Carlos III of Madrid, Spain
Jose Manuel Gómez	University of Alcalá, Spain
José R. Méndez	University of Vigo, Spain
José R. Villar	University of Oviedo, Spain
José V. Álvarez-Bravo	University of Valladolid, Spain
Joseph Giampapa	Carnegie Mellon, USA
Juan A. Botia	University of Murcia, Spain
Juan Besada	Polytechnic University of Madrid, Spain
Juan F. De Paz	University of Salamanca, Spain
Juan Gómez Romero	University Carlos III of Madrid, Spain
Juan M. Corchado	University of Salamanca, Spain
Juan Manuel Cueva Lovelle	University of Oviedo, Spain
Juan Pavón	Complutense University of Madrid, Spain
Julián Fiérrez Aguilar	Autonomous University of Madrid, Spain
Kazutoshi Fujikawa	Nara Institute of Science and Technology, Japan
Lourdes Borrajo	University of Vigo, Spain
Luis Alonso	University of Salamanca, Spain
Luis Correia	University of Libon, Portugal
Luis F. Castillo	Autonomous University of Manizales, Colombia

Contents

Artificial Intelligence Applications

User-Centric Technologies and Applications

Multiagent Systems

Distributed Computing, Grid, Cloud Computing

Bioinformatics, Biomedical Systems

Data Mining, Information Extraction, Semantic, Knowledge Representation

Image Processing, Tracking, Robotic, Control and Industrial Systems

Mobile Systems, Locating Systems

New Algorithms

Mixed Odor Classification for QCM Sensor Data by Neural Networks

Sigeru Omatu, Hideo Araki, Toru Fujinaka, Michifumi Yoshioka, and Hiroyuki Nakazumi

Abstract. Compared with metal oxide semiconductor gas sensors, quarts crystal microbalance (QCM) sensors are sensitive for odors. Using an array of QCM sensors, we measure mixed odors and classify them into an original odor class before mixing based on neural networks. For simplicity we consider the case that two kinds of odor are mixed since more than two becomes too complex to analize the classification results. We have used eight sensors and four kinds of odor are used as the original odors. The neural network used here is a conventional layered neural network. The classification is acceptable although the perfect classification could not been achieved.

Keywords: odor feature vector, neural networks, separation of mixed gasses, odor classification.

Sigeru Omatu
Osaka Institute of Technology
Faculty of Engineering, 5-16-1 Omiya Asahi-ku Osaka, 535-8585, Japan
e-mail: omatu@rsh.oit.ac.jp

Hideo Araki
Osaka Institute of Technology
Faculty of Information Science and Technology, 1-79-1 Kitayama, Hirakata, 573-0196, Japan
e-mail: araki@is.oit.ac.jp

Toru Fujinaka
Hiroshima University
Faculty of Education, 1-4-1 Kagamiyama, Higashi-Hiroshima, 599-8531, Japan
e-mail: fjnk@hiroshima-u.ac.jp

Michifumi Yoshioka
Osaka Prefecture University
Faculty of Engineering, 1-1 Gakuencho Nakaku, Sakai, 599-8531, Japan
e-mail: yoshioka@cs.osakafu-u.ac.jp

Hiroyuki Nakazumi
Osaka Prefecture University
Faculty of Engineering, 1-1 Gakuencho Nakaku, Sakai, 739-8524, Japan
e-mail: nakazumi@chem.osakafu-u.ac.jp

S. Omatu et al. (Eds.): Distributed Computing and Artificial Intelligence, AISC 151, pp. 1–8.
springerlink.com   © Springer-Verlag Berlin Heidelberg 2012

1   Introduction

Over the last decade, odor sensing systens (so-called electronic nose (EN) systems) have undergone important development from a technical and commercial point of view. The EN refers to the capability of reproducing human sense of smell using sensor arrays and pattern recognition systems [4].

We have presented a type of an EN system to classify the various odors under the various densities of odors based on a competitive neural network by using the learning vector quantization (LVQ) method in [5]. The odor data were measured by an odor sensor array made of MOGSs. We used fourteen MOGSs of FIGARO Technology Ltd in Japan. We considered two types of data for classification in the experiment. The first type was a set of four kinds of tees and the second one was a set of five kinds of coffees of similar properties. The classification results of tees and coffees were about 96% and about 89%, respectively, which was much better than the results in [3].

In this paper, we will consider the classification of mixed odors based on the sensing data by using quarts crystal microbalance(QCM) sensors. QCM sensors are sensitive to odors and we can measure the odor data precisely. Using many QCM sensors, we will try to separate the odors being mixed with two kinds of odor into the original odors based on the neural network classifier.

2   Principle of QCM Sensors

The QCM has been well-known to provide a very sensitive mass-measuring devices in nanogram levels, since the resonance frequency will change sensitely upon the deposition of a given mass on the electrodes. Synthetic polymer-coated QCMs have been studied as sensors for various gasses since QCM coated with a sensing mambrane works as a chemical sensor. The QCM sensors are made by covering the surface with several kinds of a very thin membrane with about 1 μm as shown in Fig. 1.

Since the QCM occilates with a specific frequency depending on the cross section corresponding to three axis of the crystal, the frequency will change according to the deviation of the weight due to the adsorbed odor meclulare (odorant). The membrane coated on QCM has selective adsorption rate for a moleculare and the frequency deviation show the existence of odorants and their densities. Odrants and membrane are tight relation while it is not so clear whose materials could be adsorbed so much.

In this paper we have used the following materials as shown in Table 1. The reason why fluorine compounds are used here is that the compounds repel water such that pure odorant moleculares could be adsorbed on the surface of the membrane. To increase the amount of odorants to be adsorbed it is important to iron the thickness of the membrane. In Table 1, we have tried to control the density of the solute in the organic solvent. The basic approach used here is a sol-gel method. The sol-gel process is a wet-chemical technique used for the fabrication of both glassy and ceramic materials. In this process, the sol (or solution) evolves gradually towards the formation of a gel-like network containing both a liquid phase and a solid phase. Typical precursors are metal alkoxides and metal chlorides, which undergo hydrolysis and

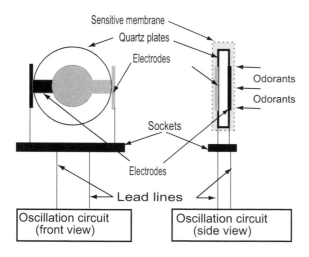

Fig. 1 Principle of QCM. The odorants attached on snsitive mambrane will make the weight of quartz plane. Thus, the original frequency of the crystal occilation will become smaller according to the density of odorants.

polycondensation reactions to form a colloid. The basic structure or morphology of the solid phase can range anywhere from discrete colloidal particles to continuous chain-like polymer networks.

Table 1 Chemical materials used as the membrane.

Sensor number	Materials of mebrane
sensor 1	Triethoxymethylsilane, ethanol(4ml), dilute nitric acid(0.023ml)
Sensor 2	Triethoxymethylsilane, water(3.13ml), ethanol(4ml), dilute nitric acid, ethylacrylate(0.043ml)
Sensor 3	Triethoxymethylsilane, water(3.13ml),ethanol(4ml), water, dilute nitric acid, ethylacrylate(0.014ml)
Sensor 4	Triethoxymethylsilane, water(3.13ml), ethanol(4ml), dilute nitric acid, ethylacrylate(0.015ml)
Sensor 5	Triethoxymethylsilane, water(0.30ml), ethanol(4ml), dilute nitric acid, ethylacrylate(0.043ml)
Sensor 6	Triethoxymethylsilane, water(0.05ml), ethanol(3.0ml), dilute nitric acid, ethylacrylate(0.043ml)
Sensor 7	Triethoxymethylsilane, water(0.30ml), ethanol(3.2ml), dilute nitric acid, ethylacrylate(0.043ml)
Sensor 8	No membrane

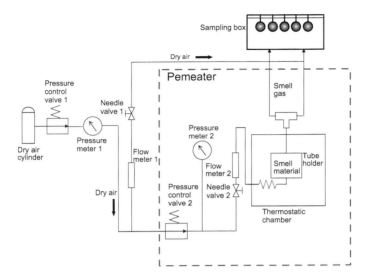

Fig. 2 Odor sensing systems. The air will be emitted from the dry air cylinder. Air flow is controlled by pressure control valves 1 and 2. By using the needle valve 2, more precise folow rate of the dry aircan be achieved and the thermostatic chamber in the pemeater can control the temperature of the dry air. Finally, the air is pull in the sampling box where the QCM sensors are attached on the ceiling of the box.

3   Odor Sensing System

Generally, it is designed to detect some specific odor in electrical appliances such as an air purifier, a breath alcohol checker, and so on. Each of QCM membranes has its own characteristics in the response to different odors. When combining many QCM sensors together, the ability to detect the odor is increased. An EN system shown in Fig. 2 has been developed, based on the concept of human olfactory system. The combination of QCM sensors, listed in Table 1, are used as the olfactory receptors in the human nose.

The odors used here are shown in Table 2. Note that the chemical properties of these odors are very similar and it has been difficult to separate them based on the measurement data by using MOGS sensors.

Table 2 Kinds of odors measured in this experimet.

Symbols	Kind of odors
A	Ethanol
B	Water
C	Methyl-salicylate
D	Triethylamine

4   Classification Method of Odor Data

In order to classify the odors we adopt a three-layered neural network based on the error back-propergation method as shown in Fig. 3.

The error back-propagation algorithm which is based on the gradient method is given by the following steps.

Step 1. Set the initial values of $w_{ji}, w_{kj}, \theta_j, \theta_k$, and $\eta (> 0)$.

Step 2. Specify the desired values of the output $d_k, k = 1, 2, \ldots, K$ correspondint to the input data $x_i, i = 1, 2, \ldots, I$ in the input layer.

Step 3. Calculate the outputs of the neurons in the hidden layer and output layer by

$$\text{net}_j = \sum_{i=1}^{I} w_{ji}x_i - \theta_j, O_j = f(\text{net}_j), f(x) = \frac{1}{1+e^{-x}}$$

$$\text{net}_k = \sum_{j=1}^{J} w_{kj}O_j - \theta_k, O_k = f(\text{net}_k).$$

Step 4. Calculate the error e_n and generalised errors by

$$e_k = d_k - O_k, \delta_k = e_k O_k (1 - O_k)$$

$$\delta_j = \sum_{k=1}^{K} \delta_k w_{kj} O_j (1 - O_j).$$

Step 5. If e_k is sufficiently small for all k, END and otherwise

$$\Delta w_{kj} = \eta O_j \delta_k, w_{kj} \Leftarrow w_{kj} + \Delta w_{kj}$$

$$\Delta w_{ji} = \eta O_i \delta_j, w_{ji} \Leftarrow w_{ji} + \Delta w_{ji}.$$

Step 6. Go to *Step* 3.

Using the above recursive procedure, we can train the odor data. The measurement data is an eight-dimensional vector which are obtained with eight sensors stated in Table 1.

5   Measurement of Odor Data

We have measured four types of odors as shown in Table 2. The sampling frequency is 1[Hz], the temperatures of odor gases are 24~26 [°C], and the humidities of gases are 6~8 [%]. To control the density of gases, we use the deffusion tubes. Odor data are measured for 600 [s]. They may include impulsive noises due to the typical phenomena of QCM sensors. To remove these impulsive noises we adopt a medean filter

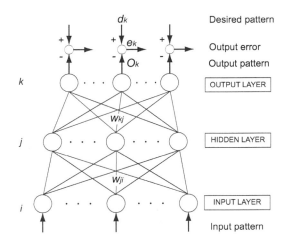

Fig. 3 Three layered neural network with the error back-propagation. The neural network consists of three layers, that is, an input layer i, a hidden layer j, and an output layer k. When the input data $x_i, i = 1, 2, \ldots, I$ are applied in the input layer, we can obtain the output O_k in the output layerwhich is compared with the desired value d_k which is assigned in advance. If the error $e_k = d_k - O_k$ occurs, then the weighting coefficients w_{ji}, w_{kj} are corrected such that the error becomes smaller based on the error back-propagation algorithm.

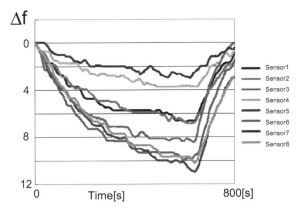

Fig. 4 Measurement data. Here, eight sensors are used and for six handred [s] the data were measured. The maximum value for each sensor among eight sensors is selected as a feature value for the sensor. Therefore, we have eight sensor values for an odor and they will be used for classification.

which replaces a value at a specific time by a median value among neighboring data around the specific time. In Fig. 4 we show the measurement data for the symbol A(ethanol) where the horizontal axis is the measurement time and the vertical axis is the frequency deviation from the standard value(9M[Hz]) after passing through a five-point medean filter.

6 Training for Classification of Odors

In order to classify the feature vector, we allocate the desired output for the input feature vector where it is nine-dimensional vector as shown in Table 3 since we have added the coefficient of variation to the usual feature vector to reduce the variations for odors. The training has been performed until the total error becomes less than or equal to $.5 \times 10^{-2}$ where $\eta = .3$.

Table 3 Training data set for ethanol (A), water (B), methyl-salicylate (C), and triethylamine (D).

Symbols	Output A	Output B	Output C	Output D
A	1	0	0	0
B	0	1	0	0
C	0	0	1	0
D	0	0	0	1

7 Testing for Classification of Odors

After training, we have tested the other data sets such that two kinds of odors are mixed with the same rate where one data set of mixed smells are {A&B, B&C, C&D, D&A, A&C, B&D}. Then the classification results are shown in Table 4 where the underlined numerals dente the top case ahrere the maximum output values are achieved. The maximum values show one of the mixed odors. But some of them does not show the correct classification for the remaining odor. Thus, we have modefied the input features such that

$$z = x - 0.9y$$

where x is the feature, y denotes the top value of each row in Table 4, and z is new feature. Using the new feature vector, we have obtained the classification results as

Table 4 Testing the mixed odors where the top and the second from the top are the classification results.

Symbols	Output A	Output B	Output C	Output D
A and B	<u>.673</u>	.322	.002	.001
B and C	.083	<u>.696</u>	.174	.001
C and D	.001	.004	.016	<u>.992</u>
D and A	.003	.003	.002	<u>.995</u>
A and C	<u>.992</u>	.006	.002	.000
B and D	.003	.003	.003	<u>.995</u>

shown in Table 5. By changing the features according to the above rlation, better classification results have been obtained. But the coefficient .9 used the abouve equation is not considered so much. The value might be replaced by the partial correlation coefficient in multi-variate analysis.

Table 5 Testing the mixed odors where except for the largest value the top is selected as the second odor among the mixed odors.

Symbols	Output A	Output B	Output C	Output D
A and B	.263	.290	.166	.066
B and C	.358	.029	.631	.008
C and D	.002	.071	.644	.163
D and A	.214	.004	.037	.230
A and C	.031	.020	.527	.026
B and D	.108	.010	.039	.325

8 Conclusions

We have presented the reliability of a new EN system designed from various kinds of QCM sensors. It has been sgown that after training the neural network for each odor, we can classify the original odor from the mixed odors in case of two odor case. More than two mixing case is open for the future research.

Acknowledgements. This research has been supported by Grant-in-Aid for Challenging Exploratory Research No. 22656089 of Japan Society for the Promotion of Science and we wish to thank JSPS for their support.

References

1. Carlson, W.L., Thorne, B.: Applied Statistical Methods. Prentice Hall International (1997)
2. General Information for TGS sensors, Figaro Engineering, http://www.figarosensor.com/products/general.pdf
3. Fujinaka, T., Yoshioka, M., Omatu, S., Kosaka, T.: Intelligent Electronic Nose Systems for Fiore Detection Systems Based on Neural Networks. In: The second International Conference on Advanced Engineering Computing and Applications in Sciences, Valencia, Spain, pp. 73–76 (2008)
4. Milke, J.A.: Application of Neural Networks for discriminating Fire Detectors. In: 10th International Conference on Automatic Fire Detection, AUBE 1995, pp. 213–222 (1995)
5. Omatu, S., Yano, M.: Intelligent Electronic Nose System Independent on Odor Concentration. In: International Symposium on Distributed Cimputing and Artificial Intelligence, Salamanca, Spain, pp. 1–9 (2011)

A Predictive Search Method of FAQ Corresponding to a User's Incomplete Inquiry by Statistical Model of Important Words Co-occurrence

Masaki Samejima, Yuichi Saito, Masanori Akiyoshi, and Hironori Oka

Abstract. We address a predictive search of FAQ corresponding to a user's incomplete inquiry that a user is inputting with important words defined in each FAQ. The important words co-occur in a user's inquiries and the rates of the co-occurrences depend on which FAQ the user's inquiry corresponds to. The co-occurrence rates of important words in inquiries are estimated from a statistical model of important words co-occurrence generated with past inquiries and FAQ corresponding to them. When the highest co-occurrence rate of them is larger than a threshold set on each FAQ, the inquiry is regarded as a corresponding FAQ. Experimental results show that the proposed method can improve the recall rate by 40% for short inquiries and the precision rate by 27% for long inquiries.

1 Introduction

Recently companies set up a web page of a help desk for answering users' inquiries. Users can send inquiry e-mails through the web form on the web page. Receiving

Masaki Samejima
Osaka University, 2-1, Yamadaoka, Suita, Osaka, Japan
e-mail: samejima@ist.osaka-u.ac.jp

Yuichi Saito
Osaka University, 2-1, Yamadaoka, Suita, Osaka, Japan
e-mail: saito.yuichi@ist.osaka-u.ac.jp

Masanori Akiyoshi
Osaka University, 2-1, Yamadaoka, Suita, Osaka, Japan
e-mail: akiyoshi@ist.osaka-u.ac.jp

Hironori Oka
Codetoys K.K., 2-6-8 Nishitenma Kita, Osaka, Japan
e-mail: oka@codetoys.co.jp

S. Omatu et al. (Eds.): Distributed Computing and Artificial Intelligence, AISC 151, pp. 9–16.
springerlink.com   © Springer-Verlag Berlin Heidelberg 2012

the inquiry e-mails, operators in the help desk need to reply the e-mails. Then, FAQ (Frequently Asked Questions) that consist of pairs of a frequent question and an answer of it are written on the web page for dealing with many inquiries. Although operators expect that users look for the answers before sending inquiries, most users send their inquiries without a glance of FAQ. If the user can get corresponding FAQ to the user's inquiry in inputting the inquiry on the web form, the user notices the answer and stops sending the inquiry. The goal of this research is to display corresponding FAQ to the incomplete inquiries that users are inputting.

In order to display corresponding FAQ, conventional methods use templates of questions [1], keyword matching [2] or Jaccard coefficient [3] between an inquiry and each question in FAQ [4, 5]. Because the methods with templates of questions can not be applied to incomplete inquiries, the others can be applied by setting "important words" as keywords on each FAQ. But, the methods with keyword matching do not display any FAQ until all important words are inputted by users. And, the methods using Jaccard coefficient display wrong FAQ when the inquiry has a few words.

We propose a predictive search method of FAQ corresponding to a user's incomplete inquiry. For example, in inquiry of "I forgot my password.", "forget" and "password" are important words and appear(co-occur) in the inquiry. Rates of the co-occurrences depend on which FAQ the user's inquiry corresponds to. So, the corresponding FAQ can be displayed when the co-occurrence rate is judged to be high by a threshold.

2 Search for FAQ Corresponding to a User's Incomplete Inquiry

2.1 The Outline of the Predictive Search of FAQ

The number of inquiries are expected to be reduced by displaying FAQ that corresponds to the user's inputting inquiry before the user completes the input of the

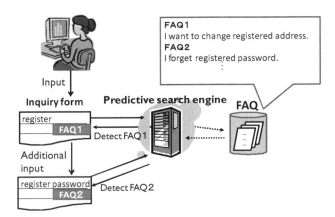

Fig. 1 Outline of the predictive search of FAQ

inquiry. So, we address a predictive search of FAQ corresponding to incomplete inquiries. Fig. 1 shows the outline of the predictive search of FAQ.

When the user inputs the inquiry in the web form, the predictive search engine searches FAQ corresponding to the inputted word. Every time another word of the inquiry is inputted by the user, FAQ that corresponds to the inquiry is displayed.

2.2   Approach and Issues

Because there are keywords that characterize the content of FAQ, operators use the important words in each FAQ as keywords for searching FAQ. The inquiry that includes all important words of a certain FAQ is likely to correspond to the FAQ. So, a typical approach of searching FAQ is to make combination patterns of important words in each FAQ, and to search the combination patterns in the inquiry. If the inquiry has a combination pattern in a certain FAQ, the FAQ is displayed to the user. Fig. 2 shows FAQ searching by the pattern matching. In the example of Fig. 2, combination patterns in FAQ1 are (address, register) and (address, change). When user inputs "I don't know my ID", the incomplete inquiry matches to the combination pattern in FAQ2. However, the typical approach has following issues:

- After the input of all important words, users submit the inquiry in a moment.
- Once the inquiry includes a combination pattern in wrong FAQ, wrong FAQ is kept displayed in spite of inputting additional words.

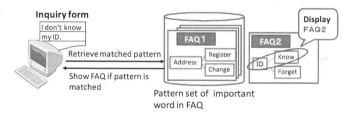

Fig. 2 Pattern matching by combination of important word

Although predictive search methods [6] have been proposed for solving the issues, the accuracy of the search by the methods is not so good. In this paper, we improve the accuracy by using important words predefined in FAQ.

3   A Predictive Search Method of FAQ by Statistical Model of Important Words Co-occurrence

3.1   Outline of a Predictive Search Method of FAQ

When some important words of a certain FAQ are inputted (co-occurred) in the inquiry, the co-occurrence rate of important words differs in each FAQ. Focusing on the difference of the rates, we propose the predictive search method by the rate of important words' co-occurrence in past inquiries. A corresponding FAQ is identified when the co-occurrence rate of important words in the FAQ is the highest. In order to prevent wrong FAQ from being kept displayed, the threshold corresponding to the number of input words is used to decide whether the FAQ is displayed or not. Fig. 3 shows the outline of the predictive search method of FAQ.

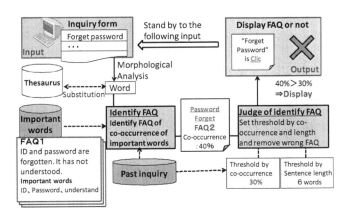

Fig. 3 Outline of FAQ presentation system that uses important word co-occurrence rate model

The inquiry sentences are processed by morphological analysis to be divided into words, and the synonyms are unified by the thesaurus. Whenever a user inputs a new important word in the inquiry, a corresponding FAQ is identified by the co-occurrence rate of important words in each FAQ. When the co-occurrence rate is larger than the threshold, the FAQ including the important words is considered to be corresponding FAQ to the inquiry.

3.2 Predictive Search by Statistical Model of Important Words Co-occurrence

Fig. 4 shows the outline of the predictive search of FAQ by the statistical model of the co-occurrence of important words. To search FAQ by some of important words, the proposed method generates the statistical model by combination patterns of important words and co-occurrence rates of the important words in past inquiries.

Co-occurrence rates are the ratios of the total number of co-occurrences of important words in past inquiries that correspond to each FAQ or "Other" that means "correspond to no FAQ". The co-occurrences of important words in each FAQi are used for searching each FAQi. Because "Other" have no important words, important words of all FAQ are used for "Other". The *co-occurrence rate*(p, i) of the co-occurrence pattern p belonging to FAQi is obtained in the following expressions:

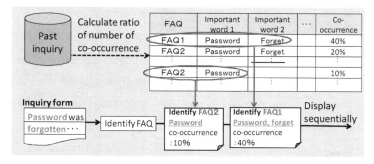

Fig. 4 FAQ search based on stochastic model of co-occurrence of important words

$$co\text{-}occurrence\,rate(p,i) = \frac{N_{p,i}}{\sum_i N_{p,i}}$$

where $N_{p,i}$ is the number of co-occurrences of p in the past inquiries correspond to FAQi.

As shown in Fig. 4, whenever the co-occurrence pattern p is inputted, FAQi where the *co-occurrence rate*(p,i) is the highest is regarded as corresponding FAQ. FAQ is not displayed when the inquiry is regarded to correspond to "Other". In Fig. 4, when a user inputs "Password", FAQ2 whose co-occurrence rate is 10% is searched. When a user inputs two words "Password" and "forget" additionally, FAQ1 whose co-occurrence rate is 40% is searched.

3.3 Judgment of Displaying FAQ

Fig. 5 shows the outline of deciding thresholds for judgment of displaying searched FAQ. The co-occurrence rate of important words concerning a specific FAQ tends to be increased as the number of input words is increased. In case that the co-occurrence rate is low in spite of a long inquiry, it is possible to judge that the displayed FAQ is wrong. The precision rate can be improved by the threshold for the co-occurrence rate according to the number of words.

First of all, past inquiries are divided for the generation of the stochastic model and for setting threshold. The proposed method calculates co-occurrence rates from past inquiries for the generation of the stochastic model. All the thresholds $Th_{p,i}(W)$ of a co-occurrence pattern p in FAQi are set according to the number of important words W. Changing the candidates of thresholds from 0% to 100% and judging whether FAQ are displayed or not for past inquiries by the thresholds, the proposed method decides thresholds for the system, when the result of the judgment indicates that F-value is maximum or the recall rate is maximum at the certain precision rate.

Moreover, because the average number of words in inquiries depends on which FAQ the inquiry corresponds to, it can be judged whether FAQ should be displayed according to the number of input words. Then, the proposed method does not display FAQ if it can be considered that the number of words is statistically different from

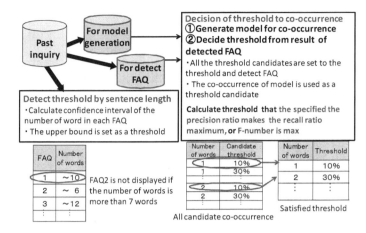

Fig. 5 Decision of threshold for FAQ presentation judgment

the number of words in past inquiries. The 95% confidence interval of the number of words in each FAQ is calculated from past inquiries, and the upper bound value of the interval is used as a threshold. The threshold of FAQ2 is six words in an example of Fig. 5. If the words are inputted more than 7 words, the inquiry is regarded to be "Other" and FAQ is not displayed. For example, the inquiry of "Because I changed my e-mail address and forgot my password, I can not access to the web page for registered members." includes a pattern (forgot, password) of FAQ2. When the pattern is inputted, the number of words in the inquiry is more than the threshold of FAQ2. So, FAQ2 is not shown to users.

4   Evaluation Experiment

We compared the proposed method to the pattern matching described in 2.2 based on inquiry e-mails sent to a help desk of the member management system of the sports association. These inquiries are classified to correct FAQ manually in advance. We use 1761 e-mails recorded by operators for a year: while 456 e-mails of them match to FAQ and 1305 e-mails belong to "Other". Because all the e-mails and FAQ are written in Japanese, Japanese morphological analysis tool "Chasen" is used. Table 1 shows the number of training data and test data. 2/3 of the inquiry e-mails is used to decide the thresholds and 1/3 of them is used to evaluate. We evaluate by the recall rate that corresponding FAQ is displayed once while an inquiry is being inputted,

Table 1 The number of supervised data and target data to be classified

FAQ	Q1	Q2	Q3	Q4	Q5	Q6	Q7	Q8
The number of inquiries as supervised data	13	166	15	9	25	13	35	27
The number of inquiries to be classified	7	83	8	5	12	6	18	14

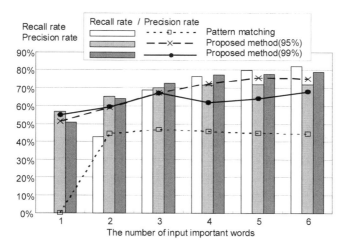

Fig. 6 Result of experiment

and by the precision rate that FAQ was correctly displayed while inquiry is being inputted. The threshold when F-value becomes the maximum is used as a threshold of the co-occurrence rate. And, the proposed method is applied with confidence levels 95% and 99% for thresholds of the number of words.

Fig. 6 shows the experiment result of precision rates and recall rates in applying the pattern matching and the propoased method with confidence levels 95% and 99%. The proposed method with the confidence level 95% improves the recall rate by 40% on the average when the number of input important words is below 2 words, and the precision rate by 27% on the average when the number of input important words is more than 3 words. In applying the method with pattern matching to long inquiries that have important words in some FAQ, wrong FAQ is often displayed.

Next, we discuss the effect of the confidence level for setting thresholds. As shown in Fig. 6, the difference of recall rates or precision rates between results with both of the confidence levels is below 10%. Because the difference of the confidence levels does not make the effect on the recall rates or precision rates, operators can set the confidence level without strictly adjusting. In this help desk, the operators want to keep the recall rate higher in inputting few words and the precision rate higher in inputting more words. So, the confidence interval 95% is appropriate for this help desk.

We show examples of inquiries where the proposed method improves the recall rate and the precision rate in the following:

- Inquiry A "Though I want to pay, the time limit of payment had passed the date yesterday…"
  In this inquiry, it is necessary to input an important word "payment", "time limit" and "pass". But, in the proposed method, corresponding FAQ can be displayed by inputting just "pass" and "time limit".

- Inquiry B "I don't know my membership code. And, the time-limit of payment . . . "
Because this inquiry includes two or more questions and is longer, it must be classified correctly into "Other". The threshold of the number of words can prevent wrong FAQ from being displayed.

Through the above discussion, it is confirmed that the proposed method is effective for the predictive search of FAQ. Now, operators must decide important words used by the proposed method. So, our future task is to decide the appropriate important words by the past inquiries.

5 Conclusion

We proposed the system that displays corresponding FAQ to a user's incomplete inquiry in the web form. The important words co-occur in a user's inquiries, and the rates of the co-occurrences depend on which FAQ the user's inquiry corresponds to. The co-occurrence rates of important words are estimated by a stochastic model that is generated with past inquiries and FAQ corresponding to them. When the highest co-occurrence rate of them is larger than a threshold set on each FAQ, the inquiry is regarded as a corresponding FAQ. Experimental results showed that the proposed method can improve the recall rate by 40% for short inquiries and the precision rate by 27% for long inquiries.

References

1. Sneiders, E.: Automated FAQ Answering with Question-Specific Knowledge Representation for Web Self-service. In: 2nd Conference on Human System Interactions (HSI 2009), pp. 298–305 (2009)
2. Sheng-Yuan, Y.: Developing an Ontological FAQ System with FAQ Processing and Ranking Techniques for Ubiquitous Services. In: Proc. of First IEEE International Conference on Ubi-Media Computing, pp. 541–546 (2008)
3. Markov, Z., Larose, D.T.: Data Mining the Web: Uncovering Patterns in Web Content, Structure, and Usage. John Wiley & Sons (2007)
4. Iwai, K., Iida, K., Akiyoshi, M., Komoda, N.: A Classification Method of Inquiry E-mails for Describing FAQ with Self-configured Class Dictionary. In: Proc. of Int. Symposium on Distributed Computing and Artificial Intelligence 2010, DCAI 2010, pp. 35–43 (2010)
5. Itakura, K., Kenmotsu, M., Oka, H., Akiyoshi, M.: An Identification Method of Inquiry E-mails to the Matching FAQ for Automatic Question Answering. In: Proc. of Int. Symposium on Distributed Computing and Artificial Intelligence 2010, DCAI 2010, pp. 213–219 (2010)
6. Yang, J., Cai, R., Jing, F., Wang, S., Zhang, L., Ma, W.: Search-based query suggestion. In: Proc. of the 17th ACM Conference on Information and Knowledge Management (CIKM 2008), pp. 1439–1440 (2008)

Decision Making for Sustainable Manufacturing Utilizing a Manufacturing Technology Database

Nozomu Mishima

Abstract. Sustainability is an important issue in manufacturing technologies in these days. However, for manufacturing engineers, quality has been the most important goal for long. Thus, in evaluating sustainability of manufacturing technologies, not only environmental aspects, but also quality of manufacturing should be taken into account. The existing work has proposed an evaluation method for manufacturing technologies which was named "Total Performance Analysis (TPA)." It can quantify the balance of the value created by the manufacturing process versus the cost and environmental impact. TPA can point out segment manufacturing processes which should be improved most. Through a simple case study, the paper concludes that the proposed method is helpful to evaluate real sustainability of manufacturing processes. In addition, by combining with manufacturing technologies database, not only pointing out the manufacturing processes to be improved, but also suggesting alternative processes will be possible.

1 Introduction

In recent world, every human activity has to be sustainable. To achieve sustainable manufacturing, it is important to reduce environmental impact of manufacturing processes. However, for manufacturing engineers, enhancing manufacturing quality had been the most significant goal. Therefore, to encourage development of sustainable manufacturing technologies, it is necessary to evaluate manufacturing quality, as well. One answer is the 'eco-efficiency [1].' Eco-efficiencies are useful indexes to evaluate environmental and economical aspect simultaneously. However, the original eco-efficiency cannot evaluate each component which composes the product, or each segment process which composes the total manufacturing process. Thus, it is difficult to suggest design improvement strategies by using eco-efficiency evaluation. In addition, it is also difficult to decide how-to-make from a long list of manufacturing technologies. In the existing work [2], a new eco-efficiency index named 'total performance indicator (TPI)' was proposed and

Nozomu Mishima
National Institute of Advanced Industrial Science and Technology, Japan

S. Omatu et al. (Eds.): Distributed Computing and Artificial Intelligence, AISC 151, pp. 17–24.
springerlink.com

a procedure of analysis based on the indicator was shown. The procedure has been applied to evaluate sustainable products [3], other manufacturing processes [4], production systems [5], sustainable businesses [6] and so on. This paper tries to apply the method to decision-making in manufacturing processes. By calculating the TPI of each segment process, bottleneck segment process in enhancing the total performance of the process can be clarified. The paper takes an actual product as an example, and estimates contribution of each segment process in creating the features of the product. If a segment process doesn't contribute much in creating value, and causes much environmental impact and cost, the segment process should be improved. One objective of this paper is to provide a decision-making method which supports manufacturing engineers to redesign sustainable manufacturing processes. And the other objective is to implement such a decision-making tool for sustainable manufacturing, by utilizing a manufacturing technologies database developed by AIST [7].

2 Basics of the Decision-Making Method

2.1 Definition of a New Eco-efficiency Index

In an existing study, the author has proposed an index to evaluate real eco-efficiency of products, by considering products' utility values, costs and environmental impacts through product lifecycle. The index is defined by (1) and named total performance indicator (TPI). To apply it to manufacturing process evaluation, it is important to take into account manufacturing qualities. To evaluate the eco-efficiencies of manufacturing processes, the same idea can be applied. The paper defines the total performance of the segment manufacturing process by (2). The equation expresses the balance of the product value created by the manufacturing process, versus cost and environmental impact of the process. The equation will be useful in evaluating the eco-efficiency of manufacturing processes.

$$TPI = \frac{UV}{\sqrt{LCC}\sqrt{LCE}} \tag{1}$$

TPI: total performance indicator, *UV*: utility value of the product
LCC: life-cycle cost of the product, *LCE*: life-cycle environmental impact

$$PROCESS \quad TPI = \frac{\Delta V_i}{\sqrt{PE_i \cdot PC_i}} \tag{2}$$

ΔV:Value increase by the segment process, *i*: number of processes
PCi: cost of a segment process, *PEi*: environmental impact of a segment process.

2.2 Decision-Making for Improvement in Manufacturing Processes

Since the purpose of the evaluation is to obtain suggestions for process improvement, it is necessary to evaluate TPI of each segment process and to determine bottleneck segment process in enhancing TPI of the total manufacturing process.

Figure 1 indicates the concept of analyzing a total manufacturing process by fo-cusing on the bottleneck segment process. The bottleneck process is shown as a segment line with a shallow slope. For example, segment process 2 in the figure has a shallow inclination. It means that this segment process does not contribute much in creating the final product value, but it generates relatively large cost and environmental impact. In other words, this process is not very efficient in enhanc-ing manufacturing quality. Therefore, there is a possibility to improve the process. In such a case, there are 3 ways to improve TPI of the total process. Those are, "1-1: reduction of environmental impact of the segment process," "1-2: enhancement of the quality," and "2: application of a new combination of processes." First two are focusing on the bottleneck process and trying to improve it. The third one is to introduce totally new processes that can replace the total manufacturing processes including the bottleneck segment process.

These approaches do not assure whether the focused segment process is actual-ly improvable. To apply the design evaluation method to an actual process and to ensure the improvement, it is indispensable to collaborate with process engineers who are aware of problems in their manufacturing process. In addition, to find an alternative segment process, integrating the knowledge database of manufacturing technologies can be a practical and appropriate method.

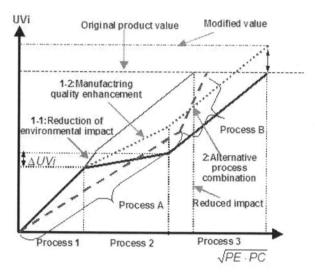

Fig. 1 Concept of evaluation and decision making in manufacturing process

3 Case Study

3.1 Case Study; A Vane Pump

To show an actual procedure of process TPA and improvement of a process, a practical example has been examined. As the target product a vane pump whose overview is shown in Figure 2 was chosen. A vane pump is a popular industrial product for liquid carriage. Roughly speaking, the main function of a vane pump is to carry liquid, but, the function can be separated into more detailed 8 functional requirements (FRs). In the 8 FRs, there are some very important functions and there are also functions that are not very important. So, in this method, every FR should be evaluated based on the relative importance. In the paper, the evaluation of the relative importance of each FR is based on QFD [8, 9] type analysis. Table 1 shows the 8 defined FRs, its' relative importance, and calculated value of each FR by assuming the price of the product as 200,000 JPY.

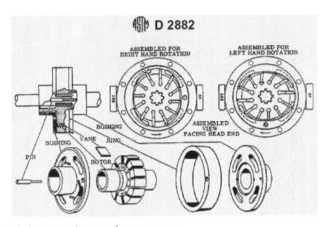

Fig. 2 Schematic image and parts of a vane pump

Table 1 Functional requirements, relative importance and monetary values

Functional requirement	Importance	Value (k JPY)
Quantity of liquid	9	39
Accuracy of liquid quantity	1	5
Cavitation free	3	13
Large exhalation pressure	9	39
Small installation space	3	13
Quiet operation sound	3	13
No liquid leakage	9	39
High reliability	9	39

3.2 Allocation of Values to Quality Characteristics

The second step of the analysis is to know how each quality characteristic is important in creating the value of the product. The calculation defines quality characteristics to express the performance of the vane pump. Then, again using the QFD approach, it tries to describe how each quality characteristic relates strongly to the functional requirements. By taking these steps, it is possible to allocate the product value to quality characteristics using the functional requirements as the intermediate step. Table 2 shows the defined quality characteristics such as 'lift,' 'rotational speed' and so on. Calculated value of each quality characteristic is also shown in the table.

3.3 Allocation of Values to Properties of Components

Products usually consist of components. As for the main structure of the vane pump, major components are vane, cum ring, rotor, housing and motor. Since the quality characteristics have relations between components and properties of components, it is possible to allocate the values of quality characteristics to the value of component properties. For example, as shown in Table 3, the 'exhalation quantity' has a strong relationship with 'shape accuracy' of 'vane.'

Table 2 Value of quality characteristics

	Importance	Value	Exhalation quantity	Lift	Rational speed	Efficiency	Inhalation pressure	Exhalation pressure	Leakage quantity from seal	Mass	Material	Volume	Expected lifetime	Total
Quantity of liquid	9	39	9	9	3	9	3	3	1	1	1			39
Accuracy of liquid quantity	1	5	9	9	3	3	3	3	1	1	1			33
Cavitation free	3	13					9						9	18
Large exhalation pressure	9	39	9	9	3	9	3	3	1					37
Small installation space	3	13								9	9	9		27
Quiet operation sound	3	13			9	1				1	3			14
No liquid leakage	9	39							9				3	12
High reliability	9	39							9					9
Value of quality characteristics (K yen)			20	20	15	21	13	7	70	6	8	4	16	200

(Leftmost vertical label: Functional requirement)

3.4 Allocation of Values to Segment Manufacturing Processes

The last step of the analysis is to know the contribution of each segment process on the value creation. By identifying the relation between each segment process composing the total manufacturing process and component properties, it is possible to calculate values of segment processes. Table 4 shows the result of the calculation of process values. Then, by using the values, costs and environmental impacts, TPI graph can be drawn. Figure 3 is the TPI graph of the manufacturing process of vanes of a vane pump. The figure shows, comparing to the TPI of the total process, segment processes 'grinding' have a relatively low TPI.

Table 3 Value allocation to component properties

		Quality characteristics											Value of component properties (k JPY)
		Exhalation quantity	Lift	Rational speed	Efficiency	Inhalation pressure	Exhalation pressure	Leakage quantity form seal	Mass	Material	Volume	Expected lifetime	
	Value of quality characteristics (K yen)	20	20	15	21	13	7	70	6	8	4	16	
Vane	Shape accuracy	9	9	1	9	9	9						18
	Surface roughness		3		3	3	3						4.6
	Weight								1				0.5
	Wear resistance									1		9	9.9
	Shape										1		0.3
Ring	Shape accuracy	9	9	1	9	9	9						
	Surface roughness	.	3		3	3	3						

Table 4 Value allocation to segment processes

Component properties		Form accuracy	Surface roughness	Weight	Wear resistance	Shape	
Value of component properties		18	4.6	0.5	9.9	0.3	
	Process name						Value
Manufacturing	Material processing			9		1	0.4
	Rough machining	3	1	3		9	5.2
	Heat treatment				9		9.9
	Finish machining	9	3	1			14.6
	Grinding		9			1	3.2
	total	11	13	13	9	11	33.3

4   Decision-Making and the Detection of Alternative Processes

The analysis shown in section 3, Figure3 can be illustrated. The figure tells that the last segment line has a relatively low slope. It suggests that "grinding" doesn't contribute much in enhancing the value but has a relatively large environmental impact and cost. Thus, "grinding" is the target of improvement. Since the suggestion doesn't tell whether there are alternative processes, combining the proposed decision making method with a knowledge database can be helpful. Figure 4 shows the index page of "cutting" of a manufacturing technologies database developed by author's research institute. Although the original database is written in Japanese, English explanations are attached. By selecting the tabs, actual data of cutting technologies can be viewed. Data of other manufacturing technologies such as electro-discharge machining, etc are also stored. The database suggested that it is possible to eliminate "grinding" by the extension of "fine machining," since surface roughness is not very important as form-accuracy, in this case.

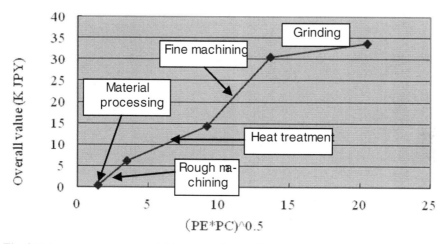

Fig. 3 Value versus environmental impact and cost of segment processes

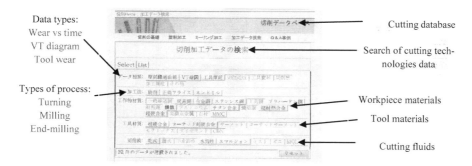

Fig. 4 Index page of the manufacturing technologies database (in Japanese)

5   Conclusion

The paper has shown a procedure to evaluate value creation efficiency of manu-facturing processes. The efficiency has the value created by the process as the nu-merator and takes square root of multiple of environmental impact and cost of the process, as the denominator. This method can be useful when a multi-criteria decision-making in establishing sustainable manufacturing is required. Through a case study, the proposed procedure can extract a segment manufacturing process to be improved, since it creates less value but has a relatively large impact and cost. In addition, only by the evaluation method, an alternative process cannot be suggested, but by combining the suggested result with knowledge database of manufacturing technologies, it is possible to determine the improvement target and the alternative processes at the same time.

Future work is to apply the procedure to more practical examples to evaluate the effectiveness of the proposed method. In addition, based on this method, it will be possible to apply a numerical method to determine the optimum set of manufacturing processes automatically. It will be a useful method of intelligent decision-making in sustainable manufacturing.

References

[1] DeSimone, L., Popoff, F.: Eco-Efficiency - the Business Link to Sustainable Development. MIT Press, Cambridge (1997)
[2] Kondoh, S., Masui, K., Hattori, M., Mishima, N., Matsumoto, M.: Total Performance Analysis of Product Life-cycle Considering the Deterioration and Obsolescence of Product Value. International Journal of Product Development 6(3/4), 334–352 (2008)
[3] Kondoh, S., Mishima, N.: Integrated Eco-design Framework of Products and Processes. In: Proc. of CIRP International Conference on LCE, Braunschweig, Germany, pp. 113–117 (2011)
[4] Kondoh, S., Mishima, N., Hotta, Y., Watari, K., Masui, K.: Total Performance Analysis of Manufacturing Processes. International Journal of Automation Technology 3(1), 56–62 (2009)
[5] Kondoh, S., Mishima, N.: Case Based Idea Generation Method for Sustainable Businesses. In: Proc. of CIRP International Conference on IPS2, Braunschweig, Germany, pp. 69–74 (2011)
[6] Mishima, N., Kondoh, S., Masui, K.: Proposal of An Efficiency Evaluation Index for Micro Manufacturing Systems. In: Proc. of IEEE ICMA 2007, Harbin, China, pp. 51–56 (2007)
[7] http://www.monozukuri.org/db-dmrc/index.html (in Japanese)
[8] Akao, K.: Quality Function Deployment. Productivity Process, Cambridge, MA (1990)
[9] Kondoh, S., Umeda, Y., Togawa, H.: Development of redesign method of production system based on QFD. Journal of Advanced Mechanical Design, Systems, and Manufacturing 1(1), 181–192 (2007)

Solving Time-Dependent Traveling Salesman Problems Using Ant Colony Optimization Based on Predicted Traffic

Hitoshi Kanoh and Junichi Ochiai

Abstract. In this paper, we propose an ant colony optimization based on the predicted traffic for time-dependent traveling salesman problems (TDTSP), where the travel time between cities changes with time. Prediction values required for searching is assumed to be given in advance. We previously proposed a method to improve the search rate of Max-Min Ant System (MMAS) for static TSPs. In the current work, the method is extended so that the predicted travel time can be handled and formalized in detail. We also present a method of generating a TDTSP to use in evaluating the proposed method. Experimental results using benchmark problems with 51 to 318 cities suggested that the proposed method is better than the conventional MMAS in the rate of search.

1 Introduction

Ant colony optimization (ACO) is a stochastic search algorithm for problem solving that takes inspiration from the foraging behaviors of ants. ACO has been formalized into a metaheuristic for combinatorial optimization problems by Dorigo et al., and many applications are now available [1]–[3].

Many studies on pheromone update methods have tried to improve the performance of ACO [1][2]. These studies focused on static optimization problems, but in practical applications, dynamic optimization problems play an important role. In this paper, we deal with the time-dependent traveling salesman problem (TDTSP), where the travel time between cities changes with time. This reflects the occurrence of traffic congestion in the real world [3]. Conventional methods for this problem repeat a search when the traffic flow changes during movement [4]–[6]. However, finding the global optimal solution by this method is difficult.

Hitoshi Kanoh · Junichi Ochiai
Department of Computer Science, Graduate School of Systems and Information
Engineering, University of Tsukuba, Tsukuba, Ibaraki, Japan
e-mail: kanoh@cs.tsukuba.ac.jp

S. Omatu et al. (Eds.): Distributed Computing and Artificial Intelligence, AISC 151, pp. 25–32.
springerlink.com   © Springer-Verlag Berlin Heidelberg 2012

Furthermore, when traffic changes rapidly, the information obtained from an old search may not be helpful.

In this paper, we propose an ACO based on the predicted traffic for a TDTSP. We previously proposed a method [7] to improve the search rate of Max-Min Ant System (MMAS) [8] for static TSPs. The method gives deviations from the initial pheromone trails by using a set of local optimal solutions calculated in advance. This method aims to build a near optimal solution at high speed, so it would be suitable for a TDTSP. In the current work, the method is extended so that the predicted travel time can be handled and formalized in detail. We also present a method of generating a TDTSP to use in evaluating the proposed method.

In the following section, we start by describing the problem. Then, we detail the algorithm of the proposed method. Finally, we present the results of experiments comparing the proposed method with MMAS using benchmark problems with 51 to 318 cities.

2 Problem Description

2.1 Time-Dependent Traveling Salesman Problem (TDTSP)

The TSP [9] can be represented by a complete graph $G = (N, A)$, where N is a set of nodes, i.e., cities, $n=|N|$ is the number of cities, and A is the set of arcs fully connecting the nodes. Each arc $(i, j) \in A$ is assigned a value $d_{i,j}$ $(=d_{j,i})$, which represents the distance between cities i and j. The TSP then is the problem of finding a shortest closed tour visiting each of the cities of G exactly once. The TSP instances used in this paper are taken from the TSPLIB benchmark library [10].

The TDTSP extends the original TSP so that traffic congestion can be included. Let $T_{i,j}$ be the travel time between cities i and j; the value of $T_{i,j}$ can be considered to be that of $d_{i,j}$. Traffic congestion can be represented by change in the travel time. In this paper, this change is defined by the following formula, where $T_{i,j}(n_{up})$ is the travel time in the n_{up}-th change, $T_{i,j}(0)$ means the original travel time of a given TSP, t is a current time, brackets [ ] indicate the Gaussian symbol, Δt is an update interval of travel time, C_f is a coefficient of fluctuation ($0 \leq C_f < 1$), and $rand \in [-1, 1]$ is a random number sequence.

$$T_{i,j}(n_{up}(t)) = T_{i,j}(n_{up}(t)-1) \times (1 + C'_f \times rand)$$

$$C'_f = \begin{cases} C_f & \text{if } rand \geq 0 \\ \dfrac{C_f}{1+C_f} & \text{otherwise} \end{cases}$$

$$n_{up}(t) = \left[\frac{t}{\Delta t} \right]$$

In this work, $T_{i,j}(n_{up})$ is considered to be a predicted value, and a set of prediction values $\{T_{i,j}(0), T_{i,j}(1), \ldots\}$ required for searching is assumed to be given in advance. The time when a salesman leaves city 1 is set to $t = 0$, and he always departs from and returns to city 1.

2.2 Greedy Algorithm for TDTSP

The greedy algorithm [1] is widely used in generating initial solutions from scratch by iteratively adding solution components to an initially empty solution until the solution is complete. In the TDTSP, a solution (tour) is built by adding the city with the shortest travel time. An algorithmic outline is given below, where i, j, and k are city numbers, Sp is a partial solution (i.e., a list of cities visited), and N' is a set of cities not yet visited.

```
Procedure Greedy(second-city)
    i = second-city;
    Sp = ( 1 i );   /* first city is always 1 */
    N' = N-{1};
    While(N' is not empty){
        N' ← N'-{i};
        k = arg min T_{i,j}(n_{up}(t));
             j∉N'
        Sp ← Sp ⊗ k;   /* k is added to Sp */
        i = k;
    }
    return Sp as a solution;
```

3 Proposed Method

3.1 Travel Time of Tour

In the following, the time required to travel from one node to another node is called the arc travel time $T_{i,j}$ (see Sect. 2.1), and the time required to travel around a tour is called the tour travel time. The arc travel time is calculated using the time t_i when a salesman reaches one of the nodes of the both ends of the arc. The tour travel time for solution S can be calculated by the following formulas:

$$T(S) = \sum_{i=1}^{n} T_{i,i+1}(n_{up}(t_i))$$

$$t_i = \begin{cases} t_1 & \text{if } i = 1 \\ t_{i-1} + T_{i-1,i}(n_{up}(t_{i-1}) & \text{otherwise} \end{cases}$$

3.2 Initial Pheromones

In MMAS, the initial pheromone trail τ_0 can be calculated by the next formula, where $S_k(\text{greedy})$ is the solution constructed by the greedy algorithm, n is the number of cities, and the number of ants is n-1.

$$\tau_0(\text{MMAS}) = \frac{n-1}{\rho \sum_{k=1}^{n-1} T(S_k(\text{greedy}))}$$

In the proposed method, we initialize the pheromone trails by the following procedure, where S_{0k} ($k=1, \ldots, n-1$) is the solution improved by a local search (such as 2-opt [1]) after construction by the greedy algorithm, and r ($0 \leq r \leq 1$) is a parameter. In addition, to use a local search is an option of the proposed method.

```
Procedure Initialize-pheromone()
    for(i=2; i<=n; i++) Greedy(i);
    /* n-1 solutions are constructed */
    Improve solutions using local search; /* option */
    Let τi,j=τ0 for all (i, j)∈A; /* see (1) */
    Update τi,j individually using (2);
```

$$\tau_{i,j} = \tau_0 = \frac{n-1}{\rho \sum_{k=1}^{n-1} T(S_{0k})} \qquad (1)$$

$$\tau_{i,j} \leftarrow (1-r)\tau_{i,j} + r\frac{1}{n-1}\sum_{k=1}^{n-1} \delta\tau_{i,j}^k \qquad (2)$$

$$\delta\tau_{i,j}^k = \begin{cases} \dfrac{1}{T(S_{0k})} & \text{if } (i,j) \in S_{0k} \\ 0 & \text{otherwise} \end{cases}$$

Here, the first term on the right side of (2) means the uniform distribution of pheromones, and the second term means centralization of the pheromones around local optimal solutions. The parameter r indicates the ratio of the decentralization of the initial pheromone trails and the centralization of those trails. When $r=0$, the proposed method and MMAS work similarly, though they differ in the quantity of trails (which increases since $T(S_{0k}) < T(S_k(\text{greedy}))$ generally). In contrast, when $r=1$, the initial pheromones are distributed only on the arcs contained in the local optimal solutions. If the local search is not used, the proposed method with $r=0$ is the same as MMAS.

3.3   General Procedure

The general procedure of the proposed method is given below, where S_{ib} is the best solution in the current iteration and S_{gb} is the best solution found since the start of the algorithm. The probability of selecting the next city can be calculated using formula (3).

$$p_{ij}^k(t) = \begin{cases} \dfrac{[\tau_{i,j}]^\alpha \cdot [\eta_{i,j}(t)]^\beta}{\sum_{l \in N^k} [\tau_{i,l}]^\alpha \cdot [\eta_{i,l}(t)]^\beta} & \text{if } j \in N^k \\ 0 & \text{otherwise} \end{cases} \tag{3}$$

$$\eta_{i,j}(t) = \frac{1}{T_{i,j}(n_{up}(t))}$$

```
Procedure Main()
   Input TDTSP;
   Set parameters;
   Initialize-pheromone();
   while(terminal condition not met) {
      for(k=1; k<=n-1; k++)
         construct solution Sₖ using (3);
      Sᵢᵦ = arg min T(Sₖ);
              k=1,n-1
      if(T(Sᵢᵦ)<T(Sgb)) Sgb=Sᵢᵦ;
      Update pheromone by the same method as MMAS[1];
   }
   Output final solution Sgb;
```

4 Experiments

4.1 Experimental Method

To evaluate how well the proposed method performs, we conducted experiments using the TSP instances eil51, eil76, eil101, kroA200, and lin318 from the TSPLIB [10]. TDTSPs were generated by the method in Sect. 2.1. Table 1 shows the number of cities, the optimal solution known in a static environment, the update interval of travel time in a dynamic environment, and the number of updates for each instance.

Table 1 TSP instances used in experiments and update of travel time.

TSP	Number of cities	Optimal solution	Update interval	Number of updates
eil51	51	426	5	86
eil76	76	538	5	108
eil101	101	629	5	129
kroA200	200	29368	300	98
lin318	318	42029	300	140

In this work, minutes and seconds are assumed as units of travel time for instances {eil51, eil76, eil101} and {kroA200, lin318}, respectively. As a result, the tour travel times obtained in the experiments were about 7 to 12 hours. In the real world, the traffic flow of a road network is measured in 5-minute intervals in many cases, and the predicted value of a traffic flow is calculated in the intervals. In the experiments, the update interval of travel time was set to 5 minutes (300 seconds); the number of updates was as shown in Table 1. Since the number of updates was 86 to 140, it can be said that the traffic flow was changed very frequently.

In this experiment, the local search which is an option of both the generic ACO and the proposed method was not used because the purpose is to evaluate the original performance of the proposed method. Consequently, the proposed method has almost no additional computation cost compared with the conventional MMAS. The other experimental conditions are as follows. The values of the ACO parameters were those generally used [1]. The degree of centralization r in formula (2) was fixed to 0.9 by preparatory experiments. The upper bound of iterations was 1000 and the coefficient of fluctuation C_f was 10%. In addition, each experimental result shown below is the average of 30 trials using different random number sequences.

4.2 Performance of Proposed Method

We first examined the property of initial solutions S_{ok} for pheromone trail initialization using TSP instance eil101. Cover ratio is defined as the ratio of the number of arcs contained in the initial solutions in the arcs that construct the final solution to the number of cities. Figure 1 shows the relationship between the cover ratio and the number of initial solutions. We can see that the large majority of arcs on the final solution can be included in the initial solutions.

Next, we compared the performance of the proposed method (r=0.9) with that of MMAS (r=0). The tour travel times of the best solutions in each iteration step for eil51 are shown in Fig. 2. The proposed method is clearly superior to MMAS in the rate of search. However the difference between the tour travel time of the best solution in the final iteration step of the proposed method and that of MMAS

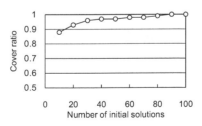

Fig. 1 Relationship between cover ratio and number of initial solutions.

was less than the standard deviation. The same results were obtained for the other problems.

Next, Fig. 3 shows the numbers of iteration steps until the tour travel time of the best solution reaches 90% of the final value. The value of 90% assumes an approximate solution in the real world. The transverse axis represents the TSP instance. We can see the following from the figure.

- The numbers of iteration steps of MMAS and the proposed methods increase slowly as the number of cities increases except for the largest instance lin318. Search for lin318 may not have converged completely.
- The time required to find approximate solutions of the proposed method is about 2.6 to 3.4 times as fast as that of MMAS.

The results suggest that the proposed method is effective for practical use problems compared with the conventional MMAS.

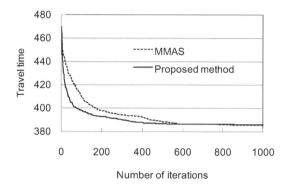

Fig. 2 Tour travel times of best solutions in each iteration step.

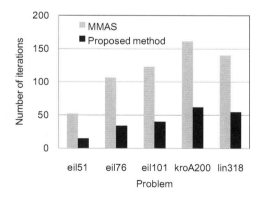

Fig. 3 Numbers of iteration steps until tour travel time of best solution reaches 90%.

5 Conclusion

In this paper, we presented a new ACO technique based on predicted traffic for TDTSPs. Experimental results using benchmark problems with 51 to 318 cities suggested that the proposed method is better than the conventional MMAS in the rate of search; the additional cost of computation for constructing local optimal solutions is negligible.

The results presented in this paper are based on five benchmark problems. Further investigation using other problems is necessary. It is important to apply the present method to real-world problems; in this case, the number of cities that a salesman can visit in one day may be less than 200. Although the proposed method is for the TDTSP, the idea of giving deviations from the initial pheromone trails can be used for other time-dependent combinatorial optimization problems.

Acknowledgments. The authors thank Mr. Yosuke Kameda for his helpful comments and suggestions. This research was partly supported by a Grant-in Aid for Scientific Research (C) of the Japan Society for the Promotion of Science (23500169).

References

[1] Dorigo, M., Stutzle, T.: Ant colony optimization. The MIT Press (2004)
[2] Dorigo, M., Birattari, M., Stutzle, T.: Ant colony optimization – Artificial ants as a computational intelligence technique. IEEE Computational Intelligence Magazine 1(4), 28–39 (2006)
[3] Dorigo, M., Stutzle, T.: Handbook of Metaheuristics. International Series in Operations Research & Management Science 146, 227–263 (2010)
[4] Mavrovouniotis, M., Yang, S.: Ant Colony Optimization with Immigrants Schemes in Dynamic Environments. In: Schaefer, R., Cotta, C., Kołodziej, J., Rudolph, G. (eds.) PPSN XI. LNCS, vol. 6239, pp. 371–380. Springer, Heidelberg (2010)
[5] Guntsch, M., Middendorf, M.: Pheromone Modification Strategies for Ant Algorithms Applied to Dynamic TSP. In: Boers, E.J.W., Gottlieb, J., Lanzi, P.L., Smith, R.E., Cagnoni, S., Hart, E., Raidl, G.R., Tijink, H. (eds.) EvoIASP 2001, EvoWorkshops 2001, EvoFlight 2001, EvoSTIM 2001, EvoCOP 2001, and EvoLearn 2001. LNCS, vol. 2037, pp. 213–222. Springer, Heidelberg (2001)
[6] Eyckelhof, C.J., Snoek, M.: Ant Systems for a Dynamic TSP. In: Dorigo, M., Di Caro, G.A., Sampels, M. (eds.) Ant Algorithms 2002. LNCS, vol. 2463, pp. 88–99. Springer, Heidelberg (2002)
[7] Kanoh, H., Kameda, Y.: Pheromone Trail Initialization with Local Optimal Solutions in Ant Colony Optimization. In: IEEE International Conference on Soft Computing and Pattern Recognition, pp. 338–343 (2010)
[8] Stutzle, T., Hoos, H.H.: MAN-MIN ant system. Future Generation Computer System 16(8), 889–914 (2000)
[9] Reinelt, G.: The Traveling Salesman: Computational Solution for TSP Applications. LNCS, vol. 840. Springer, Heidelberg (1994)
[10] Traveling Salesman Problem (TSPLIB),
http://www.iwr.uni-heidelberg.de/groups/comopt/software/TSPLIB95/

Modeling Shared-Memory Metaheuristic Schemes for Electricity Consumption

Luis-Gabino Cutillas-Lozano, José-Matías Cutillas-Lozano,
and Domingo Giménez

Abstract. This paper tackles the problem of modeling a shared-memory meta-heuristic scheme. The use of a model of the execution time allows us to decide at running time the number of threads to use to obtain a reduced execution time. A parameterized metaheuristic scheme is used, so different metaheuristics and hybridations can be applied to a particular problem, and it is easier to obtain a satisfactory metaheuristic for the problem. The model of the execution time and consequently the optimum number of threads depend on a number of factors: the problem to be solved, the metaheuristic scheme and the implementation of the basic functions in it, the computational system where the problem is being solved, etc. So, obtaining a satisfactory model and an autotuning methodology is not an easy task. This paper presents an autotuning methodology for shared-memory parameterized metaheuristic schemes, and its application to a problem of minimization of electricity consumption in exploitation of wells. The model and the methodology work satisfactorily, which allows us to reduce the execution time and to obtain lower electricity consumptions than previously obtained.

1 Introduction

Many organization, planning and logistic problems in industry, public services and different scientific disciplines, can be approached from operational research by the use of optimization models. The resolution of these models provides a solution to these complex problems and can be very useful in decision-making. But most of these models generate big combinatorial problems of type NP for which exact methods are not appropriate, and metaheuristics are used to approximate the optimum solution [7, 9]. Furthermore, the selection of an adequate metaheuristic for a particular problem is a time consuming process, which requires experimenting and tuning different metaheuristics. To alleviate this problem and so allow experiment with more metaheuristics and combinations in a shorter time we used a unified metaheuristic

Luis-Gabino Cutillas-Lozano
Aguas Municipalizadas, C/ Alona 31, 03007 Alicante, Spain
e-mail: lgabino.cutillas@aguasdealicante.es

José-Matías Cutillas-Lozano · Domingo Giménez
Departamento de Informática y Sistemas, University of Murcia, 30071 Murcia, Spain
e-mail: josematias.cutillas@um.es, domingo@um.es

S. Omatu et al. (Eds.): Distributed Computing and Artificial Intelligence, AISC 151, pp. 33–40.
springerlink.com   © Springer-Verlag Berlin Heidelberg 2012

scheme [2] and parallelized it for shared-memory [3]. But to have a parallel routine does not ensure it will be used correctly, and the execution time of the parallel routine may be far from the optimum (or even larger than the sequential time) if the number of threads used in the application of the routine is not appropriate. The autotuning problem of sequential and parallel routines has been studied in different fields [4, 5, 6, 8].

In this paper, some autotuning techniques are adapted to metaheuristic schemes. A shared-memory parameterized metaheuristic scheme is modeled, and the number of threads to reduce the execution time and to improve the obtained fitness function is selected. The application of the autotuning methodology to these schemes has some particularities which makes the adequate selection of the number of threads more difficult. As in other fields, the model of the execution time depends on the problem to be solved, the algorithmic scheme and the computational system, but in our case some new factors affect the execution time: different basic functions can be used in the metaheuristic scheme, this scheme is parameterized so that different metaheuristics and combination/hybridations are considered, and in addition to the reduction of the execution time the goodness of the value obtained can be considered. The autotuning methodology is presented and analyzed for a problem of minimization of the electricity consumption when pumping water from a set of wells.

The paper is organized in the following way. In section 2 the electricity consumption problem considered is briefly described. Section 3 shows the shared-memory parameterized metaheuristic scheme used in the solution of the problem. In section 4 the modeling of the execution time of the different basic and combined/hybridised metaheuristics in the scheme is analyzed, both theoretically and experimentally. Section 5 concludes the paper and shows some future research lines.

2 A Problem of Electricity Consumption in Exploiting Water Resources

The water system in our problem consists of a series of pumps (B) of known power, located in the corresponding wells, that draw water flow along a daily time range R. The total flow is the sum of the flows contributed by each well. The pumps may be running or out of service at a given time. The pumps operate electrically and the electricity has a daily cost which should be minimized:

$$C_e = \sum_{i=1}^{R} \sum_{j=1}^{B} T_i P_j N_i X_{ij} \qquad (1)$$

where C_e represents the cost of the electricity consumed by the combination of pumps selected in a day; T_i is the cost of the electricity in the range i; N_i is the number of hours of pump operation in the time slot i; P_j is the electric power consumed by the pump j; and x_{ij} has value 1 or 0 for pump on or off.

Using the notation for evolutionary algorithms, an individual is represented by a binary vector of size $B \cdot R$ that encodes the set of pumps distributed in different time slots. The set of individuals constitutes a population. Not all possible combinations result in feasible individuals, and each time an individual is generated or modified five constraints are evaluated: demand satisfaction, minimum flow maintenance, compliance with maximum exploitation volumes for each well, maintaining the average conductivity below the limit and compliance with maximum depths of dynamic levels. This means in some cases that obtaining a new individual is time-consuming. Furthermore, for large exploitation systems the number of wells and of time ranges can be large. So, to apply different metaheuristics efficiently a shared-memory parameterized scheme is used, and the inclusion of an autotuning methodology in the scheme is studied.

3   A Shared-Memory Parameterized Metaheuristic Scheme

When developing metaheuristics, it is possible to use a unified scheme [10], and there are different techniques to parallelize different metaheuristics [1]. In our approach, the unified scheme is parameterized to allow the application of different basic and combined/hybridized metaheuristics with the same scheme, and a shared-memory scheme is obtained by individually parallelizing each basic function in the metaheuristic scheme [3]. A scheme (Algorithm 1) is obtained with metaheuristic parameters (*ParamX*) with different values for different metaheuristics and parallelism parameters (*ThreadsX*) which should be selected to obtain low execution times.

Algorithm 1. Shared-memory parameterized metaheuristic scheme

Initialize(*S,ParamIni,ThreadsIni*)
while (not EndCondition(*S,ParamEnd*)) **do**
    SS=Select(*S,ParamSel*)
    SS1=Combine(*SS,ParamCom,ThreadsCom*)
    SS2=Improve(*SS1,ParamImp,ThreadsImp*)
    S=Include(*SS2,ParamInc,ThreadsInc*)
end while

Two basic parallel schemes are identified for the functions in algorithm 1. In the first scheme (algorithm 2) the elements of a set are treated independently. The number of threads to work in the loop (*threads − one − level*) can be selected. This scheme appears, for example, when combining elements in a genetic algorithm or when randomly generating an initial set of elements. The second scheme has two parallelism levels (algorithm 3), and the number of threads for each level can be established. This type of parallelism appears in improvement and mutation functions, where some elements are selected (first level) and each element is improved by analyzing its neighborhood (second level).

Algorithm 2. One-level parallel scheme (scheme 1)

one–loop(MetaheurParam):
    omp_set_num_threads($threads - one - level$(MetaheurParam))
    #pragma omp parallel for
      loop in elements
        treat element

Algorithm 3. Two-level parallel scheme (scheme 2)

two–level(MetaheurParam) :
    omp_set_num_threads($threads - first - level$(MetaheurParam))
    #pragma omp parallel for
      loop in elements
        second–level(MetaheurParam,$threads - first - level$)
second–level(MetaheurParam,$threads - first - level$):
    omp_set_num_threads($threads - second - level$(MetaheurParam,$threads - first - level$))
    #pragma omp parallel for
      loop in neighbors
        treat neighbor

4 Modeling and Autotuning of the Shared-Memory Parameterized Metaheuristic Scheme

The scheme in algorithm 1 can be used to apply a metaheuristic in parallel following this scheme, but to obtain reduced execution time it is necessary to appropriately select the values of the parallelism parameters (*ThreadsIni*, *ThreadsCom*, *ThreadsImp* and *ThreadsInc*), which means a model of the execution time must be obtained for each function, and the number of threads in the loop or the number of threads in the first and the second parallelism level can be established. So, the value of a large number of parameters (16 in our experiments) must be selected, and for that an autotuning methodology is systematically applied.

A scheme of the autotuning process is shown in figure 1. It is broken down into three phases, which are explained for the problem of electricity consumption:

- *First phase: Design.* The routine is developed together with its theoretical execution time. In the metaheuristic scheme, a model is obtained for each basic routine. Because two types of parallelism have been identified in these routines, two basic models can be used, one for one-level routines (algorithm 2) and another for nested parallelism (algorithm 3). For example, the generation of the initial population in function Initialize with an initial number of elements in the reference set *INEIni*, can be modeled:

$$t_{1-level} = \frac{k_g \cdot INEIni}{p} + k_p \cdot p \qquad (2)$$

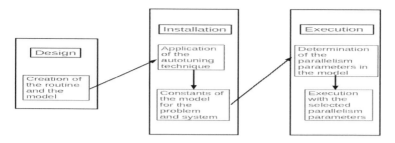

Fig. 1 Phases of the autotuning process

where k_g represents the cost of generating one individual; k_p the cost of generating one thread; and p is the number of threads. And the improvement of a percentage of the initial elements *PEIIni* with an intensification (amplitude of the considered neighborhood) *IIEIni* is modeled:

$$t_{2-levels} = \frac{k_i \cdot \frac{INEIni \cdot PEIIni \cdot IIEIni}{100}}{p_1} + k_{p,1} \cdot p_1 + k_{p,2} \cdot p_2 \tag{3}$$

where k_i represents the cost of improving one element; $k_{p,1}$ and $k_{p,2}$ the cost of generating threads in the first and second level; and p_1 and p_2 the number of threads in each level.

In our implementation, the second level is used to start more threads to work on the improvement of the fitness function (more neighbors are analyzed) but not to reduce the execution time.

For each of the other basic functions in algorithm 1, the corresponding metaheuristic parameters are determined, and the model of the execution time is obtained as a function of those parameters and the parallelism parameters (the number of threads to select to be used in each routine and subroutine).

- *Second phase: Installation.* When the shared-memory parameterized metaheuristic scheme is being installed in a particular system, the value of the parameters influenced by the system are estimated. The parameters k_g, k_i, k_p, $k_{p,1}$ and $k_{p,2}$ used in the explanation of step 1 are some of those parameters, as are the corresponding parameters for the other basic routines in algorithm 1. In addition, because in our implementation the second level of parallelism is not used to reduce the execution time but to analyze a wider neighborhood, the parameter p_2 can also be considered as a parameter of the system-algorithm. We summarize the results of the installation of the scheme in an *HP Integrity Superdome SX2000* with 128 cores of *Intel Itanium-2 dual-core Montvale* with shared-memory. The optimum number of threads varies with the number of individuals, and we are interested in the selection at running time of a number of threads close to the optimum. The model in equation 2 is used, and parameters k_g and k_p in the model are obtained by least-squares, with a reduced number of elements (in order to have low installation time). In the experiments with *INEIni* = 20 the values obtained are $k_g = 2.38 \cdot 10^{-3}$ and $k_p = 1.94 \cdot 10^{-4}$, all in seconds. By substituting

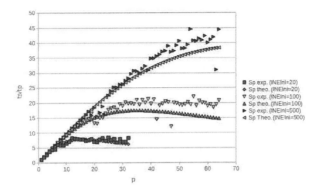

Fig. 2 Theoretical and experimental speed-up when varying the number of threads for three parameters in the one-level parallel routine

these values in the theoretical model of the execution time, the behaviour of the routine in the system is well predicted, as can be seen in figure 2 where the theoretical and experimental speed-up in the initial generation of the reference set are represented.

For a two-level routine, as for example the routine to improve elements after the initial generation or after combining or mutation, the value of p_2 is obtained by solving for a fixed number of threads in the first level and small values of the metaheuristic parameters. The value selected for this parameter is that with which the best fitness function is obtained without a substantial increment in the execution time (a maximum percentage of increment of the execution time is fixed). In the experiments for our problem and system, the value established is $p_2 = 1$. The values of the other parallelism parameters are obtained by least-squares with experiments with metaheuristic parameters $INEIni = 20$, $PEIIni = 50$ and $IIEIni = 20$. The results are $k_i = 9.10 \cdot 10^{-4}$, $k_{p,1} = 6.50 \cdot 10^{-4}$ and $k_{p,2} = 6.31 \cdot 10^{-3}$ seconds.

- *Third phase: Execution.* At execution time the number of threads in each basic function is selected from the theoretical execution time (equations 2 and 3) with the values of the metaheuristic parameters being those of the metaheuristic we are experimenting with and the values of the parallelism parameters estimated in the installation phase. The number of threads which gives the theoretical minimum execution time is obtained by minimizing the corresponding equation after substituting in it the values of the metaheuristic and parallelism parameters. For example, for the initial generation of the reference set:

$$p_{opt.} = \sqrt{\frac{k_g}{k_p} \cdot INEIni} = 3.50 \cdot \sqrt{INEIni} \qquad (4)$$

and for the improvement of the generated elements:

$$p_{1,opt.} = 1.18 \cdot 10^{-1} \cdot \sqrt{INEIni \cdot PEIIni \cdot IIEIni} \tag{5}$$

To validate the autotuning methodology the optimum execution time and the speed-up for other values of the metaheuristic parameters are compared with those obtained by applying the shared-memory parameterized metaheuristic scheme with the same metaheuristic parameters, the parallelism parameters obtained in the installation, and the number of threads selected in the execution phase. In tables 1 and 2 the optimum values and those obtained with our autotuning methodology are compared for the initial generation of the reference set and for the improvement of elements for two parameter combinations. The number of threads selected with the autotuning methodology is not far from the experimental optimum and, as a consequence, the speed-up achieved with autotuning is not far from the maximum.

Table 1 Speed-up and number of threads for $INEIni = 100$ and 500 in the one-level parallel routine. Optimum experimental values (optimum) and values obtained with autotuning (model)

	threads		speed-up	
INEIni	optimum	model	optimum	model
100	55	35	22	18
500	64	78	44	39

Table 2 Speed-up and number of threads for other parameter combinations in the two-level parallel routine. Optimum experimental values (optimum) and values obtained with autotuning (model)

			threads		speed-up	
INEIni	*PEIIni*	*IIEIni*	optimum	model	optimum	model
100	50	10	30	26	15	11
500	100	5	32	59	29	27

5   Conclusions and Future Work

An autotuning methodology has been adapted to obtain the number of threads to use in the application of a shared-memory parameterized metaheuristic scheme. Working with a problem of minimization of electricity consumption in wells exploitation and in a large shared-memory system, it has been shown the autotuning methodology provides satisfactory values for the number of threads to use in the application of the parallel metaheuristic. Due to lack of space the methodology has been shown for two basic functions in the metaheuristic scheme, but the autotuning works the same way for the other functions.

The autotuning methodology has not yet been integrated in the metaheuristic scheme, and the most immediate step in our research is to integrate it. The same ideas can be applied for message-passing or GPU parameterized metaheuristic schemes. A different direction of the research is to develop hyperheuristics which select satisfactory metaheuristics or combinations/hybridations by obtaining adequate values of the metaheuristic parameters. For that, having a scheme with autotuning would be very useful, because a large number of experiments would be necessary and the experimentation time can be reduced with a good selection of the number of threads to use in different parts of the parallel scheme.

Acknowledgements. This work has been funded in part by the Spanish MCYT under Grant TIN2008-06570-C04-02 and by the Fundación Séneca, Consejería de Educación de la Región de Murcia, 08763/PI/08. The authors gratefully acknowledge the computer resources and assistance provided by the Supercomputing Center of Fundación Parque Científico of Murcia.

References

1. Alba, E.: Parallel Metaheuristics: A New Class of Algorithms. Wiley Interscience (2005)
2. Almeida, F., Cuenca, J., Giménez, D., Llanes-Castro, A., Martínez-Gallar, J.P.: A Framework for the Application of Metaheuristics to Tasks-to-Processors Assignation Problems. The Journal of Supercomputing (2009) (published online)
3. Almeida, F., Giménez, D., López-Espín, J.J.: A parameterised shared-memory scheme for parameterised metaheuristics. The Journal of Supercomputing (2011) (published online)
4. Cuenca, J., Giménez, D., González, J.: Architecture of an Automatic Tuned Linear Algebra Library. Parallel Computing 30, 187–220 (2004)
5. Cuenca, J., Giménez, D., Martínez-Gallar, J.P.: Heuristics for work distribution of a homogeneous parallel dynamic programming scheme on heterogeneous systems. Parallel Computing 31, 717–735 (2005)
6. Clinton Whaley, R., Petitet, A., Dongarra, J.: Automated empirical optimizations of software and the ATLAS project. Parallel Computing 27, 3–35 (2001)
7. Dréo, J., Pétrowski, A., Siarry, P., Taillard, E.: Metaheuristics for Hard Optimization. Springer, Heidelberg (2005)
8. Frigo, M.: FFTW: An Adaptive Software Architecture for the FFT. In: Proceedings of the ICASSP Conference, vol. 3, p. 1381 (1998)
9. Glover, F., Kochenberger, G.A.: Handbook of Metaheuristics. Kluwer (2003)
10. Raidl, G.R.: A Unified View on Hybrid Metaheuristics. In: Almeida, F., Blesa Aguilera, M.J., Blum, C., Moreno Vega, J.M., Pérez Pérez, M., Roli, A., Sampels, M. (eds.) HM 2006. LNCS, vol. 4030, pp. 1–12. Springer, Heidelberg (2006)

Virtual Laboratory for the Training of Health Workers in Italy

Antonella Gorrino and Giovanni De Gasperis

Abstract. Virtual immersive laboratory for the training of health workers helps in the learning process of Italian as a second language for the foreign learners. Knowledge of the language is important for health workers because only through communication it is possible to build relationships between carer and user. We propose a persistent virtual environment under controlled conditions within the private network of the classroom, shared between teacher and students, which can be implemented at low cost with in a realistic situation. Few example scenarios and use cases are shown. We also introduced a frequently asked question chat-bot avatar as a language training tool to increase the attention of the learner in order to improve her/his second language skill.

Keywords: virtual learning, conversational agents, AIML.

1 Introduction

In this work, we propose technology that supports the development of skills in the Italian language for foreign students in health/social care training programs in Italy. These workers care for disabled or elderly people. Personal care is not only just to meet basic needs such as cleaning or feeding but also to establish a good relationship with the users. It is possible to properly relate with users only through a good knowledge of the language. Social health workers training in Italy is mostly done by Regions. There are courses for mother tongue trainees, separated ones for foreigners

Antonella Gorrino
Pedagogist, Independent Consultant, Alessandria, Italy
e-mail: agorrino@gmail.com

Giovanni De Gasperis
Department of Information Engineering, Computer Science and Mathematics,
University of L'Aquila, Italy
e-mail: giovanni.degasperis@univaq.it

S. Omatu et al. (Eds.): Distributed Computing and Artificial Intelligence, AISC 151, pp. 41–48.
springerlink.com   © Springer-Verlag Berlin Heidelberg 2012

were an important part of the training program is about Italian as second language. In a conventional course of study there is a large amount of time dedicated to learning of grammar which is more a scholastic way of learning and not very functional to the actual context of work. Often the person who is being assisted is usually an elderly or disabled person who could also have some difficulties understanding others and at the same making himself understood.

2 Background

In the cognitive approach to learning, many internal and external factors push or draw learners in different directions. The motivation is one of the internal factors that guides the attention process and it is crucial to handle the fatigue of learning. This work refers to the humanistic approach of Balboni [1] that emphasizes the learning actor together with its system of values and to the cognitive methods of the psycho-technologies [2] that allow a new experience of the world, mediated by information and communication artificial tools. Human thinking is not just linear; the mind explores new paths, interprets and builds up for associations, links, references and reflected forms. It is a different perspective that places the human in a position to create links, to explore, to create contact and psychological closeness, to overcome and redefine the boundaries of knowledge, to develop and strengthen old and new relationships. This work does not concentrate just about the impact of social networking on learning, but mostly studying how the use of recent advanced technology, such as the construction of three-dimensional virtual learning environments with interactive characters, can help to create the necessary conditions to experience real life like situations in a immersive modality. We aim to create a context were it is possible to develop a deep motivation in learning and that can trigger new cognitive processes to problem-solving situations [2]. Most of the students of health worker courses in Italy are foreigners; the first difficulty for them is to use of Italian as the second language in their typical working context. As reported by Content and Language Integrated Learning (CLIL) at School in Europe [3] "enabling pupils to develop: language skills which emphasize effective communication, motivating pupils to learn languages by using them for real practical purposes, knowledge and learning ability, stimulating the assimilation of subject matter by means of a different and innovative approach". The same guide lines can also be applied to the adult learner context [4] were the social care workers will be placed in situations of genuine communication at a direct contact with the user; an immersive learning technology is the most effective way to replicate this context becoming as real as possible during the learning process [5]. To learn a foreign language to be ready for a social care profession is essential to have a direct contact with people, mostly native speakers; current technology can offer low cost artificial surrogates were the second language learning experience can be mostly authentic and realistic in virtual working scenarios of interest.

3 Methodology of Intervention

In social sciences training the role playing game is a known technique [6] mainly used to construct a reality that makes possible to study and experiment with different actions on the simulated situation without suffering the irreversible, so that there is no impact on real users that health workers have to deal with. The first significant feature of an effective simulation is its analogy with reality: a model too tight and over fitted will result impossible to implement any experiment and thus obviates the achievable experience; on the other hand too far from a real situation could became too abstract and also prevents any intervention because everything seems to be random and belonging to the fantasy world. From this point of view simulation also means a deeper study, to increase the level of knowledge about a certain phenomenon, objects and situation in order to test hypotheses that would otherwise be too fragile or require a too long time of observation, entrusted to the natural course of events. The professional profile aimed by foreign learners includes essentially tasks about taking care of persons incapacitated or partially impaired. They should aim to the basic autonomy of the cared person; in this way the professional operator is held within a relationship and be part of a common context. The challenge is to create conditions by which the student can learn the written and spoken Italian language in a typical situation that could be encountered during the professional life. Therefore scenarios have to be designed so that the student can be placed o develop her/his skill. The perspective of serious games development scenarios should lead to the acquisition of skills of increasing complexity: greetings, introduce her/him self, sustain a simple conversation about daily life, include a delivery, a delivery report, manage a conflict. So, for the most significant situation under consideration we propose to construct ad-hoc three-dimensional immersive virtual environments in which learners can practice interacting within each other and with the teacher. At a highest step of difficulty the learner has to interact with a restricted artificial intelligence conversational agent [7], crafted from a frequently asked question set [8] to imitate the behavior of an elderly person in need of care in a typical private home situation .

3.1 Virtual Learning in a Metaverse

Metaverse is a term coined in 1992 by Neal Stephenson in the science fiction book "Snow Crash" [9], described as a shared virtual reality through a worldwide network, with objects persistence, in which participants are represented in three dimensions through their avatars and interact with each other and autonomous software agents. The metaverse refers to the concept of virtual reality by its very definition, namely a simulation of another reality. By the way, current technology has not yet reached a level of realism required by the original Metaverse concept. Almost all of the low cost virtual environments are predominantly rendered by means of visual and audio experiences projected on conventional equipment, such as the screen and the sound of a personal computer with 3D accelerated graphics card. However, the degree of realism is evolving rapidly, thanks to increasingly driven demands of the gaming industry. These new information technologies tools are increasingly be included in methods of teaching and learning in areas not strictly related to technical-scientific areas. Similarly, as suggested by [10] for the teaching of foreign languages it is possible to use virtual reality in the training of health and social workers, where the object of learning is the relationship and language between human beings.

3.2 Why Opensimulator

The rapid evolution of low cost Metaverses with an acceptable resolution, was triggered by the technologies offered by Linden Lab with its Second Life since 2003 [11]. Over the years the possibilities of representation of reality through the personal computer has been refined through a process of continuous release of free updates of the client software and other open source viewers, still maintaining a proprietary software on the server side (with features such as voice chat, windlight, sculpted prims, meshes, etc ...) [11]. Alongside the community of Metaverses enthusiasts have used the public specifications of the client-server communication protocol, initially called libSecondlife and now called libOpenMetaverse, to develop an open source server simulator called OpenSimulator [12], [13], or briefly OpenSim. The most important advantage of OpenSim over Second Life in a private and controlled learning environment such as the classroom can be summarized as the following:

- is not necessary to have access to a broadband connection to the Internet; it would be soon be saturated anyway by concurrent sessions of the many participants (learners) from the classroom.
- social networking interactions are limited to the participating community only
- there is no cost to create virtual environments (up to 4 whole sim in a standard laptop) without rental fees, facilitated by new tools like the open source New World Studio software package [14].

3.3 Proposed Solution

The teacher can install in her/his laptop a standalone version of sim OpenSim with at least one virtual space (65,546 square meters, about 20000 prims). She/He can easily install a a pre-packaged virtual land from among those available online, such as OpenVCE [15] available in OAR format (OpenSim Archive), install it locally and add virtual spaces, depending on the educational needs of learners and the target. The teacher can easily edit with familiar Second Life tools a series of environments and avatar-characters to be used later in the role- playing simulations adapted for foreign language learning. The teacher can later play the stereotype of assisted user by logging in the virtual world with her/his specially crafted avatar interacting with the student avatars. Through the use of PIVOTE virtual learning authoring system for virtual worlds [16] it would also be possible to design an interactive environment in which is possible to define sequences of interaction with objects and avatars in order to familiarize with the basic tasks of an environment of social care, depending on the type of users. PIVOTE has been extensively used at St. George University in Coventry (UK) in PREVIEW project [17], were a virtual patient model was used for training of paramedical staff in emergency management courses. Also robot-avatar can be equipped with scripts that links local chat to a chatbot server like Pandorabots, were a restricted artificial conversational agent can be built assembling AIML files [18] (Artificial Intelligence Markup Language). PyGenBot [8], is a Python program that can be used by the teacher to automatically generate the AIML files from a set of 5 free text files: a FAQ file (frequently asked questions), a keyword list, a multiwords list, a stop-word list and a glossary. In this experiment the input data set is referring to the conversational behavior of an elderly person in a typical private home context. In table 1 is reported a portion of the Italian FAQ text used to generate the AIML knowledge base.

Table 1 Portion of the Frequently Asked Question input set in Italian. For each question, multiple version of the question and their respective answers can be present.

Q: cosa le preparo per pranzo? A: minestra e prosciutto; A: pastasciutta e carne; A: latte e biscotti
Q: cosa le ha detto il dottore? A: mi ha detto che devo mangiare meno; A: mi ha detto che sto bene; A: mi ha detto che devo riposare
Q: vuole fare una passeggiata? A: si mi piacerebbe; A: no sono stanco; A: andiamo a fare una passeggiata dopo che ho riposato
Q: vuole andare nel letto a dormire? Q:

vuole coricarsi? A: si ho sonno e voglio andare a dormire; A no non ho sonno; A si ho sonno ma voglio vedere la TV
Q: vuole leggere una rivista, un quotidiano?; A: si voglio leggere una rivista; A: no non voglio leggere la rivista; A: non posso leggere perchè ho perso gli occhiali

The learner at the highest level of difficulty should interact in the Metaverse with the avatar controlled by the resulting FAQ-chatbot demonstrating proper knowledge of the Italian as a second language. The generated FAQ-chabot can be tested on line from Pandorabots web site [19].

3.4 The Virtual Scenarios

The virtual environment consists of a land, i.e. the artificial world, which contains several locations ad-hoc built, were in each several actions have to be made to get through the situation that simulates specific knowledge regarding the use of Italian in the context of social care from the point of view of health workers. The situations that are going to simulate through the virtual reality, will refer to the specific skill to which students have to be trained. The correct knowledge of the language it is important for the social, affective aspects of communication between health workers and users. It is necessary to build up several different scenarios and characters by which the students can experience the interaction. The difficulty of the health related work derives from the fact that every gesture can express different moods and can have different meanings. For this reason we have built different scenarios for the same situation. Feeding or talking an elderly person is different if it happens in a nursing home or in a private room. The student has to move an avatar-character (the operator), the teacher controls one of the others avatar-character, choosing the one most appropriate to the learning context. Only for expert students, it is expected the interaction with the FAQ-chatbot controlled avatar aimed to verify the level of lexical knowledge of Italian as second language.

- Virtual scene A: outside (a house with garden) inside (the various rooms - living room, bedroom, hallway)
- Virtual scene B: external (nursing home- with two-story nursing home with a garden) inside (lounge dining room, bedroom with three beds, small TV room)
- Virtual scene C: external (school with garden and games) inside (three class-rooms)

The 10 characters list: the elderly bedridden, the old man who moves and walks in the nursing home, the disabled child in wheelchair, the elderly wife, the child of school age (the grandson), the adult female, the adult male, students of a primary school, the nursing home operators, trainees (i.e. the student of the course, which later referred to as the '"avatars").

The following are some example "situations" or learning contexts:

- Situation 1 - Objective: Introduce her/himself
- Virtual scene A: It will take place in the living room. The avatars will have to present himself, knowing how to handle the situation and the type of intervention to be undertaken. Should consider the context and the speaker (user involvement and family, including the child) and will be able to use the shape of "her" for the adults and the "you" for the child, like it is used in italian language (Fig. 1)

- Situation 2 - Objective: To support a conversation about everyday life
- Virtual scene A: The same situation 1 will take place in the living room and bedroom of the elderly bedridden. The avatars will be able to cope with a conversation on general topics. Must know how to deal with matters such as weather, news reports etc.. and be able to entertain the bedridden old man with a funny short stories, (Fig. 2).
- Virtual scene B: The scene takes place inside the dining room at the nursing home. The avatars will be able to converse well with his interlocutor.

- Situation 4: Understanding a delivery of a care report
- Virtual scenes A / B: Avatar set must include a family member and / or to the client and / or other health worker related to the delivery of a report to the next worker. She/he must acquire a high number of daily jargon and technical vocabulary. The situations aims to simulate the moments of personal hygiene, ambient hygiene, meals, that all will take place both in the nursing home (Virtual scene B) is at user home (Setting A).

- Situation 5: Delivery of a care report
- Virtual scenes A / B: The situation is similar to that in step 4, but unlike the previous one, the worker avatar will be in a position that is active and you will go to the next level of text comprehension and writing of care reports.

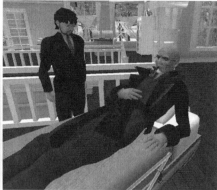

Fig. 1 Left: Sit. 2: introduce him/herself. Right: Sit. 3: support a conversation with an elderly person

- Situation 6: Managing conflict
- Virtual scenes A / B / C: A typical conflict situation shows up when the worker are willing to achieve proper goals care which can collide against the personal wishes of the user. Conflict management is the area where the involved skills are more about relationship. Experiencing the difficulty in an artificial situation of conflict, is certainly a great opportunity to learn how to handle the involved emotional aspects of dealing with a non collaborating user. In situation 6 language acquisition becomes instrumental to the achievement of the goal that involves a social care competence.

4 Conclusions

Health workers can take care of disabled or elderly persons only through a good relationship, which includes a clear communication, knowledge of words and the culture of the language. Given that learning from the experience is crucial for the learning of the language, the advent of technologies enhanced virtual learning has demonstrated an enormous potential. This work has shown an experimental low cost virtual learning project aimed at developing vocational training in the social and health care with an emphasis on acquisition of the Italian as second language; it is an essential background to enhance the "to know to be" of the learner, highlighting the need of new models for teaching and procedural innovations [20].

References

1. Balboni, P.E.: Fare educazione linguistica Attivit didattiche per italiano L1 e L2. UTET Universitá, Torino, Italy (2008)
2. De Kerckhove, D.: Connected intelligence: the arrival of the Web society. Kogan Page, London (1998)

3. Coyle, D., Hood, P., Marsh, D.: CLIL: Content and Language Integrated Learning. Cambridge University Press, Cambridge (2010)
4. Gorrino, A., De Gasperis, G.: Laboratorio Virtuale per la formazione di operatori di base stranieri nel settore socio-assistenziale. In: Proccedings of DIDAMATICA 2011, Turin, Italy, May 4-6 (2011),
   `http://didamatica2011.polito.it/content/download/259/1071/`
   `version/1/file/Full+Paper+DEGASPERIS+2.pdf`
   (accessed November 21, 2011)
5. Lucatorto, A.: Apprendere le lingue con la metodologia in "CLIL negli ambiente multimediali di apprendimento, esperienze in Second Life", Studi di Glottodidattica, Universitá degli studi di Bari, vol. 3(2), pp. 108/118 (2009),
   `http://ojs.cimedoc.uniba.it/index.php/glottodidattica/index`
   (accessed November 21, 2011)
6. Contessa, G., et al.: T-Group Storia e teoria della piú significativa invenzione sociale del secolo Clup, Turin, Italy (1990)
7. Kerry, A., Ellis, R., Bull, S.: Conversational Agents in E-Learning. In: Proceedings of AI 2008, the Twenty-Eighth SGAI International Conference on Innovative Techniques and Applications of Artificial Intelligence, Part 5, Cambridge, pp. 169–182. Springer, London (2009)
8. De Gasperis, G.: Building an AIML Chatter Bot Knowledge-Base Starting from a FAQ and a Glossary. JE-LKS. Journal of e-Learning and Knowledge Society 2, 79–88 (2010)
9. Many authors: Metaverse. English Wikipedia article (accessed November 21, 2011)
10. Garelli, P., Betti, S.: Nuove tecnologie ed insegnamento delle lingue straniere, Franco Angeli, Milan, Italy (2010)
11. Second LIfe, the commercial Metaverse from Linden Lab,
    `http://www.secondlife.com` (accessed November 21, 2011)
12. OPEN SIMULATOR, the open source virtual world server,
    `http://www.opensimulator.org` (accessed November 21, 2011)
13. Flore, G.: OpenSim, un simulatore a codice aperto per mondi virtuali in Un Viaggio dai Mondi Virtuali alla Realt Aumentata nel Segno dell'Open Source, cured by Gianpiero Moiolim, Franco Angeli, Milan, Italy (2010)
14. New World Studio, all included ready to go standalone OpenSim serve,
    `http://www.newworldgrid.com/land/free-land`
    (accessed November 21, 2011)
15. OPENVCE, the open source virtual world assets set,
    `http://openvce.net/vwassets` (accessed November 21, 2011)
16. PIVOTE, the open source virtual learning authoring system for virtual worlds,
    `http://www.pivote.info` (accessed November 21, 2011)
17. PREVIEW Project at St. George University, Coventry, UK,
    `http://www.elu.sgul.ac.uk/preview` (accessed November 21, 2011)
18. Wallace, R.S.: The Anatomy of A.L.I.C.E. in Parsing the Turing test, Part III, pp. 181–210. Springer, Netherlands (2009)
19. Elderly person FAQ-chabot online web service on Pandorabots,
    `http://www.pandorabots.com/pandora/`
    `talk?botid=d7fc01092e34da4d`
    (accessed November 21, 2011)
20. Marconi, A.: E-learning e innovazione pedagogica, competenze e certificazione - Armando Editore, Rome, Italy (2010)

The "Good" Brother: Monitoring People Activity in Private Spaces

Jose R. Padilla-López, Francisco Flórez-Revuelta,
Dorothy N. Monekosso, and Paolo Remagnino

Abstract. Population over 50 will rise by 35% until 2050. Thus, attention to the needs of the elderly and disabled is today in all developed countries one of the great challenges of social and economic policies. There is a worldwide interest in systems for the analysis of people's activities, especially those most in need.

Vision systems for surveillance and behaviour analysis have spread in recent years. While cameras are widely used in outdoor environments there are few employed in private spaces, being replaced by other devices that provide fewer information. This is mainly due to people worries about maintaining privacy and their feeling of being continuously monitored by "big brother".

We propose a methodology for the design of a multisensor network in private spaces that meets privacy requirements. People would accept video-based surveillance and safety services if the system can ensure their privacy under any circumstance, as a kind of "good brother".

1 Introduction

Demographic changes are leading to an aging population. The number of people over 50 will rise by 35% between 2005 and 2050, and those over 85 will triple by 2050.

Jose R. Padilla-López · Francisco Flórez-Revuelta
Department of Computing Technology, University of Alicante,
P.O. Box 99, 03080 Alicante, Spain
e-mail: {jpadilla,florez}@dtic.ua.es

Dorothy N. Monekosso
School of Computing and Mathematics, University of Ulster,
Jordanstown BT37 0QB, Northern Ireland
e-mail: dn.monekosso@ulster.ac.uk

Paolo Remagnino
Faculty of Science, Engineering and Computing, Kingston University,
Kingston upon Thames, KT1 2EE, Surrey, United Kingdom
e-mail: p.remagnino@kingston.ac.uk

S. Omatu et al. (Eds.): Distributed Computing and Artificial Intelligence, AISC 151, pp. 49–56.
springerlink.com   © Springer-Verlag Berlin Heidelberg 2012

In addition, birth rates are also declining [1]. The numbers of disabled people living at home are also set to grow fast. Between 2000 and 2020 they are expected to increase by 74% in Japan, 54% in France and 41% in the US [2]. For these reasons attention to the needs of the elderly and the disabled is today in all developed countries one of the great challenges of social and economic policy, which requires firm answers and innovation by both public administrations as academia, business and social organizations. Under these requirements, the question is how the information and communications technologies could be applied to the people who inhabit environments, allowing people interaction in a natural and transparent way, wherever needed, sensitive to the user and his context (situational, temporal, emotional...) and acting proactively. This is the ambient intelligence paradigm [3, 4]. Such technologies can be applied to promote and prolong independent living of the elderly, the disabled, their families or carers, which is the field of ambient assisted living (AAL).

In the last years there have been important advances in research and development of systems for monitoring the activity of daily living of people, especially those groups with more needs. For instance, there are systems for the recognition of activities of daily living (ADLs) [20, 21], fall detection [22], person-environment interaction [23], and support for people with mild dementia or Alzheimer [9, 7].

These systems are composed of networks of sensors that, under a centralised or distributed control, can analyse the environment to extract knowledge in order to detect anomalous behaviour or launch alarms to tele-care services. These advances have been possible due to the progress in of sensing, processing, and communication technologies and an associated reduction in costs. Many of these systems use different sensors embedded in the environment, such as pressure, presence or lighting sensors; state of switches, doors or windows [5, 6]. However, computer vision is essential for surveillance and safety tasks: it adds the possibility to monitor an environment and report on visual information, which is commonly the most straightforward and natural way of describing an event, a person, an object, actions and interactions [4]. Usually, in order to avoid the problem of occlusions recurrent in vision systems, it is necessary to introduce other types of sensors such as radio frequency technology (RFID) [7, 8] or microphones [9]. However, for the time being, the existing vision-based solutions are mostly laboratory prototypes [10] that have not reached deployable results and are far from marketable solutions.

Vision systems for human behaviour analysis have become widespread in recent years, mainly by security demands and the reducing costs of the devices [11, 12, 25]. However, most systems are applied to outdoor or public environments and, when they are installed indoor, they are employed in large facilities such as shopping malls, sports facilities, public transport stations, parking lots and the like [11, 13, 14, 15], only in few cases within private environments. This is mainly because of:

- Necessity of distributed vision systems, given that the area covered by each camera is very small due to the reduced size of rooms. In this domain, research is being developed in other environments [11], for human tracking and behaviour analysis with multiple cameras with or without overlapping between their vision fields. So, raw or processed data/information fusion from different devices would be needed for a number of AAL services.

- Privacy requirements. Although there are no specific surveys on the tolerance of users to have cameras watch them continuously in a private space, the first impression is that cameras are one of the devices less tolerated because they are associated with a significant loss of privacy . This fact, presumably, would be increased with elderly people who have had less contact with technology.

So, this paper deals with a methodology for the design of a multisensor network in private spaces that meets privacy requirements. This network will mainly consider vision devices in coordination also with other sensors. The remainder of this paper is organised as follows. Section 2 reviews recent approaches to preserve the privacy when taking images in private environments and our approach to tackle this issue. Section 3 presents the architecture of our system for a dynamic and hierarchic behaviour analysis.

2 Privacy Preservation

Previous works have tried to ensure privacy by applying blurring or masking filters to the area of a image where a person is detected [16, 17]. However, failing in the detection of persons, e.g. when they stand without moving, could lead to leave without blurring sensitive areas [18]. In [19] a blurring is applied in a degree according to the activity the user is performing. Results were not satisfactory because observers can make an association between the level of filtering and the on-going activity.

Other approaches try to work with binary silhouettes of people to analyse their behaviour, thus removing the main features to recognise a person. This could be done by using thermal or infrared cameras, or incorporating the appropriate filters in the pre-processing [8]. However, silhouettes may not be enough to recognise significant activities.

Our approach to privacy preservation tries to solve some of the problems these preceeding works present. Our proposal is to render an augmented virtual image of the environment displaying only the details (real and virtual elements) a carer will need to assess an activity, event or alarm. This process would involve removing from the scene all other people or activity that is not of interest for the observer. So, human activity analysis and, in some cases person identification, will be needed prior to consider the privacy issues. Therefore, the type of representation depends on the events that occur and the person who may be viewing the images, for example from a telecare centre.

In principle, the different levels of detail include (Figure 1):

- No alarm: a virtual image of the room is presented without showing any person in it (1a);
- Low-level alarm: a virtual image is displayed of both the room and the person showing position but not posture (the person would be in a standard or fixed posture, avoiding to present the real one) (1b);
- High-level alarm: a virtual image of both the room and the person (in the real posture) is presented. For instance, if a person falls in the bathroom, a carer of a telecare service would be able to see this image (1c);

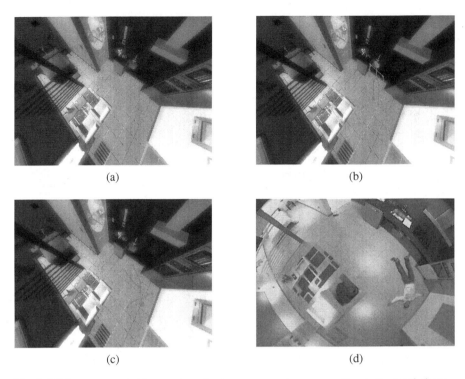

(a)   (b)

(c)   (d)

Fig. 1 Different presented images in order to preserve privacy according to the level alarm

- Very high-level alarm: the system shows the real image of the room and the person. For instance, if after a fall the person does not answer to the carer, a close relative with the appropriate permissions would be able to observe the real scene in order to assess and confirm the level of alarm (1d).

3 People Behaviour Analysis

The requisites for effective visual behaviour analysis include the modelling of the environment, the detection of motion, the classification of moving people and objects, their tracking, understanding and description of behaviours, and, in some cases, human identification. Fusion of data from multiple cameras is needed in order to cover a whole private environment with multiple rooms and overcome hard technical problems, such as occlusions. The fusion of video data and information involves the integration of information acquired by other sensors, either to increase awareness or facilitate further analysis. Figure 2 shows the general framework of the system.

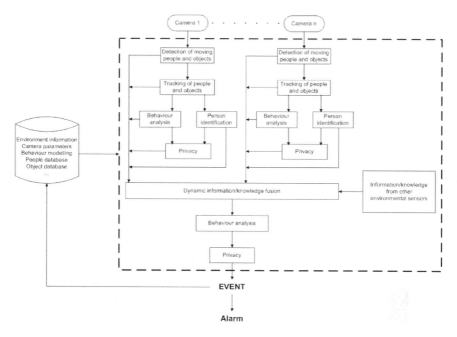

Fig. 2 Global description of the system

Almost every visual surveillance system starts with motion detection. Motion detection aims at segmenting regions corresponding to moving people and objects from the rest of an image. Subsequent processes such as tracking and behaviour analysis are greatly dependent on it. The process of motion detection usually involves background modelling, motion segmentation, and object classification, which are interwoven during processing. Although a private environment is usually quite static, unfavourable factors, such as illumination variance, shadows and occlusions, introduce many hurdles to the acquisition and updating of background images. A Gaussian mixture model is used for each pixel value and an online estimation could be carried out to dynamically update background images in order to adapt to illumination variance and disturbance in backgrounds [24]. Later, moving objects and persons are segmented by background subtraction.

Different moving regions may correspond to different moving targets. To further track people and study how they interact with objects and analyze their behaviours, it is essential to correctly classify them. The system will obtain visual features from the objects in order to recognise and track them. In some cases this will be hard to carry out due to the implicit difficulties when dealing with computer vision (occlusions, similarities between objects...). So, other characteristics or technologies could also be considered such as RFID tags or coloured markers that can be added to the objects in order to facilitate these tasks.

There are different approaches for human motion analysis depending on the requirements of the application [12, 25]: single region or blob-based tracking, multiple region tracking, articulated 2D models and articulated 3D models. Due to the well-known 3D structure of a private environment the approach is to consider a 3D articulated model modelling of the real shape of a person. However for specific use cases and scenarios, other simpler methods as blob tracking could be used. One of the key aspects is the distributed architecture of the system.

Scene analysis (behaviour analysis and people identification) will be carried out in two phases. First, the system will try to extract all the possible knowledge from each device. If activities or people are recognised, events could be generated. Later, information and knowledge from the different sensor devices (cameras and other environmental sensors that can give information about human activity, such as pressure, presence or lighting sensors; state of switches, doors or windows; energy consumptions...) can be combined to perform a higher-level behaviour analysis. This fusion will be dynamic and hierarchic according to the current state of the system determined by the previous recognised events. The system will dynamically select the inputs from the appropriate sensors to recognise actions avoiding considering the information proceeding from the whole sensor system. The system is hierarchic as sensors could be clustered with the purpose of generating local information and knowledge that, in a second stage, could be combined globally. Besides, information obtained from the multiple cameras will allow a view-invariant pose estimation in order to build the 3D representation of a person. As a result, actions performed by people will be recognised and activity will be modelled. Then, deviation from the expected will involve anomaly detection and the correspondent alarm launch.

People identification can be carried out in many ways: using complementary devices (such as RFID), classifiying morphological features obtained from previous stages, or by face recognition. As one of the requirements of this systems is unobtrusiveness, complementary devices to be wear should be discarded. That is, the system will recognize the different persons from their shapes, their clothing... If this is not possible, face recognition will be performed. In a private environment, the number of people is usually reduced, facilitating this task that in other places would be more complicated.

In regards to privacy, some of the processing could be integrated into hardware in the camera devices, assuring that no images would exit from the device if it is not necessary. This would increase the security of the system and also user confidence. However in much of the cases a fusion of data gathered from different devices would be necessary, transferring images and processing them in a central computer or in a distributed way, applying the privacy processing at a later stage.

4   Conclusions

The main goals of this PhD thesis are the design and development of distributed vision systems in coordination with other type of sensors to analyse human behaviour in private environments by considering ethical criteria to ensure privacy protection.

The use of vision systems will enable a breakthrough in the development of services and technologies for intelligent and self-adaptive environments with the aim to support independent living for everyone, but in particular for the elderly or people who require the constant attention of a carer, allowing their better quality of life and greater efficiency of their care. The maintenance of privacy will also provide greater acceptance and confidence of people when using the proposed technologies.

Currently, the research follows both lines of work, privacy preservation and behaviour analysis. We are conducting a survey to determine the acceptance of elderly people to have a vision-based system in their homes, and what services would be required and with which level of privacy. On the other hand, research on behaviour analysis is beginning with the development of a fall detection system using an omnidirectional camera located on the ceiling of the room (Figure 3).

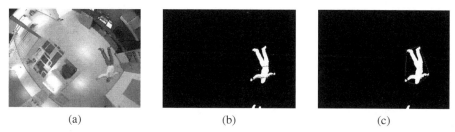

(a)   (b)   (c)

Fig. 3 Fall detection: (a) Original image, (b) Segmented image, and (c) Convex hull of the segmented person that allows to detect a fall

Acknowledgements. This work has been partially supported by the Spanish Ministry of Science and Innovation under project "Sistema de visión para la monitorización de la actividad de la vida diaria en el hogar" (TIN2010-20510-C04-02).

References

1. European Commission: Overview of the European strategy in ICT for Ageing Well (2009)
2. European Commission: Seniorwatch 2. Assessment of the Senior Market for ICT Progress and Developments (2008)
3. European Science and Technology Observatory: Science and Technology Roadmapping: Ambient Intelligence in Everyday Life (AmI@Life). In: Friedewald, M., Da Costa, O. (eds.) JRC/IPTS-ESTO Study (2003)
4. Remagnino, P., Foresti, G., Ellis, Y. (eds.): Ambient Intelligence: a Novel Paradigm. Springer, New York (2005)
5. Steinhage, A., Lauterbach, C.: Monitoring movement behaviour by means of a large area proximity sensor array in the floor. In: Gottfried, B., Aghajan, H.K. (eds.) Proceedings of the 2nd Workshop on Behaviour Monitoring and Interpretation, BMI 2008, pp. 15–27. CEUR-WS.org (2008)
6. Tapia, E.M., Intille, S.S., Larson, K.: Activity Recognition in the Home Using Simple and Ubiquitous Sensors. In: Ferscha, A., Mattern, F. (eds.) PERVASIVE 2004. LNCS, vol. 3001, pp. 158–175. Springer, Heidelberg (2004)

7. Mihailidis, A., Carmichael, B., Boger, J.: The use of computer vision in an intelligent environment to support aging-in-place, safety, and independence in the home. IEEE Transactions on Information Technology in Biomedicine 8(3), 238–247 (2004)

8. Park, S., Kautz, H.: Hierarchical Recognition of Activities of Daily Living using Multi-Scale, Multi-Perspective Vision and RFID. In: 4th IET International Conference on Intelligent Environments, Seattle (2008)

9. Hoey, J., Poupart, P., Bertoldi, A., Craig, T., Boutilier, C., Mihailidis, A.: Automated handwashing assistance for persons with dementia using video and a partially observable Markov decision process. Computer Vision and Image Understanding 114(5), 503–519 (2010)

10. European Commission: ICT & Ageing. European Study on Users, Markets and Technologies. Final Report (2010)

11. Velastin, S., Remagnino, P. (eds.): Intelligent distributed video surveillance systems. IEE (2006)

12. Moeslund, T., Hilton, A., Kruger, V.: A survey of advances in vision-based human motion capture and analysis. Computer Vision and Image Understanding 104, 90–126 (2006)

13. Ferryman, J.M. (ed.): Proceedings of PETS 2007. IEEE Computer Society (2007)

14. Shah, M., Javed, O., Shafique, K.: Automated Visual Surveillance in Realistic Scenarios. IEEE Multimedia 14(1), 30–39 (2007)

15. Tian, Y.-L., Brown, L.M.G., Hampapur, A., Lu, M., Senior, A.W., Shu, C.-F.: IBM smart surveillance system (S3): event based video surveillance system with an open and extensible framework. Machine Vision and Applications 19(5-6), 315–327 (2008)

16. Dufaux, F., Ebrahimi, T.: Scrambling for Privacy Protection in Video Surveillance Systems. IEEE Transactions on Circuits and Systems for Video Technology 18(8), 1168–1174 (2008)

17. Winkler, T., Rinner, B.: TrustCAM: Security and Privacy-Protection for an Embedded Smart Camera Based on Trusted Computing. In: Proceedings of the 7th IEEE International Conference on Advanced Video and Signal Based Surveillance, pp. 593–600. IEEE (2010)

18. Privacy enablement in a surveillance system. In: Senior, A. (ed.) Proceedings of the 15th IEEE International Conference on Computer Vision, pp. 1680–1683. IEEE (2008)

19. Neustaedter, C., Greenberg, S., Boyle, M.: Blur Filtration Fails to Preserve Privacy for Home-Based Video Conferencing. ACM Transactions on Computer-Human Interaction 13(1), 1–36 (2006)

20. Zhou, Z., Chen, X., Chung, Y.-C., He, Z., Han, T.X., Keller, J.M.: Activity Analysis, Summarization, and Visualization for Indoor Human Activity Monitoring. IEEE Transactions on Circuits and Systems for Video Technology 18(11), 1489–1498 (2008)

21. Zouba, N., Bremond, F., Thonnat, M., Anfosso, A., Pascual, É., Malléa, P., Mailland, V., Guerin, O.: A computer system to monitor older adults at home: Preliminary results. Gerontechnology 8(3), 129–139 (2009)

22. Cucchiara, R., Prati, A., Vezzani, R.: A Multi-Camera Vision System for Fall Detection and Alarm Generation. Expert Systems 24(5), 334–345 (2007)

23. Crowley, J.L., Coutaz, J., Bérard, F.: Perceptual user interfaces: things that see. Communications of the ACM 43(3), 54–64 (2004)

24. McIvor, A.: Background Subtraction Techniques. In: Proc. of Image and Vision Computing, New Zealand, Auckland (2000)

25. Poppe, A.: A survey on vision-based human action recognition. Image and Vision Computing 28(6), 976–990 (2010)

Distribution and Selection of Colors on a Diorama to Represent Social Issues Using Cultural Algorithms and Graph Coloring

Víctor Ricardo Cruz-Álvarez, Fernando Montes-Gonzalez, Alberto Ochoa, and Rodrigo Edgar Palacios-Leyva

Abstract. We present a problem know in writing about social modeling associated with the adequate distribution and color selection of societies in a diorama to specify relationships between them; and also between their principal attributes to represent the symbolic capital of a society. Our case study is related to the diversity of cultural patterns described in Memory Alpha. Thus, we use 8 principal attributes with a range of 64 colors. The purpose of this research is to apply the cultural algorithms approach with color graph to solve the proposed problem and subsequently represent the solution within a diorama. The Memory Alpha is conformed by 1087 societies, which permits to demonstrate that the matching of social issues allows correct distribution and color selection. In summary we are proposing an innovative representation for societies location.

1 Introduction

In this paper, we have focused our attention in a practical problem known in literature that is related with society modeling. In this representation the selection of visual representation includes colors and the design of features associated with their symbolic capital. Therefore we use a diorama, which facilitates setting the position that expresses the relationships between societies. The proposed problem solution is a hybridization of two techniques: color graph (for representation) and cultural algorithms (to define the distribution and social relationships), as consequence color graph is represented in the diorama. As described in [2] cultural algorithm agents are able to

Víctor Ricardo Cruz-Álvarez · Fernando Montes-Gonzalez · Rodrigo Edgar Palacios-Leyva
Universidad Veracruzana, Sebastián Camacho 5 C.P.: 91000, Xalapa, Veracruz, México
e-mail: victor.g2004@gmail.com, fmontes@uv.mx,
        roy_repl@hotmail.com

Alberto Ochoa
Universidad Autónoma de Ciudad Juárez, Ciudad Juárez, Chihuahua, México
e-mail: alberto.ochoa@uacj.mx

S. Omatu et al. (Eds.): Distributed Computing and Artificial Intelligence, AISC 151, pp. 57–64.
springerlink.com   © Springer-Verlag Berlin Heidelberg 2012

select a limited repository of colors and forms. For this work we employed societies as described in [4], which is an information repository about Artificial Societies [1]. Here societies are divided in four Quadrants. The selection of each attribute and their visual representation involves the skill evaluation of a number of individuals from diverse Quadrants. Social networking representation requires the development of some similarity measures that allows establishing locations in the dioramas [6].

2 Cultural Algorithms

The initial development of Cultural Algorithms (CAs) can be attributed to Reynolds [12] this approach is a complement to the metaphor used by evolutionary algorithms, which had focused on the concepts of genetics and natural evolution. Cultural algorithms are based on the theories of anthropologists, sociologists and archaeologists, who have tried to model the evolution as a process of cultural evolution [5]. The belief space characterizes CAs as evolutionary algorithms, which are used to store the acquired knowledge from previous generations. The information in this space must be accessible to any individual, who may use it to change their behavior and their respective proposed solution. To join the belief space and the population is necessary to establish a communication protocol, which dictates rules of the type of information to be exchanged between spaces. This protocol defines the acceptance and influence functions. The acceptance function is responsible for accepting the information or the experience that individuals have obtained in the current generation and transport into the belief space. On the other hand, the influence function is responsible for "influencing" variation operators (e.g. crossover and mutation in the case of genetic algorithms). This means that this function set some kind of pressure on resultant individuals from the application of variation operators to reach the desirable behavior, also away from undesirable results, always according to information stored in the belief space.

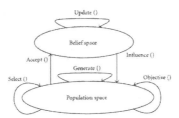

Fig. 1 Different Spaces employed by the Cultural Algorithm.

Figure 1 presents the interaction between the belief space and population space. The population space works similar to that of an evolutionary algorithm, i.e. the population consists of a set of individuals where each has an independent feature used to determine their suitability (fitness). The interaction between the two spaces makes the cultural algorithm increases the complexity in the development and computation of the evolutionary algorithm. Below we show the general pseudo-code of a cultural algorithm.

```
Begin
 t=0;
 Initialize POP(t); // Initialization of population
 Initialize BLF(t); // Initialization of believing space
 Evaluate POP(t);
 While (Do not condition of term t=t+1)
   Vote (BLF (t), Accept (POP(t))));
   Adjust (BLF (t));
   Evolve(POP(t), Influence(BLF(t)));
   t = t +1;
   Select POP(t) from POP(t-1);
 End While
End
```

3 Selection of Relationships and Colors for the Visual Representation of Related Societies

A social network is a theoretical construct useful to study social relationships. As mentioned in [3], the diorama and the graphs are common diagrams used to represent social networks. A graph is an abstract representation of a set of objects where some pairs of the objects are connected by links, for this work, each object in the graph represents one society and the link represents the relationship between societies. The representation of a social network consists of one or more graphs that represent networks. Then we set social modeling as an combinatorial problem, which is possible to resolve with the use of the cultural algithm to determine a specific solution according to the different perspectives of the the Population Space. The main strength of the CA is the capability of describing several application domains and find potential solutions in random uncertain situations. Among the different domains we have transport, race competitions, logistics and many different aspects and problems such as:

- Cooperative Interactive Based on Dynamic Knowledge and Alliance to find a solution.
- Multi-population Cooperative belief space.
- Cooperative interactive information to adopt knowledge migration.

The Cultural Algorithm employs the correlation function to measure the similarity between societies from Memory Alpha. This measure lets us relate each society with regard to the others; also to organize the societies in clusters of 64 colors. We have a weight for each attribute, which represents the importance of an individual attribute. For each pair of societies we have

$$\sum_i^n = abs[corr(W_i Attribute_i^A, W_i Attribute_i^B)] \tag{1}$$

Where:

- W_i = vector of weights that represent the importance of attribute i.
- A, B = represent the societies that we want to contrast.
- Attribute$_i$ = vector of the attributes for each society.

As a result the sum of the correlation of attributes weighted between two societies defines the similarity measure. The greater the correlation is the stronger the relationship between societies. For a threshold of 0.1, if the correlation value is less than the threshold we assume an empty relationship between the societies (zero value).The formula (1) is used as a fitness function in the cultural algorithm to measure the relationship between societies that share the same color, in other words, this formula is used for the adequate color selection of the societies.

Fig. 2 Representation of Societies presented in [4] with color selection and visual representation to distribute elements in a Diorama.

4   Distributing Elements Based on an Adequate Colored Visual Representation

From the point of view of the agents, the problem of combinatorial optimization is very complex. The best location of the individual of a society is not known with respect to the other representatives. In the proposed algorithm for the cultural interchange the individuals in the space of beliefs (Beliefspace), through their best paradigm (BestParadigm), are set to zero to represent the fact that the culture increases the amount of expectations associated with the location of a society with respect to the others. Therefore, the algorithm rewards the behavior associated with the best paradigm (BestParadigm). The social behavior of the chosen 1087 societies, described in [4], is characterized using eight attributes: emotional control, ability to fight, intelligence, agility, force, resistance, social leadership, and speed. These characteristics allow describing them as both a society and an individual, in Figure 2 an example of diorama with color distribution is shown.

5   Multiple Matching

The multiple matching is a series of seven evaluations according to different combinations of colors and a batch of 50 runs under different scenarios. In the evaluation phase social issues with more sociocultural similarities will be given a preference, then 8 social issues will be selected to compete. Each issue makes a compromise and participates in exactly seven of these evaluations. Issues must be ranked according to their preferences after tournaments end once the final list of multiple matching is evaluated. The hybrid algorithm sets the right for societies to evaluate a batch according to the organizational needs and the societies for each comparison assign the societies list before a new cycle begins. Each evaluation will have 120 societies playing over a schedule of seventeen runs.

The hybrid algorithm will be scheduled to set the timing for the comparison of different similarities using a round of multiple matching analyses based in the genre assigned to an issue. Then, societies that qualify for selection in a diorama will be chosen on the following prioritized basis.

For the first cycle of similarity, all societies in Memory Alpha (ie. Kelemane or Kioddary Societies) will be invited to participate for different comparisons. Given the organization for each society and the matches for each round in the algorithm, societies are asked to state their participation for its evaluation in each of the series. In case any of these societies decline to participate in the series, the algorithm may nominate one society to be set as a replacement, and this society has to be rated amongst the top societies in Memory Alpha. Based on an average calculation of two decimal places, the rating list in the series of comparisons, before starting a new cycle, twenty qualifiers will be selected (excluding the seven societies who will be compared in the matches). In case societies have the same average rating, the number of similarities set for the match will be used to determine its ranking.

To ensure an active participation in the future, a minimum of twenty-five games are recommended for the four included rating lists and before the main rating list. When a society does not accept to play into a Multiple Matching series, then the selection process uses the average rating plus number of games played during the rating period. The algorithm repeats this process until reaching the required qualifiers of the Multiple Matching series.

6 Experimentation

In order to obtain the most efficient arrangement of individuals in a social network, we developed a cluster for storing the data of each of the representative individuals for each society. The latter is made with the purpose of distributing an optimal form for each the evaluated societies [9]. The main experiment consisted in implementing 1087 communities in the Cultural Algorithm, with 500 agents and 200 beliefs into the belief space. The stop condition is reached after 50 runs; this allowed generating the best selection of each quadrant and their possible location in a diorama. A location is obtained after comparing the different cultural and social similarities of each community, and the evaluation of the Multiple Matching Model as in [10]. The vector of weights employed for the fitness function is $W_i=[0.6, 0.7, 0.8, 0.5, 0.6, 0.4, 0.9, 0.5]$, which respectively represents the importance of the particular attributes: emotional control, ability to fight, intelligence, agility, force, resistance, social leadership and speed. Then, the cultural algorithm will select the color of each society based on the attributes similarity. Each attribute is represented by a discrete value from 0 to 5, where 0 means absence and

5 the highest value of the attribute. The experiment design consists of an ortho-gonal array test with interactions amongst the attribute variables; these variables are studied within a color range (1 to 64). The orthogonal array is L-N(2**8), in other words, 8 times the N executions. The value of N is defined by the combina-tion of the 8 possible values of the variables, also the values in the color range. In Table 1 we list some possible scenarios as the result of combining the values of the attributes and the specific color to represent a social issue (society). The results permit us to analyze the effect of the variables in the color selection of all the possible combinations of values.

Table 1 The orthogonal array test.

Emotional Control	Ability to fight	Intelligence	Agility	Force	Resistance	Social Leadership	Speed	Color
0	1	2	2	3	3	4	5	1
0	1	2	2	3	4	5	5	1
1	1	3	2	4	4	2	1	2
1	1	3	2	5	3	2	1	2

The use of the orthogonal array test facilitates the reorganization of the differ-ent attributes. Also the array aids to specify the best possibilities to adequate cor-rect solutions (skills) for each society. Different attributes were used to identify the real possibilities of improving a society set in a particular environment, and to specify the correlations with other societies (see Figure 4b). Figure 3 presents a radar graphic repressing 36 societies (A to AI), their 8 attributes values, and the color chose based on the orthogonal test array.

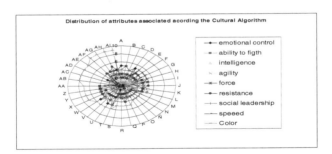

Fig. 3 The radar graphic represents the values after applying the orthogonal array for different societies.

7 Conclusions

After our experiments we were able to remark the importance of the diversity of the established cultural patterns for each community. These patterns represent a unique form of adaptive behavior that solves a computational problem that does not make clusters of the societies based only on their external appearance (the attributes for each society).

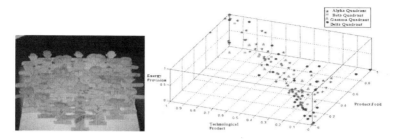

Fig. 4 The final diorama (a) is built according eight diverse attributes based on a Cultural Algorithm (Multiagents system). In (b) the visual representation of societies is analyzed according to the orthogonal array test.

The resultant configurations can be metaphorically related to the knowledge of the behavior of the community with respect to an optimization problem (to culturally select 47 similar societies, without being in the same quadrant [4]). Our implementation related each of the societies to a specific quadrant. As a result we obtained batches that included linguistic and cultural identity (see Figure 4a) for different societies. The latter, allowed us to identify changes in time related to one or another society (see Figure 4b). Here, we show that the use of cultural algorithms substantially increased the understanding in obtaining the "best paradigm". This after the classification of agent communities was made based on a relation that keeps their attributes. Therefore, we realize that the concept of "negotiation" exists based on determining the acceptance function to propose an alternative location for the rest of the communities [8]. For further implementations we intend to analyze the level and degree of cognitive knowledge for each community. Additionally, this may help to understand true similarities that share different societies based in the characteristics to be clustered and also to keep their own identity. In a related work [7], it has been demonstrated that small variations go beyond phenotypic characteristics and are mainly associate to tastes and related characteristics developed through the time. On the other hand, CAs can be used in the Evolutionary Robotic field where social interaction and decision is needed, for example in the training phase described in [11], and to organize group of robots for collaborative tasks. Another future work using CAs is related to the distribution of workgroups, social groups or social networking. Finally, CAs can be used in pattern recognition in a social database, for example: fashion styling and criminal behavior.

Acknowledgements. We thank the community of Computational Sociology that for a decade has researched about novel ways to display social topics. We also acknowledge partial support for this work from CONACyT-MEXICO grant SEP No. 0100895.

References

1. Desmond, A., Moore, J.: Darwin – la vida de un evolucionista atormentado. Generación Editorial, São Paulo, Brazil (1995)
2. Ochoa, A., et al.: Baharastar – Simulador de Algoritmos Culturales para la Minería de Datos Social. In: Proceedings of COMCEV 2007 (2007)
3. Ochoa, A., et al.: Dyoram's Representation Using a Mosaic Image. The International Journal of Virtual Reality (2009)
4. Memory Alpha, http://en.memory-alpha.org
5. Callogerodóttir, Z., Ochoa, A.: Optimization Problem Solving using Predator/Prey Games and Cultural Algorithms. In: NDAM 2003, Reykiavik, Iceland (2007)
6. Hué, T., et al.: The Emergence of Social Network Hierarchy Using Cultural Algorithms. In: VLDB 2006, Seoul, Korea (2006)
7. Vukčević, I., Ochoa, A.: Similar cultural relationships in Montenegro. In: JASSS 2005, England (2005)
8. Dennis, Z.: Culture and Organizations. McGraw-Hill, London (1991)
9. Ponce, J., et al.: Data Mining and Knowledge Discovery in Real Life Applications, Book edited by: J. Ponce, A. Karahoca, p. 438. I-Tech, Vienna (2009) ISBN: 978-3-902613-53-0
10. Ustaoglu, Y.: Simulating the behavior of a minority in Turkish Society. In: ASNA 2009, Zurigo, Svizerra (2009)
11. Nolfi, S., Floreano, D.: Evolutionary Robotic: The Biology, Intelligence, and Technology of Self-Organization Machines. MIT Press, MA (2000)
12. Reynolds, R.G., Sverdlik, W.: Problem Solving Using Cultural Algorithms. In: International Conference on Evolutionary Computation, pp. 645–650 (1994)

'Believable' Agents Build Relationships on the Web

John Debenham and Simeon Simoff

Abstract. In this paper we present the Believable Negotiator — the formalism behind a Web business negotiation technology that treats relationships as a commodity. It supports relationship building, maintaining, evolving, and passing to other agents, and utilises such relationships in agent interaction. The Believable Negotiator also takes in account the "relationship gossip" — the information, supplied by its information providing agents, about the position of respective agents in their networks of relationships beyond the trading space. It is embodied in a 3D web space, that is translated to different virtual worlds platforms, enabling the creation of an integrated 3D trading space, geared for Web 3.0.

1 Introduction

Despite the hype surrounding electronic business-to-business commerce, little business is conducted completely automatically as yet due to the complexity of the issues being negotiated and the high value of the deals involved. When human agents or artificial agents, referred to herein as 'agents' where no distinction is needed, collaborate or compete with each other, for example, in electronic business. The emergent intelligence or behaviour that did not exist prior to the interaction is commonly referred to as *collective intelligence* [1]. Research in social computing spans technologies that support activity-based interaction to theory, modelling and analysis of the variety of relationships that emerge [9].

In [6] the authors argued that *believability* is an essential feature in the next generation electronic marketplaces. [op. cit.] addressed three technological aspects of

John Debenham
Centre for Quantum Computation & Intelligent Systems, University of Technology, Sydney
e-mail: john.debenham@uts.edu.au

Simeon Simoff
School of Computing & Mathematics, University of Western Sydney
e-mail: s.simoff@uws.edu.au

S. Omatu et al. (Eds.): Distributed Computing and Artificial Intelligence, AISC 151, pp. 65–72.
springerlink.com   © Springer-Verlag Berlin Heidelberg 2012

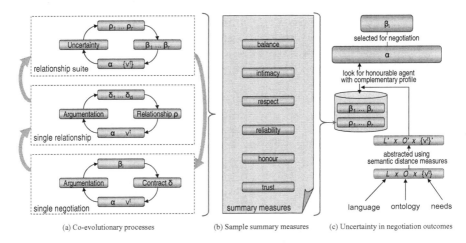

(a) Co-evolutionary processes   (b) Sample summary measures   (c) Uncertainty in negotiation outcomes

Fig. 1 The framework of the Believable Negotiator

believability of electronic marketplaces, described namely through the respective metaphors: "marketplaces where people are", "marketplaces that are alive and engaging", and "market places where information is valuable and useful". In this paper we argue that another essential feature of the next generation electronic marketplaces is the machinery that enables *believable* negotiating agents to develop and maintain relationships relevant to their business on the Web. The overall goal is to develop agents capable of establishing and reshaping their relationships with other agents to compensate for the lack of understanding of their own preferences over negotiation outcomes. According to social penetration theory [3] the establishment and growth of interpersonal relationships is a result of reciprocal exchange of information. Such exchange starts with relatively non-intimate topics and gradually progresses to more personal and private topics. These encapsulate the philosophy behind the *Believable Negotiator*.

The conceptual framework of the *believable negotiator* is shown in Figure 1. Figure 1(a) shows the fundamental co-evolutionary processes: first, an agent α interacts with other agents $\{\beta_i\}$ using argumentation to sign a contract δ in satisfaction of need ν^t, second, through repeated negotiations with β_i, α strives to form an honourable relationship with β_i, and third, agent α strives to assemble a relationship suite that respects her uncertainty profile. The superscript ^{t} denotes a function of time. Figure 1(b) lists some useful measures that summarise features of the potentially large history of illocutionary exchanges built up during many repeated negotiations — these measures are defined in [10] and [12]. Figure 1(c) shows a framework for representing uncertainty over negotiation outcomes — the potentially massive space of "language $\times$ ontology $\times$ need" is abstracted and simplified. Then using this simpler framework α looks for honourable trading partners with a complimentary profile.

2 Building Relationships on the Web

Relationships are fundamental to all but the most impersonal forms of interaction in business. Our agent summaries its relationships using 'intimacy' and 'balance' measures. Its actions are then shaped by its desired values for these two measures that represent its foreseeable social aspirations, and are called the 'target intimacy' and 'target balance'. A particular interaction with another agent is approached both with the goal of negotiating towards a satisfactory conclusion, and as an opportunity to do so in a way that gradually develops the relationship towards its target.

Relationships are built through exchanging information; this work employs *information-based agency* [11]. The intuition behind information-based agency is that all illocutionary acts give away (valuable) information. We assume that they share a common ontology and that their interactions are organised into dialogues, where a *dialogue* is a finite sequence of inter-related utterances. A *commitment* is a consequence of an utterance by an agent that contains a promise that the world will be in some state in the future. A *contract* is a pair of commitments exchanged between a pair[1] of agents. The set of all dialogues between two agents up to the present is their *relationship*.

There is evidence from psychological studies that humans seek a *balance* in their working relationships. The classical view is that people perceive resource allocations as being distributively fair (i.e. well balanced) if they are proportional to inputs or contributions (i.e. equitable). However, more recent studies [14, 15] show that humans follow a richer set of norms of distributive justice depending on their *intimacy* level: equity, equality, and need. *Equity* being the allocation proportional to the effort (e.g. the profit of a company goes to the stock holders proportional to their investment), *equality* being the allocation in equal amounts (e.g. two friends eat the same amount of a cake cooked by one of them), and *need* being the allocation proportional to the need for the resource (e.g. in case of food scarcity, a mother gives all food to her baby).

Enabling Relationships. This discussion is from the point of view of an information-based agent α in a multiagent system where α interacts with negotiating agents, β_i, information providing agents, θ_j, and an *institutional agent*, ξ, that represents the institution where we assume that all interactions happen [2]: $\{\alpha, \beta_1, \ldots, \beta_o, \xi, \theta_1, \ldots, \theta_t\}$. The institutional agent reports promptly and honestly on what actually occurs after an agent signs a contract, or makes some other form of commitment. Agents have a probabilistic first-order *internal language* $\mathcal{L}$ used to represent a *world model*, $\mathcal{M}^t$. A generic *information-based* architecture is described in detail in [11].

The architecture for managing relationships is shown in Figure 2. Agent α acts in response to a *need* that is expressed in terms of the ontology. Needs trigger α's goal/plan proactive reasoning, while other messages are dealt with by α's reactive reasoning. The *relationship strategy* determines which agent to negotiate with for a given need; it uses risk management analysis to preserve a strategic set of trading

---

[1] Sets of commitments between more than two agents are not considered here.

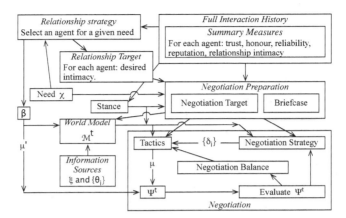

Fig. 2 The relationship architecture

relationships for each mission-critical need — this is not detailed here. For each trading relationship this strategy generates a *relationship target* that is expressed in the dialogical framework as a desired level of *intimacy* to be achieved in the long term.

Each interaction consists of a dialogue, Ψ^t, between two agents with agent α contributing utterance μ and the partner β contributing μ'. Each dialogue, Ψ^t, is evaluated using the dialogical framework in terms of the *value* of Ψ^t to both α and β. The *interaction strategy* then determines the current set of offers $\{\delta_i\}$, and then the *tactics*, guided by the *interaction target*, decide which, if any, of these offers to put forward and wraps them in argumentation dialogue. We now describe two of the distributions in $\mathcal{M}^t$ that support offer exchange.

Valuing Dialogues. Suppose that an interaction commences at time s, and by time t a string of utterances, $\Phi^t = \langle \mu_1, \dots, \mu_n \rangle$ has been exchanged between agent α and agent β. This dialogue is evaluated by α in the context of α's world model at time s, $\mathcal{M}^s$, and the environment e that includes utterances that may have been received from other agents in the system including the information sources $\{\theta_i\}$. Let $\Psi^t = (\Phi^t, \mathcal{M}^s, e)$, then α estimates the *value* of this dialogue to itself in the context of $\mathcal{M}^s$ and e as a $2 \times L$ array $V_\alpha(\Psi^t)$ where:

$$V_x(\Psi^t) = \begin{pmatrix} I_x^{l_1}(\Psi^t) & \dots & I_x^{l_L}(\Psi^t) \\ U_x^{l_1}(\Psi^t) & \dots & U_x^{l_L}(\Psi^t) \end{pmatrix} \tag{1}$$

where the $I(\cdot)$ and $U(\cdot)$ functions are information-based and utility-based measures respectively as we now describe. α estimates the *value* of this dialogue to β as $V_\beta(\Psi^t)$ by assuming that β's reasoning apparatus mirrors its own.

In general terms, the information-based valuations measure the reduction in uncertainty, or information gain, that the dialogue gives to each agent, they are expressed in terms of decrease in entropy that can always be calculated. The utility-based valuations measure utility gain are expressed in terms of "some suitable" utility evaluation function $\mathbb{U}(\cdot)$ that can be difficult to define. This is one reason why the utilitarian approach has no natural extension to the management of argumentation that is achieved here by our information-based approach.

The *balance* in a dialogue, Ψ^t, is defined as: $B_{\alpha\beta}(\Psi^t) = V_\alpha(\Psi^t) \ominus V_\beta(\Psi^t)$ for an element-by-element difference operator $\ominus$ that respects the structure of $V(\Psi^t)$. The *intimacy* between agents α and β, $I_{\alpha\beta}^{*t}$, is the pattern of the two $2 \times L$ arrays V_α^{*t} and V_β^{*t} that are computed by an update function as each interaction round terminates, $I_{\alpha\beta}^{*t} = \left(V_\alpha^{*t}, V_\beta^{*t} \right)$. If Ψ^t terminates at time t:

$$V_x^{*t+1} = \nu \times V_x(\Psi^t) + (1 - \nu) \times V_x^{*t} \tag{2}$$

where ν is the learning rate, and $x = \alpha, \beta$. Additionally, V_x^{*t} continually decays by: $V_x^{*t+1} = \tau \times V_x^{*t} + (1 - \tau) \times D_x$, where $x = \alpha, \beta$; τ is the decay rate, and D_x is a $2 \times L$ array being the decay limit distribution for the value to agent x of the intimacy of the relationship in the absence of any interaction. D_x is the *reputation* of agent x. The *relationship balance* between agents α and β is: $B_{\alpha\beta}^{*t} = V_\alpha^{*t} \ominus V_\beta^{*t}$. As a simple example, if the amount if information concerning α and β's acceptable options $\{I_\alpha^O(\Psi^{*t}), I_\beta^O(\Psi^{*t})\}$ increases, then α's greed decreases.

The notion of balance may be applied to pairs of utterances by treating them as degenerate dialogues. In simple multi-issue bargaining the *equitable information revelation* strategy generalises the tit-for-tat strategy in single-issue bargaining, and extends to a tit-for-tat argumentation strategy by applying the same principle across the dialogical framework.

Relationship Strategies and Tactics. Each dialogue has to achieve two goals. First it may be intended to achieve some contractual outcome. Second it will aim to contribute to the growth, or decline, of the relationship intimacy.

We now describe in greater detail the contents of the "*Negotiation*" box in Figure 2. The negotiation literature consistently advises that an agent's behaviour should not be predictable even in close, intimate relationships. The required variation of behaviour is normally described as varying the negotiation *stance* that informally varies from "friendly guy" to "tough guy". The stance is shown in Figure 2, it injects bounded random noise into the process, where the bound tightens as intimacy increases. The stance, $S_{\alpha\beta}^t$, is a $2 \times L$ matrix of randomly chosen multipliers, each ≈ 1, that perturbs α's actions. The value in the (x, y) position in the matrix, where $x = I, U$ and $y \in \mathscr{L}$, is chosen at random from $[\frac{1}{l(I_{\alpha\beta}^{*t}, x, y)}, l(I_{\alpha\beta}^{*t}, x, y)]$ where $l(I_{\alpha\beta}^{*t}, x, y)$ is the bound, and $I_{\alpha\beta}^{*t}$ is the intimacy.

The negotiation *strategy* is concerned with maintaining a working set of proposals. If the set of proposals is empty then α will quit the negotiation. α perturbs the acceptance machinery by deriving s from the $S_{\alpha\beta}^t$ matrix. In the early stages of the negotiation α may decide to inflate her opening offer. The following strategy uses the machinery described above. Fix h, g, s and c, set the Proposals to the empty set, let $D_s^t = \{\delta \mid \mathbb{P}^t(\text{acc}(\alpha, \beta, \chi, \delta)) > c\}$, then:

- repeat the following as many times as desired: add $\delta = \arg\max_x \{\mathbb{P}^t(\text{acc}(\beta, \alpha, x)) \mid x \in D_s^t\}$ to Proposals, remove $\{y \in D_s^t \mid \text{Sim}(y, \delta) < k\}$ for some k from D_s^t

By using $\mathbb{P}^t(\text{acc}(\beta, \alpha, \delta))$ this strategy reacts to β's history of Propose and Reject utterances.

Negotiation *tactics* are concerned with selecting some offers and wrapping them in argumentation. Prior interactions with agent β will have produced an intimacy pattern expressed in the form of $\left(V_\alpha^{*t}, V_\beta^{*t}\right)$. Suppose that the relationship target is $(T_\alpha^{*t}, T_\beta^{*t})$. Following from Equation 2, α will want to achieve a *negotiation target*, $N_\beta(\Psi^t)$ such that: $\nu \cdot N_\beta(\Psi^t) + (1 - \nu) \cdot V_\beta^{*t}$ is "a bit on the T_β^{*t} side of" V_β^{*t}:

$$N_\beta(\Psi^t) = \frac{\nu - \kappa}{\nu} V_\beta^{*t} \oplus \frac{\kappa}{\nu} T_\beta^{*t} \tag{3}$$

for small $\kappa \in [0, \nu]$ that represents α's desired *rate of development* for her relationship with β. $N_\beta(\Psi^t)$ is a $2 \times L$ matrix containing variations in the dialogical framework's dimensions that α would like to reveal to β during Ψ^t (e.g. I'll pass a bit more information on options than usual, I'll be stronger in concessions on options, etc.). It is reasonable to expect β to progress towards her target at the same rate and $N_\alpha(\Psi^t)$ is calculated by replacing β by α in Equation 3. $N_\alpha(\Psi^t)$ is what α hopes to receive from β during Ψ^t. This gives a *negotiation balance target* of: $N_\alpha(\Psi^t) \ominus N_\beta(\Psi^t)$ that can be used as the foundation for reactive tactics by striving to maintain this balance across the dialogical framework. A cautious tactic could use the balance to bound the response μ to each utterance μ' from β by the constraint: $V_\alpha(\mu') \ominus V_\beta(\mu) \approx S_{\alpha\beta}^t \otimes (N_\alpha(\Psi^t) \ominus N_\beta(\Psi^t))$, where $\otimes$ is element-by-element matrix multiplication, and $S_{\alpha\beta}^t$ is the stance. A less neurotic tactic could attempt to achieve the target negotiation balance over the anticipated complete dialogue. If a balance bound requires negative information revelation in one dialogical framework category then α will contribute nothing to it, and will leave this to the natural decay to the reputation D as described above.

Information-Based Strategies. Every communication gives away information and so has the potential to contribute to the intimacy and balance of a relationship. Information-based strategies manage the information revelation process. Let $M_{\alpha\beta}^t$ be the set of time-stamped messages that α has sent to β, and $M_{\beta\alpha}^t$ likewise both at time t. $\mathcal{M}^t$ is α's world model at time t and consists of a set of probability distributions. x^t denotes a message received at time t. $\mathbb{I}^t(\alpha, \beta, x^t)$ is the information gain — measured as the reduction of the entropy of $\mathcal{M}^t$ — observed by α after receiving message x^t. $\mathbb{I}^t(\beta, \alpha, x^t)$ is α's estimate of β's information gain after receiving message x^t from α.

The complete *information history* of both the observed and the estimated information gain, $G^t(\alpha, \beta)$, is:

$$G^t(\alpha,\beta) = \{(x^s, \mathbb{I}^s(\alpha,\beta,x^s)) \mid x^s \in M^t_{\beta\alpha}\} \cup \{(x^s, \mathbb{I}^s(\beta,\alpha,x^s)) \mid x^s \in M^t_{\alpha\beta}\}$$

respectively.

In [11] we described to the model that α constructs of β. In general α can not be expected to guess β's world model, $\mathscr{M}^t_\beta$, unless α knows what β's needs are — even then, α would only know $\mathscr{M}^t_\beta$ with certainty if it knew what plans β had chosen. However, α always knows the private information that it has sent to β — for example, in Propose($\cdot$) and Reject($\cdot$) messages. Such private information could be used by β to estimate α's probability of accepting a proposal: $\mathbb{P}^t_\beta(\mathrm{acc}(\alpha,\beta,\chi',z))$, where χ' is the need that β believes α to have.

α's information-based strategies constrain its actions, x^t, on the basis of $\mathbb{I}^t(\beta,\alpha,x^t)$ and its relation to $G^t(\alpha,\beta)$. For example, the strategy that gives β greatest expected information gain: $\arg\max_z\{\ \mathbb{I}^s_\beta(\beta,\alpha,z) \mid C^t(\alpha,\beta,z)\}$. More generally, for some function f: $\arg\max_z\{\ f(\mathbb{I}^s_\beta(\beta,\alpha,z),G^t(\alpha,\beta)) \mid C^t(\alpha,\beta,z)\}$, the idea being that the f 'optimises' in some sense the information gain taking account of the interaction history.

Ontology-Based Strategies. The structure of the ontology may be used to manage the information revelation process in particular strategic areas. For example, α may prefer to build a relationship with β in the context of the supply of particular goods only [8]. The structure of the ontology is provided by the Sim($\cdot$) function. Given two contracts δ and δ' containing concepts $\{o_1,\dots,o_i\}$ and $\{o'_1,\dots,o'_j\}$ respectively, the (non-symmetric) distance of δ' from δ is the vector: $\Gamma(\delta,\delta') = (d_k : o''_k)^i_{k=1}$ where $d_k = \min_x\{\mathrm{Sim}(o_k,o'_x) \mid x = 1,\dots,j\}$, $o''_k = \sup(\arg\min_x\{\mathrm{Sim}(o_k,x) \mid x = o'_1,\dots,o'_j\},o_k)$ and the function sup($\cdot,\cdot$) is the supremum of two concepts in the ontology. $\Gamma(\delta,\delta')$ quantifies how different δ' is to δ and enables α to "work around" or "move away from" a contract under consideration. In general for some function g; $\arg\max_z\{\ g(\Gamma(z,x^s)) \mid x^s \in M^t_{\alpha\beta} \cup M^t_{\beta\alpha} \wedge C^t(\alpha,\beta,z)\}$ the idea being that the g 'optimises' in some sense the ontological relationship with the interaction history.

3   Conclusions

Reliable relationships are fundamental to enabling interaction of value between humans. A precursor to achieving reliability for artificial agents is that the relationships should be 'believable' in the sense that they are for humans. This work aims to enable valuable interaction for human and artificial agents in a Web 3.0 environment by providing a set of measures that enable agents to gauge the strength and nature of their relationships, and then to build their relationships towards desired targets. By treating information as a valuable commodity and by managing its exchange strategically this work has been shown to enable human and artificial agents to build reliable and believable relationships in a laboratory-based Web 3.0 environment.

Acknowledgments. This research is supported by an Australian Research Council Discovery Grant DP0879789, UTS and University of Western Sydney.

References

1. Alag, S.: Collective Intelligence in Action. Manning Publications Co. (2008)
2. Arcos, J.L., Esteva, M., Noriega, P., Rodríguez, J.A., Sierra, C.: Environment engineering for multiagent systems. Journal on Engineering Applications of Artificial Intelligence 18 (2005)
3. Berscheid, E., Reis, H.: Attraction and close relationships. In: The Handbook of Social Psychology, pp. 193–281. McGraw-Hill, New York (1998)
4. Boella, G., der Torre, L.V., Verhagen, H.: Introduction to normative multiagent systems. Computational Mathematics and Organisational Theory 12, 71–79 (2006)
5. Bogdanovych, A.: Virtual Institutions. PhD thesis, Faculty of IT, University of Technology, Sydney (November 2007)
6. Debenham, J., Simoff, S.: Believable electronic trading environments on the web. In: Proceedings of the IEEE/WIC/ACM International Conference on Web Intelligence WI 2009, Milan, Italy, September 15-18, pp. 631–638. IEEE/WIC/ACM, IEEE (2009)
7. Gloor, P.A., Krauss, J., Nann, S., Fischbach, K., Schoder, D.: Web science 2.0: Identifying trends through semantic social network analysis. In: Proceedings of the International Conference on Computational Science and Engineering, pp. 215–222. IEEE Computer Society (2009)
8. Kalfoglou, Y., Schorlemmer, W.M.: IF-Map: An Ontology-Mapping Method Based on Information-Flow Theory. In: Spaccapietra, S., March, S., Aberer, K. (eds.) Journal on Data Semantics I. LNCS, vol. 2800, pp. 98–127. Springer, Heidelberg (2003)
9. King, I., Li, J., Chan, K.T.: A brief survey of computational approaches in social computing. In: Proceedings of International Joint Conference on Neural Networks, Atlanta, Georgia, USA, June 14-19, pp. 1625–1632. IEEE, IEEE Press (2009)
10. Sierra, C., Debenham, J.: Trust and honour in information-based agency. In: Stone, P., Weiss, G. (eds.) Proceedings Fifth International Conference on Autonomous Agents and Multi Agent Systems AAMAS 2006, Hakodate, Japan, pp. 1225–1232. ACM Press, New York (2006)
11. Sierra, C., Debenham, J.: Information-based agency. In: Proceedings of Twentieth International Joint Conference on Artificial Intelligence, IJCAI 2007, Hyderabad, India, pp. 1513–1518 (January 2007)
12. Sierra, C., Debenham, J.: The LOGIC Negotiation Model. In: Proceedings Sixth International Conference on Autonomous Agents and Multi Agent Systems AAMAS 2007, Honolulu, Hawai'i, pp. 1026–1033 (May 2007)
13. Sierra, C., Jennings, N., Noriega, P., Parsons, S.: A Framework for Argumentation-Based Negotiation. In: Rao, A., Singh, M.P., Wooldridge, M.J. (eds.) ATAL 1997. LNCS, vol. 1365, pp. 177–192. Springer, London (1998)
14. Sondak, H., Neale, M.A., Pinkley, R.: The negotiated allocations of benefits and burdens: The impact of outcome valence, contribution, and relationship. Organizational Behaviour and Human Decision Processes (3), 249–260 (December 1995)
15. Valley, K.L., Neale, M.A., Mannix, E.A.: Friends, lovers, colleagues, strangers: The effects of relationships on the process and outcome of negotiations. In: Bies, R., Lewicki, R., Sheppard, B. (eds.) Research in Negotiation in Organizations, vol. 5, pp. 65–94. JAI Press (1995)

Speed-Up Method for Neural Network Learning Using GPGPU

Yuta Tsuchida, Michifumi Yoshioka, and Sigeru Omatu

Abstract. GPU is the dedicated circuit to draw the graphics, so it has a characteristic that the many simple arithmetic circuits are implemented. This characteristic is hoped to apply the massive parallelism not only graphic processing. In this paper, the neural network, one of the pattern recognition algorithms is applied to be faster by using GPU. In the learning of the neural network, there are many points to be processed at the same time. We propose a method which makes the neural network be parallelized in three points. The parallelizations are implemented in neural networks which have different initial weight coefficients, the learning patterns or neurons in a layer of neural network. These methods are used in combination, but the first method can be processed independently. Therefore one of the three methods, the first method, is employed as comparison to compare with the proposed methods. As the result, the proposed method is 6 times faster than comparison method.

1 Introduction

Recently, the GPU, Graphic Processing Unit becomes popular and low cost due to developments of high quality 3DCGs, movie processing algorithms and computer games. GPU is the dedicated circuit to draw graphics, so it has many simple arithmetic circuits. These units are able to be applied to not only graphics but

Yuta Tsuchida · Michifumi Yoshioka
Department of Computer Science and Intelligent Systems, Graduate School of Engineering,
Osaka Prefecture University, Sakai, Osaka 599-8531, Japan
e-mail: tsuchida@sig.cs.osakafu-u.ac.jp,
        yoshioka@cs.osakafu-u.ac.jp

Sigeru Omatu
Department of Electronics, Information and Communication Engineering
Faculty of Engineering, Oska Institute of Technology,
Omiya, Asahiku, Osaka, 535-8585, Japan
e-mail: omatu@rsh.oit.ac.jp

S. Omatu et al. (Eds.): Distributed Computing and Artificial Intelligence, AISC 151, pp. 73–80.
springerlink.com   © Springer-Verlag Berlin Heidelberg 2012

also general parallel computations. GPU which is used in parallel computations is called General-Purpose computing on GPU (GPGPU)[5]. In order to assist developments with GPGPU, Compute Unified Device Architecture (CUDA) is distributed by NVIDIA, which optimizes parallel computations with GPU [2]. In this study, we consider a GPGPU application to intelligent signal processings especially in neural networks. As previous works, Kyoug-S et al. [4] proposed the neural network implementation for GPU. Then Daves [1] organized this method. Finally, Honghoon et al. [3] proposed the implementation with CUDA. However, these previous methods assume that networks are already trained, therefore they don't include how to train neural networks implemented in GPGPU. Under these circumstances, we propose a neural network training method which is optimized for GPGPU. At first, we propose a simple algorithm which trains neural networks with different initial weight coefficients in parallel. As next steps, we propose optimized versions with considering neural network architecture. Finally, we show the effectiveness of our proposed algorithms by some simulations.

2 Architecture of Layered Neural Network

Before the discussion about neural network implementation to GPGPU, we summarize the architecture of a simple layered neural network. Fig.1 shows a feed-forward layered neural network. It has l neurons in an input layer, m neurons in a hidden layer and n neurons in an output layer, respectively. The equations indicate the oth pattern in the P data-sets. The outputs from the hidden layer y_j^o :

$$y_j^o = f\left(\sum_{i=0}^{l} v_{ij} x_i^o\right) \quad (j = 1, \cdots m) \tag{1}$$

$$f(x) = \frac{1}{1 - \exp(-x)}. \tag{2}$$

where $x_i^o (i = 1, \cdots l)$ are input signals fed to the input layer, $x_0 = -1$ is a threshold (common by all patterns), v_{ij} are weights and (2) is a sigmoid function. The outputs from output layer z_k^o, are :

$$Z_k^o = f\left(\sum_{j=0}^{m} w_{jk} y_j^o\right). \quad (k = 1, \cdots n) \tag{3}$$

where $y_0 = -1$ is threshold (common by all patterns).

Back propagation is used as the update method for weights. The weights are updated as follows. At first, the error signals e_k^o are calculated by using supervisor signals d_k^o as follows:

$$e_k^o = d_k^o - z_k^o. \tag{4}$$

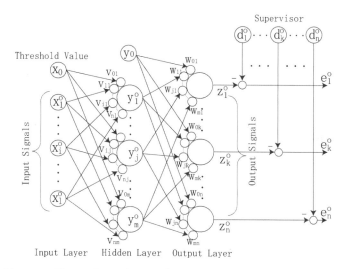

Fig. 1 Architecture of layered neural network

The weights in the output layer are calculated as follows:

$$w_{jk}^{new} = w_{jk}^{old} + \Delta w_{jk}^{o} \tag{5}$$

$$\Delta w_{jk}^{o} = \eta y_{j}^{o} \delta_{k}^{o} \quad (j = 0, \cdots m, k = 1, \cdots n) \tag{6}$$

$$\delta_{k}^{o} = e_{k}^{o} z_{k}^{o}(1 - z_{k}^{o}) \tag{7}$$

where η is the learning rate. Finally the weights in the hidden layer are calculated as follows:

$$v_{ij}^{new} = v_{ij}^{old} + \Delta v_{ij}^{o} \tag{8}$$

$$\Delta v_{ij}^{o} = \eta x_{i}^{o} y_{j}^{o}(1 - y_{j}^{o}) \sum_{k=1}^{n} w_{jk}^{new} \delta_{k}^{o} \tag{9}$$

where w_{jk}^{new} and δ_{k}^{o} are propagated values which updated in the output layer respectively. One pair of input signals and supervisor signals is defined as Pattern. The Patterns are fed repeatedly until some conditions. There are two conditions to stop the repeat. At first, the summation of e_k is less than the prescribed value. And the processes have executed the prescribed number of times. In this paper, the first condition is not defined, and the secondary prescribed number is defined as s. Therefore, the above processes are repeated Ps times, when the P Patterns are trained.

3 Proposed Method

In this section, the three training methods for neural networks in GPGPU are proposed. At first, we consider a simple implementation as the reference. Usually, in

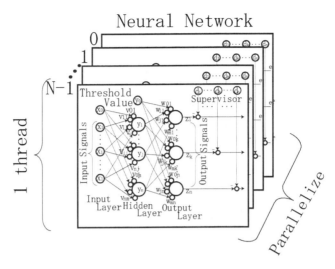

Fig. 2 Parallelization for whole network

the training process of neural networks, a trial-and-error is used with different initial weights because neural networks may converge to different local minimums according to their initial weights. By training many neural networks with different initial weights, we can reduce the number of trial and error processes. We call this method gcomparison methodh for the reference of improved methods. Fig.2 shows the method mentioned above.

As the next steps, we propose improved methods optimized for GPGPU and neural networks. From the neural network architecture, we can find some points for parallelization. As first candidate for parallelization, we adopt neurons in same layer. The calculations for neurons in same layer are performed in parallel. However, it is important that we need to wait until all calculations in same layer are completed because calculations in the next layer need the results in the previous layer. In order to guarantee this synchronization, we introduce a synchronization mechanism by using "Syncthread" function in CUDA. The number of the threads is set to the maximum of the number of the neurons of the each layer. Fig.3 shows the first method mentioned above.

As the second candidate, we adopt parallelism in training patterns. In the training of neural networks, many patterns are fed to neural networks. These calculations are able to be performed in parallel. Usually, in the neural network training, we have two types of algorithms. One is sequential and the other is batch. In the sequential type training, weights are updated with respect to each pattern, and in the batch training, all errors are accumulated and weights are updated according to the total error. However, it is difficult to use directly these algorithms in parallel case because of weights update confusion. In order to avoid the confusion, weights are fixed in the feed forward stage and updated with respect to each pattern in the back propagation

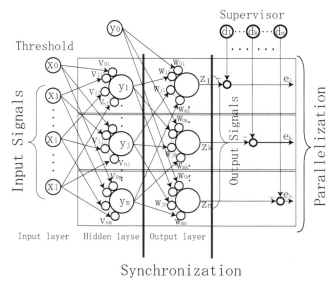

Fig. 3 Parallelization for in the network

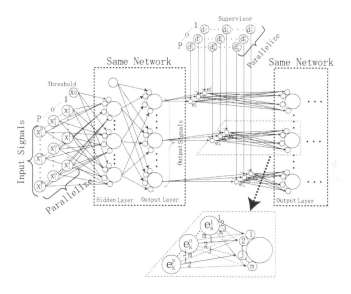

Fig. 4 Parallelization for patterns

stage. The convergence of weights in neural networks with this modified training algorithm is confirmed by some preliminary simulations. Fig.4 shows the second method mentioned above.

Table 1 GeForce GTX480

Number of cores	480
Global Memory	1.5GB
Constant Memory	64KB
Shared Memory	48KB/block
Clock rate	1.40GHz

These methods are implemented to GPU by CUDA. In CUDA, many threads are allocated in a hierarchy structure. The Grid has up to 2^{32} Blocks in two dimensions. Then each Block also has up to 1024 threads in three dimensions. In total, they have five dimensions. The each Block has a Shared memory to be shared by the threads contained in the Block. In addition, the graphic board has Global memory to be accessed from not only the all threads contained in the GPU but also the host CPU. In this study, we use the following notations. The index of the Block is (bi_x, bi_y) and the index of thread in a Block (ti_x, ti_y, ti_z) respectively where the size of Grid is $(g_x \times g_y < 2^{32})$ and the size of Block is $(b_x \times b_y \times b_z < 1024)$. The structure of the hierarchy is organized to satisfy the proposed methods as follows : $(g_x, g_y) = (N/b, p), (b_x, b_y, b_z) = (R_{max}, b, 1)$ respectively, where b is a natural number , and R_{max} is the maximum in l, m, n. The values of y, z, or δ_k are shared by the threads in same Block, so the Shared memories are used. And then the "Syncthreads" function can be operated for the threads in the same block only. The weights are allocated to Global memory in order to be accessed from the all threads in the Grid.

4 Simulation

In order to confirm the effectiveness of our proposed methods, we have two simulations. We use the comparison method as the reference, and measure calculation times of the methods. The GPU used for the inspection is GeForce GTX480. The specifications of the GPU is summarize in table 1. And the specifications of the parameter settings for this simulation is shown in table 2.

We have some simulations different in the number of the neurons in the hidden layer and patterns. Fig.5 and 6 show the results respectively. In these figures, each rectangles of the solid line depict the processing time of the proposed method, and the rectangles of the dashed line depict the comparison method.

These results show that the processing times with the proposed method don't increase though the number of hidden layer in comparison with the comparison method. It means that the proposed method is achieves much shorter processing time than the comparison method. However the processing time is increased by changing the number of pattern. It means that the effect of third method don't have enough performance. As the result, the processing time is about 16 percent shorter than that of the comparison method.

Table 2 Parameter settings used in the simulations

Symbols	Descriptions	Values
N	The number of the neural networks	40
b	The number of the networks per Block	5
l	The number of inputs	3
m	The number of the neurons in the hidden layer	9*
n	The number of output	3
η	The learning rate	0.04
P	The number of the patterns	10*
s	The number of the steps of training	2200

* is default values.

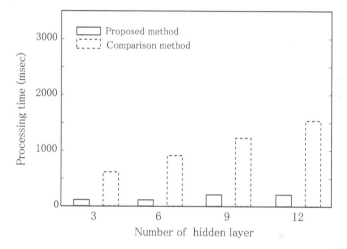

Fig. 5 Result 1: The processing time different in the number of the neurons in the hidden layer.

5 Conclusion

In this paper, we proposed the neural network training algorithm optimized for GPGPU. These methods extract parallelism in neural network training and optimize them for GPGPU. The simulation results show the effectiveness of our proposed methods combined. Our proposed method achieves about six times faster than the simple implementation, comparison method. One of the methods we discussed in this paper is that the calculations in each pattern are done in the same time. However the calculation for weight coefficients in the neuron is sequential. So, as future research, we consider that the each weights in the neuron are calculated in the same time in consideration of the calculation for the patterns. Many home computers shipped in late years have GPU. By using these systems, many applications

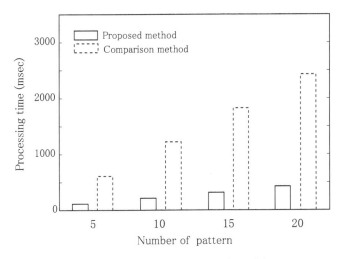

Fig. 6 Result 2: The processing time different in the number of the patterns.

developed with neural networks are available at home. As future works, we consider other machine learning algorithm implementations using GPGPU.

References

1. Davis, C.E.: Graphics processing unit computation of neural networks. M.S. thesis, Comput. Sci. Dept., Univ. New Mexico, Albuquerque, NM (2005)
2. Kirk, D.B., Hwu, W.-M.W.: Programming Massively Parallel Processors: A Hands-on Approach. Applications of GPU Computing Series. Morgan Kaufmann (2010)
3. Jang, H., Park, A., Jung, K.: Neural Network Implementation Using CUDA and OpenMP. In: Digital Image, Computing Techniques and Applications, DICTA 2008, 155–161 (2008), doi:10.1109/DICTA.2008.82
4. Oh, K.-S., Jung, K.: GPU implementation of neural networks. Pattern Recognition 37(6), 1311–1314 (2004), doi:10.1016/j.patcog.2004.01.013, ISSN: 0031-3203
5. Che, S., Boyer, M., Meng, J., Tarjan, D., Sheaffer, J.W., Skadron, K.: A performance study of general-purpose applications on graphics processors using CUDA. J. Parallel Distrib. Comput. 68(10), 1370–1380 (2008)

A Fast Heuristic Solution for the Commons Game

Rokhsareh Sakhravi, Masoud T. Omran, and B. John Oommen*

Abstract. Game Theory can be used to capture and model the phenomenon of the exploitation of the environment by human beings. The *Commons Game* is a simple and concise game that elegantly formulates the different behaviors of humans toward the exploitation of resources (also known as "commons") as seen from a game-theoretic perspective. The game is particularly difficult because it is a multi-player game which requires both competition and cooperation. Besides, to augment the complexity, the various players are unaware of the moves made by the others – they merely observe the *consequences* of their respective moves. This makes the game extremely difficult to analyze, and *there is thus no known method by which one can even understand whether the game has has an equilibrium point or not.* In the *Commons Game*, an ensemble of approaches towards the exploitation of the commons can be modeled by colored cards. In this paper, we consider the cases when, with some probability, the user is aware of the approach (color) which the other players will use in the exploitation of the commons. We investigate the problem of determining the best probability value with which a *specific* player can play each color in order to maximize his ultimate score. Our solution to this problem is an "intelligent" heuristic algorithm which determines (i.e., locates in the corresponding space) feasible probability values to be used so as to obtain the maximum average score. The basis for such a strategy is that we believe that results obtained in this manner can be used to work towards a "*expectiminimax*" solution, or for a solution which utilizes such values for a UCT-type algorithm. The solution has

Rokhsareh Sakhravi · Masoud T. Omran · B. John Oommen
School of Computer Science, Carleton University, Ottawa, Canada
e-mail: rsakhrav@connect.carleton.ca,
{mtomran,oommen}@scs.carleton.ca

* The third author is a *Chancellor's Professor*, a *Fellow of the IEEE* and a *Fellow of the IAPR*, and also an *Adjunct Professor* with the University of Agder in Grimstad, Norway. The author was partially supported by NSERC, the Natural Sciences and Engineering Research Council of Canada.

S. Omatu et al. (Eds.): Distributed Computing and Artificial Intelligence, AISC 151, pp. 81–90.
springerlink.com   © Springer-Verlag Berlin Heidelberg 2012

been rigorously tested by simulations, and the results obtained are, in our opinion, quite impressive. Indeed, as far as we know, this is a pioneering AI-based heuristic solution to the game under the present model of computation.

Keywords: Game Theory, Commons Game, Tragedy of Commons, Convergence of Commons.

1   Introduction

In Game Theory, a number of players attempt to find strategies for playing a specific game so as to lead to the best solution for a predefined goal. Although, Game Theory has been extensively studied in the field of mathematics, it also has many applications in economics, biology, engineering, politics, philosophy and computer science. In recent years, there has been a remarkable increase in research in fields that intersect game theory and computer science. Many artificial games studied in Game Theory can be used to understand the main aspects of humans using/misusing the environment. They can be tools by which we can define the aggregate behavior of humans, which, in turn, is driven by "short-term", perceived costs and benefits. We believe that humans should demonstrate a good understanding of how to protect the global environment, and in this regard Game Theory presents a potentially useful framework for the modeling and understanding of innate aggregate behaviors of multiple parties, as related to strategies for the use of resources. The *Commons Game* basically describes how a selfish player can use an unregulated resource, and instructs us of the different strategies that humans can use to cooperatively benefit from the resources found in the environment (e.g., forest, fish, water, energy, etc.)

The above scenario is modeled by the *Commons Game*, which is a simple and concise game that elegantly formulates the different behaviors of humans toward the exploitation of resources as seen from a game-theoretic perspective. In the setting of the formal *Commons Game*, an ensemble of approaches towards the exploitation of the commons can be modeled by colored cards, as explained presently. As one can understand, the game is particularly difficult because it is a multi-player game which requires both competition and cooperation. The complexity is augmented because the various players are unaware of the moves made by the others – they merely observe the *consequences* of their respective moves. Thus even the analytic techniques to analyze the game are not known, and even the existence of an equilibrium point has not been reported.

How then do we approach the problem? We consider the cases when, with some probability, the user is aware of the approach (color) which the other players will use in the exploitation of the commons. We then investigate the problem of determining the best probability value with which a *specific* player can play each color in order to maximize his ultimate score. As one can see, our solution to this problem is an "intelligent" heuristic algorithm which determines (i.e., locates in the corresponding space) feasible probability values to be used so as to obtain the maximum average score. The basis for such a strategy is that the results obtained in this manner can be used to work towards a "*expectiminimax*" solution [9], or for a solution which utilizes such values for a UCT-type algorithm [7]. This is currently being investigated.

In a commons, the individual parties using it are in competition with each other. It would, however, be more beneficial for each party (person, nation) to ensure that there will always be a harvest. But to do so, everyone in the commons must trust the others to act morally if and when an agreement to restrict exploitation is made. It will not be fair if one party restricts his exploitation, while another continues to use the resources in an unrestricted manner. The important features in a commons are thus two-fold. First, an individual's short-term selfish actions are in conflict with his long-term best interests.

The introduction of the problem to research in Game Theory dates back to 1968 when Hardin offered a simple application of Game Theory for the utilization of environmental resources in the so-called *Tragedy of the Commons* [5]. With regard to designing a computational "game" to efficiently and appropriately model this scenario, Power *et al.* [8] developed one such environmental game, the *Commons Game* in 1977 which is the theme of this study. This is a group game which has to be played cooperatively. In this game, players choose their cards individually but free communication is allowed. Players are free to act independently or interdependently in such a way that they can withdraw from the commons or take part in it. This is basically the version of the game that we use in our work.

More recently, the *Commons Game* has received a lot of attention due to its applicability to real-world issues. Baba *et al.* [1, 2, 3] suggested that the utilization of soft computing techniques (such as Genetic Algorithms (GAs) and Evolutionary Algorithms (EAs)) could contribute towards making the original game much more exciting. They also comparatively discussed three games, namely, the original *Commons Game*, the modified *Commons Game* utilizing GAs and Neural Networks (NNs), and the modified *Commons Game* utilizing EAs and NNs. They declared that the *Commons Game* which used EAs and NNs provided the best opportunity for letting players seriously consider the use of the commons. Brown and Vincent [4] analyzed the evolution of cooperation for a family of evolutionary games involving shared costs and benefits, with a continuum of strategies from non-cooperation to total cooperation. This cost-benefit game allows the cooperating parties to share in the benefit of the option to cooperate, and the recipients to be, in turn, burdened with a share of their cost.

In this paper, we shall develop a deterministic approach to maximize the average case score of a player (Player 1) when the probability by which the other players select a color is known *a priori*. While we do not claim that our solution attains the local/global optimum (which, we believe, is an intractable problem), what we endeavor to obtain is a heuristic by which a feasible solution is achieved. We analyze the game data and present various interesting properties pertaining to its performance. Finally, we illustrate the implementation results of the algorithm and provide various charts to visualize the behavior of the game in various situations.

2   State-of-the-Art of the Commons Game

The purpose of the *Commons Game* is to simulate the workings of a set of players utilizing the commons in such a way that no player has an exclusive right to the resource, while on the other hand, all the players have access to it.

In the *Commons Game*, the players can play either selfishly or cooperatively in each round. The resource is depleted gradually if the collective play is predominantly selfish, and consequently, everyone ends up with a profit less than if they had played cooperatively. On the other hand, if the play is predominantly cooperative, each player takes less than his maximum on each trial, implying that the resource is inexhaustible, and thus that they can "indefinitely" keep earning benefits from the resource. Hardin's article titled "The Tragedy of the Commons" [5] in the book by Hardin and Baden [6], and his film are excellent introductions to this game.

3   Maximum Score Computation

In this section, we develop a deterministic approach to maximize the average case score of a specific player when the probability by which other players select a color is known *a priori*. We do not claim that our solution attains the local/global optimum. What we endeavor to obtain is a heuristic by which a feasible solution is achieved. From a more philosophical perspective, one cannot draw conclusions based on best responses on how players would play. However, we are simply looking for a good playing strategy if we as a player know the probability by which other players select a color. Thus, we have proposed a heuristic which has been confirmed to have good results by simulations.

Problem Formulation: Suppose that for every player i, a probability vector P_i is given, which is the probability by which the player chooses a color. To be specific, Player i chooses color Green, Red, Yellow, Orange, and black with probability P_{G_i}, P_{R_i}, P_{Y_i}, P_{O_i}, and P_{B_i}, respectively. Let N be the number of players. Let the expected number of players that play each color be $P_X = \sum_{i=1}^{N} P_{X_i}, X \in \{G, R, Y, O, B\}$. We apologize for using the same symbol to represent different concepts. However, the context of the symbols is, generally speaking, not confusing.

As described earlier, the score of playing Green (Red) at each round is defined using the Payoff matrix, the Matrix Board, and the number of players who play Red cards. This score is in effect before invoking the consequence of playing a Black (Orange) card. For ease of explanation, we alter the Payoff matrix so that it allows not only a scalar number of players but also real numbers. For each state of the game we define a linear function that represents the score for playing a Green (Red) card based on the number of Red cards played. The score functions, denoted by *Payoff functions*, are obtained by connecting points on scalar representations of the Payoff matrices. As before, the score of playing a Yellow card is 6, and that of playing Orange and Black cards are $\frac{-6}{P_O}$ and $\frac{-6}{P_B}$, respectively.

In the original version of the game, "Replenishment" occurs in random rounds, and it is known that, on the average, it occurs every 6th round. Thus, to accommodate for the average computation, we consider it to occur every 6th round.

Based on the above problem setting, our goal is to compute a probability vector for a specific player to maximize his score when the probability vector for the other players is known *a priori*. Although we are working with the original version of the game [8], we desire a general approach which is applicable to all problems that reflect on the basics of the "Tragedy of the Commons".

The Proposed Solution: Suppose that the probability vector P_i, $i = 2, \ldots, N$, is given, and that we are planning to determine the probability vector P_1 that maximizes the average case score of Player 1. Once P_1 has been computed, the average score of playing each color at each round is available, where we use the expected number of players playing each color to compute the scores for the round. Although the scores for playing Yellow, Orange, and Black cards stay the same on different rounds, the scores of playing Green and Red cards change as a result of the change in state of the game. Once the expected numbers for all the colors have been computed, we can also compute the score for playing Green and Red cards at each round. Here, S_{G_i} and S_{R_i} are the scores for playing Green and Red cards in round i, $i = 1, \ldots, RN$, where RN is the total number of played rounds.

Our approach to determine the best probability vector for Player 1 is by considering the different cases based on the structural properties of the problem. First of all, it is straightforward to see that playing Orange and Black cards is not beneficial toward the goal of the problem. The reason for this is that for any number of Orange and Black players, the score for those who play them is negative so that they have a *contra* effect towards maximizing the score. Consequently, the maximum advantage is obtained by setting $P_{O_1} = 0$ and $P_{B_1} = 0$.

Consider the following cases:

1- $\mathbf{P_B} \geq \mathbf{1}$: Whenever the expected number of Black card players is greater than or equal to unity, playing a Green card always leads to a score of -20 points. So, playing a Green card is not beneficial, implying that $P_{G_1} = 0$. So, the best probability vector is obtained for some proportion involving P_{R_1} and P_{Y_1}, (note that $P_{R_1} + P_{Y_1} = 1$). In order to calculate the proportion, we define the average score of Red card players as $AVG_R = \frac{\sum_{i=1}^{RN}(S_{R_i} + P_0 * 10)}{RN}$, where, for sake of convenience, we compute the time average, even though, more formally, we should have computed the ensemble average. The following cases are possible:

If $(\mathbf{P_{R_1}} = \mathbf{1}) \wedge (\mathbf{AVG_R} > \mathbf{6})$: This case implies that the best probability vector is obtained by setting $P_{Y_1} = 0$ and $P_{R_1} = 1$.

Else If $(\mathbf{P_{R_1}} = \mathbf{1}) \wedge (\mathbf{AVG_R} < \mathbf{6})$: In this scenario, the best probability vector is obtained by setting $P_{Y_1} = 1$ and $P_{R_1} = 0$.

Else (i.e., $(\mathbf{P_{R_1}} = \mathbf{1}) \wedge (\mathbf{AVG_R} = \mathbf{6})$): In this scenario, either of the above solutions is the best probability vector.

The above discussion follows from the fact that decreasing P_{R_1} from its original value (equal to 1) results in a decrease in AVG_R, and so the the total score will drop more as Player 1 uses less Red cards and more Yellows cards.

2- $P_B < 1$: If the expected number of Black card players is less than unity, playing a Green card may also be beneficial. Therefore, the best probability vector is obtained for some proportion involving P_{G_1}, P_{R_1}, and P_{Y_1}, where the reader should observe that $P_{G_1} + P_{R_1} + P_{Y_1} = 1$. In order to calculate the proportion, we first consider the option of only playing Green and Red cards, and subsequently extend the solution to include playing Yellow cards.

It should be mentioned that, for the sake of convenience, instead of computing the three point convex combination of P_{G_1}, P_{R_1}, and P_{Y_1} simultaneously, we consider P_{Y_1} separate from P_{G_1} and P_{R_1}. The reasoning behind this is a consequence of the simplifying effect of assuming that changes in P_{Y_1} are independent from changes to the values of P_{G_1} and P_{R_1}. As a result, our heuristic algorithm involves computing trivial two-point convex combinations while producing reasonable results.

We define $AVG_G = \frac{\sum_{i=1}^{RN}(S_{G_i}*(1-P_B)+P_B*(-20))}{RN}$ as the average score for Green card players. The following cases are possible when a Yellow card play is not involved:

If $(P_{R_1} = 1) \wedge (AVG_G \leq AVG_R)$: In this case, the best probability vector is obtained when $P_{R_1} = 1$ and $P_{G_1} = 0$. To show that it is true, suppose that we decrease P_{R_1} and increase P_{G_1} to get a new proportion between these probabilities. The average score of playing Red and Green drops when less Reds are used because of the increasing nature of the Payoff functions. Besides this, playing more Green cards will cause the state to drop to negative values faster, which, in turn, results in a less average score for playing both Red and Green cards. Thus, the total score will drop if we decrease P_{R_1} and increase P_{G_1} not only because the average scores are decreasing, but also because we are replacing Red cards (which have a higher average score) with Green cards which have a lesser average score. Thus, the best probability vector is obtained when the most possible Red cards are played, namely $P_{R_1} = 1$. The results show that even for some values of AVG_G greater than AVG_R, $P_{R_1} = 1$ is the best probability vector for Player 1. More details of this are given in Section 4.

Else (i.e., $(P_{R_1} = 1) \wedge (AVG_G > AVG_R)$): In this case the best probability vector is obtained by finding the maximum of $(P_{G_1} * AVG_G + P_{R_1} * AVG_R)$ for all possible values of P_{R_1} and P_{G_1}, where one should observe that $P_{G_1} + P_{R_1} = 1$. Here, in order to find the maximum possible average score, we decrease P_{R_1} from its original value, 1, by some value Δ and increase P_{G_1} (initially equal to 0) by Δ and compute $(P_{G_1} * AVG_G + P_{R_1} * AVG_R)$. The highest value is the resultant maximum possible average score, namely the maximum of $(P_{G_1} * AVG_G + P_{R_1} * AVG_R)$. Because of the fact that decreasing P_{R_1} and increasing P_{G_1} will lead to a decrease in both AVG_G and AVG_R, we increase the total score by replacing Red cards (which lead to a less average score) with Green cards (with a greater average score), to work towards a maximum value. However, a more careful observation of the function at different values of AVG_G and AVG_R by decreasing P_{R_1} and increasing P_{G_1} reveals that there are some small drops and jumps in both functions. Figure 1 shows one such example. Here, we consider $\Delta = 1$ and probabilities which are given by percentages. The maximum average score in this example is obtained for $P_{G_1} = 34.32\%$ and $P_{R_1} = 65.68\%$. Note that the jumps in the functions AVG_G and AVG_R do not affect our previous discussion for the case where $(P_{R_1} = 1) \wedge (AVG_G \leq AVG_R)$ since both functions have their maxima for $P_{G_1} = 0$, and the jumps never reach to that

Fig. 1 A sample probability vector, AVG_G, AVG_R, and the function $P_{G_1} * AVG_G + P_{R_1} * AVG_R$, for increasing values of P_{G_1}.

value. Consequently, in the case that $((P_{R_1} = 1) \wedge (AVG_G > AVG_R))$, the maximum possible average score is not unique. A more detailed discussion of different cases for this scenario is presented in Section 4.

The results that we have obtained up to now do not involve the consideration of playing Yellow cards. We now consider the case where a Yellow card can also be played. We define the joint average score of playing Green and Red cards (defined earlier) as $AVG_{GR} = P_{G_1} * AVG_G + P_{R_1} * AVG_R$.

The following cases are possible when Yellow cards are also involved:

If $(\mathbf{AVG_{GR}} > \mathbf{6})$: In this case, playing a Yellow card together with Green and Red cards is not beneficial, implying that $P_{Y_1} = 0$ (see Figure 1). To prove this assertion, by way of contradiction, suppose that if we decrease P_{G_1} by some value Δ and increase P_{Y_1} by Δ, we can obtain a greater total score. If this is true we could, instead of increasing P_{Y_1}, increase P_{R_1} by Δ and get an even higher total score. But this contradicts the fact that AVG_{GR} is the maximum possible joint average score for playing Green and Red cards. Also, it is easy to see that decreasing P_{R_1} by some value Δ and adding Δ to P_{Y_1} will not increase the total score since we are replacing values greater than 6 by 6, and at the same time AVG_{GR} gets decreased as both AVG_G and AVG_R fall. So, the best probability vector would be obtained by the same probabilities as we had before.

Else If $(\mathbf{AVG_{GR}} < \mathbf{6})$: In this case, the best probability vector is obtained when $P_{G_1} = 0$, $P_{R_1} = 0$, and $P_{Y_1} = 1$ This is true because, for any combination of Green, Red, and Yellow cards, if $P_{Y_1} \neq 0$, the joint average score of playing Green and Red cards is less than AVG_{GR} obtained earlier. As a result, the best probability vector is obtained when we use all Yellow cards which yields the highest possible total score.

Else (i.e., $(\mathbf{AVG_{GR}} = \mathbf{6})$): In this setting, either of the above cases would lead to an equal total score, and this would be the best probability vector for Player 1.

The above algorithm leads the maximum average score for Player 1 when the probability vectors for the other players are given *a priori*.

4   Commons Game Implementation and Analysis

We have implemented the *Commons Game* in order to visualize and analyze the details of the game, and to test the effectiveness of our algorithm in attaining the average maximum score. Our implementation of the game is based on the manual published by Powers *et al.* [8]. We have also implemented the average maximum score computation algorithm presented in Section 3.

Our implementation illustrates some of the structural properties of the *Commons Game* and those of the new maximum score computation algorithm. The new algorithm proposed here computes the maximum possible score that Player 1 can earn (on average) by working with the average number of players that play each color in each round. The principle motivation of the new algorithm is to simplify the behavior of the game using its structural properties, which is captured by the two functions defined earlier, namely AVG_G and AVG_R. Since the Green and Red card moves are related to each other in different ways and toward the score computation, they effect the final score inter-dependently. Therefore, in our algorithm which computes the average maximum score, we first compute the joint average score of playing Green and Red cards, denoted by AVG_{GR}, and use this quantity to obtain the ultimate result.

The following properties are observable in the AVG_G and AVG_R functions:

1- Both functions are almost always decreasing, except at a few drop and jump points. The drops in the AVG_G and AVG_R functions are the result of the changes at the points at which replenishment and state changes occur.

2- Both functions decrease at about the same rate, although the rate of decrease of AVG_G is slightly more than that of AVG_R. The difference in the decrease rates is due to the fact that, in all Payoff functions, the score of playing a Green card is more than the corresponding score of playing a Red card. So, the degradation to lower states effects the score of Green card players more than that of Red card players.

3- The points at which drops and jumps occur are mostly the same in both functions, and the functions look quite similar. Note that the major reason for these drop and jump changes in AVG_G and AVG_R is the state change which affects the scores of the Green and Red cards at the same junctures.

The drops and jumps in the AVG_G and AVG_R functions also effect the quantity AVG_{GR}. Apart from this, we also record the following properties of AVG_{GR}:

1- The AVG_{GR} function starts from where the AVG_R starts and ends where it ends.

2- If both AVG_G and AVG_R were strictly decreasing, AVG_{GR} would have possessed a unique maximum value. Because of the drops and jumps in both AVG_G and AVG_R, the maximum value for AVG_{GR} is not unique. Figures 2(a), shows an example in which the maximum values for AVG_{GR} occur for two different values of P_{G_1}. In fact, there are cases when the maximum occurs for many values of P_{G_1}.

3- The higher the value of the maximum AVG_{GR}, the smoother is the shape of the functions AVG_G, AVG_R, and AVG_{GR}. Thus, the shape of the function is more like the one that is thought for. This is because in higher states, the effect of state changes and replenishment on the total score is less significant, as seen in Figure 2(b).

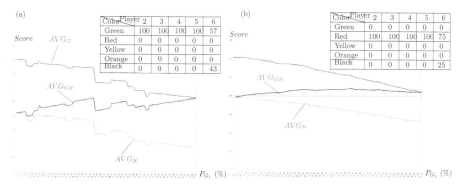

Fig. 2 (a) shows that the maximum value of AVG_{GR} is obtained for both $P_{G_1} = 42$ and $P_{G_1} = 100$. (b) An example showing the case that the maximum of AVG_{GR} occurs at higher scores.

5 Conclusions

In this paper, we have studied the *Commons Game*, the purpose of which is to simulate the workings of a set of players utilizing the commons in such a way that no player has an exclusive right to the resource, while on the other hand, all the players have access to it. In this game, the players are required to pay more attention to the so-called cost which is calculated based on the proportion of other players and their respective strategies. However, if the relative number of cheaters is higher, the perceived benefit of the others cooperating increases. Consequently, a proportion of the players must usually play cooperatively even though the others fail to, because the success of each strategy is calculated based on the proportion of the players using the various strategies in the total population. Through this work, we have developed a deterministic approach to maximize the average case score of a specific player (Player 1) when the probability by which the other players select strategies (represented by colored cards) is known *a priori*.

References

1. Baba, N., Handa, H.: COMMONS GAME Made more Exciting by an Intelligent Utilization of the Two Evolutionary Algorithms, vol. 71 (2007)
2. Baba, N., Nagasawa, K., Handa, H.: Utilization of Soft Computing Techniques for Making Environmental Games More Exciting –Toward an Effective Utilization of the COMMONS GAME. In: Lovrek, I., Howlett, R.J., Jain, L.C. (eds.) KES 2008, Part II. LNCS (LNAI), vol. 5178, pp. 411–417. Springer, Heidelberg (2008)
3. Baba, N., Sakurai, Y., Matsuda, A., Kawachi, T.: Soft computing techniques for making game playing exciting, pp. 132–135 (2005)
4. Brown, J.S., Vincent, T.L.: Evolution of cooperation with shared costs and benefits. Proceedings of the Royal Society B: Biological Sciences 275(1646), 1985–1994 (2008)

5. Hardin, G.: The Tragedy of the Commons. Science 162(3859), 1243–1248 (1968)
6. Hardin, G., Baden, J.: Managing the Commons. Freeman, San Francisco (1977)
7. Kocsis, L., Szepesvári, C.: Bandit Based Monte-Carlo Planning. In: Fürnkranz, J., Scheffer, T., Spiliopoulou, M. (eds.) ECML 2006. LNCS (LNAI), vol. 4212, pp. 282–293. Springer, Heidelberg (2006)
8. Powers, R., Duss, R., Norton, R.: THE COMMONS GAME Manual (1977)
9. Russell, S.J., Norvig, P.: Artificial intelligence: A Modern Approach, 3rd edn. Prentice-Hall, Inc., Upper Saddle River (2009)

Picture Information Shared Conversation Agent: Pictgent

Miki Ueno, Naoki Mori, and Keinosuke Matsumoto

Abstract. Recently, the various chatterbots have been proposed to simulate conversation. In this study, we propose a novel chatterbot system utilizing "pictures" of a model of situation in order to share common knowledge between users and the chatterbot by showing pictures to users. We show the basic concept and system structure of proposed system. We also describe the algorithm of estimating user interests.

1 Introduction

Recently, because of the increasing amount of information on the Internet, a conversation system called chatterbot is attracting public attention again. In order to estimate user interest, we have proposed novel chatterbot system[1].

However, there exists obvious limitation for communication between user and system with only text information, because nonverbal information like an expression or cultural background are very important in human actual conversation. Therefore, in this study, we propose a novel conversation system called *Picture Information Shared Conversation Agent* (Pictgent) in order to share conversation background knowledge by showing prepared picture to user.

There have proposed several studies to introduce pictures to communication between humans and systems. SHRDLU[2] is known as the successful model in domain specific conversation system by showing a graphical block world. In fields of psychology, *Thematic Apperception Test* (TAT)[3] utilizes ambiguous pictures to guess the mind state of subject.

Ito et al.[4] proposed the method that adds text information based on semantic ontology to usual pictograms and the tool that creates new symbolic picture.

Miki Ueno · Naoki Mori · Keinosuke Matsumoto
Department of Computer and Systems Sciences,
College of Engineering, Osaka Prefecture University,
1-1 Gakuencho, Nakaku, Sakai, Osaka 599-8531, Japan
e-mail: ueno@ss.cs.osakafu-u.ac.jp, mori@cs.osakafu-u.ac.jp,
    matsu@cs.osakafu-u.ac.jp

S. Omatu et al. (Eds.): Distributed Computing and Artificial Intelligence, AISC 151, pp. 91–94.
springerlink.com   © Springer-Verlag Berlin Heidelberg 2012

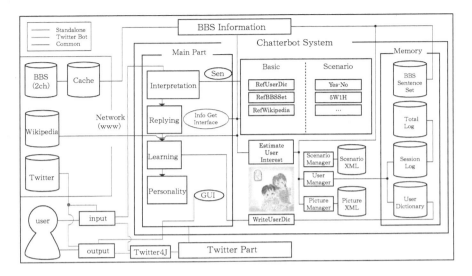

Fig. 1 Diagram of Constitution of whole System

2   Concept of Pictgent

Our proposed system called Pictgent is made of the following 3 modules.

Picture Module: This module manages the picture information that are important
items in this study. Each picture has its own model created by object oriented
modeling technique. In this module, picture model is written in XML format.

Scenario Module: This module controls transition of scenario of picture. In order
to achieve adequate transition, this module stores user's inner state as numerical
vector according to user input history.

Scenarios are written by XML which has transition map and answer example set.

Chat Module: This module replies according to user's input. There are following
two different reply modes in this module.

   Scenario mode: If user is following the scenario, the scenario mode is used. In
   this mode, Pictgent can talk with users utilizing picture's and scenario's XML.

   Chatterbot mode: If user loses interests and strays from a scenario, the mode is
   set to chatterbot mode. In this mode, Pictgent replies based on various topics
   with estimating the user interests.

3   Constitution of Pictgent

In this chapter, we describe the system outline of the Pictgent.

Fig. 1 shows the outline of the Pictgent. The system is written in Java. We introduced multi-thread programming for input and output part in order to make Pictgent

reply not only when user has finished inputs. We set an interval time to system for managing spontaneous reply.

Interpretation: Pictgent receives an input and formats it by morphological analysis.

Replying: Pictgent replies to user with an appropriate expression that is retrieved from several databases(DB) and Web information.

Learning: Pictgent will ask user about user's input sentence or any unknown keywords. It can memorize new statements and revise its memory.

Personality: Pictgent maintains a self-portrait and uses specific end of phrases.

4   Estimating Current User Interests

Generally, user interests are changing during conversation. If the Pictgent uses all logs equality, it is difficult to estimate current topics. To solve this problem, we propose the following method which estimate current user interests by emphasizing the current user input. Pictgent do not care the current user interests in scenario mode because there already exists transition map. Therefore, we focus on the "chatterbot" mode of Pictgent in following discussion.

Pictgent has a user a's interest vector c^a which represents user's internal state. Each vector element of c^a ($0 \leq c_j^a \leq 1$) relates to category j of stored BBS board information. If the degree of user interests in related board is maximum, the value of vector element becomes 1. On the other hand, use has no interest in that board, the value becomes 0.

The proposed Pictgent tries to answer based on the topic of board of larger c_j^a. However, c^a is made by all logs, there is problem that the proposed Pictgent fails to understand current topics. To avoid this, the proposed Pictgent decides the topic of answer as follows:

We define the latest input in time step T as S_T. Old inputs are represented by S_{T-1}, S_{T-2}, c, S_1. We also define the changing topic vector σ_{S_T} as follows:

$$\{\sigma_{S_T}\}_j = \sum_{x=1}^{W} N_x R_x^j \tag{1}$$

where N_i is the frequency of word i in S_T and R_i^j is the frequency word i in board j.

Next, the topic vector t^a which i-th element represents the current significance of board i is defined.

We set the default value of all elements of t^a to 0. The proposed Pictgent calculates t^a as follows whenever the proposed Pictgent obtains the newest input S_T.

$$t^a = \sum_{i=1}^{T} \gamma^{i-1} \sigma_{S_{T-i+1}} \tag{2}$$

where γ, $0 \leq \gamma \leq 1$, is discount rate. As γ decreases, the influence of past inputs decreases. We define normalized unit vector of t^a as $\hat{t}^a$.

Since we assume user interests are constant in short term, we consider that the weight of past inputs and that of current inputs are the same.

Therefore c^a is obtained by $t^a_{\gamma=1}$ of $\gamma = 1$ as follows:

$$t^a_{\gamma=1} = \sum_{i=1}^{T} \sigma s_{T-i+1}, \ c^a = \hat{t}^a_{\gamma=1} \tag{3}$$

The proposed Pictgent decides an utilizing board by referring to c^a and $\hat{t}^a$. We define board deciding vector b as follows:

$$b = \eta c^a + (1-\eta)\hat{t}^a \tag{4}$$

where $0 \leq \eta \leq 1$. The proposed Pictgent selects board i in proportion to b_i.

5 Conclusion

In this study, we proposed a novel chatterbot called Pictgent that can talk with users by sharing picture's information. We also described its system constitution. Improving each modules, especially the Chat Module and applying Pictgent to real situation like child school are important further work.

This research was supported in part by a Grant-in-Aid for Scientific Research (C), 22500208, 2010-2014 from the Ministry of Education, Culture, Sports, Science and Technology.

References

1. Ueno, M., Mori, N., Matsumoto, K.: Novel Chatterbot System Utilizing Web Information for Estimating Current User Interests. In: Proc. of the 6th IEEE International Conference on IDAACS, vol. 2, pp. 656–659 (2011) ISBN: 978-1-4577-1424-5
2. Winograd, T.: Procedures as a Representation for Data in a Computer Program for Understanding Natural Language. Cognitive Psychology 3(1) (1972)
3. Morgan, C.D., Murray, H.A.: A method of investigating fantasies. Archives of Neurology and Psychiatry 34, 289–306 (1935)
4. Ito, K., Matsuda, M., Durst, M.J., Hasida, K.: SVG Pictograms with Natural Language Based and Semantic Information, SVG Open (2007)

A Mixed Portfolio Selection Problem

Irina Georgescu and Jani Kinnunen

Abstract. The mixed portfolio selection problem studied in this paper corresponds to a situation of financial risk management in which some return rates are mathematically described by random variables and others are described by fuzzy numbers. Both Markowitz probabilistic model and a possibilistic portfolio selection model are generalized. A calculation formula for the optimal solution of the portfolio problem and a formula which gives the minimum value of the associated risk are proved.

1 Introduction

Traditionally, probability theory is the mathematical instrument to study uncertainty phenomena. On the other hand, there are types of uncertainty which cannot be approached probabilistically. Zadeh's possibility theory [16] offers an alternative to the treatment of uncertainty situations. The transition from probabilistic to possibilistic models concerns two components:

(i) random variables are replaced with possibility distributions (particularly, fuzzy numbers);

(ii) usual probabilistic indicators (expected value, variance, covariance) are replaced with adequate possibilistic indicators.

Portfolio selection is one of the crucial problems which appear in financial decision making. In Markowitz portfolio selection model [14] the security returns are considered random variables. This model uses two probabilistic indicators: mean value for return and variance for risk (see [14], [1]).

Portfolio selection problem appears naturally and various possibilistic portfolio selection models were studied in [4], [10], [11], [12], [13], [15], [17].

Irina Georgescu
Department of Economic Cybernetics, Academy of Economic Studies, Bucharest, Romania, and Department of Research, ETEA University, Cordoba, Spain
e-mail: irina.georgescu@csie.ase.ro

Jani Kinnunen
Institute for Advanced Management Systems Research, Åbo Akademi University, Turku, Finland
e-mail: jani.kinnunen@abo.fi

S. Omatu et al. (Eds.): Distributed Computing and Artificial Intelligence, AISC 151, pp. 95–102.
springerlink.com   © Springer-Verlag Berlin Heidelberg 2012

The complexity of the situations in financial field leads us to the idea of mixed models useful for possibility and probability theory. The aim of this paper is to propose a mixed portfolio selection model. Some returns will be mathematically represented by random variables and others by fuzzy numbers.

In the end of this introductory section the definition of a fuzzy number, operations with fuzzy numbers and some examples are recalled. Three indicators associated with fuzzy numbers are recalled: expected value, variance, covariance (cf. [2], [4], [5], [9]). They will be used in the next sections to build a mixed portfolio selection model. The rest of the paper is organized as follows.

In Section 2 two portfolio selection models are compared: Markowitz's and a possibilistic one derived from the former according to (i) and (ii). The difference between the two approaches consists in indicators' interpretation:

• for the first model the return is evaluated by probabilistic mean value, while for the second model the return is evaluated by the possibilistic mean value;
• for the first model the risk is evaluated by probabilistic variance, while for the second model the risk is evaluated by possibilistic variance.

Mixed portfolios are introduced in Section 3. Rentability of some assets are mathematically represented by random variables and rentability of other assets by fuzzy numbers. Two types of indicators are associated with a mixed portfolio:

• a possibilistic mean value, a probabilistic mean value, and a total mean value;
• a possibilistic variance, a probabilistic variance, and a total variance.

In Section 4 the mixed portfolio selection problem is formulated using these indicators. The main result of the section is the optimal solution of portfolio selection problem and the calculation of the minimum risk value.

In this section we recall the definition of fuzzy numbers [3], [6], [7] and some possibilistic indicators associated with them [2], [4], [5].

Let X be a set of states. A *fuzzy subset* of X is a function $A : X \rightarrow [0,1]$. For any state $x \in X$ the real number $A(x)$ is the degree of membership of x to A. The *support* of a fuzzy set A is $supp(A) = \{x \in X | A(x) > 0\}$. A fuzzy set A is *normal* if there exists $x \in X$ such that $A(x) = 1$.

In the following we consider $X = \mathbf{R}$.

Let A be a fuzzy subset of $\mathbf{R}$ and $\gamma \in [0,1]$. The γ–level set of A is defined by

$$[A]^{\gamma} = \begin{cases} \{x \in \mathbf{R} | A(x) \geq \gamma\} & \text{if } \gamma > 0 \\ cl(supp(A)) & \text{if } \gamma = 0. \end{cases}$$

cl(supp(A)) is the topological closure of the set $supp(A) \subseteq \mathbf{R}$. A is called *fuzzyconvex* if $[A]^{\gamma}$ is a convex subset of $\mathbf{R}$ for any $\gamma \in [0,1]$. A *fuzzy number* is a fuzzy set of $\mathbf{R}$, normal, fuzzyconvex, continuous, and with bounded support. Let A be a fuzzy number and $\gamma \in [0,1]$. Then $[A]^{\gamma}$ is a closed and convex subset of $\mathbf{R}$. We denote $a_1(\gamma) = \min[A]^{\gamma}$ and $a_2(\gamma) = \max[A]^{\gamma}$. Hence $[A]^{\gamma} = [a_1(\gamma), a_2(\gamma)]$ for all $\gamma \in [0,1]$.

Let A and B be two fuzzy numbers and $\lambda \in \mathbf{R}$. We define the functions $A + B : \mathbf{R} \rightarrow [0,1]$ and $\lambda A : \mathbf{R} \rightarrow [0,1]$ by $(A + B)(z) = \sup\{A(x) \wedge B(y) | x + y = z\}$;

$(\lambda A)(z) = \sup\{A(x) | \lambda x = z\}$. Then $A + B$ and λA are fuzzy numbers. If $A_1, \ldots, A_n$ are fuzzy numbers and $\lambda_1, \ldots, \lambda_n \in \mathbf{R}$ then we consider the fuzzy number $\sum_{i=1}^{n} \lambda_i A_i$.

A non–negative and monotone increasing function $f : [0,1] \to \mathbf{R}$ is a *weighting function* if it satisfies the normality condition $\int_0^1 f(\gamma)d\gamma = 1$. We fix a fuzzy number A and a weighting function f and assume that $[A]^\gamma = [a_1(\gamma), a_2(\gamma)]$ for all $\gamma \in [0,1]$. The *f–weighted possibilistic expected value* of A is defined by $E(f,A) = \frac{1}{2} \int_0^1 (a_1(\gamma) + a_2(\gamma))f(\gamma)d\gamma$. The *f–weighted possibilistic variance* of A is defined by $Var(f,A) = \frac{1}{2} \int_0^1 [(a_1(\gamma) - E(f,A))^2 + (a_2(\gamma) - E(f,A))^2]f(\gamma)d\gamma$.

Assume now that A and B are two fuzzy numbers such that $[A]^\gamma = [a_1(\gamma), a_2(\gamma)]$ and $[B]^\gamma = [b_1(\gamma), b_2(\gamma)]$ for any $\gamma \in [0,1]$. The *f–weighted possibilistic covariance* of A and B is defined by $Cov(f,A,B) = \frac{1}{2} \int_0^1 [(a_1(\gamma) - E(f,A))(b_1(\gamma) - E(f,B)) + (a_2(\gamma) - E(f,A))(b_2(\gamma) - E(f,B))]f(\gamma)d\gamma$.

2 Probabilistic and Possibilistic Portfolio Models

A Probabilistic Model of Portfolio Selection Problem

In this section we present Markowitz probabilistic portfolio selection model [14]. We consider m assets $j = 1, \ldots, m$. Assume that the returns of the m assets are random variables $X_1, \ldots, X_m$. The following elements are known: probabilistic mean returns $\mu_j = M(X_j)$, $j = 1, \ldots, m$ and probabilistic covariances $\sigma_{st} = Cov(X_s, X_t)$, $s, t = 1, \ldots, m$.

A portfolio is a vector $(y_1, \ldots, y_m) \in \mathbf{R}^m$ with $\sum_{j=1}^{m} y_j = 1$ and $y_j \geq 0$ for any $j = 1, \ldots, m$. The return of the portfolio $(y_1, \ldots, y_m)$ is the random variable $\sum_{j=1}^{m} y_j X_j$.

One associates with each portfolio $(y_1, \ldots, y_m)$: mean return $M(\sum_{j=1}^{m} y_j X_j) = \sum_{j=1}^{m} y_j \mu_j$ and probabilistic variance $Var(\sum_{j=1}^{m} y_j X_j) = \sum_{s,t=1}^{m} y_s y_t \sigma_{st}$. The probabilistic portfolio selection problem is

$$\begin{cases} \min \frac{1}{2} \sum_{s,t=1}^{m} y_s y_t \sigma_{st} \\ \sum_{j=1}^{m} y_j \mu_j = \rho, \ \sum_{j=1}^{m} y_j = 1, y_j \geq 0, j = 1, \ldots, m. \end{cases}$$

A Possibilistic Model of Portfolio Selection Problem

In this section we present a possibilistic model of [10], [11], [13]. We consider n assets $i = 1, \ldots, n$. Assume that the returns of the n assets are represented by the fuzzy numbers $A_1, \ldots, A_n$.

The following elements are known: possibilistic mean returns $\gamma_i = E(f, A_i)$, $i = 1, \dots, n$ and possibilistic covariances $\delta_{kl} = Cov(f, A_k, A_l)$, $k, l = 1, \dots, n$.

In this case a portfolio is a vector $(x_1, \dots, x_n) \in \mathbf{R}^n$. The return of the portfolio $(x_1, \dots, x_n)$ is the fuzzy number $\sum_{i=1}^{n} x_i A_i$. One associates with each portfolio: possibilistic mean return $E(f, \sum_{i=1}^{n} x_i A_i) = \sum_{i=1}^{n} x_i \gamma_i$ and possibilistic variance $Var(f, \sum_{i=1}^{n} x_i A_i) = \sum_{k,l=1}^{n} x_k x_l \delta_{kl}$. The possibilistic portfolio selection problem is

$$\begin{cases} \min \dfrac{1}{2} \sum_{k,l=1}^{n} x_k x_l \delta_{kl} \\ \sum_{i=1}^{n} x_i \gamma_i = \delta, \sum_{i=1}^{n} x_i = 1, x_i \geq 0, i = 1, \dots, n. \end{cases}$$

3 Combining Probabilistic and Possibilistic Aspects

In this section we introduce the mixed portfolio and its indicators. We consider $n + m$ assets. We make the following assumptions:

- The returns of the first n assets are fuzzy numbers $A_1, \dots, A_n$;
- The returns of the other m assets are random variables $X_1, \dots, X_m$.

We know the following elements:

- possibilistic mean returns $\gamma_i = E(f, A_i)$, $i = 1, \dots, n$;
- possibilistic covariances $\delta_{kl} = Cov(f, A_k, A_l)$;
- probabilistic mean returns $\mu_j = M(X_j)$, $j = 1, \dots, m$;
- probabilistic covariances $\sigma_{st} = Cov(X_s, X_t)$, $s, t = 1, \dots, m$.

Particularly, possibilistic variances $\delta_i^2 = \delta_{ii} = Var(f, A_i)$, $i = 1, \dots, n$ and probabilistic variances $\sigma_j^2 = \sigma_{jj} = Var(X_j)$, $j = 1, \dots, m$ are known.

A *mixed portfolio* has the form $(x_1, \dots, x_n, y_1, \dots, y_m) \in \mathbf{R}^{m+n}$, where $\sum_{i=1}^{n} x_i + \sum_{j=1}^{m} y_j = 1$ and $x_i \geq 0$, $y_j \geq 0$ for any $i = 1, \dots, n, j = 1, \dots, m$. The real numbers $x_1, \dots, x_n$ represent the investment proportions of the first n assets and $y_1, \dots, y_m$ represent the investment proportions of the other m assets.

We consider a mixed portfolio $(x_1, \dots, x_n, y_1, \dots, y_m)$. Then we define:

- portfolio's possibilistic return $A_P = \sum_{i=1}^{n} x_i A_i$;
- portfolio's probabilistic return $X_P = \sum_{j=1}^{m} y_j X_j$.

A_P is a fuzzy number and X_P is a random variable.

Now we define the following indicators of portfolio $(x_1, \ldots, x_n, y_1, \ldots, y_m)$:

- possibilistic mean return $E(f, A_P) = \sum\limits_{i=1}^{n} x_i E(f, A_i)$;

- probabilistic mean return $M(X_P) = \sum\limits_{j=1}^{m} y_j M(X_j)$;

- portfolio's mean return $R_P = E(f, A_P) + M(X_P) = \sum\limits_{i=1}^{n} x_i E(f, A_i) + \sum\limits_{j=1}^{m} y_j M(X_j)$.

We further define:

- portfolio's possibilistic variance $\delta_P = Var(f, A_P)$;
- portfolio's probabilistic variance $\sigma_P = Var(X_P)$;
- portfolio's (total) variance $Var_P = \delta_P + \sigma_P$.

We have $\delta_P = \sum\limits_{k,l=1}^{n} x_k x_l \delta_{kl}$; $\sigma_P = \sum\limits_{s,t=1}^{m} y_s y_t \sigma_{st}$; $Var_P = \sum\limits_{k,l=1}^{n} x_k x_l \delta_{kl} + \sum\limits_{s,t=1}^{m} y_s y_t \sigma_{st}$.

Var_P indicates the risk of the mixed portfolio. It comprises the possibilistic risk component δ_P and the probabilistic risk component σ_P of the portfolio.

4   Mixed Portfolio Selection Problem

In this section we will establish the mixed portfolio selection problem form, we will compute its optimal solution and the value of the associated minimum risk. We keep the notations from the previous section. The mixed portfolio problem is

$$
\begin{cases}
\min \frac{1}{2} Var_P \\
E(f, A_P) = \rho_1, M(X_P) = \rho_2, \sum\limits_{i=1}^{n} x_i + \sum\limits_{j=1}^{m} y_j = 1, x_i \geq 0, y_j \geq 0, i = 1, \ldots, n, j = 1, \ldots, m.
\end{cases}
\tag{1}
$$

To solve the mixed portfolio selection problem means to find a portfolio $(x_1, \ldots, x_n, y_1, \ldots, y_m)$ of minimum risk which ensures a possibilistic mean return ρ_1 and a probabilistic mean return ρ_2.

Taking into account the expressions of Var_P, $E(f, A_P)$ and $M(X_P)$ from the previous section, the mixed portfolio problem has the form:

$$
\begin{cases}
\min \frac{1}{2} \left[\sum\limits_{k,l=1}^{n} x_k x_l \delta_{kl} + \sum\limits_{s,t=1}^{m} y_s y_t \sigma_{st} \right] \\
\sum\limits_{i=1}^{n} x_i \gamma_i = \rho_1, \sum\limits_{j=1}^{m} y_j \mu_j = \rho_2, \sum\limits_{i=1}^{n} x_i + \sum\limits_{j=1}^{m} y_j = 1 \\
x_i \geq 0, y_j \geq 0, i = 1, \ldots, n, j = 1, \ldots, m.
\end{cases}
\tag{2}
$$

We denote $x = (x_1, \ldots, x_n)^T$, $y = (y_1, \ldots, y_m)^T$, $\gamma = (\gamma_1, \ldots, \gamma_n)^T$, $\mu = (\mu_1, \ldots, \mu_m)^T$, $e_n = (1, \ldots, 1)^T$, $e_m = (1, \ldots, 1)^T$, and $\Omega_1 = (\delta_{kl})_{k,l=1,\ldots,n}$; $\Omega_2 = (\sigma_{st})_{s,t=1,\ldots,m}$

With these notations problem (2) is written in matricial form:

$$\begin{cases} \min \frac{1}{2}[x^T \Omega_1 x + y^T \Omega_2 y] \\ x^T \gamma = \rho_1, y^T \mu = \rho_2, x^T e_n + y^T e_m = 1, x \geq 0, y \geq 0. \end{cases} \tag{3}$$

If $n = 0$, then Markowitz model is obtained from (3); if $m = 0$, then (3) is exactly the possibilistic portfolio selection model presented in Section 2 (B). It follows that that the mixed model (3) extends both Markowitz probabilistic model and the possibilistic model mentioned. Next we aim to find the optimal solution of (3).

The Lagrangian of the optimization problem (3) has the form: $L = \frac{1}{2} \sum_{k,l=1}^{n} x_k x_l \delta_{kl} +$

$\frac{1}{2} \sum_{s,t=1}^{m} y_s y_t \sigma_{st} - \lambda_1 (\sum_{k=1}^{n} x_k \gamma_k - \rho_1) - \lambda_2 (\sum_{s=1}^{m} y_s \mu_s - \rho_2) - \lambda_3 (\sum_{k=1}^{n} x_k + \sum_{s=1}^{m} y_s - 1).$

The first–order conditions are: $\frac{\partial L}{\partial x_1} = 0, \ldots, \frac{\partial L}{\partial x_n} = 0; \frac{\partial L}{\partial y_1} = 0, \ldots, \frac{\partial L}{\partial y_m} = 0; \frac{\partial L}{\partial \lambda_1} = 0, \frac{\partial L}{\partial \lambda_2} = 0, \frac{\partial L}{\partial \lambda_3} = 0.$

From the first $n + m$ conditions from above the following equations are obtained:

$$\sum_{l=1}^{n} \delta_{kl} x_l - \lambda_1 \gamma_k - \lambda_3 = 0, k = 1, \ldots, n; \sum_{t=1}^{m} \sigma_{st} y_t - \lambda_2 \mu_s - \lambda_3 = 0, s = 1, \ldots, m. \tag{4}$$

Conditions in (4) are written matricially as

$$\Omega_1 x - \lambda_1 \gamma - \lambda_3 e_n = 0; \Omega_2 y - \lambda_2 \mu - \lambda_3 e_m = 0. \tag{5}$$

Assume that the matrices Ω_1 and Ω_2 are invertible. Then we obtain from (5):

$$x = \lambda_1 \Omega_1^{-1} \gamma + \lambda_3 \Omega_1^{-1} e_n; y = \lambda_2 \Omega_2^{-1} \mu + \lambda_3 \Omega_2^{-1} e_m. \tag{6}$$

Assume that $x \in \mathbf{R}_+^n$ and $y \in \mathbf{R}_+^m$. From (6) one obtains:

$$x^T = \lambda_1 \gamma^T \Omega_1^{-1} + \lambda_3 e_n^T \Omega_1^{-1}; y^T = \lambda_2 \mu^T \Omega_2^{-1} + \lambda_3 e_m^T \Omega_2^{-1}. \tag{7}$$

By (3) we have $x^T \gamma = \rho_1, y^T \mu = \rho_2$ and $x^T e_n + y^T e_m = 1$. In these identities replacing x^T, y^T with their values from (7) one obtains:

$$\begin{cases} \lambda_1 \gamma^T \Omega_1^{-1} \gamma + \lambda_3 e_n^T \Omega_1^{-1} \gamma = \rho_1, \lambda_2 \mu^T \Omega_2^{-1} \mu + \lambda_3 e_m^T \Omega_2^{-1} \mu = \rho_2 \\ \lambda_1 \gamma^T \Omega_1^{-1} e_n + \lambda_2 \mu^T \Omega_2^{-1} e_m + \lambda_3 [e_n^T \Omega_1^{-1} e_n + e_m^T \Omega_2^{-1} e_m] = 1. \end{cases} \tag{8}$$

We denote: $A = \gamma^T \Omega_1^{-1} \gamma, B = \mu^T \Omega_2^{-1} \mu, C = e_n^T \Omega_1^{-1} \gamma, D = e_m^T \Omega_2^{-1} \mu, E = e_n^T \Omega_1^{-1} e_n + e_m^T \Omega_2^{-1} e_m.$ Then replacing in (8) we obtain the system of equations:

$$A\lambda_1 + C\lambda_3 = \rho_1, B\lambda_2 + D\lambda_3 = \rho_2, C\lambda_1 + D\lambda_2 + E\lambda_3 = 1. \tag{9}$$

We denote $F = ABE - BC^2 - AD^2$. If $F \neq 0$, then the system (3) has the solution:

$$\lambda_1 = \frac{(BE - D^2)\rho_1 + DC\rho_2 - BC}{F}, \lambda_2 = \frac{(AE - C^2)\rho_2 + DC\rho_1 - AB}{F},$$
$$\lambda_3 = \frac{-BC\rho_1 - AD\rho_2 + AB}{F}. \tag{10}$$

Replacing in (6) λ_1, λ_2 and λ_3 from (10) leads to the optimal solution of (1):

$$x = \frac{(BE - D^2)\rho_1 + DC\rho_2 - BC}{F}\Omega_1^{-1}\gamma + \frac{-BC\rho_1 - AD\rho_2 + AB}{F}\Omega_1^{-1}e_n; \tag{11}$$

$$y = \frac{(AE - C^2)\rho_2 + DC\rho_1 - AB}{F}\Omega_2^{-1}\mu + \frac{-BC\rho_1 - AD\rho_2 + AB}{F}\Omega_2^{-1}e_m. \tag{12}$$

Next we compute Var_P of portfolio $(x_1, \ldots, x_n, y_1, \ldots, y_m)$ given by (11) and (12). One notices that $\delta_P = x^T\Omega_1 x$ and $\sigma_P = y^T\Omega_2 y$. By (11) and (12) we have $x^T\Omega_1 x = \lambda_1 x^T\gamma + \lambda_3 x^T e_m$ and $y^T\Omega_2 y = \lambda_2 y^T\mu + \lambda_3 y^T e_m$, and then $Var_P = \delta_P + \sigma_P = x^T\Omega_1 x + y^T\Omega_2 y = \lambda_1 x^T\gamma + \lambda_2 y^T\mu + \lambda_3(x^T e_n + y^T e_m)$

By (3) we have $x^T e_n + y^T e_m = 1$, $x^T\gamma = \rho_1$, $y^T\mu = \rho_2$, and therefore:

$$Var_P = \lambda_1\rho_1 + \lambda_2\rho_2 + \lambda_3. \tag{13}$$

Replacing in (13) λ_1, λ_2 and λ_3 with the values from (10) it follows that

$$Var_P = \frac{1}{F}((BE - D^2)\rho_1^2 + (AE - C^2)\rho_2^2, \tag{14}$$

which can be written as $Var_P = \frac{1}{F}((BE - D^2)\rho_1^2 + (AE - C^2)\rho_2^2 + 2CD\rho_1\rho_2 - 2BC\rho_1 - 2AD\rho_2 + AB$, where Var_P is the value of the minimum risk of a portfolio which assures a possibilistic mean return ρ_1 and a probabilistic mean return ρ_2.

5 Conclusion

The mixed portfolio selection problem treated in this paper corresponds to a complex situation in financial risk management in which probabilistic aspects of risk are combined with the possibilistic aspects of risk. What we propose in this paper unifies two models known in financial management literature: Markowitz probabilistic model and a possibilistic portfolio selection model.

There might exist situations for which both Markowitz model and the posibilistic model, separately taken, cannot offer complete information. Our mixed model can give us a more complete representation of the uncertainty situation in these cases.

A formula for optimal mixed portfolio calculation and for the minimal risk associated with it is proved. The formulations and the proofs in the mixed model combine probabilistic and possibilistic indicators. The formulas for the optimal solution calculation and the minimal risk do not lead to a complicated algorithm, which makes our model easily programmable.

As for the future study we work on a portfolio selection model with three components: (i) probabilistically modeled security returns, (ii) possibilistically modeled returns, and (iii) credibilistically modeled (by fuzzy numbers defined on credibility field). The analysis of the models degree of adequacy w.r.t. real situations and real data, and the comparison with other portfolio selection models will be topics for further investigations.

Acknowledgements. I. Georgescu was upported by CNCSIS-UEFISCSU, PN II-RU 651/2010.

References

1. Altăr, M.: Teoria portofoliului. Academy of Economic Studies, Bucharest (2002)
2. Carlsson, C., Fullér, R.: On possibilistic mean value and variance of fuzzy numbers. Fuzzy Sets Syst. 122, 315–326 (2001)
3. Carlsson, C., Fullér, R.: Fuzzy reasoning in decision making and optimization. Studies in Fuzziness and Soft Computing Series, vol. 82. Springer, Berlin (2002)
4. Carlsson, C., Fullér, R., Majlender, P.: A possibilistic approach to selecting portfolios with highest utility score. Fuzzy Sets Syst. 131, 13–21 (2002)
5. Carlsson, C., Fullér, R., Majlender, P.: On possibilistic correlations. Fuzzy Sets Syst. 155, 425–445 (2005)
6. Dubois, D., Prade, H.: Fuzzy sets and systems: theory and applications. Academic Press, New York (1980)
7. Dubois, D., Prade, H.: Possibility theory. Plenum Press, New York (1988)
8. Fullér, R., Majlender, P.: On weighted possibilistic mean and variance of fuzzy numbers. Fuzzy Sets Syst. 136, 363–374 (2003)
9. Georgescu, I.: Possibility theory and the risk. Springer, Heidelberg (forthcoming, 2012)
10. Huang, X.: Portfolio analysis. Springer, Heidelberg (2010)
11. Huang, X.: Portfolio selection with fuzzy returns. J. Intelligent and Fuzzy Systems 18, 384–390 (2007)
12. Huang, X.: Minimax mean–variance models for fuzzy portfolio selection. Soft Computing 15, 251–260 (2011)
13. Inuiguchi, M., Ramik, J.: Possibilistic linear programming: a brief review of fuzzy mathematical programming and a comparison with stochastic programming in portfolio selection problem. Fuzzy Sets Syst. 111, 3–28 (2000)
14. Markowitz, H.: Portfolio selection. J. Finance 7, 77–91 (1952)
15. Tanaka, H., Guo, P., Türksen, I.B.: Portfolio selection based on fuzzy probabilities and possibility distributions. Fuzzy Sets Syst. 111, 387–397 (2000)
16. Zadeh, L.A.: Fuzzy sets as a basis for a theory of possibility. Fuzzy Sets Syst. 1, 3–28 (1978)
17. Wang, S., Zhu, S.: On fuzzy portfolio selection problems. Fuzzy Optimization and Decision Making 1, 361–377 (2002)

Towards a Service Based on "Train-to-Earth" Wireless Communications for Remotely Managing the Configuration of Applications Inside Trains

Itziar Salaberria, Roberto Carballedo, and Asier Perallos

Abstract. This paper describes a distributed service based on "train-to-earth" wireless communication, which is able to allow maintenance engineers to remotely monitor and update the configuration of the applications running inside a fleet of trains. It is a very useful tool that avoids engineers to move from their offices to trains for doing the maintenance tasks. This service is the result of five years of collaboration with a railway company in the north of Spain. Moreover, it represents the following step after the implantation of an innovative "train-to-earth" communication architecture in this company. Today, both the communications architecture and this service are being tested within the installations of the railway company with successful results.

Keywords: software configuration management, remote engineering, train-to-earth wireless communications, distributed application, railway.

1 Introduction

Advances in communication technologies and their implantation costs reduction are enabling the use of wireless technologies in the railway industry. This use is progressively growing motivated by the fact that more and more new generation services are emerging which need to exchange information between trains and ground centres [1].

Traditionally, railway services have been oriented to exploitation. However, the progressive establishment of wireless "train-to-earth" communications is becoming

Itziar Salaberria · Roberto Carballedo · Asier Perallos
Deusto Institute of Technology (DeustoTech), University of Deusto,
Avenida de las Universidades 24, 48007 - Bilbao, Spain
e-mail: {itziar.salaberria,roberto.carballedo,perallos}@deusto.es

S. Omatu et al. (Eds.): Distributed Computing and Artificial Intelligence, AISC 151, pp. 103–111.
springerlink.com   © Springer-Verlag Berlin Heidelberg 2012

a key factor for developing various digital services that will ease daily work in the rail sector, and will increase the quality of the service provided to the users [2].

One of the main targets of applications running inside the train is to provide information to facilitate the work of the driver. Usually, they need to use data generated in the ground centre. If this data changes, it needs to be updated inside the train. In addition, there are terrestrial applications that need to use data generated by some on board applications. Thus, it is necessary to design distributed solutions and tools for exchanging information between trains and the control centres in a synchronized way.

This paper presents a *Remote Applications Management Service (RAMS)* that is part of the results of the work made during the last four years alongside a regional railway company of Spain. It defines a particular service that allows the remote management of on board applications (upgrade, download and deletion of its information), and it is integrated in a general purpose wireless connectivity architecture that has been established by this company for enabling "train-to-earth" communications [3]. The main objective of RAMS is to improve the maintainability of systems and applications with are running inside the train.

The paper is organized into the following sections. The second section includes a brief description of the key functional requirements of the service developed. The third section details its architecture and design issues. The fourth section presents the results of the tests made in order to validate its performance. To close, the fifth section of the paper establishes the main conclusions of this work.

2 Functional Requirements

Traditionally, the maintenance of the train on board software systems has been completed through wired communications. Engineers had to go from train to train and connect to terminals to perform the unloading and loading of data of the applications running inside the train.

Regarding to services oriented to the driver or train staff, one of the most relevant one to develop would be a utility for downloading and/or updating documents with relevant information for them. This information would be coming from an application from the outside. This system would be very helpful because the driver could obtain updated information on any item he would like (from route itineraries to security protocols) fast, thus facilitating his work. This utility would free the driver from the burden of carrying information on paper, and even allows the on-line notification of such documentation updates.

Therefore, the main functionality of the service presented in this paper is to control the update of the information used by applications running on the train terminal (for example track flat information or supporting documentation for the driver generated by the ground information systems) as well as downloading and deleting information generated by some on board applications (for example log files) remotely from the ground centre. Thus, the main novelty of this system is the capability of schedule train systems upgrade tasks wirelessly that will be performed in an unattended way once the trains are in a coverage area.

The solution consists of two software systems, one for the ground centre (Terrestrial RAMS) and the other one to be deployed in all train terminals (On Board RAMS). Thus, the terrestrial RAMS system is installed on ground centre, and it will be responsible for managing the status of all applications in each terminal. On the other hand, on board RAMS system will handle update, download and deletion requests made by the terrestrial RAMS system. So, **terrestrial and on board RAMS functionality** involves these issues:

- Knowledge about the configuration information (files which conform it) for each application installed on each train terminal at any time (keeping a history of changes in each application's files over time). As well as about the files and/or documents currently used by these applications which can be updated, downloaded and/or deleted. This information will include: version, creation and last update date, update status (pending or not), etc. This management is done through a repository of information in a database.
- Management of information (files) of all current applications running in the train.
- Query management about the status information of the on board applications.
- Integration with the "train-to-earth" wireless communication architecture via heavy communications scheme (see section 3).

3 RAMS Architecture

To describe RAMS architecture, it is important to point out some concepts related to the "train-to-earth" communication architecture [3] in which is integrated. Thus, the adopted wireless connectivity architecture distinguishes two communication types:

(1) **"Light" communication:** this type of communication is for the transmission of small volumes of information (few Kbytes) and with high priority. In general, information that has low latency (milliseconds or a pair of seconds) and needs to be transmitted exactly when it is generated or acquired (for instance, the GNSS location of a train, or a driving order to the train diver).

(2) **"Heavy" communication:** this type of communication is tied to the transmission of large volumes of information (in the order of Mbytes) and with low priority. The importance of this information is not affected by the passage of time, so it doesn't need to be transmitted at the exact time it is generated.

Therefore RAMS application is integrated with this connectivity architecture via *heavy communication* scheme since it involves the exchange of large volumes of information that do not require real-time communications. Heavy communications purpose is to provide a way of enabling broadband communications which is suitable for the railway, due to the fact that a number of WiFi networks have been settled in places where the trains are stopped long enough to ensure the discharge of a certain amount of information (stations in the header that starts or ends a tour, workshops and garages). In this way, we can say that the WiFi coverage is not complete, but it is important to say that broadband communications are designed to update large volumes of information, which, theoretically do not need to take place in real time.

To manage heavy communications, this architecture defines the *Broadband Communications Manager (BCM)* [4]. The BCM is a system that arbitrates and distributes shifts to communicate terrestrial and train applications; in this way, the terrestrial applications request a turn when they want to establish a heavy communication with a train. This distribution shift is managed on the basis of the state of the train connection to a WiFi network (known at all times) and a system of priorities, which are allocated according to the terrestrial application that wants to communicate with the train.

Hence, when the terrestrial RAMS generates tasks which involve downloading or uploading data from and to trains, it has to communicate with BCM. In this case, BCM arbitrates communications between terrestrial and train RAMS subsystems. For proper integration with BCM, RAMS (more specifically the terrestrial one) shall be compliant with the protocol of communication established by this management entity.

At this point it is important to point out that BCM does not interfere between final applications communication, it only participates arbitrating the establishment of the communication between them. The Fig. 1 shows RAMS architecture and its integration with the "train-to-earth" wireless communication technology.

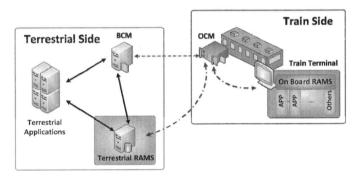

Fig. 1 RAMS and how it interoperates with the "train-to-earth" communications infrastructure: Broadband Communication Manager (BCM) in terrestrial side and On board Communication Manager (OCM) in train side.

As Fig. 1 shows, RAMS is divided into two distinct subsystems (Terrestrial RAMS and On Board RAMS in each train), which are divided into different modules that provide the previously described functionality (Fig. 2).

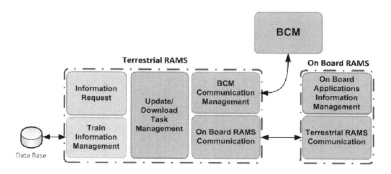

Fig. 2 RAMS architecture and its functional modules

With respect to the final communication between terrestrial and on board RAMS, there have been defined different scenarios depending on the type of tasks to be served. All of them have in common that the first message sent is a compressed and encrypted file that contains a XML document describing the task at hand (in addition to the XML, some tasks also attach all the files concerning with the task).

3.1 Terrestrial RAMS

Thus, the Terrestrial RAMS is divided into five functional modules which are described below:

– **Information Request.** This module allows the user to request different information managed by the RAMS (boarded application files' information, scheduled tasks' information, etc.).
– **Train Information Management.** Terrestrial RAMS is designed to manage the information of the applications running inside the trains. So, Terrestrial RAMS contains a similar information organization to the information loaded on each train, as a mirror and stored in a database repository, so that with each new modification made in one of the trains, such changes are reflected both on terrestrial side and in remote (train side).
– **Update/download Task Management.** This module aims to enable update and download tasks related to applications running inside the trains. These tasks could be launched in a moment or be programmed as periodical tasks.
– **BCM Communication Management.** This module is responsible for managing the interaction with BCM and implementing its protocol.
– **On Board RAMS Communication.** The wireless communication between terrestrial and on board RAMS is done following the heavy communications scheme which is included in the wireless connectivity architecture that we mentioned before. This communication will be carried out though WiFi networks placed in stations, workshops or carports. Anyway, there is also the

possibility of communication with the trains terminals on a wired mode (with a laptop for example) if there is an anomalous event that avoids communication via WiFi. This module is responsible for implementing these point to point communications based on RAMS specific protocol.

3.2 On Board RAMS

The on board RAMS is divided into two functional modules. Their functionality is described below:

- **On Board Application Information Management.** The core of this module is an embedded FTP server that will manage all existing files in the train terminal. We opted for the use of an FTP server that will be used locally (not able to access it remotely), because it is a widespread technology in file management [5].
- **Terrestrial RAMS Communication.** This module responds the terrestrial information update and downloads requests (because communication between terrestrial and on board RAMS will be always initiated by the terrestrial side). It implements the RAMS specific communications protocol, enabling point to point communication with the mirror module in terrestrial side.

4   Functional Test Scenarios

The RAMS is currently being tested within the infrastructure of EuskoTren, a railway company in the north of Spain. These tests are carried out within the installations of this company, using a new generation train (its manufacturing was finished this year) equipped with the technology needed to hold train-to-earth communications, being the BCM a part of it.

BCM has been installed on a dedicated server for its work, and is located in a local network designed to communicate it with different deployed "terrestrial" applications. Today we can find two of these services, also in test phase, relying on the BCM in the local network installed. One of them is the described RAMS. In the case of the train, it has been installed a WiFi network in the station, so the OCM can connect to it.

In this way, the terrestrial RAMS has been deployed in a PC that is in the same local network in which has been installed the BCM, whereas the on board RAMS has been installed on the terminals of the train that has been selected to realize the tests.

When running the tests, we define two basic scenarios: (a) update task and (b) download task. Their design is shown in Fig. 3.

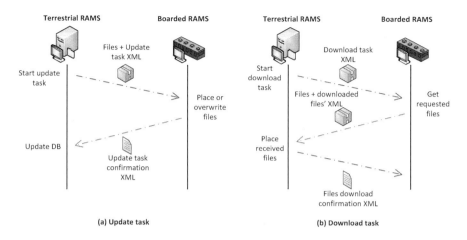

Fig. 3 Update (a) and Download (b) tasks' flowchart

4.1   Update Task

Terrestrial RAMS allows the user to configure tasks related to update information running inside the train terminals. These tasks can involve the updating of a certain application files or documents that are accessed by the train driver.

As shown in the Fig. 3, when user configures an update task indicating files to be updated (or created), the terrestrial RAMS generates a compressed and encrypted file that contains a XML document describing the update task and the files selected to be updated on the train. Then terrestrial RAMS sends this compressed file to the on board RAMS which responds with a confirmation message indicating if the task has been completed successfully.

Thus, to test the RAMS, there have been tested several scenarios: update and deletion of selected files which belong to an on board application, and remote loading of entire applications inside the train. All of them have had successful results.

4.2   Download Task

Terrestrial RAMS also allows the user to configure download tasks concerning with information of the applications running inside the train terminals. These tasks can involve file downloads (logs or other applications files) or on board application file system status requests. These status requests allow the terrestrial RAMS to identify which is the version of applications files that are in train terminals, allowing the update of them if it is considered as necessary.

As shown in the Fig. 3, when user configures a download task (file list download or status request), the terrestrial RAMS generates a XML document describing the task. This is compressed and encrypted in a file and sends it to the on board RAMS. When on board RAMS receives download task information, generates a compressed an encrypted file that contains requested files and a XML which describes the information of these attached files, and sends it as a response to terrestrial RAMS. Then, terrestrial RAMS checks the received information, and responds with a confirmation message to the on board one indicating if the task has been completed successfully.

Therefore, in order to test RAMS download functionality, there have been tested several download tasks related to different scenarios: download of log files generated by on board applications and status requests of their configuration. All the tests have been performed successfully.

Fig. 4 Utility for analysing differences between applications configurations (train / ground centre). It is launched as result of a status request made over a specific train.

The Fig. 4 shows the result of a train terminal status request. We can see two file and folder trees. The tree on the left shows the updated terminal status that is saved on the database of terrestrial RAMS and the tree on the right shows the real train terminal status. The database status is the updated one that should be in terminals on trains. So, when user configures a train terminal status request, terrestrial RAMS compare the real status of applications on terminals with the status stored on the terrestrial RAMS database. In addition, terrestrial RAMS highlights the differences between these two status trees (missing files, version differences, etc.), giving the user the option to update or delete files if it is required.

5 Conclusion and Future Work

Wireless communications are increasingly being adopted in several vertical domains. This is the case of railway industry, where these communications represent a key means to establish connectivity between trains and ground control centres. Moreover, this "train-to-earth" communications has become a requirement for developing a high number of digital railway services which could improve the way in which railways operators work.

This paper describes one of these services: a Remote Application Management Service (RAMS) based on "train-to-earth" wireless communication, which is able to allow maintenance engineers to remotely monitor and update the configuration of the applications running inside a fleet of trains. It means, to load the set of files which conform an entire application (deployment of a new application over a train), to check the status (versions) of their files and to update/delete specific files of an on board application, storing the history of all the changes made. It is a very useful tool that avoids engineers to move from their offices to trains for doing the maintenance tasks.

In this moment, our efforts are focused on adopting the web paradigm in order to become RAMS as distributed service which is accessible via a web browser. It is a very interesting challenge because it enables that the tool is hosted and administered by some technological provider company, being able to be used remotely and concurrently by several railway operators.

References

1. Aguado, M.: Railway signalling systems and new trends in wireless data communication. In: IEEE 62nd Vehicular Technology Conference, pp. 1333–1336 (2005)
2. Shafiullah, G., Gyasi-Agyei, A., Wolfs, P.J.: Survey of Wireless Communications Applications in the Railway Industry. In: 2nd International Conference on Wireless Broadband and Ultra-Wideband Communications (AusWireless), pp. 27–30 (2007)
3. Salaberria, I., Carballedo, R., Gutierrez, U., Perallos, A.: Wireless Communications Architecture for "Train-to-Earth" Communication in the Railway Industry. In: 10th International Work-Conference on Artificial Neural Networks, pp. 625–633 (2009)
4. Gutierrez, U., Salaberria, I., Perallos, A., Carballedo, R.: Towards a Broadband Communications Manager to regulate train-to-earth communications. In: 15th IEEE Mediterranean Electrotechnical Conference, pp. 1600–1605 (2010)
5. Apache MINA FtpServer, http://mina.apache.org/ftpserver

CDEBMTE: Creation of Diverse Ensemble Based on Manipulation of Training Examples

Hamid Parvin, Sajad Parvin, Zahra Rezaei, and Moslem Mohamadi

Abstract. Ensemble methods like Bagging and Boosting which combine the decisions of multiple hypotheses are among the strongest existing machine learning methods. The diversity of the members of an ensemble is known to be an important factor in determining its generalization error. We present a new method for generating ensembles, named CDEBMTE (Creation of Diverse Ensemble Based on Manipulation of Training Examples), that directly constructs diverse hypotheses using manipulation of training examples in three ways: (1) sub-sampling training examples, (2) decreasing/increasing error-prone training examples and (3) decreasing/increasing neighbor samples of error-prone training examples. Experimental results using two well-known classifiers as two base learners demonstrate that this approach consistently achieves higher predictive accuracy than both the base classifier, Adaboost and Bagging. CDEBMTE also outperforms Adaboost more prominent when training data size is becomes larger.

Keywords: Classifier Ensemble, Diversity, Training Examples Manipulation.

1 Introduction

There are many inherently different classifiers in the pattern recognition. It may be worthy to mention that usage of the broad versatility in kind of classifiers is not a trivial matter to be ignored. It means although a classification algorithm may obtain a good performance for a specific problem, it has not enough robustness for another problem. It has been always an ideal for pattern recognition communities to present an approach of how to select the best (or at least approximately the best) classifier for a specific problem. It has been shown that this desire is far from the reach; at least by employing simple classifiers as the main learner.

Fortunately there is another way to cope with the problem. It is to use many learners instead of one learner and then to employ a democracy-wise method to

Hamid Parvin · Sajad Parvin · Zahra Rezaei · Moslem Mohamadi
Nourabad Mamasani Branch, Islamic Azad University Nourabad Mamasani, Iran
e-mail: hamidparvin@mamasaniiau.ac.ir,
　　{s.parvin,rezaei,mohamadi}@iust.ac.ir

S. Omatu et al. (Eds.): Distributed Computing and Artificial Intelligence, AISC 151, pp. 113–120.
springerlink.com　　　　　　　　　　　　　　　© Springer-Verlag Berlin Heidelberg 2012

firm the final decision of the committee (classifier). One of the major advances in inductive learning in the past decade was the development of ensemble or committee approaches that learn and retain multiple hypotheses and combine their decisions during classification [2]. For example, Boosting [3]-[4] is an ensemble method that learns a series of "weak" classifiers each one focusing on correcting the errors made by the previous one; and it is currently one of the best generic inductive classification methods [6]. Therefore, recent researches are directed to the combinational methods which have more power, robustness, resistance, accuracy and generality than simple method. Combination of Multiple Classifiers (CMC) can be considered as a general solution method for pattern recognition problems. Inputs of CMC are results of separate classifiers and output of CMC is their consensus decision [7].

Melville [10] presents a new meta-learner DECORATE (Diverse Ensemble Creation by Oppositional Relabeling of Artificial Training Examples), that uses an existing "strong" learner (one that provides high accuracy on the training data) to build an effective diverse committee in a simple, straightforward manner. This is accomplished by adding different randomly constructed examples to the training set when building new committee members. These artificially constructed examples are given category labels that disagree with the current decision of the committee, thereby easily and directly increasing diversity when a new classifier is trained on the augmented data and added to the committee.

There have been many methods developed to construct an ensemble. Some of these methods, such as Bagging and Boosting are *meta-learners* i.e. they can be applied to *any* base classifier. Other methods are specific to a particular base classifier. For example, Negative Correlation Learning [8] is used specifically to build committees of Neural Networks. We only deal with construction of general ensemble, i.e. *meta-learners*. Although a special base classifier may be better suited for a particular domain than others, a *general* ensemble approach that is independent of the particular base classifier is preferred in an unknown domain. So the aim of our research is to only propose a *meta-learner*.

In an ensemble, the output of several classifiers is useful to be participated in an ensemble only when they disagree on some inputs [5] and [14]. The measure of disagreement is called as the *diversity* of the ensemble. For regression problems, *mean squared error* is generally used to measure accuracy, and *variance* is used to measure diversity.

Melville and Mooney [9] have introduced a new meta-learner DECORATE that uses an existing learner to build an effective diverse committee in a simple, straightforward manner. This is accomplished by adding different randomly constructed examples to the training set when building new committee members. These artificially constructed examples are given category labels that *disagree* with the current decision of the committee, thereby easily and directly increasing diversity when a new classifier is trained on the augmented data and added to the committee.

In this paper, a new method to obtain diverse classifiers is proposed which uses manipulation of dataset structures. The proposed method is very similar to Adaptive Learning Classification [12] in which the learner does not treat all pattern equally. In this new method it is presented to make diversity in base classifiers (of type MultiLayer Perceptron (MLP) or Decision Tree (DT)). The effect of existing or non-existing of boundary instances is evaluated. Firstly, using a simple classification, the error-prone instances and their neighbors are detected. Using these subsets, different datasets of our main dataset are created. Then, several classifiers are trained on these datasets. They may be selected into or eliminated from the ensemble based on a diversity metric. This method that we have named "CDEBMTE: Creation of Diverse Ensemble Based on Manipulation of Training Examples" is described in section 3, accurately. The used decision-tree is J48 that is a Java implementation of C4.5 [13] introduced in [15]. Cross-validated learning curves support the hypothesis that CDEBMTE generally result in greater classification accuracy.

2   CDEBMTE: Creation of Diverse Ensemble Based on Manipulation of Training Examples

In CDEBMTE, using a simple classification, the error-prone samples and their neighbors which are called respectively EPS and NS are first detected. Using EPS and NS and also the original training set, different training sets are created from the original one. Then, several classifiers are trained on these datasets. Based on a diversity metric a subset of them is selected into an ensemble. Finally, a consensus function is employed to aggregate the results of these diverse classifiers.

2.1   Preparing Different Subsets of Training Dataset

To extract different subsets out of training set, we divide the primary Training Set (TS) in K random partition. Let us denote i-th partition by TS_i. Presume that cardinality of TS, $|TS|$, is denoted by N. By training a fixed number (denoted by H) of simple base classifiers on $TS \backslash TS_i$, we reach an ensemble that is expert for TS_i. Please note that these H classifiers are produced by a Bagging mechanism named Random Forest (RF). Breiman [1] proposes a variant of bagging which he calls RF. RF is a general class of ensemble building methods that uses decision tree as its base classifier. It has a parameter α that shows the ratio of sub-sampling. To be labeled a "random forest" an ensemble of decision trees should be built by generating independent identically distributed random vectors and use each vector to grow a decision tree. In this paper RF which is one of the well known versions of bagging classifier [7] is considered as Bagging method. It means when Bagging is used it refers to RF in the paper.

Let us denote the ensemble that is expert for TS_i by E_i. We test each data point in TS_i to find out how the data point has potential to be wrongly classified. We assign a number P_j to j-th data point in TS_i indicating how many classifiers out of H ones in the ensemble E_i, wrongly classify the j-th data point. The bigger P_i, the more error-prone the i-th data point. It is also worthy to mention that P_i is an integer number between 0 to H.

This procedure is repeated K times, choosing a different part for testing, each time. When $N=K$, the method is called the leave-one-out or U-method [7].

2.2 Creating an Ensemble of Diverse Classifiers

Using more diverse classifiers is one of the most important factors controlling the performance of combinational classifiers. Based on three sets of data, TS, EPS and NS, in this step, a number of classifiers are heuristically retrained. As shown in Table 1, a number of combinations are used to create diversity in the ensemble. Also, Table 1 expresses the reasons of their usage. It is worthy to be mentioned that κ, λ and μ are three parameters that are to be set. The domains of these parameters are {0.2, 0.4, 0.6, 0.8, 1}. For example, $\kappa*TS$ means that a sub-sampling from TS that contains a κ percept of TS; and $\kappa*TS \cup \lambda*EPS \cup \mu*NS$ means that a union of three sub-samplings: (1) κ percept from TS, (2) λ percept from EPS and (3) μ percept from NS.

Retraining of classifiers, according to combinations of Table 1, results the classifiers so that each of them concentrates on special aspect of data. This can result in very good diversity in the ensemble. In other hand, although the accuracy of each classifier may not be significantly better than a simple classifier, it can yield to satisfactory diverse base classifiers.

In this study, six classes of producing a subset are defined. In first class of producing subsets, by sliding parameter κ in all possible values in its domain, we obtain 5 training subsets. Equivalently in second class, by sliding parameters κ and μ in all possible values in their domains, we obtain 5*5 training subsets. In 3rd, 4th, 5th and 6th classes, we respectively reach 125 (5*5*5), 25, 25 and 5 training subsets by sliding their parameters.

Table 1 Different data combinations and reasons of its usages

Classes of Producing Training Subset	Feature of Resultant Subsets
1. $\kappa* TS$	To create the base classifiers
2. $\kappa*TS \cup \mu*NS$	Classification by high complex boundaries with more concentration on crucial points
3. $\kappa*TS \cup \lambda*EPS \cup \mu*NS$	Classification by complex boundaries with more concentration on error prone and crucial points
4. $\kappa*(TS \backslash EPS) \cup \mu*NS$	Classification by high simple boundaries with more concentration on crucial points
5. $\kappa*TS \cup \lambda*EPS$	Classification by high complex boundaries with more concentration on error prone data samples
6. $\kappa*(TS \backslash EPS)$	Classification by very simple boundaries

2.3 Diversity of an Ensemble

To define a meaningful diversity metric first we define a between-classifier diversity metric over one example. As equation (1) shows, the between-classifier diversity metric over one example is denoted by $d(C_i, C_j, x_k)$.

$$d\big(C_i, C_j, x_k\big) = \begin{cases} 0 & C_i(x_k) = C_j(x_k) \\ 1 & C_i(x_k) \neq C_j(x_k) \end{cases}. \tag{1}$$

In the equation (1), C_i stands for i-th classifier from the ensemble and $C_i(x_k)$ stands for its output over an exemplary sample x_k. Based on the equation (1), we define a diversity metric for an ensemble according to equation (2).

$$D(E) = \frac{1}{|E|^2 * |TS|} \sum_{x_k \in TS} \sum_{C_i \in E} \sum_{C_j \in E} d\big(C_i, C_j, x_k\big). \tag{2}$$

where E stands for an ensemble of classifiers. $D(E)$ is the diversity of ensemble E.

```
TS: Training Set
FinalEnsemble={ }
Function ProduceFinalEnsemble(E)
pre_acc=0;
pre_div=0;
  For i=1 to |E|
    tempEnsemble= C_i∪FinalEnsemble;
    div=D(tempEnsemble);
    acc=TEST(tempEnsemble, TS);
    If (acc≥pre_acc)&(div≥pre_div) Then
        FinalEnsemble=tempEnsemble;
        pre_acc=acc;
        pre_div=div;
  End.
  Return FinalEnsemble
```

Fig. 1 Pseudo-code of CDEBMTE algorithm

2.4 CDEBMTE

In CDEBMTE, a rough ensemble is first generated and then pruning it into the final ensemble. A classifier is trained on each subset defined on the training data. We train each classifier on each subset defined on the training data, thereby forcing them to differ from the each other. Therefore these classifiers as an ensemble can be considered a diverse one. While enforcing the diversity into the rough initial ensemble, we still want to extract the most diverse subset of classifiers out of the rough initial ensemble. We do this by eliminating a new classifier if adding it to the existing ensemble decreases its accuracy. This process is repeated until we exceed the maximum number of iterations. The pseudo code of the CDEBMTE algorithm is presented in Fig 1.

As you can understand from Fig 1, a greedy algorithm is employed to find the most diverse subset of the classifiers existing in the ensemble. The CDEBMTE algorithm incrementally examines possibility of adding the next classifier to the pool of current ensemble by considering two factors. First it temporarily adds the next classifier to the current ensemble. If both accuracy and diversity of the current ensemble increases over TS, the search algorithm accepts the current ensemble; else the temporarily new added classifier is removed from current ensemble.

3   Experimental Results

To evaluate the performance of CDEBMTE we ran experiments on 18 representative datasets from the UCI repository [11]. The datasets are summarized in Table 2. Note that the datasets vary in the numbers of training examples, classes, numeric and nominal attributes; thus providing a diverse testbed.

The performance of each learning algorithm is evaluated using 10 complete runs of 10-fold cross-validation (except 3 monk problems). In each 10-fold cross-validation, each dataset is randomly split into 10 equal-size segments and results are averaged over 10 trials. For each trial, one segment is set aside for testing, while the remaining data is available for training.

The size ensemble in each algorithm is considered same with CDEBMTE algorithm. It means after running of CDEBMTE algorithm, the size of final ensemble is extracted and is feed as the ensemble size to other algorithms.

Table 2 Summary of datasets. The training and test sets in all three Monk's problems are predetermined.

Dataset Name	Cases	Classes	Attributes
anneal	898	6	39
audio	226	6	69
autos	205	6	25
breast-w	699	2	9
credit-a	690	2	15
glass	214	6	9
heart-c	303	2	13
hepatitis	155	2	19
colic	368	2	22
iris	150	3	4
labor	57	2	16
lymph	148	4	18
monk	432	2	6
soybean	683	19	35
splice	3190	3	62
wine	178	3	13

Fig 2-a depicts the performance of CDEBMTE comparing to the performances of DECORATE, AdaBoost and Bagging over Iris dataset. In Fig 2-a the accuracy of different ensemble methods in terms of different ratios of TS is presented.

Fig 2-b depicts the performance of CDEBMTE comparing to the performances of DECORATE, AdaBoost and Bagging averaged over all 18 datasets. In Fig 2-b the averaged accuracy of different ensemble methods in terms of different ratios of TS is presented. As it is depicted, DECORATE outperforms other methods considerably when the size of *TS* is small. It is due to adding artificial data samples to train set. It means while the number of training set is not enough to properly the other methods learn the models of the classes, DECORATE increases the size of training set to learn the models of the classes better.

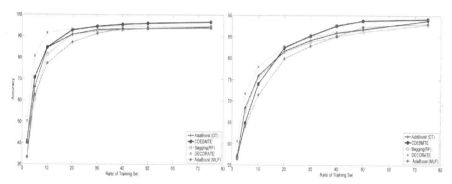

Fig. 2 Accuracy of different ensemble methods in terms of different ratios of TS. (a, Left) Accuracy over Iris dataset (b, Right) Averaged accuracy over all 18 datasets

Table 3 Experimental results.

Dataset Name	Bagging (RF)	Boosting (Arc-X4)	CDEBMTE
Wine	95.47	96.12	96.31
Iris	95.87	96.16	96.22
Monk 1	87.22	98.06	98.48
Monk 2	86.16	87.33	87.51
Monk 3	96.27	97.66	97.76

For *TS* sizes above 50 percent, DECORATE falls gradually in comparing with AdaBoost. It is finally slightly placed in rank three after AdaBoost method. While CDEBMTE sliding improves as *TS* size becomes greater. It is finally placed in rank one above all methods.

The detailed achieved results some prominent ensemble methods are presented in Table 3. In obtaining the results of Table 3, the ratio of sampling from train set in Bagging, Boosting and CDEBMTE algorithms is 70%. CDEBMTE outperforms obviously from prominent ensemble methods, Bagging and Boosting, in some of datasets according Table 3. It is again more outstanding especially in artificial dataset 2 where two of classes are composed by two clusters.

4 Conclusion

In this paper, a new method to improve performance of combinational classifier systems, CDEBMTE, is proposed. CDEBMTE is based on increasing the diversity of ensemble. First different datasets are extracted from training dataset. Then it trains a number of classifiers over them. Finally it selects a subset of the trained classifiers.

The proposed ensemble methodology, CDEBMTE, is examined on some datasets and it shows considerable improvements in comparison with some of the most competent ensemble methods.

Paper also shows that emphasizing on crucial data can cause improvement in diversity. Also we showed that usage of different datasets causes to quite diverse classifiers. It also shows that CDEBMTE can be effectively used to achieve higher accuracy and to obtain better class membership probability estimates in comparison with the most powerful methods.

References

[1] Breiman, L.: Random forests. Machine Learning 45, 5–32 (2001)
[2] Dietterich, T.G.: An experimental comparison of three methods for constructing ensembles of decision trees: Bagging, boosting, and randomization. Machine Learning 40(2), 139–157 (2000)
[3] Freund, Y., Schapire, R.E.: A decision–theoretic generalization of on-line learning and an application to boosting. Journal of Computer and System Sciences 55, 119–139 (1997)
[4] Freund, Y., Schapire, R.E.: Experiments with a new boosting algorithm. In: Saitta, L. (ed.) Proceedings of the Thirteenth International Conference on Machine Learning (ICML 1996). Morgan Kaufmann (1996)
[5] Hansen, L.K., Salamon, P.: Neural network ensembles. IEEE Transaction on Pattern Analysis and Machine Intelligence 12, 993–1001 (1990)
[6] Hastie, T., Tibshirani, R., Friedman, J.: The Elements of Statistical Learning. Springer, New York (2001)
[7] Kuncheva, L.I.: Combining Pattern Classifiers, Methods and Algorithms. Wiley, New York (2005)
[8] Liu, Y., Yao, X.: Ensemble learning via negative correlation. Neural Networks 12 (1999)
[9] Melville, P., Mooney, R.: Constructing Diverse Classifier Ensembles Using Artificial Training Examples. In: Proc. of the IJCAI, vol. I, pp. 505–510 (2003)
[10] Melville, P.: Creating Diverse Ensemble Classifiers (2006)
[11] Newman, C.B.D.J., Hettich, S., Merz, C.: UCI repository of machine learning databases (1998), http://www.ics.uci.edu/~mlearn/MLSummary.html
[12] Qiao, X., Liu, Y.: Adaptive Weighted Learning for Unbalanced Multicategory Classification. Biometrics, 159–168 (2009)
[13] Quinlan, J.R.: C4.5: Programs for Machine Learning. Morgan Kaufmann, San Mateo (1993)
[14] Tumer, K., Ghosh, J.: Error correlation and error reduction in ensemble classifiers. Connection Science 8(3-4), 385–403 (1996)
[15] Witten, I.H., Frank, E.: Data Mining: Practical Machine Learning Tools and Techniques with Java Implementations. Morgan Kaufmann, San Francisco (1999)

Enhancing the Localization Precision
in Aircraft Parking Areas of Airports
through Advanced Ray-Tracing Techniques

Antonio del Corte, Oscar Gutierrez, and José Manuel Gómez

Abstract. Parking areas in airports where the aircraft is loaded, refueled and boarded, present a high risk of accidents due to the large number of vehicles and people involved in the handling activity. For airport ground surveillance, different technologies are deployed: Radar, CCTV, GPS and Trilateration systems. All these solutions have important limitations when are using for surveillance in apron and stands areas near the terminal buildings. To solve this problem, we propose new algorithms for localization based on fingerprinting techniques that, using the available WLAN infrastructure and the ray-tracing multipath information provided by newFASANT simulation tool, allow increasing accuracy and safety in outdoor areas of the airports.

1 Introduction

Several surveillance technologies are deployed to achieve the main objective of the airports: to provide a cost-effective and efficient operation to deliver high-quality service to customers while also ensuring their safety. To detect approaching airplanes in flight, primary radar is used, while for ground surveillance, cameras, GPS, surface radar, ADS-B and trilateration systems are deployed. The ultimate goal of these systems is to improve the efficiency and the safety of airport operations. However, in the apron and stand areas near the terminal building, while the risk of accidents increases due to the large number of vehicles and people involved in the handling activity, the current surveillance systems do not

Antonio del Corte
Department of Computer Engineering

Oscar Gutierrez · José Manuel Gómez
Department of Computer Science
University of Alcalá
Alcalá de Henares. Madrid, Spain
e-mail: {antonio.delcorte,jose.gomez,oscar.gutierrez}@uah.es

S. Omatu et al. (Eds.): Distributed Computing and Artificial Intelligence, AISC 151, pp. 121–128.
springerlink.com   © Springer-Verlag Berlin Heidelberg 2012

provide adequate target localization. In this work, a new outdoor localization technique based on the signals available from the wireless devices [1] that comprise the Wi-Fi standards is presented. The localization process is done by using the fingerprinting technique [2][3]. In comparison with other techniques, such as angle of arrival (AOA) or time of arrival (TOA) that present several challenges due to multipath effects and non-line-of-sight (N-LOS) [1], the fingerprinting technique is relatively easy to implement. As cost function used inside the new fingerprinting algorithm developed, we use the information about the relative delay between rays provided by the newFASANT simulation tool. The main advantage of using the ray-tracing information is the multipath effects derived from the multiple ray-order reflections and diffractions produced between emitter and receiver [4]. Combining this technique with different similarity metric measures and with an interpolation algorithm the accuracy of the localization is increased.

2 Ground Surveillance in Airports

Apron and aircraft stand areas near the terminal building are not sufficiently protected in terms of surveillance. There are currently three normalized types of radar technologies involved in airport surveillance. The basic radar is the primary surveillance radar (PSR). PSR is a passive system because it only indicates the presence of aircraft without any information about aircraft identification. The secondary surveillance radar (SSR) is a collaborative system because the detection does not depend on the size of the object reflected, but on the response from a transponder on the aircraft. This transponder operates in a combination of transmitter-receiver mode enable of responding to the ground interrogation system to identify the plane. The passive radar is the surface movement radar (SMR). It is used to provide aircraft identification to the controller in low visibility conditions. A new radar technology for surveillance is the microwave sensor (MWS). The millimetric wave sensor (MWS) is collaborative radar that provides good target resolution in apron and stands areas but a high cost.

Conventional radar solutions do not provide coverage in apron and stands areas due to difficulty to discriminate the aircraft false detections and the corresponding to non-aircrafts (cars, trucks, trolleys, stairs). Figure 1 show the coverage provided by means of conventional radar in Madrid-Barajas International Airport. Only runway and taxiways areas (green colour) are covered by radars while apron stands areas are not under radar surveillance (yellow and magenta colors).

Alternatively, surveillance by means of television cameras in outdoor environments is a challenge due to incidence of sunlight on the lens (Fig. 2). Configure the layouts also requires expensive dedicated hardware such as Barco or Jupiter providers. GPS is another solution commonly used for fleet localization near the terminal building in airports. Three are the main problems with GPS: lack of satellite coverage, the different sources of error and the high cost for enhancing precision. Altogether GPS sources of error sums up to an error of ± 15 meters. Corrections mainly reduce ionospheric effects, but also improve orbits and clock errors. The overall error will be reduced to approximately ± 3 - 5 meters. GPS signals can also be affected by multipath issues (Fig. 3), where the radio signals reflect off surrounding terrain, buildings, canyon walls, hard ground, etc. These delayed signals can cause inaccuracy.

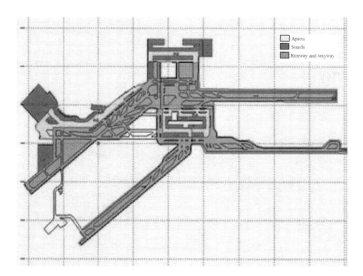

Fig. 1 Conventional radar coverage in Madrid-Barajas Airport (green colour)

Fig. 2 Panoramic view of stand areas based on CCTV system

More recently, airports have been deploying other surveillance information sources based on MLAT (multilateration) and ADS-B (Automatic Dependent Surveillance-Broadcast). ADS-B (Fig. 4) uses a conventional global navigation satellite system (GNSS) receiver to obtain precise position within the satellite constellation. Aircraft position and others flight data are sent to others ADS-B equipped aircrafts and to the ground stations. As ADS-B is GPS based, it does not work properly in ground close buildings. By the other hand, multilateration technology (Fig. 5), also known as hyperbolic positioning, is the process of locating airplanes and vehicles by calculating the time difference of arrival (TDOA) of a signal emitted by the source to three or more receivers. Shadows, multipath, reflections and garbling cause holes in multilateration coverage and degraded accuracy. Another issue is concerning the correct use of the aircraft transponder by the pilot, which will switch-off the transponder before stop the aircraft or switch it on when the aircraft is on taxi. These non-recommended actions will inhibit the detection of the multilateration signal.

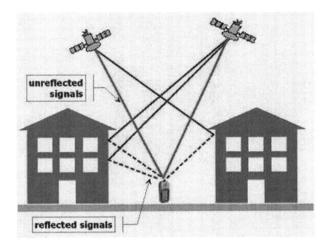

Fig. 3 Multipath effect in GPS signals

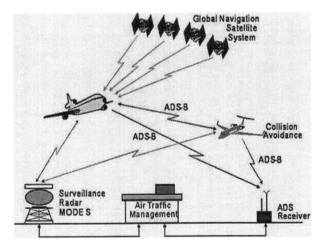

Fig. 4 ADS-B system

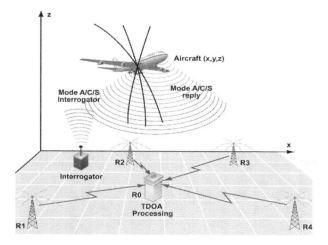

Fig. 5 Multilateration system

3 Localization Technique

The localization algorithm developed is based on fingerprinting technique. The fingerprinting (Fig. 6) can be divided in two phases [2]. In the first one, it obtains the radio map or fingerprints database. The radio-map of fingerprints is obtained by performing an analysis of the profile ray information (Fig. 7) and the power levels available from devices and multiples access points over a defined grid. The vector of ray-delays and powers information on the grid is called the location fingerprint of that point. In the second phase, it analyzes the accuracy obtained in the localization process. For this purpose, the developed technique places a significant number of mobile stations into the area covered by the radio map and it obtains the vector of received rays and powers from different APs [5] due to multipath effects. The location estimation is made by an algorithm that computes the distance between each measured mobile sample and all the fingerprints stored in the radio map. Coordinates associated with the fingerprint that results in the smallest distance are returned as the position for the mobile.

The distance metric is one of the key components used by the fingerprinting technique [6]. Therefore, it is important to explore different similarity measures to find the best distance metric that minimizes the average positioning error. Four distance metrics have been compared to find the one with the maximum localization accuracy [7]: Euclidean, Manhattan, Bray-Curtis and Mahalanobis.

In order to improve the localization precision, a set of fingerprints that present the shortest distance to the target mobile are used to interpolate lineally between fingerprinting weights based on distance metric.

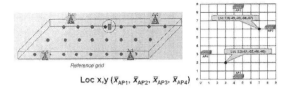

Loc x,y ($\bar{x}_{AP1}$, $\bar{x}_{AP2}$, $\bar{x}_{AP3}$, $\bar{x}_{AP4}$)

Fig. 6 Fingerprinting technique

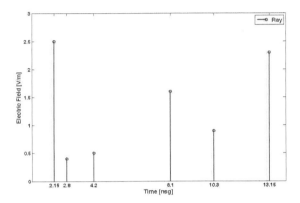

Fig. 7 Relative ray profile

4 Simulations

The geometry simulated corresponds with a section of the Madrid-Barajas International Airport parking area. The simulation area was 12 x 145 meters, where 1740
fingerprints for a grid resolution of 1 meter and 5 APs at 2.4 GHz were used (Fig.
8a). To estimate the algorithm accuracy, 100 target mobiles randomly located
were positioned (Fig. 8b). This particular section presents many problems due its
proximity to the terminal building, making existing surveillance systems impossible to work properly.

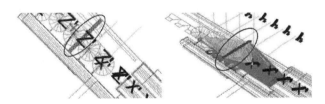

Fig. 8 Simulation area: (a) Fingerprints (left), (b) Mobiles and ray-tracing propagation
(right)

5 Results

The results obtained from the simulation are shown below. Four distance measures (Euclidean, Manhattan, Bray-Curtis and Mahalanobis) and two fingerprinting detection methods (power levels and relative ray delay between rays) were compared without interpolation (Fig. 9). It's clear that relative ray-delay detection method and the Mahalanobis distance provide the best results [8]. Finally an interpolation over the relative ray-delay detection method is showed (Fig. 10). All metrics improve the accuracy when the number of fingerprints increases between 5 and 6. Only the Bray-Curtis distance worsens due to its normalized expression.

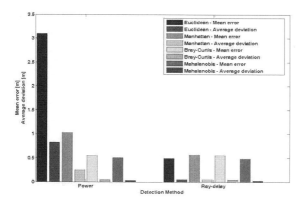

Fig. 9 Detection methods and metrics comparison

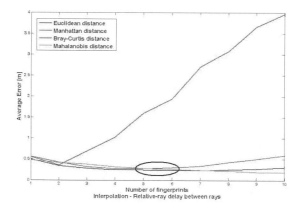

Fig. 10 Relative-ray delay detection: Interpolation effect and metrics

6 Conclusions

In this work an alternative localization method based on the fingerprinting technique has been presented. The new location algorithm is based on ray-tracing techniques over WLAN networks and was simulated with the newFASANT tool [9], providing outdoor coverage by means of a complete 3D model. The relative ray-delay detection technique presents better results in the localization precision than the traditional power detection technique. Combined with the Mahalanobis distance as metric improves the location accuracy. Finally, the interpolation algorithm is able to eliminate those fingerprints that present a high distance and therefore do not contribute to the improvement in the accuracy. The technique presented provides high localization accuracy in high risk airports areas, where other surveillance systems have important limitations.

References

[1] Cisco Wireless Location Appliance-Products. Datasheet (2006)
[2] Bahl, P., et al.: Enhancements of the Radar User Location and Tracking System. Microsoft Research Technology (February 2000)
[3] Kaemarungsi, K., Krishnamurthy, P.: Properties of Indoor Received Signal Strength for WLAN Location Fingerprinting. In: Proc. First Annual International Conf. on Mobile and Ubiquitous Systems: Networking and Services, pp. 14–23 (August 2004)
[4] Sáez de Adana, F., et al.: Propagation model based on ray tracing for the design of personal communication systems in indoor environments. IEEE Trans. on Vehicular Technology 49, 2105–2112 (2000)
[5] del Corte-Valiente, A., Gómez-Pulido, J.M., Gutiérrez-Blanco, O., Cátedra-Pérez, M.F.: Efficient Techniques for Indoor Localization based on WLAN Networks. In: Second International Workshop on User-Centric Technologies and Applications (Madrinet), Salamanca (Spain), pp. 5–15 (October 2008)
[6] del Corte-Valiente, A., Gómez-Pulido, J.M., Gutiérrez-Blanco, O.: High Precision for Indoor Localization Applications based on Relative Delay and the Fingerprinting Technique. In: Third International Workshop on User-Centric Technologies and Applications (Madrinet), Salamanca (Spain), pp. 25–32 (June 2009)
[7] del Corte-Valiente, A., Gómez-Pulido, J.M., Gutiérrez-Blanco, O.: Analysis of Fingerprinting Distance Measures for RF Indoor Location Applications Based on WLAN networks. In: Third International Workshop on User-Centric Technologies and Applications (Madrinet), Salamanca (Spain), pp. 12–25 (June 2009)
[8] del Corte, A., Gómez, J.M., Gutiérrez, O.: Increasing Airport Security with the Development of High Accuracy Location Algorithms based on Ray-tracing Techniques over WLAN Networks. In: International Conference on Electromagnetic in Advanced Applications (ICEAA), Sydney, Australia (September 2010)
[9] newFASANT Ray-tracing simulation tool, http://www.fasant.com

Variability Compensation Using NAP for Unconstrained Face Recognition

Pedro Tome, Ruben Vera-Rodriguez, Julian Fierrez, and Javier Ortega-García

Abstract. The variability presented in unconstrained environments represents one of the open challenges in automated face recognition systems. Several techniques have been proposed in the literature to cope with this problem, most of them tailored to compensate one specific source of variability, e.g., illumination or pose. In this paper we present a general variability compensation scheme based on the Nuisance Attribute Projection (NAP) that can be applied to compensate for any kind of variability factors that affects the face recognition performance. Our technique reduces the intra-class variability by finding a low dimensional variability subspace. This approach is assessed on a database from the NIST still face recognition challenge "The Good, the Bad, and the Ugly" (GBU). The results achieved using our implementation of a state-of-the-art system based on sparse representation are improved significantly by incorporating our variability compensation technique. These results are also compared to the GBU challenge results, highlighting the benefits of adequate variability compensation schemes in these kind of uncontrolled environments.

1 Introduction

Most biometric technologies are able to provide satisfactory matching performance in controlled situations where the user is cooperative and data acquisition conditions and environment can be controlled. However, in many applications, biometric data is acquired in less than ideal conditions, such as uncontrolled and unconstrained face recognition scenarios [5]. The low performance of biometrics technologies in

Pedro Tome · Ruben Vera-Rodriguez · Julian Fierrez · Javier Ortega-Garcia
Biometric Recognition Group - ATVS, Escuela Politecnica Superior
Universidad Autonoma de Madrid
Avda. Francisco Tomas y Valiente, 11 - Campus de Cantoblanco - 28049 Madrid, Spain
e-mail: {pedro.tome,ruben.vera,julian.fierrez,
        javier.ortega}@uam.es

S. Omatu et al. (Eds.): Distributed Computing and Artificial Intelligence, AISC 151, pp. 129–139.
springerlink.com   © Springer-Verlag Berlin Heidelberg 2012

these relatively uncontrolled situations has limited their deployment, therefore, a significant improvement in recognition performance in less controlled situations is one of the main challenge facing biometric technologies.

In the particular case of face recognition in uncontrolled scenarios there are numerous sources of variation, which can be known or unknown, affecting the performance. Hence, there is a need for developing methods capable of identifying and compensating/removing these variability sources in order to guarantee the robustness of the system in unconstrained and uncontrolled real environments.

In the present paper, a variability compensation approach based on Nuisance Attribute Projection (NAP) is presented for face recognition. In this field, to our knowledge only V. Štruc *et al.* in [13] have analysed such a normalization technique for illumination invariant face recognition based on NAP, which removes the illumination induced artifacts in two controlled scenarios. In our case, the proposed NAP compensation approach is used not only to compensate illumination variations, but also other variability factors. In particular, we study the uncontrolled scenario provided by the NIST - GBU still face recognition challenge, which consists of three partitions called the Good, the Bad, and the Ugly [6].

The performance of the proposed variability compensation scheme is evaluated on a state-of-the-art system based on sparse representation [14]. Results achieved show that variability compensation using NAP in combination with this system is a very interesting approach in uncontrolled face recognition environments.

The paper is structured as follows. Sect. 2 briefly describes the variability factors found in the GBU challenge. Sect. 3 describes the variability compensation approach using NAP. Sect. 3 describes the recognition system based on sparse representation. Sect. 5 presents the experimental protocol followed and the performance evaluation. Sect. 6 presents the experimental results obtained, and finally conclusions are drawn in Sect. 7.

2 Variability in Unconstrained Environments

Face recognition in unconstrained environments is a very challenging problem which has attracted increasing attention from the research community.

Some the recent studies in this field are the Multiple Biometric Grand Challenge (MBGC 2009) [7] and the Face Recognition Vendor Test (FRVT 2006) [8], whose focus of research is shifting to recognizing faces taken under less constrained conditions. As a result of the evolution of this NIST challenges a new competition called GBU has been defined, which consists of three partitions called the Good, the Bad, and the Ugly. The Good partition consists of pairs of face images of the same person that are easy to match (based on FRVT 2006 top performers); the Bad partition contains pairs of face images of a person that have average matching difficulty; and the Ugly partition concentrates on difficult to match face pairs. Fig. 1 shows an example of these three partitions and their respective histograms of match and non-match scores.

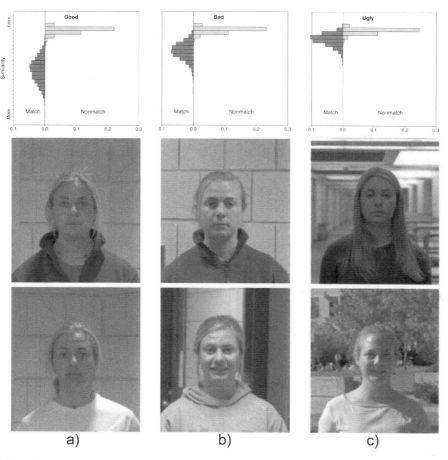

Fig. 1 GBU image samples and histograms of match and non-match distributions for the *a)* Good, *b)* the Bad, and *c)* the Ugly partitions with the relative frequency of similarity scores in horizontal axes. Extracted from [6].

Various techniques have been presented in the literature to compensate the variability present in these kind of scenarios [11, 12, 4, 15]. However, most of these techniques are focused on an isolated variability source, e.g., illumination, pose compensation, etc.

In the present paper, a variability compensation approach is presented, using the Nuisance Attribute Projection (NAP) to remove the variability induced in uncontrolled face recognition systems.

3 NAP for Variability Compensation

3.1 Nuisance Attribute Projection (NAP)

Nuisance Attribute Projection (NAP) is a powerful technique traditionally used in the field of speaker recognition for compensation of channel effects regardless of its source [10, 9], which are assumed to lie in a low dimensional variability subspace. In others fields like biometrics at a distance and unconstrained environments, the variability sources are mostly unknown and mixed, hence, we seek to understand to what extent variability compensation techniques as NAP are useful.

Consider a dataset X of n image vectors of size N pixels, where $X \in \mathfrak{R}^{n \times N}$. The NAP technique tries to remove any unwanted distortion in the images as follows:

$$X' = P(X - M), \tag{1}$$

where X' denotes the new data whose component in the variability subspace is removed, M denotes a matrix containing in each of its columns the global mean of the images in X and P stands for the $n \times n$ projection matrix:

$$P = I - VV^T = I - \sum_{i=1}^{d} v_i v_i^T. \tag{2}$$

Here, I denotes the $n \times n$ identity matrix, v_i represents the ith direction of the variability subspace base V of size d defined by NAP.

Suppose a data matrix X has n_{C_j} sample images from the jth class, whose labels of the classes are $C_1, C_2, ..., C_r$, then, for each of these images we can write:

$$x_{C_{j,k}} = x'_{C_{j,k}} + e_k, \tag{3}$$

where C_j represents the class label of the image, k denotes the index of the image in the jth class, $x'_{C_{j,k}}$ stands the variability-free part of $x_{C_{j,k}}$, and e_k represents the vector encoding the variability effects for the kth image of the jth class. Fig. 2a), describes graphically the procedure.

Assuming the unwanted variability effects inside each class coincide and they can be modelled by a Gaussian distribution, then, the base of the variability subspace defined by the matrix V, can be estimated from the first d eigenvectors (NAP directions $v_i (i = 1, 2, ..., d)$) of the matrix Σ_t:

$$\Sigma_t = \sum_{j=1}^{r} \sum_{k=1}^{n_{C_j}} (X_{C_{j,k}} - \mu_{C_j})(X_{C_{j,k}} - \mu_{C_j})^T, \tag{4}$$

where the mean value of each of the r classes μ_{C_j} ($j = 1, 2, ..., r$) represents a variability-free estimate of an image from the j class. This is typically done by using Principal Component Analysis (PCA). Fig. 2b) shows the eigenvalues of the associated eigenvectors for NAP estimated variability subspace.

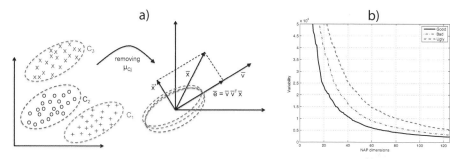

Fig. 2 *a)* Schematic illustration of the NAP technique in a 2-dimensional space. Firstly, every distribution is centred in the origin by removing the global mean of each class μ_{C_j}. The input data- and variability-dependent sample vector $(\bar{x})$ is pair up into two components $(\bar{x}')$ and $(\bar{e})$. $(\bar{e})$ stands for the component in the variability subspace, and $(\bar{x}')$ is the resulting compensated sample vector. The vector $\bar{v}$ represents the first eigenvector of the estimated variability subspace. *b) Eigenvalues of three target datasets: Good, Bad, Ugly.*

3.2   Removing Variability Effects

In the case considered in this paper, we have separated the channels YCbCr of the images and the NAP compensation scheme has been applied over the luminance (Y) component of images.

Consider an input data set in X from which we estimate the NAP directions corresponding to the unknown variability in the unconstrained scenarios. In the GBU database considered only four factors are controlled: subject aging, pose, change in camera, and variations among faces. Other factors, such as: illumination, indoor/outdoor, distance, ... are considered as unwanted variability factors. These factors do not always affect in the same level, making the problem of their compensation even more challenging. Any input image x is compensated with respect to the estimated variability effects by projecting away a number of directions in the NAP subspace. Fig. 2*a)* illustrates graphically the procedure. The compensation procedure is described by:

$$x' = P(x - \mu) = (I - VV^T)(x - \mu), \tag{5}$$

where μ represents the global mean of the images in X, I denotes the identity matrix and V stands for the NAP compensation matrix. To effectively remove the effects of annoying variability, the data matrix X must be constructed in such a way as to include the highest available number of images captured in different variability conditions.

4 Face Verification System – SRC

A system based on recent works in sparse representation for classification purposes (SRC) [2, 14] has been adopted as face verification core.

Essentially, this kind of systems span a face subspace using all known training face images, and for an unknown face image they try to reconstruct the image sparsely.

The motivation of this model is that given sufficient training samples of each person, any new test sample for this same person will approximately lie in the linear span of the training samples associated with the person.

Once a new test image y is acquired, it can be represented using samples from the database by the linear equation $y = Ax_0$, where matrix A defines our training data and x_0 represents the sparse solution.

According to the assumption that images from a given subject are sufficient to represent themselves, the solution x_0 in the linear equation $y = Ax_0$ should be very sparse. This can be approximately recovered by solving the following noise-aware l^1-minimization problem:

$$\widehat{x}_1 = \mathrm{argmin}_x \|x\|_1 \; subject\ to \; \|Ax - y\|_2 \leq \varepsilon. \tag{6}$$

To recognize a probe test image, the SRC algorithm identifies the class by computing the minimum among the residuals reconstructed per class. The robust performance of the SRC algorithm has been proved experimentally on face datasets with noises and occlusions.

The solution of equation (6), was approximated, in an efficient way, via basis pursuit using linear programming by considering L1-norm instead of L0-norm. To this end, the available package provided in [1] was used.

5 Experimental Protocol

The experiments are carried out on the The Good, the Bad, and the Ugly (GBU) database [6] included in the last still face recognition challenge from NIST - National Institute of Standard and Technology. The GBU challenge problem consists of three partitions with are called the Good (face pairs easy to match), the Bad (face pairs with average matching difficult), and the Ugly (face pairs difficult to match). Each partition consists of two sets of images, a target set and a query set, each of which contains 1,085 images from 437 distinct subjects. The distribution of image counts per person in the target and query sets are 117 subjects with 1 image; 122 subjects with 2 images; 68 subjects with 3 images; and 130 subjects with 4 images.

For the experiments in this paper, we use the segmented datasets provided by MBGC - Multiple Biometric Grand Challenge [7] compressed to 20KB with 120 pixels between the centers of the eyes. The faces were normalized following the ISO norm described in [3], from a size of 408×528 to size 168×192 pixels.

The baseline system consists on the application of the SRC algorithm with a single preprocessing stage to normalize the face illumination by histogram equalization (HQ) over the band of luminance (Y) from YCbCr color space. In the experiments described here, we have used as features the downsampled images, whose good performance combined with SRC is demonstrated in [14]. In our case the downsampling ratio is $1/8$ obtaining feature vectors of 504 dimensions.

The performance of the evaluated system is computed not using the same experimental protocol described by the GBU challenge. We also use a one-to-one matching, but using prior information of the target sets in order to compensate the variability.

In the experimental protocol we consider two experiments. In the first experiment, a NAP compensation matrix is generated for each partition of the database using only the target images. In the second experiment, a global NAP compensation matrix NAPgbu is generated for the three partitions together using all the target images. In both cases we evaluate the performance of our recognition system using two different NAP dimensions of variability (d), low compensation $d = 5$ and high compensation, $d = 125$.

6 Results

The experiments have two different goals, namely: *i*) study the benefits of variability compensation schemes in uncontrolled environments, and *ii*) show the efficiency of NAP-based variability compensation when considering multiple uncontrolled sources of variability.

Our baseline algorithm based on sparse representation achieved better results than those obtained from the LRPCA-face GBU baseline algorithm [6] at a false accept rate (FAR = 0.001). On the Good partition, the base verification rate (VR) is 0.88, for the Bad partition, the VR in 0.29, and the VR in Ugly is 0.06. Table 1 shows the comparative results.

6.1 Experiment 1: NAP over Each partition

The performance of the NAP compensation scheme is first analysed scenario by scenario. Results achieved for the *Good* partition are represented in Fig. 3*a*. In this case the compensation of few dimensions $(d = 5)$ is much better than using many dimensions $(d = 125)$. This is due to the fact that data are more or less clear of unwanted variability so the compensation of many dimensions leads to a discriminative information loss. As can be seen in Fig. 2*b*), the eigenvalues of the *Good* partition decrease faster than for the other partitions, meaning that the variability is concentrated in the first dimensions.

Results achieved for the *Bad* partition are shown in Fig. 3*b*. In this case the variability increases, implying that there are more corrupted dimensions with variability.

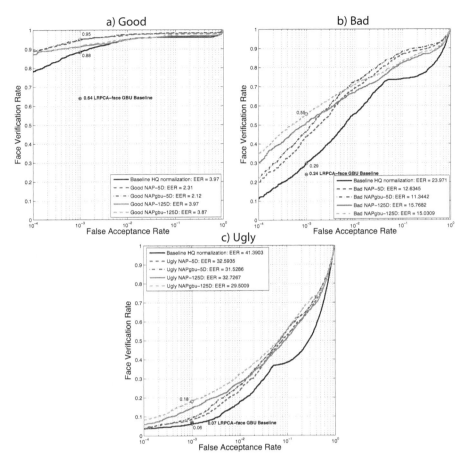

Fig. 3 ROC curves obtained for the three partitions: *a) Good* partition, *b) Bad* partition and *c) Ugly* partition. The verification rate for the LRPCA-face GBU baseline [6], our baseline system and the best NAP solution are highlighted at a FAR = 0.001.

Therefore, in this case better results are obtained for the case of compensating more dimensions ($d = 125$) with NAP working at a FAR = 0.001. On the other hand, the EER of the system is better for the case of compensating less dimensions ($d = 5$), but in this case working at a much more permissive application FAR.

Results achieved for the *Ugly* scenario are shown in Fig. 3c. In this case, as in the *Bad* one, better results are obtained when ($d = 125$) dimensions are compensated. Here, the verification rate (VR) of the baseline system versus the compensated system with $d = 125$ improves from 0.06 to 0.14 at a FAR = 0.001. Table 1 summarises all the results achieved for this experiment.

Table 1 Results achieved in Experiment 1. Performance of the LRPCA-face [6] baseline system versus our baseline and best NAP compensation results, being respectively $d = 5, 125, 125$ for Good, Bad and Ugly partitions. Also, relative improvement in the verification rate reached by NAP compensation are highlighted at a FAR = 0.001.

Partition	LRPCA face [6]	Baseline	Best NAP Comp.	Relative Improvement (%)
Good	0.64	0.88	0.94	6.8
Bad	0.24	0.29	0.51	91.6
Ugly	0.07	0.06	0.14	133.3

6.2 Experiment 2: NAP over the Whole Partitions

As mentioned before, in this experiment we generate a global NAP compensation matrix (NAPgbu) combining the three target datasets (Good, Bad, and Ugly) in order to demonstrate the potential of the proposed NAP approach on unconstrained environments. The main results are summarized in Table 2.

The difference between low and high compensation ($d = 5$ and $d = 125$) respectively, have the same behaviour than experiment 1 over the three datasets, as we can see in Fig. 3.

NAP compensation removes the intra-class variability by projecting away multiple dimensions of a low variability subspace. For this reason a subspace calculated using all possible target data is likely to improve the effect of the variability compensation. This is proved by observing who the NAPgbu compensation scheme achieves the best results over all scenarios (see Figs. 3 and 4). Note that in the *Good* partition both proposed schemes produce the same results due to the low influence of the variability source in this case.

As can be seen in Fig. 4 the *Bad* partition achieves the highest absolute improvement of VR going from 0.29 to 0.55 (at FAR = 0.001). This is possibly due to the fact that images contain a large amount of variability but still is possible to achieve reasonably good results with compensation.

Finally, in the *Ugly* partition, the balance of discriminative information against noise is very low. Fig. 4 shows how the VR improves from 0.06 to 0.18 at FAR = 0.001, reaching better results than those presented in [6]. As can be seen in Table 2, the relative improvement of the verification rate in the *Ugly* partition in this experiment is higher (200%) than in others partitions.

Table 2 Results achieved in Experiment 2. Performance of the LRPCA-face [6] baseline system versus our baseline and best NAPgbu compensation results, being respectively $d = 5, 125, 125$ for Good, Bad and Ugly partitions. Also, relative improvement in the verification rate reached by NAPgbu compensation are highlighted at a FAR = 0.001.

Partition	LRPCA face [6]	Baseline	Best NAPgbu Comp.	Relative Improvement (%)
Good	0.64	0.88	0.95	7.9
Bad	0.24	0.29	0.55	89.6
Ugly	0.07	0.06	0.18	200

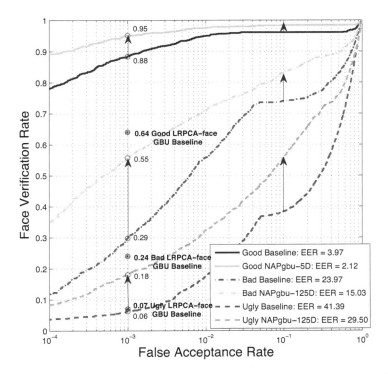

Fig. 4 ROC for the best NAP solutions vs. baseline systems on the three GBU partitions: Good, Bad and ugly. LRPCA-face GBU baseline [6] is also highlighted at a FAR = 0.001.

7 Conclusions

In the present work, a variability compensation approach based on Nuisance Attribute Projection has been presented and used to improve a state-of-the-art face recognition system based on sparse representation. The efficiency of this approach has been studied considering the three different challenge partitions designed by NIST for the still face recognition challenge "The Good, the Bad, and the Ugly" (GBU). In all cases, the baseline system performance is higher than the one achieved in the baseline algorithms from GBU challenge [6]. Furthermore, when the proposed compensation variability approach based on NAP is applied, the system performance improves significantly.

The application of NAP compensation using the whole partitions in a combined form is also analysed, highlighting the benefits of adequate variability compensation schemes in these kind of uncontrolled environments.

Acknowledgements. Pedro Tome is supported by a FPU Fellowship from Univ. Autonoma de Madrid and Ruben Vera-Rodriguez is supported by a Juan de la Cierva Fellowship from Ministerio de Ciencia e Innovacion. This work has been partially supported by projects

Bio-Challenge (TEC2009-11186), Contexts (S2009/TIC-1485), TeraSense (CSD2008-00068) and "Cátedra UAM-Telefónica".

References

1. Candes, E., Romberg, J.: l1-magic: Recovery of sparse signals via convex programming (2005), http://www.acm.caltech.edu/l1magic/
2. Huang, K., Aviyente, S.: Sparse representation for signal classification, pp. 609–616 (2006)
3. ISO/IEC JTC 1/SC 37 N 504. Biometric data interchange formats part 5: Face image (2004)
4. Li, S.Z., Chu, R., Liao, S., Zhang, L.: Illumination invariant face recognition using near-infrared images. IEEE Trans. Pattern Anal. Mach. Intell. 29(4), 627–639 (2007)
5. Li, S.Z., Schouten, B., Tistarelli, M.: Biometrics at a Distance: Issues, Challenges, and Prospects. In: Handbook of Remote Biometrics for Surveillance and Security, pp. 3–21. Springer, Heidelberg (2009)
6. Phillips, P.J., Beveridge, J.R., Draper, B.A., Givens, G., O'Toole, A.J., Bolme, D.S., Dunlop, J., Lui, Y.M., Sahibzada, H., Weimer, S.: An introduction to the good, the bad, amp; the ugly face recognition challenge problem. In: Int. Conf. on Automatic Face Gesture Recognition and Workshops (FG 2011), pp. 346–353 (March 2011)
7. Phillips, P.J., Flynn, P.J., Beveridge, J.R., Scruggs, W.T., O'Toole, A.J., Bolme, D., Bowyer, K.W., Draper, B.A., Givens, G.H., Lui, Y.M., Sahibzada, H., Scallan III, J.A., Weimer, S.: Overview of the Multiple Biometrics Grand Challenge. In: Tistarelli, M., Nixon, M.S. (eds.) ICB 2009. LNCS, vol. 5558, pp. 705–714. Springer, Heidelberg (2009), http://dx.doi.org/10.1007/978-3-642-01793-3_72
8. Phillips, P.J., Scruggs, W.T., O'Toole, A.J., Flynn, P.J., Bowyer, K.W., Schott, C.L., Sharpe, M.: FRVT 2006 and ICE 2006 large-scale experimental results. IEEE Transactions on Pattern Analysis and Machine Intelligence 99 (2009)
9. Solomonoff, A., Campbell, W.M., Quillen, C.: Nuisance attribute projection. In: Speech Communication. Elsevier Science BV, Amsterdam (2007)
10. Solomonoff, A., Quillen, C., Campbell, W.M.: Channel compensation for svm speaker recognition. In: Proceedings on Odyssey: The Speaker and Language Recognition Workshop, Toledo, Spain, pp. 41–44 (2004)
11. Tome, P., Fierrez, J., Alonso-Fernandez, F., Ortega-Garcia, J.: Scenario-based score fusion for face recognition at a distance. In: IEEE Computer Society Conference on Computer Vision and Pattern Recognition Workshops (CVPRW), pp. 67–73 (June 2010)
12. Tome, P., Fierrez, J., Fairhurst, M.C., Ortega-Garcia, J.: Acquisition Scenario Analysis for Face Recognition at a Distance. In: Bebis, G., Boyle, R., Parvin, B., Koracin, D., Chung, R., Hammoud, R., Hussain, M., Kar-Han, T., Crawfis, R., Thalmann, D., Kao, D., Avila, L. (eds.) ISVC 2010. LNCS, vol. 6453, pp. 461–468. Springer, Heidelberg (2010)
13. Štruc, V., Vesnicer, B., Mihelič, F., Pavešić, N.: Removing illumination artifacts from face images using the nuisance attribute projection. In: Proceedings of the IEEE Int. Conf. on Acoustics, Speech and Signal Processing (ICASSP 2010), Dallas, Texas, USA, pp. 846–849 (March 2010)
14. Wright, J., Yang, A.Y., Ganesh, A., Sastry, S.S., Ma, Y.: Robust face recognition via sparse representation. IEEE Trans. Pattern Anal. Mach. Intell. 31(2), 210–227 (2009)
15. Zhang, X., Gao, Y.: Face recognition across pose: A review. Pattern Recognition, 2876–2896 (2009)

Comparing Features Extraction Techniques Using J48 for Activity Recognition on Mobile Phones

Gonzalo Blázquez Gil, Antonio Berlanga de Jesús, and José M. Molina Lopéz

Abstract. Nowadays, mobile phones are not only used for mere communication such as calling or sending text messages. Mobile phones are becoming the main computer device in people's lives. Besides, thanks to the embedded sensors (Accelerometer, digital compass, gyroscope, GPS, and so on) is possible to improve the user experience. Activity recognition aims to recognize actions and goals of individual from a series of observations of themselves, in this case is used an accelerometer.

Keywords: Mobile device, Activity Recognition, Ambient Assisted Living, J48, features extraction.

1 Introduction

Ambient intelligence (AmI) could be viewed as human interface metaphor. In AmI people are empowered through a context aware environment that is sensitive, adaptive and responsive to their needs, habits, gestures and even emotions [13]. AmI sees a world where a huge mesh of sensors are integrated into daily objects, clothing, people and so on. Using this information, the environment itself can provide context-aware services to support its inhabitant.

Taking into account that people is the main actor in AmI enviroments, it is mandatory to provide a tool for user to comunicate with the enviroment. At this point, mobile phones present several advantages: they are considered essential in people life's, so they could be considered as a non-intrusive sensor; they experience the same physical forces, temperature and noise that the person who carries them out

Gonzalo Blázquez Gil · Antonio Berlanga de Jesús · José M. Molina Lopéz
Applied Artificial Intelligence Group, Universidad Carlos III de Madrid,
Avd. de la Universidad Carlos III, 22, 28270, Colmenarejo, Madrid, Spain
e-mail: {gonzalo.blazquez,antonio.berlanga,
        josemanuel.molina}@uc3m.es
        http://www.giaa.inf.uc3m.es

S. Omatu et al. (Eds.): Distributed Computing and Artificial Intelligence, AISC 151, pp. 141–150.
springerlink.com   © Springer-Verlag Berlin Heidelberg 2012

[15]. On the contrary, obtaining physical actions from smartphones presents several problems [6] [8], basically, they are not built to collect information and infer activities.

Considering these advantages, it may be possible to consider a smartphone like a non-intrusive device to obtain people activities. The ability to understand human life patterns by analyzing user mobile phone behaviour is becoming a new challenge for researchers [3].

The inference of user activities implies a large number of sensors distributed over the body and/or the environment, depending on the activities to detect [11]. Smartphones are especially well-suited to this task because they have integrated Microelectromechanical systems (MEMS) which make easier to obtain user information. They may obtain and process physical phenomena from embedded sensors (MEMS) and send this information to remote locations without any human intervention [12]. Smartphones should take advantage of mobile contextual information, such as position, user profile or device features; to offer amazing services.

GPS, Wi-Fi, Bluetooth and microphone are the most known sensors in mobile phones, however, recently, new kind sensors have been added: accelerometer, gyroscope, compass (magnetometer), proximity sensor, light sensor, etc. [5]. As a result of this, not only phone numbers and addresses are collected in the mobile phone but also location, temperature, noise, physical forces may be collected to offer user new kind of amazing applications.

Finally, a performance studio about three different ways to realize activity recognition using smartphones (Spectrogram, Continuous Wavelet Transform and mean, standard deviation and other features) is presented in this paper. Besides, a dataset is created using an HTC Magic mobile phone with Android Operating System. The quality of the given solution is measured using a J48 tree.

The paper deals with the topic of recognizing user's activities by analyzing the data produced by motion sensors embedded in mobile phones. Sensory data is collected by a mobile application made in Android and it sends to a server where prelearnt activities are recognized in real-time. Besides this study rely on the power of the GPS in order to tag every action that the mobile phone takes using speed value.

2   Related Work

There are many different methods to retrieve user activity information from raw sensor data in the literature. However, the principal steps can be categorized as preprocessing, segmentation, feature extraction, dimensionality reduction and classification [9].

Normally, raw sensor data is collected using ad-hoc accelerometers over the body. However, placing sensors in multiple body locations could be annoyed for the user. New researches try to make this more comfortable for users using a smartphone. In this section, some architectures are briefly described where a smartphone is used for that purpose.

2.1 Cenceme Architecture

Probably, Cenceme is the most known system to recognize physical activities from mobile devices (Figure 1a). This architecture aims to infer user physical activities from mobile devices and to share them on social network. The proposed architecture is split in three layer: Sense, learn and share.

Sense layer aims to collect raw sensor data from sensors embedded in the phone. In learn layer, they propose to use a variety of data mining techniques to infer user rules. These techniques are used to interpret mobile data extracted from sensor layer. Their approach is to share information in a web portal where sensor data and inferences are easily displayed.

2.2 lifeMap Architecture

Yohan Chon et all. [4] present LifeMap, an Smartphone-based Context Provider for Location-based Services. The presented architecture is split in four component: (i) All the sensor are placed on the low level, this level send the information (ii) to the Component Manager where information is processed and provide high-level information. Using high-level information from the Component Manager, (iii) the Context Generator generates a point of interest (POI) which contains the user context. The context map is stored in a database to match and aggregate user contexts. And finally, (iv) The Database Adapter is an interface to provide user context to other applications.

2.3 inContexto Architecture

Finally, in [3] was presented inContexto (Figure 1b). It is a distributed architecture to retrieve user context information from mobile phones. The architecture is split in five components. Some of them run on the cloud and some others run on the mobile phone. This paper rely on this architecture to obtain sensory data from mobile phones.

- Data Acquisition: A low-level sensing module continuously gathers relevant information about the user activities using sensors. The proposed architecture this component is located on the mobile device.
- Features Extraction: The features extraction level is also implemented in the mobile phone. The module processes the raw sensor data into features that help discriminate between activities. This level aims to process and select which features are better to identified an action.
- Activity recognition: The last layer is classification module that uses the features selected in the mobile phone to infer what activity an individual or group of individuals is engaged in, for example: Walking, running, sitting, standing. In this

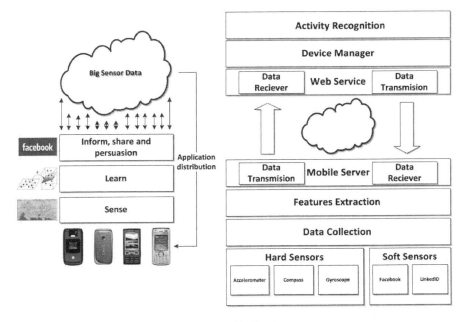

Fig. 1 Proposed Architecture by Cenceme and inContexto.

component, it will be implemented the algorithms (Supervised learning, Probabilistic classification, Model-based or instance-based learning) to figure out the taken action.

3 Experimentation

Pattern recognition answer to the description and classification of measurements taken from physical or mental processes [14]. In order to provide an effective and efficient description of patterns, pre-processing is often required improve performance, removing noise and redundancy in measurements. Then a set of characteristic measurements, which could be numerical or not, and relations between them, are extracted representing the patterns.

Collecting data is a hard task, hence in order to generate enough trajectories examples to make the training process, the data collection was made in a different way. This process has four steps: Data Collection, Trajectories generation, Features extraction and Training process.

Tagging process has been made using mobile phone GPS which distinguish every single action whith the speed value. Thus, it is not neccesary user involvement in tagging process, every time the GPS is enable, the application start to log data.

It was used three individuals who made five different activities (Running, Walking and Standing up). An HTC magic mobile phone equipped with Android OS was chosen to perform the measurement. Finally, a dataset was created for the research community and it is available online in [1].

3.1 Data Collection

In this study, the accelerations and azimuths of the pedestrian were collected using a HTC Magic. The created dataset has the following attributes: 3-axis accelerometer values in the mobile device Cartesian reference system, 3-axis Compass values, 3-axis accelerometer values in the real world reference system, GPS precision and GPS speed. Next table show the number of instances for each activity.

The sampling frequency can be adjusted according to the action studied. In this case, rely on the next study[7], the sampling frequency range requiring to obtain human actions is $0.6Hz$ to $2.5Hz$. Consequently, to prevent aliasing problem, the Nyquist-Shannon sampling theorem is followed:

$$F_E \geq 2 * F_{Max}$$

Finally, the sampling frequency was fixed to the maximum that the Android OS permits, in this case 50 Hz which is more than sufficient compared to 5Hz recognize the activities of the pedestrian.

Table 1 Number of second and samples for each activity.

	Running	Standing	Walking
Instances	5,118	7,321	24,825
Seconds	102.36	146.42	496.5

3.2 Trajectories Generation

It is necessary a big amount of trajectories to make correctly the training process. However, it is quite costly to generate enough trajectories to make this process.

In this case, the selected trajectories are made semi-automatically. First of all, we have 3 files corresponding each activity (Running, Walking, and Standing up). Subsequently, a Java program has been created to mix all the activities generated a unique trajectory. Finally all the generated trajectories have been stored to continue the pattern recognition process. However, there are some requirement to make this trajectories as real as possible:

---

[1] GIAA Web page http://www.giaa.inf.uc3m.es/

1. All the trajectories start with a Standing up action.
2. The next action could be the action besides (Figure 2) or the same action again.

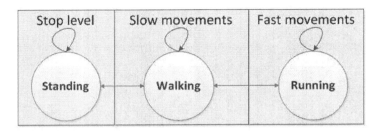

Fig. 2 Generation trajectories model.

3. The minimum duration of each action is 2 seconds and the maximum is 7 seconds.
4. Finally, each trajectory consists in 10 actions.

When the trajectories generation process is over, it is necessary to discretize the speed value due to J48 tree users nominal values. Thus, all the samples are discretized in 5 classes:

- Stop class: It is when the GPS speed measurements are less than 1 km/h.
- Walking class: Speed value from the GPS is more than 1 km/h and less than 4.
- Walking fast class: In this case, GPS speed values are among 4-6 km/h.
- Running class: It is when the GPS speed measurements are more than 6 km/h and less than 10 km/h.
- Running fast class: Finally the last class takes the GPS speed values upper than 10 km/h.

Finally, 1000 trajectories was created to infer activities. Every trajectory is different, in duration and actions, to each other. Weka[2] was used as the machine learning tool in this paper and it is necessary to transform data into arff format.

3.3 Features Extraction

Some research present different ways to obtain features in order to infer physical action. A comparison study using wavelet and frequency features (DWT, CWT, and STFT) is presented in [2]. These techniques provide several advantages, one particular advantage of frequency modulation is its resilience to signal level variations. On the contrary, in [10] is presented another features to infer activities using accelerometer values.

---

[2] Weka web page http://www.cs.waikato.ac.nz/ml/weka/

In this study, features were extracted from the raw accelerometer signals via a sliding window of 512 samples (Approximately 5 seconds), 256 of which overlap with consecutive ones. An sliding windows with 50% overlap has been defined in previous works [1]. This work uses GPS in order to obtain user's speed who is taking place the action, thus, the classifier output value is the mean of the speed in the sliding window.

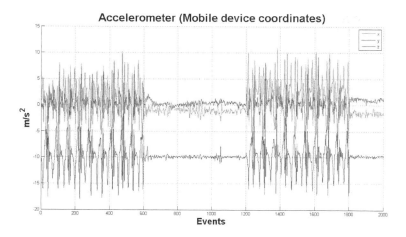

Fig. 3 Trajectory example with 4 actions (Walking, Standing up, Walking and Standing up. First graph shows the variance of the three coordinates mobile devices. The second one depicts the value of the real vertical world coordinate. The last one shows an example of spectrogram feature.)

This study is focused on compare three kind of features:

- The first one is based on Spectrogram function (STFT, Short-Time Fourier Transform). A spectrogram is a time-varying spectral representation that shows how the spectral density of a signal varies with time.
- The second one is Continuous Wavelet Transformation, is used to split a continuous-time signal into wavelets. Unlike Fourier transform, the CWT is able to construct a time-frequency representation of a signal which offers very good time and frequency localization. Both of these techniques (STFT and CWT) present several vales (higher than 150), however, they are not necessary all of them. For that reason, only the first 25 frequencies were selected such us possible feature. Besides, the signals need to be transformed from mobile device coordinate to real world coordinates following the next formula:

$$a_{realworld} = a_{mobiledevice} * R * I$$

Where I matrix is a simple rotation around the X axis and the rotation matrix R which is the identity matrix when the device is aligned with the world's coordinate system.

- The last way to make the training process is a set of different values from the raw accelerometer data, specifically, tree axes mean values, 3 axes standard deviation values, correlation between each axis and signal energy for each axis.

3.4 Results

The selected machine learning algorithm is a J48 classifier which is the Weka version from the C4.5 decision tree algorithm. J48 was chosen because gives results in tree model which can be easily transformed into real time applications.

The selected parameters for the J48 decision tree are:

- Confidence Factor = 0.25
- Minimum number of object = 2
- unpruned = false
- Test-options = 10 folds Cross-validation

After processing the training and testing sets with the J48 classifier in Weka, the results are highly accurate in vector and spectrogram features, however results are poorly accurate if CWT features extraction is used.

Table 2 Features of J4 tree generated by Weka.

	Features	Leaves	Tree size	Time(s)	Accuracy	Mean absolute error
CWT	25	8741	17481	129.32	62.85 %	0.1631
Spectrogram	25	1007	2013	41.44	95.63 %	0.0198
Vector	12	648	1295	14.57	97.20 %	0.0131

Table 2 shows result from each selected technique to extract features. The best implemented technique is features vector set, which is not only more accurate than the other ones, otherwise it provides the smallest tree generated and the minimum generation tree time. The size of the tree is very important because it will be implemented in a real application in a mobile phone. A bigger size of the tree causes more energy consumption according to the increase of CPU cycles. Another way to study the quality of the feature extraction techniques is using the confusion matrix (Figure 4).

CWT technique is the worst of all the studied technique, besides it does not present any advantage over the other ones. Secondly, spectrogram achieves great results, besides, this technique uses only one signal (vertical movement in the real world) in order to obtain the spectrogram. Although confusion matrix shows that is possible to classify an instance in a class not next to the real class. Thus, the best performance (high accurate and less tree size) is presented by Vector technique. Besides, confusion matrix figure shows that Vector features extraction just fail with the class near the one which is classified (e.g. Running instead of Running fast).

CWT

```
     a     b      c     d     e   <-- classified as
  9462  3357    269    67   132 |   a = Stopping
  3115 16957   2804   997  1852 |   b = Walking
   302  3601   2025   658  1652 |   c = WalkingFast
   114  1277    775  1100  2856 |   d = Running
   169  2144   1439  1860 20266 |   e = RunningFast
```

Spectrogram

```
     a     b      c     d     e   <-- classified as
 12815   472      0     0     0 |   a = Stopping
   473 24592    640    20     0 |   b = Walking
     0   667   7217   352     2 |   c = WalkingFast
     0    23    395  5501   203 |   d = Running
     0     0      4   212 25662 |   e = RunningFast
```

Vector

```
     a     b      c     d     e   <-- classified as
 12979   308      0     0     0 |   a = Stopping
   286 25047    392     0     0 |   b = Walking
     0   360   7639   239     0 |   c = WalkingFast
     0     1    247  5676   198 |   d = Running
     0     0      0   187 25691 |   e = RunningFast
```

Fig. 4 Confusion matrix from CWT, Spectrogram and Vector features.

4 Conclusions

In this paper, a study comparing three different techniques in order to infer activity recognition using a J48 decision tree was presented. Besides, the study rely on inContexto architecture to collect accelerometer data Overall, the presented work further demonstrates that using a mobile phone providing with accelerometers is enough to infer actions that user is taking place.

Selected features is an important field inside the Activity recognition systems. This paper aims to identify and record in real-time selected features related on user activity using a mobile device.

The best given solution obtained an overall accuracy of 97.20 % well classify instances of 79250 different actions. This solution is a vector composed by: Energy, mean, standard deviation and correlation of each axes.

The flexibility of the Android OS along with the phone's hardware capability allows this system to be extended. For example, create an application which is able to send a sms or call to your relatives if you are doing strange movements. This application may be interesting in ancient people. Another application may be a indoor GPS. Based on the user movement fingerprint, this application could follow a person who is moving indoors.

Acknowledgments. This work was supported in part by Projects CICYT TIN2011-28620-C02-01, CICYT TEC2011-28626-C02-02, CAM CONTEXTS (S2009/TIC-1485) and DPS2008-07029-C02-02.

References

1. Bao, L., Intille, S.: Activity recognition from user-annotated acceleration data. Pervasive Computing, 1–17 (2004),
   `http://www.springerlink.com/index/9AQFLYK4F47KHYJD.pdf`
2. Barralon, P., Vuillerme, N., Noury, N.: Walk detection with a kinematic sensor: frequency and wavelet comparison. In: Conference Proceedings:. Annual International Conference of the IEEE Engineering in Medicine and Biology Society. IEEE Engineering in Medicine and Biology Society. Conference, vol. 1, pp. 1711–1714 (2006),
   `http://www.ncbi.nlm.nih.gov/pubmed/17945661`,
   doi10.1109/IEMBS.2006.260770
3. Blázquez, G., Berlanga, A., Molina, J.: Incontex to: A fusion architecture to obtain mobile context. In: 2011 Proceedings of the 14th International Conference on Information Fusion (FUSION), pp. 1–8. IEEE (2011)
4. Chon, J.: LifeMap: Smartphone-based Context Provider for Location-based Services. IEEE Pervasive Computing, 1–7 (November 13, 2011),
   `http://www.computer.org/portal/web/csdl/doi/10.1109/MPRV`
5. Eagle, N., Pentland, A.: Reality mining: sensing complex social systems. Personal and Ubiquitous Computing 10(4), 255–268 (2006)
6. Ganti, R., Srinivasan, S., Gacic, A.: Multisensor Fusion in Smartphones for Lifestyle Monitoring. In: 2010 International Conference on Body Sensor Networks (BSN), pp. 36–43. IEEE (2010)
7. Henriksen, M., Lund, H., Moe-Nilssen, R., Bliddal, H., Danneskiod-Samsoe, B.: Test-retest reliability of trunk accelerometric gait analysis. Gait & Posture 19(3), 288–297 (2004)
8. Korpipaa, P., Mantyjarvi, J., Kela, J., Keranen, H., Malm, E.: Managing context information in mobile devices. IEEE Pervasive Computing 2(3), 42–51 (2003)
9. Krishnan, N., Juillard, C., Colbry, D., Panchanathan, S.: Recognition of hand movements using wearable accelerometers. Journal of Ambient Intelligence and Smart Environments 1(2), 143–155 (2009)
10. Lester, J., Choudhury, T., Borriello, G.: A practical approach to recognizing physical activities, pp. 1–16 (2006)
11. Liao, L.: Location-based activity recognition. Ph.D. thesis, Citeseer (2006)
12. Miluzzo, E., Lane, N., Fodor, K., Peterson, R., Lu, H., Musolesi, M., Eisenman, S., Zheng, X., Campbell, A.: Sensing meets mobile social networks: the design, implementation and evaluation of the cenceme application. In: Proceedings of the 6th ACM Conference on Embedded Network Sensor Systems, pp. 337–350. ACM (2008)
13. Shadbolt, N.: Ambient intelligence. IEEE Intelligent Systems 18, 2–3 (2003)
14. Verhagen, C.: Some general remarks about pattern recognition; its definition; its relation with other disciplines; a literature survey. Pattern Recognition 7(3), 109–116 (1975)
15. Want, R.: You are your cell phone. Pervasive Computing, IEEE 7(2), 2–4 (2008)

INEF12Basketball Dataset and the Group Behavior Recognition Issue

Alberto Pozo, Jesús García, and Miguel A. Patricio

Abstract. Activity recognition is one of the most prolific fields of research. For this reason, there are new fields of research that expand the possibilities of the activity recognition: Group behavior recognition. This field does not limit the number of elements in the scene, and there are a lot of new elements that must be analyzed. Each group, like each individual element, has its behavior, but this behavior depends on their elements, and the relationships between these elements. All these new elements cause that group behavior recognition was a new field of research, with some similar elements but it must be studied apart. This way, group behavior recognition is a novel field, in which there are not many researches and there are not many datasets that could be used by researchers. This situation causes the slow advance of the science in this field. This paper tries to show a complete description of the problem domain, with all the possible variants, a formal description and show a novel architecture used to solve this issue. Also describes a specific group behavior recognition dataset, and shows how it could be used.

Keywords: Dataset, Group behavior recognition, activity representation, computer vision.

1 Introduction

Behavior recognition, with humans or with other kind of elements, is one of the most prolific fields of the current research. One of the typical restriction in these research is that must be only one element in the scene. The element's behavior is analyzed and the system recognizes the activity that the element is doing. There are a lot of papers with this objective, like and.

Alberto Pozo · Jesús García · Miguel A. Patricio
Applied Artificial Intelligence Group, Universidad Carlos III de Madrid,
Avd. de la Universidad Carlos III, 22, 28270, Colmenarejo, Madrid, Spain
e-mail: {alberto.pozo,jesus.garcia,miguelangel.patricio}@uc3m.es
       http://www.giaa.inf.uc3m.es

S. Omatu et al. (Eds.): Distributed Computing and Artificial Intelligence, AISC 151, pp. 151–160.

The one-element restriction is valid for several situations, but in other cases, there are not isolated elements but these elements are joined in several groups. This could be useful in a number of situations, like group sports, animal behavior, coordinated human activities, etc. In these situation the group behavior is not only the combination of the individual behaviors, however, there are a lot of elements that must be analyzed like the internal group relationships, external group relationships (with others groups) individual behaviors, groups hierarchy, etc.

So there are several commons elements between activity recognition and group behavior recognition, but there are also several new elements so this is a novel field of research that provides a lot of new challenges and difficulties. There is an essential element in any research that cannot be missing, datasets. All the experimentation is based on this element, and it is essential to create a new system, to improve it, and to prove its results.

This new field of research has some papers that manage the problem from very different points of view, like (3) or (4). However, there are not many specific dataset that could be used for these researches, and this is a serious difficulty for the researches. This paper describes a novel representation of this problem domain, showing all the different types of group behavior recognition problems and also describes a new specific dataset, which could be used to improve the current researches and to create new group behavior recognition systems. Also this dataset could be used to prove the existing techniques with a different dataset.

The next section describes related work. Section 3 describes problem domain. Section 4 describes our architecture approach and section 5 describes our dataset, which is followed by the conclusions.

2   Related Work

As described above, there is not many dataset specializing in group behaviors for the scientific community. There is extensive literature dedicated to activity recognition, and there are a lot of dataset specializing in this area, which there is only one studied element, like KTH dataset (5) or Weizman (6); which are used in many papers like (7), (2), (8). But these dataset cannot be used to prove the group behavior recognition techniques, so we need a specializing group behavior dataset that provides the behavior of a group, with some raw information like positioning, individual actions, etc.

This is the case of the Nanez Pers work, where he has launched a three parts dataset. Each part related with one sport: Basketball, squash and European handball. Nanez Pers in (9) presents a dataset specializing in computer vision, and one of the three parts (European handball part) is focus on the group behavior recognition.

In this dataset, there is a lot of valuable information for the group behavior recognition issue. CVBASE06 (9) has ten minutes data from a European handball mach, with one team players tracking. So there are annotated all positions of each player (from one team) in each frame of video. Furthermore, CVBASE06 gather information about what are doing every player (of the tracked team) like throwing the ball, for example, and the team performed action, like aggressive attack. All this information has been annotated by an expert (European handball coach) by several views of the video. So this dataset combines both types of information, individual data like positioning and action, and group data like action performed. So CVBASE06 could be used to group behavior recognition issue, however, there are some others aspect that could be improve.

Moreover, there is scientific and educational international joint project called RoboCup (10) that could be used to test and improve the group behavior recognition systems. This competition has many different leagues divided into five main clsasses: RoboCup Soccer, RoboCup@Home, RoboRescue, RoboCupJunior and Demostratios. Inside the Soccer league there is a client-server based architecture simulator that could be gather information useful for the group behavior recognition issue.

On the other hand we could read the works from Ruonan Li and Rama Chellappa (3), (11), and Behjat Siddiquie (12) where they uses an American football dataset called GaTech Football Play. Unfortunately, there are no references to use this dataset by the other researches.

3 Group Behavior Recognition Issue: A General Overview

Group behavior recognition is a novel field of research that comes from the elimination of the one-element restriction in activity recognition issue. This field of research has a lot of potential domains like group sports, military intelligence, fauna behavior recognition, video surveillance, etc.

Group behavior recognition is composed by two steps: in the first one the features of the system should be extracted, and in the second one the features are used to recognize the behavior. The system could have a lot of types of features like position, individual action, trajectory, speed, color, etc.

In this paper we are going to focus on the second step, there is only a short description about how was the dataset construction process (feature extraction), and we try to show how this information could be used in the second step (behavior recognition).

3.1 General Description

There are a lot of types of domains and problem where we could use the group behavior recognition, for this reason it is important to define a general description that includes all of them.

In general, we have several elements that are forming groups; these elements are moving around the scene and have relationship with other elements of their group and with the other group. Also, groups have relationship with other groups. So we have a general scene where exist a number of elements (which could be fixed or not). Each element of the system has a set of features (like positioning, color, shape, etc.) The element's features could suffer changes in time. Each element of the system should belong to a group, and could belong to many groups at the same time. It is important to emphasize that any element of the system must be in a group or not, so there could be isolated elements.

Each group has an internal and an external attitude. Each attitude could be cooperative or competitive. Internal attitude defines the attitude between the members of a group, and external attitude defines the attitude of the group respecting of the rest of the groups.

Different domains could have some different descriptions, in general, there are six types of problems that's could be classify by three main features: If there are one or more groups, if there are one or more behavior on each sequence, and if the number of groups is fixed or is not.

The following image shows this classification.

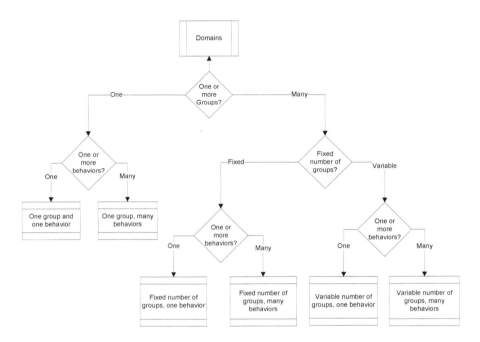

Fig. 1 Problem classification

3.2 Problem Description

In the first level the features extractor system analyze the scene and gather all the relevant information to the group behavior recognition system. So in the second level we have one sequence composed by a number of T instants, where are included a number of N elements (this number cannot change in time). The elements of the scene are distributed in a number of G groups. Each element of the scene has a set of features that depends on the specific problem domain. Some of these features could be logical (like boolean), numeric, text, etc. But in the system must be some feature that indicates the location of the element in some way.

This is because we are focused in the problems where the location and the trajectories of the elements are important, but it is important to remark that this restriction is carry out by most of the problem's domains.

The elements include in some group could not change, and the number of groups and elements could be fixed or could be not. Moreover, all the elements must be in a group, so there is not any isolated element.

So for each moment t, we have the scene information $S_t = \{G_t^1, G_t^2, ..., G_t^m, ..., G_t^M\}, 1 \leq m \leq M$ (where M is the group amount) and $G_t^m = \{I_t^1, I_t^2, ..., I_t^n, ..., I_t^{N_m}, \}, 1 < n \leq N_m$ (where N_m is the number of elements of the group m), and $I_t^n = \{f_t^1, f_t^2, ..., f_t^c, ..., f_t^C, \}, 1 \leq c \leq C$ (where C is the features amount).

There are four types of features describe above: *Positive, boolean, relative* and *enum*.

Boolean: this type of feature could be one or zero, **positive**: feature could have some value between zero and infinite, **relative**: could have some value between zero and one, and indicates how much the element fulfill with something, and **enum**: this type of features could be one of a list of possible values determinates by the problem domain.

4 Architecture

Group behavior recognition is a general research field that includes a lot of problem domains like was described above. The problem could be divided in two steps, first level must extract the features of the elements and second level must use this features to recognize the behavior of the groups.

There are two different situation depending on the number of behaviors per sequence, if there is only one, the system only have to recognize it, but if there are many behaviors the system have to cut the sequence before recognize them.

The architecture shows below tries to manage with the second situation, and it is a general approach that could be implemented with many different specific algorithms.

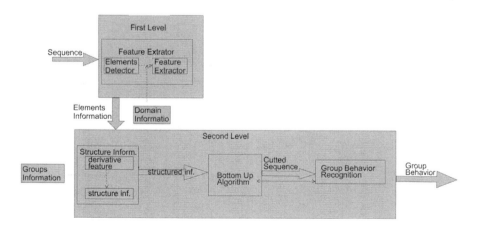

Fig. 2 Architecture

This architecture approach manages the problem of the many behaviors on one sequence with an algorithm to cut the primitive sequence, like buttom up.

The makeup group information is obtained by a external element of the system, in most case an expert.

5 INEF12Basketball Dataset

Within the field of artificial intelligence, especially in the field of activity recognition, there are plenty of papers from different authors trying to create autonomous systems that were be able to detect, classify, or simulate the insolate individual behavior.

All these papers have a lot of experiments based on a several activity recognition datasets, which allow this research field advances quickly. However, there is a novel field that emerges from the elimination of one restriction of the domain: It could be more than one element, joined in a group. This new field of research has some resemblances but also have a lot of differences so it must be manage with new approaches and techniques.

This new field of research is focus on a group of elements, these elements have a lot of relationship with others elements of the group, and with others elements of the other groups. So the number of the elements and relationships increase exponentially. This is very important, because some of the techniques use to activity recognition could be not feasible in time.

This new field still has not many researches compared to its predecessor, and it is the same with the number of specific datasets. INEF12Basketball dataset aims to contribute to this field of research, providing a set of low-level data such as location of the system's elements and other higher-level data related to the individuals and groups behavior.

So we try to create a useful dataset that could be used to improve and prove the novel techniques developed for group behavior recognition issue.

5.1 Domain

All the information of the dataset was gathered in a basketball training session, carried out by four players (two versus two). This session was recorded in 225 seconds of video composed by twenty seven sequences. Each sequence went to one basketball move.

All the moves played in the scene are composed by one pick and roll attack and some of these ways of defense: to fight over, to go below, to help and recover, to show and go, to show and recover, to switch and trap.

Group behavior information is related with the type of defense played, so all the moves have the same attack and the behavior information is focused on the defense. In a pick and roll move, the attacking team try to set a screen (pick) for a teammate handling the ball and then slips behind the defender (rolls) to accept a pass. It could be defended by several types of defenses; here we can read a short description of each of them.

Fight over: In this case, the blocked player trays to skip the screen fighting over them.

Go below: In this case, the blocked player trays to skip the screen going below, this take some extra time.

Help and recover: The player whose defense to the player that plays the screen tries to help to his partner with a short lateral movement.

Show and go: In this case the second defender goes with the ball for some time, and he gives some help to his partner.

Show and recover: This case is similar to previous one, but in this case the help is longer in time.

Switch: Both defenders change their marks.

Trap: Both defenders go to mark to the owner of the ball, in a 2vs1 situation.

So all the information is gathered in twenty seven consecutively sequences of basketball training session, with one type of attack and seven different types of defenses.

5.2 Description

INEF12Basketball dataset is composed by a video data recorded by four cameras in fixed positions on a basketball court.

There are 27 plays recorded, in three minutes and 45 seconds. At 25 frames per second the video has 5.627 frames.

Video was recorded in MPEG4 DivX avi container and no audio associated. The size of each frame is 320 pixels wide and 240 pixels high. Apart from the video, CVBASE12Basketball dataset provides useful information about the scene; like player location, group action, owner of the ball, etc. Player's location is annotated around the whole video. This information is presented in one plane text file, in two coordinated axes: screen and court. This information was captured using a modified version of ViPER program. The tracking was performed by a tracker algorithm with expert supervision, all this information is saved in the plane text file.

As well as the position, dataset contains information about which player has the ball in each frame. (There are frames in which the ball is not had by any player).

This information was annotated by an expert. The file contains the frame in which the player has the ball, and the frame in which the player louse the ball. Moreover, in order to serve as a useful research fields related to the group behavior recognition, segmentation or classification; group activity information has been incorporated.

There are information about witch type of defense is played in each turn, in the section there are the explanation of each type of defense.

Coach called Ignacio Refoyo was the expert that provides all the technical data, like defense played by the players.

All this information and much more could be found in.

5.3 Using the Dataset

All the information of INEF12Basketball dataset described above is stored in files
(.avi and .txt) but could be managing by a few set of Matlab functions described
below. These functions allow to the users manage all the information without
knowing the format and the structure of the txt files. And provide a quick and easy
way to start working on the algorithms designed, taking from the outset all the information available in the Matlab workbench.

There are seven Matlab functions:

1. CVBASE12BasketballInit
2. CVBASE12BasketballGetPos
3. CVBASE12BasketballShowTrajectories
4. CVBASE12BasketballGetFrames
5. CVBASE12BasketballShowFrames
6. CVBASE12BasketballGetTeamActivity
7. CVBASE12BasketballHasBall

The first one is used to initialize all the system, loading the information on the
workbench. Functions 2, 3; 6 and 7 allow the users manage all the information
about the location of the players, the defense played and the owner of the ball respectively.

The following image shows the information presented to the user with the function 3, that print the trajectories of the players in the couch.

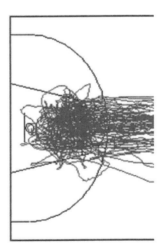

Fig. 3. Show trajectories

At last, functions 4 and 5 allow the users manage the raw video information, 4
obtains the chosen frame information provided by different cameras and the time
of the frame in several variables, and 5 show the video information as we could
see below.

Fig. 4 ShowFrames

6 Conclusions

Group behavior recognition is a field of research that emerges as an extension of the activity recognition, when the one-element restriction is eliminated. The elements and the features in group behavior recognition have several differences and it must be manage with different techniques and approaches.

So we have a novel field of research, which is still a long way to go, and it needs a lot of work to have a good approaches. One of the current difficulties on this field is that there are not many dataset to create the approaches and improve and prove it with experiments. So we need more dataset specializing in group behavior recognition issue.

This paper describes a new dataset that aims to solve this problem, using the information gather in a basketball training session.

Because in the most of the group behavior domains the location of the elements is a important feature to recognize the group behavior, INEF12Basketball dataset includes this data (the location of each player in each frame), and some other information like the current owner of the ball, and, of course, the group activity carried out by the group.

In addition, the dataset includes a set of Matlab functions that allow to the users handle the data in a quickly and easy way. So researchers only need to know some about Matlab language and could have all the gathered information in their hands.

Acknowledgments. This work was supported in part by Projects CICYT TIN2011-28620-C02-01, CICYT TEC2011-28626-C02-02, CAMCONTEXTS (S2009/TIC-1485) and DPS2008-07029-C02-02.

References

1. Liu, J., Luo, J., Shah, M.: Recognizing realistic actions from videos in the wild (2009)
2. Schindler, K., van Gool, L.: Action snippets: How many frames does human action recognition require (2008)
3. Ruonan, L., Rama, C., Shaohua, K.Z.: Learning Multi-modal Densities on Discriminative Temporal Interaction Manifold for Group Activity Recognition. S.N., New York (2009)
4. Ramos, F., Ayanegui, H.: Tracking behaviours of cooperative robots within multiagent domains. S.N., Tlaxcala (2010)
5. Schüldt, C., Laptev, I., Caputo, B.: Recognizing human actions: a local SVM approach. In: Proceedings of the 17th International Conference on Pattern Recognition, ICPR 2004, vol. 3, pp. 32–36 (2004)
6. Gorelick, L., Blank, M., Shechtman, E.: Actions as Space-Time Shapes. In: Tenth IEEE International Conference on Computer Vision, ICCV 2005, vol. 2, pp. 1395–1402 (2005)
7. Jhuang, H., et al.: A biologically inspired system for action recognition. In: IEEE 11th International Conference on Computer Vision, ICCV 2007, pp. 1–8. IEEE (2007)
8. Niebles, J.C., Fei-Fei, L.: A hierarchical model of shape and appearance for human action classification. In: IEEE Conference on Computer Vision and Pattern Recognition, CVPR 2007, pp. 1–8. IEEE (2007)
9. Pers, J.: CVBASE 06 Dataset: A Dataset for Development and Testing of Computer Vision Based Methods in Sport Environments. S.N., Ljubljana (2005)
10. Asada, M., Kitano, H.: RoboCup..: robot soccer World Cup.. Springer, Heidelberg (1999)
11. Li, R., Chellappa, R.: Group Motion Segmentation Using a Spatio-Temporal Driving Force Model. S.N., Maryland (2010)
12. Siddiquie, B., Yacoob, Y., Davis, L.S.: Recognizing Plays in American Football Videos
13. Refoyo, I., et al.: INEF12 Basketball Dataset. S.N., Madrid (2011)
14. Schindler, K., Van Gool, L.: Action snippets: How many frames does human action recognition require? In: IEEE Conference on Computer Vision and Pattern Recognition, CVPR 2008, pp. 1–8. IEEE (2008)

Development of Interactive Virtual Voice Portals to Provide Municipal Information

David Griol and María García-Jiménez

Abstract. In this paper, we describe a Voice Portal designed to provide municipal information by phone. It includes the set of modules required to automatically recognize users' utterances, understand their meaning, decide the following response and generate a speech response. The different functionalities include to consult information about the City Council, access city information, carry out several steps and procedures, complete surveys, access citizen's mailbox to leave messages for suggestions and complaints, and be transferred to the City Council to be attended by a teleoperator. The voice portal is, therefore, pioneer in offering an extensive and comprehensive range of public services accessible through speech, creating a new communication channel which is useful, efficient, and easy to use. In addition, the voice portal improves the support of public services by increasing the availability, flexibility, control and reducing costs and missed calls. The paper describes the application software, infrastructures required for its operation 24 hours a day, and preliminary results of a quality assessment.

Keywords: Voice Portals, Conversational Agents, Speech Interaction, Agent & Multiagent Systems for AmI.

1 Introduction

Technological advances currently reached by computers and mobile devices allow now its use to access information and a number of services. In addition, users want to access these services anywhere and anytime in a natural, intuitive and efficient way. This way, interfaces based on voice portals have become one of the main options to facilitate this kind of communication.

David Griol · María García-Jiménez
Computer Science Department, Carlos III University of Madrid, Avda de la Universidad, 30, 28911 - Leganés, Spain
e-mail: david.griol@uc3m.es, 100025080@alumnos.uc3m.es

S. Omatu et al. (Eds.): Distributed Computing and Artificial Intelligence, AISC 151, pp. 161–172.
springerlink.com   © Springer-Verlag Berlin Heidelberg 2012

Speech access is then as a solution to the shrinking size of mobile devices (both keyboards to provide information and displays to see the results). In addition, speech interfaces facilitate the communication in environments where this access is not possible using traditional input interfaces (e.g., keyboard and mouse). It also facilitates information access for people with visual or motor disabilities. In addition, the use of mobile technologies has been currently defined as one of the main indicators of the evolution of new information technologies.

Furthermore, Spanish Law 11/2007 on electronic access to public services[1] defines the development of multi-channel access to information as one of the main obligations of the municipalities. This Law also explicitly recognizes the right of citizens to interact electronically with Public Administration. However, a detailed study of the current situation in the provision of public services by local councils provides the main conclusion of a lack of electronic public services that meet the functions of providing information; guidance and advice; manage suggestions and complaints; as well as consult official announcements and procedures. Therefore, a public voice portal not only meets this Law, but also constitutes an efficient tool to provide speech access to public services by phone at anytime and anywhere.

The developed voice portal offers citizens different functions: consult information about the City Council (Government Team, Councils, etc.), know city information (history, geographic and demographic data, access to the city, yellow pages, movie show times, news, events, weather, etc.), carry out several steps and procedures (check lists and personal files, book municipal facilities or make an appointment), complete surveys, access citizen's mailbox to leave messages for suggestions and complaints, and be transferred to the City Council to be attended by a teleoperator.

The voice portal integrates additional technologies, such as the VoiceXML standard, databases, web and speech servers, and several programming languages (SQL, PHP, HTML), which make it more dynamic and flexible and increase its quality and efficiency.

Thanks to the developed application, speech services are automatically provided, allowing a 24 hour a day access to information in a more natural way and in environments in which using the traditional access interfaces, such as keyboard or mouse, this access would not be possible. In addition, the application facilitates the access for people with visual or motor disabilities, helping them to eliminate access barriers and enabling a more accessible technology world.

2   Conversational Agents

A conversational agent is a software that accepts natural language as an input and produces natural language as an output engaging in a conversation with the user [7, 6, 4]. To successfully manage the interaction with the users, conversational

---

[1] http://www.boe.es/aeboe/consultas/bases_datos/doc.php?id=BOE-A-2007-12352

agents usually carry out five main tasks: automatic speech recognition (ASR), natural language understanding (NLU), dialog management (DM), natural language generation (NLG) and text-to-speech synthesis (TTS).

Speech recognition is the process of obtaining the text string corresponding to an acoustic input [10, 1]. It is a very complex task as there is much variability in the input characteristics, which can differ depending on the linguistics of the utterance, the speaker, the interaction context and the transmission channel. Linguistic variability involves differences in phonetic, syntactic and semantic components that affect the voice signal. Inter-speaker variability refers to the big difference between speakers regarding their speaking style, voice, age, sex or nationality.

Once the conversational agent has recognized what the user uttered, it is necessary to understand what he said. Natural language processing is the process of obtaining the semantic of a text string [8, 1]. It generally involves morphological, lexical, syntactical, semantic, discourse and pragmatical knowledge. Lexical and morphological knowledge allow dividing the words in their constituents distinguishing lexemes and morphemes. Syntactic analysis yields a hierarchical structure of the sentences, while semantic analysis extracts the meaning of a complex syntactic structure from the meaning of its constituents. In the pragmatic and discourse processing stage, the sentences are interpreted in the context of the whole dialog.

There is not a universally agreed upon definition of the tasks that a dialog manager has to carry. Traum and Larsson [11] state that dialog managing involves four main tasks: i) updating the dialog context, ii) providing a context for interpretations, iii) coordinating other modules and iv) deciding the information to convey and when to do it. Thus, the dialog manager has to deal with different sources of information such as the NLU results, database queries results, application domain knowledge, knowledge about the users and the previous dialog history [5, 2].

Natural language generation is the process of obtaining texts in natural language from a non-linguistic representation. The simplest approach consists in using predefined text messages (e.g. error messages and warnings). Finally, a text-to-speech synthesizer is used to generate the voice signal that will be transmitted to the user.

3   Related Work

Human beings have always been interested in being able to communicate with artificial companions. In fact, one of the main challenges of AI since his early days has been to achieve the man-machine communication through natural language.

At the beginning of the XX century J.Q. Stewart built a machine that could generate vocalic sounds electrically; and during the 30s, the first electric systems covering all sounds were built. The first one was the VOCODER, an speech analyzer and synthesizer developed in Bell Laboratories that could be operated by a keyboard. At the same time appeared the first systems with very basic natural language processing capabilities for machine translation applications. During the 40s, the first computers were developed and some prominent scientists like Allan Turing pointed out their potential for applications demanding "intelligence". This was the starting point that fostered the research initiatives that in the 60s yielded the first conversational agents. For example Weizenbaum's ELIZA [12], which was based on keyword spotting and predefined templates.

Benefiting from the incessant improvements in the areas of speech recognition, natural language processing and speech synthesis, the first research initiatives related to spoken dialog systems appeared in the 80s. To some extent the origin of this research area is linked to two seminal projects: the DARPA Spoken Language Systems in the USA [3] and the Esprit SUNDIAL in Europe [9].

Among the most important research projects in the 90s with multi-domain capabilities, stands out the DARPA Communicator. This government-funded project aimed at the development of cutting-the-edge speech technologies, which could employ as an input not only speech but also other modalities. Currently experts have proposed higher level objectives to develop dialog systems, such as providing the system with advanced reasoning, problem solving capabilities, adaptiveness, proactiveness, affective intelligence, multimodality and multilinguality [4]. As can be observed, these new objectives are referred to the agent as a whole.

Related to the development of voice portals in Spain, the few existing applications currently offer access to only select between a reduced number of options. Taking into account this selection, the user is then redirected to be attended by a human operator (e.g., the voice portal developed for the Cabildo of Gran Canaria). Other applications, like VoxWebPC[2], allow speech navigation through the pages included in the website of the City, as well as to hear these contents by means of synthesized speech. Another application including these technologies is the website of the City Hall of Zaragoza[3], which provides a speech access to several sections of the webpage. A similar system has been integrated in the website of the City Hall of Alicante[4], using these technologies to convert web content into high quality digitized voice.

There are not currently voice portals in Spain offering the functionalities described for our voice portal. Then, it is pioneer in offering an extensive and comprehensive range of public services accessible through speech. This way, our voice portal creates a new communication channel which is useful, efficient, easy to use, and accessible. In addition, it improves the support of public services by increasing the availability, flexibility, control, and reducing costs and missed calls.

4   Interactive Voice Portal to Provide Municipal Information

The developed municipal voice portal has been developed following the client-server paradigm with the architecture described in Figure 1.

Regarding the VoiceXML server, the VoiceXML interpreter answers the users' calls and interpret VoiceXML documents to provide them required services. The interpreter also requests the required resources for the application, defines the logic

---

[2] http://www.voxweb.es/spanish/accesibilidad.html

[3] http://www.zaragoza.es/ciudad/servicios/voz.htm

[4] http://www.alicante.es

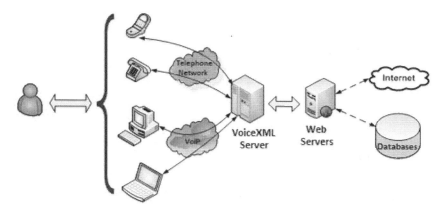

Fig. 1 Architecture designed for the Interactive Voice Portal

of the services and stores users' session state to interact accordingly. To carry out these actions, the VoiceXML interpreter include different systems to deal with users' calls, manage the communication with the servers and access the required resources, play audio files, convert text to speech, collect user data, perform voice recording, and manage sessions and events.

There are currently many VoiceXML language interpreters. One of the most important ones, given the number of functionalities provided, is Voxeo Evolution[5]. Voxeo allows creating VoiceXML applications and access them by means of a local phone number and/or a Skype number. Voxeo also allows to track calls in real time, as well as automatically create log files. These files are very useful for debugging and optimizing the application. In addition, the Voxeo platform provides a fast and efficient support system, which includes forums, support tickets and very complete documentation. Finally, Voxeo also provides the VoiceXML interpreter and the ASR and TTS components required for the voice portal. Our system uses the Prophecy 9 Multi-Language VXML implementation, which has allowed to develop the application for its use in Spanish.

Regarding the Web server, PHP and VXML files are used to implement each one of the services provided by the voice portal, in addition to access MySQL databases containing the specific information. The different functionalities and corresponding files allow users to complete more than one action in each call. To complete databases in the application with specific information, we have considered the information provided by the website of the City Hall of Alcorcón (Madrid).

---

[5] http://evolution.voxeo.com/

4.1 Static and Dynamic Information

Static information has been collected from web pages, specially from the website of the City Hall of Alcorcón, stored and well classified in the databases of the application. Each time users request this information, the system accesses the database and returns this information encapsulated into a VoiceXML file. Examples of this type of information include the history of the city, access information, or contact information of hotels and main offices in the city.

Dynamic information include local news and events, weather information, surveys, and entertainment guides for cinemas and theaters. This information is automatically updated in the application by means of a PHP-based procedure that access the required web pages, carry out a syntax processing of this information, and store the updated information the database. Each time the user requires this type of information, the system only has to access the database and return it.

All the application dialogs use voice grammars and DTMF, which means that users can access menus by speech or using the phone keys, making the application more accessible. Grammars are encoded as XML type is the standard format defined by the W3C and, therefore, supported by any VoiceXML platform. In addition, this format allows greater flexibility in terms of grammar structure and debugging.

Static grammars deal with information that does not vary over time, including a small number of options to choose from. These grammars are coded in the same file where they are used. The basic structure designed for these grammars only allows to mention exactly a word or a small set of words, besides the corresponding DTMF code according to the option that was selected. This type of grammar is used at the beginning of the interaction to inform users about how to interact with the system. To do this, system prompts include detailed information about what to say to access each of the provided functionalities.

The second structure for static grammars was designed so that the set of words that has to be uttered to to select a specific option is not unique. That is, we have defined the possible combinations of words and set of words that users might say to refer to the same option (e.g., users can access the functionality "Traffic and Transport" by saying "traffic and transport", "traffic" or "transport". This provides more natural dialogs and avoids a larger number of recognition errors. These grammars are specially useful once users already knows how to interact with the system and do not require additional information about the next action required by the system.

Dynamic grammars include information that varies with time and often deal with large amounts of data. These grammars are automatically created using PHP files to manage their contents (creation, obtain contents, modify and update information).

4.2 Modules and Functionalities

The different functionalities of the developed voice portal include to consult information about the City Council, access city information, carry out several steps and procedures, complete surveys, access citizen's mailbox to leave messages for suggestions and complaints, and be transferred to the City Council to be attended by a teleoperator.

The *Home module* implements the first dialog that is provided to the user. The options that the system provides to users are divided into 5 well-differentiated modules that takes into account the type of interaction and data that is facilitated: information, procedures and formalities, surveys, mailbox of the citizen, and human operator. Therefore, this module facilitates the access to the rest of functionalities provided by the portal.

The *Information module* provides specific information about the city. This information has been divided into 6 categories and classified so that users can easily access each one of the provided functionalities in this module:

- *City Council*: This module provides specific information related to the local government, local governing entities, teams and areas.
- *City*: By means of this module it is possible to access information related to the city (history, access, and yellow pages with bars, cafés, restaurants, shops, hostels, hotels and entertainment guides for cinemas and theaters).
- *Thematic Areas*: This module integrates the rest of information that users can consult by means of the voice portal. This information has been divided into 15 areas, which respectively provide general information, competences and contact details. Additional functionalities can be perfectly incorporated by simply adding more static information in the corresponding database of the application.
- *News*: This section provides current news of the municipality, including the date, title and corresponding description of each one of them.
- *Events*: This section reproduces the list of events of the municipality, providing its specific area, title, date, location, and detailed description.
- *Weather information*: Users can get the current weather information from the municipality and a forecast for the next two days.

An example of a dialog related to a query about a specific thematic area is shown below (dialog translation from Spanish to English).

```
S: Please select your area of interest:
   1-Social Welfare
   2-Circulation and Transportation
   3-Culture and Leisure
   4-Sports
   5-Administration
   6-Education
   7-Employment
   8-Environment
   9-Parks and Gardens
   10-Health and Consumer
   11-Security
   12-Planning
   13-Housing
   14-Finance and Equity
   15-Cooperation.
```

```
U: Social Welfare.
S: You have selected the Social Welfare area. Please select
   a specific topic:
   1-Area Expertise.
   2-General Information.
   3-Contact Information.
U: Contact.
S: The contact information of Social Welfare is:

   Social Services Centre "VALDERAS".
   Parque del Teide, 1. Alcorcón 28924.
   MetroSur: Puerta del Sur. Bus: 514, 511.
   Phone: 916 106 611 - 916 106 797 - 916 426 024.
   Fax: 916 425 311.

   Social Services Centre "EL TEJAR".
   Tiled Square, 4. Alcorcón 28921.
   MetroSur and rail: Central Alcorcón.
   Phone: 916 648 503 - 916 648 504.
   Fax: 916 425 311.

S: What do you want now?
   1 - Additional information of this area.
   2 - Menu of subject areas.
   3 - Information Menu.
   4 - Main menu.
   5 - Exit the application.
U: Exit.
S: You have chosen to exit the application. Thanks for using
   our voice portal. See you soon!
```

The *Procedures and Formalities* module provides users' access to consult procedures and steps, verify the correct incorporation into specific lists, check the status of dossiers and documents, book a municipal facility, or make an appointment related to a specific municipal service. An example of a dialog for a user asking about a specific procedure is shown below (dialog translation from Spanish to English).

```
S: Here you can access and complete procedures and
   formalities. To verify whether you are included in a list
   or not, say list or press 1. To check the status of your
   dossier, say dossier or press 2. To make a reservation
   of a sport facility, say reservation or press 3. If you
   want to schedule an appointment, say appointment
   or press 4.
U: Dossier.
S: Please tell us your ID saying or dialing the 8 digits one
   by one.
U: 47452060.
S: You have provided 47452060. The status of your dossier is
   as follows: Your certificate of registration has been
   requested.
```

```
S: What do you want to do now?
   To return to the main menu of the voice portal, say
   Alcorcón or press 1. To access or complete other
   procedures, say procedures or press 2. If you want to
   exit the application, say exit or press 3.
U: Exit.
S: You have chosen to exit the application. Thanks for
   using our voice portal. See you soon!
```

The *Surveys* module allows the City Council to easily and quickly know citizens' opinion about specific issues and queries. Surveys in the application have been designed to be completely anonymous, so the voice portal does not require users' identification. Once users have finished answering a survey, they are given the options to listen to the current stored results for the specific query. An example of a dialog for a user answering a specific survey is shown below (dialog translation from Spanish to English).

```
S: Then you can complete a municipal survey and know the
   preliminary results. I will ask you a question and
   provide you 4 possible answers.
   Which political party are you going to vote?
   1: Partido Socialista.
   2: Partido Popular.
   3: Izquierda Unida.
   4: Los Verdes.
U: Los Verdes.
S: You have selected: Los Verdes.
S: Do you want to know the results of the survey?
   Say yes or press 1, or say no or press 2.
U: Yes
S: The results of the survey "Which political party are you
   going to vote?" are as follows:
   Answer 1: Partido Socialista, 25%.
   Answer 2: Partido Popular, 25%.
   Answer 3: Izquierda Unida, 25%.
   Answer 4: Los Verdes, 25%.
   Total votes: 8.
S: What do you want to do now?
   1 - Return to the main menu of the voice portal.
   2 - Exit the application.
U: Exit.
S: You have chosen to exit the application. Thanks for
   using our voice portal. See you soon!
```

The *Citizen's mailbox* implements the functionality of recording a user' speech message and store it for further processing. Thus, citizens can provide their requests, complaints, claims or comments at anytime and anywhere. The Citizen's mailbox is then managed by a specific Office of the City Hall. In addition, if users provide their contact information (telephone, mobile phone or email), this Office would contact

them to provide a personalized response to their request. An example of a dialog
for a user accessing the Citizen's mailbox is shown below (dialog translation from
Spanish to English).

```
S: Welcome to the Citizen mailbox. You can contact us and make
   your suggestions, complaints and other comments according to
   the subjects of our voice portal.
S: You can record your message after the tone. Please provide
   first your name and phone so we can respond appropriately,
   and then provide your message. Finally, please keep waiting
   to confirm your recording. Thank you.
S: 'beep'
U: Hello, I am Juan Pérez and I want to congratulate you for
   the new voice portal service in Alcorcón. Thanks.
S: Your message is as follows: "Hello, I am Juan Pérez and I
   want to congratulate you for the new voice portal service
   in Alcorcón. Thanks."
S: Say yes or press one for sending this message. Say no or
   press 2 if you want to re-record your message.
U: Yes
S: We have saved your message. Your recording will be now
   managed by our staff, and then we will contact you. Thank
   you for collaborating with the City of Alcorcón.
S: What do you want to do now?
   To return to the main menu of the voice portal say Alcorcón
   or press 1. If you want to exit the application say exit or
   press 2.
U: Exit.
S: You have chosen to exit the application. Thanks for using
   our voice portal. See you soon!
```

Finally, the *Tele-Operator* module transfers the user's call to the switchboard aat the
City Council and, thus, the user is attended by a human operator.

5   Evaluation

The assessment, performance study and usability analysis of conversational agents
are procedures to minimize costs and optimize results in applications in which
these agents are integrated. The evaluation of this voice portal has been carried
out through quality assessments. To do this, a questionnaire has been designed to
evaluate users' subjective opinion and satisfaction with the developed voice portal,
thus obtaining a qualitative assessment of users' perception of the system.

This assessment is focused on how users appreciate that they are understood by
the system and how they understand the messages generated by the system, the
perceived interaction rate, the presence of errors, users understanding about next
actions required by the system, the similarity between the developed system and a
human operator, and the overall satisfaction with the system. In addition, additional
information from users about their knowledge level about new technologies and
previous use of dialog systems were considered as an estimator of the users' profile.
The questionnaire developed for this purpose consists of the following 10 questions:

i) *Q1*: State on a scale from 1 to 5 your previous knowledge about new technologies; ii) *Q2*: State on a scale from 1 to 5 your previous uses of speech-based interfaces iii) *Q3*: State on a scale from 1 to 5 your previous uses of voice portals; iv) *Q4*: Did the system correctly understand you during the interaction?; v) *Q5*: Did you understand correctly the messages of the system?; iv) *Q6*: Do you think that the interaction rate was adequate?; vii) *Q7*: Was it simple to obtain the requested information?; viii) *Q8*: *Set the difficulty level of the system for you.*; ix) *Q9*: Do you think that the system behaved in a similar way as human being?; x) *Q10*: In general terms, are you satisfied with the performance of the system? The possible answers to the complete set questions were the same: *Never, Rarely, Sometimes Usually* and *Always*.

The assessment test was completed by 20 professors and students of our university who were introduced on the main functionalities of the voice portal and required to complete the questionnaire once finished their interaction. Users freely chose to perform actions and select between the different functionalities, modules and sub-modules. Table 1 shows the results for each one of the questions in the subjective evaluation of the application.

Table 1 Results of the subjective evaluation of the Voice Portal

	Q1	Q2	Q3	Q4	Q5	Q6	Q7	Q8	Q9	Q10
Average value	3.3	3.0	2.9	4.1	4.5	3.1	4.2	4.1	2.9	4.3
Minimum value	1	2	2	3	4	2	3	3	2	3
Maximum value	5	4	4	5	5	4	5	5	4	5
Standard deviation	1.2	0.6	0.7	0.6	0.5	0.5	0.5	0.8	0.5	0.6

From the analysis of the results of the evaluation, it can be observed that users' knowledge about new technologies and use of dialog systems is varied. Most of the users found that the interaction with the system was very easy and the interaction rate is considered as suitable. They also considered that the system correctly understood their messages. The same fact was considered regarding the messages generated by the application. Related to the similarity between the system and a human operator, most users believed this characteristic has to be improved. Finally, users considered that they can easily obtain the required information and they very globally satisfied with the system.

6   Conclusions

This paper describes a voice portal implemented using the VoiceXML standard. The main objective is to provide users with a useful tool to access information of a specific city, carry out procedures and formalities, complete surveys, use a mailbox to make their complaints and requests, and be transferred to a PBX (Private Branch Exchange). The different services offered by the municipal voice portal are divided into modules which are accessed taking into account users decisions during each dialog. These modules are interconnected so that users can complete more than one action.

There are not voice portals in Spain offering the functionalities described. This way, it creates a new communication channel which is useful, efficient, easy to use and accessible. In addition, the voice portal improves the support of public services by increasing the availability, flexibility, control and reducing costs and missed calls. The system is also easily adaptable to the requirements of each municipality. Future works include the development of additional functionalities for each one of the described modules and the adaptation of the voice portal to each user or group of users by taking into account additional languages, previous dialogs and preferred functionalities by each one of them.

Acknowledgements. Research funded by projects CICYT TIN2011-28620-C02-01, CICYT TEC2011-28626-C02-02, CAM CONTEXTS (S2009/TIC-1485), and DPS2008-07029-C02-02.

References

1. Baker, J., Deng, L., Glass, J., Khudanpur, S., Lee, C., Morgan, N., O'Shaughnessy, D.: Developments and directions in speech recognition and understanding. IEEE Signal Processing Magazine 26(3), 75–80 (2009)
2. Bohus, D., Rudnicky, A.: RavenClaw: Dialog management using hierarchical task decomposition and an expectation agenda. In: Proc. of 8th European Conference on Speech Communication and Technology (Eurospeech 2003), Geneva, Switzerland, pp. 597–600 (2003)
3. DARPA: Speech and Natural Language Workshop. In: Book of Proceedings, San Mateo (1992)
4. Dybkjaer, L., Minker, W.: Recent Trends in Discourse and Dialogue. Springer (2008)
5. Griol, D., Hurtado, L.F., Segarra, E., Sanchis, E.: A statistical approach to spoken dialog systems design and evaluation. Speech Communication 50(8-9), 666–682 (2008)
6. López-Cózar, R., Araki, M.: Spoken, Multilingual and Multimodal Dialogue Systems. John Wiley & Sons Publishers (2005)
7. McTear, M.F.: Spoken Dialogue Technology: Towards the Conversational User Interface. Springer (2004)
8. Minker, W.: Stochastic versus rule-based speech understanding for information retrieval. Speech Communication 25(4), 223–247 (1998)
9. Peckham, J.: A new generation of spoken dialogue systems: results and lessons from the SUNDIAL project. In: Proc. of 3rd European Conference on Speech Communication and Technology (Eurospeech 1993), Berlin, Germany, pp. 33–42 (1993)
10. Rabiner, L., Juang, B.: Fundamentals of Speech Recognition. Prentice Hal (1993)
11. Traum, D., Larsson, S.:The Information State Approach to Dialogue Management. In: Current and New Directions in Discourse and Dialogue, pp. 325–354. Kluwer Academic Publishers (2003)
12. Weizenbaum, J.: ELIZA - A computer program for the study of natural language communication between man and machine. Communications of the ACM 9, 36–45 (1966)

Gesture Recognition Using Mobile Phone's Inertial Sensors

Xian Wang, Paula Tarrío, Eduardo Metola, Ana M. Bernardos, and José R. Casar

Abstract. The availability of inertial sensors embedded in mobile devices has enabled a new type of interaction based on the movements or "gestures" made by the users when holding the device. In this paper we propose a gesture recognition system for mobile devices based on accelerometer and gyroscope measurements. The system is capable of recognizing a set of predefined gestures in a user-independent way, without the need of a training phase. Furthermore, it was designed to be executed in real-time in resource-constrained devices, and therefore has a low computational complexity. The performance of the system is evaluated offline using a dataset of gestures, and also online, through some user tests with the system running in a smart phone.

1 Introduction

Mobile phones have become wearable computers equipped with various sensors due to the advance in microelectronics [1, 2], which not only increased the processing power of such devices but also made possible new forms of input interfaces, such as touch screen devices and gesture-based user interfaces [3].

Hand gesture is a powerful, natural means of communication between human beings. Now it can be a promising way to interact with computers, where gesture recognition is the core of such technique. Classically, hand gestures have been detected and recognized using camera-based computer vision algorithms. However, these techniques can be slow and require a high computational power, which leads to a significant energy consumption [2]. A more suitable approach for resource-constrained devices is to use their embedded sensors (such as magnetometers, gyroscopes, or accelerometers) to perform the recognition. Using embedded sensors has the advantage that the gesture recognition can be done in the own device and

Xian Wang · Paula Tarrío · Eduardo Metola · Ana M. Bernardos · José R. Casar
Data Processing and Simulation Group, ETSI. Telecomunicación
Universidad Politécnica de Madrid, Madrid, Spain
e-mail: {wang.xian,paula,eduardo.metola,
         abernardos,jramon}@grpss.ssr.upm.es

S. Omatu et al. (Eds.): Distributed Computing and Artificial Intelligence, AISC 151, pp. 173–184.
springerlink.com   © Springer-Verlag Berlin Heidelberg 2012

that the accuracy is not affected by lighting conditions or camera calibration. Furthermore, the cost and power consumption are lower [4].

There exist multiple challenges in hand gesture recognition for mobile phones, for example, a standardized "vocabulary" is missing, the interaction needs to be real time and the algorithms should be able to run on a platform highly constrained in terms of cost and system resources. What's more, user acceptability is also a key consideration: will they feel comfortable waving arms in a public place [2]?

Many gesture recognition systems based on accelerometers have been developed. One example is Georgia Tech Gesture Toolkit [5], that provides tools to support gesture recognition and has been used in several ongoing projects, such as an automobile gesture panel, patterned blink recognition and mobile sign language recognition. The work in [6], where 8 predefined gestures including translations, circles, and pentagram are defined and more than 98% recognition accuracy is achieved, provides a natural and intuitive way to control the browser application on a large screen. These techniques are also used in mobile gaming systems to enhance user experiences [2], where a set of 8 gestures such as single-circles, squares, triangles and double-circles are provided and an overall accuracy of 96.25% is obtained. The authors in [7] developed the uWave algorithm, which is utilized in gesture-based user authentication and interaction with a three-dimensional mobile user interface. In this work, they employ a set of 8 gestures identified by Nokia research study and the recognition accuracy is about 98%.

Gesture recognition is a type of pattern recognition. Various methods could be utilized, such as conditional Gaussian models, support vector machines [8], Bayesian networks [9], dynamic time warping (DTW) and hidden Markov models (HMMs). DTW [2, 7] and HMMs [1-3, 5, 6, 10-13] are two of the most popular approaches adopted, both of which were investigated in speech recognition area.

Only a few [2, 3, 4, 7] of the techniques mentioned above are implemented on a resource-constrained platform such as a mobile phone. What's more, most of these proposals target at user-dependent gesture recognition because of the difficulty of user-independent gesture recognition and because user-independent gesture recognition may not be that attractive as speaker-independent speech recognition, as there are no standard gestures for interaction [7]. In this work, however, our goal is to develop a user-independent system which is light, real-time, and with low consumption. To this end, we define a set of simple gestures, which are intuitive and comfortable to perform, and recognize them in a user-independent manner which is user friendly and does not require training the system.

In summary, we make the following contributions: We define a set of rotation and translation gestures and develop real-time methods to recognize them with very low computational cost. As this is a work in progress, the recognition of rotation gestures is implemented on a mobile phone and tested in real-time, whereas the recognition of translation gestures is implemented on a PC and tested offline with a dataset of gestures performed by real users. Preliminary results show a high recognition accuracy and low time complexities.

The rest of the paper is organized as follows: Section 2 enumerates possible application scenarios of our gesture recognition system. Section 3 describes the set of gestures we want to recognize. Section 4 presents the architecture of the recognition system. Section 5 reports the validation results of the rotation and translation gesture recognition systems with real experiments and simulations, respectively. Finally, we conclude in section 6 and point out future work.

2 Application Scenarios

Gesture recognition systems for mobile devices have a wide variety of applications in several scenarios, such as smart environments, teaching/learning, robot control, etc. For example, gestures performed with the user's mobile phone can be used to control the equipment of a smart room in a hotel (e.g. switching off the light when the position of the phone changes from facing up to facing down, switching on the lights by making the opposite movement, raising/lowering the blinds by doing an up/down movement while pointing to the window, opening/closing the curtains with a left/right movement, raising/lowering the room temperature by making an up/down movement while pointing to the air conditioner, calling the room service by shaking the phone, etc.). Gestures can also be used in a teaching scenario, or even in conferences or business presentations to control the computer and the slides. For example, a movement to the left/right goes to the previous/next slide, a spiral to the left/right goes to the first/last slide, moving up/down the phone will control the volume of the computer if a video/audio file is played, shaking the phone can switch on/off the projector and once it is off, an up/down movement can raise/lower the screen, etc. Another practical application of mobile gesture recognition is to control robots, toys or vehicles: turning the phone to the left or right can control the direction of the robot, turning it up and down can control the speed and shaking the phone can switch on/off the robot.

3 Selection of Gestures

The selection of the gesture set affects both the experience of user interaction and the recognition accuracy. The author of [1] argues that an extensive set of gestures becomes unpractical because too many gestures have to be learned by the users. The recognition results in [7] highlight the importance of selecting the right gesture vocabulary for high accuracy. More complicated gestures provide more features to distinguish them, resulting in higher recognition accuracy. However, complicated gestures also force users to remember how to perform them and how they are related to functions/actions. Furthermore, in [14], user studies indicate that users tend to use spatial two-dimensional gestures and that utilizing all three dimensions in one gesture is rare. In this section, we describe the gesture set selected for our recognition system, which is divided into two groups: turns and translations.

3.1 Turn Gestures

Taking into account that there are three axes in the device, we can define six possible elementary turns, as we can see in Fig. 1. In our system, we target at recognizing 45° and 90° turns of the six possible types.

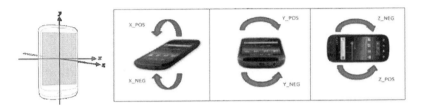

Fig. 1 Six elementary turns around the device's axes.

The complete gesture will be defined by the type of turn and by the initial or final orientation of the device. We have defined six possible orientations with respect to the ground (shown in Fig. 2).

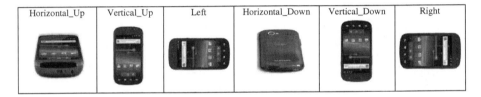

Fig. 2 Main six orientations of the device.

If we make a 90°-turn from one of the orientations in Fig. 2, we will obtain another orientation of Fig. 2. In addition, if a 45°-turn is made intermediate orientations are obtained. We classify the possible orientations in three groups:

- Type I orientations: the gravity only appears in one axis (those in Fig. 2).
- Type II orientations: the gravity appears in two axes. This happens when, from a type I position, there is a 45°-turn.
- Type III orientations: the gravity appears in all the axes. It happens when, from type II positions, there is a 45°-turn over the axis that has not been turned yet.

3.2 Translation Gestures

Following the previous research [14], in our work, we have defined six simple translations named backward, forward, left, right, up and down. The following table describes these gestures. The orientation of the device when doing the translations remains the same: in a 'Vertical_Up' orientation.

Table 1 Description of translation gestures

Gestures	Description (performed with the same fixed pose)
Backward	Pull the mobile horizontally.
Forward	Push the mobile horizontally.
Left	Move the mobile horizontally from right to left.
Right	Move the mobile horizontally from left to right.
Up	Move the mobile vertically from low position to high.
Down	Move the mobile vertically from high position to low.

4 System Architecture

The current system architecture consists of three different layers, a *sensor layer*, which acquires the information provided by the inertial sensors, a *fusion layer*, which includes the recognition algorithms to recognize the gestures, and a *communications layer*, which offers the recognition result to external devices.

4.1 Sensor Layer

The gesture detection system relies on the data provided by the gyroscope and the accelerometer embedded in the device. In our case, the acquisition process is done using a Google Nexus S smartphone running Android O.S, where data communication is done by event notifications. As a result we cannot talk about a real sampling frequency, but we have seen that the approximate sampling rates for the gyroscope and the accelerometer are, respectively, 0.0096 seconds and 0.02 seconds per measurement.

In the case of turn gestures, the information collected by the gyroscope allows us to distinguish among the six elementary turns, whereas the accelerometer gives us information about the orientation of the device. Fig. 3 shows some examples of the gyroscope and accelerometer signals for different gestures.

Fig. 3a and 3b show the values acquired from the gyroscope (angular speed around each axis) and accelerometer (acceleration on each axis), for a movement from a 'Horizontal_Up' position to a 'Vertical_Up' position. Fig. 3c and 3d correspond to a 90°-turn around the vertical axis. As it can be seen, the value of the gyroscope signal in the axis of rotation increases/decreases significantly, which allow us to recognize the type of elementary turn. The acceleration signal after the turn enables recognizing the final position (the axis with the gravity component is the one pointing to the ground).

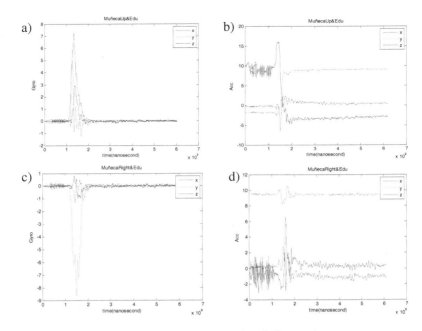

Fig. 3 Examples of gyroscope and accelerometer signals for rotation gestures.

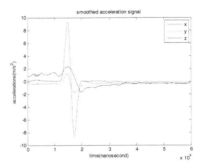

Fig. 4 Example of acceleration signals for a correctly performed 'Right' gesture

The translation gesture is a variable force movement, starting and ending with a stationary state. Essentially, the 6 translation gestures are the same but moving in different directions. After the gesture starts, the acceleration value in the direction of movement increases quickly, then changes its direction, and finally returns to zero. So ideally, the acceleration in the axis along the movement direction should look like a sinusoid in one period, and the acceleration data in the other two axes should be close to zero (see Fig. 4). The axis where this pattern appears and the order of the appearance of the curve's peak and valley indicate the direction of the movement. For example, Fig. 4 represents a 'Right' movement (sinusoid-like curve in the x axis, with the peak appearing earlier than the valley).

4.2 Fusion Layer

a. Recognition of Turn Gestures

In order to detect when a turn happens, the fusion layer is permanently gathering the data from the gyroscope sensor. The data obtained by the fusion layer is an array of three elements, each element related to one axis. If one of the elements of the array increases over a predefined threshold, a turn is detected on the corresponding axis. By also taking into account its sign, it is possible to distinguish among the six types of elementary turns. The system is able to recognize 45° and 90° turns, by comparing the initial position of the smartphone and the final position after a turn.

When the value in the axis that has increased, decreases under another threshold (set to indicate that the turn is finishing), the accelerometer sensor is enabled and a fixed number of measurements are taken to estimate the final position. These measurements are averaged, giving a higher weight to later positions in the sequence of accelerometer's data. Then, depending on the magnitude and the sign of this weighted average, a final orientation of the device is selected.

In a type I position, the gravity component relies mostly over one of the three axes. In a type II position, it will be spread over two of the three axes. And finally, a type III position has a significant component of the gravity over the three axes. So, analyzing the three components of the acceleration, if the lowest value is over 3 m/s2, a type III position is estimated. Otherwise, if the second lowest value is over 3 m/s2, a type II position is estimated. Otherwise, mobile position is included in the type I group.

For type I positions, one of the six orientations of Fig. 2 is selected (according to which axis has the gravity component and its sign). For type II/III positions, a combination of two/three type I positions is chosen, e.g. 'Horizontal_Up-Left'.

b. Recognition of Translation Gestures

For translation gesture recognition, we have developed a method based on analysis of acceleration data and feature extraction. The input of the algorithm is a time series provided by the three-axis accelerometer. Each time sample is a vector of three elements corresponding to the sampled data in the three axes.

As we said before, ideally, the curve in the axis along the movement direction should look like a sinusoid in one period (as in Fig. 4). But in practice, the signals almost never show such a clear shape because of noise and performance of the users (we should never expect the users to make gestures ideally). As a result, the sinusoid-like curves are distorted and appear in all axes with different amplitudes, as shown in Fig. 5. The algorithm of translation gesture recognition in our work consists of three steps, which are described next:

A.   *Preprocessing*

According to our definition of gestures, the mobile phone should be held vertically when doing the gesture. But in practice, a tilt of a few degrees is quite common. As gravity is a relatively large value compared to other acceleration components forced by the user, a small tilt will introduce a significant acceleration component along x and z axes, which would result in wrong recognition. In our work, the normalizing method provided in [1] is utilized to correct small amounts of tilt. Then the rectified data is passed through a Butterworth low pass filter to reduce the effect of noise. The signal is further smoothed by a moving average method where the smoothed value is the un-weighted mean of its previous n data points, where n is the size of the sample window. The order and cutoff frequency of the filter and n are all determined empirically.

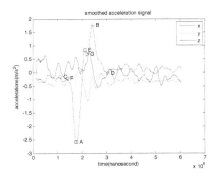

Fig. 5 Pattern generalization in the x direction. F, A, E, G are key points of the original recognized pattern. C, A, B, D are the generalized pattern.

B.   Pattern extraction for each axis

The key points of the typical translation pattern are the starting point, the ending point, the peak and the valley of the sinusoid-like curve. Because of noise or the imperfect gesture performance of users, there are more than two local extrema, so first, all the local extrema are detected, and then we find all pairs of adjacent extrema with different signs. The peak and valley of the pattern should be in the pair with the largest absolute difference of the two elements, for example, the pair A-E in Fig. 5. The starting and ending point of the pattern are the two adjacent extrema to peak and valley (F and G in Fig. 5).

Considering the noisy acceleration measurement and that this method is sensitive to fluctuations of the signal, an adjustment of the key points is employed: we look forward or backward several extrema points from the peak and valley points and check whether there are higher extrema (for the peak) or lower extrema (for the valley). This way, the method is robust to small fluctuations. In the same way, the starting and ending of the recognized pattern are the adjacent extrema points of the new peak and valley. For example, in Fig. 5, C, A, B and D are the new four key points of the pattern, which represent the shape of a sinusoid curve.

C.   Movement detection

With the pattern extraction mentioned above, the sinusoid-like curve pattern in each axis is obtained. Then we first check the symmetry of the pattern by calculating the following ratios:

$$r_1 = \begin{cases} \dfrac{C-A}{\frac{1}{2}(B-A)}, \text{if C-A} \leq \frac{1}{2}(B-A) \\ \dfrac{\frac{1}{2}(B-A)}{C-A}, \text{if C-A} > \frac{1}{2}(B-A) \end{cases} \qquad r_2 = \begin{cases} \dfrac{B-D}{\frac{1}{2}(B-A)}, \text{if B-D} \leq \frac{1}{2}(B-A) \\ \dfrac{\frac{1}{2}(B-A)}{B-D}, \text{if B-D} > \frac{1}{2}(B-A) \end{cases}$$

where A is the acceleration value at the valley, B at the peak, C at the starting point and D at the end. To be a "good" pattern, r_1 and r_2 should be both greater than a predefined threshold.

Then we select the axis where there is the largest drop from peak to valley of the pattern and that agrees with our symmetry requirement, as the dimension where the movement occurred. The appearance order of peak and valley further determines the direction of the movement along this dimension. For example, Fig. 5 represents a 'Left' gesture.

The time complexity of Butterworth low pass filtering is O(n*logn) due to the Fourier transform. The rest of processing has linear time complexity. Therefore, the method presented above has a complexity of O(n*logn), where n is the length of acceleration signal. In [3], a DTW algorithm is implemented on a mobile phone and has time complexity of O(n*n). The researchers of that paper showed that DTW implementation has lower computational load than HMMs. So hopefully, our algorithm can run faster on a mobile phone platform.

4.3 Communications Layer

The gesture recognition application runs autonomously in the device, but a communications layer has been integrated in order to transmit the detected gestures to external devices (either for a more comfortable display of the recognition results or to take the appropriate actions corresponding to the performed gestures).

To this end, a NetworkCommunication component is created in the main component of the application, so that when a gesture is detected, the information is sent through a previously created socket. This information can be gathered from an external element, like a PC, by listening to the data on an already-known IP address and port. In this way, the gesture can be shown in the screen or used to perform the appropriate action.

5 Validation

This section presents some preliminary results of the validation of the gesture recognition system. The recognition of turn gestures has been completely implemented and tested in a Google Nexus S smartphone, running Android O.S, whereas the recognition of translation gestures was implemented in Matlab and tested with a dataset of gestures, collected by several users with the same smartphone.

5.1 Online Validation of Turn Gestures with the Mobile Phone

a. Performance Statistics

The performance of the proposed system is evaluated according to its ability to correctly detect the type of gesture. The speed at which the user performs the gesture determines if it is detectable or not, as we have used an angular speed threshold to filter out random movements. In particular, we have used a threshold of 2 rad/s, which assures that usual wrist movements are detected (with detection probability nearly equal to 100%). Decreasing this threshold would allow detecting slower movements, but would increase the probability of false detections.

Supposing that the system has detected a gesture, the system quality can be described in terms of the probability of correctly recognizing the type of gesture and in terms of time delay in the recognition process. Both parameters can be tuned by modifying the number of accelerometer measurements that are analyzed. Longer data windows will increase the recognition capability, but will also increase the delay. Next table shows the delay (time interval between the instant in which the turn has finished and the instant in which the orientation is estimated) for different values of the window size:

Table 2 Time delay for different window sizes

Window size (Number of measurements)	5	10	15	20
Interval of time	0,11 s	0,21 s	0,31 s	0,41 s

After some tests, we have seen that using a window size of 10 is enough to correctly recognize most of the gestures (nearly 100% recognition accuracy) and still provides a reasonable recognition speed.

b. Computing Time

The computing time of the algorithm has been measured for type I positions. Using 15 measurements from the acceleration sensor to estimate the position of the smartphone, the average computing time was 273μs. This time is negligible compared to the acquisition time (0.02 s per measurement). So we can conclude that the computation of gesture recognition algorithm will not slow down the entire process and can be done in real time.

If the number of measurements increases, the computing time will increase almost proportionally. For 150 measurements, the computing time rises to 2.26 ms.

5.2 Offline Validation of Translation Gestures with a Dataset

a. Description of the Dataset

To test the gesture recognition method, a dataset was collected with 14 subjects (8 males and 4 females), who repeated 10 times each gesture using a Google Nexus S smartphone. The starting and ending of each gesture was marked by pressing a button of the smartphone. Before collecting the gesture data, they were given a brief introduction about how to perform each gesture.

b. Performance Statistics

Table 2 summarizes the recognition result of our approach over the collected dataset. An average of 93.2% is achieved. Most of the errors are due to the sensitivity of our approach to fluctuations of the signal. In particular, the test result demonstrates that more than half of the errors are due to the presence of a small fluctuation around zero (see an example in Fig. 6).

Table 3 Confusion matrix for the 6 translation gestures. The columns are recognized gestures and the rows are actual gestures. Average accuracy is 93.2%.

	Backward	Forward	Left	Right	Up	Down
Backward	94.3	0.7	2.9	0.7	0	1.4
Forward	0.7	94.3	1.4	2.1	0.7	0.7
Left	3.6	2.9	93.6	0	0	0
Right	3.6	0	0.7	94.3	0.7	0.7
Up	5.0	1.4	0	0	93.6	0
Right	0.7	5.0	4.3	0	0.7	89.3

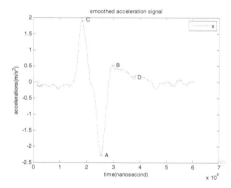

Fig. 6 Failure of pattern recognition due to fluctuation around zero between the real peak and valley (B and A are detected as peak and valley, instead of C and A)

6 Conclusion and Future Work

We focus on resource-constrained mobile phones and present algorithms to recognize turning and translation gestures in a user-independent manner. Our method utilizes embedded accelerometers and gyroscopes, which are commercially available in many mobile devices. The evaluation experiments of the turning gestures recognition system indicate that the time spent in the recognition process could be negligible compared with data sensing. The translation gesture recognition system has theoretically lower computational complexity than previous proposals implemented on mobile phones. The preliminary validation of both types of gesture recognition has shown a high recognition accuracy.

In order to further improve the recognition accuracy, we are currently making the translation gesture recognition more robust to fluctuations of the signals. Then it will be implemented on the mobile phone platform, to test its performance in real-time. As further work we are also planning to add some more gestures to our set (circles and spirals) to finally test the complete system with real users and get some feedback about the usability, accuracy, delay and energy consumption in real situations. Another possibility is to add some feedback to enhance user experience (adding sound, vibration or graphics).

References

[1] Pylvänäinen, T.: Accelerometer Based Gesture Recognition Using Continuous HMMs. In: Marques, J.S., Pérez de la Blanca, N., Pina, P. (eds.) IbPRIA 2005. LNCS, vol. 3522, pp. 639–646. Springer, Heidelberg (2005)

[2] Niezen, G., Hancke, G.P.: Gesture recognition as ubiquitous input for mobile phones. In: Proc. of the Workshop on Devices that Alter Perception (2008)

[3] Joselli, M., Clua, E.: gRmobile: A Framework for Touch and Accelerometer Gesture Recognition for Mobile Games. In: 2009 VIII Brazilian Symposium on Games and Digital Entertainment, pp. 141–150. IEEE (2009)

[4] Niezen, G., Hancke, G.P.: Evaluating and optimising accelerometer-based gesture recognition techniques for mobile devices. In: AFRICON 2009, pp. 1–6. IEEE (2009)

[5] Westeyn, T., Brashear, H., Atrash, A., Starner, T.: Georgia Tech gesture toolkit: supporting experiments in gesture recognition. In: Proc. of the 5th Int. Conf. on Multimodal Interfaces, pp. 85–92. ACM (2003)

[6] Kauppila, M., Pirttikangas, S., Su, X., Riekki, J.: Accelerometer Based Gestural Control of Browser Application. In: Int. Workshop on Real Field Identification, pp. 2–17 (2007)

[7] Liu, J., Wang, Z., Zhong, L., Wickramasuriya, J., Vasudevan, V.: uWave: Accelerometer-based personalized gesture recognition and its applications. Pervasive and Mobile Computing 5(6), 657–675 (2009)

[8] Wu, J., Pan, G., Zhang, D., Qi, G., Li, S.: Gesture Recognition with a 3-D Accelerometer. In: Zhang, D., Portmann, M., Tan, A.-H., Indulska, J. (eds.) UIC 2009. LNCS, vol. 5585, pp. 25–38. Springer, Heidelberg (2009)

[9] Cho, S.J., Oh, J.K., Bang, W.C., Chang, W., Choi, E., Jing, Y., Cho, J., Kim, D.Y.: Magic wand: a hand-drawn gesture input device in 3-D space with inertial sensors. In: 9th Int. Workshop on Frontiers in Handwriting Recognition, pp. 106–111. IEEE (2004)

[10] Kauppila, M., Inkeroinen, T., Pirttikangas, S., Riekki, J.: Mobile phone controller based on accelerative gesturing. Adjunct Proceedings Pervasive, 130–133 (2008)

[11] Hofmann, F.G., Heyer, P., Hommel, G.: Velocity Profile Based Recognition of Dynamic Gestures with Discrete Hidden Markov Models. In: Wachsmuth, I., Fröhlich, M. (eds.) GW 1997. LNCS (LNAI), vol. 1371, pp. 81–95. Springer, Heidelberg (1998)

[12] Schlömer, T., Poppinga, B., Henze, N., Boll, S.: Gesture recognition with a Wii controller. In: Proc. of the 2nd Int. Conf. on Tangible and Embedded Interaction, pp. 11–14. ACM (2008)

[13] Mäntyjärvi, J., Kela, J., Korpipää, P., Kallio, S.: Enabling fast and effortless customisation in accelerometer based gesture interaction. In: Proc. of the 3rd Int. Conf. on Mobile and Ubiquitous Multimedia, pp. 25–31. ACM (2004)

[14] Kela, J., Korpipää, P., Mäntyjärvi, J., Kallio, S., Savino, G., Jozzo, L., Di Marca, S.: Accelerometer-based gesture control for a design environment. Personal and Ubiquitous Computing 10(5), 285–299 (2006)

Towards a Lightweight Mobile Semantic-Based Approach for Enhancing Interaction with Smart Objects

Josué Iglesias, Ana M. Bernardos, Luca Bergesio, Jesús Cano, and José R. Casar

Abstract. This work describes a semantic extension for a user-smart object interaction model based on the ECA paradigm (Event-Condition-Action). In this approach, smart objects publish their sensing (event) and action capabilities in the cloud and mobile devices are prepared to retrieve them and act as mediators to configure personalized behaviours for the objects. In this paper, the information handled by this interaction system has been shaped according several semantic models that, together with the integration of an embedded ontological and rule-based reasoner, are exploited in order to *(i)* automatically detect incompatible ECA rules configurations and to *(ii)* support complex ECA rules definitions and execution. This semantic extension may significantly improve the management of smart spaces populated with numerous smart objects from mobile personal devices, as it facilitates the configuration of coherent ECA rules.

Keywords: Smart objects, smart spaces, user-object interaction, mobile middleware, embedded reasoning, ontology-based modelling.

1 Introduction

The concept of *smart object* includes any kind of device with sensing or/and processing capabilities that is capable of reacting or adapting its functionalities depending on external stimulus or users' requirements, while preserving their traditional physical interaction paradigm. Then, *spaces* become *smart* as they host several smart objects with heterogeneous functionalities. Within this scenario, smart spaces exploitation not only involves acquiring data and controlling smart objects, but a common strategy to enable the user with capabilities to coordinate them in an intelligent way. To this end, the ECA paradigm appears as a simple

Josué Iglesias · Ana M. Bernardos · Luca Bergesio · Jesús Cano · José R. Casar
Telecommunications Engineering School, Technical University of Madrid (UPM), Spain
e-mail: {josue,abernardos,luca.bergesio,
        jcano,jramon}@grpss.ssr.upm.es

S. Omatu et al. (Eds.): Distributed Computing and Artificial Intelligence, AISC 151, pp. 185–196.
springerlink.com   © Springer-Verlag Berlin Heidelberg 2012

formalism used to implement a particular perspective for user-smart object intelligent interaction.

The ECA paradigm is composed by a structure of reactive rules working over an event-driven architecture. Each ECA rule may have three kinds of 'atoms': the *event* is the signal that triggers a set of rules; the *condition* is a logical test that, if satisfied, makes the execution of the rule to continue; and, finally, the *action* identifies the execution of a process. A set of rules has the form: **ON** *event* **IF** *condition(s)* **DO** *action(s)*.

In [1] we have developed a first prototype to validate this interaction paradigm, which uses the mobile device as mediator to handle ECA rules built on the sensing and action capabilities of smart objects (details in Section 3). It happens that, the more the smart space ecosystem grows (including new smart objects), it becomes exponentially difficult to manage from a user's perspective; at the same time, the continuous rule checking process also becomes very demanding in terms of mobile processing. So, in this work we propose to extend the interaction model's capabilities from a semantic perspective, in order to enhance *(i)* user experience and *(ii)* mobile device performance.

The paper is structured as follows. Section 2 reviews the state of the art of embedded lightweight semantic tools and semantic-based ECA model approaches. Section 3 introduces the interaction scenario and the original architecture that has been extended. Section 4 presents the proposed models and addresses the enhancements obtained. Finally, Section 5 concludes the work with future lines for research.

2 State of the Art

Since 1999, when works such as the one carried out by Biegl [2] first addressed how to link and control different kinds of devices, interaction paradigms for physical interaction between devices (or digitally augmented objects) have been increasingly used to implement the smart space concept.

Trying to fill the gap between existing developments for smart spaces configuration and state-of-the-art initiatives for enhancing these paradigms (and always from a ECA model based point of view), this Section focuses on analysing *(i)* the enabling technologies for embedded context information representation and reasoning and *(ii)* some of the most up-to-date projects for semantic enhancement of ECA-based interaction models.

From a functional point of view, machine processable representations are necessary to organize, valorise and share the vast amount of information smart spaces generate (e.g., environmental conditions, users presence, available services, objects usage patterns, etc.). Within this scenario, information acquired from heterogeneous smart objects would need to be jointly processed, requiring a common and expressive enough data structure to support the smart space configuration. Personal smartphones are nowadays-common devices in people's everyday lives, becoming a potential candidate for centralizing smart spaces exploitation. However, it is still an open challenge how to optimally represent information in resource-constrained mobile devices [3].

For example, although tuple-based (or key-value) models reduce management overhead and can be easily applied to 'legacy' mobile systems (e.g., [4]), they lack from validation and scalability capabilities and they are not suitable for handling context information ambiguity. A hierarchical structure and the automatic validation are, however, some of the strong points of markup scheme modelling, although XML has a high semantic redundancy and it is not fully adapted to the limited resources of embedded devices (several works points out that a better performance can be obtained with other techniques as, e.g., JSON[1]).

Ontology-based modelling [5] combines the advantages of object-based and logic models, i.e., encapsulation, extendibility and reusability, and formalism and inference capabilities, respectively. They facilitate information fusion from heterogeneous data and knowledge sources, also providing support for automated reasoning. OWL (Web Ontology Language)[2], the standard ontology language endorsed by the W3C, enables different applications to share a common model, providing common shared domain vocabularies and a consistent mechanism for information representation. According to [6], these features are particularly important in mobile and pervasive environments, in which different heterogeneous and distributed entities must interact for exchanging users' context information.

Formal information representation facilitates automated reasoning (e.g., concept and instance classification, model and knowledge base consistency checking, etc. [30]). Although scarce compared with general context management systems, some light tools enabling reasoning in resource-constrained devices have already been described in the literature. Following the conclusions presented in [3] (whose work is focused on lightweight ontology-based data models), μJena [7] and Bossam [8] (as ontology manager and ontology rule-based reasoner, respectively) were the only ones (i) capable of dealing with ontology data, (ii) using standard formats and (iii) working in resource-constrained mobile devices. Extending this work, it is worth mentioning androJena, a new development (the first version was released on May 2010) based on a subset of the popular Jena framework migrated to Android platforms, that also fulfils these requirements. androJena has been recently used in several works; [9] and [10] can be highlighted as they include performance tests. Although these tests only measure their own particular developments, they can be used to obtain an overall idea of androJena's performance.

The state-of-the-art analysis reveals that there are still few developments of general-purpose lightweight ontological tools to be embedded in personal mobile devices with a promising success, and the existing ones are still far away from maturity. As previously said, although some performance tests can be found in the literature (e.g., for μJena [7], Bossam [8] or, more recently, androjena [9][10]), there is still a lack of experiments comparing their performances in common scenarios. Finally, it should be noted that much of these developments are discontinuous research projects. In this sense, androJena seems to be the only exception to this issue nowadays.

Focusing now on semantic technologies, they may play an important role in enhancing ECA-based interactions for smart spaces. Ontologies can be used to

---

[1] http://www.json.org/
[2] http://www.w3.org/TR/owl-features/

formally model the information offered by the smart objects and their capabilities and particularities, information that can fed different semantic reasoning mechanisms in order to offer a consistent information layer ready to be exploited.

In this line, within the Rewerse project[3], r^3 prototype (Resourceful Reactive Rules) addresses ECA rules managing from a Semantic Web perspective. It implements a rule engine capable of dealing with this kind of rules defined in different languages (even each different component –event, condition or action– may be represented using different rule languages). Every resource involved in the reasoning process (e.g., rules, engines) is described in terms of RDF triplets based on an OWL-based model: the r^3 ontology [11].

The K^{4R} project[4], maybe the most notable extension of r^3 initiative, focuses on defining a RESTful (and ontological) interface for Knowledge Resource (including Knowledge Reasoners). Authors consider this approach broad enough to include ECA rules modelling support [12].

Finally, it is worth mentioning that RIF (Rule Interchange Format), a collection of rule dialects (i.e., consistent and rigorously defined rule languages) intended to facilitate rule sharing and exchange, also consider ECA rules support as a requirement to be developed [13].

Aligned with these researches, the work presented in this paper focuses on developing real world semantic-based mobile applications for managing smart spaces. Next Section introduces the scenario where our semantic-based ECA interaction model is to be deployed and its architectural particularities.

3   ECA Interaction Model for Smart Spaces Management

3.1   ECA Model Based Smart Space Management Scenario

Our interaction scenario considers a space populated with different types of objects (with or without embedded processing capabilities). Within this scenario, the user's mobile device may be used to manage the identification, sensing, processing, etc. [14] of these smart objects, being able to act as an interaction mediator, delivering the functionality to (i) customize the responses to physical interaction with certain objects, (ii) make an object respond (physically or virtually) to a given order configured by the user, (iii) provide intuitive configuration of the smart environment through actions held in the mobile device and (iv) configure/activate features in the mobile device depending on environmental events.

The ECA interaction model aims at providing bidirectional interaction between objects and user's (mobile) devices, in order to make possible to configure the object's actions from the mobile device (note that within this general scenario, user's personal device can be also considered as a smart object). The ECA model is supported by implementing next functionalities:

---

[3] http://rewerse.net/
[4] http://code.google.com/p/k4r/

1. *Module publishing.* Smart objects are able to publish their capabilities (both events generation or actions execution). Each 'module' implements the necessary logic in order to detect an event or perform an activity.
2. *Proximity detection.* Proximity will be the starting point for interaction. After proximity detection, a mobile device will be able to download the available modules belonging to the smart object it is close to.
3. *ECA rules configuration.* Mobile mash-up tools will allow the user to easily configure ECA rules as a combination of event, condition(s) and action(s).
4. *Rule-based reasoning.* Active ECA rules will be constantly evaluated in order to detect configured events, executing the associated actions if the conditions are fulfilled.
5. *Module life-cycle management.* Installed modules will be subject to continuous updates in order to detect unused or out-of-date ones.

So, when a smart object detects a mobile phone nearby, it offers to the phone the download of a set of modules that enables the interaction with the object. After agreeing to download the modules, the user can configure ECA rules with them as described before. When an event occurs in or is detected by a smart object, the corresponding module at the mobile phone receives a notification. The mobile application then checks the conditions in the ECA rules (if any) and, if satisfied, executes the configured actions, affecting other smart objects or the device itself.

3.2 Different Architectural Approaches to ECA-Based Deployments

Our particular approach employs the user's mobile device for interacting with the smart objects deployed in the smart spaces, i.e., for module installation and ECA rules configuration. However, ECA rules evaluation (event detection, condition assessment and action triggering) may be supported by different architectural approaches:

- *Server-based centralized approach*: once an ECA rule is configured in the user's mobile device this rule is sent to a centralized server managing the whole set of smart objects of a particular smart space. This server is the one in charge of *(i)* monitoring all the smart objects (or automatically receiving state changes from them), *(ii)* detecting the configured events, *(iii)* assessing the required conditions and *(iv)* triggering the associated actions.
- *Smart object distributed approach*: in this case, each ECA rule would be distributed among the smart-objects involved in the rule. Smart objects would need to be intelligent enough to detect its own-generated events, to assess the required conditions and to execute the desired actions; they should also have communication capabilities (WiFi, Bluetooth, etc.) to automatically coordinate among themselves.
- *Mobile device approach*: this approach is equivalent to the server-based centralized one but, instead of having one server managing the ECA rules evaluation for each smart space, there would be one mobile device managing all the ECA rules configured by a particular user (regardless of the user location).

Although it has to be noted that these architectural approaches are compatible and they can coexists, this work addresses the real deployment of an architecture to be fully deployed in the mobile device.

3.3 Mobile-Based Architecture to Enable ECA Interaction Model

Before addressing the semantic extension proposed in this work (Section 4), this subsection introduces the architecture of our (non-semantic) approach for smart spaces management based on an ECA interaction model [1].

For a practical implementation of the proposed scenario, smart objects are equipped with Bluetooth or NFC tags. When detecting an object for the first time (proximity interaction), the mobile device will retrieve the object's modules from a cloud server. The mobile application is the central element to manage interaction. It is divided into three main building blocks: *core*, *application interface* and *modules*. The core manages the modules' lifecycle: it dynamically retrieves them from the infrastructure and loads them into memory, manages the interaction between events, conditions and actions and, finally, provides the GUI. The application interface defines the data structure for the communication between the core and the modules. Finally, each ECA rule aggregates module events, conditions and actions to interact with a smart object or with the mobile itself. A module offering may include none, one or more than one of each (events, conditions and actions).

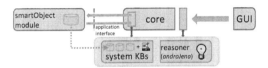

Fig. 1 ECA based architecture modules; non-semantic (red) and semantic extension (green).

This first prototype for the ECA model has been implemented on Android (v2.3, in a Google Nexus S smartphone which includes NFC technology).

Next Section shows how a lightweight ontology-based framework and rule engine (*androJena*) have been integrated inside the user's mobile device in order to exploit its semantic capabilities for *(i)* supporting the ECA rules configuration process and *(ii)* supporting the ECA rules evaluation (as depicted in Figure 1).

4   Semantic Enhancing of the ECA-Based Architecture

This work extends with semantic capabilities the ECA model based architecture for smart spaces presented in Section 3. As depicted in Figure 2, three semantic models are used:

- *Smart object ontology* ($\mathbb{O}_{so}$): is used to model the characteristics of a smart object. This ontology models *(i)* the internal (sensing and acting) capabilities of a smart object (i.e., type of capability and valid values ranges, if applicable) and *(ii)* the relationships among objects capabilities (e.g, "fixed smart objects with proximity sensing capabilities can only detect mobile objects", etc.).

- *Smart space ontology* ($\mathbb{O}_{SS}$): models the relationships between a smart space and its smart objects. In this very first approach it can be considered just as a semantic map of the environment where the smart objects are deployed (e.g., "smartObjectX is currently located in roomA", etc.).
- *ECA rule ontology* ($\mathbb{O}_{ECA}$): is a formal definition of ECA rules. It can be used to *(i)* detect inconsistencies when configuring a particular ECA rule (i.e., among smart objects capabilities) and *(ii)* identify incompatibilities among the different sets of ECA rules configured for a particular user (future works will also consider restrictions among ECA rules defined by different users).

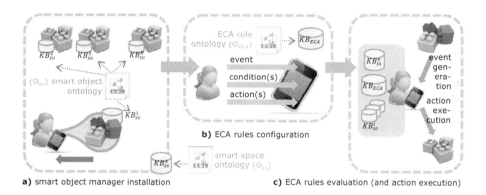

a) smart object manager installation   **c)** ECA rules evaluation (and action execution)

Fig. 2 ECA model based smart environment exploitation.

Each semantic model needs to be complemented with a rule base in order to address a comprehensive set of reasoning capabilities.

A preliminary version of these models (and rule bases) has been used in this work to *(i)* automatically detect incompatible ECA rules configurations (Section 4.1) and to *(ii)* support complex rules configuration and execution (Section 4.2).

Each smart object is defined by a smart object knowledge base KB_{so}^i shaped according a smart object ontology $\mathbb{O}_{so}$ (Figure 3.a). This ontology models a smart object (*SmartObject* class) according its particular features (*Capability*) and the set of available configurations (*CapabilityConfiguration*) modelling its valid states. Although specific capabilities are defined (e.g., *Email*, *SMS*, *Photo*, *PhoneCall*, etc.), they can be grouped in the more abstract concepts *SensingCapability* and *ActingCapability*.

$\mathbb{O}_{SS}$ models smart spaces (*SmartSpace*), Figure 3.b. A smart space is composed by several physical spaces (*Space*) hosting different smart objects. These spaces may have relative relations (*nearTo, contains, disjointSpaces*, etc.), represented in Figure 3's model as *hasSpatialRelationWith**.

ECA rules (*ECArule* class) are modelled in $\mathbb{O}_{ECA}$ ontology (Figure 3.c) as a set of atoms (*ECAruleAtom*): *ECAevent*, *ECAcondition* and *ECAaction*. Each of these atoms is associated to a smart object capability (*involvesCapability* property) and to a particular configuration for that smart object (*hasConfiguration*).

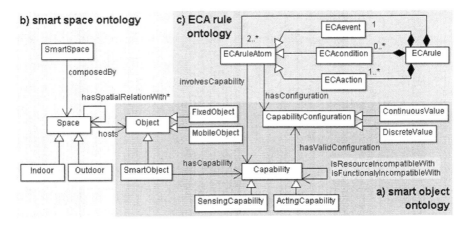

Fig. 3 ECA-based interaction conceptual model.

As stated before, this work follows a mobile centralized architecture for implementing an ECA model based smart space management system (see Section 3.2), so these semantic models are fully managed inside the user's mobile device (OWL-Lite has been used for ontology representation). Semantic models (and rule bases) management is performed using *androJena* as programmatic environment. The generic OWL-based rule engine offered by *androJena* has been also used to apply the rules bases to the semantic models defined.

4.1 Detecting Semantic Incompatibilities

Each ECA rule involves several options regarding the configuration of the available smart objects and each user may configure several ECA rules in his mobile device to personalize a particular smart space. This heterogeneity of resources available in the smart spaces (i.e., different smart objects with different capabilities) may lead to the configuration of inconsistent rules.

Semantic models described above may be exploited in order to automatically detect several kinds of inconsistencies. Although some of them may be directly detected from the formal definition of smart spaces and objects (e.g., when trying to configure certain value out of its valid range), a specific semantic has to be added to extend this for 'intra' and 'inter' ECA rules inconsistency detection (i.e., two configurations valid if isolated, may lead to an inconsistency if used inside the same ECA rule).

In the ECA rules configuration process, the information about the smart spaces modelled in this semantic way is used for:

- *ECA rules options filtering*: adapts the set of available options to be configured in the user's mobile device when creating ECA rules, just showing those smart objects whose controller module has been previously installed in the user's mobile device and adapting the configuration options of each smart object according to its features.

Additionally, four types of incompatibilities can be automatically detected using these models:

- *Exclusive resource incompatibility*: identifies those resources that cannot be employed at the same time in a ECA statement (e.g., a user cannot be located in disjoint places at the same time or s/he cannot both receive and make a call). The *isResourceIncompatibleWith* symmetric property in $\mathbb{O}_{ECA}$ ontology is used to identify these kinds of incompatibilities.
- *Smart object interaction incompatibility*: $\mathbb{O}_{so}$ ontology is used to identify inter-resource incompatibilities, as smart objects interact among them given certain restrictions. For instance, it would not be possible to set an event (or condition) like "*intelligentTray* detects *tvSet*" if the *tvSet* is defined as a fixed object and the *intelligentTray* is only able to detect mobile objects. $\mathbb{O}_{ECA}$ ontology adds the necessary semantic to extend this incompatibility detection at ECA rule level (e.g., "**ON** *userLocation=roomA* **IF** *userUses=carX* **DO** *<action>*" is an incompatible ECA statement in case *carX* cannot be inside *roomA* (this would be stated in $\mathbb{O}_{ss}$), although "*userLocation=roomA*" or "*userUses=carX*" are independently valid constructions).
- *Configuration incompatibility*: checks that each resource is configured according its valid set of values (e.g., "**ON** *<event>* **IF** *objectXtemperature<55°* **DO** *<action>*" would be a range-incompatible ECA statement if objectX temperature only range from -10° to +35°). $\mathbb{O}_{ECA}$ ontology adds the necessary semantic to extend this incompatibility detection at ECA rule level (e.g., "**ON** *<event>* **IF** *temperature<20°* **AND** *temperature>30°* **DO** *<action>*" would be also a range-incompatible ECA statement).
- *Functional incoherence*: prevents inconsistent actions to be configured (e.g., the same set of events and conditions cannot trigger opposite actions: "**ON** *event1* **IF** *condition1* **DO** *turnRadioON*" rule is functionally incompatible with "**ON** *event1* **IF** *condition1* **DO** *turnRadioOFF*"). This reasoning employs the semantic encoded in $\mathbb{O}_{ECA}$ ontology (and its associated rule base), e.g., in the *isFunctionalyIncompatibleWith* property.

Finally, it has to be noted that, as $\mathbb{O}_{ECA}$ ontology is linked with the available capabilities of the smart objects which a user's mobile device is able to configure (i.e., with $\mathbb{O}_{so}$), this information can be used in order to load in memory a filtered version of the complete knowledge base, just containing the specifications of those smart objects participating in any of the active ECA rules.

4.2 Semantic Rule-Based Reasoning Support for ECA Rules Execution

The set of ECA rules configured by a user defines those events a mobile device should be aware of. Some smart objects (those more intelligent) may filter by themselves these events, only sending to the user's mobile device those events involved in any of the configured rules. On the contrary, in other cases, the user's mobile device has to be intelligent enough to detect those events itself. So, in the worst case, user's mobile devices should be intelligent enough to *(i)* detect only those significant events involved in any of the configured ECA rules and, afterwards, *(ii)* evaluate the set of conditions associated to a particular ECA rule, in order to decide whether to launch an action or not.

The semantic models presented in the previous Section may be also used to enhance these reasoning operations. Besides, having a general purpose rule engine eases the configuration of complex rules that would only need to be semantically defined but not programmatically implemented as in previous non-semantic versions of this ECA model based smart space managing system.

4.2.1 Context-Based Dynamic Activation of ECA Rules

Each ECA rule configured in the user's mobile device involves several resources (i.e., smart objects –including the mobile device–). Context related to these resources and the particular context of the user's mobile device may cause some rules to become obsolete (not applicable), or vice versa. It is possible to use the information about the smart objects encoded in the previously presented models in order to support the dynamic activation/deactivation of ECA rules depending on:

- *Smart objects status*: an ECA rule involving a particular smart object should be deactivated if the smart object is switched off, if it loses its communication capabilities, etc. (and vice versa).
- *User's mobile device communications status*: every ECA rule requiring the mobile device to access a remote object should be deactivated if the mobile device loses communication coverage (and vice versa).

This can be achieved because smart objects (and user's mobile device) status is stored in the $\mathbb{O}_{so}$ ontology and also because the $\mathbb{O}_{ECA}$ ontology offers a mapping between each ECA rule and the smart objects involved in the rule.

Future extensions of this work have to consider multiuser scenarios, where this automatic rules activation/deactivation process should be applied when detecting contradictory rules configured by different users.

4.2.2 ECA Rules Conditions Execution Order Prioritization

ECA rules may involve evaluating 'online' conditions, i.e., those requiring to access an external resource. On the contrary, 'offline' conditions only need to access to parameters stored inside the user's mobile device. So it is quite common to configure ECA rules with both 'online' and 'offline' conditions.

Within this scenario, having information about the kind of conditions to be evaluated may be quite useful for saving mobile device resources: if the 'offline' conditions are first evaluated, once a necessary and sufficient condition is detected, no more conditions would need to be evaluated. This is shown in next simple example: "**ON** (*<offlineEvent1>* **OR** *<onlineEvent1>*) **IF** (*<offlineConditionA>* **AND** *<onlineConditionA>*) **DO** *<action>*". Having this optimization in mind, *<onlineEvent1>* and *<onlineConditionA>* would not need to be evaluated if *<offlineEvent1>* is true and *<offlineConditionA>* false, saving time and resources (e.g., processor time, battery, networking cost, etc.).

5 Conclusion and Future Works

The work presented in this paper introduces the first steps to extend a mobile-based smart space interaction paradigm with semantic capabilities. This extension implies the integration of several ontology models and a semantic and rule-based inference engine inside mobile personal devices (e.g., a smartphone).

After introducing the ECA model concept and the (non-semantic) architecture of an already developed prototype [1], the employed ontology models have been detailed, highlighting how they can be exploited in order to *(i)* automatically detect incompatible ECA rules configurations and to *(ii)* support complex rules configuration and execution.

Our current work includes a real implementation of this approach for Android-based smartphones. After validating this first version, the semantic models used should be further developed, defining particular restrictions and reusing existing ontology models. The final objective is to compare performance and usability, in order to come to conclusions that may support the choice of using compact reasoners in mobile applications with situational checking needs.

Acknowledgements. This work has been supported by the Government of Madrid under grant S2009/TIC-1485 (CONTEXTS) and by the Spanish Ministry of Science and Innovation under grant TIN2008-06742-C02-01.

References

[1] Bernardos, A.M., Casar, J.R., Cano, J., Bergesio, L.: Enhancing interaction with smart objects through mobile devices. In: 9th ACM Int. Symposium on Mobility Management and Wireless Access (MOBIWAC 2011), Miami, FL, USA, pp. 199–202 (November 2011)

[2] Beigl, M.: Point amp Click - Interaction in Smart Environments. In: Gellersen, H.-W. (ed.) HUC 1999. LNCS, vol. 1707, p. 311. Springer, Heidelberg (1999)

[3] Iglesias, J., Bernardos, A.M., Tarrío, P., Casar, J.R., Martín, H.: Design and validation of a light inference system to support embedded context reasoning. Personal and Ubiquitous Computing, 1–17 (2011)

[4] Siegemund, F.: A Context-Aware Communication Platform for Smart Objects. In: Ferscha, A., Mattern, F. (eds.) PERVASIVE 2004. LNCS, vol. 3001, pp. 69–86. Springer, Heidelberg (2004)

[5] Ye, J., Coyle, L., Dobson, S., Nixon, P.: Ontology-based models in pervasive computing systems. Knowl. Eng. Rev. 22, 315–347 (2007)

[6] Bettini, C., Brdiczka, O., Henricksen, K., Indulska, J., Nicklas, D., Ranganathan, A., Riboni, D.: A survey of context modelling and reasoning techniques. Pervasive Mob. Comput. 6, 161–180 (2010)

[7] Crivellaro, F.: μJena: Gestione di ontologiesuidispositivi mobile. MsC Thesis. Politecnico di Milano (2007)

[8] Jang, M., Sohn, J.-C.: Bossam: An Extended Rule Engine for OWL Inferencing. In: Antoniou, G., Boley, H. (eds.) RuleML 2004. LNCS, vol. 3323, pp. 128–138. Springer, Heidelberg (2004)

[9] Toninelli, A., Pathak, A., Issarny, V.: Yarta: A Middleware for Managing Mobile Social Ecosystems. In: Riekki, J., Ylianttila, M., Guo, M. (eds.) GPC 2011. LNCS, vol. 6646, pp. 209–220. Springer, Heidelberg (2011)

[10] Hachem, S., Toninelli, A., Pathak, A., Issarny, V.: Policy-Based Access Control in Mobile Social Ecosystems. In: IEEE International Symposium on Policies for Distributed Systems and Networks (POLICY 2011), pp. 57–64 (June 2011)

[11] Alferes, J.J., Amador, R.: r^3– A Foundational Ontology for Reactive Rules. In: Meersman, R., Tari, Z. (eds.) OTM 2007, Part I. LNCS, vol. 4803, pp. 933–952. Springer, Heidelberg (2007)

[12] Amador, R., Alferes, J.J.: Knowledge Resources Towards a RESTful Knowledge (2009), http://k4r.googlecode.com/files/200903_kr2rk.pdf

[13] Kifer, M.: Rule Interchange Format: The Framework. In: Calvanese, D., Lausen, G. (eds.) RR 2008. LNCS, vol. 5341, pp. 1–11. Springer, Heidelberg (2008)

[14] Siegemund, F., Floerkemeier, C., Vogt, H.: The Value of Handhelds in Smart Environments. In: Müller-Schloer, C., Ungerer, T., Bauer, B. (eds.) ARCS 2004. LNCS, vol. 2981, pp. 291–308. Springer, Heidelberg (2004)

Using GPU for Multi-agent Multi-scale Simulations

G. Laville, K. Mazouzi, C. Lang, N. Marilleau, and L. Philippe

Abstract. Multi-Agent System (MAS) is an interesting way to create models and simulators and is widely used to model complex systems. As the complex system community tends to build up larger models to fully represent real systems, the need for computing power raise significantly. Thus MAS often lead to long computing intensive simulations. Parallelizing such a simulation is complex and it execution requires the access to large computing resources. In this paper, we present the adaptation of a MAS system, Sworm, to a Graphical Processing Unit. We show that such an adaptation can improve the performance of the simulator and advocate for a more wider use of the GPU in Agent Based Models in particular for simple agents.

1 Introduction

Scientists pay more and more attention to models. They aim at simulating real complex systems to understand them, for example to prevent phenomena such as disasters. Models thus (such as Agent Based Models -ABM-) become more and more descriptive: where scientists used to built up conceptual models (KISS model), they now create well described models (KIDS Model) [1]. Tomorrow the complex system community may tend to "world model", a fully descriptive and generic model that scientists can customize according to scientific questions [2]. Making these modeling evolutions possible requires a raise of computing power as the model executions must return a result in a reasonable delay.

If a simulation can be parallelized then increasing the computing power becomes a question of cost: we need to buy more cores to increase the size of the simulation. While standard processors (CPU) are still expensive, modern Graphical Processing Units (GPU) provide good execution performance for a lower cost. These GPUs also give the possibility to execute non-graphic programs using languages such as OpenCL or CUDA.

Typical simulators are however based on a sequential design and only use one core of the main processor. So taking advantage of the computer architecture implies

Nicolas Marilleau
Institut de Recherche pour le Développement (IRD), France
e-mail: nicolas.marilleau@ird.fr

Christophe Lang · Guillaume Laville · Kamel Mazouzi · Laurent Philippe
Institut FEMTO, CNRS / Université de Franche-Comté, France
e-mail: name.surname@univ-fcomte.fr

S. Omatu et al. (Eds.): Distributed Computing and Artificial Intelligence, AISC 151, pp. 197–204.
springerlink.com   © Springer-Verlag Berlin Heidelberg 2012

to develop distributed simulators. In this context, Multi-Agent System (MAS) is an interesting way to create this kind of models and simulators. ABM are indeed often used to simulate natural or collective phenomenons whose actors are too numerous or various to provide a unified algorithm describing the system evolution.

We propose in this paper to delegate part of the ABM execution to the graphical unit of the computer, based on our experience in parallelizing and implementing part of an ABM on a GPU. We then run it on both a standard CPU processor and on a GPU. We get very good performance results and advocate for a more wider use of the GPU in ABM in particular for simple agents. In Section 2 of the paper we present the work related to agent and parallelization. We give an overview of the sworm simulator in Section 3 and we detail its GPU implementation in 4. We present the experiments and results in Section 5 then we conclude on the possible generalization of our work.

2   Related Works

To develop ABM, many frameworks are now available (e.g. Repast [3] or NetLogo [4]). Only a few of them introduces distribution in agent simulation (Madkit [5] or MASON [6]). In this context, parallel implementations are often based on threads using shared memory or on the integration of cluster libraries such as MPI. Even as multi-core or multi-socket setups become more and more readily available, these solutions stay limited to a relatively small number of parallel tasks, if not using a fully-fledged computing cluster.

Parallelizing a simulation is however complex as space and time constraints must be enforced. Time constraints are linked to the simulation execution. It is usually based on a synchronous execution of time steps by the agents and the environment. Distributing the simulation or delegating part of this execution to others processors as GPUs [7] must thus be carefully done to enforce the synchronism [8]. Space constraints are linked to the environment distribution. Using an environment on a set of computers leads to classical parallelism issues: data coherency if the environment is shared amongst all the computers or data exchanges if the environment is distributed. In the particular case of multi-scale simulations such as the Sworm simulation [9] the environment may be used at different levels. This characteristic, especially in a fractal model, could be the key of the distribution. For instance, each branch of a fractal environment could be identified as an independent area and parallelized. In addition Fractal is a famous approach to describe multi-scale environment (such as soil) and its organization [10].

3   Sworm

Sworm model aims at simulating soil functioning especially the role of macro-fauna on microbial activity and on carbon production. A first version presented in [9] focuses on soil bioturbations caused by earthworms. Soil zones, which have been modified by these earthworms, are considered by biologist as hot spots of a microbial activity (an important process that transforms organic matters into carbon and

Algorithm 1. Sequential evolution algorithm

for all $mm \in mmList$ **do**
    $breathNeed \leftarrow world.respirationRate * mm.carbon$
    $growthNeed \leftarrow world.growthRate * mm.carbon$
    if $totalAccessibleCarbon(mm) > breathNeed$ **then**
        $mm.active \leftarrow true$
        $consumCarbon(mm, breathNeed)$
        $world.CO2 \leftarrow world.CO2 + breathNeed$
        if $totalAccessibleCarbon(mm) > 0$ **then**
            $growthConsum \leftarrow max(totalAccessibleCarbon(mm), growthNeed)$
            $consumCarbon(mm, growthConsum)$
            $mm.carbon \leftarrow mm.carbon + growthConsum$
        end if
    else
        $mm.active \leftarrow false$
    end if
end for

so gives the fertility to the soil). The MIOR -MIcro-ORganisms- model tackles reproducing the microbial activity of these hot spots. MIOR focuses on a nearby 0.002 mm sized cube of soil while Sworm models 20 cm sized volume of soil.

Since these different models do not act at the same level of detail, this simulation is split into multiple, recursive levels corresponding to different scales. It uses a fractal pattern which allows lazy environment allocation: the different kinds of agents do not work at the same scale, depending of their respective sizes. As an example, worms tend to interact on a macroscopic scale (few millimeters), whereas microbial colonies work at a microscopic scale (10^{-4} millimeters). Figure 1 illustrates the fractal data representation of Sworm. White cells (in the left part of the cell) represent soil cavities, black cells represent mineral matter and gray ones represent composite cells that must be further decomposed..

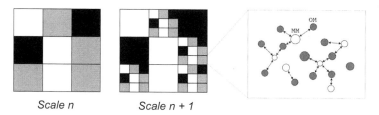

Scale n   Scale n + 1

Fig. 1 Sworm environment representation

A MIOR simulation is based on two types of agents. MMs -Meta-MIOR- represent colonies of carbon-consuming microbes naturally found in soil. OMs -Organic Matters- are deposits that represent the base resources randomly distributed in the modeled environment. Two main processes account for the model evolution, representing the MM metabolism: breathing (conversion of mineral carbon to CO_2) and growth (fixation of carbon into microbial colonies). This second process only occurs if the colony needs are covered by the environment, as described in Algorithm 1.

Even with the sparse fractal representation, simulating the microscopic-scale evolution quickly becomes time-consuming on a traditional CPU unit, where much execution resources are already required for the macroscopic simulation (nearly one week per simulation on a personal computer).

4 GPU Implementation

CUDA and OpenCL are the main programming languages for GPU scientific programming. They are based on a C-like syntax with additions to support GPU-specific features. Various open-source bindings, such as JOCL [11] and JCUDA [12] exploit GPUs with the Java language. Since our implementation must be integrated in a Java-based implementation and must be portable JOCL is a natural choice.

Sworm requires two species of agents, cognitive ones (representing worms) and reactive ones (MIOR microbial colonies). The two kinds are not as good candidates for a GPU implementation. Reactive agents tend to rely on data parallelism with few, if any, kind of branching in the code to be executed. On the opposite, cognitive agents have a variety of possible behaviors at each iteration.

Our work therefore mainly focuses on the adaptation of the MIOR part of the Sworm simulation to a GPU architecture. Since the environment modeling and evolution if often a costly aspect in many ABM, theses changes can provide a tangible benefit on the simulator executions. By using simultaneously the GPU for the environment execution and the CPU for the cognitive agents we also improve the quality of modeling (replacement of the precomputed static carbon evolution graph by dynamically computed ABMs in our case).

4.1 OpenCL Representation

Java can only be used to control the model initialization and the OpenCL executions. So a mapping of the Java Object-Oriented concepts to the C-like data structures of OpenCL is required to access and manipulate them in the GPU code. Here are some guidelines that can be used in this mapping process. The data aspect (type, attributes) can be represented as OpenCL structures. Methods can be implemented as device-local functions, and each phase of the simulation can have its own high-level function calling them. Consecutive function calls with the same dimensions and subjects may be gathered into the same kernel (set of threads) to minimize the number of context switches between host and GPU.

By applying these mapping rules, four main data structures appears in the MIOR simulator: one array of MM structures for microbial colonies, one array of OM structures for organic deposits, one neighboring matrix used to store accessibility between MM and OM and one single environment used to store global parameters such as agent numbers and output data such as the quantity of CO_2.

The design of the computing topology is of importance for the parallelism of the execution. Due to their simple behavior the reactive agents usually have limited interactions and thus allow a high level of parallelism. In our case, the action radius of each MM agent is determined by a fixed parameter of the environment, *RA*. All organic matter deposits within this distance should be accessible for the metabolism.

Since there are no dependencies between each distance calculation, we can use a two dimension kernel in OpenCL where each thread represents an (OM, MM) couple. This approach can be generalized with n dimension kernels in OpenCL depending on the modelized environment and on the GPU characteristics (usually one to three dimensions).

4.2 Data Representation and Data Dependencies

The limited interactions between reactive agents also impact the data representation as the density of interaction matrices may be low. In a typical MIOR simulation each agent is linked to a small (less than 10) number of OM on a total of some hundredth. The topology computation thus produces a low-density matrix. During the following evolution process, going through these matrices may induce numerous costly global memory accesses without actual use of the information. This issue can be addressed by creating dense continuous representations of the matrices. So on the topology stage we create two dense continuous representations of the neighboring matrix. The additional memory cost of these two data structures (one for the OM, one for the MM) can thereafter be balanced thanks to the model living steps.

Data dependency is a much difficult issue as it usually depends on the ABM characteristics. Some guidelines may be given but good results can only be obtained thanks to experience. Most of the time, as in our case, an algorithm adaptation will be needed. We present two cases of algorithm adaptation in this section.

In the MIOR simulation the data dependencies comes from the need to ensure fair access to carbon deposits for each microbial colony and thus to synchronize these accesses across the simulation. Since the colonies are randomly placed at the model initialization, no obvious geographic static locality can be extracted to split the carbon deposits in local memories for the sets of microbial colonies (MM). Therefore much data has to remain into the global slow GPU memory. A first implementation based on mutual execution and global memory synchronization thus resulted in GPU performances two order of scale slower than a sequential Java simulation. This clearly indicated the need for an algorithm adaptation. Our solution is to reduce the number of synchronization required for the execution of a given step of simulation. For that we split the living cycle process into three steps:

1. Scattering: carbon deposit resource is evenly scattered in parts across all microbial colonies accessing to it,
2. Living: each microbial colony consumes carbon for its breathing/growing process and produces CO_2,
3. Gathering: each carbon deposit content is recomputed from the remaining carbon in each part.

There are also synchronization issues in the breathing and growth part of the initial algorithm. Since multiple MM can share a OM carbon deposit, two synchronization issues must be addressed to avoid introducing a bias in the simulation. First, a synchronization on the resources is needed as multiple MM agents should not modify the OM's properties at the same time. Second, the fair access to resources from the MM must be enforced and no OM should be over-exploited.

4.3 Data Transfer between Host and GPU

Since the reference algorithm is sequentially implemented, a synchronization of the simulation steps must be implemented to prevent some agents to evolve faster than their peers. A first naive implementation may attempt to copy all modified data from and to Java objects between each simulation step. Data copy between the CPU system and the GPU-dedicated memory however uses a connexion limited to a fraction of the main memory bandwidth and with a much higher latency. So an obvious course of action to improve this is to allow the simulation to run several steps at once, or up to the termination avoiding these costly transfers. The only exception must be for values needed to detect simulation termination or convergence. The better improvements are achieved in the case where only the final state of the system is needed.

4.4 Proposed Implementations

We propose three consecutive GPU implementations of the MIOR living process with an increased adaptation level. The first implementation, further referenced as GPU v1.0, is based on a straightforward adaptation of the CPU algorithm where the relations between agents are stored as a simple two-dimensions sparse matrix located in global memory. Each simulation step iterates over the whole matrix (including empty cells). The second GPU implementation, called GPU v2.0, replaces this matrix by a compressed one using the same representation format as described in [13]: neighbors are stored in a contiguous way in the matrix along with a precomputed count of neighbor agents. This representation reduces the required number of iterations and of memory accesses. To minimize global memory access overhead, the third GPU implementation, referenced as GPU v3.0, uses the GPU private memory to store often-used data such as carbon parts and MM neighboring information.

5 Experiments

To assess the performance of our work we compare the sequential version with our MIOR implementations on two platforms, representing what a researcher could except as dedicated computing hardware or personal computer graphic card. The first platform is a GPU node with two Intel Xeon X5550 CPU at 2.67GHz and a Tesla C1060 GPU card running at 1.3GHz (240 cores organized in 30 streaming multiprocessors). The second platform is a personal computer with an Intel Q9300 CPU at 2.5GHz and a mainstream GPU card: a GeForce 8800GT at 1.5GHz (112 cores organized in 14 streaming multiprocessors).

Figures 2 and 3 give the average execution time of 50 simulations. The problem size is given by the scaling factor. A scaling factor of 6 means that the number of agents is multiplied by 6. At scale 1, the model contains 38 MM and 310 OM.

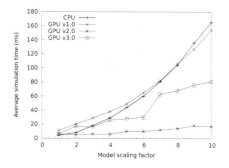

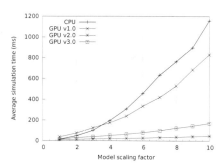

Fig. 2 CPU and GPU executions on a Tesla C1060

Fig. 3 CPU and GPU executions on a GeForce 8800GT

We can note that small sized problems result in similar execution times across all the implementations. The GPU simulations do not have enough agents to benefit from the algorithm's parallelism. From scale 5 GPU versions v2.0 and v3.0 are clearly faster than the CPU or the simple GPU implementations. At scale 10, a ratio of 10 can be observed between the v3.0 GPU implementation and the CPU one.

Several important remarks must be reported. First, the optimized GPU versions give much better performance than the non-optimized GPU one. So it really matter to work on the algorithm parallelization. Second, even a mainstream desktop graphic card provides the same order of performance gain that a much more costly solution, the Tesla card. This is rather interesting as a much larger public could benefit from GPU adaptations. Third, the same code is run on the GeForce card as on the Tesla card without any modification. So the performance does not depend on the GPU characteristics and the same benefit could be expected on other graphic cards.

6 Conclusion

In this paper we have described an adaptation of an existing ABM simulation using GPU hardware. The first result of this work is that adapting the algorithm to a GPU architecture is possible for ABMs. This adaptation may provide a significant performance improvement without so much effort, or at least not much than for a standard multi-threaded parallelization. We have shown that this approach especially suits the case of multi-scale ABMs. We also have generalized this work to show the main issues to be addressed when parallelizing an ABM and how to take benefit from the GPU architecture. Last, we show that using a mainstream card, as the GPU card of a standard computer, can even lead to a significant performance improvement and avoid the use of a costly parallel cluster.

We are currently working on defining more general guidelines for adapting SMA simulations to the GPU architecture. The aim is to use both CPU and GPU at the same time to support multi-scale ABMs. With cognitive agents being run on the CPU and reactive agents or environment being run on the GPU, we could run larger simulations thanks to the performance improvement provided by this specific architecture. In this perspective, we are working on the design of a generic library to support reactive agent simulation on graphic cards.

Acknowledgements. Computations have been performed on the supercomputer facilities of the Mésocentre de calcul de Franche-Comté.

References

1. Edmonds, B., Moss, S.: From KISS to KIDS – *An 'Anti-simplistic' Modelling Approach*. In: Davidsson, P., Logan, B., Takadama, K. (eds.) MABS 2004. LNCS (LNAI), vol. 3415, pp. 130–144. Springer, Heidelberg (2005)
2. Amouroux, E.: KIMONO: using the modelling process as an aid for research orientation. PhD thesis, UPMC, Paris, France (2011)
3. North, M.J., Howe, T.R., Collier, N.T., Vos, J.R.: A declarative model assembly infrastructure for verification and validation. In: Advancing Social Simulation: The First World Congress, FRG, Heidelberg (2007)
4. Sklar, E.: Netlogo, a multi-agent simulation environment. Artificial Life 13(3), 303–311 (2011)
5. Gutknecht, O., Ferber, J.: Madkit: a generic multi-agent platform. In: Proceedings of the Fourth International Conference on Autonomous Agents, AGENTS 2000, pp. 78–79. ACM, New York (2000)
6. Sean, L., Cioffi-Revilla, C., Panait, L., Sullivan, K., Balan, G.: Mason: A multi-agent simulation environment. Simulation: Transactions of the Society for Modeling and Simulation International 82(7), 517–527 (2005)
7. Bleiweiss, A.: Multi agent navigation on the gpu. In: Game Developers Conference, GDC 2009 (2008)
8. Marilleau, N., Lang, C., Chatonnay, P., Philippe, L.: An agent based framework for urban mobility simulation. In: PDP, France, pp. 355–361 (2006)
9. Blanchart, E., Marilleau, N., Drogoul, A., Perrier, E., Chotte, J.L., Cambier, C.: Sworm: an agent-based model to simulate the effect of earthworms on soil structure. EJSS. European Journal of Soil Science 60, 13–21 (2009)
10. Bird, N., Perrier, E.: The psf model and soil density scaling. European Journal of Soil Science 54(3), 467–476 (2003)
11. JOCL: Java bindings for OpenCL (October 11, 2011), http://www.jocl.org/
12. JCUDA: Java bindings for CUDA (October 11, 2011), http://www.jcuda.org/
13. Gómez-Luna, J., González-Linares, J.M., Benavides, J.I., Guil, N.: Parallelization of a Video Segmentation Algorithm on CUDA–Enabled Graphics Processing Units. In: Sips, H., Epema, D., Lin, H.-X. (eds.) Euro-Par 2009. LNCS, vol. 5704, pp. 924–935. Springer, Heidelberg (2009)

MultiAgent Systems for Production Planning and Control in Supply Chains

Faten Ben Hmida, Anne Seguy, and Rémy Dupas

Abstract. This paper presents a generic and extensible multiagent model for production planning and control in supply chains. The purpose of this work is to simulate different control architectures for a given supply chain in order to allow practitioners choose the control strategy that realizes the best benefits and the greater reactivity. The production planning is carried out using the material requirements planning method. Three types of control architectures are simulated on the basis of a test case application: centralized, distributed and mixed architecture.

Keywords: multiagent systems, simulation, supply chain, control, production planning.

1 Introduction

Companies are currently operating in an increasingly dynamic, complex, and uncertain environment with large fluctuations and high competitiveness. In such demanding environment, enterprises try to come together and cooperate with number of partners in order to seek a balance between profitability and customer satisfaction. These inter-enterprises collaborations define complex organizational relationships which supply chains are part of. Managing and controlling production activities of the entire supply chain are major factors of enterprises success. In this context, the organization of the supply chain decision-making system plays a very important role, however, its impact on the whole supply chain is unpredictable before its execution [1]. Therefore, there is a real need for efficient simulation tools

Faten Ben Hmida
University of Manouba, ENSI, SOIE
2010 Manouba, Tunisia
e-mail: faten.benhmida@ensi.rnu.tn

Faten Ben Hmida · Anne Seguy · Rémy Dupas
University of Bordeaux, IMS, UMR 5218
F-33400 Talence, France
e-mail: {faten.benhmida,anne.seguy,remy.dupas}@u-bordeaux1.fr

S. Omatu et al. (Eds.): Distributed Computing and Artificial Intelligence, AISC 151, pp. 205–212.
springerlink.com   © Springer-Verlag Berlin Heidelberg 2012

to analyze different supply chain control design alternatives. Agent technology is very well adapted to this issue and provides a natural way to design and implement efficient distributed manufacturing systems. Many surveys in the literature review and discuss the usage of multiagent systems in supply chains [2, 3] and consider it as very promising for next generation manufacturing [4]. The particularity of this work is not to propose an agent-based system that takes part of the decision-making process and which is integrated in the supply chain but rather a means to simulate it in order to predict and analyze the actual behaviour of the supply chain using a particular control approach. This paper is organized as follows: section 2 presents some fundamental notions related to production planning, control, and decision-making activities within supply chains. It also discusses the different control architectures to be implemented. In section 3, the proposed multiagent model for production planning and control in supply chains is detailed. Section 4 presents a test case supply chains network and the developed multiagent systems corresponding to each of the control architectures previously discussed. A conclusion and a look at future work are presented in section 5.

2   Production Planning and Control in Supply Chains

Supply chain can be defined as a network of production entities that performs the functions of raw materials procurement, transformation of these materials into intermediate or finished products, and distribution of these products to customers [5, 6]. It also includes the activities of managing supply and demand, inventory tracking, manufacturing order entry, distribution and delivery to the customers [7]. Supply Chain Management (SCM) is the management of both material and physical flows between the several facilities of the supply chain [8]. Production planning is one of the most important activities in SCM, as it addresses decisions about the acquisition, utilization and allocation of production resources to satisfy customer demands in the most efficient way over a specified time horizon [9].

To control production activity, the Material Requirement Planning (MRP) is one of the most widely practiced techniques. It consists in planning products manufacturing and determining the needs for components based on customers' orders. The needs consist in components quantities which are determined using nomenclatures and stock levels of raw material or semi-finished products [10].

In this work we focus on production planning and control of the information and material flows in supply chains. The development of procurement, production and delivery plans requires the implementation of a decision-making system defining the organization of the decision units through which the supply chain is controlled. The decisional pattern of the production planning and control activity can be organized according to different architectures which are mainly -but not only-centralized and distributed architectures.

In the distributed control there is a single decision making level [5]. In such control architecture, each enterprise of the supply chains network is totally independent and manages locally its own resources. In each enterprise, control is done through a decision making center which is responsible for production planning and resources management.

To overcome the lack of global visibility of information induced by the distributed control in supply chains, an alternative is to make information accessible to all the actors of the supply chains network and to centralize its processing. In fact, in centralized control architecture, there is a supplementary higher decision making level with a single decision center which is responsible for production planning and resources management of the overall supply chains network [5]. Thus, the different decision centers of the lower level are reduced to simple entities relaying information to the central decision making entity.

To find a balance between local autonomy induced by distributed control and global optimisation induced by centralized control, a third control strategy is proposed in [11]. Such mixed control architecture is based on many decision making centers of a higher level instead of a single one. In fact, to each supply chain of the network is associated a decision center which centralizes information and processing within that supply chain. In this case, the information exchange between decision centers of the higher und the lower levels are similar.

The purpose of this work is to present a generic agent-based solution for production planning and control in supply chains. This solution should allow simulating the different presented control architectures. In the following section, we start by introducing our multiagent model.

3   A Multiagent Model for Production Planning and Control

To define the model's specifications, the VOWELS multiagent design approach was used [13]. In this approach called also AEIO, each vowel refers to a fundamental aspect of the multiagent system: A for Agent, E for Environment, I for interaction and O for organisation.

3.1   Agents

Most definitions agree that an agent is an autonomous entity which acts in some environment and interacts with other agents. In a multiagent system, each agent has a specific role and aims to realise a local [13, 14, 15].

In this section, we specify the different active entities of our multiagent model that collaborate to ensure the production process; these entities are represented by the following agents:

- **Distributer:** this agent is responsible for the reception and the management of the customers' orders, whether there are final customers or customer enterprises. It is also responsible for managing the products' stocks and deliveries.

- **Producer:** This agent manages the material flow and ensures the effective production process; it transforms the received material into products according to composition rules defined by the nomenclature.
- **Provider:** This agent is responsible for contacting suppliers and making orders whenever some components are required.
- **Decider:** This agent synchronises the other enterprise's agents actions. It also realises the production and the material requirement planning. The distribution of the decider agents over the supply chain defines the control architecture. In the case of a distributed control architecture, each decider agent is responsible for the decision making process in its enterprise; whether in a centralized or mixed control architecture, the enterprise's decider agents rely on a common decider of a higher level.
- **Customer:** This agent represents a final customer which is wishing to buy some finished product.
- **Supplier:** This agent represents a supplier putting on sale some raw material.

3.2 Environment

The environment is the common space shared by the several agents [13, 14]. In this work, the environment is considered as the set of passive entities manipulated by the agents. Thus, two types of passive entities are identified:

- **Objects** which are products manufactured by the several enterprises and components needed for these products manufacturing.
- **Data** which describes the objects states. For each enterprise, products or components are described by the following information: manufacturing or procurement time, demands, estimated stock, expected receipt stock, production start quantities, production end quantities. All these information are needed for the production planning which is done through the MRP method [10].

3.3 Interactions

Interaction is a central aspect in multiagent systems; it ensures a coherent overall behaviour despite decentralization [14, 15]. In this work, the several system agents interact directly through messages exchange. We start first by presenting the used generic message format. Each message contains the following fields:

- The sender's and the receiver's identifiers or addresses;
- A speech act which expresses the intent of the sender and helps the receiver interpret correctly the message. It can be information, request, response or order;
- An object which allows the receiver classifying and processing the message properly. It can be about procurement, orders, deliveries, manufacturing, etc.

- A content which should be properly formatted to ease the whole message processing. This field contains the order identifier to allow orders traceability, the name of the concerned product and the list of the product's quantities with the corresponding dates (planning horizon periods).

Figure 1 shows the interaction pattern of the proposed multiagent model.

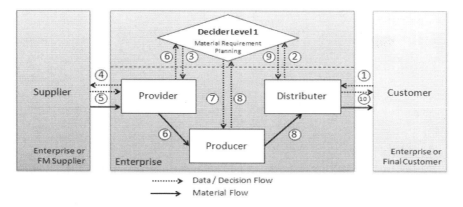

Fig. 1 Interaction Pattern

1) Procurement request	6) Order receipt information
2) Order receipt information	7) Manufacturing order
3) Purchase order	8) End of manufacturing information
4) Procurement request	9) Delivery order
5) Delivery response	10) Delivery response

3.4 Organisations

Organisations provide a means to specify and design the multiagent system's structure, as it allows defining the agents' roles and the relationships between these roles [16]. Our multiagent model defines three organisational levels:

- **Supply chain network level** which allows representing an organisation of interrelated supply chains.
- **Supply chain level** where a detailed representation of the supply chain organisation is defined. A supply chain is a set of interconnected customers, suppliers and enterprises involved in the provision, manufacturing and distribution of a specific product or category of products.
- **Enterprise level** which allows representing a detailed view of the enterprise organisation. An enterprise is defined by a set of services, namely procurement, manufacturing, distribution and decision making services.

The distribution of the decision making centers over these three organisational levels depends on the used control strategy. In fact, in the case of distributed control architecture, decision centers are defined at the enterprise level (a decider per enterprise). Whether in the case of mixed control architecture, decision centers are defined at the supply chain level (a decider per supply chain). And finally, in the case of centralized control architecture decision centers are defined at the supply chains network level (one decider for the overall network).

4 Application on a Test Case Supply Chains Network

This paper considers a test case elaborated in [11, 12]. As shown in figure 2, it is a multi-products supply chain network in the domain of furniture manufacturing. Two families of products are manufactured: tables and shelves. In a first step, wood trunks, delivered by the wood suppliers, are transformed by the sawmills into wood furniture components (trays, legs and boards). In a second step, shelves and tables are assembled by the assembly factory. Shelves are delivered to final customers, and tables are passed to the painting factory to be painted in a third and final step before being delivered to final customers. In this work, the three control architectures presented in section 2 are applied to this test case and simulated. For this purpose, our multiagent model was carried out on the JADE[1] platform.

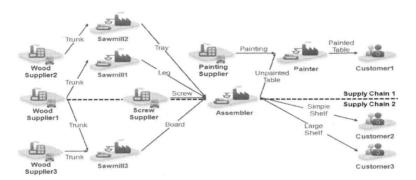

Fig. 2 Test case supply chain network

In the case of distributed control, each one of the decision centers S1.dec, S2.dec, S3.dec, ASS.dec and PNT.dec carries out locally the production planning and resources management of the enterprise to which it belongs. Whereas in the case of centralized control, there is a higher level decision center DEC which carries out by its own all the production plans for the entire supply chain network. Figure 3 shows the implemented multiagent systems in the cases of distributed and centralized controls.

---

[1] http://jade.tilab.com/

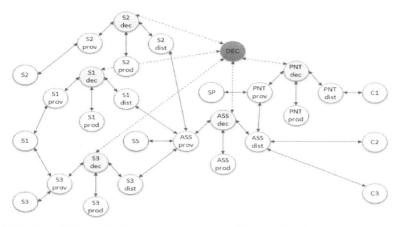

Fig. 3 Distributed / Centralized agent-based control architecture for the test case supply chains

Figure 4 shows the implemented multiagent systems in the cases of mixed control. In the case of mixed control, there is a higher level decision center for each supply chain: DEC1 and DEC2.

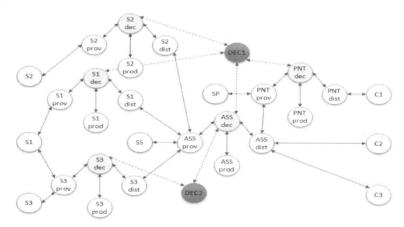

Fig. 4 Mixed agent-based control architecture for the test case supply chains

5 Conclusion

In this paper we defined and tested a generic and extensible agent-based model for production planning and control in supply chains. The multiagent model was designed using the Vowels modelling approach and allowed to simulate several control architectures for a test case supply chain network: distributed, centralized and mixed control.

As an extension to this work in progress, we are focusing on the evaluation of these three architectures, several performance measures will be used to determine the architectures that fits the best to the production planning and the decision-making process in supply chains. For this purpose we rely on the evaluation method presented in [17, 18, 19].

References

1. Chang, Y., Makatsoris, H.: Supply chain modelling using simulation. International Journal of Simulation 2(1), 24–30 (2001)
2. Moyaux, T., Chaib-Draa, B., D'Amours, S.: Supply chain management and multiagent systems: an overview. In: Multiagent Based Supply Chain Management, pp. 1–27. Springer (2006)
3. Kumar, V., Srinivasan, S.: A review of supply chain management using multi-agent system. International Journal of Computer Science Issues 7(5), 198–205 (2010)
4. Shen, W., Hao, Q., Yoon, H.J., Norrie, D.H.: Applications of agent-based systems in intelligent manufacturing: an updated review. Advanced Engineering Informatics 20, 415–431 (2006)
5. Ganeshan, R., Jack, E., Magazine, M.J., Stephens, P.: A taxonomic review of supply chain management research. Kluwer Academic Publishers, Boston (1998)
6. Ganeshan, R., Harrison, T.P.: An introduction to supply chain management (1995), http://lcm.csa.iisc.ernet.in/scm/supply_chain_intro.html (accessed October 01, 2011)
7. Lummus, R.R., Vokurka, R.J.: Defining supply chain management: a historical perspective and practical guidelines. Industrial Management and Data Systems 99(1), 11–17 (1999)
8. Thomas, D.J., Griffin, P.M.: Coordinated supply chain management. European Journal of Operational Research 94, 1–15 (1996)
9. Graves, S.C.: Manufacturing planning and control (1999), http://mit.edu/sgraves/www/ProdPlanCh.pdf (accessed September 09, 2011)
10. Mabert, V.A.: The early road to material requirements planning. Journal of Operations Management 25(2), 346–356 (2007)
11. François, J.: Planification des chaînes logistiques: modélisation du système décisionnel et performances. Ph.D dissertation. University of Bordeaux 1, France (2007)
12. François, J., Okongwu, U., Deschamps, J.C., Lauras, M.: How do the key determinants of a distributed planning process impact on the performance of a supply chain? In: 5th International Conference on Advances in Production Management Systems, Bordeaux, France (2009)
13. Demazeau, Y.: From interactions to collective behaviour in agent-based systems. In: 1st European Conference on Cognitive Science, Saint-Malo, France (1995)
14. Ferber, J.: Les systèmes multi-agents, vers une intelligence collective. Inter Editions, Paris (1995)
15. Wooldridge, M.: An introduction to multiagent systems. John Wiley & Sons, London (2002)
16. Boissier, O., Gitton, S., Glize, P.: Caractéristiques des systèmes et des applications. In: Demazeau, Y. (ed.) Systèmes Multi-Agents, Arago, OFTA. Tec & Doc, Paris (2004)
17. Ben Hmida, F., Lejouad Chaari, W., Tagina, M.: Aspect-based multiagent systems observation for performance evaluation. In: 4th International Conference on Intelligent Systems and Agents, Amsterdam, Holland (2008)
18. Ben Hmida, F., Lejouad Chaari, W., Tagina, M.: Performance evaluation of multiagent systems: communication criterion. In: 2nd International Symposium on Agent and Multi-Agent Systems: Technologies and Applications, Incheon, South Corea (2008)
19. Ben Hmida, F., Lejouad Chaari, W., Tagina, M.: Graph thoery to evaluate communication in industrial multiagent systems. International Journal of Intelligent Information and Database Systems 5(4), 361–388 (2011)

Multi-agent Bidding Mechanism with Contract Log Learning Functionality

Kazuhiro Abe, Masanori Akiyoshi, and Norihisa Komoda

Abstract. This paper addresses the agent-based bidding mechanism under trading actions from supplier sites to demand sites. Bidding includes unit price, amount, and storage cost. This trading environment assumes to be completely competitive, which means an agent cannot detect the competitive agent information. To increase the success rate of bidding, the agent must learn its bidding strategy from the past trading log. Our agent estimates the appropriate bidding price from the past "success bids" and "failure bids" by using statistical analysis. Experimental results shows an agent with such learning functionality increases its rate of "success bids" by 45.6%, compared to the agent without such functionality.

1 Introduction

Recently, networking technology has enabled us to gather and distribute information and communicate under various situations. By using such technology, applications such as electronic commerce have been developed. In electronic commerce domain, one practical example is when distributed organizations, categorized into either a supply or demand site of trading resource such as product materials, form a temporary negotiation group on the network. Each organization has its own strategy for maximizing profit and possesses its own initial supply and demand data. In case of this type of trading, the problem usually has the following issues to be solved; negotiation process is asynchronous, and the negotiation involves plans of supply chains.

Kazuhiro Abe · Masanori Akiyoshi · Norihisa Komoda
Osaka University, 2-1, Yamadaoka, Suita, Osaka, Japan
e-mail: abe.kazuhiro@ist.osaka-u.ac.jp,
    akiyoshi@ist.osaka-u.ac.jp, komoda@ist.osaka-u.ac.jp

S. Omatu et al. (Eds.): Distributed Computing and Artificial Intelligence, AISC 151, pp. 213–220.
springerlink.com   © Springer-Verlag Berlin Heidelberg 2012

Under such asynchronous and self-interested negotiation, the negotiation process involves complicated procedures such as proposal, selection, reply, and cancellation. Also in some situations, self-interested negotiation parties may collaborate to increase their profit, that is, coalition structure emerges. Given the situation mentioned so far, agent-based negotiation protocols are proposed, which aims to manage complicated tasks in the negotiation process. There are several works related to negotiation formalization, automatic negotiation mechanisms, and so forth [1][2][3][4][5][6][7].

The aforementioned negotiation protocols are intended to formalize the negotiation structure/management for establishing the agent-based trading. Under such negotiation protocols, each agent should decide its own behavior to maximize its own profit. In this paper, we design an agent-based bidding mechanism from supply sites to demand sites. Bidding includes unit price, amount, and storage cost, which are usually used in supply chains trading environment. To increase the success rate of bidding, the agent must learn its bidding strategy from the past trading log. The proposed agent estimates the most appropriate trading unit price from the past "success bids" and "failure bids" by using statistical analysis.

The rest of this paper is organized as follows. First we discuss our domain problem of bilateral trading of resource transfer. Second we introduce our negotiation process. We then introduce our bidding mechanism with estimated unit price for "success bids". Finally we shows our experimental results.

2 Asynchronous Negotiation in Supply Chains

2.1 Problem Description

When resources such as product materials are transferred from supply to demand sites, we usually consider a certain interval for transfer plans and usage of storage. For instance, a supply site may have a series of supply data from March to July. Usage of storage makes it possible to plan resource transfer even if a time difference exists for practical supply and demand. Figure 1 shows an example of resource transfer transaction that involves such an aspect of our target domain. In addition to the transfer transaction, both supply sites and demand sites pursue their own profit under independently making initial plan data such as supply capacity, demand and storage capacity, which causes asynchronous negotiation. The following is distinctive features the target domain includes.

- Asynchronous negotiation

  - Negotiation reaches mutual agreement under trading constraints.
  - Proposed transaction plans by an organization newly joined negotiation may cause more optimal cases than agreed before.
  - Previous agreement may be canceled.

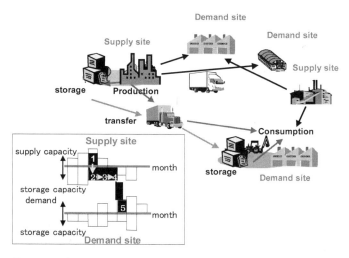

Fig. 1 An Example of Resource Transfer Transaction

- Multi-purpose optimization during the negotiation

  - Maximization is done on "sales ratio" in a supply site and "procurement ratio" in a demand site. The "sales ratio" is a ratio of quantity of agreed plans against initial supply capacity, and the "procurement ratio" is a ratio of quantity of agreed plans against initial demand.
  - Minimization is done on "transaction cost" including transportation cost and storage cost in both sites.

2.2 Negotiation Process

Concerning the above-mentioned features, we design the following negotiation process[8].

As for cancellation in asynchronous negotiation, mutually agreed transaction plans after negotiation are labeled as "pseudo-contracted", which means there is possibility of cancellation along with multi-stage negotiation. If a cancellation message on a certain transaction plan is notified from a contracted partner site, the plan is canceled immediately with some penalty payment.

As for local optimization of transaction plans, optimization technique is adopted to each site, that is, a network flow model is used for generating local optimum transaction plans, and multi-purpose optimization technique is applied when selecting transaction plans. Figure 2 shows our negotiation process with "pseudo-contracted". As shown in Figure 2, each supply site executes repetitive negotiation process of its own pace, and each demand site does so.

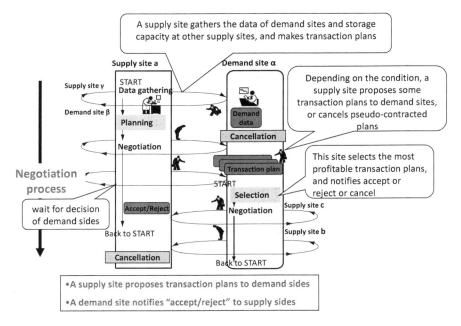

Fig. 2 An Overview of Asynchronous Negotiation

Step 1:   A supply site gathers up-to-date demand site data and storage capacity data of other supply sites that are disclosed.

Step 2:   A supply site makes transaction plans and proposes them to demand sites.

Step 3:   Each demand site selects an optimum combination of transaction plans based on its criteria. This selection causes that some of proposed transaction plans are accepted and some are rejected.

Step 4:   A supply site receiving notification of accept by demand sites sets pseudo-contracted plans and updates its own data.

2.2.1   Transfer Patterns

Supply sites basically use their own storage capacity under this self-interested environment. However, at some stage in the iterative negotiation, storage capacity is not fully used. Imagine supply "site-A" still has some not-contracted resource when its storage capacity is fully used. In that case, the "site-A" announces coalition formation on resource transfer, that is, it uses other competitive supply sites' storage not-in-use for making transaction plans to demand sites. Of course, the "site-A" pays storage cost to the coalition sites. As this case shows, it is necessary to consider transfer patterns to make transaction plans.

Transaction plans using a coalition site's storage do not involve the contracted plans of the coalition site.

Then generation of transaction plans is executed by using four types of input data such as (i) demand, storage capacity and storage cost in demand sites, (ii) storage capacity and storage cost in competitive supply sites, (iii) supply capacity, storage capacity and storage cost in a planning supply site, and (iv) transportation cost under maximization of "sales ratio" and minimization of "transaction cost".

2.2.2   Cancellation

Suppose a negotiation group consists of n supply sites and m demand sites. One of supply sites gathers demand data disclosed by a certain demand site, makes transaction plans, and finally makes contracts as pseudo-contracted plans. At this point, one other demand site starts to disclose its demand data. Then the supply site gathers demand data again and may consider that making contracts with a newly joined demand site is more profitable than a previously contracted one from transaction cost points of views. In this case, the supply site likely compares cancellation profit loss including penalty with newly expected profit. If newly expected profit is more than cancellation profit loss, the supply site likely proceeds to cancel previous contracts and propose new transaction plans to a newly joined demand site. The opposite similar situation is easily guessed that a demand site cancels pseudo-contracted plans and accepts proposed transaction plans from a newly joined supply site. These cancel operations are triggered by supply and demand sites.

Under completely competitive trading environment, an agent cannot detect the competitive agent information, which may cause self-interested bidding as a supply side resorts to end in failure at initial bid or further multistage negotiation period. To avoid such unsuccessful bids as much as possible, improved bidding mechanism is necessary. Hereafter we discuss this problem towards more successful bidding mechanism of agent-based trading.

3   Bidding Mechanism with Contract Log

Past trading log is significant information to decide the appropriate bidding attributes. For instance, even if a certain optimum transaction plan under the aforementioned resource transfer problem means no storage usage of the supply site, it is better to modify the plan to use the supply site storage for a certain period, which reduces a demand site storage cost. Initially calculated bids are modified by such estimation for achieving "success bids". As indicated before, such modification is considered as to initial bids or multistage negotiation bids. Of course the latter case is preferable, however, it is too complicated to estimate against multistage negotiation bids. Therefore this paper focuses on estimation against initial bids for achieving "success bids".

3.1 Factors Induced from a Demand Site

First of all, it is necessary to consider which factors may affect the initial bids from the supply site. As explained before as to our negotiation process, the supply site gathers information such as amount of demand, period of demand, storage capacity of demand site, storage cost of demand site and so forth. Based on these attributes, a certain proposal of transaction plans are calculated by using optimization algorithm from the supply site viewpoints. The demand site evaluates such proposed plans from many supply sites from several factors, and decides its own selected transaction plans. We assume such factors as follows.

- unit price of resource
- amount per the resource transfer plan
- period of storage usage at a demand site
- occupied rate of a demand site storage as to the proposed resource transfer plan

When using these factors for evaluation of the proposed resource transfer plan, significant factors differ in demand sites. Therefore it is inevitable to infer which factors are regarded as important ones. Figure 3 shows our idea to induce such significant factors from past trading log which includes "success bids" and "failure bids" as to a certain demand site.

Step 1: Average values of "success bids" and "failure bids" as to each factor are calculated.

Step 2: Confidence intervals are calculated as to each average value.

Step 3: Overlapped area-square measure-size of plotted graph on "successful bids" and "failure bids" as to each factor is calculated. Then the ratio of such overlapped area-square measure-size against the whole is calculated.

Step 4: Such induced ratios are sorted in ascending manner.

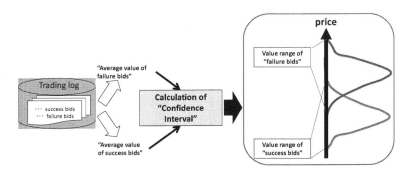

Fig. 3 Specific Factors Identification

As Figure 3 shows, the factor with the lowest ratio is considered to be the most significant factor which the demand site regards. Here we define additional attributes "success unit price range" and "failure unit price range" derived from the "Step 2" calculation as to prices.

3.2 Modification of Bidding Price

In the section 3.1, we introduce the factors affecting the result of bidding as to each demand site with the past trading log. Based on these factors, the modification is executed on the initially calculated optimum transaction plan of the supply site. If the bidding unit price is in the "failure unit price range", it should be reduced until the probable "success bids". However, it should not simply done by reducing the unit price as much as possible, because the unit price is not only a determinant factor as indicated before.

Bids with n attribute values on factors construct n dimensional vector space. Each bid is positioned in this vector space, normalized and weighted according to the inverse proportional ratio of each factor. Figure 4 shows our idea to decide which modification is desirable, using 3 dimensional vector space as an example. As indicated in Figure 4, the "cosine similarity" is used to judge the most desirable modification when reducing the unit price. The lower bound as to the reduced unit price is set to guarantee the profit according to each supply site. The modification is decided by the most minimum difference value on cosine similarity with the modified vector and "success bids" vectors.

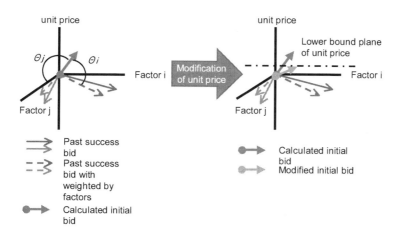

Fig. 4 Bids Modification

4 Experiment

Our proposed bidding mechanism is evaluated by 9 competitive supply site agents and 1 demand site agent. 100 past trading log with "success bids" and "failure bids" is used. There exists 46 case bids which are target ones of the proposed modification process. Table 1 shows the changes on rates of "success bids", "failure bids" and "sales account increase".

Table 1 Experimental result using modified bidding

Evaluated aspect	Improved rate
"failure bids" to "success bids"	45.6%
still "failure bids"	26.1%
increase of sales account	161.5%

5 Conclusion

This paper addresses agent-based bidding mechanism under trading actions from supplier sites to demand sites, which intends to increase the ratio of "success bids" under completely competitive environment. Past trading log is used to induce the significant factors as to "success bids" and desirable modification towards initial planned bids. Our experimental results shows the proposed bidding mechanism leads to increase the ratio of "success bids" as expected. Our further research is to extend this approach towards multistage negotiation bids modification.

References

1. Kraus, S.: Agents contracting tasks in non-collaborative environments. In: Proc. of AAAI 1993, pp. 243–248 (1993)
2. Zlotkin, G., Rosenschein, J.S.: Conflict resolution in non-cooperative domain. In: Proc. of AAAI 1990, pp. 100–105 (1990)
3. Sandholm, T.W., Lesser, V.R.: Advantages of a Leveled Commitment Contracting Protocol. In: Proc. of AAAI 1996, pp. 126–133 (1996)
4. Sandholm, T.W., Sikka, S., Norden, S.: Algorithms for optimizing leveled commitment contracts. In: Proc. of IJCAI 1999, pp. 535–540 (1999)
5. Rosenschein, J.S.: Consenting agents: Negotiation mechanisms for multi-agent systems. In: Proc. of IJCAI-1993, pp. 792–779 (1993)
6. Sandholm, T.W., Larson, K., Anderson, M., Shehory, O., Tohme, F.: Coalition structure generation with worst case guarantees. Artificial Intelligence 111, 209–238 (1999)
7. Tsvetovat, M., Sycara, K., Chen, Y., Ying, J.: Customer Coalitions in Electronic Markets. In: Dignum, F.P.M., Cortés, U. (eds.) AMEC 2000. LNCS (LNAI), vol. 2003, pp. 121–138. Springer, Heidelberg (2001)
8. Akiyoshi, M., Komoda, N.: Asynchronous negotiation protocol in supply chains. In: Proc. of IAWTIC 2005, pp. 65–69 (2005)

Bio-inspired Self-adaptive Agents in Distributed Systems

Ichiro Satoh

1 Introduction

Cellular differentiation is the mechanism by which cells in a multicellular organism become specialized to perform specific functions in a variety of tissues and organs. Different kinds of cell behaviors can be observed during embryogenesis: cells double, change in shape, and attach at and migrate to various sites. We construct a framework for building and operating distributed applications with the notion of cellular differentiation and division in cellular slime molds, e.g., dictyostelium discoideum and mycelium. It is almost impossible to exactly know the functions that each of the components should provide, since distributed systems are dynamic and may partially have malfunctioned, e.g., network partitioning. The framework enables software components, called agents, to differentiate their functions according to their roles in whole applications and resource availability, as just like cells. It involves treating the undertaking/delegation of functions in agents from/to other agents as their differentiation factors. When an agent delegates a function to another agent, if the former has the function, its function becomes less-developed and the latter's function becomes well-developed.

2 Related Work

This section discusses several related studies on software adaptation in distributed systems. One of the most typical self-organization approaches to distributed systems is swarm intelligence [1, 2]. Although there is no centralized control structure dictating how individual agents should behave, interactions between simple agents with static rules often lead to the emergence of intelligent global behavior. There

Ichiro Satoh
National Institute of Informatics
2-1-2 Hitotsubashi, Chiyoda-ku, Tokyo 101-8430, Japan
e-mail: ichiro@nii.ac.jp

S. Omatu et al. (Eds.): Distributed Computing and Artificial Intelligence, AISC 151, pp. 221–228.
springerlink.com   © Springer-Verlag Berlin Heidelberg 2012

have been many attempts to apply self-organization into distributed systems, e.g., a myconet model for peer-to-peer network [4], and a cost-sensitive graph structure for coordinated replica placement [3]. Most existing approaches only focus on their target problems or applications but are not general purpose, whereas distributed systems have a general-purpose infrastructure. Our software adaptation approach should be independent of applications. Furthermore, most existing self-organization approaches explicitly or implicitly assume a large population of agents or boids. However, since the size and structure of real distributed systems have been designed and optimized to the needs of their applications, the systems have no room to execute such large numbers of agents.

3 Basic Approach

This paper introduces the notion of (de)differentiation into a distributed system as a mechanism for adapting software components, which may be running on different computers connected through a network.

Differentiation:

When dictyostelium discoideum cells aggregate, they can be differentiated into two types: prespore cells and prestalk cells. Each cell tries to become a prespore cell and periodically secretes cAMP to other cells. If a cell can receive more than a specified amount of cAMP from other cells, it can become a prespore cell. There are three rules. 1) cAMP chemotaxically leads other cells to prestalk cells. 2) A cell that is becoming a prespore cell can secrete a large amount of cAMP to other cells. 3) When a cell receives more cAMP from other cells, it can secrete less cAMP to other cells.

   Each agent has one or more functions with weights, where each weight corresponds to the amount of cAMP and indicates the superiority of its function. Each agent initially intends to progress all its functions and periodically multicasts *restraining* messages to other agents federated with it. Restraining messages lead other agents to degenerate their functions specified in the messages and to decrease the superiority of the functions. As a result, agents complement other agents in the sense that each agent can provide some functions to other agents and delegate other functions to other agents that can provide the functions.

Dedifferentiation:

Agents may lose their functions due to differentiation as well as be busy or failed. The approach also offers a mechanism to recover from such problems based on dedifferentiation, which a mechanism for regressing specialized cells to simpler, more embryonic, unspecialized forms. As in the dedifferentiation process, if there are no other agents that are sending restraining messages to an agent, the agent can perform its dedifferentiation process and strengthen their less-developed or inactive functions again.

4 Design and Implementation

Our approach is maintained through two parts: runtime systems and agents. The former is a middleware system for running on computers and the latter is a self-contained and autonomous software entity. It has three protocols for (de)differentiation and delegation.

4.1 Agent

Each agent consists of one or more functions, called the *behavior* parts, and its state, called the *body* part, with information for (de)differentiation, called the *attribute* part.

- The body part maintains program variables shared by its behaviors parts like instance variables in object orientation. When it receives a request message from an external system or other agents, it dispatches the message to the behavior part that can handle the message.
- The behavior part defines more than one application-specific behavior. It corresponds to a method in object orientation. As in behavior invocation, when a message is received from the body part, the behavior is executed and returns the result is returned via the body part.
- The attribute part maintains descriptive information with regard to the agent, including its own identifier. The attributes contains a database for maintaining the weights of its own behaviors and for recording information on the behaviors that other agents can provide.

The agent has behaviors $b_1^k, \ldots, b_n^k$ and w_i^k is the weight of behavior b_i^k. Each agent (k-th) assigns its own maximum to the total of the weights of all its behaviors. The W_i^k is the maximum of the weight of behavior b_i^k. The maximum total of the weights of its behaviors in the k-th agent must be less than W^k. ($W^k \geq \sum_{i=1}^{n} w_i^k$), where $w_j^k - 1$ is 0 if w_j^k is 0. The W^k may depend on agents. In fact, W^k corresponds to the upper limit of the ability of each agent and may depend on the performance of the underlying system, including the processor. Note that we never expect that the latter will be complete, since agents periodically exchange their information with neighboring agents. Furthermore, when agents receive no retraining messages from others for longer than a certain duration, they remove information about them.

4.2 Function Invocation

When an agent wants to execute a behavior, it needs to select one of the available behaviors ($b_i^j, \ldots, b_i^m$), even if it has the behavior, according to the values of their weights. This involves three steps.

i) When an agent (k-th agent) wants to execute behavior b_i, it looks up the weight (w_i^k) of the same or a compatible behavior from its database and the weights ($w_i^j, \ldots, w_i^m$) of such behaviors ($b_i^j, \ldots, b_i^m$) from the database.

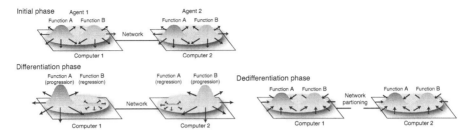

Fig. 1 Differentiation mechanism for software configuration

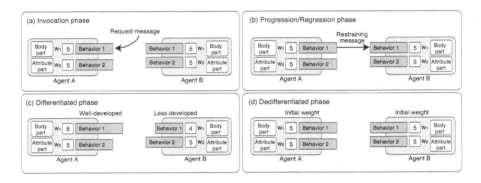

Fig. 2 Differentiation mechanism for agent

ii) If multiple agents, including itself, can provide the wanted behavior, the k-th agent selects one of the agents according to selection function ϕ^k, which maps from w_i^k and $w_i^j, \ldots, w_i^m$ to b_i^l, where l is k or $j, \ldots, m$.

iii) The k-th agent delegates the selected agent to execute the behavior b_i^l and waits for the result from the l-th agent.

There is no universal selection function, ϕ, for mapping from the weights of behaviors to at most one appropriate behavior like that in a variety of creatures. Instead, the approach permits agents to use their own evaluation functions, because the selection of behaviors often depends on their applications. For example, one of the simplest evaluation functions makes the agent that wants to execute a behavior select one whose weight has the highest value and whose signature matches the wanted behavior if its database recognizes one or more agents that provide the same behavior, including itself.

4.3 Differentiation

The approach introduces the undertaking/delegation of behaviors in agents from other agents as a differentiation factor. Behaviors in an agent, which are delegated from other agents more frequently, are well developed, whereas other behaviors, which are delegated from other agents less frequently, in the cell are less developed. Finally, the agent only provides the former behaviors and delegates the latter behaviors to other agents. Our differentiation mechanism consists of two phases. The first involves the progression of behaviors in three steps.

i) When an agent (k-th agent) receives a request message from another agent, it selects the behavior (b_i^k) specified in the message from its behavior part and dispatches the message to the selected behavior. It executes the b_i^k behavior and returns the result.

ii) The k-th agent increases weight w_i^k of the b_i^k behavior.

iii) The k-th agent multicasts a restraining message with the signature of the behavior, its identifier (k), and the behavior's weight (w_i^k) to other agents.

When behaviors are internally invoked by their agents, their weights are not increased. If the total weights of the agent's behaviors, $\sum w_i^k$, is equal to their maximal total weight W^k, it decreases one of the minimal (and positive) weights (w_j^k is replaced by $w_j^k - 1$ where $w_j^k = \mathbf{min}(w_1^k, \ldots, w_n^k)$ and $w_j^k \geq 0$). Although restraining messages correspond to the diffusion of cAMP in differentiation, they can explicitly carry the weights of the agents that send them to reduce the number of restraining messages, because they can be substituted for more than one retaining message without weights. The second phase supports the retrogression of behaviors in three steps.

i) When an agent (k-th agent) receives a restraining message with regard to b_i^j from another agent (j-th), it looks for the behaviors ($b_m^k, \ldots b_l^k$) that can have the signature specified in the received message.

ii) If it has such behaviors, it decreases their weights ($w_m^k, \ldots w_l^k$) in its database and updates the weight (w_i^j) in its database.

iii) If the weights ($w_m^k, \ldots, w_l^k$) are under a specified value, e.g., 0, the behaviors ($b_m^k, \ldots b_l^k$) are inactivated.

4.4 Dedifferentiation

Distributed systems may be damaged or stop due to disasters and problems. We need a mechanism for detecting and remedying failures in networking, computers, agents, remote computers, and other agents. To do this, each agent (j-th) periodically multicasts messages, called *heartbeat messages*, for behavior (b_i^j), which is still activated with its identifier (j). This involves two cases.

i) When an agent (k-th) receives a heartbeat message with regard to behavior (b_i^j) from another agent (j-th), it retains the weight (w_i^j) of the behavior (b_i^j) in its second database.

ii) When an agent (k-th) does not receive any heartbeat messages with regard to behavior (b_i^j) from another agent (j-th) for a specified time, it automatically decreases the weight (w_i^j) of the behavior (b_i^j) in its second database, and resets the weight (w_i^k) of the behavior (b_i^k) to the initial value or increases the weight (w_i^k) in its first database.

Note that behavior b_i^k is provided by the k-th agent and behavior b_i^j is provided by the j-th agent. The weights of behaviors provided by other agents automatically decrease without any heartbeat messages from the agents. Therefore, when an agent terminates or fails, other agents decrease the weights of the behaviors provided by the agent. If they have the same or compatible behaviors, they can then activate the behaviors, which may be inactivated. After a request message is sent to another agent, if the agent waits for the result to arrive for longer than a specified time, it selects one of the agents that can handle the message from its database and requests the selected agent. If there are no agents that can provide the behavior that can handle the behavior quickly, it promotes other agents that have the behavior in less-developed form (and itself if it has the behavior).

5 Evaluation

Although the current implementation was not constructed for performance, we evaluated that of several basic operations in a distributed system where eight computers (Intel Core 2 Duo 1.83 GHz with MacOS X 10.6 and J2SE version 6) were connected through a giga-ethernet. The cost of transmitting a heartbeat or restraining message through UDP multicasting was 11 ms. The cost of transmitting a request message between two computers was 22 ms through TCP. These costs were estimated from the measurements of round-trip times between computers. We assumed in the following experiments that each agent issued heartbeat messages to other agents every 100 ms through UDP multicasting.

The first experiment was carried out to evaluate the basic ability of agents to differentiate themselves through interactions in a reliable network. Each agent had three behaviors, called A, B, and C. The A behavior periodically issued messages to invoke its B and C behaviors or those of other agents every 200 ms and the B and C behaviors were null behaviors. Each agent that wanted to execute a behavior, i.e., B or C, selected a behavior whose weight had the highest value if its database recognized one or more agents that provided the same or compatible behavior, including itself. When it invokes behavior B or C and the weights of its and others behaviors were the same, it randomly selected one of the behaviors. We assumed in this experiment that the weights of the B and C behaviors of each agent would initially be five and the maximum of the weight of each behavior and the total maximum W^k of weights would be ten.

Figure 3 presents the results we obtained from the experiment. Both diagrams have a timeline in minutes on the x-axis and the weights of behavior B in each agent on the y-axis. Differentiation started after 200 ms, because each agent knows the presence of other agents by receiving heartbeat messages from them. Figure 3 (a) details the results obtained from our differentiation between two agents. Their weights were not initially varied and then they forked into progression and regression sides. Figure 3 (b) shows the detailed results of our differentiation between

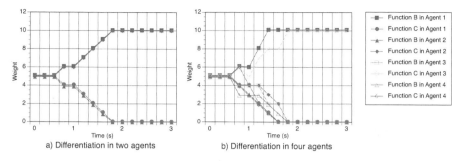

Fig. 3 Degree of progress in differentiation-based adaptation

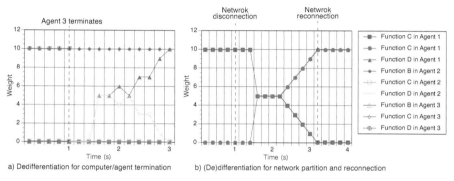

Fig. 4 Degree of progress in adaptation to failed agent

four agents and Figure 3 (c) shows those of that between eight agents. The results in (b) and (c) fluctuated more and then converged faster than those in (a), because the weights of behaviors in four are increased or decreased more than those in two agents. Although the time of differentiation depended on the period of invoking behaviors, it was independent of the number of agents. This is important to prove that this approach is scalable.

Our parameters for (de)differentiation were basically independent of the performance and capabilities of the underlying systems. For example, the weights of behaviors are used for relatively specifying the progression/repression of these behaviors.

The second experiment was carried out to evaluate the ability of the agents to adapt to two types of failures in a distributed system (4). The first corresponded to the termination of an agent and the second to the partition of a network. We assumed in the following experiment that three differentiated agents would be running on different computers and each agent had four behaviors, called A, B, C, and D, where the A behavior invokes other behaviors every 200 ms. The maximum of each behavior was ten and the agents' total maximum of weights was twenty. The initial weights of their behaviors (w_B^i, w_C^i, w_D^i) in i-th agent were $(10, 0, 0)$ in the first, $(0, 10, 0)$ in the second, and $(0, 0, 10)$ in the third.

6   Conclusion

This paper proposed a framework for adapting software agents on distributed systems. It is unique to other existing software adaptations in introducing the notions of (de)differentiation and cellular division in cellular slime molds, e.g., dictyostelium discoideum, into software agents. When an agent delegates a function to another agent, if the former has the function, its function becomes less-developed and the latter's function becomes well-developed. When agents have many requests from other agents, they create their daughter agents. The framework was constructed as a middleware system on real distributed systems instead of any simulation-based systems. Agents can be composed from Java objects.

References

1. Bonabeau, E., Dorigo, M., Theraulaz, G.: Swarm Intelligence: From Natural to Artificial Systems. Oxford University Press (1999)
2. Dorigo, M., Stutzle, T.: Ant Colony Optimization. MIT Press (2004)
3. Herrman, K.: Self-organizing Replica Placement - A Case Study on Emergence. In: Proceedings of 2nd IEEE International Conference on Self-Adaptive and Self-Organizing Systems (SASO 2007), pp. 13–22. IEEE Computer Society (2007)
4. Snyder, P.L., Greenstadt, R., Valetto, G.: Myconet: A Fungi-Inspired Model for Superpeer-Based Peer-to-Peer Overlay Topologies. In: Proceedings of 3rd IEEE International Conference on Self-Adaptive and Self-Organizing Systems (SASO 2009), pp. 40–50 (2009)

PANGEA – Platform for Automatic coNstruction of orGanizations of intElligent Agents

Carolina Zato, Gabriel Villarrubia, Alejandro Sánchez, Ignasi Barri,
Edgar Rubión, Alicia Fernández, Carlos Rebate, José A. Cabo, Téresa Álamos,
Jesús Sanz, Joaquín Seco, Javier Bajo, and Juan M. Corchado

Abstract. This article presents PANGEA, an agent platform to develop open multiagent systems, specifically those including organizational aspects such as virtual agent organizations. The platform allows the integral management of organizations and offers tools to the end user. Additionally, it includes a communication protocol based on the IRC standard, which facilitates implementation and remains robust even with a large number of connections. The introduction of a CommunicationAgent and a Sniffer make it possible to offer web services for the distributed control of interaction.

Keywords: multiagent platform, Web services, virtual organizations, IRC protocol.

Carolina Zato · Gabriel Villarrubia · Juan M. Corchado
Departamento Informática y Automática, Universidad de Salamanca, Salamanca, Spain
e-mail: {carol_zato,gvg,corchado}@usal.es

Alejandro Sánchez · Javier Bajo
Universidad Pontificia de Salamanca, Salamanca, Spain
e-mail: {asanchezyu,jbajope}@usal.es

Ignasi Barri · Edgar Rubión · Alicia Fernández · Carlos Rebate
Indra, Spain
e-mail: {ibarriv,erubion,afernandezde,crebate}@indra.es

José A. Cabo · Téresa Álamos
Wellness Telecom, Spain
e-mail: {talamos,jacabo}@wtelecom.es

Jesús Sanz · Joaquín Seco
CSA, Spain
e-mail: {jesus.sanz,joaquin.seco}@csa.es

S. Omatu et al. (Eds.): Distributed Computing and Artificial Intelligence, AISC 151, pp. 229–239.
springerlink.com   © Springer-Verlag Berlin Heidelberg 2012

1 Introduction

One of the current lines of investigation for multiagent systems aims to create an increasingly open and dynamic system. This involves adding new capabilities such as adaption, reorganization, learning, coordination, etc. Virtual agent Organizations (VOs) [1][2] emerged in response to this idea; they include a set of agents with roles and norms that determine their behavior, and represent a place where these new capabilities will assume a critical role. Possible organizational topologies and aspects such as communication and coordination mechanisms determine in large part the flexibility, openness and dynamic nature that a multiagent system can offer.

There are many different platforms available for creating multiagent systems that facilitate the work of the agent; however those that allow for the creation of VOs number much fewer, and it is difficult to find one single platform containing all of the requirements for a VO.

The remainder of the paper is structured as follows: the next section introduces some existing platforms. Section 3 presents an overview of the main characteristics of the platform. Finally, section 4 explains a case study and presents some results.

2 Related Works

All platforms for creating multiagent systems existing to date should be studied according to two principal categories: those that simply support the creation and interaction of agents, and those that permit the creation of virtual organizations with such key concepts as norms and roles. We will first present those platforms that do not incorporate organizational aspects. The FIPA-OS [4] agent platform was created as a direct derivative of the FIPA [3] standard. Another agent platform is the April Agent Platform (AAP) [5] which, unlike the majority of platforms using Java, implements the April language [6]; its development and technological support has been discontinued. One of the strong points of this platform is that it provides services to facilitate the development and deployment of agents on the Internet and is also compliant with Web Services and Semantic Web standards.

One of the most recent platforms still in development is JASON [7][8]. Its greatest contribution is the easy implementation of BDI agents [10]. The Java-developed platform contains AgentSpeak in its nucleus, an interpreter agent that acts as a language extension [9]. The platform offers two operation modes: one that runs all agents in the same machine, and another which allows distribution using SACI (Simple Agent Communication Infrastructure) [11], which in turn uses KQML [24] language instead of RIPA-ACL [23]. In practice, the most used platform for developing multiagent systems in real case studies is JADF (Java Agent Development Framework) [12]. The JADE platform focuses on implementing the FIPA reference model, providing the required communication infrastructure and platform services such as agent management, and a set of development and debugging tools. Jadex [13] is a software framework for the creation of goal-oriented agents following the belief-desire-intention (BDI) model. The Jadex project facilitates a smooth transition from developing conventional JADE agents to employing the mentalistic concepts of Jadex agents.

With the exception of JASON, these platforms follow the FIPA standard, can create agents (some with different models), and manage communication among agents and services. With VOs, however, it is necessary to consider the normative and organizational aspects that the platform itself must provide. MadKit [20] was one of the first platforms to consider basic organizational aspects. The platform architecture is rooted in the AGR (agent-group-role) model [14]; however, while it can handle the concept of role, it does not consider a role a class entity, and the behavior associated with the role is directly implemented in the agent who assumes it. Roles are strongly linked to agent architectures. This approach harms the reusability and modularity of organizations [15].

Another pioneering platform with regards to structural aspects was Jack Teams [16]. JACK Teams is an extension of JACK Intelligent Agents [17], which provides a team-oriented modelling framework. Both are extensions of the Java programming language; the implemented source code is first compiled into regular Java code before being executed.

S-MOISE+ is an organizational middleware that follows the MOISE+ model [18]. It is an extension of SACI [11] where the agents have an organizational aware architecture. Our research found systems developed in conjunction with JASON and using S-Moise+ as middleware to achieve a more complete model [19]. The result was J-Moise+ [20], which is very similar to S-Moise+ regarding overall system concepts. The main difference is how the agents are programmed: in S-Moise+ agents are programmed in Java (using a very simple agent architecture), while in J -Moise+ they are programmed in AgentSpeak.

One of the main disadvantages of VO oriented platforms is the slight loss in the concept of service and, consequently, the management of these services and the Directory Facilitator (DF) described in the FIPA standard. THOMAS was developed in response to this twofold need. THOMAS is based on the idea that no internal agents exist and architectural services are offered as web services. As a result, the final product is wholly independent of any internal agent platform and fully addressed for open multiagent systems [21].

Finally, one of the most complete and recent platforms that we found is Janus [22]. Janus is the next step towards platform organizations known as TinyMAS (no longer under development.). This platform was specifically designed to deal with the implementation and deployment of holonic and multiagent systems. Its primary focus is to support the implementation of the concepts of role and organization as first-class entities (a class in the object-oriented sense). This consideration has a significant impact on agent implementation and allows an agent to easily and dynamically change its behaviour [15].

In conclusion, it could be said that when dealing with all aspects of complex multiagent systems such as VOs, it is also necessary to deal with multiple levels of abstractions and openness, which is not the case for most solutions.

3 Architecture Overview

As we have mentioned, we are looking for a platform that can integrally create, manage and control VOs. In general terms, the proposed platform includes the following characteristics:

- Different models of agents, including a BDI and CBR-BDL architecture.
- Control the life cycle of agents with graphic tools.
- A communication protocol that allows broadcast communication, multicast according to the roles or suborganizations, or agent to agent.
- A debugging tool.
- Module for interacting with FIPA-ACL agents.
- Service management and tools for discovering services.
- Web services.
- Allow organizations with any topology.
- Organization management.
- Services for dynamically reorganizing the organization.
- Services for distributing tasks and balancing the workload.
- A business rules engine to ensure compliance with the standards established for the proper operation of the organization.
- Programmed in Java and easily extensible.
- Possibility of having agents in various platforms (Windows, Linux, MaccOS, Android and IOS)
- Interface to oversee the organizations.

Figure 1 displays the principal entities of the system, and illustrates how the roles, norms and the organizations themselves are classes that facilitate the inclusion of organizational aspects. The services are also included as entities completely separate from the agent, facilitating their flexibility and adaption.

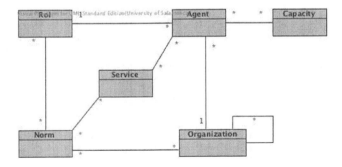

Fig. 1 Principal classes of the system

When launching the main container of execution, the communication system is initiated; the agent platform then automatically provides the following agents to facilitate the control of the organization:

OrganizationManager: the agent responsible for the actual management of organizations and suborganizations. It is responsible for verifying the entry and exit of agents, and for assigning roles. To carry out these tasks, it works with the OrganizationAgent, which is a specialized version of this agent.

InformationAgent: the agent responsible for accessing the database containing all pertinent system information.

ServiceAgent: the agent responsible for recording and controlling the operation of services offered by the agents.

NormAgent: the agent that ensures compliance with all the refined norms in the organization.

CommunicationAgent: the agent responsible for controlling communication among agents, and for recording the interaction between agents and organizations.

Sniffer: manages the message history and filters information by controlling communication initiated by queries.

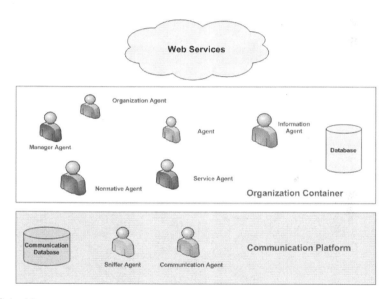

Fig. 2 Architecture

The platform examines two modes of operation. In the first mode, the agents reside in the machine itself, while in the second mode the platform allows for the possibility of initiating all agents in different machines. The latter case has the disadvantage of allowing only minimal human intervention since it is necessary to previously specify the address of the machine where each of the agents are to reside; however it has the advantage of greater system distribution.

We hope to create a service oriented platform that can take maximum advantage of the distribution of resources. To this end, all services are implemented web services. This makes it possible for the platform to include both a service provider agent and a consumer agent, thus emulating a client-server architecture. The provider agent knows how to contact the web service; once the client agent's request has been received, the provider agent extracts the required parameters and establishes the contact. Once received, the results are sent to the client agent.

Each suborganization or work unit is automatically provided with an OrganizationAgent by the platform during the creation of the suborganization. This OrganizationAgent is similar to the OrganizationManager, but is only responsible for controlling the suborganizationn, and can communicate with the OrganizationManager if needed. If another suborganization is created hierarchically within the previous suborganization, it will include a separate OrganizationAgent that communicates with the OrganizationAgent from the parent organization. These agents are distributed hierarchically in order to free the OrganizationManager of tasks. This allows each OrganizationAgent to be responsible for a suborganization although, to a certain extent, the OrganizationManager can always access information from all of the organizations. Each agent belongs to one suborganization and can only communicate with the OrganizationAgent from its own organization; this makes it possible to include large suborganizational structures without overloading the AgentManager. All of the OrganizationAgents from the same level can communicate with each other, unless a specific standard is created to prevent this.

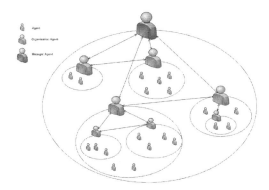

Fig. 3 OrganizationManager and OrganizationAgents

3.1 Communication Platform

This section will focus on describing the communication platform and protocol. As observed in Figure x, the communication platform includes two main agents: the CommunciationAgent and the Sniffer. The first is in charge of checking the connections to confirm that the agents are online and see which ones have disconnected. It is also in continual communication with the NormAgent to ensure that the agents respect the lines of communication and comply with the standards. The Sniffer is in charge of recording all communication, offers services so that other agents can obtain history information, and facilitates the control of information flow for programmers and users.

The IRC protocol was used to implement communication. Internet Relay Chat (IRC) is a real time internet protocol for simultaneous text messaging or conferencing. This protocol is regulated by 5 standards: RFC1459 [25], RFC2810 [29], RFC2811 [28], RFC2812 [26] y RFC2813 [27]. It is designed primarily for group conversations in discussion forums and channel calls, but also allows private messaging for one on one communications, and data transfers, including file exchanges [25]. The protocol in the OSI model is located on the application layer and uses TCP or alternatively TLS [29]. An IRC server can connect with other IRC servers to expand the user network. Users access the IRC networks by connecting a client to a server. There have been many implementations of clients, including mIRC or XChat. The original protocol is based on flat text (although it was subsequently expanded), and used TCP port 6667 as its primary port, or other nearby ports (for example TCP ports 6660-6669, 7000) [26]. The standard structure for an IRC server network is a tree configuration. The messages are routed only through those nodes that are strictly necessary; however, the network status is sent to all servers. When a message must be sent to multiple recipients, it is sent similar to a multidiffusion; that is, each message is sent to a network link only once [29]. This is a strong point in its favor compared to the no-multicast protocols such as SimpleMail Transfer Protocol (SMTP) or the Extensible Messaging and Presence Protocol (XMPP).

One of the most important features that characterize the platform is the use of the IRC protocol for communication among agents. This allows for the use of a protocol that is easy to implement, flexible and robust. The open standard protocol enables its continuous evolution. There are also IRC clients for all operating systems, including mobile devices.

All messages include the following format: prefix command command-parameters\r\n. The prefix may be optional in some messages, and required only for entering messages; the command is one of the originals from the IRC standard.

The following diagram illustrates the message flow required for an agent to enter an organization. These messages use the command PRIVMSG followed by the parameters indicated by the arrows in the diagram.

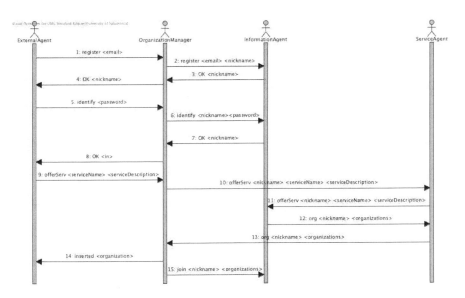

Fig. 4 Sequence of steps for an agent to enter an organization

4 Case Study and Results

The platform we have developed can create a general type of organization, and includes the possibility of creating open and highly dynamic systems. In order to test the architecture, a case study was prepared to simulate a working environment. Four organizations were created to simulate four different departments within a company: accounting (composed of 4 accounting agents, one manager and 2 secretaries); quality control (composed of 2 evaluating agents and two training specialist agents); technical services (composed of 6 technical agents); and customer service (composed of 8 telephonist agents). According to the role of each agent, there are specific services offered that allow them to resolve the queries they receive. In one possible case, the client agent contacts the telephonist agent, which simply receives the requests and redirects it to the agent qualified to resolve the request. The telephonist agent extracts the key words from the message sent by the client and contacts the Services Agent to determine which agent can address the required service. If the message contains the keyword "invoice", the query will be handled by the Accounting agent; if the keyword is "switch on" it will be handled by the Technical agent. Once the client is in contact with the appropriate agent, the agent can communicate with other agents in its organization to carry out the task.

Four 30-minute simulations were performed with 20 different types of requests randomly provided. Studying the Evaluation and Sniffer agents it was possible to study how both the simulation and message flow unfolded. Focusing specifically on the Sniffer, it is possible obtain summary charts and diagrams, and specific numbers. Once the query is made, the Sniffer consults the database, filters the data and returns a URL that displays the desired data.

It is possible to obtain the number of each type of message that a specific agent has received. Each message includes a tag that identifies the type of message, which makes it possible to filter information.

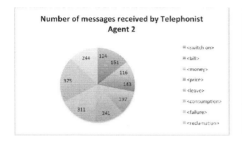 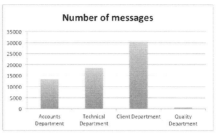

Fig. 5 Diagramas of messages.

It is also possible to obtain a diagram of messages according to organization instead of agents. Using the message identifier, it is also possible to see which agents processed a given request; using the Evaluation agents we can determine the number of requests processed by each agent.

We can conclude that the architecture we are developing has great potential to create open systems, and more specifically, virtual agent organizations. This architecture includes various tools that make it easy for the end user to create, manage and control these systems. One of the greatest advantages of this system is the communication platform that, by using the IRC standard, offers a robust and widely tested system that can handle a large number of connections, and that additionally facilitates the implementation for other potential extensions. Furthermore, the use of the Communication and Sniffer agents, offers services that can be easily invoked to study and extract message information.

Acknowledgements. This project has been supported by the Spanish CDTI. Proyecto de Cooperación Interempresas. IDI-20110343, IDI-20110344, IDI-20110345, and the MICINN TIN 2009-13839-C03-03 project. Project supported by FEDER funds.

References

1. Ferber, J., Gutknecht, O., Michel, F.: From Agents to Organizations: An Organizational View of Multi-Agent Systems. In: Giorgini, P., Müller, J.P., Odell, J.J. (eds.) AOSE 2003. LNCS, vol. 2935, pp. 214–230. Springer, Heidelberg (2004)
2. Foster, I., Kesselman, C., Tuecke, S.: The anatomy of the grid: Enabling scalable virtual organizations. Int. J. High Perform. Comput. Appl. 15(3), 200–222 (2001)
3. O'Brien, P.D., Nicol, R.C.: FIPA, Towards a Standard for Software Agents. BT Technology Journal 13(3), 51–59 (1998)
4. Emorphia, FIPA-OS, http://fipa-os.sourceforge.net/

5. Dale, J., Knottenbelt, J., Labo, F.: April Agent Platform (2011),
http://designstudio.lookin.at/research/relate%20survey/
Survey%20Agent%20Platform/April%20Agent%20Platform.htm
(accessed November 29, 2011)
6. McCabe, F.G., Clark, K.L.: APRIL—Agent PRocess Interaction Language. In: Wooldridge, M.J., Jennings, N.R. (eds.) Proceedings of the Workshop on Agent Theories, Architectures, and Languages on Intelligent Agents (ECAI 1994), pp. 324–340. Springer, New York (1995)
7. Bordini, R.H., Hübner, J.F., Vieira, R.: Jason and the Golden Fleece of agent-oriented programming. In: Bordini, R.H., Dastani, M., Dix, J., El Fallah Seghrouchni, A. (eds.) Multi-Agent Programming: Languages, Platforms and Applications, ch. 1, pp. 3–37. Springer (2005)
8. Bordini, R.H., Hübner, J.F., Wooldridge, M.: Programming Multi-Agent Systems in Agent Speak Using Jason. John Wiley & Sons, Ltd. (2007)
9. Rao, A.S.: Agent Speak(L): BDI agents speak out in a logical computable language. In: Perram, J., Van de Velde, W. (eds.) MAAMAW 1996. LNCS, vol. 1038, pp. 42–55. Springer, Heidelberg (1996)
10. Rao, A.S., Georgeff, M.P.: Modeling rational agents within a BDI-Architecture. In: Allen, J., Fikes, R., Sandewall, E. (eds.) Proceedings of the 2nd International Conference on Principles of Knowledge Representation and Reasoning (KR 1991), pp. 473–484. Morgan Kaufmann publishers, Inc., San Francisco (1991)
11. SACI – Simple Agent Communication Infrastructure,
http://www.lti.pcs.usp.br/saci/
12. Bellifemine, F., Poggi, A., Rimassa, G.: JADE – A FIPA-compliant agent framework. In: Proceedings of the Practical Applications of Intelligent Agents (1999)
13. Braubach, L., Pokahr, A., Lamersdorf, W.: Jadex: A Short Overview. Procceding Main Conference Net. Object Days, 195–207 (2004)
14. Gutknecht, O., Ferber, J.: MadKit: Organizing heterogeneity with groups in a platform for multiple multi-agent systems. Technical Report R.R.LIRMM 9718, LIRM (December 1997)
15. Gaud, N., Galland, S., Hilaire, V., Koukam, A.: An Organisational Platform for Holonic and Multiagent Systems. In: Hindriks, K.V., Pokahr, A., Sardina, S. (eds.) ProMAS 2008. LNCS, vol. 5442, pp. 104–119. Springer, Heidelberg (2009)
16. Agent Oriented Software Pty Ltd. JACK^{TM} Intelligent Agents Teams Manual. s.l. : Agent Oriented Software Pty Ltd. (2005)
17. Busetta, P., Rönnquist, R., Hodgson, A., Lucas, A.: JACK Intelligent Agents - Components for Intelligent Agents in Java. Technical report. Agent Oriented Software Pty. Ltd., Melbourne, Australia (1998)
18. Hubner, J.F., Sichman, J.S., Boissier, O.: A Model for the Structural, Functional, and Deontic Specification of Organizations in Multiagent Systems. In: Bittencourt, G., Ramalho, G.L. (eds.) SBIA 2002. LNCS (LNAI), vol. 2507, pp. 118–128. Springer, Heidelberg (2002)
19. Hübner, J.F., Bordini, R.H., Picard, G.: Using Jason and MOISE+ to Develop a Team of Cowboys. In: Hindriks, K.V., Pokahr, A., Sardina, S. (eds.) ProMAS 2008. LNCS, vol. 5442, pp. 238–242. Springer, Heidelberg (2009)
20. Hübner, J.F.: J -Moise+ Programming organisational agents with Moise+ & Jason. Technical Fora Group at EUMAS 2007 (2007)

21. Giret, A., Julián, V., Rebollo, M., Argente, E., Carrascosa, C., Botti, V.: An Open Architecture for Service-Oriented Virtual Organizations. In: Braubach, L., Briot, J.-P., Thangarajah, J. (eds.) ProMAS 2009. LNCS, vol. 5919, pp. 118–132. Springer, Heidelberg (2010)
22. Galland, S.: JANUS: Another Yet General-Purpose Multiagent Platform. Seventh AOSE Technical Forum, Paris (2010)
23. The Foundation for Intelligent Physical Agents. FIPA Standar Status Specification (2011),
   `http://www.fipa.org/repository/standardspecs.html`
   (accessed November 11, 2011)
24. Finin, T., Labrou, Y.: KQML as an agent communication language. In: Bradshaw, J.M. (ed.) Software Agents, Cambridge, MA, pp. 291–316 (1997)
25. Oikarinen, J., Reed, D.: Internet Relay Chat Protocol. RFC 1459 (May 1993)
26. Kalt, C.: Internet Relay Chat: Client Protocol. RFC 2812 (April 2000)
27. Kalt, C.: Internet Relay Chat: Server Protocol. RFC 2813 (April 2000)
28. Kalt, C.: Internet Relay Chat: Channel Management. RFC 2811 (April 2000)
29. Kalt, C.: Internet Relay Chat: Architecture. RFC 2811 (April 2000)

An OOP Agent-Based Model for the Activated Studge Process Using MATLAB

María Pereda and Jesús M. Zamarreño

Abstract. The aim of this work is to study the feasibility of using agent-based modelling and Object Oriented Programming (OOP) to study the activated sludge process. A model in MATLAB has been proposed, and experiments have been developed analyzing the behaviour of the model when some of its parameters are changed.

1 Introduction

The water and wastewater industry has undergone many changes in recent years. Of particular importance has been a renewed emphasis on improving resource management with tighter regulatory controls setting new targets on pricing, industry efficiency and loss reduction for wastewater. Meantime, the demand for water and wastewater services grows as the population increases and wishes for improved living conditions involving, among other items, domestic appliances that use water. Consequently, the installed infrastructure on the industry has to be continuously upgraded and extended, and employed more effectively to accommodate the new demands, both in throughput and in meeting the new regulatory conditions.

Sewage systems exhibit some specific characteristics that make them especially challenging from the point of view of analysis and management. These may include many complex features and behaviours: nonlinear dynamics, compositional subsystems with important delays, stochastic disturbances, etc...

Mathematical models have been shown to be useful tools for research, development and optimization of wastewater treatment processes, and designing new wastewater treatment plants (WWTP), as well as upgrading existing ones. In the recent years several kind of models have been proposed to simulate separately different processes taking place in WWTP, being the ones proposed by the IWA

María Pereda · Jesús M. Zamarreño
University of Valladolid, EII sede Mergelina, Spain
e-mail: maria.pereda@autom.uva.es, jesusm@autom.uva.es

S. Omatu et al. (Eds.): Distributed Computing and Artificial Intelligence, AISC 151, pp. 241–248.
springerlink.com   © Springer-Verlag Berlin Heidelberg 2012

(International Water Association) [3] the most accepted for the simulation of activated sludge systems.

In this work, a different approach for modelling the activated sludge Process is proposed, that is, applying the agent-based modelling (ABM) tools. Agent-based modelling [2] treats each individual component of a system as a single entity obeying its own pre-defined rules and reacting to its environment and neighbouring agents accordingly. Here, the emergent phenomena arise from the interactions of individual entities. An emergent phenomenon occurs when the system components are governed by simple rules, but the macroscopic behaviour resulting from their interaction is complex.

In the following, an ABM of the activated sludge process will be developed. The implementation will be based on the Object Oriented paradigm for better representation of the agents, and MATLAB will be used as software platform.

This paper is organized as follows. In Section 1 a brief introduction to the activated sludge process is given. Then a short introduction to agent-based modelling follows. In this section it is also described the model presented in this work. Then, Section 3 describes the possibilities for implementing agent-based Models in MATLAB. Section 4 shows the MATLAB implementation of the model proposed in this work. In Section 5 the simulation results are shown. Finally, Section 6 discusses points for further research and conclusions.

2   Description of the Agent-Based Model

An activated sludge can be defined as a microbial mass produced when the wastewater is aerated continuously. That mass is made up of microorganisms that are able to break down and metabolize the main contaminants in the wastewater. The activated sludge process is a type of biological process that takes place in the secondary treatment of sewage at a wastewater treatment plant. The activated sludge process consists of: an aeration tank (bioreactor) where the activated sludge and food (contamination or substrate water content) is mixed and aerated for a certain period of time, a system of separation of activated sludge and treated effluent (secondary clarifier or clarifier), a recirculation of activated sludge to the aeration tank, and a system of treatment and disposal of the sludge produced (purge). The aeration tank contains a sludge which is what could be best described as a mixed microbial culture, containing mostly bacteria, as well as protozoa, fungi, algae, etc. This sludge is constantly mixed and aerated either by compressed air bubblers located along the bottom, or by mechanical aerators on the surface.

In this work we are modeling a simplified version of the process in a batch reactor, i.e., an isolated reactor in which an initial charge is introduced and provided with adequate ventilation conditions for the optimal development of the reactions of substrate-biomass interaction, in order to better focus on the interaction substrate-biomass.

Under these conditions, the temporal evolution of the concentrations of substrate and biomass takes place as shown in Fig. 2. The growth curve is obtained by a count of the number of living cells (biomass) over time. The biomass curve consists of several phases which are explained below [1]. The microorganisms must first accommodate to their environment and available food. This accommodation period is called lag phase, and varies in size depending on the history of seeded microorganisms. If microorganisms are adapted to the environment, the lag phase will be very brief. Once growth has begun, it will continue rapidly. When maximum growth is occurring, the behaviour is logarithmic, this is why this phase is called the logarithmic phase. Maximum growth cannot continue indefinitely. The food available may be depleted, environmental conditions may change. Cells that are unable to obtain food from outside sources initiate endogenous catabolism, i.e., they catabolize the protoplasm stored to maintain their energy. Other cells die or break releasing their protoplasm, which is added to the food available. This stage is represented in Fig. 2 with the name of stationary phase, and represents the time along which the production of new cell material is roughly compensated by death and endogenous respiration. The endogenous respiration and death dominate the fourth phase, called endogenous phase. In this last phase, the biomass decreases slowly, asymptotically approaching the horizontal axis. The substrate curve represent the decrease of the available food in the environment.

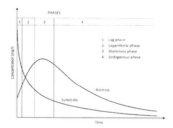

Fig. 1 Concentrations of substrate and biomass in an activated sludge batch reactor.

The proposed ABM considers the agents divided into two distinct groups: biomass agents and substrate agents. In both cases these are reactive agents. The position of each agent in the world is referenced to a Cartesian system of two axes, since it is a two-dimensional model. The agents have the properties shown at Table 1.

Besides, the model is characterized by the constants shown at Table 2.

Moreover, the model shows some degree of randomness due to different statistical arguments as Table 2 shows. Initialization of the model is as follows: assign values to the constants of the model, create the agents and place them in the world, and assign initial values of age and energy to biomass agents.

At each step the following actions are performed, as Fig. 2 shows:

1. To shake. At each step of the program, agents in the world are arranged in a random position, thus modeling a system that operates in continuous stirred-tank reactor (CSTR) conditions, obtaining homogeneous characteristics in terms of substrate and biomass concentrations.

Table 1 Agents properties.

BIOMASS	AGENTS		
Property	Variable	Description	Initial Value
Position	[x,y]	Position in the world	Uniform distribution
Size	*size_b*	Volume occupied	Normal distribution N(*initial_b_size*,1)
Age	*age*	Number of periods alive	Normal distribution N(*initial_age*, *sd_age*)
Internal energy	*energy*	Number of periods the agent can live without eating	Normal distribution N(*max_energy*, *sd_energy*)

SUBSTRATE	AGENTS		
Property	Variable	Description	Initial Value
Position	[x,y]	Position in the world	Uniform distribution
Size	*size_s*	Volume occupied	Normal distribution N(*initial_s_size*,1)

Table 2 Model constants

Constant	Description
rep_size	Size from which the biomass is able to reproduce
unit_growth	Amount by which the size of a biomass agent is increased when eating
b_density	Biomass density
s_density	Substrate density
size_percentage	Decrease of the size of a substrate agent eaten by a biomass agent
eat_radius	A biomass agent can only eat in its surrounding circular area defined by *eat_radius*
sd_death_lmax	Standard deviation for age deaths when a biomass agent is older than *max_age*
sd_death_gmax	Standard deviation for age deaths when a biomass agent is younger than *max_age*
max_age	Maximum age that a biomass agent can reach
mean_death_by_energy	Mean of an exponential distribution for energy deaths
max_energy	Maximum energy value that a biomass agent can reach
sd_energy	Standard deviation for energy at birth
i_energy_by_eating	Increase of energy of a biomass agent by eating
d_energy_by_starving	Decrease of energy of a biomass agent every iteration

Fig. 2 Stages of the model.

2. To eat. A biomass agent will look at neighbouring positions defined by a circle of radius *eat_radius* and center itself, and it will choose the closer substrate agent to eat from, and it will eat a portion of substrate agent equal to its size multiplied by *size_percentage*. The biomass agents that have eaten, increase its size by the quantity it has eaten, and its internal energy by one. The substrate agent eaten decreases its size by a percentage of the eating biomass agent (*size_percentage* parameter is used here).

3. To reproduce. Biomass agents are divided in half creating 2 individuals of the same characteristics as the parent, and half the size of the parent, according to parameter *rep_size*.
4. To die. It is checked whether biomass agents will die from one of the two cases: die because of running out of internal energy or die at a certain age. If they die, their mass becomes part of the food present in the reactor.
5. At each step of the program, the internal energy of each biomass agent is decremented by *d_energy_by_starving* and age increases by one.

This five actions are repeated iteratively until a number of predefined steps is reached.

3   Modelling Agents in MATLAB

Nowadays, there are many software options to implement agent-based models. For the first version of the model in [7], NetLogo was chosen as software platform, due to the easiness to implement agent-based models on it, the large number of examples available and the quality of the manual that is distributed with the tool. Afterwards, some disadvantages appeared: the need to do parameter optimization to adjust the model to real data and the high computational load of the model. To solve these two main disadvantages it was needed a tool that supports optimization and parallel techniques. MATLAB [4] was chosen because it is a hight level scientific and engineering computational tool, a lot of specialized toolboxes can be incorporated, a custom graphic user interface can be easily built so non-technical users can use it, it provides versatility and integration with other tools (such as CUDA, optimization algorithms, neural network toolbox...), which implies the possibility to solve the previous disadvantages and develop our future work.

The representation of Agents in MATLAB can be carried out using different alternatives:

1. Matrix representation, where the world is modeled with different matrix layers, each layer representing an agent property and the position of each element of the matrix determines the location of the agent in the world. An example of that can be found on the implementation of the Game Of Life in MATLAB [5]. This alternative has an important limitation, the agents positions in the world are restricted to the vector/matrix elements.
2. To solve the disadvantage of Matrix representation, the agents can be modeled having a property representing their position in the world, relative to a Cartesian axe. Two approaches can be used. On the one hand, data structures, where the properties of the agents can be accessed with the dot notation, e.g., *agent(number).property*. In the other hand, the agent properties can be stored in a Matrix *agent(number of agent, number of property)*.
3. Object Oriented Programing (OOP). This is the alternative selected in this work. It is a programming paradigm using *objects*, data structures consisting of data fields and methods together with their interactions. A generic agent is represented as an object, an the other agents are created inheriting their characteristics from

that agent, and adding their owns. This can be done due to the inheritance property of OOP (inheritance is a way to reuse code of existing objects, establishing a subtype from an existing object).

4   Model Implementation in MATLAB

Using the OOP paradigm, a generic agent object was created, which directly inherit from the class double of MATLAB. This object has the properties: position, size, color, shape, label; and the methods: hatch and move. From this agent, the biomass agents are created, adding two properties: age an energy. The substrate agents also inherit from the agent class. Fig.3 shows the definition of the clases agent, biomass and substrate.

```
classdef agent                           classdef Biomass < agent        classdef Sustrate < agent
    properties                               properties                      methods
        position                                 age                             function obj = Sustrate()
        size                                     energy                              obj.color = 'g';
        color                                end                                     obj.shape = 'o';
        shape                                methods                             end
        label                                    function obj = Biomass()    end
    end                                              obj.color = 'r';
                                                     obj.shape = 'x';
    methods                                      end
        function [obj,child]= hatch(obj)     end
            obj.size = obj.size/2;       end
            child = obj;
        end
        function obj = move (obj, position)
            obj.position = position;
        end
    end
end
```

Fig. 3 Definition of agent, biomass and substrate classes.

The model described in Section 2 is coded in MATLAB language and an interface is developed.

The interface is divided into two areas, as Figure 4a shows. The first area (upper left) deals with configuration. The parameter values of the model can be fixed, and also a set of preferences of the simulation can be chosen: visualizing or not the simulation progress in the graphs or only show the final result, if progress visualization is chosen, the frequency of refresh of the graphs can be selected. The rest of the interface shows the dynamic evolution of the agents such as their number, position, concentration, births and deaths. Fig. 4b shows a representation of the concentration of the agents exported from the main interface.

5   Experiments

To test the model, a set of experiments were done, analyzing the variability of behaviour when some of the parameters are changed. In the following, some of the parameters of the model will change their nominal value and the effect on the emergent macroscopic behaviour will be analyzed. Nominal behaviour for comparison is as Fig. 4b shows.

For example, if the *size_percentage* is decreased by 50%, the evolution of substrate and biomass concentrations is as shown in Fig. 5a. The biomass agents eat a

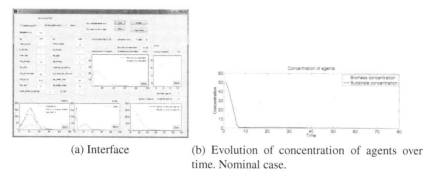

(a) Interface   (b) Evolution of concentration of agents over time. Nominal case.

Fig. 4 Interface and Nominal case.

smaller portion of substrate each time they eat, thus the food present in the environment (substrate) last longer than in the nominal case. For this reason, the biomass concentration reaches a higher maximum value than in the nominal case, being in the environment more agents (if agents reach the reproduction size, they can reproduce, so the more food, the more biomass agents). When the substrate ends, stationary phase starts, followed by the endogenous phase. The evolution of these two phases is slower than in the nominal case because there are more biomass agents.

Another experiment is a 20% of decreasing in *max_age*, so biomass agents may die younger than in the nominal case, the results are shown in Fig 5b. As it can be observed, biomass agents start dying earlier than in the nominal case, as expected. This causes the food (substrate) to last longer and the logarithmic phase ends later.

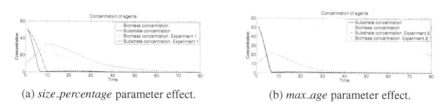

(a) *size_percentage* parameter effect.   (b) *max_age* parameter effect.

Fig. 5 Concentration of agents in experiments.

6 Conclusions and Further Research

As this paper has shown, agent-based modelling allows the researcher to develop a bottom-up approximation to the study of complex problems such as the phenomena occurring in a wastewater treatment plant, and in particular, secondary treatment by activated sludge. An Object Oriented Agent-Based Model for the activated sludge process in a batch reactor has been proposed, which allows to analyzing and a better understanding of the phenomena that occur in the system while varying some of its characteristics. The implementation environment chosen for the model, that is,

MATLAB, allows an efficient representation and provides good design tools, but the computational load is significant. In our experiments, 100 iterations of the model with 2560 biomass agents take up 90 seconds in a PC with a Core i5 3.20 GHz processor.

Thus, the following step is to reduce the computational load applying parallelization techniques, specifically, CUDA C++. CUDA (Compute Unified Device Architecture) [6] is a parallel computing architecture developed by Nvidia. CUDA is the computing engine in Nvidia graphics processing units (GPUs) that is accessible to software developers through variants of industry standard programming languages.

Another research line to follow is to model the activated sludge process in a continuous reactor and to incorporate real data to adjust model parameters, in order to be able to rehearse the various developments and techniques in a real WWTP.

Acknowledgements. The authors gratefully acknowledge the support of the CICYT through the project "Operación óptima de estaciones depuradoras de aguas residuales (EDARs)", DPI2009-14410-C02-02. This work was supported in part by the University of Valladolid under the Grant "Ayuda para la Formación de Personal Investigador".

References

1. César, E., Vázquez, A.B.: Ingeniería de los sistemas de tratamiento y disposición de aguas residuales. In: Fundación ICA, A.C. México, D.F (2003) ISBN 968-7508 05-4
2. Getchell, A.: Agent-based Modeling. Physics 22(6), 757–767 (2008)
3. Henze, M.: Activated sludge models ASM1, ASM2, ASM2d and ASM3(9). Water Science and Technology 56(8), 67–78 (2000)
4. MATLAB and Simulink for Technical Computing,
   http://www.mathworks.es/products/matlab/index.html
5. Moler, C.: Experiments with MATLAB (2011),
   http://www.mathworks.es/moler/exm/chapters.html
6. NVIDIA CUDA, http://www.nvidia.com/object/cuda_home_new.html
7. Pereda, M., Zamarreño, J.: Agent-based modeling of an activated sludge process in a batch reactor. In: 2011 19th Mediterranean Conference on Control & Automation (MED), pp. 1128–1133. IEEE, Corfu (2011)

A Case of Dictator Game in Public Finances– Fiscal Illusion between Agents

Paulo Mourão

Abstract. This paper discusses Fiscal Illusion as a special case of Agents and Multiagents Systems. Under fiscal illusion, each taxpayer does not realize how much he/she really pays to the State; therefore, he/she does not evaluate well the public actions. We will study this issue as a particular case of a 'Dictator game' with relevant applications not only for Public Finances but also for specific domains like Pervasive agents and Ambient Intelligence, or for User-centered applications and Assisting Agents.

Keywords: DicAbstract tator game, Fiscal Illusion, Agents and Multiagent Systems.

1 Introduction

Research on Agents and Multiagents systems has increased during last decade. However, when listing the set of works focused on Public (or Authoritative) Agents, we do not have yet a substantial list.

Of course, all of us recognize the large potential of exploring Agents and Multiagents Systems focused on Public Agents (like States), Societal Agents (combining Market agents and Public or Third-Sector Agents), or focused on Multiagents Systems with these sorts of identification.

This paper tries to explore a paper in this potential of scientific research.

Therefore, this paper discusses the theoretical foundations for describing the behavior of a State that tends to deceive its citizens. This is the behavior of a State classified as a 'fiscal illusion' creator.

The Multi-disciplinary discussion that evolves Fiscal Illusion (originally, fiscal illusion was discussed at the borders of Economics and Political Science) can easily be extended to specific domains of Agents and Multiagents Systems: like pervasive agents and ambient intelligence or User-centered applications and assisting agents.

Paulo Mourão
Department of Economics; University of Minho – Portugal
e-mail: paulom@eeg.uminho.pt

S. Omatu et al. (Eds.): Distributed Computing and Artificial Intelligence, AISC 151, pp. 249–254.
springerlink.com   © Springer-Verlag Berlin Heidelberg 2012

As we know, recent research on Agents and Multiagents systems (Hallé, 2005; Berger et al, 2007;Windrum and Garcia-Goñi, 2008) explores relevant topics close to fiscal illusion. These topics are the cases of Reputation, trust, privacy and security, or Evaluation, ethical and legal issues. Consequently, this paper will be also important for deepening the construction of a solid democratic system, for modeling real-time (public) multi-agent systems, and for constructing empirical simulations (not only at Agents and Multiagents systems' environments but also at Empirical Economics).

Towards these aims, all of us believe that we need solid theoretical discussions on the models' foundations. This paper provides an attempt toward this direction regarding public behavior.

2 Fiscal Illusion: A 1000-Words' Graph by Fasiani

Mauro Fasiani was an Italian economist who researched public finance in great detail (especially in his works of 1941 and of 1949) also under the influence of Vilfredo Pareto.

Puviani (1903) and Fasiani (1941) believed that in the monopolistic state, politicians manage power in the exclusive interest of the class to which they belong (Fossati, 2008). Thus, as Mourao (2007) pointed out, the rulers in this monopolistic state ask, "In order to minimize taxpayer resistance for any given level of revenues collected, how should the fiscal system be organized?" Puviani's (1903) *Teoria della Illusione Finanziaria* is the answer: Taxpayer resistance is minimized by using fiscal illusion practices (as particular practices of political illusions[1]).

Lindahl (1919) found this *Teoria* to be an "interesting" and "original" (p. 175) contribution (Fausto, 2006). However, as Buchanan (1967) observed, Puviani's *Fiscal Illusion* was an unknown work for many years[2]. In spite of this humbleness, *Teoria della Illusione Finanziaria* was substantially quoted by Fasiani (1941). As Buchanan (1968) further recognizes, Fasiani (1941) is credited with the first substantial reference to Puviani's fiscal illusion concept[3].

Although having significantly cited Amilcare Puviani, the great originality of Fasiani's (1941) work is his own attempt to model Puviani's fiscal illusion

---

[1] Mourao (2007 & 2008) details these practices in a comprehensive review of literature.

[2] It deserves to be mentioned that Buchanan was considering the anglo-saxonic literature in his remark. There are some proofs that show that Puviani's work has been discussed by Italian authors before 1941. Fasiani's German language review of 1932, which has translated into Italian and published in *Il Pensiero Economico Italiano 1850-1950*, devoted 13 pages (pp.139 to 152 of the Italian translation) to Puviani and fiscal illusion.

[3] Really, Fasiani's first analytical approach to the theory of fiscal illusion is in the paper: Riflessioni su di un punto della teoria dell'illusione finanziaria", in *Atti della Regia Accademia delle Scienze di Torino*, vol. LXIV, 1928-29, pp. 333-345 (English translation: "Some Notes on an Aspect of the Theory of Fiscal Illusion", in L. L. Pasinetti (ed.), *Italian Economic Papers*, vol. III, Il Mulino – Oxford University Press, 1998, pp. 89-97). In this paper, <<the illusion of the extent of a painful sensation>> represents a case of consumer surplus.

through the use of elements from the Paretian utility theory, such as the concepts of ophelimity[4] and utility[5].

According to Fasiani (1941), the final result of fiscal illusion is a decrease in individual utility with respect to money, as figure 1 suggests. This figure was extracted from the original book by Fasiani (1941, p. 133).

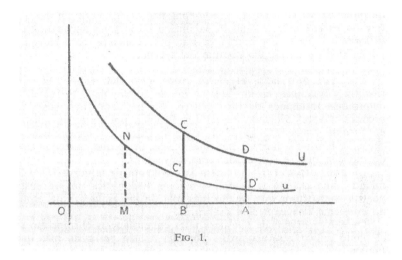

Fɪɢ. 1.

Fig. 1 Source: Fasiani (1941, p. 133)

Fasiani (1941) explained figure 1 as follows:

If we consider a single individual, we can show the [fiscal illusion] phenomenon in the following way (fig.1). Let us express the monetary units on the abscises axis, and let us express his utility on the ordinates axis. U is the utility function before the event [that generates fiscal illusion], and u is the utility function after the event. Let us assume that this individual has a fixed income equal to OA, being AB the tax. If this tax is collected when the utility function is U then it determines a burden equal to BCDA. If this tax is collected at the moment characterized by the utility function u then it determines a burden equal to BC'D'A, and the individual under the effect of the illusion has a smaller burden whose difference is equal to C'CDD.' The State can take advantage of the illusion raising the tax to AM. The increase of taxes BM determines a loss of utility equal to MNC'B. It is clear that the advantage of illusion is off when MNC'B is at a maximum, equal to C'CDD'—this is, using other words, when the tax was collected at the moment in which the utility function has been decreased, the individual did not realize the loss of utility C'CDD.' But now, if the State intends to take advantage of the

---

[4] According to Fasiani (1941, p. 241), ophelimity is the utility that is actually perceived by each taxpayer. Fasiani also states that ophelimity is equal to subjective utility.
[5] For Fasiani (1941, p.241) (objective) utility is utility *in latus sensus*, or the utility assumed for each taxpayer by the State.

[fiscal illusion] phenomenon, it may collect so many monetary units as those generating a total utility equal to C'CDD.' But beyond these limits, the illusion does not actuate and it is no longer possible to take advantage of it. (p. 133)

We can think of Fasiani's (1941) scenario (p. 132-133) as an academic simplification for a specific case of a 'dictator game'. Dictator games happen when the proposer (in this example, the State) simply allocates an endowment to the responder (in this example, the single individual). But, as a Dictator game, although the responder has a completely passive role, the proposer is influenced by a set of dimensions (Figueiredo, Carmo and Prada, 2010), also here present:

- Bias in the expected utility of the State itself(in our case, fiscal illusion cannot give the State the maximum amount that is originally expected);
- Bias in the expected gain from this strategy (fiscal illusion can generate fiscal delusion that, according to Fasiani, 1941, is a source of distrust between citizens and democratic regimens);
- And finally Bias in the utility of the responder (as argued by Fasiani, 1941, the extension of final fiscal illusion mainly depends on taxpayers' utility structure).

The Fasiani's (1941) book in question is a teaching manual and the cited extract and figure 1 represent a greatly simplified illustration to assist for public finance students to isolate one critical aspect of fiscal illusion. The influence of publicly provided goods on utility and fiscal illusion is ignored, (yet, in a previous work, Fasiani (1932) carefully discussed Puviani's "4th type[6]" of illusion under which the provision of publicly provided goods attenuate the burden of tax by increasing the enjoyment from the consumption of goods obtained privately). The absence of the expenditure side of the budget in Figure 1 avoids complexities associated with fiscal illusion, but I suggest that the simplification was deliberate and introduced for didactic purposes (and that Fasiani fully appreciated that factors other than tax and income influence fiscal illusion and utility). However, we have to recognize that Fasiani's (1941) taxonomy was greatly improved if compared to Fasiani (1932) and that Fasiani has treated the illusions related to the public expenditures (Book I, Chapter II – The Finance of the Monopolistic State) separately from the illusions regarding public revenues (Book I, Chapter III) and finally treated the limits to the different kind of illusions (Book I, Chapter IV), to which Figure 1 belongs.

Fasiani's graph refers to "the limits to the illusion arising from the association of the tax to pleasures of the taxpayer, which are originated in a private way", and its rationale is that, at the occurrence of the pleasure, the individual (erroneously) thinks to have a higher quantity of income. Thus, he perceives a higher level of utility, and then a lower level of marginal utility, because of the Fasiani's assumption of decreasing marginal utility.

Given that the sacrifice induced by the tax is the integral of the marginal utility curve between the net income and the gross income, with the introduction of the tax at the moment of the pleasure, the taxpayer receives a minor sacrifice

---

[6] Here we are using Fasiani's self taxonomy, not Puviani's.

compared to what he perceives if the event that gives rise to the pleasure has never occurred. This is owed to the fact that the marginal utility curve after the occurrence of the pleasure lies lower than the original marginal utility curve (before the occurrence of the pleasure). It follows that the government (which is constrained to produce a fixed amount of sacrifice to that taxpayer) may raise a higher tax, but only in the limit that the perceived sacrifice is no higher than the fixed amount of sacrifice.

Specifically, Fasiani represents the mentioned kind of fiscal illusion like an external shock that shifts (down) the curve of individual's marginal utility of income. From that, one can only infer that Fasiani considers a utility function $U=f(Y)$, with $UY>0$; $UYY<0$, where Y is the individual income. Perhaps, one might imagine a sort of continuum of external infinitesimal shocks. In that case, it would follow $U=f(Y, \Phi)$, where Φ is the fiscal illusion. However, since in the graph Fasiani merely shifts the utility curve, this utility function must be separable in the two arguments Y and Φ. Thus –through the illusory income increase– the role played by the fiscal illusion is to rise the level of the (total) utility perceived ($U\Phi>0$), and then to decrease marginal utility ($U\Phi \Phi <0$).

3 Conclusion

This work reviews the first model, created by Mauro Fasiani (1941, p. 132-133), that attempted to analyze fiscal illusion in public finances. Until now, literature has not appropriately analyzed the model.

Based on Pareto's class of utility functions, Fasiani (1941) created a graph to demonstrate that fiscal illusion allows states to expand more than is truly necessary. As is demonstrated by this work, this ultimate message can be summarized by three points. First, taxpayers under fiscal illusion believe that they are richer than they really are, just as electors under any political illusion (e.g., in 1903, Puviani discussed fiscal illusion as a form of political illusion) think themselves better off than they really are. Second, fiscal illusion can occur alongside economic growth. Third, an individual's utility can change, due to Paretian, non-logical actions (i.e., fiscal illusion), rather than as a result of income or pricing fluctuations.

Although their age, these claims are still relevant for us and for the construction of Agents and Multiagents Systems close to real world.

References

Berger, T., Birner, R., Díaz, J., McCarthy, N., Wittmer, H.: Capturing the complexity of water uses and water users within a multi-agent framework. Water Resour. Manage. 21, 129–148 (2007)

Buchanan, J.: Public Finance in Democratic Process: Fiscal institutions and Individual Choice. The University of North Carolina Press, Chapel Hill (1967)

Buchanan, J.: Toward a positive theory of public finance. In: The Demand and Supply of Public Goods, vol. 5, The Collected works of James M. Buchanan (edition of 1999), Liberty Fund, Indianapolis (1968)

Fasiani, M.: Der gegenwartige Stand der reine Theorie der Finanzwissenschaft in Italien. Zeitschrift fur Nationalokonomie 5, 651–691 (1932)

Fasiani, M.: Principii di Scienza delle Finanze, vol. 1. Giappichelli, Torino (1941); quoted version: Fasiani, M.: Principios de Ciencia de la Hacienda. tradução de Gabriel de Usera, Aguilar (1962)

Fasiani, M.: Contributi di Pareto alia scienza delle finanze. Giornale degli Economisti e Annali di Economia VIII (3-4), 129–173 (1949)

Fausto, D.: The Italian approach to the theory of public goods. The European Journal of the History of Economic Thought 13(1), 69–98 (2006)

Figueiredo, R., Carmo, J., Prada, R.: Do you trust me or not? – Trust games in agent societies. In: 5th European Starting AI Researcher Symposium, STAIRS 2010 (2010)

Fossati, A.: The idea of State in the 'Scienza delle Finanze' from Antonio de Viti de Marco to Mauro Fasiani. Paper Presented at History of Public Economics Conference, Paris (2008)

Hallé, S.: Automated Highway Systems: Platoons of Vehicles Viewed as a Multiagent System; Mémoire présentée pour l'obtention du grade de Maître des Sciences (M.Sc). Université Laval, Québec (2005)

Lindahl, E.: Die Gerechtigkeit der Besteuerung. Lund (1919)

Mourao, P.: The economics of illusion – a discussion based on fiscal illusion. Journal of Public Finances and Public Choice 25, 67–86 (2007)

Mourao, P.: Towards a Puviani's Fiscal Illusion Index. Hacienda Publica Espanola/Revista de Economia Publica 187(4), 49–86 (2008)

Puviani, A.: Teoria della illusione finanziaria. Sandron, Palermo (1903)

Windrum, P., Garcia-Goñi, M.: A neo-Schumpeterian model of health services innovation. Research Policy 37, 649–672 (2008)

Decentralised Regression Model for Intelligent Forecasting in Multi-agent Traffic Networks

Jelena Fiosina*

Abstract. The distributed nature of complex stochastic systems, such as traffic networks, can be suitably represented by multi-agent architecture. Centralised data processing and mining methods experience difficulties when data sources are geographically distributed and transmission is expensive or not feasible. It is also known from practice that most drivers rely primarily on their own experience (historical data). We consider the problem of decentralised travel time estimation. The vehicles in the system are modelled as autonomous agents consisting of an intellectual module for data processing and mining. Each agent uses a local linear regression model for prediction. Agents can adjust their model parameters with others on demand, using the proposed resampling-based consensus algorithm. We illustrate our approach with case studies, considering decentralised travel time prediction in the southern part of the city of Hanover (Germany).

Keywords: Regression, parameter estimation, distributed data processing and mining, multi-agent systems.

1 Introduction

Currently, travel time information plays a significant role in transportation and logistics and is applied in various fields and for different purposes. From the travellers' viewpoint, travel time information helps to select the most optimal route, which

Jelena Fiosina
Clausthal University of Technology,
Institute of Informatics,
Julius-Albert Str. 4, D-38678, Clausthal-Zellerfeld, Germany
e-mail: Jelena.Fiosina@gmail.com

* This work was supported by European Commission FP7 People Marie Curie IEF Career Development Grant. We thank also Prof. J.P. Müller for his valuable comments.

S. Omatu et al. (Eds.): Distributed Computing and Artificial Intelligence, AISC 151, pp. 255–263.
springerlink.com

minimizes delays. In logistics, accurate travel time estimation helps to reduce transport delivery costs by avoiding congested route sections, and helps also to increase the service quality of commercial delivery. For traffic managers, travel time information is an important metric of the operational efficiency of the traffic system [8].

The distributed nature of traffic networks, which are complex stochastic systems, can be suitably represented by multi-agent architecture. Centralised data processing and mining methods experience difficulties when data sources are geographically distributed and transmission is very expensive or not feasible. From practice, it is also known that most drivers rely primarily on their own experience (historical data). We use a decentralised multi-agent system to model autonomous data mining of vehicle-agents, on the based of not only the centrally processed information but also historical data.

To achieve their common goals effectively, agents should cooperate with so-called gossip scenarios. We suppose that each agent autonomously estimates its parameters and then (on demand) adjusts them, by communicating with randomly chosen neighbour agents from the transmission radius. This random selection of neighbour agents is performed several times in different combinations, thereby performing the so-called resampling procedure. The obtained weighted resampling estimators are more reliable and robust in the case of nuisance observations. This detail is important because some agents can provide unreliable estimates, and it is better to average not all possible estimates, but a random subset of local estimates [1].

In this study, we do not deal with decision-making problems as we did in the study reported in [6]. The estimated travel times serve as input data to the decision-making module, which can also be incorporated into the intelligent agents.

The contribution of this study is toward the following: 1) development of the structure of a regression model for travel time forecasting; 2) development of an decentralised resampling-based regression parameter adjustment algorithm for streaming data; 3) application of the suggested algorithm for real-time data with the objective of demonstrating its efficiency.

The remainder of this paper is organised as follows. The second section contains a description of related work. The third section formulates the problem. In the fourth section, we present the multivariate linear regression model and the iterative least square algorithm for parameter estimation. The fifth section presents the resampling-based consensus algorithm of parameter adjustment. The sixth section presents case studies. The last section presents the conclusion.

2   Related Work

The need for research in the transportation area was considered by many authors. In this area, travel time prediction is one of the important challenges. From the architectural viewpoint, centralised and distributed approaches for travel time forecasting were considered. The centralised approach was applied in various intelligent transport systems, such as in-vehicle route guidance, or advanced traffic management systems. A detailed overview is presented in [8]. The estimation of actual travel time by using vehicle-to-vehicle communication is described in [9]. In contrast to centralisation, it was demonstrated that the representation of complex systems, such as traffic networks, in the form of decentralised multi-agent systems is of fundamental importance [4]. Decision-making in multi-agent traffic systems was considered in [6]. An example of the architecture of a distributed traffic system was presented in [7].

From the algorithmic viewpoint, numerous data mining and processing techniques were suggested for travel time prediction. Statistical methods, such as regression and time series, and artificial intelligence methods, such as neural networks, are successfully being implemented for similar problems. However, travel time is affected by a range of different factors. Thus, accurate prediction of travel time is difficult and needs considerable historical data. Understanding the factors affecting travel time is essential for improving prediction accuracy [8]. Travel time prediction for bus routes using a linear regression model was employed in [10].

In this study, we propose to use a decentralised regression model for solving the travel time prediction problem for streaming data. A similar decentralised approach was suggested in [11] for the estimation of the parameters of a wireless network.

3 Problem Formulation

We consider a traffic network with several vehicles, represented as autonomous agents, which predict their travel time on the basis of their current observations and history. Each agent locally estimates the parameters of the same traffic network. In order to make a forecast, each agent constructs a regression model, which explains the manner in which different explanatory variables (factors) influence the travel time. A detailed overview of such factors is provided in [8]. The following information is important for predicting the travel time [10]: average speed before the current segment, number of stops, number of left turns, number of traffic lights, average travel time estimated by traffic management centres (TMC). We should also take into account the possibility of an accident, network overload ("rush hour") and weather conditions.

Let us consider a vehicle, whose goal is to drive through the defined road segment under specific environment conditions (day, time, city district, weather, etc.). Let us suppose, that it has no or little experience of driving in such conditions. For accurate travel time estimation, it contacts other traffic participants, which send their estimated parameters to it. The forecasting procedure of one such vehicle is shown in Fig. 1.

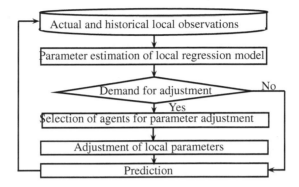

Fig. 1 Algorithm for local travel time prediction by an individual agent

4 Local Recursive Parameter Estimation

We describe the formal model, which is incorporated into each agent's local data processing and mining module. For this purpose, we first consider the classical multivariate linear regression model:

$$\mathbf{Y} = \mathbf{X}\beta + \varepsilon; \tag{1}$$

where $\mathbf{Y}$ is an $n \times 1$ vector of dependent variables (here, actual travel times); β is an $m \times 1$ vector of unknown parameters of the system to be estimated; ε is an $n \times 1$ vector of random errors; $\mathbf{X}$ is an $n \times m$ matrix of explanatory variables. The rows of the matrix $\mathbf{X}$ correspond to observations and the columns correspond to factors (here, length of the route, average speed in the system, average number of stops in the system, congestion level, etc.).

We suppose that $\{\varepsilon_i\}$ are mutually independent, have zero expectation, $E[\varepsilon] = 0$, and equal variances, $V[\varepsilon] = \sigma^2 \mathbf{I}$, where $\mathbf{I}$ is an $n \times n$ identity matrix.

The well-known least square estimator (LSE) $\mathbf{b}$ of β is:

$$\mathbf{b} = (\mathbf{X}^T \mathbf{X})^{-1} \mathbf{X}^T \mathbf{Y}. \tag{2}$$

After the estimation of the parameters β, we forecast travel time for a certain k-th future time moment:

$$E(Y_k) = \mathbf{x}_k \mathbf{b}, \tag{3}$$

where, $\mathbf{x}_k$ is a vector of observed values of explanatory variables for the kth future time moment. The described estimation procedure, (2), requires information about all observations, i.e., the complete matrix $\mathbf{X}$. In practice, for real-time streaming data, the estimation is performed iteratively, being updated after each new observation. The recurrent iterative method for the LSE was suggested in [2], [3]. This method assumes the recalculation of system parameters for each new observation.

Let us briefly describe the key aspects of this algorithm. Let $\mathbf{b}_t$ be the estimate after t first observations. After receiving the $t + 1$-th observation, we recalculate the estimates of β (Y_{t+1} - value of a dependent variable; and $\mathbf{x}_{t+1}$ - values of explanatory variables):

$$\mathbf{b}_{t+1} = \mathbf{b}_t + \mathbf{K}_{t+1}(Y_{t+1} - \mathbf{x}_{t+1}\mathbf{b}_t), \, t = 0, 1, \ldots, \tag{4}$$

where $\mathbf{K}_{t+1}$ is an $m \times 1$ vector of proportionality, smoothness, or compensation.

From (4), one can observe, that $\mathbf{b}_{t+1}$ is represented as a sum of the previous estimate $\mathbf{b}_t$ and the correction term $\mathbf{K}_{t+1}(Y_{t+1} - \mathbf{x}_{t+1}\mathbf{b}_t)$. Formula (4)is based on exponential smoothness, an adaptive forecasting method [3].

To calculate $\mathbf{K}_{t+1}$, we need values of matrices $\mathbf{A}_t$ and $\mathbf{B}_t$, obtained after the last t-th iteration. $\mathbf{A}_t$ and $\mathbf{B}_t$ are square $m \times m$ matrices. $\mathbf{B}_t$ is equal to $(\mathbf{X}^T \mathbf{X})^{-1}$ if this matrix exists, else it is equal to a pseudo-inverse matrix. Matrix $\mathbf{A}_t$ is a projection matrix, therefore, if $\mathbf{x}_{t+1}$ is the linear combination of the rows of matrix $\mathbf{X_t}$, its projection is equal to zero: $\mathbf{x}_{t+1}\mathbf{A}_t = 0$. Starting the algorithm we set the following initial values $\mathbf{A}_0 = \mathbf{I}$, $\mathbf{B}_0 = \mathbf{0}$, $\mathbf{b}_0 = \mathbf{0}$.

If the condition $\mathbf{x}_{t+1}\mathbf{A}_t = 0$ is satisfied, then

$$\mathbf{B}_{t+1} = \mathbf{B}_t - (1 + \mathbf{x}_{t+1}\mathbf{B}_i\mathbf{x}_{t+1}^T)^{-1}\mathbf{B}_i\mathbf{x}_{t+1}^T(\mathbf{B}_t\mathbf{x}_{t+1}^T),$$
$$\mathbf{A}_{t+1} = \mathbf{A}_t, \, \mathbf{K}_{t+1} = (1 + \mathbf{x}_{t+1}\mathbf{B}_t\mathbf{x}_{t+1}^T)^{-1}\mathbf{B}_t\mathbf{x}_{t+1}^T,$$

otherwise

$$\mathbf{B}_{t+1} = \mathbf{B}_t - (\mathbf{x}_{t+1}\mathbf{A}_t\mathbf{x}_{t+1}^T)^{-1}\left((\mathbf{B}_i\mathbf{x}_{t+1}^T)(\mathbf{A}_t\mathbf{x}_{t+1}^T)^T + (\mathbf{A}_t\mathbf{x}_{t+1}^T)(\mathbf{B}_t\mathbf{x}_{t+1}^T)^T\right) +$$
$$+ (\mathbf{x}_{t+1}\mathbf{A}_t\mathbf{x}_{t+1}^T)^{-2}(1 + \mathbf{x}_{t+1}\mathbf{B}_t\mathbf{x}_{t+1}^T)(\mathbf{A}_t\mathbf{x}_{t+1}^T)(\mathbf{A}_t\mathbf{x}_{t+1}^T)^T,$$
$$\mathbf{A}_{t+1} = \mathbf{A}_t - (\mathbf{x}_{t+1}\mathbf{A}_t\mathbf{x}_{t+1}^T)^{-1}\mathbf{A}_i\mathbf{x}_{t+1}^T(\mathbf{A}_t\mathbf{x}_{t+1}^T)^T,$$
$$\mathbf{K}_{t+1} = (\mathbf{x}_{t+1}\mathbf{A}_t\mathbf{x}_{t+1}^T)^{-1}\mathbf{A}_t\mathbf{x}_{t+1}^T.$$

5 Parameter Adjustment Algorithm

We are new going to introduce a notation for the local regression model of each of the s agents in the network. We use index (i,t) for the variables in formula (1), to refer to the i-th agent at time t:

$$\mathbf{Y}(i,t) = \mathbf{X}(i,t)\beta + \varepsilon(i,t), \, i = 1, \ldots, s. \tag{5}$$

Following (4), the i-th agent of s calculates the estimates $\mathbf{b}(i,t)$ of β and predicts the travel time $E[Y(i,t+1)]$ for the future time moment $t+1$, using (3).

Prior to forecasting, some agents may adjust their locally estimated parameters with other traffic participants. Let us describe this adjustment procedure more precisely.

First, the agent selects the other agents from a given transmission radius, contacts them, and requests them to send their estimated parameters. The agents can be in different situations and their observation may contain outliners. In order to make the adjustment procedure more reliable and robust to outliners, the agent perform the described selection several times in different combinations, forming so-called resamples from the available agents [1].

We implement N realisations of the following resampling procedure for the i-th agent. At the realisation q, the agent receives the parameter estimates of r randomly chosen neighbour agents. Let vector $\mathbf{L}_i^q$ contain the indices of the selected agents, $|\mathbf{L}_i^q| = r$.

The next step is the adjustment of the parameters. The agent that initialised the adjustment process considers the weighted estimates of other agents. The weights are time-varying and show the reliability level of each agent, depending on its forecasting experience as well as some other factors. Let $\mathbf{c}_i^{*q}(t)$ be a $1 \times r$ vector of the weights at the q-th realization at time t, $i = 1, \ldots, s$.

According to the logic of constructing discrete-time consensus, we assume that $\mathbf{c}_i^{*q}(t)$ is a stochastic vector for all t (the sum of its elements is equal to 1).

$$\mathbf{b}^{*q}(i, t+1) = \sum_{j=1}^{r} c_{i,j}^{*q}(t) \mathbf{b}(\mathbf{L}_{i,j}^q, t), \qquad (6)$$

where $\mathbf{b}^{*q}(i, t)$ is the adjusted estimate of β calculated by the i-th agent at the q-th resampling realisation, $i = 1, \ldots, s$, $q = 1, \ldots, N$.

Finally, the resampling estimator is obtained as an average over all N realisations:

$$\mathbf{b}^R(i, t+1) = \frac{1}{N} \sum_{q=1}^{N} \mathbf{b}^{*q}(i, t+1). \qquad (7)$$

The algorithm is a combination of the iterative LSE algorithm and resampling-based parameter adjustment. This adjustment procedure aims to increase the reliability of the estimates, especially for insufficient or missing historical data, and to contribute to the overall estimation accuracy [11].

6   Case Studies

We simulate a traffic network in the southern part of Hanover(Germany). The network contains three parallel and five perpendicular streets, creating fifteen intersections with a flow of approximately 5000 vehicles per hour. The vehicles solve a travel time prediction problem. They receive information about the centrally estimated system variables (such as average speed, number of stops, congestion level, etc.) for this city district from TMC, combine it with their historical information, and make adjustments adjust according to the information of other participants using the presented consensus algorithm. In this case, regression analysis is an essential part of the local travel time prediction process. We consider the local model (5) and implement the parameter adjustment algorithm (7).

The factors are listed in Table 1 (left). To improve the quality of the regression model, some non-linear transformations of the factors are performed. The resulting regression model remains linear by parameters, but becomes non-linear by factors. All parameters satisfy the significance test, as a measure of influence to the target variable, i.e., travel time.

We simulate ten agents and train them on the observations taken from the available dataset of size 1790. We conduct three different experiments and compare the results using analysis of variance and adjusted coefficient of determination, R^2. This provides a well-known measure of the effectiveness of prediction of future outcomes by the model [5].

Case 1: The agents transmit their observations to the central unit and regularly obtain the updated parameters of the system. This requires the transmission of a large amount of data and is therefore very expensive. Nevertheless, this model provides the best prediction results (see Table 1)(right).

Case 2: The agents estimate the parameters of the same system locally without cooperation. This case requires no transmission costs; however, it assumes that each agent has a special data processing block for performing calculations. The main problem is that each agent builds its own model using a considerably small amount of data. The data should be homogeneous because each agent estimates the parameters of the same system by using different data. These estimates should converge to the parameter values estimated in Case 1. The experiments show that for historical data sets smaller than 25, the model parameters cannot be estimated well (see Table 1(right), worth agent). To estimate the quality of the i-th agent's model, the cross-validation technique was used for testing the model with the observations of other agents.

Case 3: The agents estimate the parameters of the same system locally and use cooperation mechanisms to adjust their parameter values with other agents. The amount of transmitted information is lower than in Case 1, because not all the data but only locally estimated parameters are transmitted. This case has the same assumptions concerning intelligent module and data, as in Case 2. Agents with more experience (historical dataset) help new agents with less experience to make better estimates. The prediction experience is used as the weights (reliability level) of agents. The results show that this cooperation helps to improve performance, especially for the less experienced agents. In this case four agents have less experience (data sets smaller than $n = 15$) and six agents are more experienced. The other parameters are $r=3$, $N=10$. This case gives better results than Case 2, but slightly worse results than Case 1, owing to the loss of some information in the process of averaging and the relatively small data sets of each agent (see Table 1)(right).

Table 1 Factors and corresponding parameters **b** (left). Efficiency criteria (right)

Var.	Description	Mod.	Koef.	Est. value
Y	travel time (min);	Y		
X_1	route length(km)	X_1	b_1	0.614
X_2	avg. speed in system (km/h)	X_2	b_2	-0.065
X_3	avg. number of stops(units/min)	X_3^2	b_3	0.09
X_4	congestion level(Veh/h)	$\sqrt{X_4}$	b_4	0.159
X_5	traffic lights in the route (num)	X_5^2	b_5	0.241
X_6	left turns in the route (num)	X_6^2	b_6	-0.058

Case	R^2 of the whole system	R^2 of the worth agent
1	0.66	-
2	0.55	0.28
3	0.64	0.58

7 Conclusions

A problem of decentralised travel time forecasting was considered. A multi-agent architecture with autonomous agents was used for this purpose.

A decentralised linear multivariate regression model was developed to forecast the travel time. The iterative LSE method was used for the regression parameter estimation, which is suitable for streaming data processing. The resampling-based consensus method was suggested for coordinated adjustment of estimates between neighbour agents. We illustrate the efficiency of the suggested approach using simulation with real data from the southern part of Hanover. The experiments show the efficiency of the proposed approach.

Our future work will be continued in three directions: (a) construction of the distributed model of multiple multivariate regression, which allows forecasting of several response variables simultaneously from the same set of explanatory variables (factors), as well as an implementation of such a model for real-time traffic data; (b) application of other regression model types (i.e., general regression models using kernel estimators); (c) modification of the parameter adjustment algorithm (new strategies for the calculation of the reliability level of agents, median resampling approach, etc.).

References

1. Afanasyeva, H., Andronov, A.: On robustness of resampling estimators for linear regression models. Communications in Dependability and Quality Management: An International Journal 9(1), 5–11 (2006)
2. Albert, A.: Regression and the Moor-Penrose Pseudoinverse. Academic Press, New York and London (1972)
3. Andronov, A., Kiselenko, A., Mostivenko, E.: Forecasting of the development of Regional Transport System. KNZ UrO RAN, Sivtivkar (1991) (in Russian)
4. Bazzan, A.L.C., Wahle, J., Klügl, F.: Agents in Traffic Modelling - from Reactive to Social Behaviour. In: Burgard, W., Christaller, T., Cremers, A.B. (eds.) KI 1999. LNCS (LNAI), vol. 1701, pp. 303–306. Springer, Heidelberg (1999)
5. Draper, N., Smith, H.: Applied Regression Analysis. John Wiley and Sons, New York (1986)
6. Fiosins, M., Fiosina, J., Müller, J., Görmer, J.: Agent-based integrated decision making for autonomous vehicles in urban traffic. Adv. in Intelligent and Soft Computing 88, 173–178 (2011)
7. Lee, W., Tseng, S., Shieh, W.: Collaborative real-time traffic information generation and sharing framework for the intelligent transportation system. Inf. Sciences 180, 62–70 (2010)

8. Lin, H.E., Zito, R., Taylor, M.A.: A review of travel-time prediction in transport and logistics. In: Proc. of the Eastern Asia Society for Transportation Studies, Hamburg, vol. 5, pp. 1433–1448 (2005)
9. Malnati, G., Barberis, C., Cuva, C.M.: Gossip: Estimating actual travelling time using vehicle to vehicle communication. In: 4th Int. Workshop on Intel. Transportation, Hamburg (2007)
10. McKnight, C.E., Levinson, H.S., Kamga, C., Paaswell, R.E.: Impact of traffic congestion on bus travel time in northern new jersey. Transportation Res. Record Journal 1884, 27–35 (2004)
11. Stankovic, S.S., Stankovic, M.S., Stipanovic, D.M.: Decentralized parameter estimation by consensus based stochastic approximation. IEEE Trans. Automatic Control 56 (2009)

Multi-agent Applications in a Context-Aware Global Software Development Environment

Helio H.L.C. Monte-Alto, Alberto B. Biasão, Lucas O. Teixeira, and Elisa H.M. Huzita

Abstract. The need for skilled workers, the improving communication facilities, growing applications complexity, software development time and cost restrictions, and also the need for quality and accuracy are part of the new scenario where global software development was introduced. However, it has brought new challenges such as: communication, coordination and culture. Context information could help participants to be aware of events occurring and so improve their communication as well as their interactions. This paper presents a multi-agent mechanism for processing context information, as well as a mechanism for allocation of human resources. It is also introduced a framework to encapsulate some functionality required by a knowledge-based multi-agent system.

Keywords: context-awareness, ontology, global software development.

1 Introduction

Market competitiveness has brought new challenges that forced companies to adopt new software development strategies. The lack of skilled workers and the pursuit for cheaper employees have made these companies to embrace Global Software Development (GSD) as a new approach. Industries are reallocating their production unities in decentralized zones, where skilled labors are promptly available, taking politic and economic advantages [1].

Despite the mentioned advantages, GSD also brings different new issues such as cultural differences, geographic dispersion, communication, control and coordination, which can bring frustration, displeasure or even misunderstanding among teams. GSD also requires cooperation of individuals with different cultures, as well as the need for structure, attitudes toward hierarchy, sense of time, and communication styles.

According to these demands, context-awareness has shown up as an interesting approach to spread context information to people involved in a GSD environment.

Helio H.L.C. Monte-Alto · Alberto B. Biasão · Lucas O. Teixeira · Elisa H.M. Huzita
Informatics Department, State University of Maringá, Maringá, Paraná, Brazil
e-mail: {heliohenrique3,biasao,lucasxteixeira}@gmail.com,
    emhuzita@din.uem.br

S. Omatu et al. (Eds.): Distributed Computing and Artificial Intelligence, AISC 151, pp. 265–272.
springerlink.com   © Springer-Verlag Berlin Heidelberg 2012

Context is any information that can be used to characterize the situation of entities that are considered relevant to the interaction between a user and an application, including the user and the application themselves [2]. It is necessary that individuals who participate in a project are aware of context information while interacting [3]. Although there are researches about the characterization of context information in collaborative systems, efficient presentations of this kind of information to users, knowledge sharing and combining context-awareness techniques to achieve better communication in GSD remains a demand.

In this circumstance, a tool called Distributed Software Engineering Environment (DiSEN) is being developed by our research group, offering features to support communication, persistence and collaboration among teams geographically distributed. Then, a context-awareness based model called DiSEN-CSE (DiSEN-Context Sensitive Environment) was designed considering context information acquisition, representation, persistence, processing and sharing [3].

The purpose of this paper is to present a multi-agent context processing mechanism based on [3], called ContextP-GSD (Context Processing on Global Software Development), that transforms context information into actions to help users during their interactions. It also presents a mechanism that uses context information to support human resource allocation to software projects called DiSEN User Rating. Finally, it is also presented a framework, called DiSEN Agency, whose purpose is to structure a knowledge-based multi-agent architecture consistent with the needs and issues of performance and scalability of a context-aware distributed system. Such framework assigns roles to special agents to eliminate the load of the application agents and provides ways to extend it according to application needs.

This paper is structured in four more sessions. Section 2 introduces some related works on which we have relied for the development of the mechanisms. Section 3 is divided into two subsections, in which are presented the two mechanisms that have been implemented. In Section 4 we present the DiSEN Agency framework. Finally, in the conclusions section the contributions and future works are discussed.

2   Related Works

This section presents some related works. Some of them, such as ACAI (Agent-based Context-Aware Infrastructure) [4], CoBrA (Context Broker Architecture) [5] and PIÑAS (Platform for Interaction, Naming And Storage)/AllianceWeb [6] use the context-awareness approach and have some similarities to ContextP-GSD.

The idea of agents which maintain local knowledge bases and a central agent that would make approval of facts, persistence operations and knowledge sharing was influenced by a SemantiCore extension that provides a knowledge management process for multi-agent systems [7]. SemantiCore is a framework that provides an abstraction layer for agent-oriented application development for the Semantic Web [8].

3   DiSEN Multi-agent Applications

The DiSEN environment is still in development and since its conception the idea was that it would be mostly an agent-based system [9]. Thus, based on [3], the multi-agent applications – especially ContextP-GSD – described in the following subsections were developed. Furthermore, multi-agent systems (MAS) have been widely used to implement context-aware systems, due to their ability to make decentralized decisions about the context of the environment. Moreover, the use of agent concepts for distributed systems engineering provides several advantages for reducing complexity, including: autonomy, situatedness and high-level interactions [10]. For the implementation of agents, we have been used SemantiCore [8] platform, and for ontology manipulation we have been used Jena Ontology API [11].

3.1 ContextP-GSD

ContextP-GSD is a mechanism that focuses in sensing events generated by users in a virtual environment. Thus, ContextP-GSD's sensors acquire context information from events generated by user's interactions on DiSEN environment. Thereby, ContextP-GSD defines context processing as a set of methods and processes that are able to perform reasoning by means of derivations of environment actions and context information as consequences of inference rules.

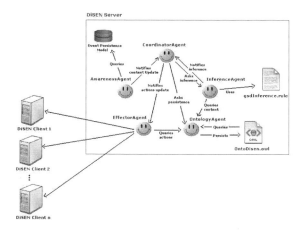

Fig. 1 ContextP-GSD's multi-agent architecture

To achieve this goal, ContextP-GSD combines features described on different approaches, creating a hybrid one that is focused on GSD challenges. Foremost, an ontology – which was called OntoDiSEN – was designed to represent the GSD's context information. Therefore, a MAS architecture was planned, to decentralize decision-making, sharing responsibilities among software agents with tasks to: (i) acquire context information from events, which are users' actions; (ii) persist this context information in the knowledge base (KB) defined by OntoDiSEN, (iii) use the context information to infer actions on behalf of participants, enhancing communication and coordination and (iv) present these actions on the workspaces of the participants. The inference is made by means of rules that can be customized according to users' needs, however, limited by the context represented in the ontology and actions implemented on DiSEN API.

Fig. 1 shows the ContextP-GSD's architecture, which was designed with five agents, each of them responsible for accomplishing a different task:

- **OntologyAgent:** acts as a persistence engine, and is the only one that can have access to the KB to update context or insert new inferred actions and facts. It is concerned with guaranteeing that OntoDiSEN remains consistent.
- **CoordinatorAgent:** this agent acts as a manager. It keeps track of the context that is being updated and inferred, notifying the interested agents.
- **AwarenessAgent:** it is a proactive agent responsible for capturing environment events from DiSEN's event model. In this model are persisted events that represent the interactions taking place in users' workspaces by means of questions in the format: what, where, when and who. After querying these questions, AwarenessAgent ask these questions for its decision engine, and it notifies the CoordinatorAgent that there is a new context to be added. At this time, CoordinatorAgent sends the new facts to the OntologyAgent to persist them.
- **InferenceAgent:** after having updated the context from the acquisition made by AwarenessAgent, InferenceAgent tries to infer actions or new context information by means of defined inference rules. If something new comes out, it notifies CoordinatorAgent that requests OntologyAgent to persist it in the KB.
- **EffectorAgent:** it is also a proactive agent that is notified by CoordinatorAgent when actions need to be called. EffectorAgent's decision engine defines which method should be called from DiSEN API and then calls it.

One of the implemented scenarios is defined by a set of time zone rules. These consist of logical rules concerned about how participants should be notified when they are allocated to a new project, taking into consideration their time zone. Thus, knowing the context of each participant, the mechanism is able to decide at runtime which and how participants must be notified. Depending on the context of each one, ContextP-GSD chooses between sending a synchronous or asynchronous message. One of these rules can be seen below. Such rule treats the case of a participant who is not present in his/her workspace but is at the same time zone as the project. The system should then send a message synchronously and asynchronously, ensuring that he/she receives the information. The rule is implemented using Jena rule notation.

```
[timeZone2: (?p od#participatesProject ?r)
    noValue(?p od#accessWorkspace ?w)
    (?r od#followsTimeZone ?t) (?l od#hasTimeZone ?t)
    (?p od#userIsInLocal ?l) ->
    (od#PopUp od#sendSynchronousMessage ?p)
    (od#Email od#sendAsynchronousMessage ?p) ]
```

Such application scenario demonstrates the flexibility of the mechanism. New rules can be implemented to transform environment context information into environment actions, supporting users' communication and coordination smartly.

3.2 *DiSEN User Rating*

In order to better exploit multi-agent and ontology reasoning capabilities, it was also implemented a prototype of a mechanism for DiSEN to support the allocation of human resources to software projects. Basically, this is a module for DiSEN that has the ability to infer the available users that can be allocated to a particular project, according to their knowledge and skills. This mechanism was called DiSEN User Rating.

As a requirement to achieve the reasoning capability, an agent responsible for rating the candidate users (RaterAgent) for a project should have access to the KB. Basically, RaterAgent must perform its inference process once it is requested. The inference process consists of the following steps: (i) load the inference rules, (ii) load the local KB and run the inference engine, (iii) extract the facts from the graph returned by the inference engine, and (iv) send the inferred facts to OntologyAgent (which is similar to ContextP-GSD's) so that they may be persisted.

The inference rule is intended to infer whether a user can be allocated to a project, considering if he/she has the required qualifications. Such qualifications include the skills and knowledge that a candidate might have. In summary, a user U can be allocated to a project P if: (i) status of U is available, (ii) U has all the skills and skill levels required for project P, (iii) U has all the knowledge and knowledge levels required for project P. A piece of this rule – using SPARQL [12] notation – can be seen below:

```
CONSTRUCT {?u :canBeAllocatedTo ?p} WHERE {
  ?u :hasResourceStatus :available .
  ?p :requiresSkill ?r . ?r :isSkillOf ?s .
  ?r :hasSkillLevel ?l . ?u :hasSkill ?ur .
  ?ur :isSkillOf ?s . ?ur :hasSkillLevel ?l .
  FILTER NOT EXISTS { ?p :requiresSkill ?r2 .
    ?r2 :isSkillOf ?s2 . ?r2 :hasSkillLevel ?l2 .
    FILTER NOT EXISTS { ?u :hasSkill ?ur2 .
      ?ur2 :isSkillOf ?k2 . ?ur2 :hasSkillLevel ?l2 . }}}
```

4 DiSEN Agency

DiSEN Agency is a framework to support the development of knowledge-based multi-agent systems. It was initially conceived as a refactoring of the reusable components of ContextP-GSD and DiSEN User Rating, and later its architecture was improved to better support high scalability and performance.

The central entity of DiSEN Agency's architecture is the Ontology Agent (OA), which was based on ContextP-GSD's OntologyAgent and ACAI's System Knowledge Base Agent [4]. In general, application agents can be divided into three groups: logical agents [13], which use knowledge and inference rules to reason about the domain (e.g. InferenceAgent and RaterAgent); context collector agents, which observe the environment via its sensors (e.g. AwarenessAgent); and generic application agents (e.g. CoordinatorAgent). There is also possible for a logical agent to be a context collector agent, and any application agent is able to act upon the environment. A graphical representation of DiSEN Agency's architecture is shown in Fig. 2.

The OA is responsible for persisting and publishing the KB in the environment, as well as ensuring KB's consistency. Therefore, OA acts as a broker between the application agents and the KB. All modifications in the KB must be performed through the OA, by means of requests for persistence and publication of new or updated facts. This aims to: (i) avoid problems with concurrent access, (ii) provide high level of flexibility and adaptation in the sense that changes in the code that handles the persistence become centralized in a single point, and (iii) standardize the process so that all logical agents remain with their local instances of the KB up to date all the time.

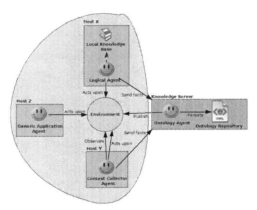

Fig. 2 DiSEN Agency's architecture.

To justify this design decision, consider a MAS where application agents are able to access the semantic repository directly. If multiple agents try to access the KB simultaneously and at least one of them wants to do a write operation, it may causes a race condition. Such problem can be avoided using some concurrence control mechanism, possibly some transactional processing. Although this solution alone seems to address the problems, it has some issues. Consider that some logical agents maintain some local knowledge, so when one agent inserts facts in the semantic repository, the other ones would have to update their local KB. Thus, without a broker this agent would have to notify or even publish the knowledge to the remaining agents. In both cases, it would imply in an increased load over the application agents, which may have other tasks to perform and may be located in a machine with low computational power. Adopting the OA centralizes such tasks into a single entity, removing these responsibilities from the application agents. Moreover, only one slice of the KB should be published, since it can be too large to be transferred through a network.

It is desirable for logical agents to maintain partial local copies of the KB, since accessing their local memory is faster than querying a remote repository. Moreover, such a repository can store a large KB, making inference and query operations slower. Another issue concerns the fact that an agent is seen as a system user, so it is more convenient for it to interact with the persistent model through interfaces than dealing with low-level details. The OA takes the role of an interface with the repository because persistence operations are made through it.

Given the presented justifications, the execution flow for the implemented OA is: (i) receive, via one of its sensors, the set of facts that must be persisted, (ii) retrieve the KB from the repository, or open a connection to it, (iii) perform three actions, each one responsible for persisting the KB according to one of the lists of facts (insert, update and remove), and (iv) publish the facts persisted to all logical agents.

The maintenance of local KBs by application agents, and the need to publish subsets of the central KB lead us to some design issues: (i) how should parts of knowledge be extracted by the OA, and (ii) how should an agent merge its KB with a new set of facts? The first one can be achieved by means of using an algorithm that extracts parts of knowledge given the application agent's interest. This extraction occurs only when an agent makes a knowledge request to OA. Once an agent retrieves knowledge at the beginning of its life cycle, it no longer needs to ask OA for knowledge. So this does not imply much additional burden over it. In addition, OA's broadcasting publications always occur in sequence with some persistence action. Thus, the sets of facts that have been just persisted are the same to be published and, therefore, it would suffice to pass them on. The second issue was simply addressed by granting the same actions to manipulate facts – insertion, update and removal – used by the OA to all logical agents of the application.

The ontology consistency checking – or approval of facts – is another important issue. Approving the facts according to the KB implies an inference process applied after an insertion of facts to check if any inconsistency was generated. If inconsistencies cannot be resolved, the facts to be persisted must be rejected. The consistency checking can be done by means of some inference engine.

5   Conclusions and Future Works

The geographic dispersion of GSD brings communication difficulties, making it harder to spread context information about activities or tasks in which the participants are involved. ContextP-GSD is a context processing mechanism, toward to ease users communication during their interactions by means of inference rules applied over a knowledge-base. The MAS approach makes it more flexible in the sense that each agent may be responsible for reasoning and collecting context about different parts of the domain and on different hosts in a distributed system. The collaboration between agents is also important to provide load balancing and smarter decision-making.

One of the main differences between ContextP-GSD and the two infrastructures ACAI and CoBrA is that the first one is a mechanism with focus on a virtual environment whereas the other ones are focused on physical environments. ContextP-GSD is more similar to ACAI once both distribute context processing among several different agents, while CoBrA centralizes it into a single one. The use of ontology to represent context distinguishes it from PIÑAS/AllianceWeb, which is an infrastructure for a virtual environment of cooperative authoring and uses relational database to represent and store context information. Ontologies have the advantage of defining semantic information (knowledge), which facilitates reasoning based on inference rules, which is one of the strengths of ContextP-GSD.

The implementation of ContextP-GSD brought up some performance and scalability issues, particularly related to how knowledge should be persisted and spread among the various agents in the MAS. Aiming at reuse, DiSEN Agency encapsulates a knowledge-based multi-agent architecture that provides solutions for these issues, based on some related works [4,7]. Its utilization in the implementation of DiSEN User Rating – which is not tightly bound to the GSD's scenario – shows that it is also applicable in many other MAS applications. For example, it is possible to use DiSEN Agency on Semantic Web applications, where agents might be distributed among Web servers and browsers, as proposed in [8]. Agents might be meant to deal with a piece of knowledge bound to a Web page (semantic annotation) and possibly keep it synchronized with a centralized KB.

Another issue, related to the OA performance, is the load that will be assigned to it. It should be located properly on the server where the semantic repository is, which must have enough computing power to handle all the processing required without generating a bottleneck on the MAS. Still, OntoDiSEN could also be modularized. In fact our algorithm for extracting subsets of the KB creates an ontology module. Our research group is already developing solutions to deal with issues related to the organization of the interfaces between modules.

References

1. Vardi, M., Mayadas, F., Aspray, W.: Globalization and Offshoring of Software. A Report of the ACM Job Migration Task Force. ACM (2006)
2. Dey, A.K., Abowd, G.D., Salber, D.: A conceptual framework and a toolkit for supporting the rapid prototyping of context-aware applications. Human Computer Interaction Journal (2001), doi:10.1207/S15327051HCI16234_02
3. Chaves, A.P., Wiese, I.S., da Silva, C.A., Huzita, E.H.M.: Um modelo baseado em context-awareness para disseminação de informações em um ambiente de desenvolvimento distribuído de software. In: CLEI 2008, Santa Fe, pp. 1365–1374 (2008)
4. Khedr, M., Karmouch, A.: Acai: agent-based context-aware infrastructure for spontaneous applications. J. Network and Computer Applications (2005), doi:10.1016/j.jnca.2004.04.002
5. Chen, H., Perich, F., Chakraborty, D., Finin, T., Joshi, A.: Intelligent agents meet semantic web in a smart meeting room. In: AAMAS 2004 (2005), doi:10.1109/AAMAS.2004.10110
6. Decouchant, D., et al.: Contextual awareness based communication and coauthoring proximity in the internet. Expert Syst. Appl. (2009), doi:10.1016/j.eswa.2008.10.082
7. Lemke, A.P.: Um framework para a organização do conhecimento de agentes de software. Pontifícia Universidade Católica do Rio Grande do Sul, Dissertation (2007)
8. Blois, M., Escobar, M., Choren, R.: Using Agents and Ontologies for Application Development on the Semantic Web. J. Braz. Comp. Soc. (2007), doi:10.1007/BF03192408
9. Pascutti, M.C.D.: Uma proposta de arquitetura de um ambiente de desenvolvimento de software distribuído baseado em agentes. Univ. Federal do Rio Grande do Sul (2002)
10. Jennings, N.: An agent-based approach for building complex software systems. Commun. ACM 44(4), 35–41 (2001)
11. Dickinson, I.: Jena Ontology API (2009), http://jena.sourceforge.net/ontology (accessed 2011)
12. Prud'hommeaux, E., Seaborne, A.: SPARQL Query Language for RDF (2008), http://www.w3.org/TR/rdf-sparql-query (accessed January 05, 2012)
13. Russel, S., Norvig, P.: Artificial Intelligence: a modern approach, 2nd edn. Pearson Ed. Inc. (2003)

SPAGE: An Action Generation Engine to Support Spatial Patterns of Interaction in Multi-agent Simulations

Kavin Preethi Narasimhan

Abstract. Space is a significant resource in human interaction. In this paper, we analyse the prospects of utilising space as an important resource in agent interaction. To do this, we created a software engine called SPAGE that generates communicative action signals for an agent based on the current state of the agent and its environment. These action signals are then evaluated against a set of conditions that are logically deduced from the literature on human face-to-face interaction. Depending upon the success or failure outcomes of the evaluation, the agent then receives a reward or a punitive signal. In either case, the states of both the agent and its environment are updated. The ultimate rationale is to maximise the number of rewards for an agent. SPAGE is incorporated into a simulation platform called the K-space in order to verify the believability of the action signals, and also to analyse the effects a sequence of actions can have in giving rise to spatial-orientational patterns of agent interaction. SPAGE is modular in nature which makes future modifications or extensions easy.

1 Introduction

Designing social behaviour for robots and avatars is very important in any application that involves interaction between humans and the AI agents [5, 25]. In the past, there have been enormous efforts to make virtual agents express human-like verbal and/or non-verbal behaviour under different contexts [1, 3, 4, 8, 9]. The implementation in these systems mirror/replicate various elements of real-life social behavioural cues. However in this paper, we specifically focus on supporting features of spatiality in agent interaction (movement, orientation etc). More precisely, we present the SPAGE system in which we abstract several basic components of

Kavin Preethi Narasimhan
Queen Mary University of London, Mile End Road, London E1 4NS
e-mail: k.p.narasimhan@eecs.qmul.ac.uk

S. Omatu et al. (Eds.): Distributed Computing and Artificial Intelligence, AISC 151, pp. 273–280.
springerlink.com © Springer-Verlag Berlin Heidelberg 2012

conventional human interaction and use them to generate different sequences of actions for virtual agents residing in a simulation platform. The system is implemented with a view to analyze the effects that a particular sequence of actions can have in giving rise to spatial-orientational patterns involving two or more agents.

2   Spatial Aspects of Human Face-to-Face Interaction

The way in which human beings manage and use physical space in all their activities accounts for the organization of their behaviour in social events. Usually, when people come together to perform any activity, they collaboratively establish and sustain a central space over which they jointly assume control for the entire duration of the activity. Such a spatial structure helps in insulating the ongoing activity from outsiders and other external disturbances. It is also the space in which the members of a conversation group have unobstructed access to carry out all their interactional activities. Over the years, several people have identified and defined the characteristics of these shared spaces in various dimensions.

In Goffman's view, participants of a conversation group tend to place themselves in a spatial-orientational arrangement such that each of them face inward, encompassing a space to which they all have immediate access [6]. Lyman and Scott defined an Interactional Territory [13] as a distinct social unit that makes implicit claims of boundary maintenance for the entire duration of an interactional activity and is constantly affected by parvenus and newcomers. Kendon's [10] observations on people's spatial arrangement in an F-formation system signifies the natural way in which people tend to position and orient themselves in social encounters. By doing this, people form a central o-space to which all the participants of a conversation group have equal, direct and exclusive access. More specifically, an individual's transactional segment is the space extending in front of a person that he uses to carry out any activity [10]. In a typical group activity (e.g. conversation), the central o-space is formed by an overlap of the individual transactional segments of participant members which they jointly control and use to carry out the intended activity. The way in which co-present individuals organize themselves into such a formation also reflects their involvement with one another [11].

From the observations of Goffman [6], Kendon [10] and Lyman and Scott [13], it is clearly evident that space is an important resource in conducting human interaction. For practical reasons rendering space as an interactional resource for AI agents could be very challenging but undoubtedly rewarding. The next section covers some of the literature that focus on modelling features of spatiality in virtual agent interactions.

3   Related Work

There has been much effort in the past to model social and communicative behaviour for virtual agents but only a select few have focused on spatiality as an important aspect of agent interaction. For example, the design of social interaction in FreeWalk3 lays specific emphasis on the importance of spatial, nonverbal signals like body orientation and position during the course of interaction [15]. However the split control interface used in FreeWalk3 poses deadlock and inflexibility issues. CADIA Populus is a social simulation platform that aims at generating believable social behaviours in avatars by drawing inspiration from human territorial behaviours during social interactions [18] but the system is not fully autonomous. The concept of automating communicative behaviours in avatars based on theories of conversation and discourse analysis [2] is another interesting approach that makes virtual interaction more effective in agent-based systems. However, at times the avatar implementation in [2] permits odd behaviour such as the avatars staring into the void or expressing repetitive/irrelevant behaviour during the planning phase. This can be a disadvantage, especially, if the avatars are designed to represent the communicative intents of their users in a virtual world. Some of the other techniques for simulating features of spatiality in agent-based systems include: an independent agent engine [23], crowd simulation engines [20], social force model [7], multimodal dialog models [22, 14] and the concept of steering layers [17, 21]. There have also been extensive efforts to analyse and design spatial behaviour for robots engaged with human users for various purposes, in various contexts [12, 16, 24].

4 SPAGE

As a more generic alternate to the works discussed in section 3, we present an action generation engine called the SPAGE which aims to simulate features of spatiality in agent-based systems based on the theories discussed in section 2. In the proposed system, we conceive of virtual agents as communicating with each other by the exchange of action signals. The conditions that impact the success or failure outcomes of these action signals are purely geometrical and derived from basic interaction rules. Our key motivation therefore, is to analyse the influence of this purely geometrical part of the success or failure conditions on how agents move and stand when trying to engage one another.

To begin with, we needed an appropriate simulation platform to verify the believability of the agents' actions generated using our system and the effect it can have in giving rise to an agent structured space arising within the constraints of a mathematical space. For this purpose, we created 'K-space', a two dimensional spatial environment in which the agents can move, position and orient themselves based on the action signals generated by our system. An overhead view of some of the scenes in the K-space simulation is shown in figure 1.

The action generation engine used in the system is modular in nature. Figure 2 illustrates the overview of the architecture of SPAGE. At any time t, the input to the system comprises of an agent's state (A_{cs}) and its environment's state (E_{cs}) at that time t. The information stored in A_{cs} comprises of an agent's ID and its sensory data (visual range of 50 pixels, vocal and auditory range of 20 pixels) calculated based on the agent's position and orientation in K-space at the time t. E_{cs} stores information regarding the individual position and orientation of each and every agent in K-space at the time t. Based on this information, the action generation engine in the system produces a random action signal (e.g. WALK, SPEECH, POINT etc) for an agent (described in section 4.1). This action signal is then verified by a set of conditions to result in two possible outcomes: success or failure (described in section 4.2).

4.1 Action Signals

Communication between the agents in K-space happens only by the exchange of action signals. At any time t in the simulation, one of the following action signals can be randomly generated for an agent in K-space.

1. **SPEECH** $\{Agent_{initiator}, Agent_{recipient}\}$ denotes a speech signal that is directed from $Agent_{initiator}$ to $Agent_{recipient}$
2. **HEAR** $\{Agent_{recipient}, Agent_{initiator}\}$ denotes a hear signal that is directed from $Agent_{recipient}$ to $Agent_{initiator}$
3. **LISTEN** $\{Agent_{recipient}, Agent_{initiator}\}$ denotes a listen signal that is directed from $Agent_{recipient}$ to $Agent_{initiator}$
   A key difference between the HEAR and LISTEN signals is that a LISTEN signal can be randomly generated for any agent at any time irrespective of whether or not there is a preceding SPEECH signal. A HEAR signal on the other hand, can be generated only when there is a corresponding SPEECH signal on the timeline of actions. An example in the real-life setting would be "I listened outside the window, but I couldn't hear what was being said."

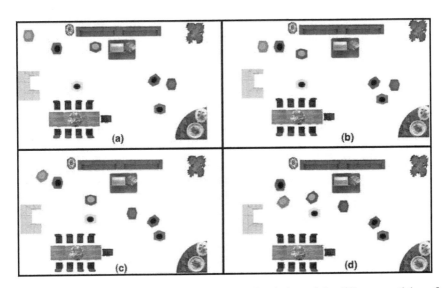

Fig. 1 Figures 1 (a), (b), (c) and (d) show an overhead view of the different spatial configurations in K-space resulting from the individual position and orientation of agents

4. **POINT** $\{Agent_{initiator}, Agent_{recipient}, Location(x_p, y_p)\}$ denotes a point signal that is directed from $Agent_{initiator}$ to $Agent_{recipient}$, wherein $Agent_{initiator}$ points $Agent_{recipient}$ to look at the $Location(x_p, y_p)$ in K-space

5. **LOOK** $\{Agent_{recipient}, Agent_{initiator}, Location(x_p, y_p)\}$ denotes a look signal that is directed from $Agent_{recipient}$ to $Agent_{initiator}$, wherein $Agent_{recipient}$ looks at the $Location(x_p, y_p)$ pointed by $Agent_{initiator}$

6. **WALK** $\{Location(x_o, y_o), Location(x_d, y_d)\}$ denotes a walk signal from a point (x_o, y_o) to another point (x_d, y_d) in K-space

7. **STAND** $Location(x_d, y_d)$ denotes a stand signal which means to remain stationary at $Location(x_d, y_d)$ in K-space

8. **TURN** $\{Direction_n\}$ denotes a turn signal which means turn to $Direction_n$ from an agent's current orientation in K-space

The first three actions (1, 2 & 3) concern agent communication, the middle two actions (4 & 5) concern agent gestures and the last three actions (6, 7 & 8) concern the movement of agents in K-space. The action generation engine in the system produces different action signals for different agents at different times. Every time we run the simulation, this non-deterministic approach results in different sequences of action signals for the agents.

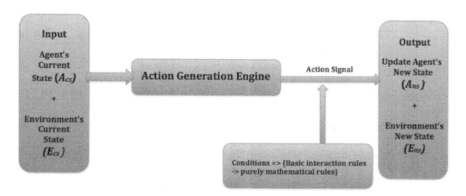

Fig. 2 System architecture of the action generation engine

4.2 Conditions for Evaluating Success or Failure of Action Signals

Every action signal generated by the system is subject to evaluation by a set of conditions that can result in two outcomes: *success* or *failure*. If an action signal succeeds by satisfying the respective conditions, the agent will get a reward or if an action signal fails, the agent will get a punishment. The ultimate aim is to maximise the number of rewards for an agent which eventually leads to having realistic sequences of actions for the agents in K-space. These conditions are derived from the literature on human face-to-face interactions [6, 10, 13].

Successful application of the SPEECH, HEAR and LISTEN action signals for
agents will depend on satisfaction of the following conditions:

1. **Condition 1**: The auditory and vocal sensory data of $Agent_{initiator}$ and
   $Agent_{recipient}$ should be in overlapping ranges. That is, [auditory and vocal range
   of $Agent_{initiator}$] - [auditory and vocal range of $Agent_{recipient}$] $\leq$ 20 pixels. The
   20 pixels value for the auditory and vocal range is an arbitrary choice made with
   respect to the total size of the K-space simulation platform.
2. **Condition 2**: $Agent_{initiator}$ and $Agent_{recipient}$ should face each other. That is,
   [Direction($Agent_{recipient}$) - Direction($Agent_{initiator}$)] $\leq$ 60°. This is because while
   standing people can usually cover a visual range of 60°on either side of their
   current orientation without having to turn their body (derived from Kendon's
   F-formation criteria [10]).

Successful application of the POINT and LOOK action signals for agents should
satisfy conditions 1 & 2 (mentioned above) along with the following condition:

3. **Condition 3**: $Location(x_p, y_p)$ should be visible to both $Agent_{initiator}$ and
   $Agent_{recipient}$. That is, $Location(x_p, y_p)$ should be within a 50 pixel radius of
   $Agent_{initiator}$ and $Agent_{recipient}$. Again, the 50 pixels value for the visual range is
   an arbitrary choice made with respect to the total size of the K-space simulation
   platform.

Successful application of the WALK and STAND action signals for agents will
depend on satisfaction of the following condition:

4. **Condition 4**: There should be no obstacle or agent in $Location(x_d, y_d)$. That is,
   if Z is a set that contains the individual location of all the agents and elements in
   K-space (set can be retrieved from E_{cs}) and D is the target $Location(x_d, y_d)$, then
   D $\notin$ Z. This is to avoid collision between two agents.

Successful application of the TURN action signal for agents should satisfy the
following condition:

5. **Condition 6**: At any time t, an agent can practically turn only 60°to the left or
   right from its current position in K-space. That is, $Direction_n$ = current orientation of the agent $\pm$ 60°. This is because while standing people can usually cover
   a visual range of 60°on either side of their current orientation without having to
   turn their body (derived from Kendon's F-formation criteria [10]).

Once an action signal generated for a particular agent is evaluated using the aforementioned conditions, irrespective of the success or failure outcomes, the agent's
state and its environment's state are respectively updated to the new states A_{ns} and
E_{ns}.

5 Conclusion and Future Work

The work presented in this paper is an introduction to the SPAGE system and we
intend to improve it in various ways. Firstly, we have to evaluate the system and

the actions generated. We are particularly interested in looking at whether or not the formations that emerge from the low-level action signals generated by our system resemble meaningful spatial configurations like Kendon's F-formation system [10]. The outcome of the evaluation could be yes or no. It would be a discovery either way: If we get F-formations, we can say that F-formations emerge from low-level action signals, and if we don't get F-formations, we can say that higher-level behavioural mechanisms should be incorporated into the system. By doing the evaluation we also aim to find the relationship between the sequences of actions and the resulting interactions. This can be done by using an appropriate machine learning algorithm on the data generated by running the K-space simulation several times. Finally, we plan to incorporate the system into an appropriate virtual character animation platform and make suitable software and hardware extensions to support real-time human agent interaction. Due to its modular nature, it will be easy to extend the functionalities of the system.

References

1. Breazeal, C.: Emotion and sociable humanoid robots. Int. J. Hum.-Comput. Stud. 59, 119–155 (2003)
2. Cassell, J., Vilhjálmsson, H.: Fully embodied conversational avatars: Making communicative behaviors autonomous. Autonomous Agents and Multi-Agent Systems 2, 4564 (1999)
3. Castelfranchi, C.: Modelling social action for ai agents. Artificial Intelligence 103(1-2), 157–182 (1998)
4. Castellano, G., Leite, I., Pereira, A., Martinho, C., Paiva, A., McOwan, P.: Its all in the game: Towards an affect sensitive and context aware game companion. In: 3rd International Conference on Affective Computing and Intelligent Interaction and Workshops, ACII 2009, p. 18 (2009)
5. Dautenhahn, K.: Socially intelligent robots: dimensions of human robot interaction. Philosophical Transactions of the Royal Society B: Biological Sciences 362(1480), 679–704 (2007)
6. Goffman, E.: Behaviour in Public Places: Notes on the Social Organization of Gatherings. Free Press, New York (1963)
7. Jan, D., Traum, D.R.: Dynamic movement and positioning of embodied agents in multiparty conversations. In: Proceedings of the 6th International Joint Conference on Autonomous Agents and Multiagent Systems, AAMAS 2007, pp. 14–114. ACM, New York (2007)
8. Jennings, R., Campos, J.: Towards a social level characterisation of socially responsible agents. Towards a Social Level Characterisation of Socially Responsible Agents 144(1), 11–25 (1997)
9. Kalenka, S., Jennings, N.R.: Socially responsible decision making by autonomous agents. In: Korta, K., Sosa, E., Arrazola, X. (eds.) Cognition, Agency and Rationality, pp. 135–149. Kluwer (1999)
10. Kendon, A.: Conducting Interaction: Patterns of Behaviour in Focused Encounters. Cambridge University Press, Cambridge (1990)
11. Kendon, A.: Spacing and Orientation in Co-present Interaction. In: Esposito, A., Campbell, N., Vogel, C., Hussain, A., Nijholt, A. (eds.) Second COST 2102. LNCS, vol. 5967, pp. 1–15. Springer, Heidelberg (2010)

12. Koay, K.L., Syrdal, D., Walters, M., Dautenhahn, K.: Living with robots: Investigating the habituation effect in participants preferences during a longitudinal human-robot interaction study. In: The 16th IEEE International Symposium on Robot and Human Interactive Communication, RO-MAN 2007, pp. 564–569 (2007)

13. Lyman, S.M., Scott, M.B.: Territoriality: A neglected sociological dimension. Social Problems 15(2), 236–249 (1967)

14. Mukai, T., Seki, S., Nakazawa, M., Watanuki, K., Miyoshi, H.: Multimodal agent interface based on dynamical dialogue model: MAICO: multimodal agent interface for communication. In: Proceedings of the 12th Annual ACM Symposium on User Interface Software and Technology (UIST 1999), pp. 69–70. ACM, New York (1999)

15. Nakanishi, H.: Freewalk: a social interaction platform for group behaviour in a virtual space. International Journal of Human-Computer Studies 60(4), 421–454 (2004)

16. Ono, T., Kanda, T., Imai, M., Ishiguro, H.: Embodied communications between humans and robots emerging from entrained gestures. In: 2003 IEEE International Symposium on Computational Intelligence in Robotics and Automation. Proceedings, vol. 2, pp. 558–563 (2003)

17. Pedica, C., Vilhjálmsson, H.H.: Social Perception and Steering for Online Avatars. In: Prendinger, H., Lester, J.C., Ishizuka, M. (eds.) IVA 2008. LNCS (LNAI), vol. 5208, pp. 104–116. Springer, Heidelberg (2008)

18. Pedica, C., Vilhjàlmsson, H.H.: Spontaneous Avatar Behavior for Human Territoriality. In: Ruttkay, Z., Kipp, M., Nijholt, A., Vilhjálmsson, H.H. (eds.) IVA 2009. LNCS, vol. 5773, pp. 344–357. Springer, Heidelberg (2009)

19. Pedica, C., Vilhjàlmsson, H.H., Làrusdòttir, M.: Avatars in conversation: the importance of simulating territorial behavior. In: Proceedings of the 10th International Conference on Intelligent Virtual Agents, IVA 2010, pp. 336–342. Springer, Heidelberg (2010)

20. Pelechano, N., Allbeck, J.M., Badler, N.I.: Controlling individual agents in highdensity crowd simulation. In: Proceedings of the 2007 ACM SIGGRAPH/Eurographics symposium on Computer animation (SCA 2007). Eurographics Association, pp. 99–108. Aire-la-Ville, Switzerland (2007)

21. Reynolds, C.W.: Steering behaviors for autonomous characters. In: Game Developers Conference, San Jose, California, pp. 763–782 (1999)

22. Traum, D., Rickel, J.: Embodied agents for multi-party dialogue in immersive virtual worlds, pp. 766–773 (2001)

23. Vilhjalmsson, H., Merchant, C., Samtani, P.: Social Puppets: Towards Modular Social Animation for Agents and Avatars. In: Schuler, D. (ed.) HCII 2007 and OCSC 2007. LNCS, vol. 4564, pp. 192–201. Springer, Heidelberg (2007)

24. Yamaoka, F., Kanda, T., Ishiguro, H., Hagita, N.: A model of proximity control for information-presenting robots. IEEE Transactions on Robotics 26(1), 187–195 (2010)

25. Zlatev, J.: The epigenesis of meaning in human beings and possibly in robots. Minds and Machines 11, 155–195 (2001), doi:10.1023/A:1011218919464

A Multi-agent Recommender System

A. Jorge Morais, Eugénio Oliveira, and Alípio Mário Jorge

Abstract. The large amount of pages in Websites is a problem for users who waste time looking for the information they really want. Knowledge about users' previous visits may provide patterns that allow the customization of the Website. This concept is known as Adaptive Website: a Website that adapts itself for the purpose of improving the user's experience. Some Web Mining algorithms have been proposed for adapting a Website. In this paper, a recommender system using agents with two different algorithms (associative rules and collaborative filtering) is described. Both algorithms are incremental and work with binary data. Results show that this multi-agent approach combining different algorithms is capable of improving user's satisfaction.

1 Introduction

Nowadays, most organizations have a Website, in order to easily deliver information to the general audience. When the size of the Website grows to a significant number of WebPages, the difficulty for users to find what they want also grows. This led organizations to become more concerned with the problem of organizing all the information efficiently, so that it may be easy to find every product or information a user is searching.

Dealing with large datasets is also the motivation for the area of Data Mining and Knowledge Discovery [1], which takes advantage of the large quantity of data

A. Jorge Morais · Eugénio Oliveira
Faculty of Engineering of the University of Porto

Alípio Mário Jorge
Faculty of Science of the University of Porto

Eugénio Oliveira
Laboratory of Artificial Intelligence and Compúter Science (LIACC)

A. Jorge Morais · Alípio Mário Jorge
Laboratory of Artificial Intelligence and Decision Support (LIAAD – INESC Porto L. A.)

A. Jorge Morais
Universidade Aberta (Portuguese Open University)

S. Omatu et al. (Eds.): Distributed Computing and Artificial Intelligence, AISC 151, pp. 281–288.
springerlink.com   © Springer-Verlag Berlin Heidelberg 2012

from previous transactions that are kept in organizations, finding useful information that is not easily visible. Considering the large number of pages in the Web, it became natural to apply this concept to the Web scope, resulting in the new area of Web Mining [2][3].

The problem of Web adaptation is not new. Recommender systems [3] have had several improvements over the last decade. One of the current solutions that are being proposed for this problem is using autonomous agents. Multi-Agent Systems [4] is a research area that has been in great development over the last decade, and has some particular characteristics that fit in this problem. In fact, it was already proposed to use a multi-agent approach, because of its flexibility and its capability of dynamic adaptation to the Web applications needs [5]. Moreover, Multi-Agent Systems are already used for automatic retrieval and update of information in Websites [6]. An architecture proposal of a recommender system using this approach was already proposed in [7].

In this paper we present a multi-agent approach for Web adaptation, where different incremental algorithms based on binary data produce item-based recommendations and make bids to provide the next set of recommendations to the user. Agents are cooperative in the sense they base their bids on client's satisfaction instead of their own revenue and they share the same data. However, their results are not combined in order to provide recommendations. Our goal is to show that this approach is able to achieve better results than the individual algorithms.

The remaining of the paper starts by presenting previous approaches and applications in the area of recommender systems and multi-agent systems, followed by the description of our approach. The results of the tests with four datasets, and some conclusions and future work complete the paper.

2 Previous Approaches and Applications

A global vision on adaptive Web sites based on user interaction analysis is given in [8]. In fact, only less ambitious approaches were proposed, such as reorganization of the Website [9], use of recommendations in the pages [10], automatic categorization of user actions [11], or seek of relevant Web sequence paths using Markov models [12].

Recommendation systems include the combination of clustering with nearest neighbour algorithms [13], Markov chains and clustering [14], association rules [15], and collaborative and content-based filtering [16]. Web dynamics has been controlled, for instance, by efficient incremental discovery of sequential Web usage patterns [17], and on-line discovery of association rules [18]. Data-driven categorization of Website usability may be done by typical usage patterns visualization [11] or with objective metrics [19].

Some platforms, like WebWatcher, use previous users' knowledge to recommend links [20]. AVANTI implements an adaptive presentation based on a model constructed from user actions [21]. WUM infers a tree structure from log records enabling experts to find patterns with predefined characteristics [22]. In [23] it was proposed an integrated tool (HDM) to discover access patterns and association rules from log records in order to automatically modify hypertext organization.

In [5] a multi-agent platform was proposed for personalization of Web-based systems, given the flexibility of this approach and its dynamic adaptation to Website needs. Multi-agent approaches for developing complex systems, like Web adaptation, were defended in [24]. Intelligent agents may also be an important contribution for autonomic computing [25]. Such systems main characteristics are being complex systems with self-administration, self-validation, self-adjustment and self-correction. Web adaptation systems should also have these characteristics, because Website environment dynamics requires either a high degree of system automation or high allocation of human resources. Another important usage of multi-agent systems in this issue is the automatic collection and actualization of information in Websites [6].

In [7] it was presented an implemented web adaption platform [26] that was the basis for this work, with the posterior adaptations to our special needs. An implementation of collaborative filtering using an incremental approach was presented in [27].

3 Multi-agent Approach

The multi-agent system recommender was implemented taking into account that agents should answer rapidly to any request from another agent and prepare in advance for the next request, and tasks that involve a large amount of time (like updating the model) should not interfere with the performance of the system.

Two recommender agents were created. The first one generates single-condition association rules and the second one uses a collaborative filtering algorithm. Since recommendations are meant to be fast in order to keep users interest and taking into account that each new response updates the recommendation models, these incremental approaches must be able to deliver a set of recommendation in a very small amount of time.

Therefore, both algorithms share a matrix $A_{n \times n}$, where n is the number of items (Webpages) and each $a_{ij} \in A$ registers the total number of co-occurrences of items i and j in the same session. The matrix is updated each time a session ends.

The single-condition association rules agent checks all possible rules $i \rightarrow j$, where i and j are items, taking into account two values (k number of sessions):

$$Support_{i \rightarrow j} = \frac{a_{ij}}{k} \qquad Confidence_{i \rightarrow j} = \frac{a_{ij}}{a_{ii}}$$

Therefore, if a set of n recommendation is requested, the n best recommendations according to the confidence that satisfy minimum confidence and support requirements are proposed.

The collaborative filtering agent uses the same matrix to compute similarity, returning the top n most similar items:

$$sim(i, j) = \frac{a_{ij}}{\sqrt{a_{ii}} \sqrt{a_{jj}}}$$

Agent biddings are based on an accumulated score for each given item obtained from previous ratings – the best N are sorted and if the next selected item was in that set it receives a score N-p+1, where p is the ordered position of the item. To this score we add the percentage of the overall score to untie equal biddings:

$$Bid_{agent,item} = Score_{item} + \frac{1}{\#requests * N} * \sum_{i \in Items} Score_i$$

The multi-agent approach was implemented in Java, using the JADE platform [28]. The communication with the browser is implemented using AJAX [29], using XMLHTTPRequest interface, so that the user can consult the Web page without losing interest. The interaction between the user and the recommender system is presented in figure 1, and the architecture of the latter is shown in figure 2. Client agents behaviour is shown in figure 3, while the behaviour of recommender agents is shown in figure 4.

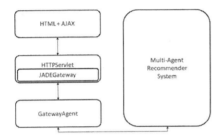

Fig. 1 Interaction with the recommender system.

Fig. 2 Multi-agent recommender system architecture.

```
// created for each client
while not end of session {
   receives request from server
   sends request to recommenders
   waits for responses
   determines winner
   sends results to providers
}
updates knowledge base
destroys itself
```

```
receives request from client
sends bid to client
builds recommendation set
if wins bid
   sends recommendations
   //directly to GatewayAgent
updates knowledge base
sends results to provider
determines winner
sends results to provider
```

Fig. 3 Client agents behaviour

Fig. 4 Recommender agents behaviour

4   Experimental Results

Experiments were undertaken offline and focused on four datasets (obtained from real Web data records). Each time recommendations are made we consider:

    a. No item was followed (discarded, no implicit knowledge – end of session).
    b. The set of recommendations was empty.

c. An item not in the recommendation set was followed.
d. One of the recommendations was followed.

For evaluation of performance there are several metrics. Since the algorithms are incremental, which means we do not have a fix split for train and test sets, the evaluation that fits better to our case is a per-user variant, where predictions are computed and the ranking metrics are calculated for each recommendation, and the average over all recommendations gives the final value [31].

There are two measures that we will use for evaluating recommendation: precision and recall [32]. Precision is the ratio of relevant items selected to number of items selected – it represents the probability that a selected item is relevant. Recall is the ratio of relevant items selected to total number of relevant items available. In our case, precision and recall are given by the following formulas (given N recommendations, and considering b, c and d of the list of possible situations above):

$$Recall = \frac{\#d}{\#b + \#c + \#d} \qquad Precision = \frac{1}{N} Recall = \frac{1}{N} \frac{\#d}{\#b + \#c + \#d}$$

This measure is also applied to the recommendation system, which combines agents' algorithms. When the recommendation set is incomplete or inexistent (because it is the first time the item appears, so there are no correlations yet), the system completes it with the most popular items. In figures 5a-d we can see the distribution of session's sizes and in table 1 the main characteristics of the datasets. In table 2, we present the results, for $N=1$ to 10 number of recommendations, with the evaluation metrics (EM) Recall (R) and Precision (P) for association rules (AR), collaborative filtering (CF), and for the winners (W) of the auctions. The best results between AR and CF are boldface, and when the mutli-agent recommender system is better than both algorithms it is also boldface.

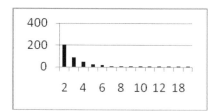

Fig. 5a e-com sessions size distribution

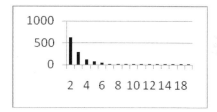

Fig. 5b e-gov sessions size distribution

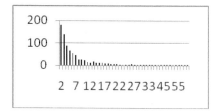

Fig. 5c pe100 sessions size distribution

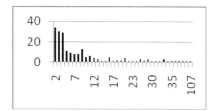

Fig. 5d pe200 sessions size distribution

Table 1 Datasets characteristics.

Dataset	#items	# records	#sessions	#records/#session	#records/#items
e-com	335	1409	413	3.411622	4.20597
e-gov	133	4047	1244	3.253215	30.42857
pe100	100	6070	803	7.559153	60.7
pe200	200	2042	200	10.21	10.21

Table 2 Results for e-com

	#R	EM	1	2	3	4	5	6	7	8	9	10
e-com	AR	R	5.22%	9.24%	11.65%	13.25%	15.46%	16.67%	17.27%	18.27%	18.78%	19.38%
		P	5.22%	4.62%	3.88%	3.31%	3.09%	2.78%	2.47%	2.28%	2.09%	1.94%
	CF	R	4.82%	7.83%	9.84%	10.84%	12.35%	13.25%	14.66%	15.26%	15.96%	16.47%
		P	4.82%	3.92%	3.28%	2.71%	2.47%	2.21%	2.09%	1.91%	1.77%	1.65%
	W	R	6.02%	10.14%	13.55%	15.36%	17.97%	19.38%	20.78%	22.09%	23.19%	24.20%
		P	6.02%	5.07%	4.52%	3.84%	3.59%	3.23%	2.97%	2.76%	2.58%	2.42%
e-gov	AR	R	13.20%	19.51%	24.22%	27.68%	31.18%	33.93%	36.46%	38.67%	40.35%	42.06%
		P	13.20%	9.76%	8.07%	6.92%	6.24%	5.65%	5.21%	4.83%	4.48%	4.21%
	CF	R	11.95%	18.27%	23.40%	27.29%	30.11%	32.36%	34.86%	36.50%	38.21%	39.56%
		P	11.95%	9.13%	7.80%	6.82%	6.02%	5.39%	4.98%	4.56%	4.25%	3.96%
	W	R	12.49%	19.27%	24.15%	27.79%	31.11%	33.61%	36.35%	38.35%	40.10%	41.96%
		P	12.49%	9.63%	8.05%	6.95%	6.22%	5.60%	5.19%	4.79%	4.46%	4.20%
pe100	AR	R	6.15%	10.14%	13.35%	16.46%	19.42%	21.95%	24.23%	26.28%	28.52%	30.43%
		P	6.15%	5.07%	4.45%	4.12%	3.88%	3.66%	3.46%	3.28%	3.17%	3.04%
	CF	R	7.23%	11.98%	15.63%	18.68%	20.96%	23.49%	25.78%	27.64%	29.49%	31.23%
		P	7.23%	5.99%	5.21%	4.67%	4.19%	3.91%	3.68%	3.46%	3.28%	3.12%
	W	R	7.50%	11.96%	15.74%	18.63%	21.28%	24.09%	26.77%	29.07%	30.91%	32.73%
		P	7.50%	5.98%	5.25%	4.66%	4.26%	4.02%	3.82%	3.63%	3.43%	3.27%
pe200	AR	R	3.26%	5.92%	7.71%	9.34%	10.53%	12.21%	13.36%	14.71%	16.02%	17.54%
		P	3.26%	2.96%	2.57%	2.33%	2.11%	2.04%	1.91%	1.84%	1.78%	1.75%
	CF	R	2.93%	5.81%	7.60%	9.28%	10.86%	12.00%	13.46%	14.98%	16.18%	16.83%
		P	2.93%	2.90%	2.53%	2.32%	2.17%	2.00%	1.92%	1.87%	1.80%	1.68%
	W	R	3.58%	6.41%	8.63%	10.37%	12.00%	13.84%	15.47%	16.99%	18.24%	19.49%
		P	3.58%	3.20%	2.88%	2.59%	2.40%	2.31%	2.21%	2.12%	2.03%	1.95%

5 Discussion and Future Work

Looking at the characteristics of the datasets, we can see that in the e-com and e-gov association rules algorithm (AR) has better results, while in pe100 collaborative filtering (CF) is the best. On the other hand, pe200 has 6 recommendation sizes where AR is better and 4 where CF is better. A possible explanation for AR success in the first two datasets is that in both cases, the percentages of 2-items sessions are around 50% (49.39% for e-com and 50.32% for e-gov), while in the others those values are below 25%.

Analysing the results, we can observe that in e-gov and pe200 datasets the multi-agent recommender system (MARS) outperforms the individual algorithms AR and CF. In pe100, the two cases where that does not happen the differences to the recall value for the best of the individual algorithms are 0.02% and 0.05%.

The only dissonant case is the e-gov dataset, where the MARS is better only once, for $N=4$. For $N=1$, we have the highest difference to the best individual algorithm (AR), 0.71%, while all the other differences vary from 0.07% to 0.32%.

As we can observe, the MARS is able to outperform the individual algorithms in most cases. In the other cases, the results have less than 1% recall difference to the best of the individual algorithms.

As future work, we will perform an in-depth analysis of the e-gov dataset to discover what characteristics are beyond its performance behaviour and we will study new improvements to the MARS in order to improve its results.

Acknowledgements. This work was funded by Fundação para a Ciência e a Tecnologia, Ref. SFRH/BD/27571/2006.

References

1. Fayyad, U.M., Piatetsky-Shapiro, G., Smyth, P., Uthurusamy, R. (eds.): Advances in Knowledge Discovery and Data Mining. AAAI/MIT Press, Menlo Park, California (1996)
2. Cooley, R., Mobasher, B., Srivastava, J.: Web mining: Information and patterns discovery on the world wide Web. In: Proceedings of the Ninth IEEE International Conference on Tools with Artificial Intelligence, Newport Beach, California, pp. 558–567 (1997)
3. Jannach, D., Zanker, M., Felfernig, A., Friedrich, G.: Recommender systems: an introduction. Cambridge University Press (2011)
4. Wooldridge, M.: An Introduction to Multi Agent Systems. John Wiley & Sons (2002)
5. Ardissono, L., Goy, A., Petrone, G., Segnan, M.: A multi-agent infrastructure for developing personalized web-based systems. ACM Trans. Inter. Tech. 5(1), 47–69 (2005)
6. Albayrak, S., Wollny, S., Varone, N., Lommatzsch, A., Milosevic, D.: Agent technology for personalized information filtering: the pia-system. In: Proceedings of the 2005 ACM Symposium on Applied Computing, SAC 2005, pp. 54–59. ACM Press, New York (2006)
7. Morais, A.J.: A Multi-Agent Approach for Web Adaptation. In: Demazeau, Y., et al. (eds.) 7th International Conference on PAAMS 2009. AISC, vol. 55, pp. 349–355. Springer, Heidelberg (2009)
8. Perkowitz, M., Etzioni, O.: Towards adaptive web sites: Conceptual framework and case study. Artificial Intelligence 118, 245–275 (2000)
9. Ishikawa, H., Ohta, M., Yokoyama, S., Nakayama, J., Katayama, K.: Web usage mining approaches to page recommendation and restructuring. International Journal of Intelligent Systems in Accounting, Finance & Management 11(3), 137–148 (2002)
10. El-Ramly, M., Stroulia, E.: Analysis of Web-usage behavior for focused Web sites: a case study. Journal of Software Maintenance and Evolution: Research and Practice 16(1-2), 129–150 (2004)
11. Berendt, B.: Using Site Semantics to Analyze, Visualize, and Support Navigation. Data Mining and Knowledge Discovery 6(1), 37–59 (2002)
12. Borges, J.L.: A Data Mining Model to Capture User Web Navigation Patterns. PhD thesis. University College London (2000)
13. Mobasher, B., Dai, H., Luo, T., Nakagawa, M.: Discovery and Evaluation of Aggregate Usage Profiles for Web Personalization. Data Mining and Knowledge Discovery 6(1), 61–82 (2002)

14. Cadez, I., Heckerman, D., Meek, C., Smyth, P., White, S.: Model-Based Clustering and Visualization of Navigation Patterns on a Web Site. Data Mining and Knowledge Discovery 7(4), 399–424 (2003)
15. Jorge, A., Alves, M.A., Grobelnik, M., Mladenic, D., Petrak, J.: Web Site Access Analysis for A National Statistical Agency. In: Mladenic, D., Lavrac, N., Bohanec, M., Moyle, S.: Data Mining And Decision Support: Integration And Collaboration. Kluwer Academic Publishers (2003)
16. Basilico, J., Hofmann, T.: Unifying collaborative and content-based filtering. In: Proceedings of ICML 2004, Twenty-first International Conference on Machine Learning. ACM Press, New York (2004)
17. Masseglia, F., Teisseire, M., Poncelet, P.: HDM: A client/server/engine architecture for real time web usage mining. Knowledge and Information Systems (KAIS) 4, 439–465 (2003)
18. Lin, W., Alvarez, S.A., Ruiz, C.: Efficient Adaptive-Support Association Rule Mining for Recommender Systems. Data Mining and Knowledge Discovery 6, 83–105 (2002)
19. Spiliopoulou, M., Pohle, C.: Data mining for measuring and improving the success of web sites. Journal of Data Mining and Knowledge Discovery, Special Issue on E-commerce 5(1-2), 85–114 (2001)
20. Armstrong, R., Freitag, D., Joachims, T., Mitchell, T.: Web Watcher: A learning apprentice for the world wide web. In: Proceedings of the AAAI Spring Symposium on Information Gathering from Heterogeneous, Distributed Environments, California, pp. 6–12 (1995)
21. Fink, J., Kobsa, A., Nill, A.: User-oriented adaptivity and adaptability in the AVANTI project. In: Designing for the Web: Empirical Studies, Microsoft Usability Group, Redmond (1996)
22. Spiliopoulou, M., Faulstich, L.C.: WUM: a tool for web utilization analysis. In: Proceedings of the International Workshop on the Web and Databases, Valencia, Spain, pp. 184–203 (1998)
23. Masseglia, F., Teisseire, M., Poncelet, P.: Real Time Web Usage Mining: a Heuristic Based Distributed Miner. In: Second International Conference on Web Information Systems Engineering (WISE 2001), vol. 1, p. 0288 (2001)
24. Jennings, N.R.: An agent-based approach for building complex software systems. Communications of the ACM 44(4), 35–41 (2001)
25. Kephart, J.O.: Research challenges of autonomic computing. In: Proceedings of the 27th International Conference on Software Engineering, ICSE 2005, 15–22. ACM Press, New York (2006)
26. Domingues, M.A., Jorge, A.M., Soares, C., Leal, J.P., Machado, P.: A data warehouse for web intelligence. In: Proceedings of the 13th Portuguese Conference on Artificial Intelligence (EPIA 2007), pp. 487–499 (2007)
27. Miranda, C., Jorge, A.M.: Item-Based and User-Based Incremental Collaborative Filtering for Web Recommendations (2009)
28. JADE (Java Agent DEvelopment Framework) Website, http://jade.tilab.com (access date October 14, 2011)
29. Asynchronous Javascript And XML (AJAX), Mozilla Developer Center, https://developer.mozilla.org/en/ajax (access date October 14, 2011)
30. Apache Derby Website, http://db.apache.org/derby/. (access date October 14, 2011)
31. Herlocker, J.L., Konstan, J.A., Terveen, L.G., Riedl, J.T.: Evaluating collaborative filtering recommender systems. ACM Transactions on Information Systems 22(1), 5–53 (2004)
32. Cleverdon, C., Kean, M.: Factors Determining the Performance of Indexing Systems. In: Aslib Cranfield Research Project, Cranfield, England (1968)

SGP: Security by Guaranty Protocol for Ambient Intelligence Based Multi Agent Systems

Nardjes Bouchemal and Ramdane Maamri

Abstract. Ambient intelligence (AmI) is an emerging multidisciplinary area based on ubiquitous computing, ubiquitous communication and intelligent user interface. AmI promises a world where people are surrounded by intelligent interfaces merged in daily life objects. However, the development of AmI systems is complex, and needs robust technologies that respond to AmI requirements such as autonomy, adaptability and context aware. One of the most prevalent alternatives is Multi Agent System (MAS) which can bring most of suitable characteristics. Yet, the success of AmI based MAS will depend on how secure it can be made. This paper presents an approach for AmI based MAS security, where each agent represents an object or a user. The set of agents is called group, and each one has a specific agent called representant. The key idea is based on encryption keys and guaranty. Members of same group share a Common Public Key (CPK), and to communicate with representant they share a Communication Key (Ck). Furthermore, if a new agent wants to communicate with the group, at least one agent belonging to this group must know it. This agent is called Guarantor Agent. The aim is to ensure that sensitive data and messages circulated among agents remain private and only trusted agents can access.

Keywords: Ambient Intelligence, Multi Agent Systems, Security.

1 Introduction

An environment endowed with ambient intelligence is capable to analyze its contexts, adapt itself to the presence of people and objects residing in it, learn from their behavior and recognize express emotion [1,21].

Nardjes Bouchemal · Ramdane Maamri
LIRE Laboratory, Computer Science Department,
Mentouri University of Constantine, Algeria
e-mail: bnarssisse@yahoo.fr, rmaamri@yahoo.fr

S. Omatu et al. (Eds.): Distributed Computing and Artificial Intelligence, AISC 151, pp. 289–296.
springerlink.com   © Springer-Verlag Berlin Heidelberg 2012

For these reasons, the development of AmI environment needs robust technologies that respond to AmI necessities such as autonomy, adaptability and context aware [5, 7].

On the other hand, agents have a set of characteristics such as autonomy, reasoning, reactivity, social abilities, pro-activity, mobility, organization, etc., which allow them to cover several needs for AmI environments.

Agent and multi-agent systems have been successfully applied to several AmI scenarios such as education, culture, entertainment, medicine, robotics, home [2, 15, 16, 18]. An agent can be defined as a computational system situated in an environment and is able to act autonomously in this environment, to achieve its design goals [8].

Several systems already exist we mention ALZheimer multi-agent system (ALZ-MAS) [6, 10] and Telemonitoring Homecare [3].

But, the success of these systems will depend on how secure it can be made, how privacy [19, 20] and other rights of individuals can be protected.

This article addresses security issues and presents an approach to guaranty that only trusted agents/users can communicate with AmI based MAS. In section 2 we present some related work on protection of MAS and we discuss the need of new robust protection approaches. Section 3 presents our proposition based on a Guarantor Agent and cryptography keys. In section 4 we present some implementation issues using Jade for agents and AES algorithm for encryption keys. Finally, a conclusion summarizes the paper and future works.

2   Related Work

We mention in this section some approaches to protect classical MAS, then we discuss the need of more powerful mechanisms which response to AmI requirements.

2.1   Static Mutual Protection

Mana et al. in [13], proposed an approach to protect a society of collaborating agents, by making every agent collaborate with one or more remote agents running in different hosts. These agents act as secure coprocessors for the first one. Likewise, these agents are in turn protected by other agents.

2.2   Dynamic Mutual Protection

Mana et al. in [12] proposed a new strategy called Dynamic Mutual Protection where each agent is able to execute arbitrary code sections on behalf of other agents in the society. Each agent includes a public part, an encrypted private part and a specific virtual machine.

2.3   Trusted Computing Platform

Trusted Computing Platforms take the advantage of the use of a hardware element in order to provide a secure environment. These hardware elements are called

TPM (Trusted Platform Modules) [14]. It is not easy to apply it in ubiquitous environments, where there is a high level of device heterogeneity due to the different physical requirements of these devices.

For the provision of appropriate security in the context of multi agent systems, it is not enough that agent platform provides a set of standard security mechanisms such as sandboxing, encryption and digital signatures [9, 13]. But the most interesting question to ask is: *"in what way does AmI make a difference to security in this setting"?* [11, 20]. AmI environments present new challenges such as energy devices management, devices physical theft or alteration, and identity stealing. Furthermore, users move to a free and arbitrary leaving or joining the group at any time.

Moreover, agents embedded in AmI devices have limited capacities because of energy problems of devices, so we mustn't endow agents with complex cryptography concepts or historic data.

3 Security by Guaranty Protocol

In this section we present our approach to protect AmI based MAS environment. First, we present principal departure points, and then we summarize the protocol.

3.1 Presentation

We consider an AmI based MAS system, where each agent is embedded in a device, and represents an object and/or a user.

1. Agents are fixed into mobile devices and the set of these agents is called **Group.**
2. Each group has a specific and trust agent called **representant.** Its goal is the management of group members: joining and/or leaving, moving elsewhere then rejoining the same group, verifing integrity and trust of an element by checking its identity and/or asking other representants.
3. Group representant has the instantly list of all agents of group and the instantly list of representant of adjacent groups.
4. An agent can belong to two groups at once, or more.
5. Agents belonging to the same group know each other and cooperate to complete various tasks.
6. All members of the group have communication keys (Ck_i for an agent i) to establish secure channels with the representant. Ck_i are frequently changed by the representant.
7. All members of the group share a common public key, called CPK (Common Public Key) and known only by group members, to communicate with privacy. If we have several groups, each one must possed a CPK different from other. CPK is frequently changed by the representant.
8. Each agent has a unique identification (**ID**), the list of all agents' identities of the group (LID), and the identity of group representant.

3.2 The Protocol

We summarize the proposal into four points: Initialization of the group, identification between agents of the same group, arrival of new agent, departure of an agent and departure of a representant.

Initialization of the Group. Initially, the representant establish a secure channel with agents of group, by sending a Communication Key to each agent (Cki to agent i). Then, representant sends Cki(CPK) to all members, where CPK is the Common Public Key of the group.

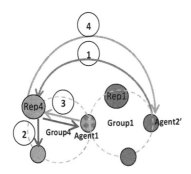

1. Agent2' asks for communication to Rep4.
2. Rep4 ask agents of group if they know it.
3. Agent1 responds positively to Rep4.
4. Rep4 verifies and accord to Agent2' its demand.

Fig. 1 Case I. One agent knows Agent2

Identification between Agents. In figure 1, when we can say that Agent1 from group4 knows Agent2' from group1? Agent1 belongs to both groups, so it has CPK and identity list of both groups. To identify Agent2', Agent1 sends CPK1(ID_Agent1) to Agent2', who decrypts the message and finds that ID_Agent1 belongs to its liste of identity(LID). Agent2' sends CPK1(ID_Agent2') to Agent1 to confirm its identity. After that proceeding, Agent1 and Agent2' communicates using CPK1.

Arrival of a New Agent. In this section we will discuss three cases when an agent arrives into a group and request communication. First, at least one agent belonging to the group knows the new agent. Secondly, no agent knows the new agent, in this case the representant group will ask adjacent groups. Finally, the new agent claims that it was to a given group, so the representant must verify.

At least one agent of the group knows the new agent. In figure 1, Agent1 belongs to group4 and group1, so it knows agents of both systems: (Agent1 ∈ group4 and Agent1 ∈ group1) and (Agent2' ∈ group1). Agent1is called ***Guarantor Agent*** and sends a confirmation message about Agent2' to representant of the group.

Representant verifies that Agent2' is not in its list of malicious agents (where Agent2' has already made unsuccessful demands to group4).

If Agent1 guaranties Agent2', and it's not in malicious agents list, representant integrates Agent2' into the group:

1. It establishes a secure channel by sending a communication key (Ck2') to Agent2'.
2. It sends a new group key Ck2' (new_ CPK), and the list of identities of group members to Agent2'.
3. It sends this new key to the other members of the group: Old_CPK (New_CPK) and the identity of the new agent: New_CPK (Id_Agent2').

An intruder can not access the new key because it does not know the old group key and communication key of the new member.

No agent knows the new agent. If no agent knows the new one, representant will ask the representants of adjacent groups (fig.2). If the answer is positive, representant will proceed to the integration (previous section); otherwise the request of integration is sent to a human administrator.

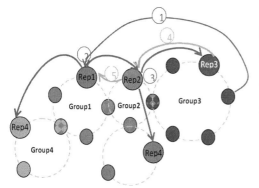

1. Agent2' asks rep1 to communicate.
2. Rep1 asks Representative Agent of adjacent groups.
3. Representative of adjacent groups ask their adjacent groups.
4. Positive response comes from Rep3 to Rep2.
5. Positive response is sent to Rep1 from Rep2.

Fig. 2 Case II: No agent knows Agent2', but adjacent groups do

The new agent claims that it belonged to the group at a given time. In this case the representant will check in its history. If it finds that agent was not in the group, it denies the request. If it finds that it really belonged to this group, it proceeds to a questionnaire. The goal is to ensure that the agent was not altered or his identity was not stolen. We propose some questions such as: What is the creation date of the group? In the date on which you pretend belonging, who was the representant? Give me the identity of at least two members belonging to the group at that date. Representant changes frequently the questionnaire to avoid the case of stealing.

Departure of an Agent. If an agent leaves the group, representant sends a New CPK: Cki (New_CPK) to group members, where Cki is a new communication key of an agent i. Agent who left the group cannot obtain new key (New_CPK), because it is encrypted with communication keys of remaining members.

Departure of the representant. If the representant leaves the group, it must delegate a trusted agent from the group to replace it and gives it all necessary information: communication keys, CPK and the list of members. The new representant will in turn change communication keys, CPK, and sends it to all remaining members. We suggest that representant chooses a member with less tasks and more time to stay within the group.

4 Implementation

In this section we describe some implementation issues using JADE platform (Java Agent DEvelopment framework) [4]. We choose to implement the proposed encryption keys (CPK, Cki) AES (Advanced Encryption Standard) encryption algorithm [22]. We use basic AES encryption/decryption functions in Java, because agents mustn't be endowed with complex algorithms.

Figure 3.a shows how a representant encrypts CPK using Ck1 and sends it to Agent1, who receives encrypted data and decrypts it to obtain CPK (fig 3.b).

```
byte[] Ck1 = //...secret sequence of
             // bytes
byte[] CPK = ...

Cipher c = Cipher.getInstance("AES");
SecretKeySpec K =
     new SecretKeySpec(Ck1, "AES");
     c.init(Cipher.ENCRYPT_MODE, K);
     byte[] encryptedData =
c.doFinal(CPK);
     // now send encryptedData to
//Agent1...
```

```
byte[]Ck1 = //we know secret!
byte[] CPK =//...received from rept

Cipher c = Ci-
pher.getInstance("AES");
     SecretKeySpec K =
     new SecretKeySpec(CK1, "AES");
     c.init(Cipher.DECRYPT_MODE, K);
     byte[]data=
c.doFinal(encryptedData);

     // decrypted data is the CPK
```

Fig. 3a RepAgent sends.   **Fig. 3b** Agent1 receives and decrypts

We summarize some principles of Agents implementation using JADE:

1. We create three containers, everyone represents a group: main container (Groupe1, RepAgent, and Agent1), container1 (Groupe2, RepAgent1, Agent1 and Agent2) and container2 (Groupe3, RepAgent2, Agent2 and Agent3) (Fig4).
2. Each container has a representant agent.
3. Main container and container1 have Agent1 as a common agent.
4. Container1 and Container2 have Agent2 as a common agent.

If Agent3 from container2 wants to communicate with Agent1 from Main container, RepAgent from

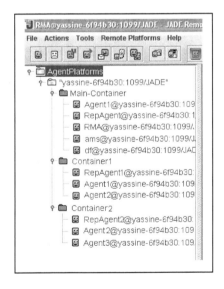

Fig. 4 Implementation using JADE

Main container asks adjacent groups. A positive response comes from RepAgent2 (container2), where Agent2 is the guarantor of Agent3, and Agent1 is the guarantor of Agent2.

5 Conclusion

The goal of this paper is to present an approach to protect AmI based Multi Agents Systems, where a set of agents represents dynamic devices, objects or users. This set is called group and have a specific agent called representant.

If a new agent wants to communicate and share data with agents of this group, at least one agent belonging to the group must know it. This agent is called Guarantor Agent. Otherwise, the representant of the group contacts adjacent groups, and asks them if they know the agent.

We have detailed the protocol and introduced a set of encryption keys: CPK (Common Public Key) known by all members of a group and Cki (Communication Key for Agent i) sent by the representant to an agent i. Finally, we presented an implementation using JADE framework to represent Agents.

Our ongoing work is focused on how privacy property can be verified in a context of AmI based MAS. Moreover, we would like to simulate our approach with an AmI based Multi Agents System platform.

References

[1] Augusto, J.C.: Ambient Intelligence: Basic Concepts and Applications. In: Computer and Information Science: Software and Data Technologies, vol. 10, Part 1. Springer, Heidelberg (2008)

[2] Acampora, G., Loia, V.: A proposal of ubiquitous fuzzy computing for ambient intelligence. Inf. Sci. 178(3), 631–646 (2008)

[3] Alonso, R.S., Saavedra, A., Tapia, D.I., de Paz, J.F., Corchado, J.M.: Heterogeneous wireless sensor networks in a telemonitoring system for homecare. In: Proceedings of IWANN, vol. (2), pp. 663–670 (2009)

[4] Bellifemine, F., Caire, G., Trucco, T., Rimassa, G.: Jade Programmer's Guide, version 3.2, Livre, édition Juillet (2004)

[5] Casert, R.: Workshop ambient intelligence: in the service of man? In: Societal Aspects of Ambient Intelligence in the Field of Health and Surveillance. Rep. RP-DIV-167, Rathenau Institute, The Hague (December 2004)

[6] Corchado, J.M., Bajo, J., De Paz, Y., Tapia, D.I.: Intelligent environment for monitoring Alzheimer patients, agent technology for health care. Decision Support Systems, 382–396 (2008)

[7] Duman, H., Hagras, H., Callaghan, V.: Intelligent association exploration and exploitation of fuzzy agents in ambient intelligent environments. J. Uncertain Syst. 2(2), 133–143 (2008)

[8] Ferber, J.: Les Systèmes Multi-Agents. Informatique Intelligence Articielle. Inter. Edition (1995)

[9] Gong, L., Mueller, M., Prafullchandra, H., Schemers, R.: Going Beyond the Sandbox: An Overview of the New Security Architecture in the Java Development Kit 1.2. In: Proceedings of the USENIX Symposium on Internet Technologies and Systems, Monterey, California (December 1997)

[10] IST Advisory Group, Scenarios for Ambient Intelligence in 2010. European Commission (2010)

[11] Ko, H., Ramos, C.: A Study on Security Framework for Ambient Intelligence Environment (ISyRAmISF: ISyRAmI Security Framework. In: Fifth International Conference on Wireless and Mobile Communications, IEEE (2009)

[12] Maña, A., Muñoz, A., Serrano, D.: Towards Secure Agent Computing for Ubiquitous Computing and Ambient Intelligence. In: Indulska, J., Ma, J., Yang, L.T., Ungerer, T., Cao, J. (eds.) UIC 2007. LNCS, vol. 4611, pp. 1201–1212. Springer, Heidelberg (2007)

[13] Maña, A., Muñoz, A.: Mutual Protection for Multiagent Systems. In: Proceedings of the Third International 3rd International Workshop on Safety and Security in Multiagent Systems, SASEMAS 2006 (2006)

[14] Maña, A., López, J., Ortega, J., Pimentel, E., Troya, J.M.: A Framework for Secure Execution of Software. International Journal of Information Security 3(2) (2004)

[15] Miguel, A., Perez, L., Susperregi, L., Maurtua, I., Ibarguren, A.: Software Agents for Ambient Intelligence based Manufacturing. In: Proceedings of the IEEE Workshop on Distributed Intelligent Systems: Collective Intelligence and Its Applications, DIS 2006 © (2006), doi:0-7695-2589-X/06 $20.00

[16] Pecora, F., Cesta, A.: Dcop for smart homes: a case study. Comput Intell. 23(4), 395–419 (2007)

[17] Roovers, R., Aarts, E.: Embedded system design issues in ambient intelligence. In: Basten, T., Geilen, M., Groot, H.D. (eds.) Ambient Intelligence: Impact on Embedded System Design, pp. 11–29. Kluwer, Norwell

[18] Riva, G.: Ambient intelligence: The evolution of technology. In: Communication and Cognition Towards the Future of Human_Computer Interaction, IOS Press (2005); Sellami, Z.: Intelligence ambiante: étude d'une approche par auto-organisation coopérative et mise en oeuvre d'une plate-forme de conception générique. Master thesis. Université Paul Sabatier, Toulouse (2008)

[19] Stajano, F.: Security Issues in Ubiquitous Computing, Handbook of Ambient Intelligence and Smart Environments. Security Issues in Ubiquitous Computing 281.Springer Science+Business Media, LLC (2010), doi:10.1007/978-0-387-93808-0_11

[20] Weber, W., Rabaey, J.M., Aarts, E.: Ambient Intelligence. Springer, New York (2005)

[21] http://awr.free.fr/javadoc/fr/free/awr/crypto/kls/algo/AES.html

Discrete Dynamical System Model of Distributed Computing with Embedded Partial Order

Susmit Bagchi

Abstract. The traditional models of distributed computing systems employ the mathematical formalisms of discrete event dynamical systems along with Petri Nets. However, such models are complex to analyze and computationally expensive. Interestingly, the evolving distributed computing systems closely resemble the discrete dynamical systems. This paper constructs and analyzes the model of distributed computing systems by applying mathematical formalisms of the discrete dynamical systems. The proposed model embeds the partial ordering of states under happened-before relation in the set of globally observable states. The stability of the proposed model is analyzed using the first order linear approximation.

Keywords: distributed computing, partial order, discrete dynamical systems, Petri Nets.

1 Introduction

The present day distributed computing systems are large scale systems, which are difficult to construct, analyze and, control with reliability. The processes in a distributed computing system compute with incomplete view of the entire system and the processes coordinate by using synchronous or asynchronous messages through FIFO or non-FIFO channels [1, 5, 9, 10]. Often, the formal methods are employed to construct and analyze distributed computing systems in order to understand controllability and observability of the systems. The mathematical models are generally used to describe and analyze the dynamic behavior of computing systems [2]. Interestingly, the distributed computing systems evolve over time and closely resemble the discrete dynamical systems (DDS). In general, the dynamic systems are modeled using finite state automata having partially observable events coupled with state transitions [4, 15]. However, the observation of global predicates in a distributed computing system is NP-complete due to the

Susmit Bagchi
Department of Informatics, Gyeongsang National University, Jinju, South Korea
e-mail: susmitbagchi@yahoo.co.uk

S. Omatu et al. (Eds.): Distributed Computing and Artificial Intelligence, AISC 151, pp. 297–304.
springerlink.com   © Springer-Verlag Berlin Heidelberg 2012

combinatorial explosion of states of the set of processes. The DDS model is a formal mechanism to capture, observe and predict the behaviour of the trajectory of an evolvable dynamic system [16]. The advantages of constructing DDS model of distributed computing systems are as followings:

- Formal verification of concurrent distributed computing systems.
- Analyzing the structural and functional stability of the system architecture.
- Understanding the trajectory of computations of the evolvable distributed systems.
- Constructing the linear and non-linear models of higher order computing systems.
- Timed DDS model implements logical-clock based distributed computing.

This paper constructs and analyzes the timed DDS model of a large scale distributed system. The stability of the system is verified through analysis. It is explained that, the timed DDS model embeds reflexive partial order of the states of processes under Lamport's happened-before relation of the traditional model. The rest of the paper is organized as follows. Section 2 explains the DDS model construction of a distributed computing system. The analytical comparisons between partial ordering of events and states under the happened-before relation in traditional partial ordered set (poset) model and DDS model are explained in section 3. The evaluations of system stability and fault-tolerance of the model are described in section 4. Section 5 depicts the related work. Section 6 concludes the paper.

2 Distributed Computing as DDS

The dynamical behaviour of DDS closely resembles the evolvable large scale distributed computing systems. A distributed computing system may have infinitely many processes but each of the state transitions can only be finitely many [1]. In the logical clock based distributed systems, detection of a global predicate is NP-complete and, there exists no bound on the number of processes for all valid state transitions. According to finite arrival model, no communication protocol can be designed using a defined upper bound on the number of processes in a distributed computing system [6, 14]. The space-time diagram of a distributed computing is illustrated in Fig. 1.

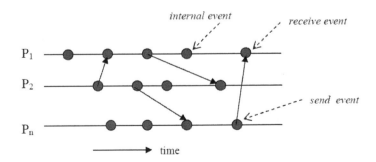

Fig. 1 Standard event-based model of distributed computing.

Let, a distributed computing system D is comprised of a set of nodes $N = \{n_a : a = 1, 2, 3, \ldots m\}$. Any failed node $n_f \in N$ generates a series of faulty events represented by $\{e_\phi\}$ and the unique halting or terminal state of a node in D is $\{s_\phi\}$. The distributed system D is synchronized by a monotonically increasing integer clock C, $C \in I^+$. The set of events generated by distributed processes is $\mathcal{C}$ given by, $\mathcal{C} = (\cup_{a = 1,2,..m} E_a)\cup\{e_\phi\}$, where E_a is a series of events generated by node $n_a \in N$. Again, $\forall n_a \in N$, the internal-states of the node is given by $S_a = \{s_a[x] : x \in C\}\cup\{s_\phi\}$. Thus, the set of states in D is, $S_D = \cup_{a = 1,2,..m} S_a$. Accordingly, the nodes in D periodically converge to a set of the globally observable states given by, $S_\omega = \cap_{a = 1,2,..m} S_a$, $S_\omega \neq \phi$ and, $|S_\omega| \geq 1$. The internal computational outputs at a node $n_a \in N$ is a relation R_a, $R_a \subset (\mathcal{C} \times C)$ such that, $\{\phi\} \notin R_a$. The intermediate local state transitions at any node in D is a function τ_a, which can be defined as, $\forall n_a \in N$, $\tau_a : (S_D \times R_a) \rightarrow S_D$. The transition to a stable state is controlled through a global transition function T defined on S_D at every node in N as, $T : S_D \rightarrow S_D$. Hence, a distributed system can be formally represented as $D = \langle N, S_D, \mathcal{C}, T \rangle$. The representation of D is illustrated in Fig. 2, where n_f is the halting node in D.

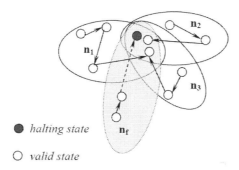

halting state n_f

valid state

Fig. 2 Concept of a distributed computing system as DDS.

Let, $f = (T \circ \tau_a)$ is a non-commutative composition at the node n_a. Thus, $\exists s_a[x]$, $s_a[x+1] \in S_D$ and $\exists r_a = (e_j \in \mathcal{C}, x) \in R_a$ such that, $s_a[x+1] = f(s_a[x], r_a)$. Initially, at $x = 0$, $s_a[x] = s_a[0]$ be the initial stable internal-state of a node $n_a \in N$. Let, $\forall s_a[x] \in S_\omega$, $x = 0, 1, 2, \ldots$, the globally observable stable states are, $s_a[1] = T(s_a[0])$ and $s_a[x] = T^x(s_a[0])$. The globally observable transitions can be functionally defined as,

$$\forall n_a \in N, T^x(s_a[0]) = \begin{cases} s_a[x] & \text{if } s_a[x] \in S_\omega \\ s_a[x-1] & \text{if } s_a[x] \notin S_\omega \\ s_\phi & \text{if } T^{x-1}(s_a[0]) = s_\phi \end{cases} \tag{1}$$

Hence, the valid execution trajectory of a node n_a in D at logical clock x is represented as, $\wp_a[x] = \langle s_a[k] : k = 0, 1, 2, \ldots x; s_a[k] \in S_\omega - \{s_\phi\}\rangle$. It is evident from the definition of $T^x(s_a[0])$ that, it will restrict the final transition states of the nodes in globally observable domain in the system D irrespective of any intermediate transitions executed by individual nodes in the system D due to internal (local) computations. The DDS model of distributed computing systems allows divergence in trajectories of the nodes due to internal local-computations. However, the DDS model implements the deterministic convergence of the distributed computation to globally observable stable states.

3 Comparison between $(\mathcal{C}, \rightarrow)$ and $\langle N, S_D, \mathcal{C}, T \rangle$

Traditionally, the distributed computing systems are modeled using Lamport's happened-before relation $(\rightarrow)$ in combination with partially ordered set (poset) structure [5, 18]. In this section, the embedding of $\rightarrow$ and configuring consistent cuts in DDS model are explained along with comparison to standard model.

3.1 Embedding $\rightarrow$ in DDS Model

According to standard poset model of distributed computing, $(\mathcal{C}, \rightarrow)$ is a poset of distributed events $\mathcal{C}$ in D such that, $\forall e_j, e_k \in \mathcal{C}$, either $e_j \rightarrow e_k$ or $e_k \rightarrow e_j$. Additionally, $(\mathcal{C}, \rightarrow)$ maintains that, $\neg((e_j \rightarrow e_k) \vee (e_k \rightarrow e_j)) \Rightarrow e_j \| e_k$. Hence, the $(\mathcal{C}, \rightarrow)$ is an event-based poset model of distributed computing system D. On the other hand, the DDS model of distributed computing system D employs the dynamic state transition structure, where state transitions are induced by the events (i.e. internal events, message send events or message receive events). However, the Lamport's happened-before relation is maintained in $\langle N, S_D, \mathcal{C}, T \rangle$. One of the main distinctions of DDS model is that, non-commutative composition f at n_a allows repetitive transitions to a stable state at monotonically increasing different clock values in special cases during the convergence cycles, while maintaining the relation $\rightarrow$. Such repetitive states are considered as different timed-states in DDS model by using the logical clock (C) values. Thus, $\forall s_a[x], s_a[y], s_b[y] \in S_D$, and $y \in C$,

$$\left.\begin{array}{l} (x < y) \Rightarrow (s_a[x] \rightarrow s_a[y]) \oplus (s_a[x] \rightarrow s_b[y]) \\ \\ \neg((s_a[x] \rightarrow s_b[y]) \vee (s_b[y] \rightarrow s_a[x])) \Rightarrow s_a[x] \| s_b[y] \end{array}\right\} \quad (2)$$

and,

In Eq. 2, the symbol $\oplus$ represents XOR operation. Hence, the Lamport's happened-before relation is preserved in the DDS model of distributed computing systems.

3.2 Irreflexive $(\mathcal{C}, \rightarrow)$ and Reflexive $(S_\omega, \rightarrow)$

In traditional model, the poset of distributed computing systems $(\mathcal{C}, \rightarrow)$ has following property, $\forall e_j \in \mathcal{C} : \neg(e_j \rightarrow e_j)$. Thus, the happened-before relation induces an irreflexive partial order. However, in case of DDS model the non-commutative composition f induces a reflexive partial order at a node $n_a \in N$. Suppose, a distinct state of a node n_a is represented by $\chi \in S_\omega$ so that, $\chi = s_a[v]$ for some value $v \in C$. The following can be derived from Eq. 1 for a computational output $r_a = (e_j, x) \in R_a$,

$$\exists s_a[v] \in S_\omega : ((x > v) \wedge (T^x(s_a[0]) \notin S_\omega)) \Rightarrow (f(s_a[x], r_a) = \chi) \quad (3)$$

Thus, for all $x > v$ in the distributed computing system D, $s_a[x] = \chi$. However, following Eq. 2, $(v < x) \Rightarrow (s_a[v] \rightarrow s_a[x])$. Hence, the happened-before relation in DDS model is reflexive as, $(\chi,\chi) \in \rightarrow$. This proves, $(S_\omega, \rightarrow)$ is a reflexive poset under non-commutative composition f in D.

3.3 Consistent Cuts in DDS Model

In general, in the event-based poset model of a distributed computing system a consistent cut is $H_e \in L((\mathcal{C}, \rightarrow))$, where $L((\mathcal{C}, \rightarrow))$ is a lattice such that, $\forall e_j, e_k \in H_e$, $e_j \| e_k$ and $|H_e| = |N|$. However, a consistent cut in DDS model is a state-based approach where a consistent cut $H_D \subseteq S_\omega$ and $|H_D| = |N|$. In other words, $\exists s_a[x] \in H_D$ such that, if $s_a[x] \notin S_\omega$ or $|H_D| < |N|$ then H_D is the inconsistent cut in D.

4   System Stability and Fault-Tolerance

4.1   Analyzing System Stability

A distributed system is stable if there exists a global converging state or a set of global states in the system under consideration. A monotonic divergence of all the nodes of a distributed computing system from the stable state(s) may lead to unreliability and instability of the system. Let, a globally observable equilibrium of a distributed computing system is $s_a[\beta] \in S_\omega$ for the node $n_a \in N$ of D, $\beta \in C$. The Taylor expansion of $T(s_a[x])$ considering first-order derivative at $s_a[\beta]$ is given by,

$$T(s_a[x]) = T(s_a[\beta]) + (s_a[x] - s_a[\beta]).\Delta_x T(s_a[\beta]) \tag{4}$$

Where, $\Delta_x T(s_a[\beta])$ is the discrete derivative of T in discrete time domain of logical clock at β and $(s_a[x] - s_a[\beta])$ represents shortest linear distance between the two states in $\wp_a[x]$ of the node n_a. It is clear that, $T(s_a[x])$ represents a discrete dynamical system in simple form. However, due to the stable equilibrium of D at logical clock β, the drift around β should be converging i.e., $\Delta_x T(s_a[\beta]) < 1$. Thus, around stability point the predicate P will hold true in D, where P is represented as, $P \Rightarrow ((s_a[x+1] = s_a[\beta]) \wedge (T(s_a[\beta]) = s_a[\beta]))$. As, $T(s_a[x]) = s_a[x+1]$ hence, at $s_a[\beta]$, $\Delta_x T(s_a[\beta]) = 0$ for the distributed computing system D. This indicates that Taylor expansion (Eq. 4) can be reduced to, $T(s_a[x]) = T(s_a[\beta])$. Deriving further following Eq. 1 and P, it can be said that, $T^x(s_a[0]) = s_a[\beta]$, $x \in C$. As $s_a[\beta] \in S_\omega$ thus, $s_a[\beta]$ is locally observable by all nodes in D. On the other hand, $\exists n_b \in N$ such that, stability point $s_b[\beta] \in S_\omega$, $s_b[\beta] \neq s_a[\beta]$ then, nodes n_a and n_b will be stable in D following Eq. 1 and Eq. 4, respectively. Hence, the distributed computing system D will be highly stable at $s_a[\beta]$, $\forall n_a \in N$, which is a globally observable stability in D. This indicates, the discrete dynamical behaviour of a distributed computing system would eventually lead to a globally observable and stable consistent state irrespective of any local intermediate state-transitions at the nodes. The convergence property of a discrete dynamical system makes a distributed computing system model globally stable.

4.2 Analyzing Fault Tolerance

In general, there are two broad classifications of failures of a distributed computing system such as, Byzantine failure and non-Byzantine failure. In this section, these two are analyzed by considering individual failure models.

4.2.1 Non-Byzantine Failure Stabilization

Let, $a = h$ is a halting node at $x \in C$ during any arbitrary computation in D. The non-Byzantine may either be fail-stop or may generate a series of internal faulty events $\{e_\phi\}$ without any inter-node communications. As, $\{\phi\} \notin R_h$, hence $R_h(e_\phi, y)$ $\neq \phi$ for the node n_h. So, at $y = x+1$, $y \in C$, $s_h[y] \in \tau_h$. Thus, from Eq. 1, it can be said that at n_h, either $T^y(s_h[0]) = s_h[x]$ or $T^y(s_h[0]) = s_\phi$. If $T^y(s_h[0]) = s_h[x]$ is valid at the halting node then, it is a fail-stop state of the corresponding failed node. Otherwise, at the halting node n_h, $T^{y+1}(s_h[0]) = T^y(s_h[0]) = s_\phi$, where $x < y$. Hence, combining both the conditions, $s_\phi = s_h[y] = s_h[x]$ for $n_h \in N$. Now, let D' is the distributed system transformed from D such that, $S_{D'} = (S_D - S_h) \cup \{s_\phi\}$, where $|S_{D'}|$ $= |S_D|$. Again, the transformed event set of D' is given by, $\mathcal{C}' = (\mathcal{C} - E_h) \cup \{e_\phi\}$, where $|\mathcal{C}'| = |\mathcal{C}|$. Let, $\exists n_a \in N$, $a \neq h$ such that, $\tau_a(S_{D'}, R_a) = s_a[x]$ at $x \in C$ and $s_a[y]$ $= T(s_a[x])$. According to Eq. 1, if $s_a[y] \in S_\omega$ then, n_a is stable even after non-Byzantine failure of n_h if $s_a[y] \neq s_\phi$. Otherwise, if $s_a[y] = s_\phi$ then n_a halts at logical clock y because, $T^{y+1}(s_a[0]) = s_\phi$. On the other hand, if $s_a[y] \notin S_\omega$ then, $s_a[y] = s_a[x]$. Thus, the node n_a will be blocked from further computation due to halting node n_h if they have mutual dependencies in carrying out distributed computation. Hence, $\forall n_a \in N$, $s_a[y] = T(s_a[x])$ will enforce deterministic behaviour in the distributed computing system D, where $x < y$.

4.2.2 Byzantine Failure Stabilization

Let, $n_h \in N$ is a node under Byzantine failure in D at $y \in C$, $y \geq x+1$ and, it generates a series of $\{e_\phi\}$ at y. Thus, the trajectory of n_h is $\wp_h[x]$. As $\{\phi\} \notin R_h$, hence, $s_h[y] \in \tau_h$. According to Eq. 1, $T^y(s_h[0]) = s_h[x]$. Thus, from the definition of trajectory one can derive, $s_h[y] \notin \wp_h[y]$. So, $T^y(s_h[0]) = s_h[x]$ is a blocking state for the corresponding halting node in D if $s_h[y] \notin \wp_h[y]$. Hence, the halting node, n_h, will be effectively stabilized in trajectory $\wp_h[x]$ through the functional composition of non-commutative discrete transformations given by $(T^o \tau_h)$ in the distributed computing system D.

5 Related Work

The characteristics of distributed computing systems resemble the dynamic systems. The dynamic model of an asynchronous message-oriented distributed system is proposed in [1]. In general, the process model having finite arrival rate is employed to design a distributed computing architecture [6, 14]. In the process model of distributed systems, the processes communicate using persistent and reliable broadcast semantics [8]. However, it is often not desired to design a distributed computing architecture based on such ideal or near to ideal operational environments in practice. The distributed consensus mechanisms based upon the failure-detectors are explained in [11]. In static systems, the weak failure-detectors can be successfully employed [11]. However, the weak failure-detector model may not stabilize a dynamic system. In static systems, the leader election in a distributed system model assumes synchrony between process speed and message communication delays [7]. Hence, such synchronous model would fail in a dynamic asynchronous distributed computing environment. In another direction, the concurrent computing systems are modeled based on the DDS formalism [2]. However, the DDS modeling of a large scale distributed computing system is not considered in the construction of the concurrent systems. Researchers have proposed to design load balancing mechanisms in distributed systems considering the dynamical discrete-time domain in the presence of time delays [3]. Apart from dynamic load-balancing in discrete-time, the mechanism does not model the overall characteristics of a distributed system as a single entity. The applications of DEDS as a tool for modeling and controlling a distributed system are well researched [13, 15]. For example, the analysis of stability of any DEDS is explained in [4]. The DEDS model is applied to implement distributed diagnosis in a distributed computing system [12]. The DEDS model is often used in association with timed Petri Nets. The one of the main difficulties of DEDS model with timed Petri Nets is the indeterminism of stability of the systems [17]. In addition to the existing models, the DDS formalism [16] can be very effective and reliable mechanism to design and analyze the characteristics as well as execution trajectories of any distributed computing system. The DDS model successfully captures the uncertainties of large scale distributed computing systems and can serve as a concrete model to design fault-tolerant distributed systems architectures.

6 Conclusion

The dynamics of large scale distributed computing systems can be modeled by using the formalisms of discrete dynamical systems (DDS). The DDS model of a distributed computing system facilitates the analysis and evaluation of overall system stability and observability. The global state transition function defined in the DDS model of a distributed computing system implements controllability of the dynamics of computations. The Lamport's logical clock based happened-before relation can be integrated within the DDS model of distributed computing systems. However, the happened-before relation and globally observable stable states of DDS model induce a reflexive partial order.

References

1. Mostefaoui, A., Raynal, M., Travers, C., Patterson, S., Agrawal, D., Abbadi, E.A.: From Static Distributed Systems to Dynamic Systems. In: Proceedings of the 24th IEEE Symposium on Reliable Distributed Systems (SRDS). IEEE CS Press (2005)

2. Pelayo, L.F., Valverde, C.J., Pelayo, L.M., Cuartero, F.: Discrete Dynamical Systems for Encoding Concurrent Computing Systems. In: Proceedings of the 9th IEEE International Conference on Cognitive Informatics (ICCI). IEEE CS Press (2010)

3. Dhakal, S., Paskaleva, B.S., Hayat, M.M., Schamiloglu, E., Abdallah, C.T.: Dynamical Discrete-Time Load Balancing in Distributed Systems in the Presence of Time Delays. In: Proceedings of the 42nd IEEE International Conference on Decision and Control, USA (2003)

4. Ozveren, M.C., Willsky, A.S., Antsaklis, P.J.: Stability and Stabilizability of Discrete Event Dynamic Systems. Journal of the Association of Computing Machinery (ACM) 38(3) (1991)

5. Lamport, L.: Time, Clocks and the Ordering of Events in a Distributed System. Comm. of the ACM 21(7) (1978)

6. Aguilera, M.K.: A Pleasant Stroll through the Land of Infinitely Many Creatures. ACM SIGACT News, Distributed Computing Column 35(2) (2004)

7. Aguilera, M.K., Delporte-Gallet, C., Fauconnier, H., Toueg, S.: Communication-efficient Leader Election and Consensus with Limited Link Synchrony. In: Proceedings of the 23rd ACM Symposium on Principles of Distributed Computing (PODC). ACM Press (2004)

8. Friedman, R., Raynal, M., Travers, C.: Two Abstractions for Implementing Atomic Objects in Dynamic Systems. In: Proceedings of the 24th ACM Symposium on Principles of Distributed Computing (PODC). ACM Press (2005)

9. Chandy, K.M., Misra, J.: How Processes Learn, Distributed Computing, vol. 1(1). Springer (1986)

10. Flocchini, P., Mans, B., Santoro, N.: Sense of Direction in Distributed Computing. In: Kutten, S. (ed.) DISC 1998. LNCS, vol. 1499, pp. 1–15. Springer, Heidelberg (1998)

11. Raynal, M.: A Short Introduction to Failure Detectors for Asynchronous Distributed Systems. ACM SIGACT News, Distributed Computing Column 36(1) (2005)

12. Fabre, E., Benveniste, A., Jard, C.: Distributed Diagnosis for Large Discrete Event Dynamic Systems. In: Proceedings of the IFAC World Congress (2002)

13. Capkovic, F.: Modelling and Control of Discrete Event Dynamic Systems, BRICS Technical Report, RS-00-26 (2000) ISSN: 0909-0878

14. Merritt, M., Taubenfeld, G.: Computing with Infinitely Many Processes. In: Herlihy, M.P. (ed.) DISC 2000. LNCS, vol. 1914, pp. 164–178. Springer, Heidelberg (2000)

15. Sobh, M.T.: Discrete Event Dynamic Systems: An Overview, Technical Report (CIS), MS-CIS-MS-CIS-91-39. University of Pennsylvania (1991)

16. Galor, O.: Discrete Dynamical Systems, JEL Classification numbers: C62, O40. Brown University (2005)

17. Konigsberg, R.Z.: The Stability Problem for Discrete Event Dynamical Systems Modeled with timed Petri Nets Using a Lyapunov-Max-Plus Algebra Approach. International Mathematical Forum 6(11) (2011)

18. Chandy, K.M., Lamport, L.: Distributed Snapshots: Determining Global States of Distributed Systems. ACM Transactions on Computer Systems 3(1), 63–75 (1985)

Evaluation of High Performance Clusters
in Private Cloud Computing Environments

J. Gómez, E. Villar, G. Molero, and A. Cama

Abstract. In recent years, an increasing number of organizations — including universities, research centers, and businesses — have begun to use Cloud Computing technology as an essential and promising tool for optimizing existing computing resources and increase their efficiency. Among the most important advantages it offers is a scalable and low-cost computing system which is adapted to the needs of the client, who only pays for the resources used. On the other hand, the use of High Performance Clusters for solving complex problems is increasing. If we unify both technologies, we will be able to produce flexible, scalable, and low-cost High Performance Clusters. This paper analyzes the operation and performance of a High Performance Cluster on a Cloud Infrastructure.

Keywords: Cloud Computing, Cluster, Performance Evaluation, MPI.

1 Introduction: Cloud and High Performance Computing

In recent years, Cloud Computing has become the technological benchmark for the optimization of organizations use of computing resources, as it allows a significant increase in their efficiency in comparison to previous solutions [1].

The general idea on which Cloud Computing is based is providing users, transparently, with the resources of a computer system through a network. Depending on the location of the system resources, different topologies are defined – including Public Clouds, Private Clouds and Hybrid Clouds [2].

Cloud Computing refers to the demand for services rendered over the Internet as well as the hardware and software that enable the delivery of these services. This view is encompassed by the concept of Anything as a Service (XaaS).

J. Gómez · E. Villar · G. Molero
Department of Computer Science and Languages, University of Almería, Spain
e-mail: jgomez@ual.es, evf989@alboran.ual.es,
    meg501@alboran.ual.es

A. Cama
Department of Architecture and Computer Technology, University of Granada, Spain
e-mail: acamap@correo.ugr.es

S. Omatu et al. (Eds.): Distributed Computing and Artificial Intelligence, AISC 151, pp. 305–312.
springerlink.com   © Springer-Verlag Berlin Heidelberg 2012

Assuming that virtually any computing resource can be offered as a service through the application of Cloud Computing, today and in practice, the current trends in providing services are classified as: Infrastructure as a Service (IaaS), Platform as a Service (PaaS) and Software as a Service (SaaS).

The Infrastructure as a Service (IaaS) model provides users the necessary infrastructure to use the desired software and execution environment. This infrastructure, the provision of which is usually the role of virtualization techniques, mainly consists of hosting applications and providing the necessary resources in terms of computing, connection and storage capacity as a service; it is therefore sometimes called HaaS (Hardware as a Service). We can mention, among the most notable IaaS solutions, GoGrid [3], Amazon Web Services EC2 [4] and S3 [5], Joyent [6], Enomaly´s Elastic Computing Platform [7], Chef [8], Puppet [9], OpenNebula [10], Eucalyptus [11], Ubuntu Enterprise Cloud [12], and OpenStack [13].

The service rendered is particularly affected by the infrastructure model used. If we use a private Cloud, control over services is total; this includes control of access, infrastructure and customization. On the other hand, if the model used is a public Cloud, the service is owned and controlled by its supplier, thus reducing the possibilities for customization and personalization of the service.

IaaS is mainly aimed at avoiding the underutilization of resources, and the advantages for users are many: no investment is necessary in the infrastructure, there are no operation and maintenance costs – only the cost of actual consumption – and, moreover, there is no usually a minimum-term contract to use the service.

Cloud Computing introduces new challenges and opportunities, but creates problems of management, resource-planning [14], legality, security, privacy, and confidentiality [15]. In the case of public Clouds, systems, applications and personal data are stored in data centers belonging to third parties, the resources of which are shared by multiple users.

On the other hand, High Performance Computing (HPC) uses supercomputers and computer clusters to solve advanced computing problems. At present, the use of clusters is the most popular option for the deployment of HPC infrastructures. These systems are usually programmed using MPI [16], OpenMP [17], and, additionally, PGAS languages (Partitioned Global Address Space) such as Unified Parallel C (UPC) [18].

A cluster is a grouping of two or more interconnected computers, referred to as nodes, based on any operating system, and which usually operate jointly with the aim of solving a problem. The main objective of their implementation is usually to expand the functionality of a server or to improve its performance and/or availability. The disadvantages of clusters are high cost and limited flexibility.

If we take advantage of the flexibility provided by Cloud Computing to deploy High Performance Clusters, we could create flexible clusters that are adaptable to our needs at a low cost. This type of hybrid computing environment – called Elastic Cluster – [19], allows us to obtain architectures that support functionalities not otherwise available, in a self-adjusting (in size and resources), scalable and fault-tolerant manner. However, not all are advantages; the creation of a Cloud infrastructure inherently involves a loss of performance, which must be assessed in order to decide whether or not its use is convenient.

An abstraction layer is created from the physical infrastructure in order to deploy the virtual machines run in the Cloud (see Figure 1). The virtual machines are managed by a virtualization solution such as XEN [20] or KVM [21]. Finally, the different virtual machines make up the High Performance Cluster.

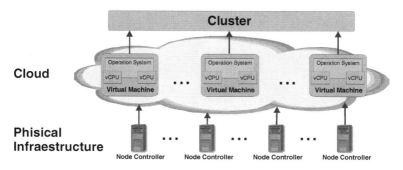

Fig. 1 Cluster diagram used by a Cloud infrastructure.

Related work has been published previously in order to study the Public Cloud Computing infrastructures –such as Amazon EC2- performance and suitability for High Performance Scientific Computing Clusters [22], as well as to check and optimize the overhead introduced, in general, by virtualization techniques [23] (particularly on the Xen Virtual Machine Monitor).

This paper analyses the use of Private Cloud Computing Techniques as support for the creation of High Performance Clusters. Section 2 shows the process of creating the Elastic Cluster, the tools and infrastructures used, etc. Section 3 analyzes the performance of a physical High Performance Cluster compared with a cluster using Cloud Computing.

2 Testing Environment: Physical Cluster vs. Elastic Cluster

To analyze the advantages and disadvantages of deploying clusters in a Cloud infrastructure, we should analyze, using the same equipment, the performance of a physical cluster compared with a Cloud cluster –Elastic Cluster. Table 1 shows the characteristics of the physical servers used in conducting the tests.

Table 1 Technical characteristics of the servers.

Model	CPU	Memory	Hard Disk	Units
Dell R210	4 x INTEL XEON X340 – 2.40 GHz	4GB	250 GB HDD	5
HP DL120 G5	2 x INTEL XEON E3110 – 3.00 GHz	4GB	250 GB HDD	1

The following configurations have been created in order to analyze the performance of the system:

- **Physical Host Cluster (PHC).** The cluster has been created using 5 Dell R210 servers with the operating system Ubuntu 10.10 Server.
- **Virtual Machine Cluster (VMC).** First, the Cloud infrastructure has been established using Ubuntu Enterprise Cloud 10.10 Maverick Meerkat as Cloud Computing software solution, which it is known, it is powered by Eucalyptus, employing last to date KVM version (qemu-kvm 0.14) as virtualization layer on its default configuration. Taking into account Eucalyptus architecture, the HP DL120 server has been used as Cloud Controller and 5 Dell R210 servers as Node Controller. After creating the Cloud infrastructure, 10 instances of 2 processors have been created to form the cluster. Although it is possible to create a different configuration in terms of instances (for example, five instances of four processors), the results are practically the same.

In both cases, the cluster has a total of 20 processors and the physical infrastructure is made up of 5 Dell R210 servers.

For the configuration of the systems, we have used the distribution Ubuntu 10.10 Maverick Meerkat (2.6.35 linux kernel) in the servers, Ubuntu 10.04 Lucid Lynx (2.6.32 linux kernel) in the instances of the virtual machines in the Cloud, and MPICH2 1.2.1 for the cluster.

3   Execution of the Performance Testing

The objective of the performance testing is to determine how creating the cluster in the Cloud infrastructure affects the performance of the system and thus to quantify its influence and evaluate whether it is convenient to apply these technologies to HPC application. The following benchmarks have been used to analyze the performances of the PHC and VMC:

- **Basic Benchmark.** The objective is to analyze the performance of the basic elements of the system: filesystem, RAM and processing performance. A benchmark has been created in order to measure the performance of the filesystem and RAM, while a MPI Benchmark has been used to measure the power of the system [24].
- **NAS Parallel Benchmarks (NPB).** The NPB [25] have been developed by NASA to study the performance of parallel supercomputers. NPB consists of eight benchmarks, each of which focuses on solving problems in aerospace engineering, which require great processing power. These benchmarks, formed by a set of computational kernels (such as MG, IS, FT or CG) and applications (BT, LU or SP), represent the parts with the highest computing workload on CFD (Computational Fluid Dynamics) simulations used by the NASA to evaluate the current and future computers performance taking into account its supercomputation needs.

3.1 Basic Benchmark

In order to measure the performance of the filesystem and RAM, we have compared the performance of a physical server with one instance of the Cloud using a C-coded benchmark. To make this a 4-core physical Dell R210 Server has been used, used as well to instantiate an image on the cloud sharing its hardware characteristics. Results for the filesystem test were obtained from a 100Mb-size file sequential read and write operation on both physical and virtual machines, while the RAM performance test consisted on 1Mb array creation timing, read and written 1000 times. As shown in Table 2, the system virtualization causes a 23.3% loss of performance in the filesystem and a 1.11% performance in the use of RAM memory.

Table 2 PHC vs. VMC: RAM and filesystem performance.

	PHC	VMC	%
Filesystem	52.87 Mbyte/s	43.22 Mbyte/s	23.3%
RAM	332.63 Mbyte/s	328.86 Mbyte/s	1.11%

To measure the processing power of the system in both environments, we have run the MPI benchmark, varying the number of processors in the cluster and calculating the efficiency grade of the system in each case.

Efficiency is usually a value between zero and one that allows us to determine the degree of usable processing time that processors employ in solving a problem, compared to the effort carried out in communication and synchronization tasks. The ideal efficiency of a system is 1, meaning that 100% of the processing time of the system is used to solve the problem.

Figure 2 shows a comparison between the degrees of efficiency of the PHC and VMC methods, respectively. Logically, the efficiency for p=1 in the case of VMC has been calculated over the efficiency for p=1 in PHC.

The efficiency results for both PHC and VMC were fairly good; using 20 processors, they achieved efficiency ratings of 0.991 and 0.917, respectively. At any given time, PHC has a higher efficiency and its performance presents a greater stability (smaller mean deviation).

Specifically, VMC registers an 8.07% loss of performance, which is caused by the penalty that the virtualization layer of the Cloud produces.

A fact to consider is that efficiency sometimes increases in comparison with previous execution and can even exceed the ideal value, which is 1. This occurs when we introduce a server into the cluster that performs better than the average of the previous servers that form the cluster.

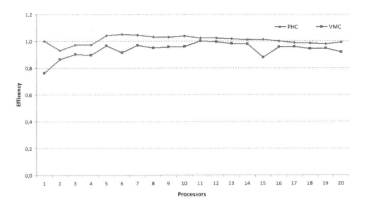

Fig. 2 PHC vs. VMC: Efficiency.

3.2 NPB Benchmark

The NAS Parallel Benchmarks (NPB) were developed by NASA to study the performance of parallel supercomputers. In particular, we will use the LU and SP Simulated CFD Application benchmarks. The LU algorithm resolves 3-dimension Navier-Stokes equations using finite differences discretization and the lower-upper matrix decomposition method (LU) whereas SP algorithm does that using Beam-Warming factorization to find out the solution on 5x5-sized fully diagonalized Jacobians. Figure 3 show the results of the execution of the LU and SP benchmarks, respectively, in the PHC and VMC environments. The LU benchmark is run for 2^n processors while the SP benchmark is run for n^2 processors. On both cases, results of execution are expressed on seconds.

VMC shows worse performance than PHC. Specifically, the performances in running the LU and SP benchmarks in the VMC environment are, respectively, 27.29% and 19.89% slower. Therefore, the execution of the benchmark presents a 23.59% average performance penalty.

The results show a worsening in the performance of the VMC system as the number of processes increases. This is logical, as increasing the number of processes naturally increases the system´s communication load, and differences in performance become more apparent due to the existence of the virtualization layer of the Cloud.

We can observe, from the results obtained in the previous performance tests, that creating the cluster in a Cloud environment results in a performance penalty of between 8.07% and 23.59%, depending on the degree of parallelization of the application.

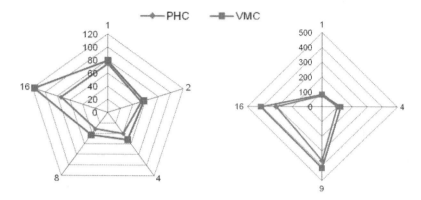

Fig. 3 NPB benchmark results by number of processors in seconds a) LU   b) SP

4 Conclusions

Since its inception, the use of Cloud Computing to improve the utilization of computing resources has become widespread. The main characteristic of Cloud Computing is the ability to create powerful systems adapted to the needs of the client, at a low cost. At the same time, there is an increasing use of high-performance processing systems (HPC) to solve many types of problems. However, the drawback of traditional HPC systems is cost. If we unify both technologies, we will be able to produce flexible, scalable, and low-cost High Performance Clusters.

In this paper, we have shown the advantages of creating such HPC systems on a Private Cloud Computing Infrastructure based on the nowadays most extended solution, Ubuntu Enterprise Cloud using KVM as virtualization layer. Once the system is implemented, we have analyzed its performance in order to compare it with a physical cluster that uses the same resources. Different benchmarks have been used to analyze the performance of clusters, and they have been applied to a computationally expensive problem, cryptanalysis of hash functions. Although the creation of clusters using the Private Cloud Computing infrastructure causes a loss of performance due mainly to communication issues and patterns on HPC applications, this technology, despite showing an affordable overhead, is ideal for implementing high-performance clusters to run applications that require large computing resources taking advantage of the benefits brought by these technologies.

Acknowledgements. Spanish Ministry MICINN and Ingenieros Alborada IDI under grant TRA2009-0309.

References

[1] Zhang, S., Zhang, S., Chen, X., Huo, X.: Cloud Computing Research and Development Trend. In: Second International Conference on Future Networks (2010)
[2] Sun Microsystems. Introduction to Cloud Computing Architecture (June 2009)

[3] GoGrid, http://www.gogrid.com/ (last accessed November 2011)

[4] Amazon, http://aws.amazon.com/ (last accessed November 2011)

[5] Amazon S3, http://aws.amazon.com/s3/ (last accessed November 2011)

[6] Joyent, http://www.joyent.com/ (last accessed November 2011)

[7] Enomalys Elastic Computing Platform, http://www.enomaly.com/ (last accessed November 2011)

[8] Chef, http://wiki.opscode.com/display/chef/Home (last accessed November 2011)

[9] Puppet, http://www.puppetlabs.com/ (last accessed November 2011)

[10] OpenNebula, http://opennebula.org/ (last accessed November 2011)

[11] Eucalyptus, http://www.eucalyptus.com/ (last accessed November 2011)

[12] Ubuntu Enterprise Cloud, http://www.ubuntu.com/business/cloud/overview (last accessed November 2011)

[13] OpenStack http://www.openstack.org/ (last accessed November 2011)

[14] Jackson, K.R., Ramakrishnan, L., Muriki, K., Canon, S., Cholia, S., Wasserman, J.S.H.J., Wright, N.J.: Performance Analysis of High Performance Computing Applications on the Amazon Web Services Cloud. In: 2nd IEEE International Conference on Cloud Computing Technology and Science (2010)

[15] Vecchiola, C., Pandey, S., Buyya, R.: High Performance Cloud Computing: A View of Scientific Applications. In: 10th International Symposium on Pervasive Systems, Algorithms, and Networks (2009)

[16] MPI Forum, http://www.mpi-forum.org (last accessed November 2011)

[17] OpenMP, http://openmp.org (last accessed November 2011)

[18] Unified Parallel C, http://upc.gwu.edu (last accessed November 2011)

[19] Mateescu, G., Gentzsch, W., Ribbens, C.J.: Hybrid Computing – Where HPC meets grid and Cloud Computing. Elsevir B.V. (2010)

[20] XEN, http://www.xen.org/ (last accessed November 2011)

[21] KVM, http://www.linux-kvm.org/ (last accessed November 2011)

[22] Walker, E.: Benchmarking Amazon EC2 for high-performance scientific computing. USENIX; Login: Magazine 33(5), 18–23 (2008)

[23] Cherkasova, L., Gardner, R.: Measuring CPU Overhead for I/O Processing in the Xen Virtual Machine Monitor. In: Proceedings of the Annual Conference on USENIX Annual Technical Conference, ATEC 2005, pp. 24–24. USENIX Association, Berkeley (2005)

[24] MPI Benchmark, http://www.generacio.com/cluster/mflops.c (last accessed November 2011)

[25] Bailey, D.H., et al.: The NAS Parallel Benchmarks. International Journal of Supercomputer Applications 5(3) (Fall1991), 63–73 (1996)

Power Conservation in Wired Wireless Networks

John Debenham and Simeon Simoff

Abstract. A joint university / industry collaboration has designed a system for conserving power in LTE (Long Term Evolution) wireless networks for mobile devices
such as phones. The solution may be applied to any wireless technology in which all
stations are wired to a backbone (e.g. it may *not* be applied to an 802.11 mash). This
paper describes the solution method that is based on a distributed multiagent system
in which one agent is associated with each and every station. Extensive simulations
show that the system delivers robust performance: the computational overhead is
within acceptable limits, the solution is stable in the presence of unexpected fluctuations in demand patterns, and scalability is achieved by the agents making decisions
locally.

1 Introduction

In mobile radio networks, over 50% of the total power is consumed by base stations [6], thus base stations can be targeted for reducing energy consumption. The
Alcatel-Lucent lightRadio^{TM} cube is LTE technology [1] claimed to reduce energy consumption of mobile networks by up to 50% over current technology. The
lightRadio^{TM} cube is designed to complement or replace the existing large base stations that currently supply mobile signal. It covers multiple frequency bands and
supports 2G, 3G and LTE networks. Due to its small size, the cube is very easy to
be installed. The cubes support beamforming in both the horizontal and vertical dimensions. Alcatel-Lucent claims a capacity improvement of 30% through vertical
beamforming alone [2].

John Debenham
Centre for Quantum Computation & Intelligent Systems, University of Technology, Sydney
e-mail: john.debenham@uts.edu.au

Simeon Simoff
School of Computing & Mathematics, University of Western Sydney
e-mail: s.simoff@uws.edu.au

S. Omatu et al. (Eds.): Distributed Computing and Artificial Intelligence, AISC 151, pp. 313–320.
springerlink.com   © Springer-Verlag Berlin Heidelberg 2012

A joint project between The University of Technology, Sydney and Alcatel Lucent Bell Labs has investigated techniques for power conservation [8]. The solution derived may be applied to any network in which all stations are wired to a backbone, and is described in this paper. The solution is based on a distributed multiagent system in which one agent is associated with each and every station. The downlink capacity of the system [4] has been tested in extensive simulations and delivers robust performance: the computational overhead is within acceptable limits, the solution is stable in the presence of unexpected fluctuations in demand patterns, and scalability is achieved by the agents making all decisions locally. The system is a distributed autonomous system in the sense that incapacitating some of the agents will not prevent those that remain from doing their job.

2 The Generic Network Model

The basic elements of the model are: a set of *mobiles*, $\mathcal{M}$, (mobile devices), a wired network of *stations*, $\mathcal{C}$, and intelligent *agents* (in a cloud).

A *mobile* is an abstraction of a handheld device — the set of mobiles at any time is denoted by $\mathcal{M}$. There are a random number of mobiles that we assume are somehow randomly distributed in a *service region* $\mathcal{S}$ that is considered as two-dimensional. The mobiles seek only download capacity from a wireless network of wired stations distributed in and around $\mathcal{S}$. We assume that for every mobile $m \in \mathcal{M}$, β_{op} is the universally acceptable operational bandwidth.

Each *station* $c \in \mathcal{C}$ transmits in a radial *beam* within which it delivers service to mobiles. A station c's beam is specified as a pair, $(\underline{a}(c), \overline{a}(c))$, where $\underline{a}(c), \overline{a}(c) \in [0°, 360°]$, $\underline{a}(c) < \overline{a}(c)$. The angle of c's beam is: $a(c) = \underline{a}(c) - \overline{a}(c) \leq a_{max}(c)°$ the greatest angle that the station is capable of — angles being measured with respect to an arbitrary, fixed standard direction. The stations are assumed to have on-board intelligence to manage their beam angle and direction. A stack of stations is also assumed to intelligently coordinate the beam management of the stations in the stack.

Each station $c \in \mathcal{C}$ has range of transmission that is determined by the transmission power of a station, the angle of its beam, the distance-power gradient, and by any associated interference. The term *range of effective transmission*, $\rho(c)$, means that any mobiles connected to c receive at at least β_{op} bits/s/Hz when interference effects are ignored. That is, $\rho(c)$ will be determined by: the transmit power of c, c's beam angle, the distance-power gradient and the characteristics of c's antenna.

A station c's *settings* are: $c_g = (P_t(c), \underline{a}(c), \overline{a}(c))$ where $P_t(c)$ is the transmit power of station c. The *set of settings* of a set of stations $\mathcal{C}$ is the Cartesian product of the settings of the individual stations: $\times_{c \in \mathcal{C}} c_g$ — let $\mathcal{C}_g$ denote the set of all possible sets of settings of the set of stations $\mathcal{C}$.

Given a station $c \in \mathcal{C}$ at some location $l(c)$, its beam $(\underline{a}(c), \overline{a}(c))$ and its range of effective transmission $\rho(c)$ determine a segment of a circle centred at c within which c may service mobiles. This segment consists of a set of points in two-dimensional space and is called the *segment* of c; it is specified by $s(c_g) = (l(c), \rho(c), \underline{a}(c), \overline{a}(c))$.

$l(c)$ can be specified as Cartesian coordinates with respect to some arbitrary origin. A station's *maximal segment* is: $s^*(c_g) = (l(c), \overline{\rho}(c), 0°, 360°)$ where $\overline{\rho}(c)$ is the maximal range that c is capable of servicing with an isotropic antenna. We expect $\rho(c) > \overline{\rho}(c)$.

For the network to be performing satisfactorily in the *service region* $\mathscr{S}$: $\forall m \in \mathscr{M}, \exists c \in \mathscr{C} \cdot b(m,c) > \beta_{op}$, where $b(m,c)$ is the bandwidth delivered by station c to mobile device m. If $\mathscr{C}_G \in \mathscr{C}_\mathscr{G}$ is a set of settings for the stations $\mathscr{C}$ such that $\mathscr{S} \subseteq \cup_{c \in \mathscr{C}} s(c_g)$ then the set of settings $\mathscr{C}_G$ for $\mathscr{C}$ is said to *cover* $\mathscr{S}$. It is assumed that: $\mathscr{S} \subseteq \cup_{c \in \mathscr{C}} s(c_g)$ at all times; otherwise additional stations are required to cover $\mathscr{S}$.

3 Topology

Any automatic adjustment of either the power or the beam of a station should ideally ensure that the adjustment does not jeopardise the signal strength at any mobile. This poses the problem of how a set of stations is to determine whether or not their combined segments deliver satisfactory bandwidth to the entire service region. That is, how are the stations to understand the topology of the space that they are intended to service?

In this work the question is addressed by providing a finite set of "reference points" $\mathscr{R}$, $\mathscr{R} \subset \mathscr{S}$, in the service region. A *reference point* is a point $r \in \mathscr{R}$ is such that *if* $\forall c \in \mathscr{C}, r \in s(c_g)$ *then* c knows the sub-quadrant in which it lies. The idea being that there are sufficiently many reference points so that if station c reduces its power then it will know the name of a reference point that it could previously service but can now no longer do so.

If a station adjusts its power or beam then this may well effect the desirable power and beam of neighbouring stations. So stations need some understanding of the concept of 'neighbourhood'. The set of stations $\mathscr{C}_1(c)_g = \{c' \in \mathscr{C} \mid \exists r \in \mathscr{R}, r \in s(c_g) \wedge r \in s(c'_{g'})\}$ is the *one-hop set* of c, and the set of stations $\mathscr{C}_1^*(c)_g = \{c' \in \mathscr{C} \mid \exists r \in \mathscr{R}, r \in s^*(c_g) \wedge r \in s^*(c'_{g'})\}$ is the *maximal one-hop set* of c. The set of stations $\mathscr{C}_2(c)_g = \{c' \in \mathscr{C} \mid \exists r \in \mathscr{R}, \exists c'' \in \mathscr{C}_1(c)_g, r \in s(c'_{g'}) \wedge r \in s(c''_{g''})\}$ is the *two-hop set* of c. In general, the set of stations $\mathscr{C}_{n+1}(c)_g = \{c' \in \mathscr{C} \mid \exists r \in \mathscr{R}, \exists c'' \in \mathscr{C}_n(c)_g, r \in s(c'_{g'}) \wedge r \in s(c''_{g''})\}$ is the *$(n+1)$-hop set* of c. If every station has at least one reference point in its segment for all of its settings then: $c \in \mathscr{C}_1(c)$ and $\mathscr{C}_n(c) \subset \mathscr{C}_{n+1}(c)$.

4 Agents

Each station $c \in \mathscr{C}$ is uniquely associated with an agent, $\gamma_c \in \Gamma$, that exists in a cloud. An agent is responsible for adjusting its station's power settings and for interacting with nearby agents.

An agent can make the following adjustments to its station's settings:

- adjusting (i.e. increasing or decreasing) its transmission power, including:
- turning the station's transmitter on and off, i.e. putting it in standby mode.

Each station's beam angle and direction is managed by its beam management algorithms that we assume are on-board as described generally in Section 2. If a station's power is adjusted then we expect this to trigger adjustments to the station's beam.

Agent Interaction. The only actions that software agents can make is to send and receive messages. To do this an agent needs an illocutionary communication language such as that described in [3]. Those languages typically contain illocutionary constructs such as "inform" and "propose". Agent interaction is often called negotiation and hopefully terminates with the participating agents each *committing* to perform some action. For cooperative agents with a common goal the situation is simpler as long as they share both a common utility function and a common method for evaluating utility. Armed with knowledge of the agents' utility, an agent γ is able to propose a set of commitments whose enactment increases utility from its perspective and from the perspective of the other agents in the negotiation. This is how the deliberative autonomous agents cooperate in Section 6.

If an autonomous agent is to spontaneously propose to adjust the power of their station and other nearby stations, it is necessary to ensure that another proposal is not under way in the same region at the same time. To manage this potential problem we adopt the following simple "locking" procedure. Suppose that agent γ_c for station c wishes to propose that the settings for $\mathscr{C}_1(c)$ be changed from: $\mathscr{C}_1(c)_G \in \mathscr{C}_1(c)_{\mathscr{G}}$ to: $\mathscr{C}_1(c)_{G'} \in \mathscr{C}_1(c)_{\mathscr{G}}$. Agent α_c communicates with the agents for stations $\mathscr{C}_2(c)$ as follows:

- send message "request lock by c" to all agents in $\mathscr{C}_2(c)$
- if all agents in $\mathscr{C}_2(c)$ respond "accept lock from c" then:
    - send message "propose change $\mathscr{C}_1(c)_G$ to: $\mathscr{C}_1(c)_{G'}$" to all agents in $\mathscr{C}_1(c)$
    - when all agents in $\mathscr{C}_1(c)$ respond "$\mathscr{C}_1(c)_G$ changed to: $\mathscr{C}_1(c)_{G'}$" then
        - send message "request unlock by c" to all agents in $\mathscr{C}_2(c)$
        - if all agents in $\mathscr{C}_2(c)$ respond "accept unlock from c" then done

- else abandon

5   Power Conservation

The power conservation problem is to reduce power consumption in networks without compromising 'unduly' the quality of service delivery. It has the following characteristics:

- the network is large — so a centralised algorithm is not feasible — distributed algorithms are indicated and the issue of stability of the solution must be addressed,
- the set of mobiles is large, moving and changing rapidly — so there is no virtue in seeking an optimal solution, rather to try to *improve quickly* the network settings, and
- the environment is chaotic and unpredictable — so no solution can guarantee to maintain 'good' network settings all the time.

To formalise the *power conservation problem*, given a set of stations $\mathscr{C}$ we define a two-dimensional *utility function* across the space of the settings of $\mathscr{C}$, $\mathscr{C}_{\mathcal{G}}$. $\mathbf{U} : \mathscr{C}_{\mathcal{G}} \to \mathbb{R}^2_+$, the two-dimensional space of positive reals. Given a set of settings $\mathscr{C}_G$ for a set of stations $\mathscr{C}$, the first component of $\mathbf{U}$: $U_1(\mathscr{C}_G) \triangleq \sum_{c \in \mathscr{C}} P_{v,\iota}(c)$, where $P_{v,\iota}(c)$ is the received power at the "close-in distance", v, when c is using an isotropic antenna. The second component is: $U_2(\mathscr{C}_G) \triangleq I(\mathscr{C}_G)$, where $I(\mathscr{C}_G)$ is an estimate of the interference between the stations in $\mathscr{C}$ in setting $\mathscr{C}_G$. An *ordering* is defined for $\mathbf{U}$ as follows: suppose $\mathbf{U}(\mathscr{C}_G) = (p, i)$ and $\mathbf{U}(\mathscr{C}_{G'}) = (p', i')$ then: $p < p' \to \mathbf{U}(\mathscr{C}_G) > \mathbf{U}(\mathscr{C}_{G'})$, and $(p = p') \wedge (i < i') \to \mathbf{U}(\mathscr{C}_G) > \mathbf{U}(\mathscr{C}_{G'})$. This ordering of $\mathbb{R}^2_+$ gives first priority to saving power, and second priority to reducing interference. These two priorities are interlinked in that reducing interference, perhaps by narrowing a beam angle, may then reveal new opportunities for power saving. Further adjusting power may then lead of adjustment of the beams.

The *power conservation problem* is: given a set of stations $\mathscr{C}$, and a set of mobiles $\mathscr{M}$, to find the settings $\mathscr{C}_{G^*}$ for $\mathscr{C}$:

$$\mathscr{C}_{G^*} = \arg\max_{\mathscr{C}_G}\{\mathbf{U}(\mathscr{C}_G) :$$

$$||\{m \in \mathscr{M} : \forall c \in \mathscr{C} \cdot b(m,c) < \beta_{op}\}|| \text{ is minimal}\}$$

where $||\{m \in \mathscr{M} : \forall c \in \mathscr{C} \cdot b(m,c) < \beta_{op}\}||$ is the number of mobiles that are receiving less than β_{op} — hopefully this will be zero. In other words, for all settings that deliver satisfactory bandwidth, to find the setting with highest utility.

The criterion for a satisfactory solution, $||\{m \in \mathscr{M} : \forall c \in \mathscr{C} \cdot b(m,c) < \beta_{op}\}||$ is minimal, is not the only suitable criterion.

6 Agent Reasoning

An approach to autonomic network management that was extensively tested on a variety of problems on 802.11 wireless mesh is described in [7]. Its strength is its ability to deal quickly with changing circumstances, to improve the network settings and to exhibit stable performance. [7] does not address the power problem directly although the essential characteristics of the problems addressed in [7] and the power problem are so similar that we are compelled to believe that it shows a way to effective algorithms for the power problem.

Our agents are hybrid agents whose logic consists of two opposing components:

- a *reactive component* that deals with unexpected problems — in our case this is limited to a station being overloaded. The reactive logic acts quickly, without consultation of deliberation, to attempt to deal with station overload.
- a *proactive component* that aims to achieve the system goal — in our case this is to move the settings towards G^* the optimal setting.

where, as is usually the case for autonomous agents, the reactive logic overrides the proactive logic.

Reactive Logic. A station is *overloaded* when it is at, or 'near', capacity. When a station is overloaded it attempts to shed some of its load. This is achieved by reducing power. Adjusting the beam angle will alter the region of effective transmission of the station and so may increase interference as well as the level of overload.

If a station acts to reduce its transmit power it may find that some reference points that it previously covered are no longer covered; we say that such reference points have been *abandoned* by the action. The reactive logic repeats the following simple procedure: while station c with settings c_g is overloaded it reduces its transmit power by one click *and* for each reference point that c will abandon by this action c sends an inform message to the agent of each station in the set: $\{\gamma : \gamma_c \in \mathscr{C}_1(c_g)\}$ — those agents may alyready be aware of the overload siuation via their listening capabiliy — in any case, if an agent in $\{\gamma : \gamma_c \in \mathscr{C}_1(c_g)\}$ is aware of an overload at c and is switched off then it switches on, and commences its 'warm up' cycle.

Reactive logic is normally expressed as: event-condition-action rules. The above logic then is:

if observe overload
    and not at minimum power
then reduce power by one click
    and inform all in $\{\gamma : \gamma_c \in \mathscr{C}_1(c_g)\}$ "overload at c"
if receive "overload at c"
    and switched off
then switch on

Proactive Logic. The agent γ_c for station c desires to change the settings of $\mathscr{C}_1(c)$ from: $\mathscr{C}_1(c)_G \in \mathscr{C}_1(c)_{\mathscr{G}}$ to: $\mathscr{C}_1(c)_{G''} \in \mathscr{C}_1(c)_{\mathscr{G}}$ if γ_c estimates that: $\mathbf{U}(\mathscr{C}_2(c)_{G''}) > \mathbf{U}(\mathscr{C}_2(c)_G)$. But agents only adjusts their agent's power. The general idea is that agent γ_c will propose to adjust the power as long as the proposal is expected to result in a net increase in utility for $\mathscr{C}_2(c)$. This is the 'trick' that ensures stability [7].

As the adjustment process occurs very quickly we assume that the state of the mobiles is unchanged during the process. The process then consists of two steps: first agent γ_c proposes to adjust the power for stations in $\mathscr{C}_1(c)$, and second the beam management algorithms adjust the stations' beams — when this is all complete the agents in $\mathscr{C}_1(c)$ confirm that the change is completed as described in Section 4. For this process to be (fairly) certain of yielding a net increase in utility for $\mathscr{C}_2(c)$: (1) the power adjustment step needs to be significant and, (2) the beam adjustment step needs to be 'intelligent' as we define below. In detail these two steps are:

1. γ_c proposes to adjust the power settings for stations in $\mathscr{C}_1(c)$ from $\mathscr{C}_1(c)_G$ to $\mathscr{C}_1(c)_{G'}$ if:

$$\gamma_c \text{ estimates that } \mathbf{U}(\mathscr{C}_2(c)_{G''}) > \mathbf{U}(\mathscr{C}_2(c)_G), \text{ and}$$

$$U_1(C_1(c)_G) > U_1(C_1(c)_{G'}) \times \kappa$$

where κ is a constant, $\kappa > 1$, to ensure that the improvement in power consumption is significant, and

2. the beam management algorithms change the station settings from $\mathscr{C}_2(c)_{G'}$ to: $\mathscr{C}_2(c)_{G''}$.

This will lead to a net increase in utility for $\mathscr{C}_2(c)$ if: $||\{m \in \mathscr{M} : \forall c \in \mathscr{C} \cdot b(m,c) < \beta_{op}\}||$ is not increased. We hypothesise that this should occur if: $I(\mathscr{C}_2(c)_{G'}) \leq I(\mathscr{C}_2(c)_{G''})$ which we conveniently adopt as the meaning of *intelligent beam forming*.

Adjusting the Power. It remains to discuss how γ_c will determine G' that starts the proactive process in the first step above. If station c reduces its transmit power then it will in general abandon references points at which it delivers the strongest signal. So c needs to know which reference points will be abandoned if its transmit power is reduced by some amount. It uses the reference point calibration settings to do this — described in Section 3. Suppose reference point r is in the range of station c's beam. c accesses the data for r and identifies all of r's neighbouring stations including c. For each of these neighbours, c', c interpolates an estimate of c''s received power at r given c''s actual transmit power, $P_t(c')$, or $P_{v,t}(c')$. c then identifies the c' for which the received power at r is greatest.

Station c now knows the set of reference points at which it delivers maximum power. Mobiles at those reference points will connect to c given that c has available capacity. Station c also knows how this set of reference points will change if it, or any of r's neighbours, change their power settings. Now stations are connected to mobiles and not to reference points. The number of mobiles connected to any station is common knowledge, but we assume that their precise location is known to no one. We deal with this lack of knowledge as follows: we assume that the mobiles connected to a station are somehow randomly distributed over the reference points to which it delivers the strongest signal.

So c now knows which reference points will be abandoned if it reduces power and the load (i.e. number of mobiles) associated with those reference points. c also knows which stations could service those potentially abandoned reference points, and how much additional power they would use to do so. And, given that another station may have to increase power to service an abandoned reference point it may mean that yet another station can reduce power.

7 Conclusions

The distributed solution to power management was developed in tandem with simulation systems that were designed around the characteristics of the Alcatel-Lucent lightRadio^{TM} cube except the maximum number of mobiles that each cube could service was reduced to twenty to make the display easier to follow in real time. For the Alcatel-Lucent lightRadio^{TM} cube the power model is: $P_c = 4 * P_t + 10$, where P_t is the transmission power, and P_c is power consumed in watts. P_t varies between 2 and 5 watts. $a_{max}(c) = 120°$ and $\delta a = 20°$. We believe that the assumptions that underly this approach are not unduly restrictive [5]. We are confident that the algorithms described will perform well in the field.

Acknowledgment. This collaborative research project is supported by an ARC Linkage Grant LP0883580, University of Technology, Sydney and Alcatel-Lucent Bell Labs (Paris).

References

1. Dahlman, E., Parkvall, S., Sköld, J., Beming, P.: 3G Evolution: HSPA and LTE for Mobile Broadband. Academic Press (2008)
2. Fehske, A.J., Richter, F., Fettweis, G.P.: Energy efficiency improvements through micro sites in cellular mobile radio networks. In: Globecome Workshops (2009)
3. FIPA. Communicative act library specification. Technical Report SC00037J, Foundation for Intelligent Physical Agents, Geneva, Switzerland (2002)
4. Jorgušeski, L., Prasad, R.: Downlink Resource Allocation in Beyond 3G OFDMA Cellular Systems. In: The 18th Annual IEEE International Symposium on Personal, Indoor and Mobile Radio Communications (PIMRC 2007), Athens, pp. 1–5. IEEE Computer Society (2007)
5. Lin, K., Debenham, J., Simoff, S.: Managing Power Conservation in Wireless Networks. In: Cao, L., Zhong, J., Feng, Y. (eds.) ADMA 2010, Part II. LNCS, vol. 6441, pp. 314–325. Springer, Heidelberg (2010)
6. Lorincz, J., Capone, A., Bogarelli, M.: Energy savings in wireless access networks through optimized network management. In: Proceedings of the 5th IEEE International Conference on Wireless Pervasive Computing, pp. 449–454. IEEE Press, Piscataway (2010)
7. Prodan, A.: A Multiagent System for Topology Control on Multi-Radio Wireless Mesh Networks. PhD thesis. University of Technology, Sydney (2010)
8. Vadgama, S., Hunukumbure, M.: Trends in green wireless access networks. In: 2011 IEEE International Conference on Communications Workshops (ICC), pp. 1–5 (2011)

Risk Assessment Modeling in Grids at Component Level: Considering Grid Resources as Repairable

Asif Sangrasi and Karim Djemame

Abstract. Service level agreements (SLAs), as formal contractual agreements, increase the confidence level between the End user and the Grid Resource provider, as compared to the best effort approach. On the other end, SLAs fall short of assessing the risk in acceptance of the SLA, risk assessment in Grid computing fills that gap. The current approaches to risk assessment are based on node level risk assessment. This work is differentiated by that it provides risk assessment information at the granularity level of components. A risk assessment model at the component level based on Non-Homogeneous Poisson Process (NHPP) model is proposed. Grid failure data is used for the experimentation at the component level. The Grid risk model selection is validated by using a goodness of fit test along with graphical approaches. The experimental results provide detailed risk assessment information at the component level which can be used by Grid Resource provider to manage and use the Grid resources efficiently.

Keywords: Risk Assessment, Repairable Grid resources, Probability of Failure.

1 Introduction

Grid technologies enabled coordinated resource sharing between distributed resources in dynamic and heterogeneous environments [1, 10, 11]. A good way of defining the Grid is presented in [9] using Foster's three point checklist.

Service Level Agreements (SLAs) provide an opportunity to improve on the deficiencies of the best effort approach and in turn increase the Grids commercial uptake [13, 14]. The best effort approach presents with scenarios where the Resource provider provides no guarantees for the successful running of a job. An SLA includes the requirements of the End user that the Resource provider agrees to deliver. It also specifies the reward or amount that the End user pays to resource

Asif Sangrasi · Karim Djemame
School of Computing,University of Leeds, Leeds, UK
e-mail: {scas,scskd}@leeds.ac.uk

S. Omatu et al. (Eds.): Distributed Computing and Artificial Intelligence, AISC 151, pp. 321–330.
springerlink.com   © Springer-Verlag Berlin Heidelberg 2012

provider when the SLA is successfully fulfilled. In case that the SLA is not successfully fulfilled by the resource provider, a penalty is paid to the End user [5-8, 12]. The introduction of SLAs in Grid business transactions provided a means for expressing the requirements of End users and the assurances of the resource providers in terms of QoS. However, resource providers were still skeptical about agreeing and fulfilling the SLA due to the probability of failing to fulfilling the SLA due to any hardware, software and network failure. The introduction of risk brings new opportunities and options for both the End user and the Resource provider [18-20].

Risk is defined as the sum of uncertainty and the outcome (damage) [33]. The work in Risk Assessment and Management in Grid computing has been limited to the work of [18, 21]. However, these risk assessment efforts were based on the node or machine level risk assessment. As a result the risk information is limited to node level and the component level risk information is not available to the resource provider. As a node may contain a number of components such as memory, CPU, software, disk, etc, the failure of node may in turn be a failure of any of these components. As a result of this risk assessment at the component level was introduced in [22], considering the grid resources as replaceable.

The contributions of this work are:

- A Non-Homogeneous Poisson Process based risk assessment model, to allow risk assessment at the component level. The risk model uses the grid failure data, once the grid failure data is analyzed, to work out the point estimations of the risk model parameters.
- A goodness of fit test along with the graphical approaches validates the risk model selection.
- A risk model which provides risk estimations for time in future at the granularity level of components for a number of nodes as required in the SLA, by means of the Probability of Failure (PoF). The risk is expressed by the PoF in this work.

The rest of this work is structured as follows: section two describes the related work. Section three encompasses the analysis of Grid failure data. Repairable system models along with the goodness of fit tests used in this work for model selection are discussed in section four. Section five presents the risk model and point estimations of the NHPP model parameters. Experimental results along with the discussion of the limitations of model are elucidated in Section six. Section seven describes the future work and concludes the work.

2 Related Work

Assessgrid introduced the concept of risk assessment in Grid Computing, by providing risk models [18]. The risk modeling framework in Assessgrid is aimed at two levels i.e. the provider and the broker level. These probabilistic and possibilistic set of models, at the provider level, provide risk information at the granularity level of nodes. These risk models at the provider level are in the same context as this work and our past work [22]. The risk assessment models in Assessgrid at the resource provider layer provide risk assessment by estimating the risk associated with the acceptance of an SLA to a resource provider that has been submitted by the End user. The probabilistic risk models proposed in AssessGrid are based on Bayesian models operating on Poisson Processes with parameter estimates on Gamma distribution [18, 23].On the other hand the possibilistic modeling in AssessGrid is based on the work of Zadeh [16]. In AssessGrid a broker was introduced to search and negotiate with a number of resource providers on behalf of the End user [24]. The risk assessment models in the AssessGrid at the resource provider level do not account for the nature of the data i.e. repairable or replaceable unlike the current work and [22]. The risk assessment models at the resource provider layer in AssessGrid are aimed at the granularity level of nodes, unlike this work that is aimed at the granularity level of components. The broker level scenarios addressed in AssessGrid are out of scope of this work.

Risk Assessment in Grids is addressed in [21]; however the work falls short of introducing any risk assessment model, the future work intentions in [21] are aimed at the node level but not at the granularity level of components. A number of scenarios closely related to risk assessment and management in Grids are addressed in [26-30].The issues mentioned in these works vary from rates of failure of resources to usage of distributions to estimate future reliability estimations. However none of these address the scenario of risk assessment modeling in Grids at the granularity level of components or consider the Grid resources repairable.

3 Data Analysis

The risk modeling methodology for this work consists of the steps of data analysis, risk modeling and experimentation. The failure data from the Los Alamos National Laboratory (LANL) [3] used in this research contains about 23700 failure events. The Grid failure data was divided into the categories of Hardware, Software, Facilities, Network, Human and Undetermined. The hardware failure events were further subcategorized in CPUs, Memory, Disks, Interconnect, power supply failure events. For a set of nodes, failure data was inspected, sorted and transformed to extract the time to failure information at granularity level of components. A more detailed analysis of the failure data is available in our past work [22].

4 Repairable Systems

Repairable systems are categorized according to the repair nature, minimal or perfect repair. Minimal repair brings a failed system back to the condition to which the system was just before the failure, also termed as "as bad as old". Perfect repair or renewal repair brings the failed system back to the perfect condition of as if new [15, 2]. In the case of the current scenario the repairable system observes repairs that are of minimal nature. Consequently the system is brought back to the state after repair, in which the system was before the failure.

One of the potential ways to model the risk at the component level was to use the Bayesian reliability however this depends on the estimations of the prior and the posterior distributions. For the estimation of the prior distribution, the Mean Time to Failure (MTTF) information from the hardware vendors was required, which was not available. As a result of it, instead of Bayesian reliability, Poisson Process based modeling was utilized to assess the risk in future time at the component level. The Poisson Process can have two shapes, i.e. Homogeneous Poisson Process (HPP) and Non-homogeneous Poisson Process. HPP accounts for the systems that do not change where as NHPP models the systems that improve or deteriorate. There are a number of ways by using which it can be determined whether the Grid failure data is governed by a HPP or NHPP based model. One set of informal approaches are Duane graph and trend plots, however, the interpretation of these plots is subjective. As a result of which one of the more formal Goodness of fit test called the ratio-power transformation and Cramer Von Mises Test [15] was used in this work. The interpretation of the variables used in this work is as follows.

t is the time age of the system/component in time units
θ is the scale parameter of the NHPP model
$\hat{\beta}$ is the parameter of the NHPP model that exhibits system deterioration/ improvement
N is the total number of failure events
t_i is the time of current failure
d is the time in future
R(t) is the probability of the component will not fail in time (t,t+d)
P(t) is the probability of the system/component will fail in time (t,t+d)

The ratio-power transformation is given by

$$\hat{R}_i = \left(\frac{t_i}{t}\right)^{\bar{\beta}}$$

The Cramer-von Mises test statistic will be given as

$$C_R^2 = \frac{1}{12n} + \sum_{i=1}^{n}\left(\hat{R}_i - \frac{2i-1}{2n}\right)^2$$

The table of critical values for Cramer-von Mises goodness-of fit test [15] provides the critical values for C_R^2 considering the value of M, i.e the number of failure events. The null hypothesis is. H_0: The failure times of the Grid failure data are governed by power law/NHPP Process. The results of the ratio-power transformation and Cramer Von Mises Test, when tested against the Grid Failure data, presented in Figure 1 suggest the governance of NHPP based model for most of the Grid resources. Consequently we do not reject the null hypothesis and conclude that the Grid failure data is governed by NHPP based model. In this work,, clusters represented by C; nodes represented by N and the component name form the identity trio for any component and C_iN_j denote Cluster i and node j.

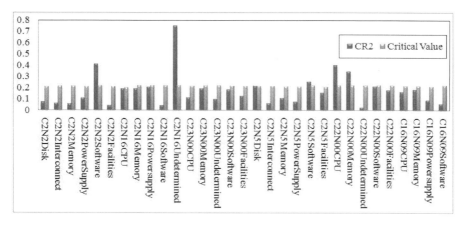

Fig. 1 Goodness of Fit Test

5 Risk Modeling

The power law process or non homogeneous poison process has an intensity function of [20, 15].

$$\lambda(t) = \frac{\beta}{\theta} \left(\frac{t}{\theta}\right)^{\beta-1}$$

The improvement and the deterioration patterns are effected by β parameter. If $\beta = 1$ the system is reduced to HPP while if $\beta < 1$ the system is improving and intensity function is decreasing. If $\beta > 1$ the system is deteriorating and intensity function is increasing. [15, 25]. The value of θ is directly related to the scale of time and is directly related to the time of occurrence of the latest failure event of system. Point estimations of β and θ were conducted by using the following equations and shown in Table 1.

The MLE (Maximum Likelihood Estimators) of β and θ for N>0 are equal to [15],

$$\hat{\beta} = \frac{N}{\sum_{I=1}^{n} \log\left(t/t_i\right)}$$

$$\hat{\theta} = \frac{t}{N^{\frac{1}{\hat{\beta}}}}$$

While the conditionally unbiased estimator of β is given by

$$\bar{\beta} = \frac{N-1}{N}(\hat{\beta}) = \frac{N-1}{\sum_{I=1}^{n} \log\left(t/t_i\right)}$$

Table 1 The point estimates of the β and θ.

Component	β	θ	Component	β	θ
C2N2Disk	2.812	335072	C2N5Disk	3.818	269970
C2N2Interconnect	1.793	2970215	C2N5Interconnect	1.261	484473
C2N2Memory	1.440	1453246	C2N5Memory	2.878	2000693
C2N2PowerSupply	2.260	1577369	C2N5PowerSupply	1.631	923267
C2N2Software	0.452	19056	C2N5Software	0.397	752.6256
C2N2Facilities	0.934	181977	C2N5Facilities	0.929	282989
C16N09CPU	1.984	1152284	C22N00CPU	1.301	227423
C16N09Memory	1.511	571665	C22N00Memory	1.040	305681
C16N09Powersupply	1.793	872349	C22N00Undetermined	0.589	23104
C16N09Undetermined	3.754	1359911	C22N00Software	1.959	653118
C16N09Software	4.398	1389167	C22N00Facilities	1.1306	183099
C23N00CPU	1.055	252069	C16N09CPU	1.3469	776651
C23N00Memory	0.440	4900.573	C16N09Memory	2.5208	1181907
C23N00Undetermined	1.023	159887	C16N09Powersupply	1.8322	783943
C23N00Software	1.058	222043	C16N09Software	2.3564	72985
C23N00Facilities	0.914	234004	C16N09Software	0.7297	32236

The Probability of Failure Estimation was carried out using the following equations given below based on [15, 4].

$$\lambda = \frac{1}{\theta^\beta}$$

$$R(t) = e^{-[\lambda(t+d)^\beta - \lambda(t)^\beta]}$$

$$P(t) = 1 - R(t)$$

$$P(t) = 1 - [e^{-[\lambda(t+d)^\beta - \lambda(t)^\beta]}]$$

6 Experimental Results and Discussion

The objective of the experiments is to demonstrate the usage of this risk model to obtain risk assessment information at the granularity level of components. The experimental results for different components of the nodes are given in figures 2-6. In figures, y-axis represents PoF while x-axis represents time (t) in million minutes. The figures show that the convergence of the estimated PoF to one in future time t varies for all the components. This variation of convergence shows that components will fail at different timings in future. For some nodes the component CPU or Disk is missing, this due to the nature of the available Grid failure data, where there were failures for either the Disk or the CPU component. The figures reflect that there is no constant pattern for the estimations of PoF for the individual nodes, either within the same cluster or within different clusters. However there is constant pattern shown for the nodes, even the most reliable component within these nodes is expected to fail within about 2 million minutes.[1]

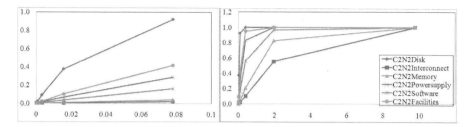

Fig. 2 PoF estimations of components of C2N2 against future Time (t) in million minutes

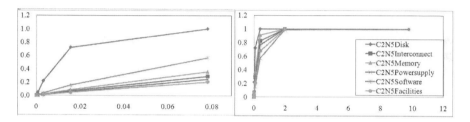

Fig. 3 PoF estimations of components of C2N5 against future Time (t) in million minutes

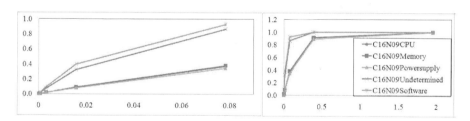

Fig. 4 PoF estimations of components of C16N09 against future Time (t) in million minutes

Following are set of assumptions and limitations for this work.

• It is assumed that the Grid resources are repairable and not replaceable.
• The Grid failure data should be available, as this risk model utilizes the Grid failure data to estimate the risk of failure in future time.
• The failure of the data should either available at the granularity level of the components or it should be possible to extract that information at the granularity level of components.

---

[1] 1 million minutes are approximately equivalent to 694 days.

- The repairable Grid resources after repair come to a working condition close to the state before failure but the repair will not change the condition of the resource to" like new" condition. That means that the Grid resources are considered as repairable but not renewal systems.
- The failures of the components in Grids are not independent and identically distributed (i.i.d.).
- The cause-effect relationship or the effects of a failure of a component on another component are not considered in this study.

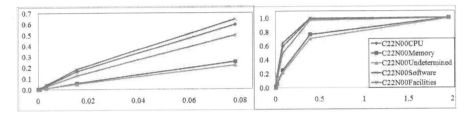

Fig. 5 PoF estimations of components of C22N00 against future Time (t) in million minutes

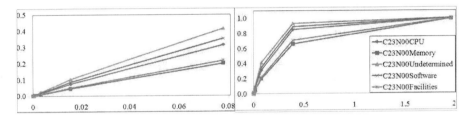

Fig. 6 PoF estimations of components of C23N00 against future Time (t) in million minutes

7 Conclusion and Future Work

Risk Assessment at component level provides more detailed risk information to the Grid resource provider. This risk information can be used by the resource provider to use and schedule the Grid resources more efficiently. The aim of this work was to propose risk model at the component level considering resources as repairable. Although this model does not take reparability extrinsically nevertheless provides risk information about Grid resource failure in future at component level by means of PoF. The PoF values of the components show that the PoF values of the resources do not follow a consistent pattern within and outside the clusters. In our future work, component level risk will be modeled by approaches that take reparability extrinsically rather than intrinsically. We will also look towards a comparison of the results of this work and our previous work [22] where Grid resources are considered replaceable.

References

[1] Foster, I., Kesselman, C., Tuecke, S.: The anatomy of the grid: Enabling scalable virtual organizations. Int. J. High Perform. Comput. Appl. 15(3), 200–222 (2001)

[2] Krivtsov, V.: Practical extensions to NHPP application in repairable system reliability analysis. Reliability Engineering & System Safety 92(5), 560–562 (2007)

[3] (Los Alamos National Laboratories),
http://institute.lanl.gov/data/lanldata.shtml

[4] Crow, L.H.: Reliability analysis for complex, repairable systems. Tech. Rep. (December 1975)

[5] Burchard, L.O., et al.: A Quality-of-Service architecture for future grid computing applications. In: Proceedings of the 19th IEEE IPDPS 2005, p. 132a (2005)

[6] Leff, A., Rayfield, J.T., Dias, D.M.: Service-level agreements and commercial grids. IEEE Internet Computing 7(4), 44–50 (2003)

[7] Ludwig, H., et al.: A service level agreement language for dynamic electronic services. In: Advanced Issues of E-Commerce and Web-Based Information Systems, pp. 25–32 (2002)

[8] Padgett, J., Djemame, K., Dew, P.: Grid-Based SLA Management. In: Sloot, P.M.A., Hoekstra, A.G., Priol, T., Reinefeld, A., Bubak, M. (eds.) EGC 2005. LNCS, vol. 3470, pp. 1076–1085. Springer, Heidelberg (2005)

[9] Foster, I.: What is the grid? - a three point checklist. GRIDtoday 1(6) (July 2002)

[10] Schopf, J.M., Nitzberg, B.: Grids: The top ten questions. Sci. Program. 10, 103–111 (2002)

[11] Schwiegelshohn, U., et al.: Perspectives on grid computing. Future Gener. Comput. Syst. 26(8), 1104–1115 (2010)

[12] Talia, D., et al.: Using SLA for resource management and scheduling - a survey grid middleware and services. In: Grid Middleware and Services, ch. 22, pp. 335–347. Springer, Boston (2008)

[13] Djemame, K., et al.: Brokering of risk-aware service level agreements in grids. Concurrency Computat.: Pract. Exper. 23(13), 1558–1582 (2011)

[14] Sahai, A., et al.: Specifying and monitoring guarantees in commercial grids through SLA. In: Proceedings of the CCGRID 2003. IEEE Computer Society, Washington, DC (2003)

[15] Rigdon, S.E., Basu, A.P.: Statistical Methods for the Reliability of Repairable Systems. Wiley-Interscience (2000)

[16] Zadeh, L.A.: Fuzzy Set as a Basis for a Theory of Possibility. Fuzzy Sets and Systems 1, 3–28 (1978)

[17] Haimes, Yacov, Y.: Risk Modeling, Assessment, and Management, 3rd edn. Wiley series in system engineering (1998)

[18] Djemame, K., et al.: Introducing risk management into the grid. In: Second IEEE International Conference on e-Science and Grid Computing, p. 28. The Netherlands (December 2006)

[19] Hovestadt, M., et al.: The first step of introducing risk management for prepossessing slas. In: IEEE International Conference on Services Computing, SCC 2006, pp. 36–43 (2006)

[20] Hamada, M.S., et al.: Bayesian Reliability. Springer series in statistics. Springer (2008)

[21] Alsoghayer, R., Djemame, K.: Probabilistic risk assessment for resource provision in grids. In: Djemame, K. (ed.) Proceedings of the 25th UK Performance Engineering Workshop, Leeds, UK, pp. 99–110 (July 2009)

[22] Sangrasi, A., Djemame, K.: Component level risk assessment in grids: A probablistic risk model and experimentation. In: Proceedings of the 5th IEEE International Conference on Digital Ecosystems and Technologies Conference (DEST), pp. 68–75 (May 2011)

[23] Carlsson, C., Fullér, R.: Risk Assessment of SLAs in Grid Computing with Predictive Probabilistic and Possibilistic Models. In: Greco, S., Pereira, R.A.M., Squillante, M., Yager, R.R., Kacprzyk, J. (eds.) Preferences and Decisions. SFSC, vol. 257, pp. 11–29. Springer, Heidelberg (2010)

[24] Gourlay, I., et al.: Reliability and risk in grid resource brokering. In: 2nd IEEE DEST (2008)

[25] Bunday, B.D.: Statistical methods in reliability theory and practice, ser. Maths and its applications. Statistics, Operational Research, and Computational Maths. Ellis Horwood (1991)

[26] Krautsevich, L., et al.: Risk-Aware Usage Decision Making in Highly Dynamic Systems. In: International Conference on Internet Monitoring and Protection, vol. 0, pp. 29–34 (2010)

[27] Iosup, A., et al.: On the dynamic resource availability in Grids. In: Proceedings of the 8th IEEE/ACM International Conference on Grid Computing, USA, September 19-21, pp. 26–33 (2007)

[28] Hovestadt, M., et al.: The first step of introducing risk management for prepossessing slas. In: IEEE International Conference on Services Computing, SCC 2006, pp. 36–43 (2006)

[29] Pinheiro, E., et al.: Failure trends in a large disk drive population. In: Proceedings of the 5th USENIX Conference on File and Storage Technologies (February 2007)

[30] Lingrand, D., et al.: Optimization of Jobs Submission on the EGEE Production Grid: Modeling Faults Using Workload. Journal of Grid Computing 8, 305–321 (2010)

R & D Cloud CEIB: Management System and Knowledge Extraction for Bioimaging in the Cloud

Jose Maria Salinas, Maria de la Iglesia-Vaya, Luis Marti Bonmati,
Rosa Valenzuela, and Miguel Cazorla

Abstract. The management system and knowledge extraction of bioimaging in the cloud (R & D Cloud CEIB) which is proposed in this article will use the services offered by the centralization of bioimaging through Valencian Biobank Medical Imaging (GIMC in Spanish) as a basis for managing and extracting knowledge from a bioimaging bank, providing that knowledge as services with high added value and expertise to the Electronic Patient History System (HSE), thus bringing the results of R & D to the patient, improving the quality of the information contained therein. R & D Cloud CEIB has four general modules: Search engine (SE), manager of clinical trials (GEBID), anonymizer (ANON) and motor knowledge (BIKE). The BIKE is the central module and through its sub modules analyses and generates knowledge to provide to the HSE through services. The technology used in R & D Cloud CEIB is completely based on Open Source.

Within the BIKE, we focus on the development of the classifier module (BIKE-Classifier), which aims to establish a method for the extraction of biomarkers for bioimaging and subsequent analysis to obtain a classification in bioimaging available pools following GIMC diagnostic experience.

1 Introduction

The bioimaging has now become one of the most innovative multidisciplinary fields of medical research given the important role it plays in the diagnosis of diseases. New needs and improved technology require us to look for the best proposals to promote diagnostic based on medical imaging with the help of innovative technologies and signal analysis through optimization of our platforms. That is why,

Jose Maria Salinas · Miguel Cazorla
Dpto. Ciencia de la Computacion e I.A. Universidad de Alicante, Spain

Maria de la Iglesia-Vaya · Luis Marti Bonmati · Rosa Valenzuela
CEIB in Agencia Valenciana de Salud, Spain
e-mail: {salinas,miguel}@dccia.ua.es,
        {delaiglesia_mar,marti_lui,valenzuela_ros}@gva.es

S. Omatu et al. (Eds.): Distributed Computing and Artificial Intelligence, AISC 151, pp. 331–338.
springerlink.com

among other display systems for medical imaging diagnosis is considered cutting-edge medical devices, which are due to incorporate advanced analysis techniques to meet the expectations that society expected of public services. Among others, the standardization of imaging biomarkers, providing increasing value to aid imaging by obtaining objective measures to identify, measure and monitor those underlying pathophysiologic processes not detectable by the observer's subjectivity .

In Valencian Community, the Health Information System of the Valencian Health Service (AVS) is a particularly large portfolio that offers an assortment of highly specialized solutions. The centralization of bioimaging through Valencian Biobank Medical Imaging (GIMC), will support R & D to the scientific community through the implementation of logic services and retrieval determined bioimaging sets. For the exploitation of the valuable information stored in the GIMC, there was designed a logical bus of R & D services in the cloud based on open source technologies, which we call R & D Cloud CEIB. This is intended to provide high service expertise and excellence in biomedical imaging as a portfolio in this area through service-oriented architecture (SOA) which is being implemented in the AVS within the project epSOS - Smart Open Services for European Patients to provide the highest quality and information of the patient's clinical history (HSE).

Within R & D Cloud CEIB we can find services that enable data mining on the information entered from traditional capture devices of medical bioimaging through the headings of the DICOM [12] standard format; advanced post processing bioimaging techniques through open source libraries (FSL [1], etc.); and different tools for the diagnosis, among which is the bioimaging classifier from the optimal selection of visual biomarkers which will be discussed later on.

The article continues with an overview of the centralized imaging system in AVS (GIMC) and a short description of the proposed system R & D Cloud CEIB defining each of the modules to better understand the global system. Then, it comes the classifier bioimaging, establishing the general features of it. In the last section, conclusions and future work will be outlined the main objectives of the system R & D Cloud CEIB.

2 System Overview

In the field of bioimaging, centralized storage systems are a reference within the strategic framework of the AVS and the European Community (EuroBioimage, The Euro-Bioimaging Vision "to provide a clear path of access to imaging technologies for every biomedical scientist in Europe"), creating a Europe-wide plan for this type of infrastructure that are harmonized and coordinated among all the nodes involved.

As a result of the creation of GIMC, images from patients from the entire population of the Valencian Community through archiving systems and departmental image transmission (PACS) will form the basis of knowledge of the future community science in our society through R & D services that are presented. The architecture defined in the R & D Cloud CEIB is defined as the following elements: bioimaging

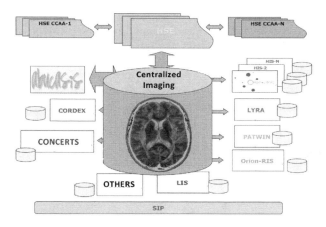

Fig. 1 Centralized imaging system of AVS.

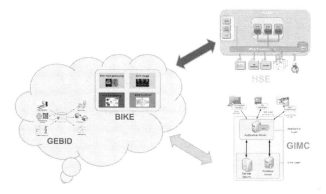

Fig. 2 Structure of R & D Cloud CEIB

bank, scientific community, search engine, anonymizer, clinical trials manager and knowledge engine. In the following sections we describe each of those elements.

2.1 Bioimaging Bank (GIMC)

The Valencian Medical Imaging Biobank (GIMC) is the system in charge of centralized storage of all the bioimaging of the AVS, having as sources all the bioimaging generated in different health centres across the Valencian Community through the synchronized copy of their internal PACS. The GIMC is comprised in three blocks: The storage system, which manages the optimized storage of all the images collected in DICOM format; the database system, which manages DICOM headers of images received through a relational database (storage and index); the application server, which allows abstracting the system of these two previous blocks from an application layer that facilitates the management of biobank image information. The GIMC provides access to all of the AVS corporate applications, such as Orion Clinic (specialized care management), Abucasis management (primary care) and other applications through DICOM web access services (WADO). GIMC also forms the storage basis for the retrieval systems (search engines) implemented in the R & D Cloud CEIB that enables the provision of indexed image blocks to the knowledge engines and clinical trial manager.

2.2 Scientific Community (CC)

One of the main goals pursued by the proposed system is to be able to offer to the scientific community a basis for clinical trials from subsets of images from the GIMC. Scientific community can make structured requests to the GIMC system (e.g. 40-50/female/ms/alicante will return a set of images from women from 40-50 years old with multiple schlrerosis diagnostic from Alicante province) that will provide to the scientific community tools to manage this information as well as a set of advanced bioimaging post processing tools.

2.3 Anonymizer (ANON)

In compliance with the Data Protection Law, all images provided from the GIMC must be provided in an anonymised form, always preserving the anonymization of the patient information. System allows different types of anonymity, from the alteration of the existing text information in DICOM headers up to image-level deformation of parts that can identify the patient (especially in neuroimaging obtained by magnetic resonance). This part provides a restricted module, in which the information needed to reverse the anonymization process is stored, so that, given the event that specific needs of a trial, if more information is needed of the patient under study, the system will be able to provide more information about the patient.

2.4 R & D Bioimage Trials Manager System (GEBID)

The R & D bioimage trials manager system is designed to provide to the scientific community a platform to help them to manage information from clinical trials. The GEBID is based on the implementation of a customized instance of XNAT [8] (eXtensible Neuroimaging Archive Toolkit). XNAT is an open source platform designed to facilitate the management of image sets and associated data (assessments, reconstructions and any other information). Initially it is designed to work with neuroimaging, but the open data model and customizable XML-based technologies allow to adapt the platform for any type of bioimaging. XNAT follows a three-tier architecture that includes a data file, an user interface and a middleware engine. The data file can be incorporated into the platform through different ways, such as XML files, web forms, DICOM transfers from image capture devices or image viewers like Oxiris and so on. Among its most important features are the personalized safe access to information, quality control processes of data and image information, classification and storage of data, ability to run custom searches, communication with bioimaging generating systems, programmability of process flows using scripts (pipelines), the incorporation of intermediate results and conclusions to the study, recording of all actions taken to control and monitor quality, etc. All these features make XNAT an ideal platform for the management of clinical trials.

2.5 Bioimaging Knowledge Engine (BIKE)

The knowledge engine of the R & D Cloud CEIB (BIKE) consists of a series of modules: bioimaging post-processing, defining and quantifying biomarkers aid, study of DICOM header and bioimaging classifier. We describe this modules in the next sections.

2.5.1 Module of Bioimaging Post-processing Aid (BIKE-Post-processing)

The digital processing of data obtained by the medical imaging acquire machines is a field that can extract information which is beyond the simple observation of images on film or on monitors of the diagnostics services. The digital bioimaging processing allows to precise the anatomy of the area of study and obtain functional, and even molecular, information. With this service, the BIKE equips the system with a set of tools based in open source graphics libraries (FSL) that helps the bioimaging post-processing in clinical trials through GEBID. These tools may be used individually or grouped sequentially through process management applications such as LONI Pipelines. Given the complexity of many of the post-processing techniques required for the calculation of results, the system will leverage the cloud architecture to enhance parallel processing. BIKE-post-processing serves as a basis for all necessary bioimaging analysis in other modules of BIKE such as bioimaging classifier and the module of defining and quantifying biomarkers.

2.5.2 Module of Defining and Quantifying Biomarkers (BIKE-Image)

Image biomarkers define objective features extracted from medical images, related to normal biological processes, diseases or therapeutic responses. In recent years it has been shown that imaging biomarkers provide useful complementary information to traditional radiologic diagnosis to establish the presence of a disturbance or injury; to measure biological status; to define its natural history and progress; stratify the abnormal phenotypes and to evaluate the effects of treatment.

To develop an imaging biomarker it must be performed a series of steps designed to validate their relationship to the reality studied and checking its reliability, both clinical and technical. BIKE-Image module provides all the necessary tools to carry out effectively from simple measurements of size or shape to the implementation of complex models. This facilitates the definition of proof of concept and mechanism, standardized and optimized acquisition of anatomical images, functional and molecular, analysis of data using computer models, adequate visualization of the results, obtaining appropriate statistical measures, and testing of principle, efficacy and effectiveness. BIKE-Image module used as the basis for these processes, tools provided by the BIKE-Post-processing.

2.5.3 Module of Study of DICOM Headers (BIKE-Datamining)

Within the world of medical imaging, DICOM is the standard format used. This format, file-level, also includes the image information obtained from the radiological procedure, includes in its header information in text format such as patient demographic information, clinical information, quality control data of the image, technical data of the capture device and image type, and many more features. BIKE-Datamining module provides tools to exploit that information in the DICOM headers for creating dashboards for the analysis of various indicators of quality, radiation, etc. . Using these data, specialized statistical reports can be generated, which allow the quantification and control of processes, and reporting corporate structured format using the DICOM-SR.

2.5.4 Bioimaging Classifier Module (BIKE-Classifier)

Starting from the definition of clinical decision support system, an image decision support system (SADI) is a computer system that provides specific knowledge for the interpretation of medical imaging for the diagnosis purpose, prognosis, treatment or management processes of care. The features of SADI search may include findings associated with the diagnosis or prognosis of the patient, therapy planning and control and operations, quality control of biomedical signals multicenter biobanks anomalous pattern matching. The use of these systems can enhance the medical skills in the management of multiple variables in biomedical care processes and help achieve balance in the health service through the optimal use of resources and knowledge available.

As experiences in other communities, there is a system of computer-aided diagnosis (CAD) for mammography, already implemented in some hospitals in Castilla La Mancha community, which processes images from mammography generating the same analysis in which indicate the possible injuries that may exist, thus helping the radiologist in their diagnosis. BIKE includes among its modules the BIKE-classifier, a classification system that allows multiple classification in a number of existing diagnostic groups. This classification is based on an optimal selection of biomarkers and visual characteristics. This selection of biomarkers (Feature Selection (FS [19])) can be performed using mutual information. FS is a combinatorial computational complexity problem. FS Methods must be oriented to find suboptimal solutions in a feasible number of iterations.

The BIKE-Classifier uses the BIKE-Image to extract biomarkers and other visual indicators quantified, and the BIKE-Post-processing for the extraction of visual features that are not based on biomarkers. This system performs a supervised learning.

In supervised classification, models or algorithms are capable of learning from a set of instances or cases. Each instance is a vector of features labelled with a class variable. Formally, the problem of supervised classification is to assign a set value of the variable class to a new instance. A classifier can be viewed as a class assignment to each of the instances. In the first phase we have a set of training or learning (for designing the classifier) and another called test or validation (for classification), these will serve to build a model or rule for classification. In the second phase is the actual process of classifying objects or samples of the class is unknown to which they belong.

The methods that have been used so far are: neural networks, Bayesian classifiers, support vector machines. In our case, we propose the use of the Support Vector Machines (SVM). The entry of this process is the optimal selection of biomarkers and visual markers that allow classification into diagnostic groups in the system.

3   Conclusions and Future Work

The GIMC generated within the AVS is an ideal data source for analysing the images acquired with all bioimaging modalities. The proposed system, R & D Cloud CEIB, will provide these bioimages available to the community science through different tools like search engine (SE), clinical trials manager (GEBID) and knowledge engine (BIKE). BIKE provides services to perform data mining activities at the DICOM header (BIKE-Datamining), image post-processing (BIKE-Post-processing), definition and quantification of biomarkers (BIKE-Image) and classification (BIKE-classifier).

The main goal of R & D Cloud CEIB is that all knowledge acquired in the system will be move via web services to HSE to improve patient's data. The possibility of querying real-time visual analysis and extraction of information through the bioimaging and get specific reports comparing the experience of the biobank allows much more information available to the doctor, improving quality in patient's diagnosis process.

References

1. FSL Group, http://www.fmrib.ox.ac.uk/fsl/
2. Jenkinson, M., Beckmann, C.F., Behrens, T.E.J., Woolrich, M.W., Smith, S.M.: FSL. NeuroImage (2011) (in press)
3. Woolrich, M.W., Jbabdi, S., Patenaude, B., Chappell, M., Makni, S., Behrens, T., Beckmann, C., Jenkinson, M., Smith, S.M.: Bayesian analysis of neuroimaging data in FSL. NeuroImage 45, S173–S186 (2009)
4. Smith, S.M., Jenkinson, M., Woolrich, M.W., Beckmann, C.F., Behrens, T.E.J., Johansen-Berg, H., Bannister, P.R., De Luca, M., Drobnjak, I., Flitney, D.E., Niazy, R., Saunders, J., Vickers, J., Zhang, Y., De Stefano, N., Brady, J.M., Matthews, P.M.: Advances in functional and structural MR image analysis and implementation as FSL. NeuroImage 23(S1), 208–219 (2004)
5. caBIG Community Website, https://cabig.nci.nih.gov/

6. Rex, D.E., Ma, J.Q., Toga, A.W.: The LONI Pipeline Processing Environment. Neuroimage 19(3), 1033–1048 (2003)
7. Dinov, I.D., Lozev, K., Petrosyan, P., Liu, Z., Eggert, P., Pierce, J., Zamanyan, A., Chakrapani, S., Van Horn, J.D., Parker, D.S., Magsipoc, R., Leung, K., Gutman, B., Woods, R.P., Toga, A.W.: Neuroimaging Study Designs, Computational Analyses and Data Provenance Using the LONI Pipeline. PLoS ONE 5(9), e13070 (2010), doi:10.1371/journal.pone.0013070
8. XNAT - Open source informatics for biomedical imaging research, http://www.xnat.org
9. Marcus, D.S., Olsen, T., Ramaratnam, M., Buckner, R.L.: The Extensible Neuroimaging Archive Toolkit (XNAT): An informatics platform for managing, exploring, and sharing neuroimaging data. Neuroinformatics 5(1), 11–34 (2007)
10. Manjon, J.V., Marti-Bonmati, L., Robles, M., Celda, B.: Postproceso en Imagen Medica: morfologia, funcional y molecular
11. Marti Bonmati, L., Alberich-Bayarri, A., Garcia-Marti, G., Sanz Requena, R., Perez Castillo, C., Carot Sierra, J.M., Manjon Herrera, J.V.: Biomarcadores de imagen, imagen cuantitativa y bioingenieria. In: Radiologia 2011 (2011)
12. DICOM, http://dicom.nema.org/
13. Downing, G.: Biomarkers Definitions Working Group. Biomarkers and surrogate endpoints. Clin. Pharmacol. Therap. 69, 89–95 (2001)
14. Schuster, D.: The opportunities and challenges of developing imaging biomarkers to study lung function and disease. Am. J. Respir. Crit. Care Med. 176, 22–30 (2007)
15. Van Beers, B., Cuenod, C.A., Mart-Bonmat, L., Matos, C., Niessen, W., Padhani, A.: European Society of Radiology Working Group on Imaging Biomarkers. White paper on Imaging Biomarkers. Insights Imaging. 1, 42–45 (2010)
16. Campbell, C.: Kernel methods: a survey of current techniques. Neurocomputing 48, 63–84 (2002)
17. Mavroforakis, M.E., Georgiou, H.V., Dimitropoulos, N., Cavouras, D., Theodoridis, S.: Mammographic masses characterization based on localized texture and dataset fractal analysis using linear, neural and support vector machine classifiers. Artif Intell Med. 37(2), 145–162 (2006)
18. Siuly, Li, Y., Wen, P.P.: Clustering technique-based least square support vector machine for EEG signal classification. Comput Methods Programs Biomed. 104(3), 358–372 (2011)
19. Bonev, B., Escolano, F., Cazorla, M.: Feature selection, mutual information, and the classification of high-dimensional patterns: Applications to image classification and microarray data analysis. Pattern Analysis and Applications 11(3-4) (August 2008)

Performance Comparison of Hierarchical Checkpoint Protocols Grid Computing

Ndeye Massata Ndiaye, Pierre Sens, and Ousmane Thiare

Abstract. Grid infrastructure is a large set of nodes geographically distributed and connected by a communication. In this context, fault tolerance is a necessity imposed by the distribution as any node can fail at any moment and the average time between failures highly decreases. To improve the robustness of supercomputing applications in the presence of failures, many techniques have been developed to provide resistance to these faults of the system. Fault tolerance is intended to allow the system to provide service as specified in spite of occurrences of faults. To meet this need, several techniques have been proposed in the literature. We will study the protocols based on rollback recovery classified into two categories: checkpoint-based rollback recovery protocols and message logging protocols. However, the performance of a protocol depends on the characteristics of the system, network and applications running. Faced with the constraints of large-scale environments, many of algorithms of the literature showed inadequate. Given an application environment and a system, it is not easy to identify the recovery protocol that is most appropriate for a cluster or hierarchical environment, like grid computing. Hence there is a need to implement these protocols in a hierarchical fashion to compare their performance in grid computing. In this paper, we propose hierarchical version of these protocols. We have implemented and compare their performance in clusters and grid computing using the Omnet++ simulator.

Ndeye Massata Ndiaye
Gaston Berger Univsersity of Saint-Louis Senegal and Regal team Paris Jussieu
e-mail: ndeye-massata.ndiaye@lip6.fr

Pierre Sens
Regal team Paris Jussieu
e-mail: Pierre.Sens@lip6.fr

Ousmane Thiare
Gaston Berger University of Saint-Louis Senegal
e-mail: ousmane.thiare@ugb.edu.sn

S. Omatu et al. (Eds.): Distributed Computing and Artificial Intelligence, AISC 151, pp. 339–346.
springerlink.com © Springer-Verlag Berlin Heidelberg 2012

1 Introduction

Today, grid computing technologies make it possible to securely share data and programs for multiple computers, whether desktop or personal supercomputers. These
resources are networked and shared through software solutions. Many grids are appearing in the sciences, production grids are now being implemented in companies and among agencies: Grid'5000, TeraGrid, Sun Grid, Xgrid ... Grid computing
will allow dynamic sharing of resources among participants, organizations and businesses in order to be able to pool, and thus run compute-intensive applications or
treatment of very large volumes of data. Since the probability of failure increases
with a rising number of components, fault tolerance is an essential characteristic of
massively parallel systems. Such systems must provide redundancy and mechanisms
to detect and localize errors as well as to reconfigure the system and to recover from
error states. A fault tolerant approach may therefore be useful in order to potentially
prevent a faulty node affecting the overall performance of the application. Fault tolerance appears then as an indispensable element in grid computing. Many protocols
for distributed computing have been designed [1]. These protocols are classified
into four different groups, namely, coordinated checkpointing, communication induced checkpointing, independent checkpointing and log-based protocols. We have
implemented and compared the performance of these protocols in clusters and grid
computing using the Omnet++ simulator [7]. Section 2 describes the protocols implemented in Omnet++. The experimental setup and results obtained by executing
these protocols are presented in Section 4. In section 5, we present the related work
and finally section 6 concludes.

2 Checkpoint and Rollback-Recovery Protocols

Checkpointing is a standard method for the repairing of faults in systems. The idea is
to save the state of the system on a stable period to prevent breakdowns (Figure. 1).
That way, when you restart after a power failure, the state last saved is restored and
the execution before the crash resumes. The overall status of a distributed system is
defined by the union of local states of all processes belonging to the system.

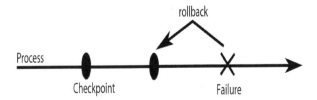

Fig. 1 Rollback recovery

There are two main classes of protocols: coordinated checkpointing and message logging.

2.1 Coordinated Checkpointing

The protocol requires processes coordinate their checkpoints to form a consistent global state. A global state is consistent if it does not include any orphan messages (i.e a message received but not already sent). This approach simplifies the recovery and avoids the domino effect, since every process always restarts at the resume point later. Also, the protocol requires each process to maintain only one permanent checkpoint in stable storage, reducing the overhead due to storage and release of checkpoints (garbage collection) [1].

Its main drawback however is the large latency that require interaction with the outside world; in this case the solution is to perform a checkpoint after every input / output. To improve the performance of the backup coordinated, several techniques have been proposed. We have implemented as non-blocking coordinated checkpointing.

- Non-blocking coordinated checkpointing: the example of coordinated checkpoint non-blocking is that of Chandy and Lamport algorithm [2]. This algorithm uses markers to coordinate the backup, and operates under the assumption of FIFO channels. In [3], a comparison of protocols between a blocking and a non-blocking coordinated checkpoint has been made. Experiments have shown that the synchronization between nodes induced by the blocking protocol further penalize the performance of the calculation with a non-blocking protocol. However, using frequencies of taken checkpoints usual performance of the blocking approach is better on a cluster to high-performance communications.
- Communication induced checkpointing: this protocol defines two types of checkpoints [1]: local checkpoints taken by processes independently, to avoid the synchronization of coordinated backup and forced checkpoints based on messages sent and received and dependency information carried 'piggyback' on these posts, so to avoid the domino effect of uncoordinated backup, ensuring the advancement of online collection. Unlike coordinated checkpoint protocols, the additional cost due to the medium access protocol disappears because the protocol does not require any exchange of message to force a checkpoint: this information is inserted piggyback on the messages exchanged.

2.2 Message-Logging Protocols

The logging mechanism uses the fact that a process can be modeled as a sequence of deterministic state intervals, each event begins with a non-deterministic. An event may be the receiving or the issuing of a message or other events in the process. It is deterministic if from a given initial state, it always occurs at the same final state [1]. The principle of Logging is to record on a reliable storage support any occurrences of non-deterministic events to be able to replay them in recovering from a failure. During execution, each process performs periodical backups of their states, and records log information about messages exchanged between processes. There are three message-logging categories: optimistic, pessimistic and causal.

- Pessimistic message-logging: this protocol was designed under the assumption that a failure may occur after any nondeterministic event (i.e. message reception). It is often made referring to the synchronized because when logging process logs an event of non-deterministic stable memory, it awaits for an acknowledgment to resume its execution. In a pessimistic logging system, the status of each process can be recovered independently. The main drawback is the high latency of communications which results in degradation of the applications response time.

- Optimistic message-logging: this protocol uses the assumption that the logging of a message on reliable support will be complete before a failure occurs. Indeed, during the execution of the process, the determinants of messages are stored in volatile memory, before being saved periodically on stable support. The storage of stable memory is asynchronous. Induced latency is then very low. However, a failure may occur before the messages are saved on stable storage support. In this case, the information stored in volatile memory of the process down are lost and the messages sent by this process are orphaned.

- Causal message-logging: this protocol combines the advantages of both previous methods. As optimistic logging, it avoids the synchronized access to the stable support, except during the input / output. As pessimistic logging, it allows the process to make interactions with the outside world independently, and does not create process orphan. Causal logging protocols piggyback determinants of messages previously received on outgoing messages so that they are stored by their receivers.

3 Hierarchical Checkpointing for Grids

The architecture of a grid can be defined as a set of clusters connected by a WAN-type network. The cluster consists of multiple nodes connected by a broadband network. We adopt a hierarchical scheme. In each cluster, there is one leader connected to all other nodes of its cluster. All leaders are connected together (Figure. 2). The leader assumes the role of intermediary in the inter-cluster communications. The backup takes place in four phases:

- Initialization: an initiator sends a checkpoint-request to its leader,
- Coordination of leaders: the leader transfers the checkpoint request to the other leaders
- Local checkpointing : Each leader initiates a checkpoint inside its cluster
- Termination: When local checkpoint is over, each leader sends an acknowledgement to the initial leader.

The recovery follows the same rules as the backup: coordination phase of the leaders, and a phase of recovery limited to the cluster.

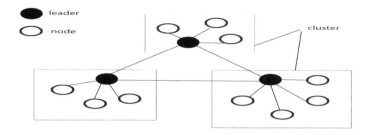

Fig. 2 Hierarchical checkpointing for grids

4 Performance Evaluation

In most of the previous studies, fault tolerance algorithms were tested in flat architectures, namely in a cluster. The aim of our study is to determine which algorithm
best suits the architectural grid. For that purpose, we implement the seven checkpoint algorithms described in Section 2: the 3 main messages logging protocols
(represented as ML in the figures), Chandy-Lamport, Communication induced protocol (CIC in figures), and blocking coordinated checkpointing (CheckpointCoord).
We compare the performance of these algorithms in cluster and grid environments.
We use the Omnet++ simulator [7]. The cluster is configured with 25 nodes. For the
grid configuration, 25 nodes were uniformly spread in 5 clusters. The intra-cluster
delay is fixed to 0.1 ms and the inter-cluster delay is fixed to 100ms. Our tests were
carried out with 50 application processes. Messages between processes were randomly generated.

4.1 Failure Free Performance

Figure. 3 presents the performance of the algorithms in both configurations. It is obvious that the time taken to run an application with checkpointing is longer than the
time it takes to run it without checkpoint. Protocol overhead checkpoint coordinated
non-blocking is less compared to other approaches to that phase synchronization is
limited to the cluster and the second concerns only the leaders of each cluster. The
additional cost of communications-driven approach is due to the forced checkpoints
during execution. Logging protocols are sensitive to characteristics of the application, especially in communications-intensive applications. Indeed, they produce a
large overhead due to the backup of messages on stable storage and the increasing
size of messages to piggyback determinants.

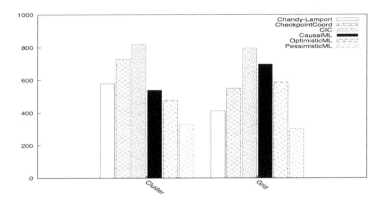

Fig. 3 Failure free performance, Checkpoint interval=180s, Execution time=900s

4.2 Recovery Time

The recovery time depends on the number of checkpoints maintained by the protocol and the number of rollbacks. In coordinated checkpointing and pessimistic logging, recovery is simplified because the system is rolled back only to the most recent checkpoint. In the grid approach, the additional cost of recovery decreases slightly. Indeed, if the faulty node has no dependencies with nodes of other cluster nodes, the fault is confined to the cluster node's fault. So all the nodes of the grid do not perform the recovery procedure. By cons, if the inter-cluster communications are intensive, the overhead increases as in the case of causal and optimistic logging.

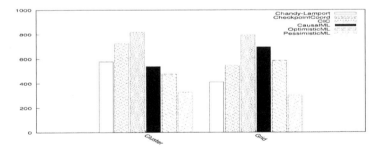

Fig. 4 Overhead of recovery,checkpoint interval=180s,execution time=900s,numbers of fault=10

4.3 Number of Rollbacks

For coordinated checkpoint protocols, all processes must resume during recovery. The logging protocol reduce the number of rollback. This number is minimal in pessimistic approach since only faulty processes need to be rolled back. For the

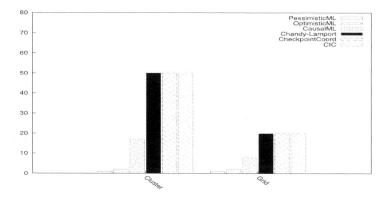

Fig. 5 Number of process, Checkpoint interval=180s, Execution time=900s, Numbers of fault= 1

other logging protocol, this number depends on the information stored in backups and in the main memory of correct processes

5   Related Work

Paul and al. [4] proposes a hierarchical protocol based on coordinated checkpoint. The hierarchical checkpoint protocol is in two phases. The first phase is the execution of the blocking coordinated checkpoint algorithm limited to the cluster. The second phase is a coordinated checkpoint but the leaders of the clusters are the only participants, with the initiator, which acts as a coordinator. The experiments showed that the overhead of checkpointing in the hierarchical approach is lower than in the standard flat coordinated protocol. However the protocol hierarchy is sensitive to the frequency of messages between clusters. Indeed the extra cost of checkpoint increases progressively as the frequency of messages increases, and tends towards that of the checkpoint protocol standard.

Bhatia and al. [5] propose a hierarchical causal logging protocol that addresses the scalability problems of causal logging. Authors propose a hierarchical approach using a set of proxies spread on the network that act as a distributed cache. This approach highly reduces the amount of information piggybacked on each messages. However, the use of proxies decreases the performance of recovery since the recovery information is spread on the proxies.

Monnet and al. [6] propose a hierarchical checkpointing protocol which combines coordinated checkpointing inside clusters and a checkpoints induced by communications between clusters. Simulation of the protocol shows that it generates a high number of forced checkpoints when the communication rate between clusters increases. Then, this approach is more suitable for code coupling applications where communications are mainly local inside clusters.

6 Conclusion

In this paper, we compared checkpoint protocols and message-logging in grid computing. We propose a hierarchical approach to combine different algorithms. We found that the protocols that require the recovery of all processes in case of single failure are poorly suited to systems with many processes. The message logging protocols are more suitable for large configuration with the exception of some causal logging approach which induces communications to all processes during the recovery. Non-blocking coordinated checkpoint are not sensitive to the rate of communications. They therefore represent an attractive solution for applications and highly interconnected grid architectures by reducing the number of markers sent during the synchronization phase.

References

1. Elnozahy, E.N., Alvisi, L., Wang, Y.-M., Johnson, D.B.: A Survey of Rollback- Recovery Protocols in Message-Passing Systems. ACM Computing Surveys 34(3), 375–408 (2002)
2. Chandy, M., Lamport, L.: Distributed snapshots: Determining global states of distributed systems. ACM Trans. Computing Systems 3(1), 63–75
3. Coti, C., Herault, T., Lemarinier, P., Pilard, L., Rezmerita, A., Rodriguez, E., Cappello, F.: Blocking vs. non-blocking coordinated checkpointing for large-scale fault tolerant MPI. In: SC 2006: Proceedings of the 2006 ACM/IEEE Conference on Supercomputing, New York, USA, p. 127 (2006)
4. Himadri, S.-P., Gupta, A., Badrinath, R.: Hierarchical Coordinated Checkpointing Protocol. In: International Conference on Parallel and Distributed Computing Systems, pp. 240–245 (2002)
5. Bhatia, K., Marzullo, K., Alvisi, L.: Scalable causal Message Logging for Wide- Area Environments. Concurrency and Computation: Practice and Experience 15(3), 873–889 (2003)
6. Monnet, S., Morin, C., Badrinath, R.: Hybrid Checkpointing for Parallel Applications in cluster Federations. In: Proc. 4th IEEE/ACM International Symposium on Cluster Computing and the Grid, Chicago, USA, pp. 773–782 (2004)
7. http://www.omnetpp.org

A Scientific Computing Environment
for Accessing Grid Computing Systems
Using Cloud Services

Mariano Raboso, José A. de la Varga, Myriam Codes, Jesús Alonso,
Lara del Val, María I. Jiménez, Alberto Izquierdo, and Juan J. Villacorta

Abstract. This paper shows how virtualization techniques can be introduced into
the grid computing infrastructure to provide a transparent and homogeneous
scientific computing environment. Today's trends in grid computing propose a
shared model where different organizations make use of a heterogeneous grid,
frequently a cluster of clusters (CoC) of computing and network resources. This
paper shows how a grid computing model can be virtualized, obtaining a simple
and homogeneous interface that can be offered to the clients. The proposed system
called *virtual grid*, uses virtualization support and is developed from integration of
standard grid and cloud computing technologies. Furthermore, a Scientific
Computing Environment (SCE) has been developed to provide uniform access to
the virtual grid.

Keywords: grid computing, cloud computing, virtualization, scientific computing
environment, message passing interface.

1 Introduction

Increasing demand of computer resources for scientific research has been a strong
motivation for the community to develop a wide variety of high performance
computing infrastructures (HPC). Huge supercomputer resources are not always

Mariano Raboso · José A. de la Varga · Myriam Codes · Jesús Alonso
Facultad de Informática, Universidad Pontificia de Salamanca. Compañía 5,
37002 Salamanca, Spain
e-mail: mrabosoma@upsa.es

Lara del Val · María I. Jiménez · Alberto Izquierdo · Juan J. Villacorta
Departamento de Teoría de la Señal y Comunicaciones e Ingeniería Telemática,
Universidad de Valladolid, E.T.S.I. Telecomunicación, Paseo Belén 15,
47011 Valladolid, Spain

S. Omatu et al. (Eds.): Distributed Computing and Artificial Intelligence, AISC 151, pp. 347–353.
springerlink.com   © Springer-Verlag Berlin Heidelberg 2012

available to small research groups, usually limited by restrictive budgets, deploying tasks not always suitable for these systems. Initiatives, such as the European Grid Infrastructure (EGI) [1], aim to develop a sustainable grid infrastructure for all European researchers.

A grid computing system is a grid of parallel and distributed computing resources, working together towards a single goal and providing high global computational power if convenient parallelization and concurrency issues are attained. A cloud computing system is an approach to computing, based on on-demand efficient use of aggregate resources, self-managed and consumed as a service.

We have developed an infrastructure combining grid and cloud services that provides transparent virtual grids to the clients. Although the grid is shared, the clients will use their own virtual grids with physical computing nodes through common services provided by the SCE (Scientific Computing Environment). Classic grid environments provide shared resources as CPU time and queues for batch processes but virtual grids, as they provide full access to their own infrastructure, allow the user to perform their own administration tasks.

Other solutions [2][3][4][5] use customized or proprietary software for sharing implementation, creating their own interfaces and middleware, or use cloud technology as a part of the HPC system, affecting the overall system performance. Our solution uses a standard and open cloud technology based on EC2 Amazon technology that allows not only a private cloud model but also a public or even hybrid. Cluster performance is not influenced by cloud technology, as once the virtual grid is configured and assigned, clients get direct and full access to the computing resources.

Furthermore, virtual clusters assignment can take in account cluster location so computing assignments can be done depending on site location. This is very useful to avoid network delays or low-bandwidth link issues when several organizations shared resources over a network. We have designed the cloud infrastructure to integrate resources at the Computer Science Faculty (Pontifical University of Salamanca) and the Array Processing Group (GPA) at the Valladolid University. Both universities are connected through RedIris network.

Integrating grid and cloud technologies has interesting advantages as efficiency and security improvements. Furthermore, providing a common interface by an SCE, makes the system more accessible and hides the clients from the technical details for customizing their own grid.

- Uniformity. Virtualizing the grid implies that every client sees their own grids and so, collateral effects can be minimized or even null. Users access through a common interface provided by the SCE using an authentication service. Cloud services are in charge of running virtual machines providing the master nodes of every virtual grid, but no participate from computation jobs.
- Customizable grids. It is possible to offer customized grids by deploying specific images for the virtual master machines. The cloud infrastructure and configuration server provide the images that can be customized with user-defined parameters such CPU resources, memory, nodes number or time limit.
- Efficient assignment of local grid infrastructure allowing users to run jobs from different sites, avoiding bottlenecks caused by bandwidth or latency issues. As

grid nodes can be located on different sites connected through WAN links, the configuration server can select only local nodes to deploy the grid.

- Flexible accounting. It is possible to extend general accounting plans from the cloud computing system, to offer different quality of service. This feature is provided by the cloud service.
- Improved LAN security. Independent grids isolate user jobs running in the grid. VLAN management is implemented on the cloud system, so different grids can be assigned to different VLANs.
- User-level security. Cloud technology by Eucalyptus uses credentials and certificates to login into the system and manage the virtual machines.

The next sections describe how virtualization process can be applied to the grid. Virtualization techniques used are described in detail in section 2. In order to provide a common access, a SCE is developed. It is described in section 3. Finally, some conclusions are made.

2 Virtualizing the HPC System

In order to batch their jobs, HPC users must usually connect to a master node. Some effort has been made to offer a standard and uniform interface, so hardware and software component details are transparent for users [6], providing a middleware that enable users to solve complex scientific problems, while using simple interfaces from their Scientific Computing Environments (SCE) [7].

Another solution for sharing the grid computing system can be addressed. The main goal is to manage local virtual grids to serve client requests. This virtual grid contains a master (virtual machine) and several computing nodes. The system runs a new virtual machine using a previously configured master server image. This master machine is responsible of running jobs in the new grid and collecting statistics for the grid.

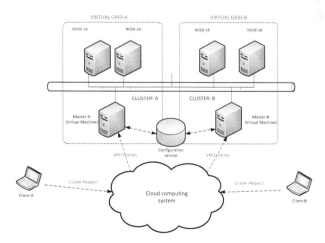

Fig. 1 Virtual grid system architecture and VM creation

Once the new master is running, the clients will manage their own grid with the computing resources (nodes) assigned through the master and the SCE. The infrastructure developed is called *virtual grid*. Figure 1 shows the virtual machine creation process.

When clients want to gain access to the grid (virtual), they first make a request to the cloud system through the SCE, a web frontend that integrates the cloud computing and configuration services. The cloud service then runs an instance of a master virtual machine that downloads and self-configures the grid with the subnet, nodes and policies assigned. As a result, the clients get their own grids with a master node and a set of computing nodes assigned.

The configuration is dynamically loaded by the virtual machine from the configuration server and includes node assignment, job policy algorithms and time limit if configured. This configuration is shown to the client through the SCE.

Figure 2 shows a sequence diagram describing relationships between clients, the SCE and the configuration server.

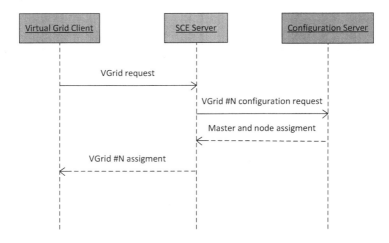

Fig. 2 Sequence diagram showing Virtual Grid creation and assignment.

The grid is based on a powerful technology for grid implementation called MPI (Message Passing Interface). Open-MPI middleware [8], which is used worldwide, provides a cluster of associated machines performing a grid. Each grid has a master node controlling the program execution. This special node is here performed by a virtual machine using the cloud technology. The cloud services are provided integrating Eucalyptus [9] technology and Ubuntu server. The rest of the nodes are not virtualized and are only assigned to the corresponding virtual grid.

Once the virtual grid is assigned to a client, it is accessible through a ssh session to the master node. Programs, usually written in C, C++ or Fortran then runs using the middleware with software MPI wrappers as mpicc and mpirun.

The configuration server stores on a database all the details for the system, the cloud and grid computing nodes available, client sessions details and node assignments.

3 Scientific Computing Environment for Virtual Grids

Using a scientific computing environment makes easier, for the scientific groups, to use HPC systems. Complex hardware and software architectures are convenient encapsulated as researchers, from a variety of science fields, are not usually concerned to such details. Therefore, users can access by a common and uniform interface.

It hides virtual grid implementation providing a direct link with the machine that is responsible of batching their jobs. The grid can be also easily configured using a web based frontend, and monitored using a platform called Ganglia.

Configuration tasks through the SCE allow the users of each virtual grid, to resize the infrastructure to match their requirements. Resources aggregated are measured in terms of number of slots, a typical measurement for MPI systems related to the number of processor cores. Figure 3 shows the configuration interface.

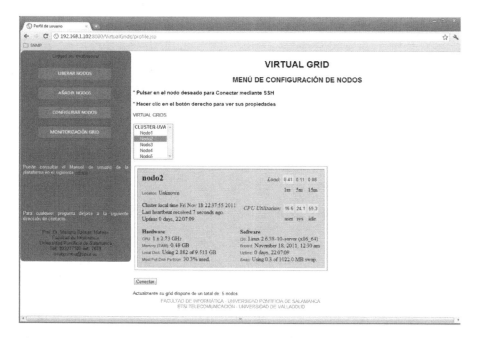

Fig. 3 SCE node configuration view for *CLUSTER-UVA* virtual grid.

When the client connects to the master machine a new window with a ssh terminal appears, providing full access to the grid. The client can connects to every node, but it usually only needs to batch the jobs in the master node. File systems are configured and connected through NFS, and Ganglia configuration files are already set by the configuration service when the master machine is started.

Performance is measured integrating Ganglia [10], a distributed monitoring system that uses a large variety of sensors that acquire information and send it to the management station (see Figure 4).

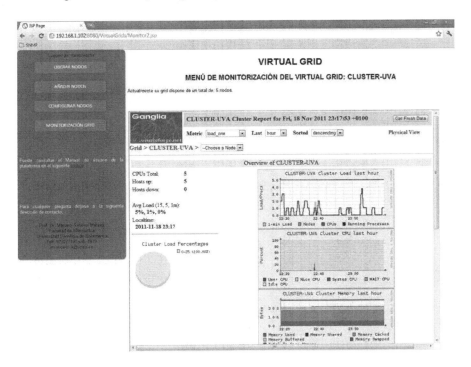

Fig. 4 Ganglia integrated performance view for *CLUSTER-UVA* virtual grid.

Ganglia monitoring system uses daemons running on every computing node that deliver statistics to the master node. A report can be downloaded from the master node where Ganglia is also installed. Through the SCE, the client views the statistics accessing via http protocol to the master machine.

4 Conclusions

Traditional high performance computing systems based on grid computing can be improved using cloud computing infrastructures. A new infrastructure for HPC called *virtual grid* has been developed, virtualizing the master node responsible of running user jobs. This virtualizing process improves security, flexibility and efficient use of resources.

Tests made have shown that delay when the cloud virtual machines start is too high. This issue does not affect to the grid performance as it only occurs once, when the virtual machine starts. It is being considered to use VMware virtualization API to deploy the master machines for faster serving the virtual grids.

Performance for the whole system is determined by the cluster node capacity: computing node hardware, network and MPI. The virtualization service only introduces delays when preparing the virtual grid upon client requests, so it can be measured independently. Current hardware configuration introduces delays from three to five minutes, depending on the workload.

This research has been supported by projects: 10MLA-IN-S08EI-1 (Pontifical University of Salamanca), and PON323B11-2 (Junta de Castilla y León).

References

[1] European Grid Infrastructure, http://www.egi.eu
[2] Walker, E., Gardner, J.P., Litvin, V., Turner, E.L.: Creating personal adaptive clusters for managing scientific jobs in a distributed computing environment. In: Challenges of Large Applications in Distributed Environments, pp. 95–103. IEEE (2006)
[3] Lin, L., Decker, K.M., Jognson, M.J., Domain, C., Souffez, Y.: ISCN: towards a distributed scientific computing environment. In: High Performance Computing on the Information Superhighway, April 28-May 2. HPC Asia 1997, pp. 157–162 (1997)
[4] Chine, K.: Scientific Computing Environments in the age of virtualization toward a universal platform for the Cloud. In: Chine, K.: 2009 IEEE International Workshop on Open-source Software for Scientific Computation (OSSC), pp. 44–48 (Septtember 2009)
[5] Li, X., Palit, H., Foo, Y.S., Hung, T.: Building an HPC-as-a-Service Toolkit for User-Interactive HPC Services in the Cloud. In: 2011 IEEE Workshops of International Conference on Advanced Information Networking and Applications (WAINA), March 22-25, pp. 369–374 (2011)
[6] Innovative Computing Laboratory, GridSolve: A system for Grid-enabling general-purpose scientific computing environments,
http://icl.cs.utk.edu/netsolve/
[7] Zhihui, D., Lingjiang, W., Haili, X., Hong, W., Xuebin, C.: A Lightweight Grid Middleware Based on OPENSSH - SCE. In: Sixth International Conference on Grid and Cooperative Computing, August 16-18, pp. 16–18 (2007)
[8] Open MPI: Open Source High Performance Computing.,
http://www.open-mpi.org
[9] Eucalyptus system,
http://www.eucalyptus.com/products/
ubuntu_enterprise_cloud
[10] Massie, M., Chun, B., Culler, D.: The Ganglia Distributed Monitoring System: Design, Implementation, and Experience. Parallel Computing 30(7) (July 2004)

Grid Computing and CBR Deployment: Monitoring Principles for a Suitable Engagement

Luis F. Castillo, Gustavo Isaza, Manuel Glez Bedia, Miguel Aguilera, and Juan David Correa

Abstract. This paper presents a mathematical technique for modeling the generation of Grid-solutions employing a Case based reasoning system (CBR). Roughly speaking, an intelligent system that tries to be adapted to highly dynamic environment needs an efficient integration of high-level processes (deliberative and time-costly) within low-level (reactive, faster but poorer in quality) processes. The most relevant aspect of our current approach is that, unexpectedly, the performance of the CBR-system do not get worse any time that it retrieves worse cases in situations even when it has enough time to generate better solutions. We concentrate on formal aspects of the proposed Grid-CBR system without establishing which should be the most adequate procedure in a subsequent implementation stage. The advantage of the presented scheme is that it does not depend on neither the particular problem nor a concrete environment. It consists in a formal approach that only requires, on one hand, local information about the averaged-time spent by the system in obtaining a solution and, on the other hand, an estimation

Luis F. Castillo
Universidad Nacional de Colombia, Industrial Engineering Dpt.
Universidad de Caldas. Manizales (Colombia), Computer Science and
Systems Engineering Dpt
e-mail: lfcastil@autonoma.edu.co

Gustavo Isaza
Universidad de Caldas. Manizales (Colombia), Computer Science and
Systems Engineering Dpt.
e-mail: gustavo.isaza@ucaldas.edu.co

Manuel Glez Bedia · Miguel Aguilera
Universidad de Zaragoza (Spain), Computer Science Dpt
e-mail: {mgbedia,maguilera}@unizar.es

Juan David Correa
Universidad Autónoma de Manizales, Systems Engineering Student
e-mail: jdcorrea@autonoma.edu.co

S. Omatu et al. (Eds.): Distributed Computing and Artificial Intelligence, AISC 151, pp. 355–361.
springerlink.com   © Springer-Verlag Berlin Heidelberg 2012

about their temporal restrictions. The potential use of industry standard technologies to implement such a Grid-enabled CBR system is discussed here too.

Keywords: Case base reasoning, Grid computing, Artificial Intelligence.

1   Introduction

Grid technologies consist in a collection of geographically dispersed computing, storage and networking resources. They are able to provide many advantages over conventional computing because they can be coordinated to improve the performance and scalability of computing activities. These technologies allow us designing collaborative decision support across virtual organizations because it enables the sharing of applications and data in a heterogeneous environment. The Grid can be thought of as a distributed system with non-interactive workloads that involve a large number of files.

This paper describes how Grid and CBR tools are brought together to deliver a substantial further improvement in facilitating the performance of an appropriated system for real-time environments. In general, a real-time system is the one that accomplishes a sequence of processes with temporal restrictions on a dynamic environment. Getting to a most technical point, an appropriated system for real-time environments must achieve its goals and tasks with the added difficulty of temporal restrictions. Therefore, efficient integration of high-level deliberative processes and reactive structures in lower layers seems to be necessary in order to achieve a proper tradeoff between goals and available time. In our case, the use of a CBR method must be bounded in order to enforce the whole system to behave with temporal restrictions. We can find in the literature several approaches in the CBR community to deal with time-constraints.

The structure of the paper is as follows: in section 2, a brief overview about the link between CBR models and Grid technologies is presented; in section 3, the proposed mathematical model and empirical results will be shown ; in section 4, conclusions and future work will be introduced.

2   Grid Computing and CBR Strategies

A CBR model provides a framework to display solutions for solving a current problem by reusing or adapting other solutions that were applied in similar previous problems. It offers a reasoning paradigm that is similar to the way people routinely solve problems. Faced with a new problem, a human often relates the problem to one or more memory episodes and composes a solution from these episodes. As a technological tool, CBR-based programs have been successfully applied to a variety of problems in several fields and disciplines [Ontanon, 2010], [Julian, 2004], [Cheng, 2004], [Dean, 1998], [Bereni, 2005]. However, one of the main problems in deploying these systems is that have been designed to "static" domains (domains in which the system has unlimited amount of time to solve each problem and during this time, the state does not change) but most realworld (and most interesting) domains are dynamic . In the real-world, CBR-systems have time-constraints and must deal with dynamic conditions that changes over time. On the other side, the ability to capture resources and instantly upload them to the Grid-based system greatly increases the accuracy of the CBR systems. The advantage in rapidly sharing the information across the entire system is that it enables several other engine experts, at different geographical locations, to collaborate in solving the problem. Although Grid techniques can be very useful– they enable the sharing of applications and data in an open, heterogeneous environment and applications previously considered to be host-centric can now be distributed– we should solve previously fundamental problems of temporal restrictions on a dynamic environment.

In this paper, we have explored how stated the upper time-limit in the retrieval stage, expressed in mathematical terms. The aim is focused on formal aspects without establishing which should be the most adequate procedure in a subsequent implementation stage. The advantage of the presented scheme is that it does not depend on neither the problem nor the model of the environment. It will be illustrated in the following sections.

3   A New Sight on the CBR Methodology: Better Solutions Do Not Always Involve Optimal Performance

Case-based reasoning has been defined as a problem-solving technique based on the hypothesis that reasoning is reminding. The generic CBR cycle consists of the following steps [Aamodt et al. 1994]: (1) retrieval of relevant cases from the repository based on cues derived from problem requirements, (2) reuse of applicable cases to suggest solutions to a new problem, (3) testing-based verification and rule-based validation to ensure correctness, and (4) retention of past solutions and failures to enable learning. For simplicity, we do not consider stage 3 in our model. Moreover, stages 1 and 2 will be integrated together in a new stage called "generation of solution". Therefore, in our approach, the cycle of a CBR will consist in: (1) a generation stage, that covers retrieving and adapting processes, and (2) an execution stage, that ends in a retain process. It is obvious that execution and problem solving in a CBR system are decoupled.

In connection with that assumption, it is also accepted that in a dynamic environment where the system does not know how changes on its surrounding are going to be, the system is built considering that the optimal solution to a problem should be composed by optimal intermediate solutions obtained in every cycle. In other words, it assumed that given a time period T, resulting from a temporal windows series $\sum_{i=1}^{n} \Delta t_i$, the best strategy is a greedy strategy, the one that holds $optimal(T) = \sum_{i=1}^{n} \text{optimal}(\Delta t_i)$.

That statement is assumed but no demonstrated, and we feel that it embodies a *naive* and extended error related to the design of artifacts in dynamical environments. It will be questioned if that assumption is right, i.e. if the optimal strategy, under conditions of uncertainty, must be a *greedy strategy*. If it is represented the suitability (or the quality) of a solution from the CBR in an instant t and denoted by $a(t) \in [0,1]$, we will assume that:

1. It is assumed that the system shows an exponential relation between the quality of a solution and time (the effort in obtaining better results grows in relative terms with time, $a(t) = K(1 - e^{-t/\tau})$, where τ is the adjustment speed.
2. It is also assumed that the solution degrades throughout time as the environment changes, and the exponential functional dependency between quality of a solution and time will be $a(t) = K(e^{-t/\varepsilon})$, where ε stands for the degradation rate.

The resolution structure of the system can be captured with a single variable denoted by $\gamma(t) \in \{\gamma_0, \gamma_1\}$, that is, as the binary exclusive choice of the system over time, γ_1 representing adjustment and ãl deployment. Now, the following equations to describe the behavior of the system result from the previous formalization:

- *Generation phase*: $a(t) = (1 - e^{-t/\tau}), \gamma(t) = \gamma_0$
- *Execution phase*: $a(t) = (e^{-t/\varepsilon}), \gamma(t) = \gamma_1$

The structure of the dilemma can thus be reduced to finding the strategy (i.e. the $\gamma(t)$ values) that obtain the better results. In order to compute the quality of the results obtained by a specific $\gamma(t)$ we will define the evolution of the fitness over time:

$$\frac{d}{dt}a(t) = \begin{cases} \frac{1}{\tau}(1 - a(t)), & \gamma(t) = \gamma_0 \\ -\frac{1}{\varepsilon}a(t), & \gamma(t) = \gamma_1 \end{cases}$$

The agent performance will be obtained only integrating the suitability of the system during the deployment periods (the ones in which the agent is obtaining a benefit from the world, so we will take $\gamma_0 = 0$, and $\gamma_1 = 1$). Both previous functions can be combined, obtaining the global behavior equation:

$$\frac{d}{dt}a(t) = -\gamma(t)\frac{1}{\varepsilon}a(t) + \frac{1}{\tau}(1 - \gamma(t))(1 - a(t))$$

And the quality of the obtained results will be defined by the performance of the system, $\overline{p}(T)$, evaluated in an interval $(0, T)$ and defined by:

$$\overline{p}(T) = \frac{1}{T}\int_0^T \gamma(t) \cdot a(t)\, dt$$

The optimal solution of the system, $a_{opt}(t)$, is the one that maximizes $\overline{p}(T)$, so it is wanted to find the set $\{\gamma_k(t_k)\}$ that maximizes $\overline{p}(T)$. Discretizing,

$$\begin{cases} a_{k+1} - a_k = -h\left(\frac{1}{\tau}(1 - a_k) - \gamma_k \cdot \left(\frac{1}{\tau} + a_k \cdot \left(\frac{1}{\varepsilon} - \frac{1}{\tau}\right)\right)\right) \\ p_{k+1} - p_k = h(\gamma_k \cdot a_k) \end{cases}$$

where h is a temporal step, $= 0, 1, 2, \dots, N$, so $a(0) = a_0$, p$(T) = p_N$, given $T = \{t_1, t_2, \dots, t_N\}$.

Case a:

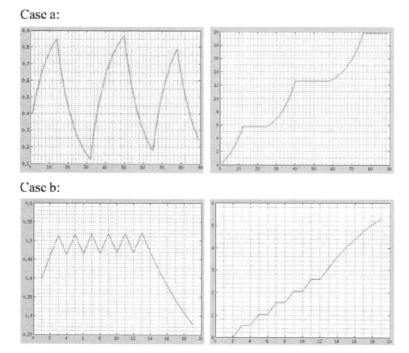

Case b:

Fig. 1 Comparison of strategies and representation of the adjustment function a(t) and the adjustment value p(t) for: (a) the greedy strategy where every window of opportunity is optimized, and (b) the coupled solution CBR-environment (with $\tau = 1$, $\varepsilon = 1$) which offers the optimal solution.

Under these considerations, the problem can be reformulated by the following statement (knowing that h is constant): "Find the set of decisions $\{\gamma_k(t_k)\}$ that maximize $\sum_{k=0}^{N} \gamma_k a_k$". The classical model propels us to compute the better solution in the available time (every window of opportunity) and execute it until it stops being good enough (as it is illustrated in Figure 1.a). When solving the system [Bellman 1957], the results forces us to change that point of view (see Figure 1.b): it is shown that the global result do not depend on the quality of the chosen solution, but in how that solution is coupled in the available time window.

In a nutshell, the optimal solution can be captured under the following dictum: *"Using a CBR when the environment changes, the best behavior is the one that maximizes the number of cycles with the world, being the optimal level fitness level determined by the dynamics of the environment characterized by the relation between τ and ε".* In other words, the obtained result tells us the following: when the environment changes, quick and fast generation solutions are preferred than the ones that spent more time in being generated. Much to our surprise, there has been shown that in dynamical environments, solutions locally considered as "bad" give the optimal response with time. And there has also been shown that, when analyzing dynamical systems, does not matter how simple they are, the solution can be surprising when the global picture is took.

4 Overview and Conclusions

With the emergence of Grid computing, computing is becoming less confined to traditional computing platforms. Grid technology promises access to vast computing power and large data resources across geographically dispersed areas.

This capability can be significantly enhanced by establishing relationships between these kind of techniques and traditional and successful tools as CBR systems. But these promising technological symbiosis require solving fundamental problems as the adjustment of a CBR strategy in an dynamical environment. In this paper, a model recently developed is presented.

CBR steps are implemented helping the translating to a distributed and scalable problem on the Grid that can deliver increased value and performance. In this paper, a traditional perspective about the use of CBR in dynamic environments has been criticized. It is characterized for:

- Decoupling the generation and execution phases with time
- Considering the presence of uncertainty (about the future in the environment) only as 'noise', ignoring it as a source of opportunities.
- Since the system has not information about how the world is going to change, it is maximized the adjustment of particular solutions in the available time every window of opportunity.

Our mathematical and experimental study, reveal us:

- It is possible to develop a system where the generation and execution of solutions are coupled in time without explicitly knowing how the world is going to change.
- In dynamic environments, solutions that maximize the local adjustment do not necessarily find the global optimal strategy.

Acknowledgments. This work was supported in part by Colciencias (Learning Management System -LMS- Proyect, cod. 121948725660, num. 47, 2009). Miguel Aguilera currently holds a predoctoral fellowship from Gobierno de Arag´on, Programa de Formación de Personal Investigador 2010.

References

[1] Aamodt, A., Plaza, E.: Case-Based Reasoning: Foundational Issues, Methodological Variations, and System Approaches. AI. Commun. 7(1), 39–59 (1994)
[2] Bellman, R.E.: Dynamic Programming. Princeton University Press, Princeton (1957); Republished 2003: Dover, ISBN 0486428095
[3] Berenji, H., Wang, Y., Saxena, A.: Dynamic Case Based Reasoning in Fault Diagnostics and Prognostics. In: FUZZIEEE, Reno (May 2005)

[4] Cheng, D., Thawonmas, R.: Case-based plan recognition for realtime strategy games. In: Proceedings of the International Conference on Computer Games: Artificial Intelligence, Design and Education (CGAIDE 2004), pp. 36–40. University of Wolverhampton, Reading, UK (2004)

[5] Dean, T., Boddy, M.: An analysis of time-dependent planning. In: Proc. of the 7th National Conference on Artificial Intelligence, pp. 49–54 (1988)

[6] Eremeev, A., Varshavskiy, P.: Case-based Reasoning Method for Real-time Expert Diagnostics Systems. Institute of Information Theories and Applications FOI ITHEA 15(2) (2008) ISSN: 1313-0463

[7] Garvey, Lesser, V.: A survey of research in deliberative Real-Time Artificial Intelligence. The Journal of Real-Time Systems 6, 317–347 (1994)

[8] Goldman, R.P., Musliner, D.J., Krebsbach, K.D.: Managing Online Self-Adaptation in Real-Time Environments. In: Proc. of 2nd Int. Workshop on Self Adaptive Software (2001)

[9] Howe, E., Hart, D.M., Cohen, P.R.: Addressing real-time constraints in the design of autonomous agents. The Journal of Real-Time Systems 2, 81–97 (1990)

[10] Julian, V.J., Botti, V.: Developing real-time multiagent systems. ICAE 11(2), 150–165 (2004)

[11] Marling, M.T., Gillen, M., Alexander, D., Chelberg, D.: Case-based reasoning for planning and world modeling in the robocup small size league. In: Workshop on Issues in Designing Physical Agents for Dynamic Real-Time Environments, IJCAI (2003)

[12] Ontanon, S., Mishra, K., Sugandh, N., Ram, A.: Line Case-Based Planning. Computational Intelligence, 84–119 (2010)

[13] Vina, Seiver, A.: Guardian: A prototype intensive-care monitoring agent. Artificial Intelligence in Medicine 4, 165–185 (1992)

A Low-Cost Solution to Distribute Repetitive and Independent Tasks by Clustering

Ignacio Traverso Ribón, Mª Ángeles Cifredo Chacón,
Ángel Quirós-Olozábal, and Juan Barrientos Villar

Abstract. This paper evaluates three different solutions to distribute a list of independent tasks among the nodes of a cluster of computers. Operative systems are unable to take advantage of modern multicore or multiprocessor based computers in the execution of this kind of work when it is not paralellized, thus, the availability of a software tool that allows the concurrent use of a set of computers can be very useful in terms of reduction of the execution time and efficiency.

The three solutions have been evaluated using a set of tasks related to the implementation of digital electronic circuits in Field Programmable Gate Arrays (FPGAs), but they can be applied to any other set of independent tasks.

The results have been obtained using a cluster composed by four relatively old and low performance computers, and have been compared with the results obtained by a quad-core processor based computer executing the same task list in a purely sequential way. This comparison clearly shows the power of concurrent execution, that allows a reduction of execution time by a factor between three and four.

1 Introduction

Nowadays cluster's topologies are included in the following categories [1]:

- High performance: to resolve problems those require a intensive computing capacity.
- High availability: with the aim of offering high security, reliability and availability services.
- High efficiency: built to execute the maximum number of tasks in a given period of time.

Ignacio Traverso Ribón · Mª Ángeles Cifredo Chacón · Ángel Quirós-Olozábal ·
Juan Barrientos Villar
Grupo de Diseo de Circuitos Microelectrónicos, Universidad de Cádiz, Cádiz, Spain
e-mail: mangeles.cifredo@uca.es

S. Omatu et al. (Eds.): Distributed Computing and Artificial Intelligence, AISC 151, pp. 363–370.
springerlink.com   © Springer-Verlag Berlin Heidelberg 2012

The work we introduce in this paper fits in the third topology, a set of computers working as a team to complete a long list of independents tasks in the shortest time and without repetition. It aims to improve the computing performance distributing a large list of independent tasks that, currently, are executed in series.

Initially the cluster must have a list with the set of tasks to execute in order to be distributed among its nodes. Because of the heterogeneity of the computers and the different elapsed time to finish each task, it is necessary a coordinator that will assign tasks to the nodes when they enter in an idle state and that will check that no task is repeated.

This kind of cluster can be very useful when non-parallels and no open source applications are used and it is necessary to execute the same programs many times with different parameters. These applications do not take advantage from multicore or multiprocessors systems. They are not able to distribute tasks among the different cores to achieve a reduction of the elapsed time either. The proposed solution avoids the manual and error prone execution of the applications, and automatically distributes the multiple tasks among the cluster's nodes. Using our algorithm the user's work is just to create the task list and to wait for the tasks to be completed. The results are organized and stored in a single hard disk with the help of a Network File System (NFS).

2 Tasks Distribution Strategies

All the solutions aim to reduce the global execution time of the task list without repeating any of them.

C++ and Python have been the languages considered to code this application. C++ maturity makes it a very polished and tested language, besides being one of the most widespread, which let use a large list of libraries. However it is less legible than Python. Pointers disappear in Pythton so it is possible to get a higher abstraction level, making easier the inclusion of modifications. This has been the main reason to choose Python instead of C or C++ to do this work.

Cluster's features are showed in the table 1, where it can be noticed that its nodes are rather obsolete and very economic computers, acting the master as a NFS server for the remaining nodes. The first objective is to centralize the information (input and output) of the different tasks that will be executed by each node. Cluster's connections are based on Ethernet technology using a domestic router-switch.

The closed but free application ISEWebpack[1] was installed in each node as test program. It requires two parameters for its execution: the HDL[2] description of a digital circuit to be synthesized and the FPGA target device. A test bench composed of 16 HDL projects and 105 Xilinx's[3] FPGA was used. In previous version [2] of

---

[1] Digital circuit development environment of the FPGAs manufacturer Xilinx Inc. This version is free and can be downloaded from its web site.

[2] Hardware Description Language.

[3] A Xilinx's tool called ISE is mandatory to implement a digital electronic circuit in a Xilinx FPGA, isn´t possible to use third-party or free software.

this tool the elapsed time to complete the task list was 84 hours, using a quad core computer equipped with 4Gbytes of RAM.

To monitor the cluster, a tool ,called Ganglia, was installed in the master computer. This tool shows in a web-frontend module a collection of data such as used RAM, network traffic, load-average, etc. using graphics and tables. One of the most important parameters obtained from this tool is the load-average, measured as the number of process in the scheduler's queue of the operative system, waiting to be executed.

Table 1 Cluster's features

	Master	Node1	Node2	Node3
Processor	Athlon 3500+	Athlon XP 2000+	Athlon 3500+	Athlon 3500+
Memory	1002MB	465MB	1002MB	1002MB

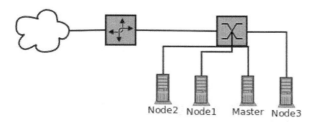

Fig. 1 Cluster's topology

2.1 Concurrent Scripts: First Attempt

MPI library [6] is used in this solution. The algorithm determines the number of task and divides them among many nodes as the cluster has. This way guarantees that the load-average for each node is equivalent.

MPI library lets you know how many nodes make up the cluster. In addition it assigns an identifier to each computer, easier to distribute tasks equitably in quantity.

Adding computers to this cluster requires file modifications. Furthermore, appending them during execution is not possible because tasks are distributed at the very beginning of the process.

Figure 2 corresponds to the flow diagram of this solution.

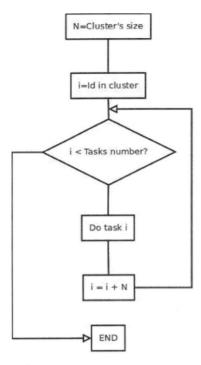

Fig. 2 First solution diagram flow

2.2 Concurrents Scripts: Second Attempt

The performance of the first attempt was better than the serial execution but improvable. The reason was the process was not completed until the slowest node finished. Thus, the efforts were focused in a new solution. It consists in leaving a mark that lets you know which tasks have been done in order to not repeat them.

NFS was the key in this method because acts as marks manager and displays, on the request of cluster's nodes, the remaining task in the list. Paths and sub-paths are used as marks, if a path exists, it means that the job is assigned and the node must look for another one.

Every node checks, before executing a task, if there is a mark in the NFS. If that mark exists means that another node is doing or has done the task and the node will choose a new task, otherwise the node will execute it. This process is repeated until there are no tasks to be done.

Checking process let add new nodes easily, even during execution. A new node just has to find a task whose directory or mark does not exist. It does not matter if other nodes have already executed several tasks.

Figure 3 corresponds to the flow diagram of this solution.

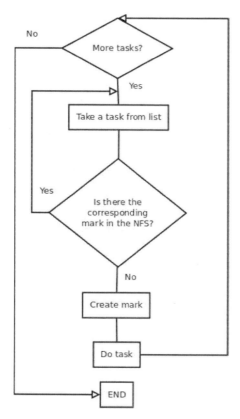

Fig. 3 Second solution diagram flow

2.3 Concurrent Scripts: Third Attempt

Previous solution was not refined enough because there was a strong dependence with NFS and file checks required a lot of disk accesses. Consequently, a new solution based in remote object technology [4] was tested.

Remote objects let methods being called from remote computers. It is easier to implement mutual exclusion of a critical resource with a local outline (shared variables, semaphores, etc [5]) in a computer acting as a server than doing it with a distributed outline.

The server computer will be responsible for providing tasks to the clients when they ask for one, implementing a task distributor on demand. This way we avoid polling and there is no need to ask to the NFS server. This solution is smarter, more legible and safer than previous because it has reduced the NFS dependence.

In this solution when a client wants to execute a task, it sends a request to the server. The server, which owns the task lists, unlocks it, responses to the client with a new task and locks the list again. This way, it is impossible for two or more clients to access to the list at the same time. Clients must wait until the list is unlocked to get a new task. Just for that we are sure that a task will not be done more than once.

The server is responsible of providing tasks correctly to the clients. Thus it is possible to add a new node during the execution of a job. The new node asks for a task to the server, which will reply the client with a not done task. The user just has to start the process in the new node, the server will take care of the rest. This is the reason because we can say that the process to add a node is automatic.

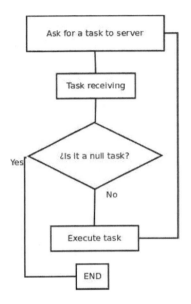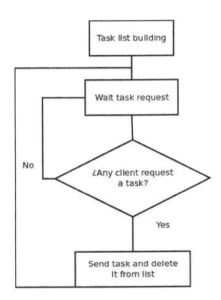

Fig. 4 Client's and Server flow diagram

3 Experimental Results

To check the efficiency of each solution they were implemented in a new open-source application called THOr (Tool Hardware analyatiOn). [7]

THOr is a tool designed to automate logic synthesis, implementation and exhaustive analysis of a batch of digital circuits using FPGAs as target technology. The tool simplifies the automatic and unattended logic synthesis and implementation of one or more digital circuits projects into a user selectable set of FPGA devices. Thereby, a user could know in a short time which is the most optimal FPGA to host their design. THOr has been developed to interact with Xilinx ISEWebpack and, although it has more features, it has been used in this work only to generate the task list with the HDL circuit and FPGA model as parameters to invoke ISEWebpack. When the task list is built, THOr will distribute the jobs among cluster's nodes.

Several files are generated in each execution of ISEWebpack for every pair of HDL circuit and FPGA model. These files include relevant data for choosing the best target FPGA to implement the tested digital circuit. There are other necessary files to deal each stage in the process. Furthermore, THOr must generate a file system hierarchy to orderly store all these files. When the task list is done the designer will have all the results in a single file system thanks to the NFS.

To check the different proposed strategies we used a set of 16 circuits, described in VHDL and Verilog languages, available in Opencore repository. As target technologies we used 105 Xilinx's FPGA models from Spartan3 serie. THOr's first task is to create a hierarchy of 1680 folders to store the results. After that, it distributes the jobs among the cluster's nodes until the task list is completed.

This test was also executed in a single computer whose features are showed in the figure 2, taking 89 hours and 19 minutes to complete.

Table 2 Previous computer's features

	Previous computer
Processor	QuadCore 2.4GHz
Memory	4096MB

Table 5 shows information about the minimum time to finish the task list, the load average of the cluster and the scalability.

Table 3 Elapsed time to finish the task list

Algorithm	Language	Time	Load Average	Scalability
No cluster	Bash	84 hours	-	No
Tasks preassignment	Python	27 hours	90%	Requires files modifications
File existence	Python	21 hours 45 minutes	90%-110%	Automatic
Remote objects	Python	21 hours	120%	Automatic

3.1 Conclusions

In table 3 and figure 5 is showed that the last two solutions have similar elapsed times. If we focus on the main objective (reduce the execution time) the third solution is the best one. If we suppose that the work is not critical, server reliability is not necessary

Load averages that are showed by the tables are determined by the monitoring tool Ganglia. It considers the load average as the number of process ready for being executed in the operative system scheduler queue.

Although we have performed the test with ISEWebpack, this tool can be used to run other applications that require many executions with different parameters each time.

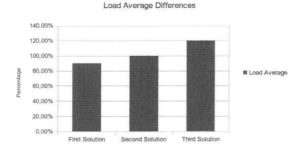

Fig. 5 Comparative between three solutions

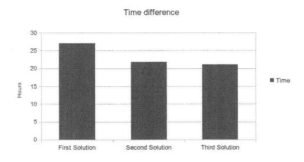

Fig. 6 Comparative between three solutions

References

1. Lucke, R.W.: Building Clustered Linux Systems. Prentice Hall (2004)
2. Mª Ángeles Cifredo Chacón: Un procedimiento para la verificación y clasificación priorizada de FPGAs de bajo coste ante aplicaciones sectoriales electrónicas. PhD Thesis (2010)
3. Gaj, K., Kaps, J.-P., et al.: ATHENa Automated Tool for Hardware EvaluatioN: Toward Fair and Comprehensive Benchmarking of Cryptographic Algorithms using FPGAs
4. Irmen de Jong, Pyro Documentation (2011),
   http://packages.python.org/Pyro/
5. Ben-Ari: M. Concurrent and Distributed Programming. Prentice Hall (1990)
6. Message Passing Interface Forum (2011), http://www.mpi-forum.org/
7. Thor Program Web, https://forja.rediris.es/projects/thor/

Identifying Gene Knockout Strategies Using a Hybrid of Bees Algorithm and Flux Balance Analysis for *in Silico* Optimization of Microbial Strains

Yee Wen Choon, Mohd Saberi Mohamad, Safaai Deris, Chuii Khim Chong, Lian En Chai, Zuwairie Ibrahim, and Sigeru Omatu

Abstract. Genome-scale metabolic networks reconstructions from different organisms have become popular in recent years. Genetic engineering is proven to be able to obtain the desirable phenotypes. Optimization algorithms are implemented in previous works to identify the effects of gene knockout on the results. However, the previous works face the problem of falling into local minima. Thus, a hybrid of Bees Algorithm and Flux Balance Analysis (BAFBA) is proposed in this paper to solve the local minima problem and to predict optimal sets of gene deletion for maximizing the growth rate of certain metabolite. This paper involves two case studies that consider the production of succinate and lactate as targets, by using *E.coli* as model organism. The results from this experiment are the list of knockout genes and the growth rate after the deletion. BAFBA shows better results compared to the other methods. The identified list

Yee Wen Choon · Mohd Saberi Mohamad · Safaai Deris · Chuii Khim Chong · Lian En Chai
Artificial Intelligence and Bioinformatics Group, Faculty of Computer Science and Information Systems, Universiti Teknologi Malaysia, 81310 Skudai, Johor, Malaysia
e-mail: ywchoon2@live.utm.my, saberi@utm.my, safaai@utm.my, ckchong2@live.utm.my, lechai2@live.utm.my

Zuwairie Ibrahim
Department of Mechatronics and Robotics, Center for Artificial Intelligence and Robotics (CAIRO), Faculty of Electrical Engineering, Universiti Teknologi Malaysia, 81310 UTM Skudai, Johor Darul Takzim, Malaysia
e-mail: zuwairiee@fke.utm.my

Sigeru Omatu
Department of Electronics, Information and Communication Engineering, Osaka Institute of Technology, Osaka 535-8585, Japan
e-mail: omatu@rsh.oit.ac.jp

S. Omatu et al. (Eds.): Distributed Computing and Artificial Intelligence, AISC 151, pp. 371–378.
springerlink.com   © Springer-Verlag Berlin Heidelberg 2012

suggests gene modifications over several pathways and may be useful in solving challenging genetic engineering problems.

Keywords: Evolutionary Programming, Metabolic Engineering, Bees Algorithm, Gene Knockout, Optimization.

1 Introduction

Microbial strains are strains of microorganisms which have become popular for genome-scale metabolic networks reconstructions in recent years [1]. Reconstructions of the metabolic networks are found to be very useful in health, environmental and energy issues [2]. A vast numbers of high-throughput experimental data has expedited the development of computational models for simulating the actual processes inside the cell. One of the main goals in system biology is to construct an efficient and accurate pathway models that may be useful in predicting cellular responses and providing better understanding of complex biological functions.

Many algorithms were developed in order to identify the gene knockout strategies for obtaining improved phenotypes. Maranas *et al.* [3, 4] developed the first rational modeling frameworks (named OptKnock) for introducing gene knockout leading to the overproduction of a desired metabolite. OptKnock identifies a set of gene (reaction) deletions to maximize the flux of a desired metabolite without affecting the internal flux distribution such that growth is optimized.

OptKnock is implemented by using mixed integer linear programming (MILP) to formulate a bi-level linear optimization that is very promising to find the global optimal solution. OptGene is an extended approach of OptKnock which formulates the *in silico* design problem by using Genetic Algorithm (GA). These meta-heuristic methods are capable in producing near-optimal solutions with reasonable computation time, furthermore the objective function that can be optimized is flexible. SA is then implemented to allow the automatic finding of the best number of gene deletions for achieving a given productivity goal. However, SA faces the problem of falling into local minima far from the global optimum solution.

In this paper, a hybrid of Bees Algorithm and Flux Balance Analysis (BAFBA) is proposed to predict the gene knockout strategies. Bees Algorithm (BA) is a typical meta-heuristic optimization approach which was introduced by [5]. The search process of BA is based on the intelligent behaviors of honey bees. BA is proven to be efficient in solving optimization problems in the previous studies [5]. While the Flux Balance Analysis (FBA) approach which is used to calculate the fitness function is based on a steady state approximation to concentrations of the internal metabolites, which reduces the corresponding mass balances to a set of linear homogeneous equations. There are two advantages of BAFBA. First, BAFBA requires less computational time and thus it capable to solve larger size problems. Secondly, BA is capable of performing local and global search simultaneously and thus it works out the local minima problem. This paper presents the results obtained by BAFBA to two case studies where *E.coli* is the target microorganism. This paper also evaluates the performance of BAFBA for identifying gene knockout strategies with existing tools and compares the performance of BA with the existing methods within experimental approaches.

2   A Hybrid of Bees Algorithm and Flux Balance Analysis

In this paper, BAFBA is proposed to predict the gene knockout. Fig 2.1 shows the flow of a basic BA. The flow of BAFBA is presented in Fig 2.2. The important steps are explained in the following subsections.

2.1   *Bee Representation of Metabolic Genotype*

In the metabolic model, one or more genes can be found in each reaction.   In this proposed method, each of those genes is represented by a binary variable indicating its absence or presence (0 or 1), these variables form a 'bee' representing a particular mutant that lacks some metabolic reactions when compared with the wild type (Fig 2.3.)

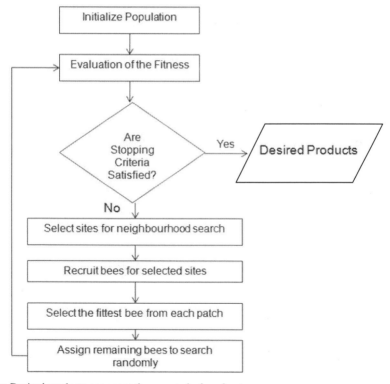

Note: Desired products represent the gene to be knockout.

Fig. 2.1 Flowchart of a basic BA.

2.2 Initialization of the Population

The algorithm starts with an initial population of n scout bees. Each bee is initialized as follows: assume that a reaction with n genes. Bees in the population can be initialized by assigning present or absent status to each gene randomly.

2.3 Scoring Fitness of Individuals

The fitness computation process is carried out for each site visited by a bee through FBA (Fig 2.4.). Cellular growth is defined as the objective function Z, vector **c** is used to select a linear combination of metabolic fluxes to include in the objective function, **v** is the flux map and i is the index variable (1, 2, 3, …, n).

Maximize Z, where

$$Z = \sum_i c_i v_i = \mathbf{c.v}$$

(2.1)

where c = a vector that defines the weights for of each flux.

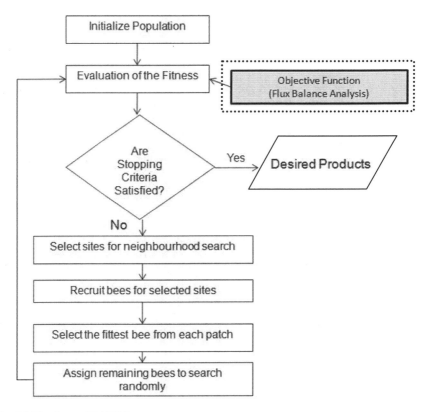

Fig. 2.2 The flow of BAFBA.

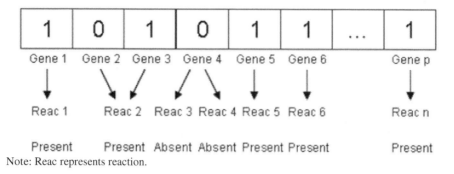

Note: Reac represents reaction.

Fig. 2.3 Bee representation of metabolic genotype

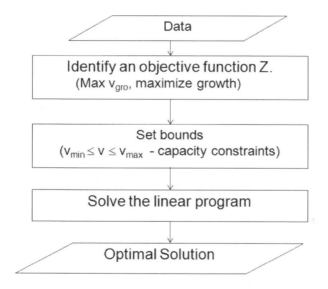

Fig. 2.4 Steps in FBA

2.4 Neighbourhood Search

The algorithm carries out neighbourhood searches in the selected sites, assigning more bees to search near the best sites. The bees can be chosen directly according to their fitnesses associated with the sites they are visiting. Searches in the neighbourhood of the best sites which represent more promising solutions are further more detailed by recruiting more bees to follow them than other selected bees.

2.5 Randomly Assigned and Termination

The remaining bees in the population are assigned randomly around the search space scouting for new potential solutions. These steps are repeated until a stopping criteria is met. The stopping criteria are either the maximum loop value is met or the fitness function has converged. At the end of each iteration, the colony produces two parts to its new population – representatives from each selected patch and other scout bees assigned to conduct random searches.

3   Experimental Results

In this paper, the *E.coli* dataset is used to test on the operation of BAFBA. All simulations were performed for aerobic minimal media conditions. The glucose uptake rate was fixed to 10 mmol/gDW/hr while a set non-growth associated maintenance of 7.6 mmol ATP/gDW/hr. The results obtained are compared to the previous works reported in the literature studies [3, 6]. Millimole (mmol) is the unit of concentration whereas millimoles per hour (mmol/hr) is used as the unit measurement in the experiments.

Table 3.1 and Table 3.2 summarize the results obtained from BAFBA. As shown from the results, this method has produced better results to the previous works.   In this paper, potential genes which can be removed are identified.

BAFBA suggests the removal of three reactions from the network results in succinate growth rate reaching 0.91665 which is better than the other two methods. The list obtained is to disable the phosphotransferase system which causes the network to rely solely on glucokinase for glucose uptake [3].

Table 3.1 Comparison between different methods for production of Succinate

Method	Growth Rate (mmol/hr)	List of knockout genes
BAFBA	0.91665	ACALD, ACKr, ATPM
SA + FBA [6]	0.35785	MALS, ORNDC, FUM, GLYCL, GHMT2, ADPT, DCYTD, DUTPDP, URIDK2r, NTD8, PUNPI, THD2, GND, PFL, SUCFUMt
OptKnock [3]	0.31	PYK, ACKr, PTAr, Phosphotransferase system

Note: The shaded column represents the best result.

Next, BAFBA is applied to identify knockout strategy for producing lactate. Table 3.2 shows the best result is obtained from this method is 0.91665. From the list of knockout genes, it can be concluded that the flux toward lactate at the maximum biomass yield is redirected by blocking acetate and ethanol production [3]. BAFBA produced the best results in both cases, and this is because BA performs local and global search simultaneously to avoid being trapped at locally optimal solutions.   BA splits the search into exploration and exploitation, which are then executed parallely rather than serially like SA. Thus, BA performs better than SA where it solves the local minima problem faced by SA.

In addition, Table 3.3 and Table 3.4 show the results of three of the identified gene knockout strategies for succinate and lactate overproduction.

Table 3.2 Comparison between different methods for production of Lactate

Method	Growth Rate (mmol/hr)	List of knockout genes
BAFBA	0.91665	ACALDt, ALCD2x, ATPM
SA + FBA [6]	0.39850	ACLD19, DRPA, GLYCDx, F6PA, TPI, LDH_D2, EDA, TKT2, LDH_D-
OptKnock [3]	0.28	ACKr, PTAr, ACALD

Note: The shaded column represents the best result.

Table 3.3 Result of different knockout strategies for production of Succinate

Mutants	Growth Rate (mmol/hr)	List of knockout genes
A	0.87392	ACALD, ACKr
B	0.91665	ACALD, ACKr, ATPM
C	0.87392	ACALD, ACALDt, ACKr, ALCD2x

Table 3.3 shows three of the identified gene knockout strategies (i.e., mutants A, B, and C). For the production of succinate, acetate kinase (ACKr) which contributes to the phosphotransferase system for all three mutant A, B, and C is disabled, this causes the network to rely exclusively on glucokinase for glucose uptake [3]. The deletion of acetaldehyde dehydrogenase (ACALD) results earlier coupling of succinate with biomass yields. For mutant C, the additional deletion of alcohol dehydrogenase (ALCD2x) eliminated the production of ethanol.

Table 3.4 Result of different knockout strategies for production of Lactate

Mutants	Growth Rate (mmol/hr)	List of knockout genes
D	0.87392	ACALD, ACKr
E	0.91665	ACALDt, ALCD2x, ATPM
F	0.91665	ACALD, ACALDt, ACKr, ATPM

Table 3.4 shows the result of different knockout strategies for the production of lactate, phosphotransferase system for mutant D and mutant F are disabled. The additional deletion of ACALD results earlier coupling of lactate with biomass yields. For mutant E, the knockout strategy eliminated the competing byproduct (i.e, ethanol). In conclusion, the phosphotransferase system and ethanol affect greatly to both production of succinate and lactate.

4   Conclusion and Future Works

In this paper, BAFBA is proposed to predict optimal sets of gene deletion in order to maximize the production of certain metabolite. This method is based on BA, which is capable of performing local and global search simultaneously where the local minima problem faced by SA is worked out. The FBA approach is used as a fitness function whereby it is based on a steady state approximation to concentrations of the internal metabolites, which reduces the corresponding mass balances to a set of linear homogeneous equations.

Experimental results on *E.Coli* core model dataset obtained from literature [4] showed that BAFBA is effective in generating optimal solutions to the gene knockout prediction, and is therefore a useful tool in Metabolic Engineering.

In regard to further improve the performance of BAFBA, we are interested in applying an automated pre-processing operation in BAFBA to simplify the genome-scale metabolic model. Another interesting feature is the development of multi-objective optimization algorithms in a single run to achieve two goals, for example, maximizing the biomass and the desired product. Lastly, as BA employs many tunable parameters which are difficult for the user to select, it is important to find ways to help the user choose appropriate parameters.

Acknowledgments. This work is financed by Institutional Scholarship MyPhD provided by the Ministry of Higher Education of Malaysia. We also would like to thank Universiti Teknologi Malaysia for supporting this research by UTM GUP research grants (vot number: Q.J130000.7123.00H67 and Q.J130000.7107.01H29).

References

[1] Feist, A.M., Herrgård, M.J., Thiele, I., Reed, J.L., Palsson, B.O.: Reconstruction of biochemical networks in microorganisms. Nat. Rev. Microbiol. 72, 129–143 (2009), doi:10.1038/nrmicro1949
[2] Chandran, D., Copeland, W.B., Sleight, S.C., Sauro, H.M.: Mathematical modeling and synthetic biology. Drug Discovery Today Disease Models 5(4), 299–309 (2008), doi:10.1016/j.ddmod.2009.07.002
[3] Burgard, A.P., Pharkya, P., Maranas, C.D.: Opt. Knock: A bilevel programming framework for identifying gene knockout strategies for microbial strains optimization. Biotechnol. Bioeng. 84, 647–657 (2003), doi:10.1002/10803
[4] Pharkya, P., Burgard, A.P., Maranas, C.D.: Opt. Strain: a computational framework redesign of microbial production systems. Genome. Res. 14, 2367–2376 (2004), doi:10.1101/gr.2872004
[5] Pham, D.T., Ghanbarzadeh, A., Koç, E., Otri, S., Zaidi, M.: The bees algorithm – a novel tool for complex optimization problems. In: Proceedings of the Second International Virtual Conference on Intelligent Production Machines and Systems, pp. 454–461 (2006)
[6] Rocha, M., Maia, P., Mendes, R., Pinto, J.P., Ferreira, E.C., Nielsen, J., Patil, K.R., Rocha, I.: Natural computation meta-heuristics for the in silico optimization of microbial strains. BMC Bioinformatics 9, 499 (2008), doi:10.1186/1471-2105-9-499
[7] Segre, D., Vitkup, D., Church, G.M.: Analysis of optimality in natural and perturbed metabolic networks. Proceedings of National Academy of Sciences of the United States of America 99, 15112–15117 (2002), doi:10.1073/pnas.232349399

Inferring Gene Regulatory Networks from Gene Expression Data by a Dynamic Bayesian Network-Based Model

Lian En Chai, Mohd Saberi Mohamad, Safaai Deris, Chuii Khim Chong,
Yee Wen Choon, Zuwairie Ibrahim, and Sigeru Omatu

Abstract. Enabled by recent advances in bioinformatics, the inference of gene regulatory networks (GRNs) from gene expression data has garnered much interest from researchers. This is due to the need of researchers to understand the dynamic behavior and uncover the vast information lay hidden within the networks. In this regard, dynamic Bayesian network (DBN) is extensively used to infer GRNs due to its ability to handle time-series microarray data and modeling feedback loops. However, the efficiency of DBN in inferring GRNs is often hampered by missing values in expression data, and excessive computation time due to the large search space whereby DBN treats all genes as potential regulators for a target gene. In this paper, we proposed a DBN-based model with missing values imputation to improve inference efficiency, and potential regulators detection which aims to lessen computation time by limiting potential regulators based on expression changes. The performance of the proposed model is assessed by using time-series expression data of yeast cell cycle. The experimental results

Lian En Chai · Mohd Saberi Mohamad · Safaai Deris · Chuii Khim Chong ·
Yee Wen Choon
Artificial Intelligence and Bioinformatics Research Group, Faculty of Computer Science
and Information Systems, Universiti Teknologi Malaysia, Skudai, 81310 Johor, Malaysia
e-mail: lechai2@live.utm.my, saberi@utm.my, safaai@utm.my,
  ckchong2@live.utm.my, ywchoon2@live.utm.my

Zuwairie Ibrahim
Department of Mechatronics and Robotics, Center for Artificial Intelligence and Robotics,
Faculty of Electrical Engineering, Universiti Teknologi Malaysia,
Skudai, 81310 Johor, Malaysia
e-mail: zuwairie@fke.utm.my

Sigeru Omatu
Department of Electronics, Information and Communication Engineering,
Osaka Institute of Technology, Osaka 535-8585, Japan
e-mail: omatu@rsh.oit.ac.jp

S. Omatu et al. (Eds.): Distributed Computing and Artificial Intelligence, AISC 151, pp. 379–386.
springerlink.com   © Springer-Verlag Berlin Heidelberg 2012

showed reduced computation time and improved efficiency in detecting gene-gene relationships.

Keywords: Dynamic Bayesian Network, Gene Regulatory Networks, Gene Expression Data, Inference.

1 Introduction

The development of microarray technology has enabled researchers to facilitate new experimental methods for understanding gene expression and regulations. The output, usually referred as gene expression data or microarray data, contains vast information such as the behaviors revealed by the system under normal conditions; abnormalities of the system if certain parts cease to function; the robustness of the system under extreme conditions [1], hence providing a holistic viewpoint of gene expression to the researchers instead of only a few genes as in the classical experiments.

Motivated by the need of researchers to understand the complex phenomena of gene regulations, gene expression data have obtained significant importance in the inferring of GRNs to explain the phenotypic behaviors of a specific system. The traditional trial and error method of inferring GRNs from gene expression data is obviously not feasible in handling large-scale data due to the time-consuming nature of repeating the routine to achieve accurate results [2]. To analyze and utilize the massive amount of gene expression data, researchers have already developed numerous computational methods to automate the inferring procedure [2, 3]. In particular, Bayesian network (BN), which models conditional dependencies of a set of variables via probabilistic measure, was widely utilized by researchers in inferring GRNs from gene expression data.

BN's effectiveness in inferring GRNs is mainly due to its ability to work on locally interacting components with a relatively small number of variables; able to assimilate other mathematical models to avoid the overfitting of data; allows the combination of prior knowledge the strengthen the causal relationship. Despite the advantages stated above, BN has two critical limitations in which it does not allow feedback loops and is unable to handle the temporal aspect of time-series microarray data.

In view of the fact that feedback loops represent the importance of homeostasis in living organisms, researchers have developed the dynamic Bayesian network (DBN) as a promising substitute. Since the pioneering work of Murphy and Mian [4], DBN has attracted particular attention from numerous researchers [5, 6, 7, 8, 9]. Nevertheless, normal DBN usually assumes all genes as potential regulators against target genes, and consequently causes the excessive computational cost which inhibits the efficiency of DBN on large scale gene expression data [8, 9]. In addition, the missing values commonly found in expression data may influence up to 90% of the genes [10], thus affecting the inference results. To tackle the two problems, we proposed a model of DBN with missing values imputation to improve the inference efficiency, and potential regulators selection which reduce computation time by limiting the numbers of potential regulators for each target gene. The details of our model are discussed in the following section.

2 Methods

In this section, we describe the details of the proposed DBN-based model for inferring GRNs from gene expression data. In essence, the proposed model consists of three main steps: missing values imputation, potential regulators selection and dynamic Bayesian network. The following sub-sections (2.1 – 2.3) discuss in detail for each of the three main steps. Table 2.1 shows the overview of our proposed model and existing DBN-based models.

Table 2.1 Overview of our proposed model and existing DBN-based models for inferring gene regulatory networks from gene expression data.

Our proposed model	Previous work [8, 9]	Previous work [15]
Missing values imputation	Potential regulators selection	Dynamic Bayesian network
Potential regulators selection	Dynamic Bayesian network	
Dynamic Bayesian network		

2.1 *Experimental Data and Missing Values Imputation*

The experimental study is based on the *S. cerevisiae* cell cycle time-series gene expression data from Spellman *et al.* [11] This dataset contains two short time series (cln3, clb2; both 2 time points) and four medium time series (alpha, cdc15, cdc28 and elu; 18, 24, 17 and 14 time points). However, the dataset contains missing values which must be processed. Conventional methods of treating missing values include repeating the microarray experiment which is not economical feasible, or simply replacing the missing values by zero or row average. A better solution is to use imputation algorithms to estimate the missing values by exploiting the observed data structure and expression pattern. In view of this, we applied the Bayesian principle component analysis (BPCA) imputation algorithm [12] due to its ability to assumes a global covariance structure of the dataset by iteratively estimating the posterior distribution of the missing values until convergence is achieved, and its effectiveness on large-scale data (9 minutes and 13 seconds for the experimental data on a Core i3 PC).

2.2 Potential Regulators Selection

The work of Yu *et al.* [13] showed that most transcriptional factors (TFs) experience changes in expression level before or simultaneously with their target genes. Following this fact, it is possible to derive an effective algorithm to reduce the search space by limiting the potential regulators of each target genes. Firstly, we determined the cutoff threshold for up-regulation (≥ 1.4) and down-regulation (< -1.1) based on the distribution of the gene expression values. Then, we discretized the dataset into three classes (up-, down-regulation and normal) and search only for the data located in the upper and lower bound classes. A time gap of two time points width is created to slide through the data to group regulation pairs. Thus, each target gene contains a subset of potential regulators which exhibit earlier or simultaneously expression changes. These are used as the input for the subsequent network inference step using DBN.

2.3 A Dynamic Bayesian Network

The network inference step is done by applying DBN, which is actually an extension of BN to describe the stochastic evolution of a network against time. This is mainly because BN is limited to steady-state data (static data), and DBN readily handles time-series data to identify the causal relationships among a set of variables. It also enables the modeling of cyclic structure while inheriting the advantages of BN. In essence, in modeling time-series data, values of a set of random variables are observed at different points in time. Assuming each time point as single variable Y_i, the simplest causal model for asequence of data $\{Y_1,...,Y_t\}$ would be a *first-order Markov chain*, in which the state of the next variable is dependent on the previous variable only. By applying the chain rule of probabilities and conditional independencies based on Bayes theorem, the joint probability distribution (JPD) of the graph has the general form of $P(Y_1, Y_2, ..., Y_t) = P(Y_1)P(Y_2|Y_1) ... P(Y_t|Y_{t-1})$. DBN consists of two stages: the parameter learning stage followed by the structure learning stage. In the parameter learning stage, we used the results from the previous step to create the data matrices of all target genes with their subsets of potential regulators. We then updated the data matrices by calculating the conditional probabilities of each target gene against each of its respective potential regulators. In light of recent work [14] showed that learning DBN structure is not definitely NP-hard, we employed a globally optimal search strategy [15] instead of using local search strategy in the structure learning stage. Figure 2.1 illustrates the overview of our proposed DBN-based model.

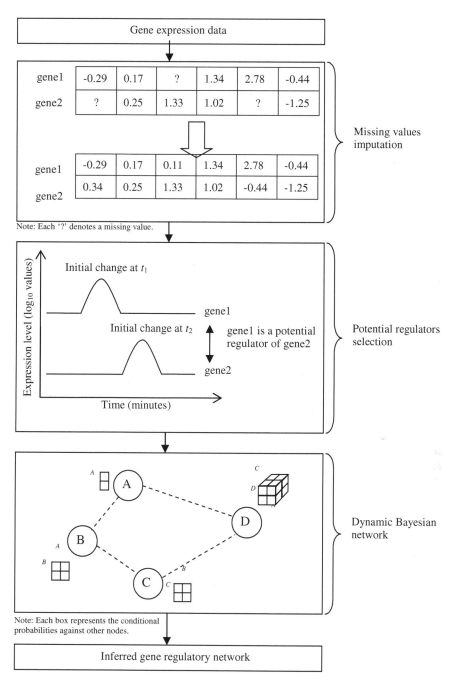

Fig. 2.1 Overview of our proposed DBN-based model with missing values imputation and potential regulators selection.

3 Results

In this study, we compared the efficiency and computation time of our DBN-based model against normal DBN [15]. The experiment results are evaluated by comparing with the yeast cell cycle pathway compiled at KEGG (Figure 3.1) and summarized in Table 3.1. In Table 3.1, row 1 represents the network inferred by our proposed model and row 2 represents the network predicted by normal DBN. Our proposed model used 45 minutes and 9 seconds against normal DBN which in turn used 1 hour 38 minutes and 23 seconds on a Core i3 PC with 4GB main memory. This is due to the reduced search space by applying potential regulators selection prior to DBN learning. Each target gene has a limited number of potential regulators instead of assuming all genes as potential regulators. Our proposed model was able to identify 14 gene-gene relationships against normal DBN which identified 12 gene-gene relationships. The missing values imputation helped to improve the efficiency of our proposed model by making use of the data structure and pattern to impute missing values. Interestingly, our proposed model also incorrectly identified more relationships (6 against 3). Putting aside that, the results of this study proved that the performance of DBN in inferring GRNs can be improved by imputing missing values and potential regulators selection.

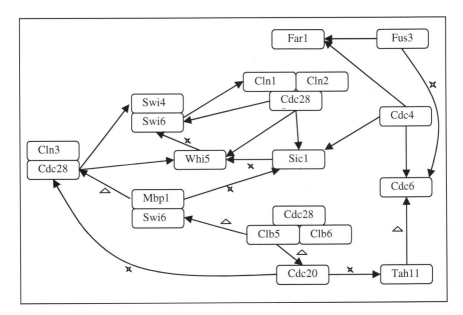

Fig. 3.1 Predicted cell cycle pathway for *S. cerevisiae* dataset using our proposed DBN-based model. A cross represents an incorrect inference; a triangle represents a misdirected relationship; an edge without any attachment is a correct inference. Genes that are grouped closely together represent a complex.

Table 3.1 The results of experiment study

Inference model	Correctly identified relationships	Misdirected relationships	Incorrectly identified relationships	Computation time (HH:MM:SS)
DBN_prs	14	4	6	00:45:09
DBN_norm [15]	12	8	3	01:38:23

Note: Shaded row represents the network inferred by our proposed model (DBN_prs) and unshaded row represents the network predicted by normal DBN (DBN_norm). Relationships refer to the gene-gene relationships.

4 Conclusion and Future Work

As a conclusion, our proposed DBN-based model is showed to perform better than normal DBN in terms of computation time and efficiency. However, it should be noted that our proposed model could only deals with inter-time slice edges. To learn DBN with both inter- and intra-time slice edges remains an interesting point of research. It is suggested by Vinh *et al.* [16] to learn intra-time slice edges separately before combining with the inter-time slice edges and post-processing as an alternative to describe gene-gene interactions. Additionally, we are also interested in taking account of the transcriptional time lag which is commonly found in GRNs. As Zou and Conzen [8] pointed out, the lack of an algorithm to handle transcriptional time lag is one of the main factors that contribute to the relatively low accuracy of inferring GRNs using DBN. Researchers have implemented time lag mechanism in the potential regulators selection algorithm [8, 9]. Lastly, despite the extensive usage of DBN on gene expression data to infer GRNs, it is by no means to replace gene intervention experiments completely. The resultant networks should be treated as a guideline or framework of the studied biological pathways for future hypotheses testing and intervention experiments.

Acknowledgments. This work is financed by Institutional Scholarship MyPhd provided by the Ministry of Higher Education of Malaysia. We also would like to thank Universiti Teknologi Malaysia for supporting this research by the UTM GUP research grants (Vot number: QJ130000.7107.01H29 and QJ130000.7123.00H67).

References

[1] Karlebach, G., Shamir, R.: Modelling and analysis of gene regulatory networks. Nature Reviews Molecular Cell Biology 9(10), 770–780 (2008), doi:10.1038/nrm2503

[2] Lee, W.P., Tzou, W.S.: Computational methods for discovering gene networks from expression data. Briefing in Bioinformatics 10(4), 408–423 (2009), doi:10.1093/bib/bbp028

[3] Bansal, M., Belcastro, V., Ambesi-Impiombato, A., di Bernado, D.: How to infer gene networks from expression profiles. Molecular Systems Biology 3, 78 (2007), doi:10.1038/msb4100120

[4] Murphy, K., Mian, S.: Modelling gene expression data using dynamic Bayesian networks. Technical Report. Computer Science Division, University of California, Berkeley (1999)

[5] Perrin, B.E., Ralaivola, L., Mazurie, A., Bottani, S., Mallet, J., d'Alche-Buc, F.: Gene networks inference using dynamic Bayesian networks. Bioinformatics 19(suppl.2), 138–148 (2003), doi:10.1093/bioinformatics

[6] Kim, S.Y., Imoto, S., Miyano, S.: Inferring gene networks from time series microarray data using dynamic Bayesian networks. Briefing in Bioinformatics 4(3), 228–235 (2003), doi:10.1093/bib/4.3.228

[7] Yu, J., Smith, V.A., Wang, P.P., Hartemink, A.J., Jarvis, E.D.: Advances to Bayesian network inference for generating causal networks from observational biological data. Bioinformatics 20(18), 3594–3603 (2004), doi:10.1093/bioinformatics/bth448

[8] Zou, M., Conzen, S.D.: A new dynamic Bayesian network (DBN) approach for identifying gene regulatory networks from time course microarray data. Bioinformatics 21(1), 71–79 (2005), doi:10.1093/bioinformatics/bth463

[9] Jia, Y., Huan, J.: Constructing non-stationary dynamic Bayesian networks with a flexible lag choosing mechanism. BMC Bioinformatics (11) S27(11) (2010), doi:10.1186/1471-2105-11-S6-S27

[10] Ouyang, M., Welsh, W.J., Geogopoulos, P.: Gaussian mixture clustering and imputation of microarray data. Bioinformatics 20, 917–923 (2004), doi:10.1093/bioinformatics/bth007

[11] Spellman, P.T., Sherlock, G., Zhang, M.Q., Iyer, V.R., Anders, K., Eisen, M.B., Brown, P.O., Botstein, D., Futcher, B.: Comprehensive identification of cell cycle regulated genes of the yeast Saccharomyces cerevisiae by microarray hybridization. Molecular Biology Cell 9, 3273–3297 (1998)

[12] Oba, S., Sato, M., Takemasa, I., Monden, M., Matsubara, K., Ishii, S.: A Bayesian missing value estimation method for gene expression profile data. Bioinformatics 19(16), 2088–2096 (2003), doi:10.1093/bioinformatics/btg287

[13] Yu, H., Luscombe, N.M., Qian, J., Gerstein, M.: Genomic analysis of gene expression relationships in transcriptional regulatory networks. Trends in Genetics 19(8), 422–427 (2003), doi:10.1016/S0168-9525(03)00175-6

[14] Dojer, N.: Learning Bayesian Networks Does Not Have to Be NP-Hard. In: Proceedings of International Symposium on Mathematical Foundations of Computer Science, pp. 305–314 (2006), doi:10.1007/11821069_27

[15] Wilczynski, B., Dojer, N.: BNFinder: exact and efficient method for learning Bayesian networks. Bioinformatics 25(2), 286–287 (2009), doi:10.1093/bioinformatics/btn505

[16] Vinh, N.X., Chetty, M., Coppel, R., Wangikar, P.P.: GlobalMIT: Learning Globally Optimal Dynamic Bayesian Network with the Mutual Information Test (MIT) Criterion. Bioinformatics (2011), doi:10.1093/bioinformatics/btr457

A Hybrid of SVM and SCAD
with Group-Specific Tuning Parameter
for Pathway-Based Microarray Analysis

Muhammad Faiz Misman, Mohd Saberi Mohamad, Safaai Deris,
Raja Nurul Mardhiah Raja Mohamad, Siti Zaiton Mohd Hashim, and Sigeru Omatu

Abstract. The incorporation of pathway data into the microarray analysis had lead to a new era in advance understanding of biological processes. However, this advancement is limited by the two issues in quality of pathway data. First, the pathway data are usually made from the biological context free, when it comes to a specific cellular process (e.g. lung cancer development), it can be that only several genes within pathways are responsible for the corresponding cellular process. Second, pathway data commonly curated from the literatures, it can be that some pathway may be included with the uninformative genes while the informative genes may be excluded. In this paper, we proposed a hybrid of support vector machine and smoothly clipped absolute deviation with group-specific tuning parameters (gSVM-SCAD) to select informative genes within pathways before the pathway evaluation process. Our experiments on lung cancer and gender data sets show that gSVM-SCAD obtains significant results in classification accuracy and in selecting the informative genes and pathways.

Muhammad Faiz Misman · Mohd Saberi Mohamad · Safaai Deris ·
Raja Nurul Mardhiah Raja Mohamad
Artificial Intelligence & Bioinformatics Research Group, Faculty of Computer Science and
Information Systems, Universiti Teknologi Malaysia, Johor, Malaysia
e-mail: faizmisman@gmail.com, {saberi,safaai}@utm.my,
        mardhiah.mohamad@gmail.com

Siti Zaiton Mohd Hashim
Soft Computing Research Group, Faculty of Computer Science and Information Systems,
Universiti Teknologi Malaysia, 81310, Skudai, Johor Darul Takzim, Malaysia
e-mail: sitizaiton@utm.my

Sigeru Omatu
Department of Electronics, Information and Communication Engineering,
Osaka Institute of Technology, Osaka 535-8585, Japan
e-mail: omatu@rsh.oit.ac.jp

S. Omatu et al. (Eds.): Distributed Computing and Artificial Intelligence, AISC 151, pp. 387–394.
springerlink.com   © Springer-Verlag Berlin Heidelberg 2012

1 Introduction

In order to obtain further biological information, researchers in recent years have begun to incorporate the microarray data with biological prior knowledge such as pathway data. Currently there are two approaches used in pathway-based microarray analysis, enrichment analysis approaches (EA) and supervised machine learning approaches (ML) [1, 2].

Beside the advantages, this pathway-based microarray analysis also provides some challenges to researchers. One of the challenges is the quality of the pathway data. When the pathway data is curated from the literature or other resources, the informative genes may be excluded while uninformative genes may be included [1]. Chen *et al.* [3] stated that since the pathway data are defined from the biological context free, when dealing in the specific biological context (e.g. cancer development), typically only a subset of genes within pathway are responsible for the corresponding cellular process. In order to deal with these challenges, we used the ML approaches since it have an advantage compared to EA, where ML can select informative genes within pathways by including the gene selection method while EA tends to consider all the genes within pathways are equally important [1]. This is because, gene selection methods provide several advantages such as improves the classification accuracy, remove uninformative genes, and it can reduce computational time [4]. Therefore, we proposed a hybrid of support vector machines and smoothly clipped absolute deviation with group-specific tuning parameter method (gSVM-SCAD) with aim to effectively select the informative genes and pathways that related to a specific biological context.

2 The Proposed Method and Experimental Data

Given a data set $\{(x_i,y_i)\}$, $y_i \in \{-1,1\}$ is the sample tissue with possible two classes $y_i = -1$ and $y_i = 1$ for each data set used in this paper, while $x_i = (x_{i1,...} x_{id}) \in R^d$ represents the input vector of expression levels of d genes of the i-th sample tissue. SVM is a large margin classifier which separates classes of interest by maximizing the margin between them [5]. This has been widely used especially in microarray classification area [6]. SVM distinguish input variables into its classes by a margin of

$$min_{\beta,c}\Sigma[1-y_if(x_i)]_+ + pen_\lambda(\beta) \qquad (1)$$

where $[1-y_if(x_i)]_+$ is the SVM convex hinge loss function, while $pen_\lambda(\beta)$ is the penalty function with parameters λ, where $\beta = (\beta_1,...., \beta_i)$ are the coefficients of the hyperplane, while c is the intercept of the hyperplane. Even though SVM has proven its superior ability in classifying high dimensional data, the standard SVM can suffer from irrelevant data, since all the variables are used for constructing the classifier [5]. This is due to the usage of the L_2 penalty in a soft-thresholding function for the common SVM. The detailed applications of L_2 penalty in a soft-thresholding function and its drawbacks in identifying noises can be obtained from [5].

2.1 SVM-SCAD

A penalty function is usually used as a variable selection in the statistics, in bioinformatics it is called as gene selection. SCAD is different from other popular penalty functions such as LASSO, also called as the L_1 penalty [7]. This is because SCAD provides nearly unbiased coefficient estimation when dealing with large coefficients. This is contrary to other penalty functions that usually increase the penalty linearly as the coefficient increases [8]. SCAD penalty has the form of

$$\text{pen}_\lambda(\beta) = \Sigma_{j=1}^d P_\lambda(\beta_j) \tag{2}$$

where $P_\lambda(\beta_j)$ is a penalty function with tuning parameter λ for β_j. For providing nearly unbiased, sparsity, and continuity estimate of β, the continuous differentiable penalty function is defined as

$$\text{pen}_\lambda(\beta_j) = \begin{cases} \lambda|\beta| & \text{if } |\beta| \le \lambda \\ -(|\beta|^2 - 2a\,\lambda|\beta| + \lambda^2)/(2(a-1)) & \text{if } \lambda < |\beta| \le a\lambda \\ ((a+1)\,\lambda)/2 & \text{if } |\beta| > a\lambda \end{cases}$$

where a and λ are tuning parameters with $a > 2$ and $\lambda > 0$ [8]. For a tuning parameter a, Fan and Li [8] suggested the parameter $a = 3.7$ due to the minimal achievement in a bayes risk while λ is a tuning parameter obtained using general approximate cross validation (GACV) method (as discussed latter).

In order to surmount the limitations of the SVM due to its inability to distinguish between noise and informative data, Zhang et al. [5] proposed the SVM-SCAD by replacing the L_2 penalty in Equation (1) with Equation (2), which takes the form

$$\min_{\beta,c} \frac{1}{n}\Sigma[1 - y_i f(x_i)]_+ + \Sigma_{j=1}^d P_\lambda(\beta_j) \tag{3}$$

and thus the SVM-SCAD can simultaneously provide gene selection and classification. In order to select the informative genes, SVM-SCAD have to minimize the Equation (3) using the successive quadratic algorithm (SQA) and repeated for kth times until convergence. During the procedure, if $\beta_j^k < \epsilon$, the gene is considered as uninformative. Where β is the coefficient for the gene j in the kth iteration and ϵ is a preselected small positive thresholding value with $\epsilon = y_i - f(x_i)$.

2.2 Tuning Parameter Selection Method

In SCAD there are two tuning parameters namely a and λ that plays an important role in determining an effective predictive model. The tuning parameter selector method in SVM-SCAD is only used to estimate the nearly optimal λ in order to identify the effective predictive model for SCAD. In this paper, a GACV by Wahba et al. [9] is used in order to select the nearly optimal λ. The formula on calculating the GACV as given below:

$$GACV_\lambda = \frac{1}{n} \sum_{i=1}^{n} [1 - y_i f(x_i)_\lambda]_+ + DF_\lambda \tag{4}$$

where n is a total number of samples, DF_λ is a degree of freedom where

$$DF_\lambda = \frac{1}{n} \left[2 \sum_{y_i f(x_{i\lambda}) < - -1} \frac{\alpha_{\lambda i}}{2n\lambda} \cdot \|K(.,x_i)\|_{Hk}^2 + \sum_{y_i f(x_{i\lambda}) \epsilon [-1,1]} \frac{\alpha_{\lambda i}}{2n\lambda} \cdot \|K(.,x_i)\|_{Hk}^2 \right]$$

where $\frac{\alpha_{\lambda i}}{2n\lambda} = \frac{f(x_{i\lambda}) [y_i] - f(x_{i\lambda})[x]}{y_i - x}$ and $\|K(.,x_i)\|_{Hk}^2$ is the reproducing kernel hilbert space (RKHS) with SVM reproducing kernel K (refer [10] for further explanations on RKHS). If all samples in microarray data are correctly classified, then $y_i f(x_{i\lambda}) > 0$ and sum following 2 in DF_λ does not appear and $DF_\lambda = K(0,0)/n\gamma^2$ where γ is the hard margin of an SVM [9]. The nearly optimal tuning parameter λ is obtained by minimizing the error rate from the GACV.

2.3 The Proposed Method (gSVM-SCAD)

Since parameter a in SVM-SCAD has been setup as 3.7 [8], there is only parameter λ that play an important role. In order to incorporate pathway data, the gSVM-SCAD used group-specific parameters λ_j estimation, using the framework proposed by Tai and Pan [11]. In this paper, there are k groups of genes where k = 1...n, each gene is able to be in one or more pathways. We grouped the genes based on their pathway information from the pathway data. In order to provide the group-specific tuning parameters, we modified Equation (2) to the form of

$$pen_{\lambda k}(\beta_j) = \sum_{j=1}^{d} P\lambda_k (\beta_j) \tag{5}$$

by allowing each pathway to have it own parameter λ_k as in (5) instead of general λ in Equation (2), the genes within pathways can be selected and classified more accurately. Figure 1 illustrates the procedure of gSVM-SCAD.

There are several main differences between gSVM-SCAD and other current methods in ML. First, it provides the genes selection method to select the informative genes within a pathway that related to the phenotype of interest. Second, the penalty function SCAD is more robust when dealing with a high number of genes, and it selects important genes more consistently than popular L_1 penalty function [5]. And lastly, with group-specific tuning parameters, the gSVM-SCAD provides more flexibility in choosing the best λ for each pathway. Therefore, by selecting the informative genes within a pathway, the gSVM-SCAD can be seen as the best method in dealing with pathway data quality problems in pathway-based microarray analysis.

```
Algorithm: gSVM-SCAD
Input:    GE: Microarray data
          PD: Pathway data
          λ : Tuning parameter
Output:   SP: Informative pathways
          IG: Informative genes
Begin
Step 1: Grouping genes based on their pathway information
          For j=1 to max number of pathways in PD do
          Find and assign genes from GE that have same pathway info as group
          End-for
Step 2: Evaluate the pathways
     For j=1 to max number of pathways in PD do
       Step 2.1: Estimation of TP using a GACV
          For TP = 0.001 to 0.009 ,0.01 to 0.09 and TP = 0.1 to 1 do
```

$$GACV_\lambda = \frac{1}{n} \sum_{i=1}^{n} [1 - y_i f(x_i)_\lambda]_+ + DF_\lambda$$

```
          End-for
          λ = argmin_λ{GACV(λ)} // best λ produces minimum GACV error
       Step 2.2: Select the informative genes using the SVM-SCAD
          Let β^k as the estimate of β at step k where k = 0, … , n
          The value of β^0 set by an SVM
          While β^k not converge do
```
$$\text{Minimizing the } \frac{1}{n}\Sigma[1\text{-}y_i f(x_i)]_+ + \Sigma_{j=1}^{d} P_\lambda(\beta_j)$$
```
                    k = k + 1
          If β_j^k ≤ ε then
                    The gene j considered as non-informative and discarded
          End-if
          End-while
       Step 2.3: Classify the selected genes using an SVM
Step 3: Calculate the classification error using a 10-fold cross validation
     End-for
End
```

Fig. 1 The gSVM-SCAD procedure

2.4 Experimental Data

The performance of the gSVM-SCAD is tested using two types of data, microarray and pathway data. The role of pathway data is as a metadata or prior biological knowledge. For the pathway data, there are a total of 480 pathways with 168 taken from KEGG and the other 312 pathways from BioCarta. The information of the microarray data sets is shown in Table 1. Both data can be downloaded at http://bioinformatics.med.yale.edu/pathway-analysis/datasets.htm.

Table 1. Microarray data sets

Name	Total samples	Total genes	Class	Reference
Lung	86	7129	2 (normal and tumor)	[13]
Gender	32	22283	2 (male and female cells)	unpublished

3 Results and Discussion

3.1 Performance Evaluation

In order to evaluate the performance of gSVM-SCAD, we used a 10-fold cross validation (10-fold CV) classification accuracy. The selected gene and pathways are validated with the biological literatures and databases. The biological validation results can be obtained in our supplementary page (http://www.utm.my/aibig/people/mohd-saberi-mohamad/research/supplementary-information.html).

For the performance evaluation of SCAD penalty function, SCAD was compared with L_1 penalty function by hybridizing it with an SVM classifier (L_1 SVM), obtained from R package penalizedSVM [14]. The L_1 SVM also applied with group-specific tuning parameters to determine λ. Then, the gSVM-SCAD was compared with the current SVM-SCAD with respect to one general tuning parameter for all pathways, the tuning parameter $\lambda = 0.4$ as used by Zhang et al. [5]. For comparison with other classification methods without any gene selection process, the gSVM-SCAD was compared with four classifiers that are without gene selection method. The classifiers are PathwayRF [12], multi layer perceptron neural networks with 3 layers (MLP), k-nearest neighbor with one neighbours (kNN), and linear discriminant analysis (LDA). The results of the experiment were shown in Table 2.

In comparing gSVM-SCAD with L_1-SVM and SVM-SCAD, it is interesting to note that gSVM-SCAD outperforms the other two penalized classifiers in both data sets with gSVM-SCAD is 18.63% higher than L_1-SVM for lung cancer data set and 6.57% higher in gender data set. This is due to the SCAD as a non-convex penalty function is more robust to biasness when dealing with a large number of coefficients β in selecting informative genes compared to the L_1 penalty function [5]. In contrast to L_1 penalty, SCAD produces sparse solution by thresholding small estimated β to zer (Please refer [5] and [8] for further information of the robustness of non convex penalty in microarray data). Therefore, the proposed method with SCAD penalty function selected more informatively genes within a pathway than the LASSO penalty. Table 2 further shows that the gSVM-SCAD had better results than the SVM-SCAD, with 20.27%, 9.37% higher in lung cancer and gender data sets respectively. It is demonstrated that group tuning parameters in the gSVM-SCAD provided flexibility in determining the λ for each pathway compared to the use of a general λ for whole pathways. This is because usually the genes within pathway have a different prior distribution.

Table 2 further shows that result in lung cancer data set outperformed compared to gender data set. This is because one feature selection method may find many different subsets of features (in this research, features are referred as gene and pathway) that can achieve similar or different classification accuracy [15, 16]. It is believed that, this is related to the instability of the SVM-SCAD as a gene selection method in selecting the informative genes within pathway, since this research focuses only on accuracy-based strategy in analyzing the performance of the gSVM-SCAD. By using the accuracy-based strategy the stability in feature selection method may not be fully reliable in selecting the true informative genes [15].

Table 2 A comparison of averages of 10-fold CV accuracy from the top ten pathways with other methods

Method	Lung Cancer (%)	Gender (%)
gSVM-SCAD	**73.77**	**87.33**
L₁-SVM	55.14	80.76
SVM-SCAD	53.50	77.96
MLP	70.39	81.54
kNN	61.73	82.44
LDA	63.24	75.81
PathwayRF [13]	71.00	81.75

Note:
The texts in **Bold** are the highest 10-fold CV accuracy.
The texts in *italic* are the methods from the self-running experiment.

In order to show that not all genes in pathways are contributed to the development of specific cellular processes, the gSVM-SCAD is compared with four classifiers. The results are also shown in Table 2. For the lung cancer data set, it shows that the gSVM-SCAD outperformed all the classifiers, with 2.77% higher than PathwayRF, 3.38% higher than MLP, 10.53% higher than LDA, and lastly 12.04% higher than kNN. While for the gender data set, the gSVM-SCAD obtained 5.58% higher than PathwayRF, 5.79% higher than MLP, 4.89% higher than kNN one neighbour and 11.52% higher than LDA. From the results in Table 2, the gSVM-SCAD shows a better performance when compared to almost four classifiers for all two data sets. This is because the standard classifiers built a classification model using all genes within pathways. If there are uninformative genes inside the pathways, it reduced the classification performance. In contrast, the gSVM-SCAD does not include all genes in the pathways into the development of a classification model, as not all genes in a pathway contribute to cellular processes, due to the quality of pathway data.

4 Summary

This paper focuses on to identify the informative genes and pathways that relate to phenotypes of interests by proposing the gSVM-SCAD. From the experiments and analyses, the gSVM-SCAD was shown to outperform the other supervised machine learning methods in almost all three data sets. In comparison of penalty functions, gSVM-SCAD has shown its superiority in selecting the informative genes within pathways compare to L₁ SVM. By providing group-specific tuning parameters, gSVM-SCAD had shown a better performance compare to an SVM-SCAD that provides a general penalty term for all pathways. The proposed method also had shown its ability in identifying the informative genes and pathways.

Acknowledgments. This work is financed by Institutional Scholarship MyPhD provided by the Ministry of Higher Education of Malaysia. We also would like to thank Universiti Teknologi Malaysia for supporting this research by UTM GUP research grants (vot number: Q.J130000.7123.00H67 and Q.J130000.7107.01H29).

References

[1] Wang, X., Dalkic, E., Wu, M., et al.: Gene module level analysis: identification to networks and dynamics. Curr. Opin. Biotechnol. 19, 482–491 (2008), doi:10.1016/j.copbio.2008.07.011

[2] Misman, M.F., Deris, S., Hashim, S.Z.M., et al.: Pathway-based microarray analysis for defining statistical significant phenotype-related pathways: a review of common approaches. In: Int. Conf. Inf. Manag. Eng. (2009), doi:10.1109/ICIME.2009.103

[3] Chen, X., Wang, L., Smith, J.D., et al.: Supervised principle component analysis for gene set enrichment of microarray data with continuous or survival outcome. Bioinformatics 24, 2474–2481 (2008), doi:10.1093/bioinformatics/btn458

[4] Mohamad, M.S., Omatu, S., Deris, S., et al.: A modified binary particle swarm optimization for selecting the small subset of informative genes from gene expression data. IEEE Trans. Inf. Technol. Biomed (2011), doi:10.1109/TITB.2011.2167756

[5] Zhang, H.H., Ahn, J., Lin, X., et al.: Gene selection using support vector machines with non-convex penalty. Bioinformatics 22, 88–95 (2006), doi:10.1093/bioinformatics/bti736

[6] Guyon, I., Weston, J., Barnhill, S., et al.: Gene selection for cancer classification using support vector machines. Mach Learn 46, 389–422 (2002), doi:10.1093/bioinformatics/btl386

[7] Tibshirani, R.: Regression shrinkage and selection via the lasso. J. R. Stat. Soc. Ser. B (Methodol.) 58, 267–288 (1996), doi:10.1.1.35.7574

[8] Fan, J., Li, R.: Variable selection via nonconcave penalized likelihood and its oracle properties. J. Am. Stat. Assoc. 96(456), 1348–1360 (2001), doi:10.2307/3085904

[9] Wahba, G., Lin, Y., Zhang, H.: GACV for support vector machines, or, another way to look at margin-like quantities. In: Smola, A.J., Bartlett, P., Schoelkopf, B., Schurmans, D. (eds.) Advances in Large Margin Classifiers. MIT Press, Cambridge (2000)

[10] Wahba, G.: Support vector machines, reproducing kernel Hilbert spaces, and randomized GACV. In: Schoelkopf, A.B., Burges, C.J.C., Smola, A.J. (eds.) Advances in Kernel Methods – Support Vector Learning. MIT Press, Cambridge (1999)

[11] Tai, F., Pan, W.: Incorporating Prior Knowledge of Predictors into Penalized Classifiers with Multiple Penalty Terms. Bioinformatics 23, 1775–1782 (2007), doi:10.1093/bioinformatics/btm234

[12] Pang, H., Lin, A., Holford, M., et al.: Pathway analysis using random forest classification and regression. Bioinformatics 16, 2028–2036 (2006), doi:10.1093/bioinformatics/btl344

[13] Battacharjee, A., Richards, W.G., Satunton, J., et al.: Classification of human lung carcinomas by mRNA expression profiling reveals distinct adenocarcinoma subclasses. Proc. Natl. Acad. Sci. 98, 13790–13795 (2001), doi:10.1073/pnas.191502998

[14] Becker, N., Werft, W., Toedt, G., et al.: PenalizedSVM: A R-package for feature selection SVM classification. Bioinformatics 25, 1711–1712 (2009), doi:10.1093/bioinformatics/btp286

[15] He, Z., Yu, W.: Stable feature selection for biomarker discovery. Computational Biology and Chemistry 34, 215–225 (2010), doi:10.1016/j.compbiolchem.2010.07.002

[16] Zucknick, M., Richardson, S., Stronach, E.A.: Comparing the Characteristics of Gene Expression Profiles Derived by Univariate and Multivariate Classification Methods. Statistical Application in Genetics and Molecular Biology 7, 1–28 (2008), doi:10.2202/1544-6115.1307

A Reliable ICT Solution for Organ Transport Traceability and Incidences Reporting Based on Sensor Networks and Wireless Technologies

Asier Moreno and Ignacio Angulo

Abstract. This paper describes an ICT solution based on an intelligent onboard system which is able to trace the organs inside a medical van during delivery routes to the hospitals, without altering the carriers daily tasks. The intelligent onboard system is able to ensure the safety of the cargo by means of a sensor network who permanently evaluates their status. The van understands its environment, including: its location, the temperature and the humidity of the transported organs; and can report incidences instantly via wireless communications to anyone interested. It is a non-intrusive solution which represents a successful experience of using smart environments in a viable way to resolve a real social and healthcare necessity.

Keywords: medical traceability, transportation of organs, incidence management, sensor networks, wireless technologies.

1 Introduction and Related Work

Traceability provides the historical location and trajectory of a product or batch of products along the supply chain. Traceability commercial systems are adapted to areas in which governments establish minimum traceability requirements: mainly food and pharmaceutical industry, or those sectors in which large industries are those than indicate the parameters that determine traceability to suit their demanding quality systems, such as electronics or automotive [1].

The importance of tracking goods has reached such a point that is being regulated by law in some countries. For example in Europe, Regulation 178/2002 requires the traceability of all food from farms to the end of the supply chain. This is one of the reasons because most of the projects about tracking are focused on the

Asier Moreno · Ignacio Angulo
Deusto Institute of Technology - DeustoTech, University of Deusto,
Avenida de las Universidades 24, 48007 - Bilbao, Spain
e-mail: {asier.moreno,ignacio.angulo}@deusto.es

S. Omatu et al. (Eds.): Distributed Computing and Artificial Intelligence, AISC 151, pp. 395–402.
springerlink.com

food industry, as for example the Trace FP6 project (http://www.trace.eu.org) or those ones based on RFID or ZigBee technology and described in [2] [3] [4].

There are other critical sectors as the pharmaceutical one. Thus, the PharmaX initiative is proposed to shed light on the pharmaceutical traceability and over-all-process regulation. This system ensures that all pharmaceutical supply chain participants can integrate with each other, resulting in information sharing, consistency checking and anti-counterfeit [5]. All these projects are seeking the following benefits: processes automation, turnover increase, and stock management improvement [6].

This paper describes an ICT-based solution that meets the needs of a specific scenario: transportation of organs and biological samples. In this environment, the decisive factors in industrial environments such as the cost of deployment and return of investment are replaced by others that lead to system reliability. Maintain organs in optimum conditions since the certification of brain death of the donor to its reintroduction in compatible receivers is a critical process. Any incident that affects the environmental conditions of the packaging must override the transplant process to the enormous risk involved in the receiver [7]. The proposed system constantly monitors the environmental variables immediately reporting any change in them which endangers the transport process.

The use of this kind of ICT tracking solutions in the medical sector is innovative due to the lack of previous references adapted to this scenario. Despite that, the proposed architecture adapts to any other sector that requires sensitive freight.

The result of our work is an innovative ICT solution with real-time tracking for improving the incidence recovery timing in the transportation of organs and biological samples.

2 Functional Description

There are two well differentiated parts in the proposed solution: the onboard system and the control software solution. An architecture schema of the complete system is shown in the graph below.

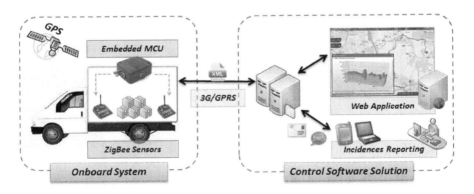

Fig. 1 System Arquitecture

2.1 Onboard System

Responsible for managing the proper transport and delivery of organs to the hospitals, is a key element for the functionality of the whole system. Its function is to monitor by humidity, brightness and temperature sensors the state of the refrigerators in which organs are transported. Likewise, and by geolocation technologies, it will send the position of the cargo to guarantee traceability at all times.

Refrigerators, specifically designed to the organ transportation, will be provided with wireless sensors able to measure its temperature, brightness level and humidity. With the ZigBee technology used, whose actual operation is detailed in Section 3.1 of this paper, the solution is capable of generating a mesh network autonomously so that the addition of a new refrigerator in the van with its corresponding sensor is performed transparently to the communications system.

Another component present in the onboard system is an embedded microcontroller that is connected to the ZigBee coordinator. With the processing power and communication interfaces of this device the solution will be capable of collecting and processing data which will be subsequently sent to the central server. This data shall consist of the values of the set of sensors present in the medical van providing the transport, as well as geolocation values obtained through the GPS capabilities of the microcontroller module.

By processing the raw data received by the embedded system from the ZigBee coordinator, the system will generate relevant alerts or incidents, taking into account the maximum or minimum values established for that specific transport.

2.2 Control Software Solution

Installed on the servers of the control center, it is composed of two different applications, integrated and designed as part of the solution. The applications complement each other for the treatment, management and dissemination of the data collected and received from the onboard systems installed on the medical transports.

Communications Management Server
The communications manager is the server side of the distributed system that comprises the solution. The clients (the medical transports on route) will request access to the server so it will generate dedicated processing threads to meet these requests dynamically increasing the QoS offered by the application in important aspects such as reliability, performance, transparency and scalability.

Traceability Web Application
It is used to track the medical transport vans and the transported organs as well as monitor and keep track of their status at all times. The information received will be displayed through a web application designed to facilitate usability and present data in an attractive and complete way.

The functionality offered by the web application can be subdivided into three distinct areas according to their purpose. These are: the traceability of the medical transports, the freight status monitoring and the incident management.

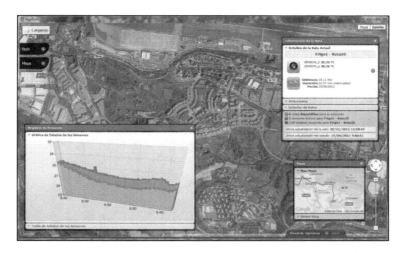

Fig. 2 Web Application Snapshot

1. Freight Condition Monitoring

The system has a robust database where all route information is stored. It keeps a record for both finished and real-time routes with all the information collected on the status of the sensors that monitor each refrigerator. That is, once the interval of data sending is defined in the application itself , the system stores the status of sensors and GPS position data during all the course of the route.

2. Medical Transports Traceability

The fleet management system provides trace and tracking of moving medical vehicles using a web-map. The computed routes with corresponding times (including stopovers) are recorded to assist in the prediction of appropriate time intervals for successive freight distributions which is of great importance considering that time is a critical factor in the field of medical transportation. This leads to optimized routes with improved delivery time, traffic monitoring and a preventive transport distribution.

3. Incidences Management

Although you can perform a complete monitoring of the state of the transport and its cargo via web, it is also necessary to include a complete incidences management system to notify to anyone interested about the possible deviations when they occur. All this incidences are managed from the incidences reporting tab, where incidences are displayed at the instant they occur in the form of warning. There is also an historical record of incidents in which we can seek them ex post using search filters for transport, date, route, etc.

Another required functional feature that belongs to the incidences system is the ability to subscribe to alerts for a particular route or transport and for a certain time. This subscription allows the user to receive notifications via email or SMS depending on the configuration chosen. Upon receipt of an incidence the system takes care of consulting the list of subscribers and then sends via SMTP server or SMS server the appropriate notice to all of them.

The relevance of the scenario encourages the time constraint in reporting incidences. Being a distributed architecture, the delay is additive between all elements that collaborate in the incidence transmission. The Zigbee End Device integrates a microcontroller continuously monitoring environment conditions of embedded refrigerator. If any sensor reading is out of the programmed scope, an incidence is immediately reported to the Zigbee Coordinator, which sends an alert through serial communication to the MTX microcontroller. The alert is formatted into an XML frame and sent through a TCP socket to the Central Server. Delay in this process depends on the complexity of the Zigbee network and conditions of the 3G communication. All the tests performed showed a time window of 3 to 7 seconds since the detection of the alert until the incidence is notified.

3 Technical Solution

3.1 Onboard System ZigBee Sensor Network

ZigBee is a low-cost, low-power, wireless mesh network standard. The low cost allows the technology to be widely deployed in wireless control and monitoring applications. Low power-usage allows longer life with smaller batteries. Mesh networking provides high reliability and more extensive range [8].

The ZigBee mesh sensor network that conform the devices that are arranged both in the conditioned refrigerators and in the medical transport allows the system to know the status of the organs transported at all times.

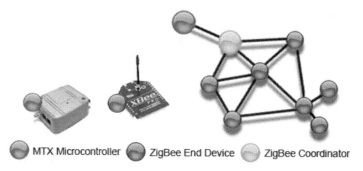

MTX Microcontroller   ZigBee End Device   ZigBee Coordinator

Fig. 3 Onboard Sensor Network

As shown in Figure 3, the network is composed of two types of ZigBee devices (the End Devices or ZEDs and the Coordinator) and a Microcontroller unit.

- ZigBee End Device (ZED): Contains just enough functionality to talk to the parent node (either the coordinator or a router); it cannot relay data from other devices. This relationship allows the node to be asleep a significant amount of the time thereby giving long battery life.

- ZigBee coordinator (ZC): The most capable device, the coordinator forms the root of the network tree and might bridge to other networks. There is exactly one ZigBee coordinator in each network since it is the device that started the network originally.

The use of the proposed ZigBee network will allow the system to meet the specific requirements of the healthcare field as the reliability, as well as desirable non-functional requirements such as low cost, high durability or auto configuration. All these features encourage the use of this type of networks in the proposed field.

3.2 Embedded Microcontroller and Communications Device

MTX-65+G embedded microcontroller is connected to the ZigBee coordinator. This terminal is a GPS receiver combined with a GSM-GPRS for data, voice and SMS connectivity. It is made with a SIEMENS XT65 modem, with a high sensitivity and low consumption GPS receiver integrated into the GSM module that can be controlled with AT commands, not being necessary to know the NMEA plot.

The embedded microcontroller is responsible for collecting and transforming the data which will be subsequently sent to the central server. Fig. 4 illustrates the XML schema generated and sent to the central server.

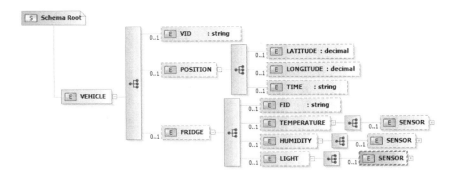

Fig. 4 Status Message XML Schema

As shown in the graph, the file is composed of the values of the set of sensors present in each of the fridges carried in the medical transport (temperature, humidity and light levels), as well as geolocation values (latitude, longitude and time) obtained through the GPS.

3.3 Control Software Solution

All the functionalities described in Section 2.2 have been developed on a Rich Internet Application (RIA). Below are listed the technical characteristics of the business logic, data management and graphical components of the web application.

Business Logic and Services
In this module is contained all the logic needed to meet the functional requirements set by the application proposal, outlined in points 2.2.1 to 2.2.3. Thus, we find the necessary logic for the treatment and management of the routes or buckets, as well as the user and incidences management. It also has all the required data about the geolocation of the routes, obtained through external services, in this case using the Google Maps Javascript API.

Data Management
"Microsoft SQL Server 2008" was used for the data storage of the applications related to the ICT Solution. The data is stored during the entire transport distribution by generating the necessary entries in the Database Management System (DBMS) of the server.

Web Application
The development of the control solution is completed with the web application which offers features beyond the typical application of fleet management. This application has been developed taken into account two fundamental characteristics: (1) to maintain a friendly and attractive interface, (2) without prior installation or further configuration. This Web application approaches usability to a modern desktop application but with all the advantages that such implementation offers, being available globally via the Internet to a vast number of devices supported.

The Control Panel, based on asp.NET development framework, has been made extensive use of technologies designed for creating Rich Internet Applications (RIA): JavaScript, CSS3, HTML5, Ajax and jQuery, along with the use of the tools offered by Google for displaying and processing of geographic and positioning information. This feature improves not only the final visual aspect of the application but also the overall usability. The whole site is based on an asynchronous behavior, so interaction eliminates the sense of loading data and responds instantly. All kind of choices as routes, transports or incidences, represent a dynamic and transparent loading of data and an almost immediately response to their interaction.

4   Conclusion

The work presented in this paper provides a comprehensive solution that can detect environmental anomalies that affect the organs and biological samples during the transport process. Temperature, humidity and light are continuously monitored since organs are collected in any of the extractors centers network of organ and tissue until they are deposited in the immunology service of the nearest certified laboratory or in the center linked to the selected receiver. The system can detect trends that may lead to the loss of an organ, generating autonomously a warning before jeopardizing its integrity. Furthermore, the system provides real-time geolocation for transported organs allowing the medical teams to plan the intervention with maximum accuracy.

The inclusion of the ZigBee motes, responsible for the acquisition of environmental variables on the inside of the refrigerator where organs are transported, so as the installation of on-board system in the medical transport vehicles allows carriers to run their transport actions in a non-intrusive way without causing any alteration in the transport process, fully established by the protocol for the transport of biological samples and transfer of organs and/or tissues.

Acknowledgments. This work has been funded by the Ministry of Science and Innovation of Spain under INNPACTO funding program (Trazamed project, IPT-090000-2010-7).

References

1. Cleland-Huang, J., Settimi, R., Romanova, E., Berenbach, B., Clark, S.: Best Practices for Automated Traceability. Computer 40(6), 27–35 (2007)
2. Schwägele, F.: Traceability from a European perspective. In: 51st International Congress of Meat Science and Technology (ICoMST), Meat Science, vol. 71(1), pp. 164–173 (2005)
3. Kelepouris, T., Pramatari, K., Doukidis, G.: RFID-enabled traceability in the food supply chain. Industrial Management & Data Systems 107(2), 183–200 (2007)
4. Manikas, I., Manos, B.: Design of an integrated supply chain model for supporting traceability of dairy products. International Journal of Dairy Technology 62(1), 126–138 (2009)
5. Huang, G.Q., Qin, Z., Qu, T., Dai, Q.: RFID-enabled pharmaceutical regulatory traceability system. In: 2010 IEEE International Conference on RFID-Technology and Applications (RFID-TA), Guangzhou, China, pp. 211–216 (2010)
6. Bertolini, M., Bottani, E., Rizzi, A., Volpi, A.: The Benefits of RFID and EPC in the Supply Chain: Lessons from an Italian pilot study. The Internet of Things, Part 4, 293–302 (2010)
7. Moreno, A., Valls, A., Ribes, A.: Finding Efficient Organ Transport Routes using Multi-Agent Systems. In: Proceedings of the IEEE 3rd International Workshop on Enterprise Networking and Computing in Health Care Industry (Healthcom), L'Aquilla, Italy (2001)
8. Gutierrez, J.A.: IEEE 802.15.4, Low-Rate Wireless Personal Area Networks: Enabling Wireless Sensor Networks. Institute of Electrical & Electronics Enginee (2003) ISBN: 0738135577

Applying Lemur Query Expansion Techniques in Biomedical Information Retrieval

A.R. Rivas, L. Borrajo, E.L. Iglesias, and R. Romero

Abstract. The increase in the amount of available biomedical information has resulted in a higher demand on biomedical information retrieval systems. However, traditional information retrieval systems do not achieve the desired performance in this area. Query expansion techniques have improved the effectiveness of ranked retrieval by automatically adding additional terms to a query. In this work we test several automatic query expansion techniques using the Lemur Language Modelling Toolkit. The objective is to evaluate a set of query expansion techniques when they are applied to biomedical information retrieval. In the first step of the information retrieval searching, indexing, we compare the use of several techniques of stemming and stopwords. In the second step, matching, we compare the well-known weighting algorithms Okapi and TF-IDF BM25. The best results are obtained with the combination of Krovetz stemmer, SMART stopword list and TF-IDF. Moreover, we analyze the document retrieval based on Abstract, Title and Mesh fields. We conclude that seems more effective than looking at each of these fields individually. Also, we show that the use of feedback in document retrieval results a improvement in retrieving. The corpus used in the experiments was extracted from the biomedical text Cystic Fibrosis Corpus (CF).

Keywords: Query expansion, Biomedical information retrieval, Lemur, MEDLINE.

1 Introduction

Nowadays there is a need to improve information search systems in order to increase their performance [9]. When a user formulates a query on an Information Retrieval (IR) system, typically enters short queries without some words that, if provided,

A.R. Rivas · L. Borrajo · E.L. Iglesias · R. Romero
Univ. of Vigo, Computer Science Dept.
e-mail: arrivas@correo.ei.uvigo.es, lborrajo@uvigo.es,
        eva@uvigo.es, rrgonzalez@uvigo.es

S. Omatu et al. (Eds.): Distributed Computing and Artificial Intelligence, AISC 151, pp. 403–410.
springerlink.com   © Springer-Verlag Berlin Heidelberg 2012

would be very useful search terms. In order to achieve reasonable retrieval effectiveness on these short queries, query expansion techniques are used.

Query expansion techniques have improved the effectiveness of ranked retrieval by automatically adding additional terms to a query. It consists on the building of a new query from an old one by adding synonyms or hierarchical related terms or other kinds of semantically related terms [5]. Such terms may come from documents identified during the search, or by semantic tools such as thesauri. It may be done automatically or intellectually.

In this study some automatic query expansion techniques using Lemur Language Modelling Toolkit [3, 10] are implemented. Our objective is to test the Lemur query expansion techniques [7, 2] when they are used in biomedical information retrieval, and to show which is the best combination of techniques. Lemur is a software tool designed to facility research in language modelling and IR providing methods to parse queries, indexing and retrieving documents related to queries using weighting algorithms.

The structure of this article is the following: In the next section the IR model is showed. In section 3 the different tests developed in this research are described. The section 4 presents the statistical analysis and, in the last section, the conclusions and future work are described.

2   The IR Model

Information Retrieval is composed of two main processes, indexing and matching. The first step indexes a corpus composed by three elements: documents, queries and relevance judgements given by the experts, for obtaining keywords to be used in the process. These keywords represent relevant terms in documents and queries. Matching is the process of computing a measure of similarity between documents and queries by weighting algorithms of terms, being TF-IDF and Okapi BM25 the most applied. In this research, both are used.

In the indexing process, stemming [6, 17] and stopword lists and acronyms are usually used. There are two major stemmers in use for English IR, the Porter stemmer and the Krovetz stemmer. These two stemming algorithms within Lemur are tested in this work.

Besides, in Information Retrieval, a document is indexed by the frequency of words in it. Statistical analysis shows that some words have quite low frequency, while others have high frequency [18]. For example, *and*, *of* and *the* appear frequently in the documents without significant information to the document. This set of words is referred as stopwords. Elimination of stopwords could significantly reduce the size of the indexing structure, speed up the calculation and increase the accuracy. Up to now, a lot of stopword lists have been developed for English language. For example, the U. S. National Library of Medicine (NLM) oficial stopword list [1] and the stopword list built by Gerard Salton and Chris Buckley for the experimental

---

[1] NLM stopword list http://www.netautopsy.org/umlsstop.htm

SMART information retrieval system at Cornell University [2]. In this research, NLM and SMART stopword lists are used.

The tests performed, showed in the next section, are also based on an analysis of the benefits produced by the use of acronyms in the searching. Acronyms are widely used in biomedical literature. The names of many clinical diseases and procedures, and of common entities such as genes or proteins, have widely used acronyms. To recognize acronyms is important in order to not be affected by stemming and stopwords processes and thus make a direct mapping between them in documents and queries [12, 13] [8, 16]. In this research the Lemur acronym recognition module [3] is tested.

At the end of the indexing process, the keywords that represent relevant words in documents and queries are obtained.

The next step on IR model is the matching process. This process computes a measure of similarity between documents and queries by weighting terms, being TF-IDF and BM25 the most important weighting algorithms. In this research, both algorithms are used.

Parameterize weighting algorithms is an important point to improve results with the collection used. This was the next test in this research. Besides realizes several tests, the best parameters to the text collection are obtained.

Also, following previous studies by other authors, we test the benefit of working with a set of fields of the documents compared to only work with one of them (the Abstract).

To improve the results of the above processes, relevance feedback was made. Most retrieval systems return a ranked list of documents in response to a query. These documents are ordered such that the documents more similar to the query are first on the list. Once obtained the first documents ranked for a query, expansion techniques can be applied again. Retrieved documents are analyzed taking into account new keywords representing them. And these keywords can be added to initial query in order to reranking the documents. Lemur toolkit implements a simplified Rocchio feedback algorithm, and we use this algorithm.

Many authors have worked with MeSH and abstract fields to retrieve information, so we focuse the last part of our research to improve the expansion obtained by the abstract with the MeSH field [11]. MeSH terms index journal articles for subject analysis of biomedical literature NLM. It imposes uniformity and consistency to the indexing of works. MeSH field in the documents is considered one of the most important fields in retrieving documents that are appropriate to the queries.

In our tests, firstly we make the expansion with the queries and the expanded queries leaving only the main MeSH terms (MeSH Descriptors) that are used to cataloge the publications indicating what article is about. Secondly, we use synonyms, alternate forms, and other closely related terms to a given MESH record (usually known as Entry Terms or Terms).

---

[2] SMART http://www.lextek.com/manuals/onix/stopwords2.html
[3] Acronym
http://www.acronymslist.com/cat/human-genome-acronyms-p3.html

The possibility of feedback the Abstract results with retrieved documents by searching in the MeSH terms [11] is also explored.

3 Experimental Results

In this section we are going to describe the tests achieved in this research. The corpus used in the experiments was extracted from the Cystic Fibrosis Corpus (CF). It consists of 1239 documents discussing Cystic Fibrosis Aspects in the Medline composed by Abstract, Title and MeSH fields -between others- and 100 queries with the respective relevant documents as answers belong to experts [1].

In order to evaluate results, trec_eval [4] program inside the TREC conference is used. This program allows getting measures like total number of documents over all queries (Retrieved, Relevant and Rel-ret (relevant and retrieved)) or MAP, R-prec and Interpolated Recall-Precision Averages.

Firstly, different combinations of stemming functions and stopwords lists are studied (see Table 1). Although weak (Krovetz) and strong (Porter) stemming methods performances are usually comparable [4], in our experiments the results are similar. In more detail, Porter stemmer is the best based on the MAP measure, and in terms of R-prec Krovetz stemmer is slightly better.

Different stopword removal methods are compared, and the results show that removing stopwords improves the performance. From our experiments, using the SMART stopword list results are better than using lists with less stopwords (NLM).

Table 1 Improvements using different stemming and stopword techniques

Test	Technique	MAP	RPrec	Gm_MAP
1	Baseline	0,1263	0,2098	0.0703
2	Porter stemmer	**0,1253**	0,2154	0.0806
3	Krovetz stemmer	0,1334	**0,2231**	0.0847
4	NLM stopword	0,1391	0,2242	0.0784
5	SMART stopword	0,1395	0,2243	0.0788
6	Porter stemmer + NLM stopword	0,1408	0,2332	0.0895
7	Porter stemmer + SMART stopword	0,1433	0,2342	0.0912
8	Krovetz stemmer + NLM stopword	0,1530	0,2333	0.0935
9	Krovetz stemmer + SMART stopword	**0,1544**	0,2324	0.0939

Thus, we conclude that applying the Krovetz stemmer and SMART stopword gives the best results for this test corpus, so the remaining experiments are performed using this combination. It is also statistically showed. In Table 2 significance tests corresponding to the test 6 to test 9 included in Table 1) are presented. Taking into account those values we conclude that all general tests based on p-values are

---

[4] trec_eval http://trec.nist.gov/trec_eval/

extremely significant. However if we take a look to χ^2 distribution values, the best results are obtained by tests 8 (Krovetz stemmer + NLM stopword) and 9 (Krovetz stemmer + SMART stopword). Besides both got the same value, suggesting that there is not difference between them and we can use either indifferently. Since, moreover, the Krovetz stemmer and SMART stopword combination results in the best MAP, we have finally chosen that option.

Table 2 Statistical significance

Tests Queries	One hundred queries	
	χ^2	P-value
Porter stemmer + NLM stopword	330.080	less than 0.0001
Porter stemmer + SMART stopword	332.170	less than 0.0001
Krovetz stemmer + NLM stopword	336.371	less than 0.0001
Krovetz stemmer + SMART stopword	336.371	less than 0.0001

The following test was to study the advantages of recognizing acronyms in biomedical queries and documents. The Lemur acronym recognition module is used, without stemming nor stopword list. Experiments show that expanding the query using acronyms and their long forms is not beneficial for this collection. In Table 3 (row 1) result of MAP using only acronyms is shown.

Subsequently, the behavior of weighting algorithms in the matching process was analyzed. In our tests we obtain best results optimizing the parameters of the Okapi and TF-IDF algorithms based on the abstracts (AB) of the collection and the type of queries (see rows 2 and 3 of Table 3). The values showed in rows 4 and 5 (TF-IDF LogTF and TF-IDF RawTF formulas without parameters) were the worst.

Many researchers use the weighting algorithms to retrieve information in several sections in the documents, not only in the Abstract field. By this assumption, we also test how the MAP increases if we look for documents related to the queries in the Abstract (AB), Title (TI) and MeSH (MH) fields using the BM25 formulas. Results (see Table 3, rows 6 to 9) are comparable to the obtained by others authors [4, 14, 15].

Row 10 (Table 3) shows the best values obtained for Abstract Relevance feedback for TF-IDF BM25. We test how the MAP measure is increased looking for documents based on queries applied to Abstract, Title and MeSH fields using the TF-IDF BM25 relevance feedback algorithm (row 11, Table 3). These results are comparable with those obtained by Kwangcheol Shin and Sang-Young Han [11] in other biomedical text corpus.

Evaluating the use of MeSH fields to improve the expansion querys, the advantage of using the MeSH field respect to the Abstract field was studied (see Table 3, row 3 and 12). It can be concluded that search MeSH field does not benefit the retrieval, and the feedback obtained is worse than the original result (see Table 3, row 13).

Table 3 Krovetz stemmer and SMART stopword list results

	Technique	MAP
Acronyms		
1	with Acronyms	0,1808
Weighting algorithms in Abstract field		
2	Okapi BM25 in AB	0,1824
3	TF-IDF BM25 in AB	0,1866
4	TF-IDF RawTF in AB	0,1436
5	TF-IDF LogTF in AB	0,1731
Weighting algorithms in Abstract, Title and MeSH fields		
6	Okapi BM25 in AB+TI+MH	0,2737
7	TF-IDF BM25 in AB+TI+MH	0,2904
8	TF-IDF LogTF in AB+TI+MH	0,2654
9	TF-IDF RawTF in AB+TI+MH	0,2228
Using relevance feedback and MeSH field		
10	TF-IDF BM25 feedback AB	0,2033
11	TF-IDF BM25 feedback AB+TI+MH	**0,3435**
12	TF-IDF BM25 MH	0,1826
13	TF-IDF BM25 feedback MH	0,1782
14	MH in Queries	0,1863
15	MH feedback in Queries	0,1819
16	AB feedback with MH	0,2005

Then, a new form of expansion was tested consisting in the searching of MeSH descriptors of the Entry Terms in queries and launching against the MeSH field of documents. This expansion method is based on the work of Kwangcheol Shin and Sang-Young Han [11]. We prove the advantage of using descriptors to expand the query instead of working with terms that are not related to the MeSH field on documents (see Table 3, row 14). In this aproximation, feedback is non-relevant (see Table 3, row 15).

Finally, documents retrieved with the MeSH process were selected to make feedback. We found that there is not relevant variation between the Abstract feedback with documents previously retrieved in the same field or feedback with the documents retrieved from the MeSH process (see Table 3, row 16).

4 Conclusions and Future Work

In this study we develop and evaluate query expansion techniques for retrieving documents in several fields of biomedical articles belonging to the corpus Cystic Fibrosis, a corpus of MEDLINE documents.

We test the benefit of using several techniques of indexing and matching in the query expansion following the investigations of other authors. After evaluating the obtained results, we can conclude:

1. The best results are obtained with the combination of Krovetz stemmer and SMART stopword list.
2. Comparing the weighting algorithms Okapi BM25 and TF-IDF BM25, TF-IDF is superior in its results.
3. Document retrieval based on Abstract, MeSH and Title fields seems more effective than looking at each of these fields individually.
4. The use of feedback in document retrieval results a great improvement in retrieving the relevant documents.
5. The results obtained are similar to the presented by other authors with different corpus.

As future work, stemming could be used when working with expansion techniques based on Entry Terms following the research of K. Shin and S.Y. Han [11]. Working with MeSH Headings and MeSH fields in documents, the mapping between terms should be enough. Also it would be interesting to test whether to reduce the root of MeSH Headings in queries and documents could improve the results.

Finally, we want to implement query expansion techniques applying different scientific dictionaries used in many investigations as: EntrezGene, HUGO, Eugenes and ARGH for gene/protein names; GO for molecular functions, biological processes and cellular components; UMLSKS for related names and symbols or WordNet for general words/phrases.

Acknowledgements. This work has been partially funded by the Spanish Ministry of Science and Innovation, the Plan E from the Spanish Government and the European Union from the ERDF (TIN2009-14057-C03-02).

References

1. Baeza-Yates, R.A., Ribeiro-Neto, B.: Modern Information Retrieval. Addison-Wesley Longman (1999)
2. Chen, J., Yu, P., Ge, H.: Unt 2005 trec qa participation: Using lemur as ir search engine. In: Voorhees, E.M., Buckland, L.P. (eds.) Proceedings of the Fourteenth Text REtrieval Conference, TREC 2005, volume Special Publication 500-266, National Institute of Standards and Technology, NIST (2005)
3. Eckard, E., Chappelier, J.C.: Free Software for research in Information Retrieval and Textual Clustering. Technical report, Ecole Polytechnique Federale de Lausanne (2007)
4. Fan, Y., Huang, X., An, A.: York university at trec 2006: Enterprise email discussion search. In: Voorhees, E.M., Buckland, L.P. (eds.) Proceedings of the Fifteenth Text REtrieval Conference, TREC 2006, volume Special Publication 500–272, National Institute of Standards and Technology, NIST (2006)
5. Gauch, S., Wang, J., Rachakonda, S.M.: A corpus analysis approach for automatic query expansion and its extension to multiple databases. ACM Transactions on Information Systems 17, 250–269 (1999)

6. Leveling, J., Jones, G.F.: Sub-word indexing and blind relevance feedback for english, bengali, hindi, and marathi ir. ACM Transactions on Asian Language Information Processing (TALIP) 9, 12:1–12:30 (2010)

7. Mitra, M., Singhal, A., Buckley, C.: Improving automatic query expansion. In: Proceedings of the 21st Annual International ACM SIGIR Conference on Research and Development in Information Retrieval, pp. 206–214. ACM Press (1998)

8. Pustejovsky, J., Castanho, J., Saur, R., Rumshinsky, A., Zhang, J., Luo, W.: Medstract: creating large-scale information servers for biomedical libraries. In: Proceedings of the ACL 2002 Workshop on Natural Language Processing in the Biomedical Domain, pp. 85–92. Association for Computational Linguistics, Morristown (2002)

9. Ramampiaro, H., Li, C.: Supporting biomedical information retrieval: The biotracer approach. T. Large-Scale Data- and Knowledge-Centered Systems 4, 73–94 (2011)

10. Shi, Z., Gu, B., Popowich, F., Sarkar, A.: Synonym-based query expansion and boosting-based re-ranking: A two-phase approach for genomic information retrieval. In: Voorhees, E.M., Buckland, L.P. (eds.) Proceedings of the Fourteenth Text REtrieval Conference, TREC 2005, volume Special Publication 500-266, National Institute of Standards and Technology, NIST (2005)

11. Shin, K., Han, S.-Y.: Improving Information Retrieval in MEDLINE by Modulating MeSH Term Weights. In: Meziane, F., Métais, E. (eds.) NLDB 2004. LNCS, vol. 3136, pp. 388–394. Springer, Heidelberg (2004)

12. Si, L., Lu, J., Callan, J.: Combining multiple resources, evidence and criteria for genomic information retrieval (2006)

13. Stokes, N., Li, Y., Cavedon, L., Zobel, J.: Exploring criteria for successful query expansion in the genomic domain. Inf. Retr. 12, 17–50 (2009)

14. Trotman, A.: An artificial intelligence approach to information retrieval (2004)

15. Trotman, A.: Learning to rank. Information Retrieval 8, 381 (2005)

16. Yu, H., Kim, W., Hatzivassiloglou, V., Wilbur, W.J.: Using medline as a knowledge source for disambiguating abbreviations and acronyms in full-text biomedical journal articles. Journal of Biomedical Informatics 40(2), 150–159 (2007)

17. Zazo, A.F., Figuerola, C.G., Berrocal, J.L.A., Rodríguez, E.: Term expansion using stemming and thesauri in Spanish. In: Peters, C., Braschler, M., Gonzalo, J. (eds.) CLEF 2002. LNCS, vol. 2785, pp. 177–183. Springer, Heidelberg (2003)

18. Zipf, G.K.: Human Behavior and the Principle of Least Effort. Addison-Wesley, Reading (1949)

Genetic Evaluation of the Class III Dentofacial in Rural and Urban Spanish Population by AI Techniques

Marta Muñoz, Manuel Rodríguez, Mª Encarnación Rodríguez,
and Sara Rodríguez

Abstract. The etiology of skeletal class III malocclusion is multifactorial, complex and likely results from mutations in numerous genes. In this study, we sought to understand genotype correlation of the class III dentofacial deformity in rural and urban spanish population of more than one generation. The genetic analyze was made using a Genome-wide scan. It will hold a novel classification using Artificial Intelligence techniques highlighting the difference between the two groups at the level of polymorphism. Our phenotypic and genetic analysis highlights that each group is unique.

Keywords: Skeletal class III malocclusion, genetic etiology, genome-wide scan, extraction of knowledge, artificial intelligence.

1 Introduction

Skeletal Class III malocclusion is a general morphological description of a diverse group of dentofacial conditions in which the mandibular teeth are forward in relationship to the maxillary teeth, resulting in an anterior crossbite or underbite. The term skeletal implies that the positions of the teeth are the result of underlying jaw relationships. This type of skeletal occlusal pattern is also referred to as true Class III or true mesiocclusion. These conditions are developmental to the extent that they are not recognizable at birth and by definition, until the individual is dentate, it is not possible to make a diagnosis of skeletal Class III malocclusion.

Marta Muñoz · Manuel Rodríguez · Sara Rodríguez
University of Salamanca, Salamanca, Spain
e-mail: {mmbruguier,mrrvm,srg}@usal.es

Mª Encarnación Rodríguez
Dentistry and Stomatology, Hnos. García Noblejas, Madrid
e-mail: maritarodrod@hotmail.com

S. Omatu et al. (Eds.): Distributed Computing and Artificial Intelligence, AISC 151, pp. 411–419.
springerlink.com   © Springer-Verlag Berlin Heidelberg 2012

As one would expect there is a higher incidence of this condition in the transitional and adult dentitions than there is in the primary dentition.

A Class III deformity can be an attribute of a syndrome, as in achondroplasia with associated midface deficiency resulting from a failure in the development of the cartilaginous nasal capsule or can merely be a manifestation of normal morphologic variation. Where a growth effect is responsible for the skeletal Class III problem, the affect can be primary and active, such as in acromegaly where an increased production of pituitary growth hormone acts on the condylar cartilage creating exuberant mandibular growth. Recent gene mapping and linkage analysis of individuals with achondroplasia and acromegaly have identified some of the responsible genes.

The Class III dentofacial deformity is clinically and genetically heterogeneous presenting with a distinct subphenotype and genotype in 2 cohorts. The main goal of this study is to conduct genome-wide scans followed by linkage analysis to identify the geneticloci associated with the Class III trait in rural and urban Spanish populations. The populations of individuals chosen to carry out the study are two groups of individuals of second generation rural and urban environments in Spain. It will hold a novel classification using Artificial Intelligence techniques highlighting the difference between Class III and normal individuals at the level of polymorphism.

The article is structured as follows: Section 2 makes a review of the problem description and the reasons that led to the realization of this research. Sections 3 introduces the most important requirements and functionalities of the AI (Artificial Techniques) used. Finally, some experimental results and conclusions are given in Sections 4 and 5.

2 Background

The genetic etiology of Class III malocclusion has been demonstrated in several studies. In [1], Bui et al. demonstrated that the Class III trait was inherited in an autosomal dominant fashion in the 12 families that they studied. This has been previously suggested by other studies [7][8]. Certain syndromic conditions with a genetic etiology, such as Crouzon syndrome, acromegaly and achondroplasia, have been described as presenting with skeletal Class III malocclusion [10][11][18].

In recent studies is reported results from the mutation on specific chromosomes and mutations in genes [16] [13], results in achondroplasia and amelogenesis. These recent advances have fallen on the heels of the Human Genome Project (HGP) that began in 1990. As a result of the HGP comprehensive genetic maps have been created that locate and identify genes underlying susceptibility to disease. Increasingly detailed knowledge of the human genome at the DNA level forms the basis of our understanding of genetic transmission and gene action. The HGP has mapped 30,000 genes thus far, and therefore provides the basis for genetic diagnosis and therapy.

The field of molecular genetics continues to improve and advance, it should be possible to identify relevant genetic markers for such traits. Skeletal Class III malocclusion or mandibular prognathism has been analyzed genetically [4][15][9].

Human studies have played a major role in the developing hypothesis that Class III malocclusion is at least in part due to genetic factors. Orofacial structures are significant in the development of the craniofacial complex and have been shown to be under genetic control, hence they should be considered in the etiology of the development of skeletal Class III malocclusion [7][8][3][14][12].

The existence of familial aggregation of mandibular prognathism (MP) suggests that genetic components play an important role in its etiology. A genetic etiology of class III malocclusion is suggested by many lines of evidence [5][2][6]. In Jena's report, a pair of monozygotic female twins were presented. The girls exhibited a marked similarity in facial appearance. They both had a similar dentition, but their occlusions were dissimilar to some extent. Twin 1, reverse overjet, overbite and class III molar relations were more severe than twin 2. Both twins had bilateral posterior crossbite. The cephalometric parameters did not reveal a very significant difference in skeletal morphology. Height of the anterior face was similar in both the twins, but posterior facial height was more in twin 2. Position of mandible in relation to anterior cranial base and Frankfort-horizontal plane was significantly different among the twins [5]. In this study, the concavity of the face (Angle of convexity) in twin 1 was more compared to twin 2.

Relatively more backward position of the maxilla (Angle SNA, N Perpendicular to point-A) and forward position of the chin (Angle SNB, N Perpendicular to Pog) contributed to such difference in the severity of the facial concavity. The antero-posterior position of the mandible in the twin study was influenced significantly by environmental factors. However, in a previous study undertaken, a report was made that the anterior-facial posterior position of the mandible is genetically determined. Anterior facial height of both twins was apparently equal. It showed that the height of the anterior face is genetically determined and did not play any role in the discordance of class III malocclusion. This is in agreement with the result from a study done by Townsend and Richards [17]. Another study investigated the role of genetic influences in the etiology of class III malocclusion [2]. In this study, a segregation analysis of 37 families of patients that were treated for mandibular prognathism, was performed. A study conducted by Yamaguchi, et al, in 2005 [19], utilized a genome-wide linkage analysis to identify loci susceptible to MP with 90 affected sibling-pairs in 42 families, comprised of 40 Korean sibling-pairs and 50 Japanese sibling-pairs. Two non-parametric linkage analyses, GENEHUNTER-PLUS and SIBPAL, were applied and detected nominal statistical significance of linkage to MP at chromosomes 1p36, 6q25, and 19p13.2. The best evidence of linkage was detected near D1S234 (maximum Zlr = 2.51,P = 0.0012). In addition, evidence of linkage was observed near D6S305 (maximum Zlr = 2.23, P = 0.025) and D19S884 (maximum Zlr = 1.93, P = 0.0089). This study while helpful relied on sibling pairs, which is less powerful than the family studies that we report in this publication.

The identification of the susceptible genes in the linkage regions will pave the way for insights into the molecular pathways that cause MP, especially overgrowth of the mandible, and may lead to the development of novel terapeutic tools.

3 AI Techniques

Conduct genome-wide scans followed by linkage analysis to identify the geneticloci associated with the Class III trait in the rural and urban populations. The GeneChip Mapping 10K 2.0 AssayR version by Affymetrix (Santa Clara, CA 2004), is a mapping tool designed to identify regions of the genome that are linked to or associated with a particular trait or phenotype. It is also useful for determination of allele frequencies in various populations and for mapping regions with chromosomal copy. This array system provides genotypes for 10,000 human single nucleotide polymorphisms (SNPs) on a single array.

The extraction of knowledge that is presented to the human expert is carried out using the J48 algorithm [21]. The J48 algorithm is the Java implementation of the C4.5 algorithm, an evolution of the original ID3 [20], whose main advantage is that it allows incorporates numerical attributes into the logical operations carried out in the test nodes. There are other alternatives for the generation of decision rules which operate similar to the decision trees, including RIPPER [22] and PART [23]. The J48 [20] algorithm attempts to minimize the width of the decision tree by using heavy search strategies. In summary, the algorithm defines two terms: gain and rate of gain with respect to the information $I(S)$ contained in a node S. Using only the gain criteria, attributes with multiple values are more highly favored given that they can more easily divide the elements into numerous subsets. To avoid the effect of favoring attributes with multiple values, the concept of gain rate is added.

$$I(S) = -\sum_{j=1}^{n} f_j^S \cdot \log(f_j^S) \qquad (1)$$

Where f_j^S represents the relative frequency of class j in the node S.

The gain function is defined as follows:

$$G(S, B) = I(S) - \sum_{i=1}^{t} \frac{|S_i|}{|S|} I(S_i) \qquad (2)$$

B represents the test that separates the modes and | | the number of elements of the set.

To avoid favoring the partitions with branches containing few elements, the rate of gain term is introduced in the following manner:

$$P(S, B) = -\sum_{i=1}^{t} \frac{|S_i|}{|S|} \log\left(\frac{|S_i|}{|S|}\right) \qquad (3)$$

Finally test B is selected from the previous tests since it maximies the following criteria:

$$G(S,B)/P(S,B) \qquad (4)$$

The system uses the equation 4 for calculating the nodes of the decision tree, the nodes represent the relevant SNPs that allow classifying the patients.

4 Experimental Results

The sample consists in 48 individuals in 4 families in rural environments and 25 individuals in 3 families urban environment were genotyped using 500 microsatellite markers prior to the refinement of the Single Nucleotide Polymorphism (SNP) genotyping methods. Some individuals belong to skeletal Class III malocclusion and other normal individuals. In order to amplify the DNA at these loci, polymerase chain reaction (PCR) was carried out using primers surrounding a previously identified locus. The initialization step took place at 95°C for 15 minutes and the DNA denaturation step took place at 95°C for 30 seconds. The annealing step took place at62°C for 30 seconds. The elongation step took place at 72°C for 1 minute. Chromatography was then used to analyze the gene fragment. A chromatogram was analyzed to determine the size of the fragments (bp) as well as whether subjects were heterozygous or homozygous for a given allele. A chromatogram of a heterozygous individual typically exhibits two defined peaks with smaller leading and lagging peaks, while that of a homozygous individual typically exhibits one defined peak. The GeneChip Mapping 10K 2.0 AssayR version by Affymetrix (Santa Clara, CA 2004), is a mapping tool designed to identify regions of the genome that are linked to or associated with a particular trait or phenotype. It is also useful for determination of allele frequencies in various populations and for mapping regions with chromosomal copy.

An analysis by inspection revealed an autosomal dominant mode of inheritance among all the families. Rural family #1, is composed of 3 generations of both affected and unaffected individuals. Each generation exhibits an approximately equal number of affected and unaffected family members. In this family, more females than males exhibit the phenotype. This family appears to exhibit autosomal dominant mode of inheritance of the Class III trait. Rural family #2, comprises 4 generations of individuals affected and unaffected with the skeletal Class III trait. This family has twice as many females exhibit the phenotype. This family also appears to exhibit an autosomal dominant mode of inheritance. Rural family #3, is composed of 4 generations with approximately males and females affected the trait equally, exhibiting an autosomal dominant mode of inheritance. Rural family #4, comprises 2 generations with twice as many female affected as males and all are affected with the trait. The mode of inheritance is autosomal dominant. AI family #19 and #33 are both composed of 6 and 7 generations respectively, with an equal number of males and females affected with the skeletal Class III trait, thereby exhibiting an autosomal dominant mode of inheritance.

Urban family #1, is made up of three generations of both affected and unaffected individuals. Each generation exhibits an approximately equal number of affected and unaffected family members. This family has twice as many males as females exhibiting the phenotype and appears to exhibit an autosomal dominant mode of skeletal Class III malocclusion. Urban Family #2 in the urban cohort, is composed of 2 generations with an equal amount of males affected with the skeletal Class III trait as females. This family appears to exhibit autosomal dominant inheritance of the skeletal Class III malocclusion. Urban family # 3, has 5 generations with 4 generations affected. Males seem twice as likely to inherit the condition as females. This family appears to exhibit autosomal dominant inheritance of this trait. Urban Family #4, has 2 generations with both generations affected with skeletal Class III malocclusion. Twice as many males are likely to be affected than females. This family appears to exhibit an autosomal dominant mode of inheritance of the skeletal Class III phenotype. Urban Family # 5, has 3 generations with 1 generation of affected individuals. More males were affected than females. There was some uncertainty regarding the skeletal Class III affection status of some of the individuals, hence a statement regarding the mode of inheritance was not made for this family.

Urban family #6, had 3 generations with 2 generations of affected individuals and an equal number of affected females as males. This family also had a few individuals where the affection status of the skeletal Class III trait was uncertain, however, with the information currently available, this family appears to have an autosomal dominant mode of inheritance. Urban Caucasian family #7 had 2generations with only females affected with the trait. Urban Family #8 in the, had 2 generations of affected individuals with twice as many females affected as males, revealing an autosomal dominant mode of inheritance.

The relevant regions obtained using J48 are between markers D1S2865 to D1S435, D1S435 to D1S206, D3S3725 to D3S3041, and D12S368 to D12S83 yielded on chromosomes 1, 3 and 12. Furthermore, single nucleotide polymorphism chip technology (SNP) has been used to identify regions associated with skeletal Class III malocclusion in the urban families.

The Affymetrix GeneChip Mapping Assay (Affymetrix, Inc., Santa Clara, CA) has been used to analyze SNP's in both affected and unaffected individuals. We anticipate the results from the urban linkage analysis will reveal other genes in addition to those found from the results of the rural analysis.

Table 1 Linkage Analysis

Markers	Chromosome
D1S2865 to D1S435	1
D1S435 to D1S206	1
D3S3725 to D3S3041	3
D12S368 to D12S83	12

Although the findings of this study are suggestive of linkage at chromosome 1 for the skeletal Class III trait, the answers regarding the genetic etiology of true mandibular prognathism might be found in the IGF system which would suggest that the mutation could be on chromosome 15. If the somatomedin hypothesis introduced by Daughaday almost a half-century ago is correct, the regulation of the growth of the skeleton would be the key to the entire genetic study on skeletal Class III malocclusion.

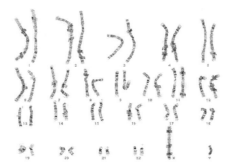

Fig. 1 Diagrammatic Representation of 23 Chromosomes and Relative Location of Markers D1S2865 – D1S435 using Parametric Linkage Analysis for rural cohort

5 Conclusions

A visual inspection of the pedigrees suggests an autosomal dominant mode of inheritance of skeletal Class III malocclusion. Results from the linkage analysis in this study suggest that chromosomes 1, 3 and 12 are suggestive of linkage to the skeletal Class III trait. In light of other genetic studies currently being done with improved technology, these results are not consistent with some of the other previous studies which would suggest that the IGF-1 gene located on chromosome 15, is involved in the regulation of growth hormone and hence the development of the skeleton.

Once the skeletal Class III trait is phenotypically characterized according to type, i.e., maxillary hypoplasia, mandibular prognathism, or a combination of both, it may be possible to utilize the candidate genes identified for other syndromes that have a skeletal Class III component in the identification of the genes involved in the development of this trait.

Acknowledgments. This work has been supported by the Spanish JCyL project SA225A11-2.

References

[1] Bui, C., King, T., Proffit, W., Frazier-Bowers, S.: Phenotypic Characterization of Class IIIPatients, A Necessary Background for Genetic Analysis. Angle Orthodontist, 564–569 (2006)

[2] El-Gheriani, A.A., et al.: Segregation Analysis of Mandibular Prognathism in Libya. J. Dent. Res. 82(7), 523–527 (2003)

[3] Horowitz, et al.: A cephalometric study of craniofacial variation in adult twins. Angle Orthodontist 30, 1–5 (1960)

[4] Huang, C.M., et al.: Mandibular prognathism in the rabbit: discrimination between singlelocus and multifactorial models of inheritance. The Journal of Heredity 72(4), 296–298 (1981)

[5] Jena, A., et al.: Class III malocclusion: Genetics or environment? A twins study. J. of Indian Society of Pedodontics and Preventive Dentistry 23(1), 27–30 (2005)

[6] Litton, S.F., Ackeramn, L.V., Isaacson, R.J., Shapiro, B.L.: A genetic study of Class III malocclusion. Am. J. Orthod. 58, 565–577 (1970)

[7] Mossey, P.A., et al.: The Heritability of Malocclusion: Part 1. Genetics, Principles and Terminology. British Journal of Orthodontics 26(3), 103–113 (1999)

[8] Mossey, P.A., et al.: The Heritability of Malocclusion: Part 2. The Influence of Genetics in Malocclusion. British Journal of Orthodontics 26(2), 195–203 (1999)

[9] Pereira, A., et al.: Association of GH and IGF-1 polymorphisms with growth traits in a synthetic beef cattle breed. Genetics and Molecular Biology 28(2) (2005)

[10] Preising, M., et al.: On the effect of mutations of the fibroblast growth factor receptors as exemplified by three cases of craniosynostoses. Klin. Monatsbl Augenheilkd 220(10), 669–681 (2003)

[11] Machicek, S.L., et al.: Mandible Size and Prognathism of MEK1 Transgenic Achondroplastic Mice. Presentation at AADR Meeting (March 2007)

[12] Nakasima, A., Ichinose, M., et al.: Hereditary factors in the craniofacial morphology of Angle's Class II and Class III malocclusions. Am. J. Orthod. 82(2), 150–156 (1982)

[13] Ravassipour, D.B., Powell, C.M., et al.: Variation in dental and skeletal open bite malocclusion in humans with amelogenesis imperfecta. Archives of Oral Biology 50, 611–623 (2005)

[14] Singh, G.D., et al.: Craniofacial heterogeneity of prepubertal Korean and European-American subjects with Class III malocclusions: Procrustes, EDMA andcephalometric analyses. Int. J. Adult Orthod. Orthog. Surg. 13, 227–240 (1998a)

[15] Sutter, N., et al.: A single IGF1 allele is a major determinant of small size in dogs. Science 316(5821), 112–115 (2007)

[16] Tamura, Y.: A kindred of familial acromegaly without evidence for linkage to MEN-1 locus. Endocr J. 49(4), 425–431 (2002)

[17] Townsend, G.C., Richards, L.C.: Twin and twinning, dentists and dentistry. Aust. Dent. J. 35, 317–327 (1990)

[18] Yagi, T., et al.: Surgical Orthodontic Correction of Acromegaly with Mandibular Prognathism. Angle Orthodontist 74(1), 125–131 (2004)

[19] Yamaguchi, T., et al.: Genome-wide Linkage Analysis of Mandibular Prognathism in Korean and Japanese Patients. J. Dent. Res. 84(3), 255–259 (2005)

[20] Pinzón, C.I., De Paz, J.F., Navarro, M., Bajo, J., Julián, V., Corchado, J.M.: Real-time CBR-agent with a mixture of experts in the reuse stage to classify and detect DoS attacks. Applied Soft. Computing 11(7), 4384–4398 (2011)

[21] Saravanan, N., Cholairajana, S., Ramachandran, K.I.: Vibration-based fault diagnosis of spur bevel gear box using fuzzy technique. Expert Systems with Applications 36(2-2), 3119–3135 (2009)

[22] Holte, R.C.: Very simple classification rules perform well on most commonly used datasets. Machine Learning, 63–91 (1993)

[23] Frank, E., Witten, I.H.: Generating accurate rule sets without global optimization, pp. 144–151. Morgan Kaufmann, San Mateo (1998)

[24] Vogiatzis, D., Tsapatsoulis, N.: Active learning for microarray data. International Journal of Approximate Reasoning 47(1), 85–96 (2008)

A Methodology for Learning Validation in Neural Cultures

V. Lorente, F. de la Paz, E. Fernández, and J.M. Ferrández

Abstract. In this paper a bio-hybrid system based on neural cultured is described and the learning processes for programming this biological neuroprocessor are revised. Different authors proposed many different learning techniques for managing neural plasticity, however it is necessary to provide a formal methodology for verifying this induced plasticity and validating this bio-hybrid programming paradigm.

Keywords: Hybrid systems, cultured neural network, induced plasticity, learning.

1 Introduction

Using biological nervous systems as conventional computer elements is a fascinating problem that permits the hybridization between Neuroscience and Computer Science. This synergic approach can provide a deeper understanding of natural perception and may be used for the design of new computing devices based on natural computational paradigms. Classical computational paradigms consist in serial and supervised processing computations with high-frequency clocks silicon processors, with moderate power consumption, and fixed circuits structure. However the brain uses millions of biological processors, with dynamic structure, slow commutations compared with silicon circuits, low power consumption and unsupervised learning. This kind of computation is more related to perceptual recognition, due to the natural variance of the perceptive patterns and the a priori lack of knowledge about the perceptual domain.

E. Fernández · J.M. Ferrández
Instituto de Bioingeniería, Universidad Miguel Hernández, Alicante, Spain

J.M. Ferrández · V. Lorente
Departamento de Electrónica, Tecnología de Computadores y Proyectos,
Universidad Politécnica de Cartagena, Spain

F. de la Paz
Departamento de Inteligencia Artificial, UNED, Spain
e-mail: jm.ferrandez@upct.es

S. Omatu et al. (Eds.): Distributed Computing and Artificial Intelligence, AISC 151, pp. 421–428.
springerlink.com   © Springer-Verlag Berlin Heidelberg 2012

A wetware computer may be built from leech neurons, capable of performing simple arithmetic operations and simple pattern recognition tasks. Another wetware computation is based on cellular cultures. Cells could be neurons from dissociated hippocampus or cortical tissue, or even neuroblastoma cells. Some problems to solve within this subgroup are: What are the requirements for wetware computations? Which are the new programming paradigms or learning techniques for wetware computing? What are the methods for validating learning in neural cultures?

In this paper a bio-hybrid system based on neural cultured is described and the learning processes for programming this biological neuroprocessor are revised. Different authors proposed many different learning techniques for managing neural plasticity; however it is necessary to provide a formal methodology for verifying this induced plasticity and validating this bio-hybrid programming paradigm.

2 Learning in Neural Cultures

Hippocampal cells are dissociated from the brains of embryonic rats and seeding onto planar Multi-Electrode Arrays (MEAs). The MEA is filled with a conventional cell culture medium containing nutrients, growth hormones and antibiotics, which is replaced weekly. Within the first hour after seeding, neurons start to extend connections no nearby cells and, within 24 hours, a thick layer of neuronal extensions is visible across the seeded area. The connectivity between seeded cells increases rapidly over subsequent days. After a week, electrophysiological spontaneous activity is observed in form of actions potentials, which transform into dense bursts of simultaneous electrical activity across the entire network over the following week. This bursting activity continues through the maturation phase (towards 20 days in vitro).

Cultures usually remain active until approximately 40 days approximately, but they may be active until 3 months of age if they are maintain in proper conditions (e.g. with Potter rings). During this time they are maintained in a humidified, 37°C, 5% CO_2 incubator to preserve its electrophysiological properties.

The first studies demonstrating functional plasticity in cultured networks began 1990s. The research group of Akio Kawana at NTT in Japan reported that tetanic stimulation through one or several electrodes resulted in plasticity [1]. They observed a change in the probability of evoking bursts by test pulses, as well as a change in the rate of spontaneous bursting, as a result of repeatedly evoking bursts using strong tetanic stimulation. Jimbo et al. reported similar results with a different tetanic stimulation and used voltage clamp to observe inward currents associated with evoked bursts [2]. The following year, Jimbo et al. reported that tetanizing a single electrode resulted in changes in the responses to test pulses to other electrodes [3]. The responses along the culture to a particular stimulation electrode were either all upregulated or all downregulated, a phenomenon they called 'pathway-dependent plasticity'. Individual responses throughout the array to stimuli on one particular electrode or pathways were upregulated or downregulated depending on the correlation between responses to stimuli applied to the test electrode and to the tetanization electrode. In another paper, Jimbo et al. used simultaneous tetanization through a pair of electrodes to induce more precise forms of plasticity, expressed in detailed spike patterns evoked by electrical (probe) pulses [4]. Since then, a few other groups have reported on other forms of plasticity in MEA neural cultures.

Typically, these later papers have focused on more abstract plasticity results, more related to the network level than to the synaptic level. For instance, Shahaf and Marom reported that networks could be learn to respond in specific ways to test pulses, by repeatedly stimulating until the desired response was obtained [5], while Ruaro et al. reported that cultured networks could learn to extract a specific pattern from a complex image that had been presented repeatedly as spatial patterns of multielectrode stimulation [6].

In the following years, researchers have tried using more complex stimulation patterns in order to induce plasticity in neural cultures. Wagenaar et al. [7] looked for plasticity expressed in changes in spontaneous burst patterns, and in array-wide response patterns to electrical stimuli, following several induction protocols related to the previous ones, as well as some novel ones. Madhavan et al. [8] investigated patterns of spontaneous multi-single-unit activity to study the potential role of bursts of action potentials in memory mechanisms. Their analysis revealed spatiotemporally diverse bursts occurring in well-defined patterns, which remained stable for several hours. Chao et al. [9] compared five established statistical methods to one of their own design, called center of activity trajectory (CAT), to quantify functional plasticity at the network level. Stegenga studied the possibility of changing the spatio-temporal structure of spontaneous bursts using different configurations of tetanic stimulation. They obtained a profile of the array-wide spiking rate, a *burst profile* (BP) and also calculated the per-electrode spiking rate profile, the *phase profiles* (PPs). None of their stimulation methods had a measurable effect on of the specific burst statistics (peak firing rate, rise and fall times). However, they found many PP changes in their experiments, which can be seen as a confirmation that the analysis is sensitive to changes in the network.

Other researchers have focused in changing some stimulation parameters (voltage vs. current, frequency, amplitude...) to achieve learning. Brewer et al. [10] used chronic stimulation for getting an increase in evoked spike counts per stimulus and in spiking rate. The results obtained suggested that plastic network changes induced by chronic stimulation enhance the reliability of information transmission and the efficiency of multi-synaptic network communication. In turn, Martinoia et al. [11] applied low-frequency stimulation constantly applied over weeks. They found that the stimulation had a delayed effect modulating responsiveness capability of the network without directly affecting its intrinsic *in vitro* development. Table 1 summarizes an overview of these protocols and the principal results achieved by each of the above-mentioned studies on neural cultures.

Table 1 Overview of studies on changes in network activity by electrical stimulation. Tetani are trains of stimuli. A volley is like a train in which a different electrode is stimulated with each pulse.

Study	N. electr.	Stimulation Parameters	Evaluated Aspect	Effect
Maeda et al. (1998)	5	Tetani: (20x, 20 Hz; ITI 10-15 s) 5-10x	Burst rate and size, spontaneous and induced by test stimuli.	Increase in burst rate and spikes per burst. Both spontaneous and as induced by test stimuli.
Jimbo et al. (1998)	1 or 8	Tetani: (11x, 20Hz; ITI 5s) 10x	Cell voltage	Increase in EPSC currents.
Jimbo et al. (1999)	1	Tetani: (11x, 20Hz; ITI 5s) 10x	Number and timing of spikes induced per neuron in response to test stimuli.	Increase or decrease of EPSC's and number of induced spikes depending on pathway.
Tateno et al. (1999)	1 or 2	Tetani: (10x, 20Hz; ITI 5s) 20x 100 us pulses	Reliability of action potentials after test stimuli, time between test stimuli and first spike.	Shortening of latency and reduction of jitter of first spike in some neurons. Increase of effect from 1 to 2 electrodes.
Shahaf et al. (2001)	2	Bipolar stimulation, ISI 1-3s until desired response, 10min max.	Increase in number of spikes on 1 evaluation electrode.	Increase in spikes induced in window.
Ruaro et al. (2005)	15	Tetani: (100x, 250 Hz, ITI 2s) 40x	Responses to test stimuli.	Increase in setFR between 1 and 50 ms.
Wagenaar et al. (2006)	2	Tetani: 20x, 20Hz; ITI 6s, 150x, shift 5ms	Responses to test stimuli, 10-50 ms. after test stimuli.	Increase in number of spikes induced by leading electrode; decrease for following electrode.
Madhavan et al. (2007)	2	Tetani: 18000x, 20 Hz (15 min), shift 10 ms	Spontaneous expression rate of burst types (BT).	Generation of new BTs, increase or decrease in expression rate of BTs.
Chao et al. (2007)	2	Tetani: 18000x, 20Hz (15 min), shift 10 ms	Center of Activity Trajectory (CAT) of evoked activity.	CAT's change significantly on some of the test electrodes.
Stegenga et al. (2008)	1	SESP: single pulse, 0.1 to 1Hz	Burst profiles (BPs) and phase profiles (PPs).	No measurable effect on Increase in effect on non-stimulated PPs.
Brewer et al. (2009)	30	Biphasic paired pulses, 50ms ISI, 5s ITI, 0-3 hours 30uA, 100us phase	Evoked responses, spike activity.	Increase of evoked responses and in spike frequency.
Martinoia et al. (2010)	8	Biphasic pulse 36x, 0.2 Hz sequentially to each electrode, 200us/phase, 1.5Vpp	Spontaneous firing rate, burst rate.	Stimulation significantly changes the mean firing rate of the networks.

3 A Methodology for Learning Validation in Neural Cultures

Learning validation in Artificial Neural Networks is usually achieved by computing the learning curve that is the error produced by the network during the learning phase. This curve needs to converge below a global minimum over certain number of epochs, avoiding local minimum in order to validate the learning process. A typical learning curve shows the number of stimuli required to evoke the desired response in every stimulation cycle. When learning is achieved, the learning curve is stabilized below the convergence criterion indicating that the desired response is elicited with a less number of stimuli. Fixing this convergence criterion is critical for avoiding overlearning and providing generalization. We propose the following methodology:

1) Impose a learning paradigm over the neural culture until a convergence criteria is reached.
2) Check that the network has impressed the desired function over its functional structure and this change is persistent.
3) Validate that the temporal n-responses from the individual neurons change from the initial spontaneous firing to a controlled temporal firing pattern induced by the functional connections.

The first step is necessary for the cells behaviour programming, while the second one is essential for analysing the subjacent neural structure and checking that from an original and random connectivity matrix, a controlled and desired stabilized matrix is reached. The third step is related with the functional response of aisle cells to the provided stimuli. Initially, each neuron has a temporal spiking response elicit by the provided stimulation. The whole network has unique and original n-temporal responses to the stimuli. For validating the modified structure in the network, a contrasted changed in the n-temporal unit responses is also required. These changes capture the induced connectivity matrix observed in the prior procedure.

3.1 Functional Connectivity

Functional connectivity [12,13] captures patterns of deviations from statistical independence between distributed neurons units, measuring their correlation/covariance, spectral coherence or phase locking. Functional connectivity is often evaluated among all the elements of a system, regardless whether these elements are connected by direct structural links; moreover, it is highly time-dependent (hundreds of milliseconds) and model-free, and it measures statistical interdependence (e.g. mutual information) without explicit reference to causal effects.

Correlation and information theory-based methods are used to estimate the functional connectivity of in-vitro neural networks: Cross-correlation, Mutual Information, Transfer Entropy and Joint Entropy. Such methods need to be applied to each possible pair of electrodes, which shows spontaneous electrophysiological activity. For each pair of neurons, the connectivity method provides an estimation of the connection strength (one for each direction). The connection strength is supposed to be proportional to the value yielded by the method. Thus, each method is associated to a matrix, the Connectivity Matrix (CM), whose elements (X, Y) correspond to the estimated connection strength between neuron X and Y.

High and low values in the CM are expected to correspond to strong and weak connections. By using such approach, inhibitory connections could not be detected because they would be mixed with small connection values. However, non-zero CM values were also obtained when no apparent causal effects were evident, or no direct connections were present among the considered neurons.

In our experiments *Connectivity maps* (Fig. 1) offered a visualization of the connectivity changes that occur in the culture. Connectivity maps were generated using the connectivity matrix (CM) obtained after applying the analysis and Cross-Correlation or Mutual Information. By setting thresholds in the CM (Fig. 1 right), it is possible to filter out some small values that may correspond to noise or very weak connections. In consequence, these maps show the strongest synaptic pathways, and can be used for visualizing the neural weights dynamics, and validate the achieved learning.

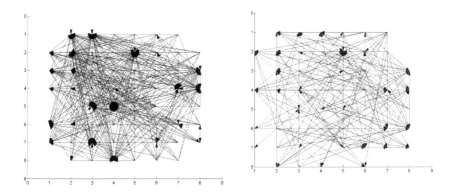

Fig. 1 Initial and Final Connectivity Maps with thresholds.

3.2 Post-Stimulus Time Histogram (PSTH) and Potentiation Index (PI)

To investigate the neuronal activity evoked by stimulation, the PSTH needs also to be computed. It represents the impulse response of each site of the experimental preparation to the electrical stimulation. The PSTH is calculated by considering time windows from the recordings that follow each stimulus. Each time window is divided into bins and then counted the number of spikes occurring in each time bin. Finally, the histogram obtained is normalized dividing the computed number of spikes by the number of stimuli times the bin size. Figure 2 shows the PSTH of the overall units from one of our experiments. In this case, an early response to a test stimulus is obtained in some neurons, while others have a delayed response according with their intrinsic connectivity.

If an electrode presented a PSTH with an area below '1', that channel must be eliminated from the statistics because it is considered to be inactive (i.e. not able to evoke a minimal response to stimulation).

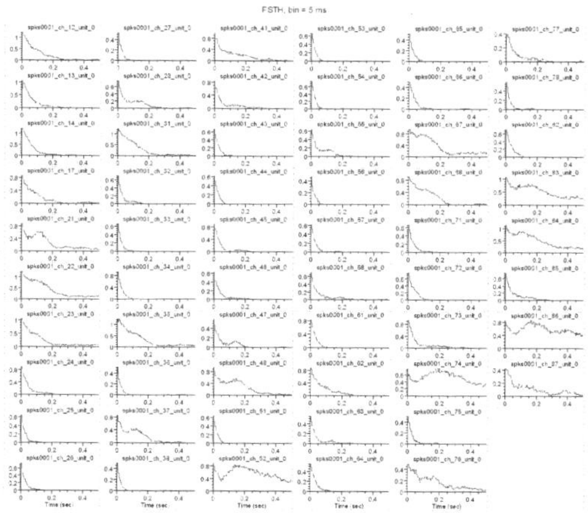

Fig. 2 PSTH with 5ms bins. An early response to the stimuli is observed in some units.

One of the problems in dealing with multi-unit extracellular recordings is to give a clear definition of 'potentiation' at a specific recording site. A long-term change in the excitatory post-synaptic potential amplitude (usually recognized as the evident sign that a change in connectivity has occurred) must produce a change also in the evoked firing rate or in the number of evoked spikes in a single or in a group of cells. Thus, to check if a stimulus was able to induce potentiation, a parameter, potentiation index (PI), related to changes in the PSTH area (i.e. number of evoked spikes) for each recording channel in two experimental conditions: pre- and post-stimulus is used. A 'potentiated response' means an increase of the PSTH area (given as a percentage) to a value equal or higher than a settled threshold. In this way, we are able to ensure as significant only those changes actually due to the application of the stimulation and not to the spontaneous fluctuations of the activity level. The PI was simply calculated as the ratio between the number of electrodes belonging to the stimulating session S → A presenting a change of the PSTH area above the threshold and the total number of active electrodes in the same session. For validating the learning the Potentiation index must exceed a threshold defined using the culture physiologic response.

4 Conclusions

Learning in biological neural cultures is a complex task. Different authors have proposed different methods for inducing a desired and controlled plasticity over the biological neural structure. For normalizing the heterogeneity of these learning procedures it is necessary to propose a formal methodology for validating the achieved learning.

In addition to the learning curve for determining the convergence criterion it is necessary to visualize the initial connectivity matrix, and the final neural connections for checking that the network has impress the desired function over its functional structure. It is also essential that the temporal n-responses from the individual neurons change from the initial spontaneous firing to a controlled temporal firing pattern induced by the functional connections. Validating the temporal stability of the network, how last the induced functional structure, or determining the plasticity required for normal operation are still open question that need to be addressed for using neural cultures as bio-hybrid computing elements.

Acknowledgments. This work is being funded by grant 12361/FPI/09 from Séneca Foundation, Science and Technology Agency from the region of Murcia and by the project 2010V/PUNED/0011 from Universidad Nacional de Educación a Distancia.

References

[1] Maeda, E., Kuroda, Y., Robinson, H.P., Kawana, A.: Modification of parallel activity elicited by propagating bursts in developing networks of rat cortical neurones. Eur. J. Neurosci. 10(2), 488–496 (1998)

[2] Jimbo, Y., Robinson, H.P., Kawana, A.: Strengthening of synchronized activity by tetanic stimulation in cortical cultures: application of planar electrode arrays. IEEE Trans. Biomed. Eng. 45(11), 1297–1304 (1998)

[3] Jimbo, Y., Tateno, T., Robinson, H.P.C.: Simultaneous induction of pathway specific potetiation and depression in networks of cortical neurons. Biophys. J. 76(2), 670–678 (1999)

[4] Tateno, T., Jimbo, Y.: Activity-dependent enhancement in the reliability of correlated spike timings in cultured cortical neurons. Biol. Cybern. 80(1), 45–55 (1999)

[5] Shahaf, G., Marom, S.: Learning in networks of cortical neurons. J. Neurosci. 21(22), 8782–8788 (2001)

[6] Ruaro, M.E., Bonifazi, P., Torre, V.: Toward the neurocomputer: image processing and pattern recognition with neuronal cultures. IEEE Trans. Biomed. Eng. 52(3), 371–383 (2005)

[7] Wagenaar, D.A., Pine, J., Potter, S.M.: Searching for plasticity in dissociated cortical cultures on multi-electrode arrays. Journal of Negative Results in Biomedicine 5(16) (2006)

[8] Madhavan, R., Chao, Z.C., Potter, S.M.: Plasticity of recurring spatiotemporal activity patterns in cortical networks. Phys. Biol. 4(3), 181–193 (2007)

[9] Chao, Z.C., Bakkum, D.J., Potter, S.M.: Region-specific network plasticity in simulated and living cortical networks: comparison of the center of activity trajectory (CAT) with other statistics. J. Neural. Eng. 4(3), 294–308 (2007)

[10] Ide, A.N., Andruska, A., Boehler, M., Wheeler, B.C., Brewer, G.J.: Chronic network stimulation enhances evoked action potentials. J. Neural. Eng. 7(1), 16008 (2010)

[11] Bologna, L.L., Nieus, T., Tedesco, M., Chiappalone, M., Benfenati, F., Martinoia, S.: Low-frequency stimulation enhances burst activity in cortical cultures during development. Neuroscience 165, 692–704 (2010)

[12] Sporns, O., Tononi, G.: Classes of network connetivity and dynamics. Complexity 7, 28–38 (2002)

[13] Friston, K.J., Frith, C., Frackowiak, R.: Time-dependent changes in effective connectivity measured with PET. Human Brain Mapping 1, 69–79 (1993)

Intelligent Working Environments, Handling of Medical Data and the Ethics of Human Resources

Céline Ehrwein Nihan

Abstract. The development of Ambient Intelligence (AmI) will radically transform our everyday life and social representations. These transformations will notably impact the working environment. The objective of this paper is to offer a first survey of the main ethical issues raised by the development of intelligent working environments (IWEs). It especially focuses on the capacity of such environments to collect and handle personal medical data. The first section describes the features of intelligent environments in general and presents some of their applications at the workplace. The second section of this paper points out some of the main ethical issues raised by these environments and their capacity to collect and handle medical data. The third and final section attempts to offer some elements of reflection regarding the ethical principles that should guide the development of IWEs in the future.

1 Introduction

Over the last decades we have witnessed not only the growth of the so-called new technologies but also the constant interaction between these technologies. This phenomenon has led to what is often termed the "converging" technologies, or NBIC technologies (in reference to Nanotechnology, Biotechnology, Information Technology and Cognitive sciences). The so-called "intelligent environments" are representative of this phenomenon as well as of the transformations that it implies. As the authors of the report on "Ambient Intelligence" of the Rathenau Instituut in the Netherlands state (Schuurman 2009), in the next few years, the AmI is going to radically transform the organization of our everyday life, of our relation to the world, to the others and to ourselves as well as our global perception of reality.

In recent years some scholars have begun to show interest in the ethical issues implied in the development of intelligent environments. However, very few among them have been interested in the questions raised by the use of such environments at work.

Céline Ehrwein Nihan
School of Business and Engineering Vaud, Switzerland
e-mail: celine.ehrwein@heig-vd.ch

S. Omatu et al. (Eds.): Distributed Computing and Artificial Intelligence, AISC 151, pp. 429–436.
springerlink.com   © Springer-Verlag Berlin Heidelberg 2012

This article partly attempts to bridge this gap. Nevertheless its focus is limited: it does not tackle the question of IWEs in the perspective of economic ethics at large nor does it provide any definitive solution to the ethical issues raised by such environments. It tries more modestly to 1) *offer a first survey of the possible positive and negative ethical impacts on the management of human resources about the capacity of IWEs to collect and handle data containing personal medical information* and, 2) *question some elements of reflection on the ethical principles that should guide the development of IWEs in the near future.*

2 What Are We Talking about? Origins, Features and Applications

It is current to attribute the paternity of the AmI vision to the American researcher Mark Weiser (1952-1999) who was Head of Laboratory and then Chief Technologist at the Xerox Palo Alto Research Center (PARC) between 1987 and 1999 (Bohn et al. 2004). The basic idea was apparently to develop "wall-sized, flat panel computer displays [which could also] function as input devices for electronic pens" (Weiser et al. 1999). Very quickly, a further idea was introduced, namely, that computers should be spread ubiquitously in the environment (soil, walls, etc.) and in the objects of our everyday life (furniture, clothes, accessories, etc.). This is why Weiser himself used the notion of "ubiquitous computing" to describe his project (on the use of concepts, see Ronzani 2009).

One of the aims of Weiser and his colleagues was to push the computer in the background and to bring back the human being at the center of the interactions:

> We wanted to put computing back in its place, to reposition it into the environment background, to concentrate on *human-to-human* interface and less on *human-to-computer* onces. By 1992, when our first experimental "ubi-comp" system was being implemented, we came to realize that we were, in fact, actually redefining the *entire* relationship of humans, work, and technology for the post PC-era (Weiser et al. 1999).

Ubiquity is not the only striking feature of AmI. AmI is also characterized by invisibility or, more exactly, by its *non-perceptibility*. As Nijholt states it, the AmI does not only remove the computer from our field of vision, but also provokes "the mental disappearance of the computing device" (Nijholt 2004).

Sensitivity is another important feature of AmI. Miniaturized biosensors enable the computer to measure and communicate information on its environment. It can be chemical, biological or physiological data, such as the temperature of a room, the presence of a substance in the atmosphere, one's heart rate or emotional state.

Now the intelligence of the ubiquitous computing is not restricted to its context awareness. The intelligent machine should also be capable of adapting itself to its environment by reacting to that environment in a suitable way. In this regard some scholars speak about the *personalization capacity* of the AmI, as far as it is able to respond to the particular needs of its users (Stefani et al. 2007).

This last capacity combines with what can be termed the *"anticipation skill"* of intelligent environments. In other words, it is expected from the AmI to meet the user's needs even before these needs become manifest.

The *large memory capacity* is another central feature of the AmI. Thus, the latter is able for instance to record and collect in the long term an extensive set of physiological data of workers in a company, and to spot on this base variations which affect their health.

Finally, AmI is often associated to, or even regarded as equivalent to, the so called "persuasive" technologies. Since it is able to anticipate the user's behaviour and to adapt itself to it, the computer becomes a powerful *instrument of persuasion* (van den Broeck et al. 2006; Kaptein et al. 2010).

Therefore, as it has already been raised by some authors, the AmI reflects and contributes to the *humanization of the machine* and, even more significantly, to the humanization of the material world of the objects which surround us. Endowed with ultrasensitive computers capable of adaptation and anticipation our environment gradually acquires a large set of typically human features. "Things, so as Hildebrand puts it, have become 'actants'" (Hildebrand 2008).

Let me say now a few words about the applications of AmI, and more precisely, the application of AmI at the place of work. Such applications are indeed manifold, and it is not possible to present all of them. In the context of this paper, I will limit myself to a short enumeration of applications which includes the collection of data containing personal medical information.

- a system which is able to detect the temporary memory loss of a worker, to remind him or her of the tasks that have been done and of those that still need being done, and, if necessary, to contact automatically a support person (Bühler 2009);
- a smart computer which is able to assess the fatigue of its user by measuring his/her blink rate and to react by "increasing font size or screen contrast in order to ease" reading (Allanson and Fairclough 2004).
- a computational interface which is capable of evaluating the level of frustration of a person by measuring his or her heart rate and blood pressure, and which, on this base, may adapt itself to the user's needs (Allanson and Fairclough 2004).
- smart clothes provided with sensors which allow them to monitor the level of stress and hydration of a person, and to activate an alarm in case of great medical risk. This kind of application may bring an advantage for people working in extreme situation like the firemen for instance (Cf. BIOTEX project).

3 Main Ethical Issues

As I said I do not want to offer a detailed overview of all the ethical issues raised by the development of IWEs. Such a project would deserve much more than a few pages in an article. In the context of this presentation I would rather focus on the ethical issues raised by the capacity of IWE to collect and handle personal medical information. I have tried, to this purpose, to draw up a list with the advantages and disadvantages that are implied by this capacity. As far as I know, such a survey has never been conducted so far.

Ehrwein Nihan - Table 1

These questions point to even more fundamental issues that are related 1) to our conception of the individual, 2) our definition of social roles and relationships, as well as 3) to the configuration of the different social spaces that coexist in society. Let me briefly comment on each of these issues.

1) Regarding the individual, the development of IWEs seems to carry the *risk of increase in reification*. Thanks to AmI, the machine or the person who controls it (namely, the technician, the employer, or even – why not? – the shareholders), does not merely acquire the capacity to watch the worker's *actions* in a panoptical perspective (Foucault 1975). Nowadays, in fact, this sort of monitoring is already possible through video surveillance, mobile phones or RFID access cards. AmI technology's strength over the other forms of monitoring mentioned, lies in the possibility of watching and having an impact on a worker's physical and mental health, and even more widely to influence his or her behaviors in real time and in a way that is not perceptible for that person. The risk here, if we do not establish clear institutional and legal rules, is to favor the worker's instrumentation, namely, that the worker becomes a mere instrument whose function is to improve the performance of the company. In other words, by pushing the computer in the background the AmI vision does not only bring back the human being at the center of the interactions, it also leads to a radical transformation of the machine and of the human being. On one hand it contributes to the anthropomorphization or humanization of the computer (as Hildebrand says it). On the other hand, it tends to reinforce the trend towards the *reification of the human being who gets gradually transformed not only into a thing* (see the work of the Frankfurt School) *but even into a machine*. This transformation questions profoundly the comprehension of the human being as free (co-)agent of the world (Arendt 1958).

2) Besides that, AmI is likely to have an *important impact on the definition of social roles and relationships*. There is no doubt that a working environment that can measure the workers' physiological data and react to them, will profoundly affect the employer's function and power. Indeed, in the world of AmI, through the knowledge and the skills given by the machine, the employer is potentially entitled to take an active part in the preservation of the worker's health. In this way, the employer's function becomes something else, moving closer towards the function of a doctor or a therapeutic advisor. Moreover, the blurring of lines that is involved in the attribution of medical competences to the head of a company is also likely to change the role of doctors and other healthcare specialists. More than ever, healthcare experts will become economic agents who, because of their medical knowledge, may contribute to the performance of companies. If this happens, we may legitimately fear that the professional autonomy of a doctor, or of a medical researcher, which is already subject to important budgetary pressure, will be more dramatically reduced.

3) Furthermore, this issue concerns not only the definition of social roles but also the delimitation of institutional spaces. The fusion of social roles and functions implies a *radical reconfiguration of the different institutional spaces*. (In a similar but slightly different way, Bohn et al. 2004, following Gary T. Marx, point out the

border crossing capacity of AmI). Consequently, if the (con)fusion of roles that I have just described really occurs, it will mean that the line of demarcation between the world of business and work, on one hand, and the world of medicine and care, on the other hand, will deeply evolve as well. By making the permanent monitoring of workers' health possible, IWEs transfer to the company some of the fundamental characteristics of a therapeutic service. To say it in other words, AmI should indeed be able to transform the workplace in a clinic. How should this new medical service match the kind of services which already exist today, such as doctors' offices, hospitals, etc.? And how can we guarantee, in this context, fairness in the domain of care (among the workers of one company, but also between people in employment and non-working people) and quality of care (which could be seriously threatened by the application of a purely economic logic)? In addition to that, the confusion of social roles and institutional spaces implied by the IWEs raises some important problems if we take it for granted that we need a minimum of delimitation between our different activities and the spots in which they take place in order to construct our identity and to live with the others (Arendt 1958).

4 Propositions for an Ethical Framework

Against the background of these and similar questions, it becomes urgent not only to make known to the public the developments in AmI technology, but also to encourage the debate within companies and within society in general on the ethical, social and human challenges which these developments represent.

The first question which needs being raised is the question of the relevance of the development of IWEs. Do we share their vision? And can we accept it?

Whatever answer we personally give to this question, we should take care not to deny that, like other technologies, AmI carries a profound ambivalence. As we have shown it, IWEs are not without risks and they raise fundamental ethical problems. Yet they also bring promises and opportunities: the opportunity of a better integration of people with disabilities in work, a better prevention of accidents, and so on. Their strong power of attraction is another element which must be taken into account. Having said so, it is rather obvious that the total social and political rejection of IWEs seems unlikely to happen. Thus we should anticipate the high probability of their development and the concrete use which could be made of them. We will also need to develop the means creating a new framework of research on AmI technology and related issues, so that the development of such technology may contribute to shape tomorrow's world *in the most positive way*.

I would like to suggest, to this effect, some elements of reflection on the ethical principles that should guide the development of IWEs in the near future. In doing so I do not intend to close the discussion nor to impose my own point of view on how best to deal with IWEs. My intent is to initiate, in the perspective of the *discourse ethics* (Habermas 1994), an *open and public debate on the ethical issues raised by the IWEs and the common social rules that should be established in order to frame the risks they carry with*.

In my point of view and considering the risks, the collection and handling of medical data through IWEs may be admitted provided that the following rules are observed:

- Protection of privacy

Medical data measured in the workplace are properties of the employee. The employer (or any other person) may have access to them only if some other preponderant good requires it.

I am not entitled to say, on my own, what this preponderant good might be. It must be defined democratically through an open and public discussion. Nevertheless, it seems obvious to me that some good which will bring advantages to a restricted portion of the population cannot be considered as preponderant.

  ➤ Privacy over productivity
The improvement of the company's productivity may not be considered as preponderant good. It does not justify the recording and handling of workers' medical data.

  ➤ Balance between privacy and protection against poverty
The case is slightly different if the existence of the company is threatened. Nevertheless, in my opinion, the risk of the financial bankrupt of a company does not t allow us to harm the workers' privacy. *From the employee's point of view*, we may understand that he or she may be willing to renounce to his or her privacy in order to secure his or her existence and the existence of his or her family. *But from a political ethical perspective*, the infringement of the workers' privacy for the sake of a company survival, or even, for the sake of economic interests is hard to defend. As a society we have the power to set a legal framework which may prevent companies from being dependent on the medical information delivered through IWEs. To say it in other words, nowadays companies can survive without these kinds of environments and without storing any special information on their employee's health. Possessing such environments and such information does not constitute a competitive advantage. Yet, this aspect can change in future. If we want to protect privacy, we must assume our political responsibility and prohibit the collection of workers' medical data in order to gain a competitive advantage.

Balance between privacy and population health.

A major epidemiologic risk could justify the gathering and handling of a worker's medical data through IWEs. In this case, it will be necessary to take into account the probability as well as the gravity of the risk (degree and speed of contagion, degree of morbidity and mortality, gravity of the disease, existence of preventive and therapeutic means, etc.).

- Health protection

The medical data collected through IWE should not be harmful to the workers' health. Only some preponderant good may justify health damage.

> ➢   Best possible sanitary situation

A major epidemiologic risk could justify the use of IWEs causing health damage. In this case, however, it will be necessary, to take into account the probability as well as the gravity of the risk (see above).

> ➢   Workers' health over productivity

The improvement of the company's productivity may not be considered as preponderant good. It does not entitle anyone to cause damage to the workers' health.

- Free and informed consent

Workers must receive complete and comprehensive information about the data that will be collected at their place of work (nature and extent of the data collected, way they will be used, etc.) before the activation of the AmI system. They must also have the possibility to give their free consent to the collection and the handling of such data (item mentioned in the employment contract for instance).

- Control of data handling

The way in which the data is used by the company must be submitted to regular control by an external institution. Disregard for the legal and ethical rules must be submitted to sanction.

- Protection of the weakest

The medical data collected through IWEs must not be used in order to justify managerial measures causing a negative discrimination among the workers of a company. Yet, we may accept as legitimate the use of such data in order to improve the working conditions of a person experiencing physical or psychological deficiency.

- Protection against false data

Everything must be done in order to guarantee the validity of the (medical) data collected at the workplace (fight against forgery, viruses, etc.).

Conclusion

As I have tried to show, IWEs and their ability to collect and handle medical data raise fundamental ethical issues. These issues need being tackled urgently. AmI research moves fast. According to the scientific literature, the first applications should be accessible to a large public in about 10 years and, as stated at the start of this paper, such applications are likely to have long-term implications in regards to the organization of our everyday life and to our relation to the world. If we want to take an active part in this foreseeable change, we have to open the debate now.

References

Allanson, J., Fairclough, S.H.: A research agenda for physiological computing. Interacting with Computer 16, 857–878 (2004)

Arendt, H.: The Human Condition. The University of Chicago Press, Chicago (1958)

Bühler, C.: Ambient intelligence in working environments. In: Stephanidis, C. (ed.) Universal Access in HCI, Part II, pp. 143–149. Springer, Heidelberg (2009)

Bohn, J., et al.: Living in a world of smart everyday objects – social, economic, and ethical implications. Journal of Human and Ecological Risk Assessment 10(5), 763–786 (2004)

Foucault, M.: Surveiller et punir. Naissance de la prison. Gallimard, Paris (1975)

Gandy Jr., O.H.: Engaging rational discrimination: exploring reasons for placing regulatory constraints on decision support systems. Ethics Inf. Technol. 12, 29–42 (2010)

Habermas, J.: Justification and application: remarks on discourse ethics. MIT Press, Cambridge (1994); Original German version (1991)

Hildebrandt, M.: Ambient intelligence, criminal liability and democracy. Crim. Law and Philos. 2, 163–180 (2008)

Kaptein, M.C., et al.: Persuasion in ambient intelligence. Journal of Ambient Intelligence and Humanized Computing (2010), doi:10.1007/s12652-009-0005-3

Ronzani, D.: The battle of concepts: Ubiquitous computing, pervasive computing and ambient intelligence in mass media. UbiCC Journal 4(2), 9–19 (2009)

Schuurman, J.G., et al.: Ambient intelligence. Viable future or dangerous illusion? Rathenau Instituut (2009),
http://www.rathenau.nl/uploads/tx_tferathenau/
Ambient_Intelligence_ENG.pdf (accessed October 03, 2011)

Spiekermann, S., Pallas, F.: Technology paternalism – wider implications of ubiquitous computing. Poiesis and Praxis (2006), doi: 10.1007/s10202-005-0010

Stefani, O., et al.: Ambient intelligence in working environments. In: Stephanidis, C. (ed.) Universal Access in HCI, Part II, pp. 708–714. Springer, Heidelberg (2007)

van den Broek, E.L., Schut, M.H., Tuinenbreijer, K., Westerink, J.H.D.M.: Communication and Persuasion Technology: Psychophysiology of Emotions and User-Profiling. In: IJsselsteijn, W.A., de Kort, Y.A.W., Midden, C., Eggen, B., van den Hoven, E. (eds.) PERSUASIVE 2006. LNCS, vol. 3962, pp. 154–157. Springer, Heidelberg (2006)

Weiser, M., Gold, R., Brown, J.S.: The origins of ubiquitous computing research at PARC in the Late 1980s. IBM System Journal 38(4), 693–696 (1999)

Wieland, M., et al.: Methods for conserving privacy in workflow controlled smart environments – A technical and philosophical enquiry into human-oriented system design of ubiquitous work environments. In: Proceedings of the The Third International Conference on Mobile Ubiquitous Computing, Systems, Services and Technologies (2009), doi: 10.1109/UBICOMM.2009.44

An Interpretable Guideline Model to Handle Incomplete Information

Tiago Oliveira, João Neves, Ângelo Costa, Paulo Novais, and José Neves

Abstract. Healthcare institutions are both natural and emotional stressful environments; indeed, the healthcare professionals may fall into practices that may lead to medical errors, undesirable variations in clinical doing and defensive medicine. On the other hand, Clinical Guidelines may offer an effective response to these irregularities in clinical practice, if the issues concerning their availability during the clinical process are solved. Hence, in this work it is proposed a model intended to provide a formal representation of Computer-Interpretable Guidelines, in terms of the extensions of the predicates that make their universe of discourse, as well as a Decision Support System framework to handle Incomplete Information. It will be used an extension to the language of Logic Programming, where an assessment of the Quality-of-Information of the extensions of the predicates referred to above is paramount.

1 Introduction

The healthcare institutions exposes the practitioners to situations where their decisions are subjected to greater scrutiny, which reinforces the need for the development of tools that may support a good clinical practice [1], being the Clinical Guidelines (CGs) a good basis to fulfill such rationale [2]. Undeniably, the cumulative effect of these features diminishes their reaction capability, which may lead to the development of occupational stress [1, 3, 4, 5]. It is necessary to address these issues, by promoting evidence-based medicine and disseminating standards for a good clinical practice [6, 7]. CGs are evidence based statements that provide recommendations to patients and healthcare professionals about the suitable clinical procedures within specific circumstances [8]. However, the current formats of CGs are usually available as long textual documents with a high degree of complexity, which makes them difficult to consult and complex to update. Such static documents are unable to keep up

Tiago Oliveira · Ângelo Costa · Paulo Novais · José Neves
CCTC, Department of Informatics, University of Minho, Braga, Portugal
e-mail: `tiago.jose.martins.oliveira@gmail.com,`
`{acosta,pjon,jneves}@di.uminho.pt`

João Neves
Centro Hospitalar de Vila Nova de Gaia/Espinho, EPE, Portugal
e-mail: `joaocpneves@gmail.com`

S. Omatu et al. (Eds.): Distributed Computing and Artificial Intelligence, AISC 151, pp. 437–444.
springerlink.com   © Springer-Verlag Berlin Heidelberg 2012

with the fast growth of scientific knowledge [9]. Moreover, they should interact with the clinical staff by providing information about the patients, being placed in context or not, but on the fly. On the other hand, they are unable to handle Incomplete Information [10], concerning the description of the condition of a patient, a key issue in all this process. In fact, the objective of this work is to present a computer guideline representation model, embedded in a clinical Decision Support System, which is simple and encompasses all the dimensions of the information in the clinical process, including the incomplete one. We will focus on the knowledge representation and inference mechanisms, regarding the decision making facets of the model.

2 Computer-Interpretable Guidelines

Computer-Interpretable Guidelines (CIGs) [11] are computer representations of CGs that are incorporated in clinical Decision Support Systems (DSSs), allowing immediate appliance. On the other hand, studies have shown that the implementation of current CIGs have effectively reduced the occurrence of medical errors and the expenses with unnecessary complementary exams, thus improving the attendance of patients [12, 13], and a reduction of costs. Moreover, a formal model of representation of CGs provides a better understanding of the clinical process and enables the development of automatic means for validating their logic and syntactic structures. One of the critical issues in the implementation of CIGs is the depiction model. However, there are several approaches to tackle the problem, with different description models in use. These approaches define guidelines as instances of building blocks, called primitives, which represent the different tasks to bring to a close, in order to conclude a clinical process [14, 15, 16, 17]. The current approaches use models of human task execution, reflected in the primitives of their interpretation models, such as actions, decisions and patient states. The model proposed in this work merges this type of account with formal logic in decision making, to handle Incomplete Information, which is an issue that none of the current approaches deals with.

3 A Guideline Description Model

The proposed symbolic formalism draws its inspiration from PROforma [18], that represents guidelines as flowcharts. Its objective is to offer an intuitive representation of clinical concepts and procedures. *Tasks, Decisions, Actions, Enquiries*, and *Plans* are the constituints of this formalism [19]. The model is depicted in Fig. 1. Each primitive has a set of attributes, relevant to the type of task to which they stand for. These attributes are aimed at controlling the execution of such tasks (e.g., time contraints, clinical objectives). The case study depicted in Fig. 2, corresponds to a fragment of the ATPIII Hypercholesterolemia Guideline. It refers to the selection of the appropriate treatment for high levels of cholesterol, based on the presence of coronary heart disease, risk factors and a ten year risk of coronary heart disease. Going from the left to the right, the guideline is inserted in a *Root Task* for a management of guidelines, and the guideline itself is rep-resented as *Plan P1*. Inside *Plan P1*, it is found *Plan P2*, which embraces the necessary enquiries in order to obtain the parameters used in *Decision D1*. On the other hand, *D1* is responsible for choosing the appropriate treatment based on the current state of the patient. The details about the way the decision is made, will be discussed latter on in this work. *Plans P3, P4* and *P5* hold back the procedures for treatment, being their execution bounded by clinical goals expressed in terms of blood levels of *LDL*. The support format for the guideline model is based on the *XML*. The structure of the *XML* files is described in *XML Schema Definition (XSD)*.

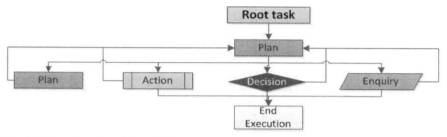

Fig. 1 Schematic view of the primitives of the representation model, where each primitive denotes a task of the clinical process.

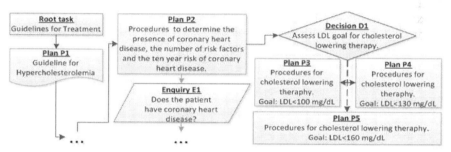

Fig. 2 Representation of a fragment of the ATPIII Hypercholesterolemia Guideline in the proposed model.

4 Knowledge Representation and Quality-of-Information

The image of the state of a patient within the clinical *DSS* resorts to *Extended Logic Programming (ELP)* [10, 20]. *ELP* encompass one of the few account mechanisms that cover the classic cases of *Incomplete Information*, such as *Uncertainty*, *Incompleteness*, *Inaccuracy*, and *Incoherence*. *ELP* uses default negation (*not p*) along with classic negation ($\neg p$), to explicitly represent negative information, which is useful to distinguish what is false because it cannot be proven, from what is false because its negation can be proven. Indeed, many approaches for knowledge representation and reasoning have been proposed using the *Logic Programming (LP)* paradigm, namely in the area of *Model Theory* [21], and *Proof Theory* [20, 21, 22]. We follow the proof theoretical approach and an extension to the *Language* of *Logic Programming* [22], to knowledge representation and reasoning. An *ELP* denotes a finite set of clauses in the form:

$$q \leftarrow p_1 \wedge \dots \wedge p_n \wedge not\ q_1 \wedge \dots \wedge not\ q_m$$
$$?\ p_1 \wedge \dots \wedge p_n \wedge not\ q_1 \wedge \dots \wedge not\ q_m\ (n,\ m \geq 0)$$

where *?* is a domain atom denoting falsity, the p_i, q_j, and p are classical ground
literals. In this representation formalism, every program is associated with a set of
abducibles [10, 22], given here in the form of exceptions to the extensions of the
predicates that make the program. To reason about the body of knowledge presented
in a particular set, that considers *Incomplete Information* on the base of the formalism referred to above, let us consider a procedure given in terms of the extension of
a predicate denoted as *demo* [10, 22]. This meta predicate is given by the signature
demo:T,V →{true, false, unknown}, according to the following set of terms:

```
demo(T, true) ← T.
demo(T, false) ← ¬T.
demo(T, unknown) ← not T, not ¬T.
```

Indeed, under this scenery, the first clause establishes that a theorem to be proved is
put to a knowledge base of positive information returning the truth-value *true* (*1*);
the second clause denotes that the theorem to be proved recurred to the negative information presented in the knowledge base, returning the truth-value *false* (*0*); the
third clause stands for itself, associating the theorem to be proved with a truth-value
in the interval *]0,1[*, i.e., a measure of system confidence in the proof process. As an
example, let us consider the case depicted in Figure 2, with respect to a first scenario
where a patient does not have coronary heart disease. However, the clinical staff is
unable to detect exactly how many risk factors he/she shows, but they believe the
number to be in the set {1, 2, 3}. The ten year risk of coronary heart disease is also
impossible to determine with accuracy, given the uncertainty about the risk factors,
yet the healthcare professionals know it is either 19% or 21%. In the context of *ELP*,
the information about the presence of coronary heart disease, the number of risk factors and the ten year risk of coronary heart disease can be represented with the following extensions of predicates *chd*, *risk_factors* and *ten_year_risk*, in the form:

```
¬chd(X) ← not chd(X), not abducible_chd(X).
abducible_chd(X) :- chd(unknown).chd(unknown).
¬risk_factors(X) ← not risk_factors(X), not abducible_risk_factors(X).
abducible_risk_factors(1). abducible_risk_factors(2). abducible_risk_factors(3).
¬ten_year_risk(X) ← not ten_year_risk(X), not abducible_ten_year_risk(X).
abducible_ten_year_risk(19). abducible_ten_year_risk(21).
? (ten_year_risk(X) ∨ ten_year_risk(Y))∧ ¬ (ten_year_risk(X) ∧
ten_year_risk(Y)).
```

where the first clause of each predicate denotes its closure. The hypotheses in terms
of the number of *risk factors* and *ten year* risk are represented here as *abducibles*. The
last clause in the extension of the predicate *ten_year_risk* stands for an invariant,
denoting that the abducibles are not disjunct, making that one as to consider one of
the values *19* or *21* for the *ten year risk*, but not both, i.e., the values in the set
{19,21}. On the other hand, the *abducibles* for the predicate *risk_factors* denote a disjunction, making that one has to consider the values in the set *{1, 2, 3, {1,2}, {1,3},
{2,3}, {1,2,3}}*. In order to establish a link between the parameters of the patient state
and the ones of the guideline recommendations, one has to look to an attribute, in
each *Decision Task*, named *Argument Rules*. *Argument Rules* point to production
rules that state the conditions that must hold in order to choose one or more of the
treatment options available at a given *Decision Task*. The use of production rules

confers transparency to the clinical process, since they are interpretable by both people and machines, i.e., they stand for themselves. For example, the set of rules that may enforce *Decision D1* might be given in the form:

```
if chd(yes) or ten_year_risk(X) > ten_year_risk(20) then
ldl_goal_less(100).
if ten_year_risk(X) =< ten_year_risk(20) and chd(no) and
risk_factors(X) >= risk_factors(2) then    ldl_goal_less(130).
if (risk_factors(0) or risk_factors(1)) and chd(no) then
ldl_goal_less(160).
```

To make a decision about the *LDL* goal of the cholesterol lowering therapy in the previously described cases with *Incomplete Information*, it is necessary to measure the reliability of the available information used in each rule. *ELP* appears here associated with an *Quality-of-Information (QoI)* evaluation method [10, 22], which is defined in terms of truth values taken from the interval *[0,1]*. If the information is *known*, the *QoI* for the term of the extension of the predicate *p* under consideration is *1*. For cases where the information is *unknown*, the *QoI* for a term in the extension of predicate *p* is given by:

$$QoI_p = \lim_{N \to \infty} 1/N = 0 (N >> 0) \tag{1}$$

where *N* is the number of terms for *p*. This is the case in the extension of predicate *chd* referred to above. For cases such as the one that is found in the extension of predicate *ten_year_risk*, where the extension of the predicate is unknown, but can be taken from a set of mutually exclusive values, the *QoI* is given in the form:

$$QoI_p = 1/Card \tag{2}$$

where *Card* corresponds to the number of abducibles of *p*. If the set of abducibles is disjoint, the QoI is depicted as follows:

$$QoI_p = 1/(C_1^{Card} + ... + C_{Card}^{Card}) = 1/(\sum_{j=0}^{Card} C_j^{Card} - 1) = 1/(2^{Card} - 1) \tag{3}$$

where C_{Card}^{Card} is a card-combination subset with Card elements. The next element to be considered is the relative importance that a predicate assigns to each of its attributes under observation, i.e., w_i^k, which stands for the relevance of attribute k in the extension of predicate i [10, 22]. This parameter spawns from the fact that some guidelines use in diagnostic situations look at symptoms that have a different weight in the detection of a disease. It is also assumed that the weights of all the attribute predicates are normalized, that is to say:

$$\sum_{1 \le k \le n} w_i^k = 1, \forall i \tag{4}$$

where $\forall$ denotes the universal quantifier. It is now possible to define a predicate's scoring function $V_i(x)$ so that, for a value $x = (x_1, ..., x_n)$ in the multi-dimensional space, defined in terms of the attributes of predicate i, one may have:

$$V_i(x) = (\sum_{1 \le k \le n} w_i^k \times QoI_i(x))/n. \tag{5}$$

It is now possible to engender all the possible scenarios that denote the universe of discourse, in terms of the extensions of the predicates *chd*, *risk_factors* and *ten_year_risk* given above.

The last, but not the least, it must be rendered attention to the fact that in a conjunction of terms, all terms will contribute to the score; on the other hand, if we are faced with a disjunction of terms, only the term with the highest *truth value* will make a payment to the score. However, this is not always the case, since a symptom can be more denotative of a disease than others that may also occur along with it. The evaluation scheme may evolve towards distinguishing conditions in terms of the symptoms relative weight, which is the case when there is a profound knowledge of the clinical domain by its part-owners.

5 The Decision Model

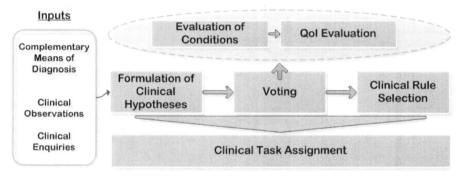

Fig. 3 A schematic view of the decision model based on the CGs recommendations.

To reason about the knowledge base of a patient, it was thought-out a logic inference engine that mimics the state of affairs of a diagnosis, i.e., once it is required to execute a given assignment, it must consider the guideline recommendations for that particular patient. Such model is depicted in Figure 3. Initially, the *Formulation of Hypotheses* is considered. At this stage it is carried out a survey on the available options with respect to a *Decision Task*. Subsequently, for each rule, there is a *Voting Stage*. The *Voting Stage* starts with the *Evaluation of Conditions,* at the knowledge base level; under a successful outcome, the rule is considered to be inferable and the *QoI Evaluation* is the next phase in the decision process. Otherwise, the process goes back to the initial phase and a subsequent rule will be considered. Indeed, in terms of the scenarios that were described above, and according to the rules of *Decision D1*, we may come to a conclusion. At the *Evaluation Stage*, the *QoI* method is used to determine the truth or falsity of the logic conditions and the scores of each rule. Considering the first scenario, the truth-values that denote the *QoI* of each of the extensions of the predicates that make the universe of discourse of a patient state and the scores of the rules would be the following: $QoI_{chd}=1$, $QoI_{ten\_year\_risk}=0.50$ and $QoI_{risk\_factors}=0.14$. Thus, the scores of the available rules would be: $V_{ldl\_goal\_less(100)}=0.50$, $V_{ldl\_goal\_less(130)}=0.55$ and $V_{ldl\_goal\_less(160)}=0.57$, so all rules are valid. For the second scenario, only the truth value of *chd* would be different, with $QoI_{chd}=0$, resulting in the following scores: $V_{ldl\_goal\_less(100)}=0.50$, $V_{ldl\_goal\_less(130)}=0.21$ and $V_{ldl\_goal\_less(160)}=0.07$, with only one valid rule. The *Clinical Rule Selection* stage requires that all the scores of the valid rules have been evaluated. The options are put in a preference list, according with the decreasing order of their scores and an information dashboard, like the one depicted in Figure 4, containing the contributions of the conditions of each rule to the scores, is shown to the user. The selected rule is the one with the highest score and will determine the next task in the clinical process at the *Clinical Task Assignment* stage. The distribution of tasks after a pronouncement is assured by trigger conditions placed at the subsequent tasks that match the rules offered at the *Decision Task*.

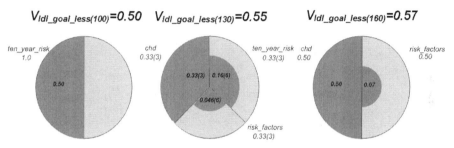

Fig. 4 A dashboard of the available rules for the first scenario.

6 Conclusions and Future Work

The proposed decision model summarizes the state of the information available to the user at every decision step, of an inclusive decision task. It is a good asset to clinical decision making, since it effectively reproduces the cognitive processes involved in real life diagnosis and is supported by the recommendations of *CGs*. The work done was focused on the development of an intuitive guideline representation model and of a decision one that supports the processing of *Incomplete Information*. The future plans for this work include a constant work towards the improvement of the architecture, and the application of the technical solutions that will make the implementation of these decision models possible.

References

1. Maslach, C., Schaufeli, W.B., Leiter, M.P.: Job Burnout. Annual Review of Psychology 52, 397–422 (2001)
2. Farquhar, C.M., Kofa, E.W., Slutsky, J.R.: Clinicians' Attitudes to Clinical Practice Guidelines: A Systematic Review. Medical Journal of Australia 177, 502–506 (2002)
3. West, C., Tan, A., Habermann, T., Sloan, J., Shanafelt, T.: Association of Resident Fatigue and Distress with Perceived Medical Errors. JAMA 177, 502–506 (2002)
4. Hampton, J.R.: Evidence-based Medicine, Practice Variations and Clinical Freedom. Journal of Evaluation in Clinical Practice 3, 123–131 (1997)

5. Studdert, D.M., Mello, M.M., Sage, W.M., Des Roches, C.M., Peugh, J., Zapert, K., Brennan, T.A.: Defensive Medicine among High-risk Specialist Physicians in a Volatile Malpractice Environment. JAMA 293, 2609–2617 (2005)
6. Woolf, S.H., Grol, R., Hutchinson, A., Eccles, M., Grimshaw, J.: Potential Benefits, Limitations and Harms of Clinical Guidelines. BMJ 318, 527 (1999)
7. Sackett, D.L.: Evidence-based Medicine. Seminars in Perinatology 21, 3–5 (1997)
8. Mead, P.: Clinical Guidelines: Promoting Clinical Effectiveness or a Professional Minefield? JAN 31, 110–116 (2000)
9. Rosenbrand, K., Croonenborg, J., Wittenberg, J.: Guideline Development. In: Computer-based Medical Guidelines and Protocols: A Primer and Current Trends, pp. 3–21. IOS Press (2008)
10. Neves, J., Ribeiro, J., Pereira, P., Alves, V., Machado, J., Abelha, A., Novais, P., Analide, C., Santos, M., Fdez-Delgado, M.: Evolutionary Intelligence in Asphalt Pavement Modeling and Quality-of-Information. In: Progress in Artificial Intelligence. Springer (to appear, 2012)
11. De Clercq, P., Kaiser, K., Hasman, A.: Computer-Interpretable Guideline Formalisms. In: Computer-based Medical Guidelines and Protocols: A Primer and Current Trends, pp. 22–43. IOS Press, Amsterdam (2008)
12. Sánchez, H., Albala, C., Dangour, A., Uauy, R.: Compliance with Guidelines for the Management of Community Acquired Pneumonia at Primary Healthcare Centres. Revista Médica de Chile 137, 1575–1582 (2009)
13. Cannon, D., Allen, S.: A Comparison of the Effects of Computer and Manual Reminders on Compliance with a Mental Health Clinical Practice Guideline. JAMIA 7, 196–203 (2000)
14. Kim, S., Haug, P.J., Rocha, R.A., Choi, I.: Modeling the Arden Syntax for medical Decisions in Xml. Medical Informatics 77, 650–656 (2008)
15. Peleg, M., Boxwala, A.A., Ogunyemi, O., Zeng, Q., Tu, S., Lacson, R., Bernstam, E., Ash, N., Mork, P., Ohno-Machado, L., Shortlife, E.H., Greenes, R.A.: Glif3: The Evolution of a Guideline Representation Format. Journal of Biomedical Informatics 37, 147–161 (2004)
16. Shahara, Y., Mikschb, S., Johnsonc, P.: The Asgaard Project: A Task-specific Framework for the Application and Critiquing of Time-oriented Clinical Guidelines. Artificial Intelligence in Medicine 14, 29–51 (1998)
17. Ram, P., Berg, D., Tu, S., Mansfield, G., Ye, Q., Abarbanel, R., Beard, N.: Executing Clinical Practice Guidelines using the SAGE Execution Engine. Medinfo 11, 1–5 (2004)
18. Fox, J., Johns, N., Rahmanzadeh, A.: Disseminating Medical Knowledge: The PROforma Approach. Artificial Intelligence in Medicine 14, 157–182 (1998)
19. Oliveira, T., Costa, A., Neves, J., Novais, P.: Digital Clinical Guidelines Modelling. In: Modelling and Simulation 2011, pp. 392–398. EUROSIS (2011) ISBN: 978-9077381-66-3
20. Neves, J.: A Logic Interpreter to handle Time and Negation in Logic Databases. In: ACM 1984 Proceedings of the 1984 Annual Conference of the ACM on The Fifth Generation Challenge, pp. 50–54. ACM, New York (1984)
21. Gelfond, M., Lifschitz, V.: The Stable Model Semantics for Logic Programming. In: Kowalski, R., Bowen, K. (eds.) Logic Programming: Proceedings of the Fifth International Conference on Logic Programming, pp. 1070–1080 (1988)
22. Neves, J., Machado, J., Analide, C., Abelha, A., Brito, L.: The Halt Condition in Genetic Programming. In: Neves, J., Santos, M.F., Machado, J.M. (eds.) EPIA 2007. LNCS (LNAI), vol. 4874, pp. 160–169. Springer, Heidelberg (2007)

Modeling a Mobile Robot Using a Grammatical Model

Gabriel López-García, Javier Gallego-Sánchez, J. Luis Dalmau-Espert,
Rafael Molina-Carmona, and Patricia Compañ-Rosique

Abstract. Virtual Worlds Generator is a grammatical model that is proposed to define virtual worlds. It integrates the diversity of sensors and interaction devices, multimodality and a virtual simulation system. Its grammar allows the definition and abstraction in symbols strings of the scenes of the virtual world, independently of the hardware that is used to represent the world or to interact with it.

A case study is presented to explain how to use the proposed model to formalize a robot navigation system with multimodal perception and a hybrid control scheme of the robot.

1 Introduction

Autonomous robots are physical agents that perform tasks by navigating in an environment and by manipulating objects in it. To perform these tasks, they are equipped with effectors to act on the environment (wheels, joints...) and with sensors that can perceive it (cameras, sonars...) [6].

The growing disparity of available sensors adds complexity to systems, but it also allows the control of robots to be more accurate. For example, humans and other animals integrate multiple senses [7]. There are other reasons of mathematical nature: combining multiple observations from the same source provides statistical advantages because some redundant observations are obtained for the same estimation.

In this paper we deal with the integration of multimodal inputs in the sense stated by Signhal and Brown [8], that is, the use of data of different nature for decision-making in high-level tasks performed by a robot. However, the proposed system can also deal with the concept of fusion.

Gabriel López-García · Javier Gallego-Sánchez · J. Luis Dalmau-Espert ·
Rafael Molina-Carmona · Patricia Compañ-Rosique
Grupo de Informática Industrial e Inteligencia Artificial, Universidad de Alicante,
Ap.99, E-03080, Alicante, Spain
e-mail: {glopez,ajgallego,jldalmau,rmolina,patricia}@dccia.ua.es

S. Omatu et al. (Eds.): Distributed Computing and Artificial Intelligence, AISC 151, pp. 445–452.
springerlink.com   © Springer-Verlag Berlin Heidelberg 2012

The Virtual Worlds Generator (VWG), our proposal, is a grammatical model, which integrates the diversity of interaction and sensing devices and the modules that make up a Graphics System (Graphics, Physics and AI engines). The scene definition is separated from the hardware-dependent characteristics of the system devices. It uses a grammar definition which integrates activities, visualization and interaction with users. The hypothesis is that it can be used as a formal framework to model a robot navigation system, including several multimodal inputs, sensor fusion and integration, and behaviour strategies.

2 Model for Virtual Worlds Generation

In the VWG model, a virtual world is described as an ordered sequence of primitives, transformations and actors. A primitive is the description of an object in a given representation system (typically, they are graphical primitives but they could also be sounds or any other primitive in a representation space). Transformations modify the behaviour of primitives, and actors are the components which define the activities of the system in the virtual world. The actors may be finally displayed through primitives and transformations. To model the different actor's activities, the concept of an event is used. Events cause the activation of a certain activity that can be processed by one or more actors.

Each element in the scene is represented by a symbol from the *set of symbols of the scene*. The symbols make up strings that describe the scenes, in accordance with a language syntax, which is presented as a grammar [1].

A grammar M is a tuple $M = < \Sigma, N, R, s >$, where Σ is the finite set of terminal symbols, N is the finite set of non-terminal symbols, R is the finite set of syntactic rules (a syntactic rule is an application $r: N \rightarrow W^*$, where $W = \Sigma \cup N$) and $s \in N$ is the initial symbol of the grammar. In out case, M is defined as:

1. $\Sigma = P \cup T \cup O \cup A_{ATTR}^D$, where:

   - P: set of symbols for primitives.
   - T: set of symbols for transformations.
   - $O = \{\cdot ()\}$: symbols for indicating the scope () and the concatenation $\cdot$.
   - A_{ATTR}^D: set of symbols for actors, where D is the set of all the types of events generated by the system and $ATTR$ is the set of all the attributes of actors which define all the possible states. For example, the actor a_{attr}^H will carry out its activity when it receives an event e^h, where $h \in H$, $H \subseteq D$ and $attr \in ATTR$ is its current state.

2. $N = \{$WORLD, OBJECTS, OBJECT, ACTOR, TRANSFORM., FIGURE$\}$.
3. Grammar rules R are defined as:

   - Rule 1. **WORLD** $\rightarrow$ OBJECTS
   - Rule 2. **OBJECTS** $\rightarrow$ OBJECT | OBJECT $\cdot$ OBJECTS
   - Rule 3. **OBJECT** $\rightarrow$ FIGURE | TRANSFORM. | ACTOR
   - Rule 4. **ACTOR** $\rightarrow a_{attr}^H$, $a_{attr}^H \in \mathbf{A}_{ATTR}^D, H \subseteq D$

- Rule 5. **TRANSFORMATION** $\to t(\text{OBJECTS})$, $t \in T$
- Rule 6. **FIGURE** $\to p^+$, $p \in P$

4. $s = \text{WORLD}$ is the initial symbol of the grammar.

M is a context-free grammar. $L(M)$ is the language generated by the grammar M:
$L(M) = \{w \in \Sigma^* \mid \text{WORLD} \xrightarrow{*} w\}$

Apart from the language syntax, it is necessary to define the semantics of $L(M)$. It will be defined with a denotational method, that is, through mathematical functions.

Rule 6 defines a figure as a sequence of primitives. Primitive's semantics is defined as a function $\alpha : P \to G$. Each symbol in the set P carries out a primitive on a given geometric system G. So, depending on the definition of the function α and on the geometry of G, the result of the system may be different. G represents the actions to be run on a specific visual or non-visual geometric system (e.g. the actions on OpenGL or on the system of a robot). The function α provides the abstraction needed to homogenize the different implementations of a rendering system. Therefore, only a descriptive string is needed to run the same scene on different systems.

In **Rule 5**, two functions are used to describe the semantics of a transformation, whose scope is limited by the symbols "()": $\beta : T \to G$ (carried out when the symbol "(" is processed) and $\delta : T \to G$ (run when the symbol ")" is found). These two functions have the same features that the function α, but they are applied to the set of transformations T, using the same geometric system G.

Rule 4 refers to actors, which are the dynamic part of the system. The semantics of the actor is a function which defines its evolution in time. For this reason, the semantic function is called *evolution function* λ and it is defined as: $\lambda : A_{ATTR}^D \times E^D \to L(M)$, where E^D is the set of events for the set of all event types D. The function λ has a different expression depending on its evolution. However, a general expression can be defined. Let $H = \{h_0, \ldots, h_n\} \subseteq D$ be the subset of event types which the actor a_{ATTR}^H is prepared to respond to. The general expression for λ can be seen at (e1), where $u_0, \ldots, u_n$ are strings of $L(M)$. This equation means that an actor a_{ATTR}^H can evolve, that is, it is transformed into another string u_i when it responds to an event e^h which the actor is prepared to respond to. However, the actor remains unchanged when it is not prepared to respond.

As well as dynamic elements, actors can also have a representation in the geometric space G. To be displayed, an actor must be converted to a string of primitives and transformations. This visualization function is defined as: $\theta : A_{ATTR}^D \times E^V \to L(M')$, where $V \subseteq D$, $E^V \subseteq E^D$ are events created in the visualization process, and $L(M')$ is a subset of the language $L(M)$, made up of the strings with no actors. Let $H \cap V = \{v_0, \ldots, v_n\} \subseteq D$ be the subset of visual event types which the actor a_{ATTR}^H is prepared to respond to. The expression of θ can be seen at (e2).

$$\lambda(a_{ATTR}^H, e^h) = \begin{cases} u_0 \in L(M) & \text{if } h = h_0 \\ \cdots \\ u_n \in L(M) & \text{if } h = h_n \\ a_{ATTR}^H & \text{if } h \notin H \end{cases} \quad \text{(e1)}$$

$$\theta(a_{ATTR}^H, e^v) = \begin{cases} z_0 \in L(M') & \text{if } v = v_0 \\ \cdots \\ z_n \in L(M') & \text{if } v = v_n \\ \varepsilon & \text{if } v \notin H \cap V \end{cases} \quad \text{(e2)}$$

The semantic function of these **Rules 1, 2, and 3** breaks down the strings and converts them into substrings, executing the so called *algorithm of the system*, which performs the complete evolution of the system and displays it in the current geometric system. It performs several actions, which are described in the following paragraphs.

To display the scene on the geometric system G, the function φ is defined, for the set of symbols that can directly be displayed: primitives and transformations. Given a string $w \in L(M)$ and using only symbols of P and T, φ is defined as:

$$\varphi(w) = \begin{cases} \alpha(w) & \text{if } w \in P \\ \beta(t); \varphi(v); \delta(t) & \text{if } w = t(v) \wedge v \in L(M) \wedge t \in T \\ \varphi(u); \varphi(v) & \text{if } w = u \cdot v \wedge u, v \in L(M) \end{cases}$$

In the case of strings including both displayable elements, and actors, two functions must be defined. The first one is the so called *function of the system evolution* η, which requires a sequence of sorted events $S = e^1 \cdot e^2 \dots e^n$, where every $e^i \in E^D$ and a string of $L(M)$ including actors, and implements a set of recursive calls to the function λ to perform the evolution of all the actors in the system at a given frame:

$$\eta(w, S) = \begin{cases} w & \text{if } w \in P \\ t(\eta(v, S)) & \text{if } w = t(v) \\ \prod_{e^i \in S} \lambda(a_{attr}^H, e^i) & \text{if } w = a_{attr}^H \\ \eta(u, S) \cdot \eta(v, S) & \text{if } w = u \cdot v \end{cases}$$

The operator $\prod_{e^i \in S} \lambda(a_{attr}^H, e^i)$ concatenates the strings of the function λ.

For the actors to be displayed in the system, they must be converted to displayable elements, that is, primitives and transformations. The second function, returns a string of the language $L(M')$ given a string $w \in L(M)$ and a sequence of ordered visualization events $S' = e^1 \cdot e^2 \dots e^n$, where every $e^i \in E^V$ and $S' \subseteq S$. This function is called *function of system visualization* π and it is defined as:

$$\pi(w, S') = \begin{cases} w & \text{if } w \in P \\ t(\pi(v, S')) & \text{if } w = t(v) \\ \prod_{e^i \in S} \theta(a_{ATTR}^H, e^i) & \text{if } w = a_{ATTR}^H \\ \pi(u, S') \cdot \pi(v, S') & \text{if } w = u \cdot v \end{cases}$$

The **events** are the mechanism to model the activity in the system. The actors activity is carried out when a certain type of event is produced. The following event definition is established: e_c^d is *defined as an event of type $d \in D$ with data c*.

A new function called **event generator** is defined as: Let $C^d(t)$ be a function which creates a sequence of ordered events of type d at the time instant t, where $d \in D$ and D is the set of event types which can be generated by the system. This function is: $C^d : Time \rightarrow (E^D)^*$

Different event generators can create the same type of events. So, a priority order among event generators must be established to avoid ambiguities.

Once all the elements involved in the model have been defined, the **System Algorithm** can be established. It defines the system evolution and its visualization at every time instant 't' or frame:

1. $w = w_0$; $t = 0$
2. **while** $w \neq \varepsilon$ **do**
    - S = collect events from generators C^* in order of priority.
    - Z = extract visual events from S.
    - $w_{next} = \eta(w, S)$
    - $v = \pi(w, Z)$; $g = \varphi(v)$
    - $w = w_{next}$; $t = t + 1$
3. **end while**

Where w_0 is the initial string, $C^* = \{$ All the event generators which generate events of type $D \}$, $D = \{$ Set of all the types of possible events in the system $\}$, g is the output device, S is a sequence of all the events generated by the system at instant t, Z is a subsequence of S, and it includes all the events from visual devices. These events are the input of the visual algorithm π.

3 Case Study

Let us consider a robot with several sensors that provide information about the environment. It is programmed to autonomously navigate in a known environment, and to transport objects from one place to another. The input data are the data from a range sensor, the image from a camera to identify objects and places using markers and a humen supervisor that he is controlling the robot. The information is combined using a multimoldal algorith based on priorities.

A system like this can be modeled using a classical hybrid scheme (figure 1). This hybrid scheme can be adapted using the VWG introduced in the previous section.

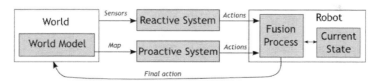

Fig. 1 Hybrid scheme for a robotic system.

In this picture the world is the real environment. The world model is a map containing the static elements of the environment. The reactive system is made of several generators, for the sensors and for the user's orders. The proactive system is the AI of the robot. The robot is the only actor in the system. The current state is the set of robot attributes. The multisensorial integration process is the evolution function of the robot. The final action is the result of the process of sensor integration and the final action carried out by the robot.

Only one primitive is needed, the robot, and it is modified by two possible transformations: move and rotate (table 1). When the system is executed in a real environment, the transformations correspond to the actual operations performed by

Table 1 Primitives and transformations of the robotic system.

	Real environment	Simulator
PRobot	No action	Draw the robot in the GS
TMove$_{<dist>}$	Move a distance *dist*	Move a distance *dist* in the GS
TRotate$_{<angle>}$	Rotate an angle *angle*	Rotate an angle *angle* in the GS

the robot. If it is executed in a simulator, the primitive and the transformations will represent the operations carried out in the graphics system (GS).

Events are used to define the activity in the system. Each event is defined by its identifier and some attributes. They produce changes on the actors through their evolution functions. These events are produced by generators. There is a generator for each event type. In the robotic system, five generators are needed:

- *gLaser*: It generates an *eLaser* event when the laser detects an obstacle, by obtaining the laser data and processing them to find the possible obstacles. It is defined as: *gLaser = eLaser*$_{<dist,angle>}$ if obstacle, where *dist* is the distance to the obstacle and *angle* is the angle to the obstacle.
- *gCamera*: It generates an *eCamera* event when a marker is detected in the camera image. Markers are used to identify the rooms in the environment. It is defined as: *gCamera = eCamera*$_{<marker>}$ if a marker is detected.
- *gDecide*: It generates an *eDecide* event each frame to indicate to the robot to make a decision. Is is defined as: *gDecide = eDecide* each frame.
- *gExecute*: It generates an *eExecute* event to indicate the system to execute the robot actions in the current representation space. If the representation space is the real environment, the real operations will take place (move the robot, rotate the robot...). If the current space is the simulator, the operations will take place in the graphics system. It is defined as: *gExecute = eExecute* each frame.
- *gObjective*: It generates an *eObjective* event to set a new objective marker. This generator is connected to the user orders. Users can specify a new target room simply by selecting its associated marker. It is defined as: *gObjective = eObjective*$_{<marker>}$ if user order.

An order relation must be defined to establish an execution priority among generators. In the robotic system, the order relation is: *gLaser, gCamera, gObjective, gDecide, gExecute*. Therefore, events related with the acquisition of data have the highest priority, compared with the events of decision and execution.

The only actor in our robotic system is the robot which is defined as: $ARobot_{<grid,row,column,angle,objective,action>}^{eLaser,eCamera,eDecide,eExecute,eObjective}$, where the superscript are the events which it is prepared to respond to, and the subscript are the attributes, whose meanings are: the *grid* represents the environment where the robot moves in. Each cell stores the registered data obtained from the sensors (the detected obstacles and markers). *Row and column* are position occupied by the robot in the grid. *Angle* is the robot orientation. *Objective* is the objective room, represented by its marker. And *action* is the string of primitives and transformations which indicates the next command to be executed. To simplify, in the following equations this actor will be referred as $ARobot^E_{<g,r,c,an,o,ac>}$.

The evolution function defines the way the robot behaves in the environment. Let e be an event that is received by the actor, the evolution function is defined as:

$$\lambda(ARobot^E_{<g,r,c,an,o,act>}, e) = \begin{cases} ARobot^E_{<g',r,c,an,o,ac>} & if \ e = eLaser_{<dist,angle>} \\ ARobot^E_{<g',r,c,an,o,ac>} & if \ e = eCamera_{<marker>} \\ ARobot^E_{<g,r',c',an',o,ac'>} & if \ e = eDecide \\ \alpha(ARobot^E_{<g,r,c,an,o,ac>}) & if \ e = eExecute \\ ARobot^E_{<g,r,c,an,o',ac>} & if \ e = eObjective_{<marker>} \\ ARobot^E_{<g,r,c,an,o,ac>} & otherwise \end{cases}$$

(1)

where the symbol apostrophe (') indicates that it has changed as follows:

- If $e = eLaser_{<dist,angle>}$, the grid (g) must be updated to indicate that an obstacle has been detected. The cell to mark is the one in position $(r + dist \ \cos(ang + angle), c + dist \ \sin(ang + angle))$.
- If $e = eCamera_{<marker>}$, the grid must be updated to indicate that a marker has been detected. The cell to mark is $(r + dist \ \cos(ang), c + dist \ \sin(ang))$.
- If $e = eDecide$, the current position and orientation of the robot (r, c, ang), must be updated, as well as the actions to be executed.
- If $e = eExecute$, the actions of the robot must be executed in the representation space, through the use of the α function.
- If $e = eObjective_{<marker>}$, a new objective has been set by the user, so the objective (o) must be changed to the new one $(marker)$.
- In any other case, the actor must remain unchanged.

The initial string in our system is defined as: $ARobot^{eLaser,eCam.,eDecide,eExec.,eObjct.}_{<grid,row,column,angle,\varepsilon,\varepsilon>}$, where the attribute grid is initialized to a set of empty cells, the attributes row, column and angle are the initial position, and the objective and the actions are empty.

3.1 Analysis

A set of tests has been designed to prove the features of our model: The aim of the first test is to prove the suitability of the evolution function to introduce new AI algorithms. This test is not to obtain the best AI algorithm to achieve the goal. Two simple decision algorithms have been used to decide how the robot should move. The first algorithm makes decisions randomly to find the target position. The second algorithm is A* [4]. These two AI algorithms were introduced without making any changes in other parts of the system, just by changing the evolution function.

The aim of the second test is to prove that the input devices can be replaced without changing the system. We can change the laser to a Kinect to detect obstacles. To change this device, we have just designed a new event generator (gKinect) that creates events of the same type that the ones generated by the laser generator.

In the third test we want to test the extensibility of the system. New instances of the actor symbols (representing robots) have been added to the definition string to extend the system and create a multi-robot system in an almost immediate way. The updating of the definition string supposes the extension of the model and the addition of new features. Moreover, most elements can be reused in new definition strings to obtain new behaviours with little effort.

In the last experiment we tested the flexibility to work under different conditions. To prove this feature, the system has been tested with different maps, in the case of the simulated robot, and in different real environments, in the case of the real robot.

4 Conclusions

A new model to formally define virtual worlds, independently from the underlying physical layer, has been presented. Is has been used to model the control of a mobile robot, navigating in a given environment, and using a set of multimodal inputs from different types of sensors.

Taking into account the diversity of virtual worlds available nowadays and the wide variety of devices, this model seems to be able to provide interesting features. Firstly, it is a formal model that allows to abstract and represent the states of the system in a general way by avoiding specific features. It is a device-independent model, therefore, is not linked to the implementation. It allows the replacement of physical devices by simulated ones, and the easy addition of new ones.

In conclusion, it has been achieved the main objective of defining a new formal and generic model that is able to model general virtual worlds systems by avoiding the specific peculiarities of other models existing today.

References

1. Davis, M., Sigal, R., Weyuker, E.J.: Computability, Complexity, and Languages, Fundamentals of Theoretical Computer Science, 2nd edn. Elsevier Science, San Diego (1994)
2. Ernst, M.O., Bülthoff, H.H.: Merging the senses into a robust percept. TRENDS in Cognitive Sciences 8(4) (2004)
3. Ingrand, F., Chatila, R., Alami, R.: An Architecture for Dependable Autonomous Robots. In: IARP-IEEE RAS Workshop on Dependable Robotics (2001)
4. Luo, R., Lin, Y.-C., Kao, C.-C.: Automous mobile robot navigation and localization based on floor paln map information and sensory fusion approach. In: IEEE MFI (2010)
5. Posadas, J.L., Poza, J.L., Simó, J.E., Benet, G., Blanes, F.: Agent-based distributed architecture for mobile robot control. In: Engineering Applications of Artificial Intelligence, pp. 805–823 (2008)
6. Russell, S.J., Norvig, P.: Artificial intelligence: a modern approach. Prentice Hall (2010) ISBN: 0136042597
7. Sharma, R., Pavlovic, V.I., Huang, T.S.: Toward Multimodal Humar-Computer Interface. Proceedings of the IEEE 86(5), 853–869 (1998)
8. Singhal, A., Brown, C.: Dynamic bayes net approach to multimodal sensor fusion. SPIE (1997)

A Classification Method of Knowledge Cards in Japanese and Chinese by Using Domain-Specific Dictionary

Xiaopeng Liu, Li Cai, Masanori Akiyoshi, and Norihisa Komoda

Abstract. This paper addresses a classification method to classify the knowledge cards written in Japanese and Chinese by using domain-specific dictionary at off-shore software development company. The method has two phases in classification process; pre-process for morphological analysis, translation or filtering original cards, and sample-based categorization process with statistical information. Finally, we take the classification experiment to verify the feasibility of our method and also discuss the experimental results.

1 Introduction

Nowadays, as the offshore software development between Japan and China is growing rapidly, a large amount of related knowledge cards that mixed with Japanese and Chinese have been accumulated in the database of many offshore software development companies. The contents of these knowledge cards consist of software development technology, employee management and language learning, which are very helpful in technical training and self-study for the employee. So, in order to take use of these knowledge cards and know-how of the companies, an efficient knowledge classification system which can handle the knowledge cards that mixed with Japanese and Chinese is strongly expected.

Towards the situation that mentioned above, the knowledge management system, so called RKMS [1] was developed to store the knowledge cards that collected in

Xiaopeng Liu · Masanori Akiyoshi · Norihisa Komoda
Osaka University, 2-1, Yamadaoka, Suita, Osaka, Japan
e-mail: liu.xiaopeng@ist.osaka-u.ac.jp,
    akiyoshi@ist.osaka-u.ac.jp, komoda@ist.osaka-u.ac.jp

Li Cai
Jinan Ryouka Science & Technology Co., Ltd., Qilu Soft Park, No. 1 Shunhua Road,
Jinan Hi-tech Zone, Shangdong, 250101, China
e-mail: cai@ryouka.com.cn

S. Omatu et al. (Eds.): Distributed Computing and Artificial Intelligence, AISC 151, pp. 453–460.
springerlink.com   © Springer-Verlag Berlin Heidelberg 2012

the form of Q&A (Question and Answer) from BBS (Bulletin Board System) of the company. These collected knowledge cards are used by references when users face problems, and the BBS search engine helps a user to get necessary information. In order to get the required reference more quickly, RKMS should also classify the knowledge cards. Such cards are written in Japanese, Chinese, or both of them sometimes. The existing method [2] is able to classify the Japanese knowledge cards well, but cannot handle the knowledge cards that mixed with Chinese.

So, this paper discusses how to classify the knowledge cards that mixed with Japanese and Chinese by using a domain-specific dictionary.

2 Word-Based Classification

2.1 Outline of Existing Card Classification System

Fig.1 shows the outline of existing classification method for Japanese knowledge cards [2].

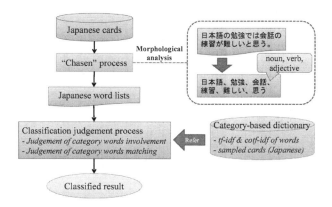

Fig. 1 Outline of existing opinion classification method

As the process flow shown in Fig.1, the Japanese cards are divided into a set of Japanese words under the "Chasen" [3] process. "Chasen" is a morphological analysis tool for Japanese, and it can split a Japanese sentence into noun, verb, adjective words and something else, just like the example which is shown in the right side of Fig.1 above. The category-based dictionary consists of some sampled Japanese cards that judged by our tool maintenance team members, and also provides two types of characterizing indices on each category from word statistical information points of view. One is $tf - idf$ (term frequency inverted document frequency), and the other one is $cotf - idf$ (co-occurrence $tf - idf$). $cotf - idf$ is an extension of $tf - idf$ and it focuses on the two words co-occurrence to characterize each category. Referring to the category-based dictionary, the involvement degree of category words and common words between the target card and a category, R_t and R_{cot} can be calculated respectively. Based on the R_t, R_{cot}, summing the weighted parameter α, the category words involvement degree R between the target knowledge card and a category can be calculated by the Formula 1

$$R = \alpha \times R_t + (1 - \alpha) \times R_{cot} \tag{1}$$

If the involvement degree R of the target knowledge card is larger than the preset threshold, the system judges the content of the target knowledge card is quite similar to the sample cards in a category, and intends to classify the target card to this category prospectively. Then, the judgement of category words matching degree is going on. In this process, the average value of *Jaccard* coefficient between the target knowledge card and the other cards in a category will be calculated as the category words matching degree of the target knowledge card. If the category words matching degree is larger than the preset threshold, the system determines that the category words matching degree between the target knowledge card and a category is high, and classifies the target knowledge card to this category. Besides, the *Jaccard* coefficient is calculated by the Formula 2 below.

$$Jaccard\ coefficient = \frac{N_{sx}}{N_s + N_x - N_{sx}} \tag{2}$$

In the Formula 2, N_{sx} means the number of common words between a knowledge card S and a knowledge card X, N_s does the number of words in a knowledge card S, and N_x does the number of words in a knowledge card X. S means a sampled card stored in a category, and X does a target card under judgement processing.

2.2 Problems of Classifying Knowledge Cards Involving Chinese

Knowledge management systems of many offshore software development companies always have the knowledge cards that written in two languages or more. For example, in the RKMS that mentioned above, 66% of the knowledge cards are written in Japanese, 29% are in Chinese, and the 5% left are mixed with Japanese and Chinese.

The existing classification method referring to Japanese category-based dictionary can classify Japanese knowledge cards effectively; however it cannot handle the classification of Chinese cards, let alone the mixed ones.

In order to solve the problem above, there are two kinds of proposals. One is to translate all the Chinese knowledge cards into Japanese by hand or software. But that way is always inefficient or low quality. The other way is to convert the Chinese word lists that generated by Chinese morphological analysis tool into Japanese by using a domain-specific dictionary. Although we cannot provide a comprehensive dictionary as the fast-changing IT technology, there is still a possible way to compensate the conversion loss.

So, in the following, based on the existing card classification method for Japanese, we address our classification system which can handle the knowledge cards that mixed with Japanese and Chinese by using a domain-specific dictionary.

3 Classification System of Knowledge Cards in Japanese and Chinese Referring to Domain-Specific Dictionary

3.1 Approach

Fig.2 indicates the process flow of our classification method of knowledge cards that represented in Japanese and Chinese.

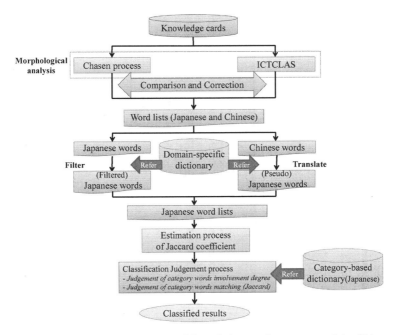

Fig. 2 Outline of classification system of knowledge cards represented in Chinese and Japanese

In the beginning, we skip the language judgement of knowledge card, and directly take morphological analysis of the knowledge cards with "Chasen" and ICTCLAS [4] As "Chasen" for Japanese, ICTCLAS is a morphological analysis tool for Chinese. Then, the system compares the two analysis results morpheme by morpheme, and generates the correctly analyzed word lists. After that, in order to make use of the existing card classification method and reduce the unimportant Japanese words which may worsen the classification results. We translate the Chinese words of the word lists into Japanese by using the preset domain-specific dictionary [5]. Then, referring to the same dictionary, we also filter the Japanese words left of the word lists. The new word lists after translating and filtering has fewer words than the original ones, because all of the words that do not included in the dictionary will be deleted. If we still take the word lists to the classification judgement process, we will get a wrong *Jaccard* coefficient, which will cause an incorrect classification result. So, before the classification judgement, the estimation process of *Jaccard* coefficient should be added. Finally, knowledge cards will be classified by using the existing classification method that mentioned in Section 2.1.

3.2 *Morphological Analysis Process*

As the input knowledge cards of our system can be written either in Japanese, Chinese, or both of them, the general way of classification is to judge the language of all knowledge cards at first. Instead of such judgement, in our method, we take another way to solve the language problem for the classification. That is taking advantage of the feature of two morphological analysis tools. Both "Chasen" and ICTCLAS can divide the knowledge cards in their original language into morphemes correctly. However, when facing the knowledge cards that written in other language, the two morphological analysis tools intend to divide every morpheme into a few smaller segments or even individual characters [7, 8]. For example, "勉強" (English word is "Study") is a Japanese unique word. If ICTCLAS take morphological analysis of it, as there is no such word in Chinese, it will be divided into "勉" and "強" the two individual characters. Similarly, when "Chasen" analyzes a Chinese word, we will get the same result. So, referred to such feature of the two morphological analysis tools, if we contrast the two analysis results morpheme by morpheme, choose the correct parts and assemble all of them, we will get the correctly analyzed word list.

The criterion for system to judge which analysis result is correct is the number of characters in a morpheme. As the two morphological analysis tools will divide every morpheme into a few smaller segments or even individual characters when facing the knowledge cards that written in other language. The number of characters in these smaller segments and individual characters should be less than the one of correctly analyzed morphemes. For example, the Japanese unique word "勉強" can be correctly analyzed by "Chasen" and the number of characters of it is 2. Then, when it is analyzed by ICTCLAS, it will be divided into "勉" and "強" the two individual characters, and either of the two characters has only one character, which is less than the number of characters of "勉強". So, the system judges that the analysis result provided by "Chasen" is correct.

Fig.3 shows the morphological analysis process of the input data in different languages. Firstly, when the input data is written in Japanese only, "Chasen" provides better analysis accuracy than ICTCLAS. So, the analysis result of "Chasen" is set as the final output. Then, when the input data is written in Chinese only, the analysis result of ICTCLAS is more correct than "Chasen" and the system chooses that as the final output. However, when the input data is mixed with the two languages above, both "Chasen" and ICTCLAS have mistakes in their analysis results. In this time, the system will contrast the two analysis results from one morpheme to another, chooses every correct part and assembles them into the final output.

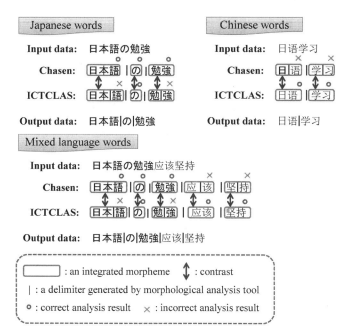

Fig. 3 Samples of morphological analysis process

3.3 Loss Correction Process

When translating the Chinese words of the word lists into Japanese, the Chinese words that do not exist in the domain-specific dictionary will be deleted. Moreover, in the process of filtering, the Japanese words which are not in the dictionary will also be deleted. So, if we directly take the word lists that after translating and filtering to calculate the *Jaccard* coefficient by the Formula 2, as the values of parameters in the denominator and numerator of the formula are not correct, we will get a wrong *Jaccard* coefficient, which will lead to an incorrect classification result. So, in the estimation process of *Jaccard* coefficient, we should add the loss correction for the parameters that may have loss [9].

Referring to the Formula 2, we find that N_s is a regular value, N_x and N_{sx} will be affected by translation and filtering. The original value of N_x can be obtained easily and directly from the word list before translation and filtering process. So, we should discuss the loss correction for N_{sx} only. We set a parameter "w" to compensate the loss correction for N_{sx} and its definition is shown by the Formula 3 below.

$$w = \frac{1}{1 - loss\,rate\,of\,common\,words} \qquad (3)$$

When knowledge cards are written in Chinese, w is a conversion loss correction rate for Chinese words common to the cards. When knowledge cards are in Japanese, w means a filtering loss correction rate for Japanese words common to the cards.

Moreover, if knowledge cards are mixed with Chinese and Japanese, w is a loss correction rate for the words of word lists common to the cards.

$$estimated\,Jaccard\,coefficient = \frac{w \times N_{sx}}{N_s + N_x' - w \times N_{sx}} \tag{4}$$

Formula 4 shows the calculation method of the estimated *Jaccard* coefficient. Besides, N_x' means the number of words in a knowledge card X before translation and filtering.

3.4 Experimental Result

570 knowledge cards in Chinese from the database of RKMS are set as the input data of the following experiments. The category-based dictionary consists of 36 sample Japanese cards. Besides, the domain-specific dictionary has 1587 conversion word pairs and most of them are specialized vocabulary about information system. Our experiments and the calculation of loss correction rate w proceed as follows.

Firstly, we get 100 knowledge cards from the 570 Chinese cards randomly, and take the value of w when the F value is highest as the loss correction rate by using the former 100 knowledge cards. F value is set as the evaluation criterion of classification, which is the weighted harmonic mean of recall rate and precision rate. The recall rate means the fraction of the documents that are relevant to the query that are successfully retrieved, and the precision rate does the fraction of the documents retrieved that are relevant to the user's information need. Then, based on the 470 knowledge cards left, we make 3 types of dataset such as Chinese cards, Japanese cards (translated by hand) and the knowledge cards mixed the two languages (made by hand). After that, referring to the w that obtained above, the 3 datasets are classified by our classification system. The process of the experiment above repeats 10 times, and the average values of recall rate, precision rate and F value for the 3 datasets are calculated at last.Although the 3 datasets are written in different languages, as the contents of them are the same, we can predict that the classification results of the 3 datasets are in the same level.

Table 1 Comparison of experimental results

Num. of knowledge cards	Language	Ave. of recall rate(%)	Ave. of Precision rate(%)	Ave. of F value(%)
470	C&J	66.49	70.76	68.54
	CHN	67.29	70.75	68.74
	JPN	67.80	70.85	69.32

Table 1 shows the comparison of the results of the three classification experiments. In the Table 1, C&J means the knowledge cards that mixed with Chinese and Japanese; CHN means the knowledge cards in Chinese and JPN does the knowledge cards that written in Japanese.

Referring to the Table 1, the average F value of the Chinese cards and the knowledge cards mixed with Chinese and Japanese are 68.74% and 68.54%. Both of them are very close to the one of Japanese cards which gets 69.32%. So, it is proved that our system is able to handle the classification of Japanese cards, Chinese cards and the knowledge cards mixed with Chinese and Japanese effectively.

4 Conclusion

This paper addresses a classification method of knowledge cards represented in Japanese and Chinese at offshore software development. The classification is based on the existing card classification method for Japanese and is able to classify the knowledge cards that mixed with Chinese and Japanese effectively. Referred to the experimental results, the feasibility of our proposed method is well proved. In the future, we will take wider experimental sample to testify the reliability of our method.

References

1. Cai, L., Wang, Z., Jiao, Y., Akiyoshi, M., Komoda, N.: Prototype of Knowledge Management System in Chinese Offshore Software Development Company. WSEAS Transactions on Information Science & Application 5(3), 252–257 (2008)
2. Negoro, K., Oiso, H., Akiyoshi, M., Komoda, N.: A Classification Method of Users Opinions using Category-based Dictionary generated from Answers in Open-ended Questionnaire System. In: IEE J. Workshop on Information System IS-07-16 (2007) (in Japanese)
3. Matsumoto, Y., et al.: Japanese Morphological Analysis System "Chasen" version 2.2.7, http://Chasen.aistnara.ac.jp
4. Chinese Academy of Sciences; ICTCLAS (Institute of Computing Technology, Chinese Lexical Analysis System), http://ictclas.org/
5. Cai, L., Wang, Z., Akiyoshi, M., Komoda, N.: A Knowledge Cards Classification Method with Conversion Loss Correction for Incomplete Translation Dictionary. IEE J. Transactions on Electrical and Electronic Engineering 6(6), 566–570 (2011)
6. Graham, T.: Unicode: a primer, Unicode Consortium and Mulberry Technologies, Inc. IDG Books Wold Wide, Inc., Foster City (2000) ISBN: 0-7645-4625-2
7. Zhang, H.-P., Yu, H.-K., Xiong, D.-Y., Liu, Q.: Hhmm-based chinese lexical analyzer ictclas. In: Proceedings of the Second SIGHAN Workshop on Chinese Language Processing, pp. 184–187 (2003)
8. Yin, D., Shao, M., Jiang, P., Ren, F., Kuroiwa, S.: Rule-based Translation of Quantifiers for Chinese-Japanese Machine Translation. In: Proceedings of the 10th WSEAS International Conference on Computers, Vouliagmeni, Athens, Greece, pp. 559–564 (2006)
9. Pirkola, A., Hedlund, T., Keskustalo, H., Jarvelin, K.: Dictionary-based cross-language information retrieval: problems, methods, and research findings. Inf. Retrieval 4, 209–230 (2001)

A Taxonomy Construction Approach Supported by Web Content

Ana B. Rios-Alvarado, Ivan Lopez-Arevalo, and Victor Sosa-Sosa

Abstract. The growth of unstructured information available inside organizations and on the Web has motivated the building of structures for represent and manipulate such information. Particularly, an ontology provide a structural organizational knowledge to support the exchange and sharing. A crucial element within an ontology is the taxonomy. For building a taxonomy, the identification of hypernymy/hyponymy relations between terms is essential. Lexical patterns have been used in analysis of text for recovering hypernyms. In addition, the Web has been used as source of collective knowledge and it seems a good option for finding appropriate hypernyms. This paper describes an approach to get hypernymy relations between terms belonging to a specific domain knowledge. This approach combines WordNet synsets and context information for building an extended query set. This query set is sent to a web search engine in order to retrieve the most representative hypernym for a term.

Keywords: Knowledge representation, knowledge management, search heuristics.

1 Introduction

At the beginning of the 21st century the easy way to access to digital information resources has motivated an exponential growth in the available unstructured information. This growth is not only present on web resources, but it also can be seen inside organizations, institutions, and companies. In an organization, for example, documents represent a significant source of collective expertise (*know how*). In order to store, retrieve, or infer knowledge from this information, it is necessary represent it using a conceptual structure. This can be achieved by means of taxonomies or

Ana B. Rios-Alvarado · Ivan Lopez-Arevalo · Victor Sosa-Sosa
Information Technology Laboratory, Cinvestav - Tamaulipas
Scientific and Technological Park. Victoria, Mexico
e-mail: {arios,ilopez,vjsosa}@tamps.cinvestav.mx

S. Omatu et al. (Eds.): Distributed Computing and Artificial Intelligence, AISC 151, pp. 461–468.
springerlink.com   © Springer-Verlag Berlin Heidelberg 2012

ontologies. Taxonomy is the science that deals with the study of identifying, grouping, and naming organisms according to their established natural relationship. In current usage within Knowledge Management, taxonomies are considered narrower than ontologies since ontologies apply a larger variety of relation types. According to Gruber [1], *"ontologies are often equated with taxonomic hierarchies of classes"*; thus, it can be said that the key component in the ontology is the taxonomy. Such taxonomies, as the main component for an ontology, provide an organizational model for a domain (domain ontology), or a model suitable for specific tasks or problem solving methods (ontologies of tasks and methods). Nevertheless, constructing taxonomy is a very hard task.

The identification of hypernymy/hyponymy relations between terms (in this work only nouns are considered as terms) is mandatory for building a taxonomy. A hyponym can be defined as: a word of more specific meaning than a general or superordinate term applicable to it. By contrast, a hypernym is a word with a broad meaning constituting a category under which more specific words fall. For example, *dog*, *cat*, and *turtle* are hyponyms of *Animal* whereas *Animal* is a hypernym of *dog*, *cat*, and *turtle*. Other names for the hyponym relationship are *is-a*, *parent-child*, or *broader-narrower* relationships. Caraballo [2] claimed that according to WordNet[1], *"a word A is said to be a hypernym of a word B if native speakers of English accept the sentence B is a (kind of) A"*.

On the other hand, the Web has become a source of collective knowledge, reason why it seems a good option for finding suitable hypernyms. In addition to using Web and lexical patterns, some works [3, 4, 5] identify new lexical patterns making possible to obtain more specific hyponyms; but it is necessary rely on the known hyponymy relationships for training a classifier, which is not always possible. In this paper, an approach to find hypernym relations between terms from text belonging to a domain knowledge is presented. Particularly, this approach combines WordNet synsets and contextual information for building an extended query set. Whit this query set, a web search is executed in order to retrieve the most representative hypernym for a term.

The rest of this paper is structured as follows. In Section 2, a brief description of the related work about automatic discover of hypernyms is given. In Section 3 the approach and the method to find hypernyms are described. Later, in the Section 4, the experiments and preliminary results are presented. Finally, Section 5 gives some conclusions and the further work.

2   Related Work

One of the first ideas in automatic discovering hypernyms from text was proposed by Hearst [6]. She proposed a method to identify a set of lexico-syntactic patterns occurring frequently in the text. Caraballo [2] proposed to automatically build a noun hierarchy from text using data on conjunctions and appositives appearing in the Wall Street Journal corpus. Both methods are limited by the number of patterns

---

[1] http://wordnet.princeton.edu/

used. Pantel *et al.* [7] showed how to learn syntactic patterns for identifying hypernym relations and binding them with clusters that were built from co-occurrence information. Blohm and Cimiano [3] proposed a procedure to find lexico-syntactic patterns indicating hypernym relations from the Web. Snow *et al.* [4] generated hypernym patterns and combined them with noun clusters to generate high-precision suggestions for unknown noun insertion into WordNet. They also described a variant of their classifier including evidence from *coordinate terms* (terms with common ancestor classes) increasing precision. Ritter *et al.* [8] presented a method based on lexical patterns that find hypernyms on arbitrary noun phrases. They used a Support Vector Machine classifier to find the correct hypernyms from matches to the Hearst patterns. Kozareva and Hovy [9] proposed a semi-supervised algorithm to learn automatically hyponym-hypernym pairs from the Web. They presented a graph algorithm that derives from scratch the integrated taxonomy structure of all the terms. Most of these studies are limited due to the hand selection of pairs of terms that have hypernymy relationship, which represent the initial seed for discovering new patterns. On the other hand, Cimiano and Staab [10] showed that a potencial way to avoid the *knowledge acquisition bottleneck* is acquiring collective knowledge from the Web using a search engine. This idea was used by Sánchez [11], using the Web for acquiring taxonomic and non-taxonomic relationships. The use of lexical patterns and the Web as knowledge resource permits to get taxonomic relations supported by lexical evidence and domain independent.

3   The Method: Taxonomy Construction

For building a taxonomy, is important consider terms and relationships. These elements should be relevant in the domain of the input corpus. This section presents a method for extracting relevant hypernyms taken from the information given by a corpus and issued from the knowledge contained in the Web.

3.1   The Representation Model

Typically text is represented using the *bag of words* model. This model assumes that the order of words has no significance. However, current applications consider that a semantic representation focused on *Natural Language Proccessing* (NLP) has a major potential for new developments. Thus, word-context matrices and pair pattern matrices are most suited for measuring the semantic similarity of word pairs and patterns [12]. In the approach presented in this paper, the proposal is to use a syntactic parser to extract the grammatical context where each word occurs. It is of special interest the focus on two dependency relationships *<subject, verb>* and *<verb, object>*. A pair-term matrix is used as representation model. Thus, by means of mutual information is possible to find two related terms. The Pointwise Mutual Information (PMI) is the measure used for the association strength between two words (w_1, w_2). By using the Equation 1, the values of PMI for each verb-noun pair are calculated. The representation model is obtained on the overall corpus.

$$PMI(w_1, w_2) = log_2 \frac{p(w_1 AND w_2)}{p(w_1) * p(w_2)} \tag{1}$$

3.2 Querying the Web

For obtaining close results to the domain of the input corpus, it is proposed the construction of a query set that considers the most representative terms in the input corpus and in the WordNet synsets. The discovering hypernyms consists of the following phases.

- Pre-processing: It is performed to identify dependencies between nouns sharing a verb in the same context. These dependencies are obtained using the Minipar[2] parser. The pair-pattern matrix is used as representation model. In the pair pattern matrix, the pairs correspond to the terms appearing in a triple term structure *<subject> verb <object>*. Due to a noun can be a subject or an object within a sentence, the representative nouns are obtained by pairs like *<subject-verb>* and *<verb-object>*. The corresponding PMI (Equation 1) between a noun and a verb is used to calculated the distributional similarity between any two nouns, and so to extract the topics from the corpus.
- Topic extraction: The topics from the corpus are inferred using an adaptation of the CBC algorithm proposed by Pantel [13].
- Discovering hypernyms: For each noun in the topic, a set of queries is generated; this process considers the following:

1. The Hearst patterns have shown good evidence identifying that entity A (noun) is a hyponym of B. Other patterns considered in this work are shown in Table 1.

Table 1 Lexical patterns

Hearst's patterns	Other patterns
A, and other B	B, called A
A, or other B	B, particularly A
A is a B	B, for example A
B, such as A	B, among which A
B, including A	
B, especially A	

2. A general query on the Web like *such as <hyponym>* is not enough to obtain interesting and precise information. In order to get useful information,

---
[2] http://webdocs.cs.ualberta.ca/ lindek/minipar.htm

the query needs to be more specific [14]. This is the reason why related information is added to the query: 1) contextual information and 2) supervised information. The contextual information is given to the terms with the higher frequencies in the corpus (without stopwords and after a lemmatization process). The supervised information is given to the most representative terms in the WordNet synset corresponding with the term. For extracting terms from WordNet, the gloss of the term is tagged and the words (three words) labeled as noun are considered as supervised information.

3. Query sets are constructed using the lexical patterns and the related information. Each query is sent to a web search engine for using the Web as a source of knowledge.

4. For each query in the hypernym query set, the n first pages are retrieved. The text for each n page is cleaned and parsed avoiding non-essential information (eliminating images, videos, banners, etc.). Each sentence is POS-tagged using the Stanford tagger[3], thus the lexical pattern of the query and their candidate hypernym are identified. A term is selected as hypernym if it is a noun but it is not a stopword.

5. The list of candidate hypernyms is evaluated using a new query set, where each possible hypernym will be replaced in the lexical pattern. Using its query set and the number of hits obtained in the web search each candidate hypernym (CH) is evaluated by means of the following measure to score candidate hyperonym (SCH) [10] (Equation 2):

$$SCH = \frac{hits(LexicalPattern(term, CH))}{hits(CH)} \tag{2}$$

where the *LexicalPattern(term, CH)* corresponds to build a query like: *<term>, + and + other + <CandidateHypernym>*; *and other* corresponds to some lexical pattern. The total score for a CH is given by the sum of scores obtained for each lexical pattern. Thus, the hypernym with the highest total score in the result for the query will be the hypernym associated to the term.

4 Experiments and Preliminary Results

A sample of the Lonely Planet[4] corpus was used in the experiments. To illustrate the experiment, the term *museum* was considered. The terms with the higher frequencies in the sample corpus were: *cash*, *travel*, and *product*. The web query set for term *museum* is constituted by *<museum, + lexicalPattern + cash + travel + product>*. Other query set was composed by the extracted words from the WordNet

---

[3] http://nlp.stanford.edu/software/tagger.shtml
[4] The Lonely Planet corpus is related to travels. It is part of an initiative for automated construction and population of ontologies. http://olc.ijs.si

synset for term *museum* (*collection*, *object*, and *display*): <*museum,+ lexicalPattern + collection + object + display*>, where *lexicalPattern* corresponds to one of the patterns in Table 1.

Using the query set with only lexical patterns, the list of candidate hypernyms was: <*site, place, attraction, department of history*>. Using a query with added information, the new candidate hypernyms were: <*depository, institution*>. A new lexical pattern query set was created using each one. Then, using the number of obtained hits in the web search, the corresponding score was computed for each candidate hypernym. Following this, the best candidate hypernym is chosen.

In Table 2 can be seen that the best hypernym to *museum* is *attraction* and into the tourist context could be a good option, but it is important to note that the second best candidate is *institution*.

According to different authors, the definitions of *museum* are:

...a museum is a building or institution which houses and cares for a collection of artifacts and other objects of scientific, artistic, or historical importance and makes them available for public viewing through exhibits that may be permanent or temporary...[5]

Museums enable people to explore collections for inspiration, learning and enjoyment. They are institutions that collect, safeguard and make accessible artefacts and specimens, which they hold in trust for society...[6]

The museum is an empowering institution, mean to incorporate all who would become part of our shared cultural experience...[7]

According to the added information to queries, the term *institution* is a good candidate hypernym to *museum*. The Fig. 1 shows the created taxonomy for the group of terms <*art, culture, library, science, book, travel*> related to *museum*. It is important remark that the taxonomy was generated taking into account exclusively the corpus Lonely Planet and the terms into the topic where term *museum* appears.

Table 2 Total score of candidate hypernyms for term *museum*

Candidate Hypernym	Total score
attraction	3.74220
institution	3.65833
depository	1.50125
department of history	0.82055
place	0.21463
site	0.09794

[5] Edward Porter Alexander, Mary Alexander. Museums in motion: an introduction to the history and functions of museums. Rowman & Littlefield, 2008 ISBN 0-7591-0509-X.

[6] http://www.museumsassociation.org/about/frequently-asked-questions

[7] Mark Lilla. The Great Museum Muddle. New Republic, April 8, 1985. pp.25-29.

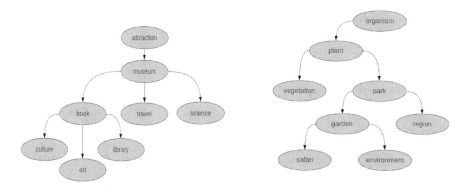

Fig. 1 Taxonomy created for the group of terms related with *museum*

Fig. 2 Taxonomy created for the group of terms related with *plant*

Following the experiments in the Biology domain[8], a query set was constructed for the term *plant* and their group of related terms using the WordNet synsets terms: *flora*, *botany*, and *organism*. The found hierarchical structure is shown in Fig. 2. This taxonomy was generated taking into account exclusively the corpus of news from Biology domain. It does not represent the universal knowledge about Biology.

The resulting taxonomies do not represent the universal knowledge on an specific domain because the structure depends strongly on the input corpus. At this moment, for illustration, only small corpora were used. However, the found structure can be used in applications where the content of input corpus is significant. The taxonomic relations are better when the input corpus contains broader documents (for example, journal articles and technical reports). Therefore, the result is directly associated to the size and quality on content of the input corpus.

5 Conclusions and Future Work

In this paper, an approach to construct taxonomies from unstructured text has been presented. The use of lexical patterns support the evidence lexical of the relationships in the text. The related information used for extending a query seems to be a good approximation to narrow the search results. Furthermore, the extended query set indicates that 1) there is a relation between terms and 2) the terms and their hypernym are in the same context. The method can be applied to any domain knowledge. The generated taxonomies are consistent with the input corpus, which allows to do search, location and retrieval tasks in efficient way. The obtained results for this approach can be improved adding new lexical patterns to queries and extending the search over other kind of resources (Wikipedia or Frequently Questions Blogs).

---

[8] A sample of Biology news (Biology subtopic) from the Biology News Net website http://www.biologynews.net/

Also, in the future experiments will be consider different sizes and domain knowledge corpus.

Acknowledgements. This work was partialy funded by project number 176656 from "Fondo Mixto Conacyt-Gobierno del Estado de Tamaulipas".

References

1. Gruber, T.: A translation approach to portable ontology specifications. Knowledge Acquisition 5(2), 199–220 (1993)
2. Caraballo, S.: Automatic construction of a hypernym-labeled noun hierarchy from text. In: Proceedings of the 37th Annual Meeting of the Association for Computational Linguistics, Association for Computational Linguistics, pp. 120–126 (1999)
3. Blohm, S., Cimiano, P.: Learning Patterns from the Web-Evaluating the Evaluation Functions-Extended Abstract. In: OTT 2006, vol. 1, p. 101 (2007)
4. Snow, R., Jurafsky, D., Ng, A.: Learning syntactic patterns for automatic hypernym discovery. In: Advances in Neural Information Processing Systems, vol. 17, pp. 1297–1304 (2005)
5. Ortega-Mendoza, R.M., Villaseñor-Pineda, L., Montes-y-Gómez, M.: Using Lexical Patterns for Extracting Hyponyms from the Web. In: Gelbukh, A., Kuri Morales, Á.F. (eds.) MICAI 2007. LNCS (LNAI), vol. 4827, pp. 904–911. Springer, Heidelberg (2007)
6. Hearst, M.: Automatic acquisition of hyponyms from large text corpora. In: Proceedings of the 14th Conference on Computational Linguistics, vol. 2, pp. 539–545. Association for Computational Linguistics (1992)
7. Pantel, P., Ravichandran, D., Hovy, E.: Towards terascale knowledge acquisition. In: Proceedings of the 20th International Conference on Computational Linguistics, p. 771. Association for Computational Linguistics (2004)
8. Ritter, A., Soderland, S., Etzioni, O.: What is this, anyway: Automatic hypernym discovery. In: Proceedings of AAAI 2009 Spring Symposium on Learning by Reading and Learning to Read, pp. 88–93 (2009)
9. Kozareva, Z., Hovy, E.: A semi-supervised method to learn and construct taxonomies using the web. In: Proceedings of the 2010 Conference on Empirical Methods in Natural Language Processing, EMNLP 2010, pp. 1110–1118. Association for Computational Linguistics, Stroudsburg (2010)
10. Cimiano, P., Staab, S.: Learning by googling. ACM SIGKDD Explorations Newsletter 6(2), 24–33 (2004)
11. Sánchez, D.: Domain ontology learning from the web. The Knowledge Engineering Review 24(04), 413–413 (2009)
12. Turney, P.D., Pantel, P.: From frequency to meaning: vector space models of semantics. J. Artif. Int. Res. 37, 141–188 (2010)
13. Pantel, P.: Clustering by committee. PhD thesis. University of Alberta (2003)
14. Sang, E.: Extracting hypernym pairs from the web. In: Proceedings of the 45th Annual Meeting of the ACL on Interactive Poster and Demonstration Sessions, Association for Computational Linguistics, pp. 165–168 (2007)

Semantic Graph-Based Approach for Document Organization

Erika Velazquez-Garcia, Ivan Lopez-Arevalo, and Victor Sosa-Sosa

Abstract. Actual document search engines base searches on the file name or syntactic content, which means that the word or part of the word to search must exactly match. This article proposes a semantic graph-based method for document search. The approach allows to organize, search, and display documents or groups of documents. Groups are formed according to topics contained in documents.

1 Introduction

Nowadays text files (documents) are the most common means used to capture information and knowledge in all subjects of modern life. Current information retrieval techniques commonly execute syntatic and lexical search. Typical users try to write their own words as they want to search, instead of the exact set of present in the text. The large amount of documents available in hard drives becomes another problem, considering that this type of information is stored using different formats and could have different ways to be organized. These problems cause that some important results can be discarded in a search.

The recent growth of the amount of documents in hard disks or in online storage units requires a better organization that makes it easier to find and retrieve information. As a way to organize their personal information, users adopt their own schemes. By definition, this way of organizing information involves users to know the location of each document, or at least to have an idea about when they were created or what information they contain. However, studies by Ravasio Golemati *et al.* [1] and Ravasio *et al.* [2] show that file names or paths do not provide significant help to searches and existing tools do not facilitate the management and retrieval of such information.

Erika Velazquez-Garcia · Ivan Lopez-Arevalo · Victor Sosa-Sosa
Information Technology Laboratory, Cinvestav - Tamaulipas
Scientific and Technological Park.Victoria, Mexico
e-mail: {evelazquez,ilopez,vjsosa}@tamps.cinvestav.mx

S. Omatu et al. (Eds.): Distributed Computing and Artificial Intelligence, AISC 151, pp. 469–476.
springerlink.com   © Springer-Verlag Berlin Heidelberg 2012

This paper proposes a semantic graph-based method to organize, search, and display groups of documents according to topics they contain, based on the collection of synonyms, hypernyms and hyponyms of each term. The paper is organized as follows. Section 2 presents a brief review of related work. The strategy used to solve the search and document representation by using graphs is presented in section 3. Preliminar results are presented in section 4. Finally, some concluding remarks and future work ideas are given in section 5.

2 Related Work

The use of semantic in search and information retrieval is not a new tendency. Some approaches consider, for example, the representation of a document using ontologies such as in Yao *et al.* [3] and Ocampo [4], among others.

Some researchers have addressed this problem in different ways. Shi *et al.* [5] represent semantic documents connecting concepts as a graph and calculates the similarity of documents to measure the distance between the connections in the graph of concepts. They used only two semantic relationships: hypernym and the relationship *is-part-of*, and tested their approach using RDFs semantic document and ontologies. Wang and Taylor [6] propose a *concepts-forest* by using an ontology and the WordNet[1] lexical database. Another interesting work was developed by Malo *et al.*[7] where the authors attempt to relate the topic of a document with other issues associated with the content of Wikipedia and not the actual content of the document unlike as in our approach.

3 Proposal

The stages of the proposed method are illustrated in Figure 1 and explained below.

1. Transform the content of the documents, which are in formats like *HTML*, *XML* and derived formats, *Microsoft Office* document formats, *OpenDocument formats*, *PDF*, *EPUB*, *RTF*, compression formats, and text formats.
2. Remove stopwords. In this step verb-noun relations are extracted. For this, the dependencies of each sentence are obtained using grammatical-probabilistic parser and tagger. Thus, a vector representation based on the weight of a verb-noun ratio is obtained, calculating for this the value of *tf-idf* [8].
3. Find the most significant terms of each document using LDA (Latent Dirichlet Allocation) [9]. Labeled vector representation is the gateway to a network SOM (Self-Organizing Map) [10] that is used to determine the number of topics by document and thus weigh the number of topics throughout the corpus. These data serve as input for the LDA algorithm.

---

[1] http://wordnet.princeton.edu

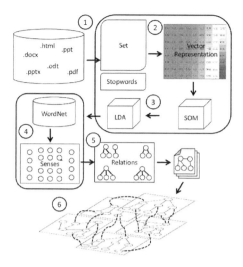

Fig. 1 Graph-based approach for document search

4. Find the meaning of each relevant term in WordNet. This is achieved using synonyms, hyponyms, and hypernyms of WordNet. With all the returned words of WordNet, a graph is constructed in which each word represents a node and each node has information of the document to which it belongs.
5. Determine the relationships of terms in the graph structure by considering relationships of words with similar meaning. That means, how strong or weak is the relationship between each term in the documents. This is done by assigning values based on whether a word is in the document, as synonym, hypernym or hyponym; the values were 1.0, 0.9, 0.75, and 0.6 respectively.
6. Build a semantic meta-graph (graph of graphs) for all documents. First a graph is generated for each document taking into account the relationships (synonyms, hyponyms, and hypernyms) of its most relevant terms. Thus, a meta-graph is generated relating all the relationships for terms in all documents.
7. Compute a local score for search and a global score for document search.

Search in the Graph

The search of documents in the graph involves to compute a local score to find terms in documents and compute a global score to find related topics to documents. This is explained below.

Local Score. With the built meta-graph, it is possible to represent documents and identify the relationship between them. The next step is to access the information. When receiving a query, it is converted to an input vector, called *search vector*, represented as:

$$q = \{q_1, q_2, \ldots q_m\} \equiv \mathbf{q} = [1.0, 1.0, \ldots, 1.0]$$
$$d_1 = \{q_1, q_2, \ldots, q_m\}$$
$$d_2 = \{q_1, q_2, \ldots, q_m\}$$
$$\vdots$$
$$d_n = \{q_1, q_2, \ldots, q_m\}$$
$$q \cdot d_i = \|q\| \|d_i\| \cos \theta$$

where $q_1 \ldots q_m$ are the lexemes of the words entered in the search.

In order to build the graph, three types of terms have been considered: verbs, nouns, and compound nouns. The search vector is compared against the *matching vectors of documents*, where each vector d_j is the j-th vector formed by the coincidence of the lexeme or word found in the document j. It is decided whether or not the q_i term represent in each document in the graph. For example, given the query: *"migrate if it is based on download"*, extracted lexemes are *download, migrate, base and be*, whose vector representation is equivalent to $[1.0, 1.0, 1.0, 1.0]$

$$q = \{\text{migrate, be, base, download}\} \equiv \mathbf{q} = [1.0, 1.0, 1.0, 1.0]$$
$$d_1 = [1.0, 1.0, 0.9, 0.0]$$
$$d_2 = [1.0, 1.0, 1.0, 0.9]$$
$$d_3 = [0.75, 1.0, 0.9, 0.0]$$
$$d_4 = [1.0, 1.0, 0.9, 0.0]$$

This query has returned four documents, which means each of these documents contains at least one lexeme that matches the query. Looking at the vector of document d_2, the first three values correspond to 1.0, this means that the first three lexemes are written in the document two and the last value, 0.9 tells us that the word "download" is a synonym of a word in d_2. In the case of d_1 it has no lexeme related to "download".

The way to calculate how similar is the search vector against each vector of documents is getting the cosine of the angle between vectors[11].

Global Score. It may happen that verbs or nouns be very common in the graph. This could cause that the local score gives more importance to common terms, whose frequency is higher. To avoid this problem, the PageRank (PR) algorithm is used to measure a global score. For this, a graph with verb-noun connections among documents has been used. These relationships are obtained by the LDA (Latent Dirichlet Allocation) algorithm and represent the most significant relationships of each set of topics for a document. This calculation does not consider all the documents, but only those who obtained the highest score in the local score. Considering only the set of documents that have the highest score, the search is executed for what is related the query documents and how many connections have between them. The idea of the global score is to obtain the weight of a document with respect to others, so, documents related to the desired topics will be more connected and have a higher weight. At this step the isolated nodes (which are more specialized terms within document) become more important. The initial PageRank algorithm is shown in Equation 1.

$$PR(D) = (1 - c) + c * \sum_{i=1}^{n} \frac{PR(d_i)}{N(d_i)} \qquad (1)$$

where $PR(D)$ is the value of PageRank for the document D, $PR(d_i)$ means the value of PageRank for the i-th document and a link of d_1 to D, $N(d_i)$ is the total number of outbound links on document (meta-node) , c is a value between 0 and 1 (usually 0.85) and n is the number of documents in D.

The total value assigned to each document d is the result of the sum of local weights determined by the appearance of terms in the graph (local score) and the number of connections to other documents (global score), as shows Equation 2.

$$similarity = \cos(\theta) + PR(D) \qquad (2)$$

From the perspective of the overall calculation, it may be that all or most of the documents were connected. Given this additional consideration can not be taken into account all the documents overall return calculation. According to the results of the experiments was chosen the Equation 3 as the better number of returned documents for a query.

$$retrieved\ documents = n/rcl \qquad (3)$$

where n is the number of lexemes in a query and rcl is the amount of documents retrieved by the local score.

4 Experiments

An initial implementation in Java has been obtained for the approach, which has been tested with a set of PDF documents taken from *www.springerlink.com*; these files have been classified manually by the editors of Springer. It has been considered a scenario of five topics with 50 files per topic belonging to the subject of Computing. This topic has been chosen because it contains highly technical terms, and the use of technical terms is considered the worst case for the application given that WordNet contains only non-specialized terms. The topics are:

1. *Algorithmica*
2. *Knowledge and Information Systems*
3. *Multimedia Tools and Applications*
4. *Software Quality*
5. *Supercomputing*

Examples of the 31 executed queries are:

1. Knowledge and Information Systems
2. Multimedia Tools and Applications
3. supercomputing
4. minimizing the symmetric difference
5. The problem of similarity search on high dimensional data

6. substitution of improvisation with the common rules that can ensure acceptable quality levels
7. reducing the total power consumption in a significant amount
8. vertex does not disconnect the graph
9. . . .

The results of these tests have been measured with the *precision* measures [13]. To perform these tests were used 31 queries (according to the Central Limit Theorem[12]). The proptotype was compared against Google Desktop[2].

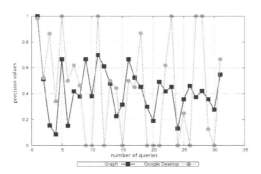

Fig. 2 Graph comparing precision values of the proposal and *Google Desktop*.

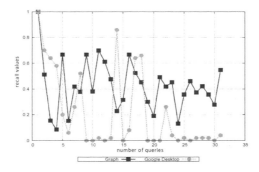

Fig. 3 Graph comparing recall values of the proposal and *Google Desktop*.

Figure 2 shows that the implementation of Google Desktop in 10 of the 31 cases can not retrieve relevant even though the sentences are related to the issues and that our application behaves more stable than the Google application. The graph in Figure 3 shows that in cases where Google Desktop seems to have better precision, the number of documents retrieved is very poor, as seen for example in queries 6, 9, 11, 12, 13, 15, 16, 19, 20, 21, 22, 23, 24, 25, 26 27, 28, 29, 30, and 31.

---

[2] http://desktop.google.com/

Table 1 Comparison of precision and recall values, for *Google Desktop* and our proposal

Query	Proposal Precision	Google Desktop Precision	Proposal Recall	Google Desktop Recall
1	1.0000	0.9803	0.9750	1.0000
2	0.5121	0.5223	0.4200	1.0000
3	0.1547	0.8648	0.2600	0.7400
4	0.0857	0.3411	0.0600	1.0000
5	1.6666	1.0000	0.1000	0.2000
6	0.1520	0.5000	0.3800	0.1200
7	0.4189	0.6190	0.6200	0.4200
8	0.3783	0.4642	0.5800	1.0000
9	0.6666	0.0000	0.4800	0.0000
10	0.3809	0.0000	0.6400	0.0000
11	0.6984	1.0000	0.8800	0.0200
12	0.6111	0.0000	0.4400	0.0000
13	0.4761	0.5000	0.6000	0.0400
14	0.2272	0.4432	0.1000	1.0000
15	0.3157	0.0000	0.2400	0.0000
16	0.6666	0.5000	0.2800	0.1600
17	0.5238	0.4507	0.6900	1.0000
18	0.4523	0.8684	0.3800	0.7600
19	0.3000	0.0000	0.1800	0.0800
20	0.1904	0.0000	0.1600	0.0000
21	0.4920	0.0000	0.6200	0.0000
22	0.4189	0.6190	0.6200	0.4200
23	0.4523	1.0000	0.3800	0.0400
24	0.1309	0.0000	0.2200	0.0600
25	0.3571	0.2500	0.6000	0.0800
26	0.4603	0.0000	0.5800	0.0260
27	0.3734	1.0000	0.6200	0.0200
28	0.4216	1.0000	0.7000	0.0200
29	0.3571	0.1250	0.1000	0.1600
30	0.2777	0.0000	0.2000	0.0000
31	0.5476	0.6666	0.9200	0.0600

5 Conclusions

This paper presents an alternative way to represent and retrieve documents that not only allows search by keywords but also it finds terms semantically similar and related topics. From the initial evaluation about performance of the proposed approach, it is concluded that Google Desktop retrieves fewer relevant documents for all queries. In terms of general performance, the implementation of the approach retrieves more relevant documents compared to Google Desktop. An important feature of the implementation is that it is able to link all the input documents, so the user can see which documents are more general. However, from a user perspective, it is not feasible to review all files. Although the time response for queries is small, the time for semantic processing of documents is the major drawback. As a future work, the time for semantic processing will be reduced and to set a threshold for the number of documents returned by a query. It has also been observed that despite the large number of documents returned, the first positions are always the most relevant, so to defining a threshold would not affect significantly the success of the method.

References

1. Golemati, M., Katifori, A., Giannopoulou, E.G., Daradimos, I., Vassilakis, C.: Evaluating the Significance of the Windows Explorer Visualization in Personal Information Management Browsing tasks. In: 11th International Conference on Information Visualization, IV 2007, pp. 93–100 (2007), doi:10.1109/IV.2007.46, ISSN 1550-6037
2. Ravasio, P., Schär, S.G., Krueger, H.: In pursuit of desktop evolution: User problems and practices with modern desktop systems. ACM Trans. Comput.-Hum. Interact. 11(2), 156–180 (2004) ISSN 1073-0516
3. Yao, Y., Lin, L., Dong, J.: Research on Ontology-Based Multi-source Engineering Information Retrieval in Integrated Environment of Enterprise. In: Int. Conference on Interoperability for Enterprise Soft. and Applications, China, pp. 277–282 (2009)
4. Guzmán, I.O.: Enfoque basado en datos para el aprendizaje de ontologías, Master Thesis. Cinvestav - Tamaulipas. Cd. Victoria, Tamaulipas, México (2010)
5. Shi, B., Fang, L., Yan, J., Wang, P., Dong, C.: Classification of Semantic Documents Based on WordNet. In: International Conference on E-Learning, E-Business, Enterprise Information Systems, and E-Government, pp. 173–176. IEEE Computer Society (2009)
6. Wang, J.Z., Taylor, W.: Concept Forest: A New Ontology-assisted Text Document Similarity Measurement Method. In: Proceedings of the IEEE/WIC/ACM International Conference on Web Intelligence, pp. 395–401. IEEE Computer Society (2007)
7. Malo, P., Siitari, P., Ahlgren, O., Wallenius, J., Korhonen, P.: Semantic Content Filtering with Wikipedia and Ontologies. In: IEEE International Conference on Data Mining Workshops, pp. 518–526 (2010)
8. Zhou, G.-S., Li, H.-L., Wang, Z.-N., Liu, G.-H.: An efficient intra prediction method for H.264 based on probability and statistics. In: International Conference on Computational Problem-Solving, pp. 413–415 (2010)
9. Blei, D.M., Ng, A.Y., Jordan, M.I.: Latent dirichlet allocation. J. Mach. Learn. Res. 3, 993–1022 (2003), JMLR.org
10. Kohonen, T.: Self-Organized Formation of Topologically Correct Feature Maps. Biological Cybernetics 43(1), 241–254 (1982)
11. Manning, C.D., Raghavan, P., Schütze, H.: An Introduction to Information Retrieval. Cambridge University Press (2009)
12. Fischer, H., Fischer, H.: Conclusion: The Central Limit Theorem as a Link Between Classical and Modern Probability Theory. In: A History of the Central Limit Theorem, pp. 353–362. Springer, New York (2011)
13. Hersh, W.R.: Information retrieval: a health and biomedical perspective. Springer (2009)

Identifying Concepts on Specific Domain by a Unsupervised Graph-Based Approach

Franco Rojas-Lopez, Ivan Lopez-Arevalo, and Victor Sosa-Sosa

Abstract. This paper presents an unsupervised approach to Word Sense Disambiguation on a specific domain to automatically to assign the right sense to a given ambiguous word. The approach proposed relies on integration of two source information: context and semantic similarity information. The experiments were carried on English test data of SemEval 2010 and evaluated with a variety of measures that analyze the connectivity of graph structure. The obtained result were evaluated using precision and recall measures and compared with the results of SemEval 2010 the approach is currently under test with another semantic similarity measures, preliminary results look promising.

1 Introduction

In Natural Language Processing (NLP) and recently in Computational Linguistics literature, the problem of assign concepts to words in texts has been called Word Sense Disambiguation (WSD); which is defined as a task that consists on selecting the correct sense for a given ambiguous word in a given context. A word is ambiguous when its meaning varies depending on the context in which it occurs. There are several approaches that have been proposed for WSD. In general, in the literature there are two main approaches: supervised and unsupervised. Supervised approaches rely on the availability of sense labeled data from which the relevant sense distinctions are learned, while unsupervised approaches typically refer to disambiguating word senses without the use of sense-tagged corpora. Most of the unsupervised approaches proposed in the WSD literature are knowledge based, i.e. they

Franco Rojas-Lopez · Ivan Lopez-Arevalo · Victor Sosa-Sosa
Information Technology Laboratory, Cinvestav - Tamaulipas
Scientific and Technological Park.Victoria, Mexico
e-mail: {frojas,ilopez,vjsosa}@tamps.cinvestav.mx

S. Omatu et al. (Eds.): Distributed Computing and Artificial Intelligence, AISC 151, pp. 477–484.
springerlink.com   © Springer-Verlag Berlin Heidelberg 2012

exploit only the information provided by a Machine Readable Dictionary (MRD). Some unsupervised WSD system also use unlabeled data together with the dictionary information to perform an unsupervised approach to disambiguate words. An effective approach to this task would be useful for a number of NLP applications: for example Information Retrieval, Content Analysis, Information Extraction, etc. Over the last years the interest on WSD has been motivated by WSD competition such as SemEval[1] where different system may evaluate their performance. The purpose of SemEval is to perform a comparative evaluation of WSD systems in several kinds of tasks. Particularly the obtained results in the task #17 (All-words WSD on a Specific Domain) are reported in this paper. It is organized as follows, Section 2 presents relevant works on WSD. Section 3 describes the graph-based approach. Section 4 give the carried out experiments, and finally, the conclusions and further work are given in Section 5.

2 Background

The last advances in WSD report that graph-based methods have been applied in the network analysis area, such as centrality models, and recently applied to linguistic knowledge bases, including unsupervised WSD. These methods explore the structure and link of the graph underlying a particular lexical knowledge base. Some important work in this area are presented by Navigli and Mirella [1], Rada and Sinha [2], Reddy *et. al.* [3], and Navigli [4]. In these approaches a graph representation for senses (vertices) and relation (edges) is first build from a lexical knowledge base. According to its performance, similar works reported in the literature are based on clustering techniques. For example, Aguirre and López [5] proposed a method to group senses of words of fine granularity within one of coarse granularity to reduce the polysemy[2]. Pedersen *et. al.* [6] proposed and unsupervised approach that solves name ambiguity by clustering the instances of a given name into groups, each of which is associated with a distinct underlying entity. In this approach, given a name, the actual contexts are grouped to represent the meanings of a word. In this paper we describe and evaluate a graph-based approach to assign the right sense to an ambiguous word by obtaining and merging context information and semantic similarity information; the main idea is mutually reinforcing between both techniques. The preliminary experiments carried out show promising response on WSD.

3 Approach

The graph-based representation relies on the combination of two techniques to select the right sense for a given ambiguous word: the context and semantic similarity

---

[1] http://semeval2.fbk.eu/semeval2.php?location=
[2] The association of one word with two or more distinct meanings.

using a specific domain corpus, both techniques use information from WordNet (a lexical database). Figure 1 illustrates the proposed methodology, the complete description of the involved procedures is given in the following sections.

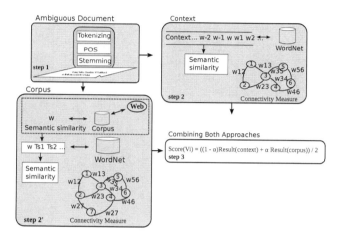

Fig. 1 Proposal

3.1 Pre-processing

Given an ambiguous word and its context,[3] a graph is constructed recovering senses of second order vectors[4] from WordNet as well as semantically related words extracted from external knowledge sources. The experiments were performed using the *all-words* dataset on specific domain of SemEval 2010. The input file consist of several instances of the ambiguous word, each instance is a context in which a particular ambiguous word appears. So, the content of this file firstly is tagged[5] (step 1 of the methodology), for this task, the Stanford parser is used to Part-of-Speech tag the test data. The steps 2 and 2' are performed in parallel. In step 2, the context window size is defined, different window sizes were tested in the experiments to determine how many words before and after a ambiguous word w must be included in the context, so, the better resulting window size was $2\beta + 1$, with $\beta = 1$.

On the another hand in step 2' an untagged corpus from the environment domain provided by SemEval 2010 was used to extract keywords in the domain based on their frequency of occurrence. For example *species, biodiversity, conservation*, etc.

---

[3] The parts that immediately precede and follow a word or passage and clarify its meaning.

[4] Given an ambiguous word the senses are retrieved from WordNet, each recovered sense again is tagged with the Part-Of-the-Speech to recover the additional senses for each word within the first sense.

[5] The assignment of parts of speech to each word in the document.

appear frequently in the environment domain. The first 20 words, in descending order according to their frequency, were selected and combined in pairs to create a web query of length two according to Iosif and Potamianos [11]. For example, for "*specie and biodiversity*" the web querys were sent to several search engines (Google, Yahoo, Bing, HotBot, and MetaCrawler) according to the study of Aguilar [12]. Also the corpus is Part-Of-Speech tagged and stemming by using the Stanford parser. After the pre-processing phase, the semantically similarity terms for each ambiguous word are retrieved using Mutual Information (MI) [10, 9] (see Equation 1). In this case the window size was defined as $2\beta + 1$, $\beta = 5$, according to Islam and Inkpen [9]. MI compares the probability of observing X and Y together ($f(X,Y)$) with the probabilities of observing X and Y independently ($f(X), f(Y)$).

$$IM(X,Y) = log_2 \frac{f(X,Y)}{f(X)f(Y)} \tag{1}$$

3.2 Graph Construction

Some semantic similarity measures have been implemented to quantify the degree of similarity between two words using information drawn from WordNet hierarchy (see Ted Pedersen et. al. [8]). Particularly the Lin and Vector measures were taken into account because they have a good performance on WordNet hierarchy and results in the conducted research. Once contexts are recovered, the senses for each word in the context are retrieved from WordNet and weighted by a semantic similarity score using the WordNet::Similarity[6] score between the senses of word w and the senses for each word in the context. These measures return a real value indicating the degree of semantic similarity between a pair of concepts.

Formally let $C_w = \{c_1, c_2, \cdots, c_n\}$ the set of words in the context related to an ambiguous word w. Let *senses(w)* be the set of senses of w and let *senses(c_n)* be the set of senses for a word in the context, a list ranked is returned in order descending of semantic similarity between w and c_n, the items that maximize this score are filtered according to the statistical mean. These items constitute the named first order vectors. For each ambiguous word, two graph are built (see Figure 1). In this representation, $G = (V, E, W)$ where V are the vertices (concepts), E are the edges (semantic relations) and W (a strong link between two concepts or vertices). So, each recovered sense again is tagged with the Part-Of-the-Speech to recover the additional senses for each word within the first sense. These semantic relations for senses constitute the connections in the graph. Once the semantic graph is built, its structure and links are analyzed applying the algorithms described in the section 3.3.

---

[6] This is a Perl module that implements a variety of semantic similarity and relatedness measures based on information found in the lexical database WordNet.

3.3   Graph-Based Measures

Vertex-based centrality is defined in order to measure the importance of a vertex in the graph; a vertex with high centrality score is usually considered more highly influential than other vertex in the graph. In the experiments, four algorithms have been implemented to determine which node is the most important examining the graph structure: in-degree, Key Problem Player, Jaccard, and Personalized PageRank, which are described bellow.

Indegree [1], the simplest and most popular measure is degree centrality. In a undirected graph the degree of the vertex is the number of its attached links; it is a simple but effective measure of nodal importance. A node is important in a graph as many links converge to it. In the implementation, V is the set of vertices on the graph and v a vertex, see Equation 2.

$$score(v) = \frac{indegree(v)}{|V|-1} \qquad (2)$$

Key Problem Player [1], consists in find a set of nodes that is maximally connected to all other nodes. Here, a vertex (denoted by v and u, V is the set of vertices) is considered important if it relatively close to all other vertices, see Equation 3.

$$kpp(v) = \frac{\sum\limits_{u \varepsilon V: u \neq v} \frac{1}{d(u,v)}}{|V|-1} \qquad (3)$$

Jaccard coefficient computes the probability that two vertex i and j will have a common neighbor k. According to Granovetter [7], the link strength between two vertex depends on the overlap of their neighborhoods. If the overlap of neighborhoods between the vertex i and vertex j is large, it is considered that i and j have a strong tie. Otherwise, they are considered to have a weak link, see Equation 4.

$$Jaccard(i,j) = \frac{|N_i \cap N_j|}{|N_i \cup N_j|} \qquad (4)$$

where, N_i and N_j indicate the neighborhoods of the vertex i and j respectively.

PageRank is a link analysis algorithm traditionally applied on directed graphs, this algorithm can be also applied to undirected graphs, in which case the outdegree of a vertex is equal to the in-degree of the vertex. For this, an adaptation to the PageRank algorithm has been proposed, Personalized PageRank (PPRank) algorithm [14]. After running the algorithm, a score is associated with each vertex as shows the Equation 5.

$$PR(v_i) = (1 - \alpha) + \alpha * \sum_{v_j \varepsilon In(v_i)} \frac{w_{ji}}{\sum_{v_k \varepsilon Out(v_j)} w_n k} PR(v_j) \qquad (5)$$

According to the literature, the α is a factor which is usually set as 0.85 that is the value used in the evaluation of the implemented WSD prototype.

Finally the context and semantic similarity are combined (see step 3 in Figure 1) using the Equation 6 to get a ranked list in order descending according to their relevance so, the node with the highest value is selected as the right sense for the ambiguous word in question. Several experiments were carried out with different values for δ so, the better result was $\delta = 0.6$, thus we give more importance to semantic similarity because surprisingly the best results were obtained using the background documents.

$$Score(v_i) = \frac{(1 - \delta)Result(context) + \delta Result(corpus)}{2} \qquad (6)$$

4 Experiments and Results

The purpose of this evaluation is show the relevancy of the hypothesis on the combination of contextual semantic relationships and semantic similarity of a domain contributes to WSD in an unsupervised manner, usually only the context or expanded context has been used to WSD. Therefore in this approach the context and semantic similarity information were integrated and used afterwards to assign the right sense to an ambiguous word. So, To evaluate the performance of the WSD approach and to be able to compare it with others algorithms, the experiments were carried on English test data of SemEval 2010 [15]. Precision (percentage of words that are tagged correctly, out of the words addressed by the system) and Recall (percentage of words that are tagged correctly, out of all words in the test set) were used as evaluation measure. The dataset is a file with 1398 ambiguous words, 366 verbs, and 1032 nouns. The WSD approach was performed by using WordNet 3.0 as lexical database. Table 1 and 2 shows the results using the context and semantic similarity alone. The table 3 shows the gain of the combination of the both techniques (context and semantic similarity) table 4 the results obtained in the WSD competition, the results shown that the proposed approach is low, equal to the Yoan's system and far from Anum's system when is evaluated using the PPRank algorithm this is because unlike the other ranking algorithms, PPRank takes into account edge weights when computing the score associated with the vertex. The other algorithms only make use of the content or links information, that could explain the worse performance. The results obtained by our approach were worse with those reported in the literature but the preliminary results of these algorithm are promising if we retrieve the semantically most similar words for an ambiguous word, this could help improve the process of disambiguation.

Table 1 Performance using context alone

Algorithm	Precision (%)	Recall (%)	Nouns (%)	Verbs (%)
KPP	15.1	14.73	15.21	13.38
Indegree	15.61	15.23	15.98	13.11
Jaccard	17.37	16.95	18.02	13.93
PPRank	17.15	16.73	18.02	13.11

Table 2 Performance using semantic similarity information alone

Algorithm	Precision (%)	Recall (%)	Nouns (%)	Verbs (%)
KPP	15.61	15.23	18.99	4.64
Indegree	15.68	15.3	18.89	5.19
Jaccard	16.58	16.16	19.57	6.55
PPRank	17.15	16.73	18.89	7.92

Table 3 Gain in the combination the both techniques (context and semantic similarity)

Algorithm	Precision (%)	Recall (%)	Nouns (%)	Verbs (%)
KPP	33.94	33.11	33.52	36.33
Indegree	33.87	33.04	31.68	36.88
Jaccard	34.38	33.54	32.94	35.24
PPRank	35.11	34.26	31.78	36.88

Table 4 Overall results for the domain WSD of SemEval 2010

Algorithm	Precision (%)	Recall (%)	Nouns (%)	Verbs (%)
Anup Kulkarni	51.2	49.5	51.6	43.4
Andrew Tran	50.6	49.3	51.6	42.6
Andrew Tran	50.4	49.1	51.5	42.5
Aitor Soroa	48.1	48.1	48.7	46.2
⋮	⋮	⋮	⋮	⋮
Radu Ion	35.1	35.0	34.4	36.8
Yoan Gutierrez	31.2	30.3	30.4	30.1
Random baseline	23.2	23.2	25.3	17.2

5   Conclusions and Future Work

This paper describes an approach aimed to tackle the WSD problem on specific domain. The adaptation and integration of the tested techniques have been implemented in a first prototype. With this prototype, a semantic graph is obtained by using *second order vectors* of senses recovered from WordNet; which corresponds to a specific ambiguous word. Thus, two semantic graph are obtained and evaluated given the context and words related to an ambiguous word. The approach have been only tested on a standard benchmark dataset released by SemEval 2010 in all-words domain specific WSD task. Tests on other datasets are desirable in the future, for example medicine and tourism. As further work, another measure for semantic similarity will be integrated, for example distributional similarity measure [13], which allows to get semantic terms with more accuracy.

References

1. Navigli, R., Lapata, M.: Graph connectivity measures for unsupervised word sense disambiguation. In: Veloso, M.M. (ed.) IJCAI, pp. 1683–1688 (2007)
2. Sinha, R., Mihalcea, R.: Unsupervised graph based word sense disambiguation using measures of word semantic similarity. In: ICSC, pp. 363–369. IEEE Computer Society (2007)
3. Kilgarriff, A., Siva, R., Jan, P.: A Corpus Factory for many languages. In: Proc. LREC, Malta (2010)
4. Navigli, R.: Using cycles and quasicycles to disambiguate dictionary glosses. In: Proc. of EACL 2009, pp. 594–602 (2009)
5. Agirre, E., de Lacalle, O.L.: Clustering wordnet word senses. In: Recent Advances in Natural Language Processing III (2004)
6. Pedersen, T., Purandare, A., Kulkarni, A.: Name discrimination by clustering similar contexts. In: Proceedings of the Sixth International Conference on Intelligent Text Processing and Computational Linguistics, Mexico City, pp. 220–231 (February 2005)
7. Granovetter, M.: The strength of weak ties. American Journal of Sociology 78(6), 1360–1380 (1973)
8. Pedersen, T., Patwardhan, S., Michelizzo, J.: Wordnet:similarity - measuring the relatedness of concepts. In: Proceedings of the Nineteenth National Conference on Artificial Intelligence (AAAI 2004), pp. 1024–1025 (2004)
9. Islam, A., Inkpen, D.: Second order co-occurrence PMI for determining the semantic similarity of words. In: Proceedings of the International Conference on Language Resources and Evaluation, Genoa, Italy, pp. 1033–1038 (2006)
10. Church, K.W., Hanks, P.: Word Association Norms, Mutual Information, and Lexicography. Computational Linguistics 16(1), 22–29 (1990)
11. Iosif, E., Potamianos, A.: Unsupervised Semantic Similarity Computation Between Terms Using Web Documents. IEEE Transactions on Knowledge and Data Engineering 22(11), 1637–1647 (2009)
12. Aguilar, D., Ivan Lopez, I., Sosa, V.: Web search based on domain ontologies. Polish Journal of Environmental Studies 17(4C), 197–200 (2008) ISSN 1230-1485
13. Rychlý, P., Kilgarriff, A.: An Efficient Algorithm for Building a Distributional Thesaurus (and other Sketch Engine Developments). In: Proceedings of the 45th Annual Meeting of the Association for Computational Linguistics Companion Volume Proceedings of the Demo and Poster Sessions, pp. 41–44. Association for Computational Linguistics, Prague (2007)
14. Agirre, E., Soroa, A.: Personalizing pagerank for word sense disambiguation. In: Proc. of EACL, pp. 33–41 (2009)
15. Agirre, E., Lopez de Lacalle, O., Fellbaum, C., Hsieh, S., Tesconi, M., Monachini, M., Vossen, P., Segers, R.: Semeval-2010 task 17: All-words word sense disambiguation on a specific domain. In: Proceedings of the 5th International Workshop on Semantic Evaluations (SemEval 2010). Association for Computational Linguistics (2010)

The Problem of Learning Non-taxonomic Relationships of Ontologies from Text

Ivo Serra, Rosario Girardi, and Paulo Novais

Abstract. Manual construction of ontologies by domain experts and knowledge engineers is a costly task. Thus, automatic and/or semi-automatic approaches to their development are needed. Ontology Learning aims at identifying its constituent elements, such as non-taxonomic relationships, from textual information sources. This article presents a discussion of the problem of Learning Non-Taxonomic Relationships of Ontologies and defines its generic process. Three techniques representing the state of the art of Learning Non-Taxonomic Relationships of Ontologies are described and the solutions they provide are discussed along with their advantages and limitations.

Keywords: Ontology, Ontology learning, Non-taxonomic relationships, Natural Language Processing.

1 Introduction

Manual construction of ontologies by domain experts and knowledge engineers is a costly task, thus automatic and/or semi-automatic approaches to their development are needed. Ontology Learning (OL) [2] [3] aims at identifying the constituent elements of an ontology, such as non-taxonomic relationships from textual information sources. Some techniques have been proposed for Learning Non-Taxonomic Relationships of Ontologies (LNTRO). All of them use Natural Language Processing (NLP) techniques [1] [4] to annotate the corpus with the information needed for the subsequent processing. Information Extraction (IE) techniques [7] are used to extract from the annotated corpus possible relationships and Machine Learning (ML) [9] or Statistic Techniques (ST) to make a refinement of

Ivo Serra · Rosario Girardi
Federal University of Maranhão, Computer Science Departament, São Luís, Brazil
e-mail: `ivocserra@gmail.com, rosariogirardi@gmail.com`

Paulo Novais
University of Minho, Computer Science Departament, Braga, Portugal
e-mail: `pjon@di.uminho.pt`

S. Omatu et al. (Eds.): Distributed Computing and Artificial Intelligence, AISC 151, pp. 485–492.
springerlink.com   © Springer-Verlag Berlin Heidelberg 2012

the relationships outputted from the previous phases. This article discusses the problem of LNTRO, identifying its phases and what kind of techniques can be used to perform the activities of each phase. Three techniques of the state of the art on LNTRO are also described and the advantages and limitations of the solutions they adopt for each phase of LNTRO are discussed.

The paper is organized as follows. Section 2 introduces the lexical realizations of non-taxonomic relationships. Section 3 defines the problem of LNTRO, its phases and what techniques can be used to approach each one. Section 4 describes three representative techniques of the state of the art on LNTRO and which solutions they adopt for each of its phases described in section 3. Finally, section 5 presents the conclusions discussing general and open research topics on LNTRO.

2   Non-taxonomic Relationships

Non-taxonomic relationships can be classified as domain independent or domain dependent. Domain independent relationships are of two subtypes ownership or aggregation. Aggregation is the "whole-part" relationship. For example, in the sentence "The car's wheel is out of order." there is a non-taxonomic relationship of aggregation between "car" and "wheel". The linguistic realization of the relationship of aggregation occurs in two forms: the possessive form of English (apostrophe) and the verb "to have" in any conjugation. However, the converse is not true, that is, the occurrence of such linguistic accomplishments does not imply a relationship of aggregation as will be explained in the next case. Ownership relationships are held as in the example: "Father and mother will wait for the court's decision." in which there is a relationship of ownership between "court" and "decision". The linguistic realization of this kind of relationship occurs in two forms: the possessive form of English (apostrophe) and the verb "to have" in any conjugation. However, the converse is not true, that is, the occurrence of such linguistic accomplishments does not imply a relationship of possession. Domain dependent relationships are expressed by particular terms of an area of interest. For example, the sentence "The court will judge the custody in three days." holds the relationship "judge" between "court" and "custody" which is characteristic of the legal field.

3   The Problem of LNTRO

LNTRO is an approach to automate or semi-automate the extraction of these relationships from textual information sources. An example of this kind of relationship is "represents" between "lawyer" and "client" in the legal domain.

LNTRO can generally be accomplished through the tasks of "Corpus construction", "Extraction of candidate relationships" (which in turn consists of the subtasks of "Corpus annotation" and "Extraction of relationships") and "Refinement". The task of "Corpus construction" consists of selecting documents on the domain we expect to extract relationships from. This is usually a costly task and the outcome of any LNTRO technique depends on its quality. The "Extraction of candidate relationships" task aims at identifying a set of possible relationships. It has the corpus built in the previous phase as input and candidate relationships as its product. It is composed of two sub-activities: "Corpus annotation" and "Extraction of relationships". The "Corpus annotation" task consists of applying tags to the text with NLP techniques that are necessary for the next steps in LNTRO. The "Extraction of relationships" task consists of searching in the annotated corpus for evidences suggesting the existence of relationships. For example, Maedech [8] considers the existence of two instances of ontology concepts in a sentence as evidence that they are non-taxonomically related. For Villaverde et al. [13] a relationship is identified by the presence of two concepts of an ontology in the same sentence with a verb between them. This sub-task can also receive the concepts of the ontology as input. In this case the search space for relationships is reduced and there is a potential for achieving greater precision in the extraction of relationships.

Relationships from the previous task should not be recommended to the specialist, since there is usually a substantial amount of them that do not correspond to good suggestions. For this reason Machine Learning (ML) or Statistic Techniques (ST) can be used in the "Refinement" phase. The ontology taxonomy can also be given as input. In this case the LNTRO technique is able to suggest to the specialist the best possible level in the hierarchy where to add the relationship. This functionality is explained in section 4.2.

4   Techniques for LNTRO

In the following sections, three state of the art techniques for LNTRO are presented. The solutions adopted to approach the generic phases of LNTRO are highlighted and their positive aspects and limitations are discussed.

4.1   LNTRO Based on the Extraction of Association Rules

This technique described in [13] has two phases "Identification of occurrences of relationships" and "Mining associations". The "Identification of occurrences of relationships" receives a corpus and a set of concepts of an ontology and outputs a set of tuples in the form $<c_1, v, c_2>$, where c_1 and c_2 are the ontology concepts and v is a verb. Initially using Wordnet [6] each ontology concept is extended with its synonyms to increase the recall of the search. Then the POS-tagging is performed in order to identify the verbs. For sentences that satisfy the following two conditions a tuple (c_1, v, c_2) is generated: (a) sentences that have exactly two concepts and a verb between them and (b) the two concepts are at a maximum distance of D terms. "D" is a parameter whose value is defined experimentally by the specialist and corresponds to the maximum number of terms that must exist between two concepts for them to be considered related. For example if D = 3 then for the sentence " The court judged the custody in three days." a tuple <court, judge, custody> is generated since there are two terms between the concepts. However, for the sentence " The court of North Dakota will judge the custody in three days." no tuple is generated.

Once a set of candidate relationships (set of tuples outputted from the previous phase) is obtained, "Mining associations" can be performed that aim at refining the results of the previous phase before they are suggested to the specialist. For this purpose an algorithm to extract association rules [12] is used. The product of this phase are non-taxonomic relationships represented by association rules in the form $\langle c1 \wedge c2 \rangle \rightarrow \langle v \rangle$, which have values of support and confidence greater than the minimum defined experimentally by the specialist.

For example, in the sentence "Our data suggests that lipoxygenase metabolites activate ROI formation which then induce IL-2 expression via NF-kappa B activation", Lipoxygenase (Li) and Reactive Oxygen Intermediates (ROI) are concepts and Activate (Ac) is a verb. In the first phase the tuple $\langle Li, ROI, Ac \rangle$ is generated representing the fact that the two extraction conditions described previously were satisfied. In the second phase if the rule $\langle Li, ROI \rangle \rightarrow \langle Ac \rangle$ has values of support and confidence greater than or equal to the minimum support and confidence, it is recommended to the specialist. A positive aspect of this proposal is that it labels with verbs the relationships between two concepts found in each sentence. In addition, the search space for relations is restricted since ontology concepts are given as input to the technique, thus potentially leading to better results. The technique only extracts concepts from the text; concept instances are ignored. Furthermore, it doesn't use stemming, a NLP technique that could lead to better recall values. Moreover, one restriction is the fact that no treatment is given to the possessive form "'s" that is one of the linguistic realizations of non-taxonomic relationships which can be present in the corpus with reasonable frequency. In addition, the authors refer to the verbs as single words when in fact, in most of the cases, they appear in the form of verbal phrases. In Genia [10], the corpus used to illustrate and evaluate the technique, coincidentally most of the verbal phrases are composed of a single term, which is a uncommon fact. Therefore, to be applied to corpora without this characteristic, the technique should be updated either to work with verb phrases or with the information of which verb, among those of the verb phrase, should be used as the label of the relationship. For the evaluation of the technique, the recall and precision measures are used, fact that we consider too restrictive for a noisy area like AO. Table 1 shows which solutions have been adopted for each one of the generic phases for LNTRO as defined in section 3.

Table 1 Solutions for LNTRO based on the Extraction of Association Rules.

Phase	Adopted solution
Corpus construction	A corpus already available in the medical field (Genia) was used in its experiment.
Corpus annotation	POS-tagging
Extraction of relationships	Uses the algorithm already described to extract candidate relationships in the form of tuples ($\langle c_1, v, c_2, \rangle$)
Refinement	Uses a technique known as the "Extraction of Association Rules" to suggested non-taxonomic relationships in the form of rules ($\langle c_1 \wedge c_2 \rangle \rightarrow \langle v \rangle$)

4.2 LNTRO Based on the Extraction of Generalized Association Rules

Maedech and Staab [8] propose a process similar to that of Villaverde et al. [13], with the difference that it uses an algorithm of generalized association rules [8] to suggest the possible most appropriate hierarchical level for the relationship and works with texts in German. The technique has two phases "Text processing" and "Mining associations". In the first phase, the objective is to extract pairs of concepts from the text that correspond to candidate relationships. For this purpose, the title and the sentence heuristics are used. The first one says that a pair of related concepts should be created for every concept in the text with every concept in the title. This heuristic is based on the intuition that the concepts that appear in the text body are related to the concepts that appear in the title. The second one sets up a tuple for each pair of concepts that are present in the same sentence. In the second phase relationships in the form of pairs of concepts from the previous phase are submitted to an algorithm for Mining Generalized Association Rules [8]. The goal is to extract non-taxonomic relationships in the form of association rules and suggest the best possible level in the hierarchy where to add the relationships. After applying the Extraction of Generalized Association Rules, the rule *area → hotel* is discarded because *area -> accommodation* is an ancestral rule (its concepts are in the same or higher levels in the ontology taxonomy) and has values for support or confidence greater or equal than the descendent rule. The same happens to the rules *room -> television* and *room → furnishing* (Table 2). The solutions adopted for each one of the generic phases for LNTRO are shown in Table 3.

A positive aspect of this proposal is the use of the algorithm for the Extraction of Generalized Association Rules that suggests the best possible level in the ontology taxonomy where the relationship should be added. On the other hand, a limitation is the fact that the technique does not label the relationships but, only indicates what classes are related.

Table 2 Extracted Relationships [8].

Discovered relations	Confidence	Support
(area → accommodation)	0,38	0,04
~~(area → hotel)~~	~~0,1~~	~~0,03~~
(room → furnishing)	0,39	0,03
~~(room → television)~~	~~0,29~~	~~0,02~~
(accommodation → address)	0,34	0,05
(restaurant → accommodation)	0,33	0,02

Table 3 Solutions for LNTRO based on the Extraction of Generalized Association Rules.

Phase	Adopted solution
Corpus construction	A corpus already available in the touristic domain (Lonely Planet) was used in its experiment.
Corpus annotation	Uses chunking, stemming and NER.
Extraction of relationships	Uses sentence and title heuristics to extract candidate relationships.
Refinement	Uses a technique known as mining generalized association rules [8] to recommend relationships as rules in the form $c_1 \rightarrow c_2$.

4.3 LNTRO Based on Queries on Web Search Engines

Sanchez and Moreno [11] propose an automatic technique for LNTRO that is able to learn verbs from a domain, extract related concepts and label them using the Web instead of a traditional corpus as a source for the construction of an ontology. Despite being diverse and unstructured, according to the authors, the redundancy of information in an environment as vast as the Web is a measure of its relevance and veracity. The first phase is the extraction and selection of verbs that express relationships characteristic of the domain. Based on morphological and syntactic analysis, verbs that have a relationship with the domain keyword are extracted. Then, the degree of relationship between each verb and the domain is measured. To do so, statistical measures are made about the term distribution on the web. The obtained values are used to rank the list of candidate verbs. This lets one choose the labels of non-taxonomic relationship that are closely related to the domain. The domain related verbs are used to discover non-taxonomic related concepts. To do so it queries the web with the patterns "domain-keyword verb" or "verb domain-keyword " that returns a corpus related to the specified query. The goal is to search the content of documents to find concepts that proceed ("High sodium diets are associated with hypertension") or succeed ("Hypertension is caused by hormonal problems") the constructed patterns. These concepts are candidate to be non-taxonomically related to the original keyword. Table 4 shows which solutions have been adopted for each one of the generic phases for LNTRO as defined in section 3.

Table 4 Solutions for LNTRO based on Queries on Web Search Engines.

Phase	Adopted solution
Corpus construction	Based on documents returned by a Web search engine.
Corpus annotation	Chunking.
Extraction of relationships	Extracts verb phrases and noun phrases as labels and concepts of relationships respectively.
Refinement	Statistical processing based on the result of queries in a web search engine.

A positive aspect in this proposal is that specialists do not have to deal with the construction or selection of corpora, a generally laborious task. They are automatically created with the help of a web search engine. On the other hand, one limitation is that learning relationships is dependent of learning concepts and vice versa which makes the process less flexible.

5 Concluding Remarks

This work approached the LNTRO problem, its phases and the knowledge areas which provide solutions to them. Three techniques of the state of the art on LNTRO were presented and the solutions each one adopted for the phases of LNTRO were highlighted. Advantages and limitations of each of the techniques were also discussed. To end our considerations on LNTRO, we now discuss some relevant issues and point out a line of research.

The corpus used for LNTRO may contain classes, instances of classes or both. For the first case, a search is performed for classes in the text. This search can include the synonyms of the concepts and/or their stems, thus increasing the recall of the extracted concepts from the corpus. If the corpus has only instances of classes it is necessary to use Named Entity Recognition (NER). If the corpus has both classes and instances, all these solutions can be used together. Non-taxonomic relationships are generally represented by a pair of concepts and optionally a label. The first representation has the disadvantage of being semantically poorer because we know which classes are related but do not have a name giving a meaning to the relationship. The second is the representation that has the highest semantics since the relationships are constituted by a pair of concepts and a label. The label is generally a verb phrase found between the two concepts in a sentence.

LNTRO techniques that use an ontology taxonomy as input can suggest the best level in the hierarchy where to insert the relationship. Those that receive only the ontology concepts have the search space for relationships reduced and have the potential of obtaining better results when compared to those that don't receive this input. Techniques that don't receive any of these sets as input often consider noun phrases as concepts. Techniques on LNTRO are usually evaluated comparing their results against reference ontologies [5]. However, comparing them when executed under similar conditions is a work that still has to be done.

References

1. Allen, J.: Natural Language Understanding. The Benjamin/Cummings Publishing Company, Inc., Redwood City (1995)
2. Buitelaar, P., Cimiano, P., Magnini, B.: Ontology Learning from Text: An Overview. DFKI, Language Technology Lab. AIFB, University of Karlsruhe. ITC-irst (2003)
3. Buitelaar, P., Cimiano, P., Magnini, P.: Ontology Learning from Text: Methods, Evaluation and Applications. IOS Press, Amsterdam (2006)
4. Dale, R., Moisl, H., Somers, H.L.: Handbook of natural language processing. CRC (2000)

5. Dellschaft, K., Staab, S.: On How to Perform a Gold Standard Based Evaluation of Ontology Learning. In: Cruz, I., Decker, S., Allemang, D., Preist, C., Schwabe, D., Mika, P., Uschold, M., Aroyo, L.M. (eds.) ISWC 2006. LNCS, vol. 4273, pp. 228–241. Springer, Heidelberg (2006)
6. Fellbaum, C.: WordNet: An Electronic Lexical Database. MIT Press, Cambridge (1998)
7. Freitag, D.: Information extraction from HTML: Application of a general machine learning approach. In: Proceedings of the 15th Conference on Artificial Intelligence, pp. 517–523 (1998)
8. Maedche, A., Staab, S.: Mining Ontologies from Text. In: Dieng, R., Corby, O. (eds.) EKAW 2000. LNCS (LNAI), vol. 1937, pp. 189–202. Springer, Heidelberg (2000)
9. Mitchell, T.: Machine Learning. McGraw Hill (1997)
10. Rinaldi, F., et al.: Mining relations in the GENIA corpus. In: Proceedings of the Second European Workshop on Data Mining and Text Mining for Bioinformatics (2004)
11. Sanchez, D., Moreno, A.: Learning non-taxonomic relationships from web documents for domain ontology construction. Data and Knowledge Engineering 64(3), 600–623 (2008)
12. Srikant, R., Agrawal, R.: Mining generalized association rules. In: Proc. of VLDB 1995, pp. 407–419 (1995)
13. Villaverde, J., Persson, A., Godoy, D., Amandi, A.: Supporting the discovery and labeling of non-taxonomic relationships in ontology learning. Expert Syst. Appl. 36(7), 10288–10294 (2009)

Improving Persian Text Classification and Clustering Using Persian Thesaurus

Hamid Parvin, Atousa Dahbashi, Sajad Parvin, and Behrouz Minaei-Bidgoli

Abstract. This paper proposes an innovative approach to improve the classification performance of Persian texts. The proposed method uses a thesaurus as a helpful knowledge to obtain more representative word-frequencies in the corpus. Two types of word relationships are considered in our used thesaurus. This is the first attempt to use a Persian thesaurus in the field of Persian information retrieval. Experimental results indicate the performance of text classification improves significantly in the case of employing Persian thesaurus rather the case of ignoring Persian thesaurus.

Keywords: Persian Text, Persian Thesaurus, Semantic-Based Text Classification.

1 Introduction

In the current century Information Technology is considered as one of the most important scientific fields (if not the most important field) among the researchers. Ever-increasing growth pace of data makes its appropriate and efficient management significantly important and also its appropriate usage inevitable. Indeed proper responding to user queries is considered as a crucial challenge in the Information Technology [1]. Two of the most important challenging problems in the field of Information Technology include:

- How can one handle information retrieval problem in a huge number of texts efficiently?
- How can one extract useful information out of a huge mass of data efficiently?

From this perspective, usage of text keywords has been considered as a very promising approach for researchers to handle two mentioned challenges.

This paper proposes to use existing between-word-relationships to help us build an automatic thesaurus-based indexing approach in Persian language.

Rest of the paper is organized as follows. Section 2 deals with related works. In section 3, the proposed method is explained. Section 4 demonstrates the results of

Hamid Parvin · Atousa Dahbashi · Sajad Parvin · Behrouz Minaei-Bidgoli
Nourabad Mamasani Branch, Islamic Azad University Nourabad Mamasani, Iran
e-mail: hamidparvin@mamasaniiau.ac.ir,
　　　{dahbashi,s.parvin,b_minaei}@iust.ac.ir

S. Omatu et al. (Eds.): Distributed Computing and Artificial Intelligence, AISC 151, pp. 493–500.
springerlink.com　　　　　　　　　　　　　　　　© Springer-Verlag Berlin Heidelberg 2012

our proposed method against traditional one comparatively. Finally, the paper is concluded in section 5.

2 Related Works

In 1999, Turney showed that keyword extraction field is one of the most important factors accelerating and facilitating the information retrieval applications, but until then there is no attempt to improve the quality of extracted keywords [5].

Simultaneously in 1999, Frank et al. who worked in the field of artificial intelligence tried to improve the quality of extracted keywords by presenting machine processing algorithm. Their work was based on a Simple Bayes algorithm. Their system is named "KEA". In the KEA method, although the quality of extracted keywords significantly increased, linguistic issues were not taken into considerations during keyword extraction process [6]. The general process of keyword extraction was introduced by Liu et al. in 2005. They first elect a number of candidate words as potential keywords, then assign a weight to each potential keyword, and finally consider potential keywords with the highest weights as the final extracted keywords [7]. Franz in 2002 combined statistical analysis and linguistic analysis [8]. He believed that without considering information about linguistic knowledge, statistical analysis considers disadvantageous and non-keywords [8].

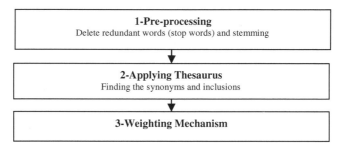

Fig. 1 Proposed indexing framework.

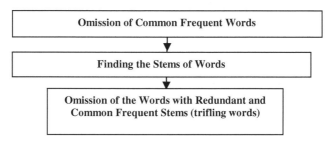

Fig. 2 Pre-processing phase of proposed framework.

Along with previous researches, to solve drawbacks of the traditional keyword extraction approaches (that extract disadvantageous and non-key words instead of the keywords), Freitas et al. modeled process of the keyword extraction into a classification problem in 2005 [9]. Zhang et al. used a decision tree as classifier to recognize the keywords among all words [10]. Halt used the features based on N-gram concept in the context of information retrieval [11]. In the first attempt, Deegan used thesaurus concept in 2004 to improve information retrieval efficacy [12]. After that Hyun tried to use a specialized thesaurus for special-formatted queries [13]. There are some successive works that try to improve information retrieval efficacy after then [14]-[16].

There are some related works done in the field of Persian language. While there are many methods in Persian language, there is a lack of employing a thesaurus in Persian so far. The curious reader is referred to [4] and [17]-[20] for more detail. The only work that employs a thesaurus is Parvin et al. work that is a very simple and immature one [21].

3 Proposed Framework

Fig. 1 depicts the proposed framework. The first step in Fig. 1 is expanded in Fig. 2. As seen in Fig. 2, in preprocessing step, Persian texts are refined into useful texts to get rid of the trivial words that are unnecessary for keyword extraction phase.

Indeed the pre-processing step of proposed framework consists of three phases (sub-steps). In first phase the common frequent words like prepositions are omitted. Then the stem of each word is found. Third the common frequent stems, like "*be*", are also omitted from the text.

Fig. 3 A typical text with three words that are synonyms.

To clarify second step, please consider Fig. 3. In Fig. 3 assume that the $word_1$, $word_2$ and $word_3$ are synonyms of each other. Using a thesaurus these three words, i.e. $word_1$ and $word_2$ and $word_3$ are considered as the single word that is first observed, i.e. $word_1$ with a frequency as many as sum of their frequencies, here 3. Here $word_1$ is head word of those three words and two words, $word_2$ and $word_3$, are children of head word $word_1$.

So after second step a table of words is obtained from the input text that depicts the words next to their frequencies; for example the table of words for the text presented in Fig. 3 is like Table 1. In the table of words, words are partitioned into two types: (a) *head* type and (b) *child* type. Only words with *head* type are considered in the final step. Consider the table of words extracted from the previous example and presented in Table 1. It contains three words, $word_1$, $word_2$ and $word_3$. Only the word $word_1$ is considered as *head* type and its frequency is equal to 3. Two other words are considered as *child* type.

So in obtaining a table of words, weight for a synonym/antonym relationship is considered by a one, i.e. occurring the synonym/antonym of a word is equal to occurring the original word. Another relationship that is taken into consideration is inclusion. For example a word like *animal* includes a *wolf*. So in a text that has a word *animal* as a head type word, occurring a word *wolf* is equal to occurring a word *wolf* and also occurring the *head* type word *animal* with weight α, where α is less than one and vice versa. It means if an inclusion word has been occurred as a *head* type word so far, occurring an included word is to occur the included word by weight one, and including word by a weight α, where α is a real number below one. For example consider text presented in Fig. 4. Assume that $word_5$ is a special kind of $word_4$ and $word_4$ is a special kind of $word_3$. As before, $word_1$, $word_2$ and $word_3$ are synonyms/antonym of each other.

Table 1 Table with frequencies of words of Fig. 3.

word	frequency	Type
.	.	.
.	.	.
.	.	.
word1	3	headi
word2	3	childi
word3	3	childi
.	.	.
.	.	.
.	.	.

---------------word$_1$--word$_5$--------
---------word$_2$--word$_2$---
---------word$_4$--word$_3$...

Fig. 4 A typical text with five words.

Now a table is extracted from the text presented in Fig. 4 that the frequencies of its words are like Table 2. For simplicity we assume that α is 1/4 for this example.

Table 2 Table with frequencies of words of Fig. 4.

word	frequency	type
.	.	.
.	.	.
.	.	.
$word_1$	4+1/4+1/4*1/4	$head_i$
$word_2$	4+1/4+1/4*1/4	$child_i$
$word_3$	4+1/4+1/4*1/4	$child_i$
$word_4$	1+4*1/4+1/4	$head_{i+1}$
$word_5$	1+1/4+4*1/4*1/4	$head_{i+2}$
.	.	.
.	.	.
.	.	.

In Table 2, word *word₁* is the *head* for three words, *word₁*, *word₂* and *word₃*. Because those words, *word₁*, *word₂* and *word₃*, are occurred 4 times, their frequencies are considered 4 at least. Besides, due to occurring the word *word₄* that is a special kind of *word₃*, a 1/4 (α) is added to their frequencies. Due to occurring the word *word₅* that is a special kind of word *word₄*, a 1/4*1/4 (α^2) is added to their frequencies. From another side, the frequency of the word *word₄* is at least 1, due to its one direct appearance. Because of four appearances of the word *word₁*, 4 times 1/4 (4*α) is added by its one appearance. Besides because of one appearance of word *word₅* another 1/4 (α) is added to its frequency. This scenario is valid for the word *word₅*. It means that one appearance of the word *word₅*, plus 1/4 (α) due to appearance of the word *word₄* plus 4 appearances of the word *word₁* that has inclusion relationship with length 2, i.e. 4*1/4*1/4 (4*α^2), is considered as frequency of the word *word₅*.

4 Experimental Studies

Employed criteria based on which an output of a classifier or a clustering algorithm are evaluated, are discussed in the first part of this section. The details of the used dataset are given in the subsequent part. Then the settings of experimentations are given. Finally the experimental results are presented.

We have two different parts of experimentations. In the first part of experimentations we use a simple classifier to show the effectiveness of the proposed method. We employ confusion matrix to visually show the distribution of articles in different classes. Each row in the confusion matrix represents the instances in a predicted class, while each column of the confusion matrix represents the instances in an actual class. One benefit of a confusion matrix is that it is easy to see if the system is confusing two classes. To evaluate the performance of the classification, the accuracy, entropy and purity measures are taken as the evaluation metrics throughout all the paper.

Table 3 Details of used dataset.

Row	Topic	# of articles	Average # of words	Average # of words after refinement phase
1	Sport	146	204	149
2	Economic	154	199	135
3	Rural	171	123	76
4	Adventure	89	160	115
5	Foreign	130	177	124

All the classification experiments are done using 4-fold cross validation. The results obtained by 4-fold cross validation are repeated as many as 10 independent runs. The averaged accuracies over the 10 independent runs are reported. Confusion matrix of 1-nearst neighbour classifier with leave-one-out technique is presented as a comprehensive study of performance of classification.

In the second part of experimentations k-means clustering algorithm is applied over dataset. Here the normalized mutual information (NMI) between the output partition and the real labels of dataset is considered as the main evaluation metric of the final partition [2]. Second alternative to evaluate a partition is the accuracy metric, provided that the number of clusters and their true assignments are known.

In order to test the proposed method five different categories have been collected from Hamshahri newspaper [3]. The detail of the dataset is presented in the Table 3.

After refinement of dataset, the average number of words in each category is reduced as the last column of Table 3. And then after applying refinement phase, we produce a feature space as illustrated in Table 4.

In Table 4, parameter n is the number of the words which are considered as *head* word type in one article at least. The entity j-th column of i-th row in Table 4 is equal to frequency value of *head* word j in i-th article. The parameter m that shows the number of articles in dataset is 400. It means 75 articles per class. The averaged number of features in dataset, n, is 171.5.

By filling values of Table 4 by using thesaurus and without using thesaurus we obtain two different datasets.

The used thesaurus is produced considering the manual presented by Hori [22].

Table 4 Dataset after refinement.

	Head Word$_1$	Head Word$_2$	Head Word$_3$	....	Head Word$_n$
Article$_1$					
Article$_2$					
00000					
Article$_m$					

We use 1-neareast neighbour classifier as base classifier and averaged on 10 independent runs each of which obtained by 4-fold cross validation is reported. Parameter α is considered 1/4 throughout all the experimentations. The true labels of this dataset are employed for obtaining the accuracy metric. For reaching confusion matrices (Table 6 and Table 7) we use 1-neareast neighbour and leave-one-out technique.

Table 5 Performances of 1-NN classifier and k-means clustering with and without thesaurus.

	Without thesaurus	With thesaurus
1-NN Accuracy	68.31%	81.45%
1-NN Entropy Measure	58.81%	42.71%
1-NN Purity Measure	70.49%	81.16%
k-means Accuracy	60.03%	72.39%
k-means NMI Measure	27%	42%

In clustering the real number of cluster (here 5) is feed to k-means algorithm. The similarity measure to reach similarity matrices is based on normalized Euclidean distance.

Table 6 Confusion Matrix of Documents with thesaurus.

Cluster	Sport	Economic	Rural	Adventure	Foreign	Entropy	Purity
1	2	1	2	123	4	20.86	93.18
2	13	120	3	6	4	42.63	82.19
3	132	14	5	2	2	35.85	85.16
4	19	7	73	11	8	72.22	61.86
5	5	12	6	4	112	46.16	80.58
Total	171	154	89	146	130	42.71	81.16

Table 5 shows the main results of first part of experimentations. The table shows in first row Accuracy measures of 1-NN classifier with and without thesaurus. It then shows the Entropy and Purity measures of the classifier in the two subsequent rows. Then it presents k-means clustering accuracy and NMI measures in the two subsequent rows. The confusion matrix for 1-NN classifier on features obtained by help of thesaurus is shown in the Table 6. The confusion matrix for 1-NN classifier on features obtained without help of thesaurus is shown in the Table 7.

Table 7 Confusion Matrix of Documents without thesaurus.

Cluster	Sport	Economic	Rural	Adventure	Foreign	Entropy	Purity
1	17	99	4	11	9	60.82	70.71
2	119	24	7	2	3	47.48	76.77
3	9	18	7	8	94	63.50	69.12
4	3	2	7	109	15	44.70	80.15
5	23	11	64	14	9	81.60	52.89
Total	171	154	89	144	130	58.81	70.49

5 Conclusion and Future Works

In this paper, we have proposed a new method to improve the performance of Persian text classification. The proposed method uses a Persian thesaurus to reinforce the frequencies of words. With a simple classifier, it is shown that using thesaurus can improve the classification of Persian texts. We consider two relationships: synonyms and inclusion. We use a hierarchical inclusion weighting, and linear synonym weighting. As it is concluded the text classification and clustering both outperforms significantly in the case of applying a thesaurus.

As a future work, one can turn to research on the different weighting methods. For another further future work it can be studied how further relationships, like contradiction, can affect the text classification performance.

Acknowledgments. This research is partially supported by Iran Communication Research Center, Tehran, Iran.

References

[1] American Society of Indexers. Frequently Asked Questions Indexing. Index review in Books, Ireland, http://www.asindexing.org/site/indfaq.shtml

[2] Strehl, A., Ghosh, J.: Cluster ensembles - a knowledge reuse framework for combining multiple partitions. Journal of Machine Learning Research 3, 583–617 (2002)

[3] Hamshahri newspaper, http://www.hamshahrionline.ir

[4] Yousefi, A.: Principles and methods for computerized indexing. Journal Books 9(2) (2010) (in Persian)

[5] Turney, P.D.: Learning Algorithms for Keyphrase Extraction. Information Retrieval 2(4), 306–336 (1999)

[6] Frank, E.: Domain-Based Extraction of Technical Keyphrases. In: International Joint Conference on Artificial Intelligence, India (1999)

[7] Liu, Y., Ciliax, B.J., Borges, K., Dasigi, V., Ram, A., Navathe, S.B.: Comparison of two schemes for automatic keyword extraction from MEDLINE for functional gene clustering. In: Computational Systems Bioinformatics Conference, Stanford (2005)

[8] Frantzi, K., Ananiadou, S., Mima, H.: Automatic Recognition of Multi-word Terms: the C-value/NC-value Method. Digital Libraries 3(2), 115–130 (2002)

[9] Freitas, N., Kaestner, A.: Automatic text summarization using a machine learning approach. In: Brazilian Symposium on Artificial Intelligence (SBIA), Brazil (2005)

[10] Zhang, Y., Heywood, N.Z., Milios, E.: World Wide Web Site Summarization Web Intelligence and Agent Systems. Technical Report, CS-2002-8 (2006)

[11] Hult, A.: Improved automatic keyword extraction given more linguistic knowledge. In: 8th Conference on Empirical Methods in Natural Language Processing (2003)

[12] Deegan, M.: Keyword Extraction with Thesauri and Content Analysis, http://www.rlg.org/en/page.php?Page_ID=17068

[13] Hyun, D.: Automatic Keyword Extraction Using Category Correlation of Data, Heidelberg, pp. 224–230 (2006)

[14] Witten, W., Medley, I.H.: Thesaurus based automatic keyphrase indexing. In: 6th ACM/IEEE-CS JCDL 2006 (Joint Conference on Digital Libraries) (2006)

[15] Klein, M., Steenbergen, W.V.: Thesaurus-based Retrieval of Case Law. In: 19th International JURIX Conference, Paris (2006)

[16] Martinez, J.L.: Automatic Keyword Extraction for News Finder, Heidelberg, pp. 405–427 (2008)

[17] Shahabi, A.M.: Abstract construction in Persian literature. In: Second International Conference on Cognitive Science, Tehran, p. 56 (2002) (in Persian)

[18] Bahar, M.T.: Persian Grammar, ch. IV, p. 111 (1962) (in Persian)

[19] Khalouei, M.: indexing machine. Journal Books 6(3) (2009) (in Persian)

[20] Karimi, Z., Shamsfard, M.: Automatic summarization systems Persian literature. In: 12th International Conference of Computer Society of Iran (2005) (in Persian)

[21] Parvin, H., Minaei-Bidgoli, B., Dahbashi, A.: Improving Persian Text Classification Using Persian Thesaurus. In: Iberoamerican Congress on Pattern Recognition, pp. 391–398 (2011)

[22] Hori, E.: A Manual to make and develop a multilingual thesaurus, Scientific Documentation Center (2003) (in Persian)

Towards Incremental Knowledge Warehousing and Mining

Habiba Drias, Asma Aouichat, and Aicha Boutorh

Abstract. In this paper, we propose new ideas around the concepts of knowledge warehousing and mining. More precisely, we focus on the mining part and develop original approaches for incremental clustering based on k-means for knowledge bases. Instead of addressing the prohibitive amounts of knowledge, the latter is gradually exploited by packets in order to reduce the problem complexity. We introduce original algorithms named ICPK/k-means for *Incremental Clustering by Packets of Knowledge*, ICPKG/k-means for *Incremental Algorithm by Packets of Knowledge and Grouping of clusters* for determining the number of desired clusters, LICPK/k-means for *Learning Incremental Clustering by Packets of Knowledge* and LIGPKG/k-means for *Learning Incremental Clustering by Packets of Knowledge and Grouping of clusters* for handling the clustering of large amount of knowledge. Experimental results prove the effectiveness of our algorithms.

Keywords: knowledge warehouse, knowledge mining, packets of knowledge, k-means, incremental knowledge clustering, learning incremental knowledge clustering.

1 Introduction

Warehousing and mining are two technologies that are exploited in the systems of decision support[4,5]. Currently, they are relatively well mastered for the data, in a sense that data warehousing and mining are *midgame* activities. They help creating knowledge from a tremendous available volume of data. Knowledge is considered nowadays as one of the most valuable assets of a company. In this paper we introduce a new paradigm called *knowledge warehousing and mining* that we consider as the *end game* of a whole process starting from data warehousing and mining and ending at knowledge warehousing and mining. We will

Habiba Drias · Asma Aouichat · Aicha Boutorh
USTHB, LRIA, Department of Computer Science
e-mail: hdrias@usthb.dz, asma_aouichat@yahoo.fr,
    bout_aicha@hotmail.fr

S. Omatu et al. (Eds.): Distributed Computing and Artificial Intelligence, AISC 151, pp. 501–510.
springerlink.com   © Springer-Verlag Berlin Heidelberg 2012

present more precisely the part of the process that deals with knowledge mining and show how to apply data mining tasks on knowledge. Let us point out that this concept is different from the one we found in the literature [7,8]. The issue is to mine knowledge instead of data and the result of the desired task is therefore a meta-knowledge.

Mining tasks such as classification, prediction and clustering have been for a long time applied to a great number of domains. In this work, we will focus especially on clustering knowledge objects. Like for data mining, the major difficulty in the development of techniques for knowledge mining resides in the large amounts of knowledge, which are created everyday for various purposes. However there exists also a more complicated issue for such question and it consists in modeling knowledge similarity for the purpose of classifying, clustering or even extracting associations or any mining treatment. The task considered in this paper is clustering which is an important concern of knowledge mining. Due to the very large size of the considered knowledge bases, it is infeasible to perform this task without undertaking the process in an incremental way [3]. In a mount of this work, we propose four different incremental clustering algorithms based on k-means.

2   Knowledge Rules Clustering

The paradigm in which we are interested belongs to the domain of knowledge acquisition, which has been already known while essential, as difficult and costly. Undertaking mining tasks on a knowledge base is even harder. In this section we will attempt to perform Clustering operations on knowledge base rules.

2.1   The Paradigm of Knowledge Mining

In the literature [7, 8], the concept of knowledge mining is defined as the mechanism of extracting new knowledge from a set of data when integrating domain knowledge in the process of data mining, and in a summarized way as follows: *Data + Prior Knowledge + Goal → New Knowledge.*

This definition is inconsistent with our idea and the aim of the present work is the study of a new paradigm, which consists in mining a set of knowledge. Our vision about knowledge mining is the extraction of knowledge from knowledge and hence as a result from this process, we acquire meta-knowledge. It is then described as: *Knowledge + Prior Knowledge + Goal → Meta-Knowledge*

In this case, knowledge mining features literally *the exploitation of knowledge,* the goal being wealth acquisition and in our case wealth is meta-knowledge. Consequently, we define knowledge mining as: "*a set of techniques used to generate a meta-knowledge base from a huge amount of knowledge in order to facilitate handling the latter in a rapid way*".

2.2   Knowledge Representation

As knowledge representation is a wide field, we cannot browse all the possible representations and study knowledge mining for each of them. We have primarily considered a knowledge base in order to extend this study to large scale reasoning since rules can be easily used to design reasoning systems. Another subsequent extension is the development of super agent technology. Many other enhancements such as studying mining knowledge represented by hierarchies and associations may be launched from the ideas developed in this work.

2.3 Knowledge Rules Clustering

In this section, we will present the necessary tools for developing techniques for rules clustering according to their similarity. An example of a simple rule is the following derivation rule:

If (k is a constant) and (f(x) = k*x) then (f'(x) = k).

In this rule, the clauses *(k is a constant) and (f(x) = k*x)* belong to the premise part while the clause *(f'(x)= k)* to the consequence one. As shown in the example, the pair (variable = value) form a clause. Other relational operators such as > and < can be used besides equality.

2.3.1 Distance Measure for Knowledge Rules

In general, the distance separating two objects is a way of quantifying their similarity according to some semantic. Two knowledge entities are similar or close if they record the minimum difference computed as the smallest distance. To evaluate the similarity between knowledge rules, we propose the following distance measure:

If **Ri, Rj** represent two production rules where *Ri* is different from *Rj* then:

$$D(R_i, R_j) = \frac{a(|C_i \cup C_j| - |C_i \cap C_j| + |V_i \cup V_j| - |V_i \cap V_j|)}{(a+1)(|R_i| + |R_j|)}$$

Where a is a weight assigned to the clauses and set by experiments with the condition that a>1, the weight of the variables being set to 1. Ci and Cj are respectively the set of the clauses of Ri and Rj. The union of these two sets gives a set containing all the clauses of both rules and the intersection is the set of clauses that are common to both rules. Vi and Vj include respectively the set of variables of rules Ri and Rj. The union and the intersection between Vi and Vj are defined the same way as for the clauses. Finally the quantities $|Ri|$ and $|Rj|$ represent respectively the number of clauses in Ri and Rj. Let's look at the following rules:

R1 : if x=1 and y=2 then z=3
R2 : if y=2 and x=2 and f=3 then h=0
then:
(C1 U C2) = {(x, 1), (y, 2), (z, 3), (x, 2), (f, 3), (h, 0)}, |C1 U C2| = 6, (C1 ∩ C2) = {(y, 2)}, |C1 ∩C2| = 1,
(V 1 U V 2) = {x, y, z, f, h}, |V 1 U V 2| = 5, (V 1 ∩ V 2) = {x, y}, |V 1 ∩ V 2| = 2, | R1 | = 3, | R2 | = 4

If we choose ($a = 2$) and apply the distance formula, then: $D(R_1, R_2) = \frac{2(6-1)+(5-2)}{3(3+4)} = 0.62$.

2.3.2 Centroid Computation

Finding the exact centroid of a cluster containing knowledge rules is not obvious, nevertheless, an approximate calculation is proposed. The centroid of a cluster is defined as a set of clauses built from the cluster rules and that verify the following conditions:

- The size of the centroid is equal to the average size of all the cluster rules.
- If different clauses have the same frequency, those containing the most frequent variable in the rules set are selected.
- The distance between the centroid and each rule is approximately equivalent.
- Moving knowledge from one cluster to another one disturbs the position of the centroid, which hence has to be updated.

Let illustrate this concept with the following rule added to *R1* and *R2*:

R3 : If i z=3 and y=2 and x=3 and f=0 then h=1

The size of each rule is successively: $|R1| = 3$, $|R2| = 4$, $|R3| = 5$. The size of the centroid is computed as:

$$|G| = \frac{|R_1| + |R_2| + |R_3|}{number\ of\ rules} = \frac{3 + 4 + 5}{3} = 4$$

The four most frequent clauses in the centroid G are computed as shown in the following steps:

- The most frequent clauses are *y=2, z=3*, the other clauses have the same frequency.
- The choice of the other two clauses depends on the variables frequency. The most frequent variables are *x* and *y*. The variable *y* has only one value in the rules, in addition it was already taken into account. The variable *x* takes two different values *x=1* or *x=2*, in this case we choose the most frequent one. If they have the same frequency, as is the case of the example, one of these clauses will be selected randomly (*x=1*), then the set of clauses so far is *y = 2, z = 3, x = 1*.
- For the last clause, and given that the variables *f* and *h* have the same frequency, and clauses containing these variables have the same frequency, the fourth clause will be drawn randomly, let it be *f= 0*.
- At the end we get *G = (y = 2, z = 3, x = 1, f = 0)*.

$(Ri,Rj) = 0$ then Ri and Rj represent the same rule and if distance $(Ri,Rj) = 1$ then Ri is the furthest from Rj. The serial and parallel approaches are described in the next subsections.

4.2 Serial Incremental Clustering Approach

The serial proposed approach consists in handling the incremental mining in a sequential way that is, clustering the first packet then integrating the rules of the current packet in the result of clustering of the previous packets. To perform such task, we propose two techniques. The first one called ICPK/k-means for *Incremental Clustering by Packets of Knowledge based on k-means* consists of a classical k-means with a fixed number of clusters. The second one called LICPK/k-means is for *Learning Incremental Clustering by Packets of Knowledge based on K-means* with a number of clusters to learn. The difference between both techniques resides in the way of inserting rules into clusters. In the first one, k is fixed by the input whereas in the second one, k is learnt. After having inserted all the rules of one packet, we perform a k-means in order to update the clusters that is, recalculate the centroids and relocate rules in the clusters. The number of execution of k-means is then equal to the number of packets. The insertion of each packet rules in the clusters takes advantage of the knowledge learned during the insertion of the previous packets.

4.2.1 ICPK/k-Means

As mentioned previously, the technique ICPK/k-means has a different insertion function from the one used in the second technique. The insertion function in this case consists in introducing a rule into the cluster having the nearest centroid from the rule according to the distance measure.

4.2.2 LICPK/k-Means

The insertion of new rules for this second technique depends on the threshold value T and is performed according to the following constraints :

- if distance-min $(Ri, G) <= T$ then insert Ri in the cluster of the centroid G.
- iIf distance-min $(Ri, G) > T$ then create a new cluster with Ri as its centroid.

distance-min is a function that calculates the minimum distance that separates Ri from G.

4.3 Parallel Incremental Clustering Approach

The parallel approach consists in handling the incremental mining in a parallel way that is, clustering each packet independently from the others then combine the results of each packet in the same structure by regrouping the clusters. To perform such task, we propose two techniques, the first one called *ICPKG/K-means* for *Incremental Clustering by Packets of Knowledge and Grouping of clusters based on K-means* consists in grouping the nearest cluster according to the distance with a fixed number of clusters and a second one called *LICPKG/K-means* for *Learning Incremental Clustering by Packets of Knowledge and Grouping of clusters based on K-means* also consists in grouping clusters but with a number of clusters to determine. The difference between both techniques resides in the way of regrouping clusters. Concerning the rules that must be relocated in the clusters, we propose two methods:

- Assign the rules using k-means after regrouping clusters pair by pair of packets
- distribute the rules among clusters using k-means once the combination of clusters for all packets is done.

4.3.1 ICPKG/K-Means

To join two clusters $Cl1$ and $Cl2$ it is necessary to calculate the distance between each centroid of the initial packet with each centroid of the new packet. Clusters that express a minimum distance between their centroids will be grouped to generate a new cluster containing the rules of both clusters. An update of the centroid of the result cluster will be necessary to unify the two centroids.

4.3.2 LICPKG/K-Means

The merging of clusters for this second technique of parallel approach depends on the threshold value T, we start by calculating the distance between each centroid of the first packet and the one of the second packet and considering the two clusters of minimum distance according to the following condition: if distance-min <= T then combine both clusters into one cluster and recalculate the centroid else do not perform grouping.

5 Experimental Results

The algorithms are applied on 1000 rules drawn at random and the results are compared in terms of execution time and performance. A general criterion for evaluating the results of clustering consists in comparing the calculated partition with The "pertinent" partition. A traditional measure to evaluate the conformity between two partitions of n elements is the value called "Rand"[6]. If C = {C1,C2, ...,Cu} is the calculated structure of clustering and P = {P1, P2, ... , Pv} is a predefined partition then each pair of items can be assigned to the same cluster or two different clusters. Let assume the following hypotheses:

- a is the number of pairs in the same cluster C and in the same cluster P.
- b is the number of pairs whose elements belong to two different clusters of C and two different clusters of P.

Then the conformity between C and P can be estimated using the formula:

$$Rand(C,P) = \frac{a+b}{N(N-1)/2}$$

A good partition should have a value of "Rand" close to 1. We use this index to calculate the precision in our experiments.

5.1 Clustering by k-Means

To tune the parameter of a distance formula, several tests were performed, $a = 2$ is the value that generates the best results for *Rand and* the best value of k is equal to 5. The main observed drawback of the k-means is the negative impact of the initial partition (which is often drawn at random) on the final result.

5.2 ICPK/K-Means versus ICPKG/K-Means

The series of experimentations shown in fig.1 aims at comparing the behavior of ICPK/k-means with that of the IAPKG/k-means, in the case of fixed number of clusters. The second essential criterion, after k, in our incremental clustering approach is the size of packet. The results are presented by setting k=5.

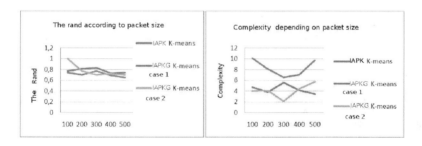

Fig. 1 Rand and Complexity for IAPK/k-means and IAPKG/k-means.

The best packet size for both techniques that yields a reasonable complexity and a good efficacy is around "100 rules" for 1000 rules.

5.3 LICPK/K-Means versus LICPKG/k-Means

Fig. 2 illustrates sequentially the values of *Rand* and number of clusters achieved by the methods described above for different threshold values. The packet size was fixed using the previous results to 100 rules.

Based on the achieved results, it appears that the learning strategy is the most efficient and that in general the results of the experimentations are encouraging for rules clustering. In addition, the conducted experiments have shown that the designed approaches yield a good performance score, by comparing generated partitions with correct ones, in an optimal time of complexity.

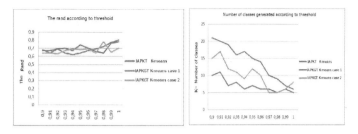

Fig. 2 Rand and Number of clusters for LICPK/k-means and LICPKG/k-means.

6   Conclusions

In this paper, we have proposed several incremental clustering algorithms based on k-means for mining a huge knowledge base. In our design of incremental clustering, we have identified two techniques with respect to the number of clusters to be generated. In a first implementation, the number of clusters corresponds to the parameter *k* of the k-means algorithm, which can be set by the user or by experimentations. Therefore we have developed the algorithms *ICPK/k-means* and *ICPKG/k-means*. In a second implementation, the number of clusters is learnt by using a minimum distance threshold and in this context we have developed the algorithms *LICPK/k-means* and *LICPKG/k-means*.

Extensive experiments have been performed and the achieved results are very promising. Besides they showed that the incremental clustering approaches we have designed are robust and capable to handle large scale knowledge.

References

1. Drias, H., Mosteghanemi, H.: Bees Swarm Optimization Based Approach for Web Information Retrieval. Web Intelligence, 6–13 (2010)
2. Drias, H., Sadeg, S., Yahi, S.: Cooperative Bees Swarm for Solving the Maximum Weighted Satis_ability Problem. In: IWANN, pp. 318–325 (2005)
3. Fisher, D.: Knowledge Acquisition Via Incremental Conceptual Clustering. In: Fisher, D., Shavlik, Dietterich (eds.) Readings in Machine Learning, pp. 267–283. Morgan Kaufmann (1990)
4. Han, J., Gonzalez, H., Li, X., Klabjan, D.: Warehousing and Mining Massive RFID Data Sets. In: ADMA, pp. 1–18 (2006)
5. Han, J., Kamber, M., Pei, J.: Data Mining: Concepts and Techniques, 3rd edn. Elsevier (2011)
6. Hartigan, J.A., Wong, M.A.: A K-means Clustering Algorithm. Journal of the Royal Statistical Society 28(1) (1979)
7. Kaufman, K., Ryszard, S.M.: From Data Mining to Knowledge Mining, Handbook in Statistics. In: Rao, C.R., Solka, J.L., Wegman, E.J. (eds.) Data Mining and Data Visualization, vol. 24, pp. 47–75. Elsevier/North Holland (2005)
8. Ryszard, S.M.: Knowledge mining: A proposed new direction, School of Computational Sciences George Mason University and Institute for Computer Science Polish Academy of Sciences (2003)

Analysis of Sequential Events for the Recognition of Human Behavior Patterns in Home Automation Systems

Adolfo Lozano-Tello and Vicente Botón-Fernández

Abstract. Learning users' frequent patterns is very useful to develop real human-centered environments. By analyzing the occurrences of events over time in a home automation system, it's possible to find periodic patterns based on action-time relationships. However, the human behavior could be better defined if it's related to chained actions, creating action-action relationships. This work presents IntelliDomo's learning layer, a data mining approach based on ontologies and production rules that aims to achieve those objectives. This module is able to acquire users' habits and automatically generate production rules for behavior patterns to anticipate the user's periodic actions. The learning layer includes new features looking for the adaptability and personification of the environment.

1 Introduction

In recent years, the design of smart environments is one of the research areas which are rapidly gaining importance in fields such as health care, energy savings, etc where the interaction among the users and the environment is a fundamental factor. An important feature that these environments need to possess is the ability to adapt themselves to the residents' whims and have the versatility to make decisions in a variety of situations. In this sense, finding behavior patterns in a sequence of events in order to predict future actions can lead us to the natural interaction we are looking for. Therefore, the system will be able to recognize human behavior and anticipate to the needs and preferences of the inhabitants. Obviously, discovering these habits requires a previous task of learning. In a smart environment, learning means that the environment has to gain knowledge about the preferences and common behavior of the user in an unobtrusive and transparent way.

Adolfo Lozano-Tello · Vicente Botón-Fernández
Telefónica Chair of Extremadura University, and Quercus Software Engineering Group.
Escuela Politécnica, Campus Universitario s/n, 10003, Cáceres, Spain
e-mail: {alozano,vboton}@unex.es

S. Omatu et al. (Eds.): Distributed Computing and Artificial Intelligence, AISC 151, pp. 511–518.

Using ontologies [1] to classify the types of devices and their functionality can be an appropriate way to understand users' behavior. Ontologies and SWRL (Semantic Web Rule Language) rules provide a precise definition of a smart home taxonomy and are reusable, so other users can get them to classify their own domotic components and to build rules that will allow inferring new information.

The present paper describes a new behavior analysis of IntelliDomo's learning layer, a data mining approach for the management of ontology-based AmI systems and the learning of human behavior, including the recognition of action-action relationships. The rest of the paper is as follows: Section 2 identifies publications using data mining in intelligent systems. Section 3 sums up the main features of IntelliDomo. Section 4 shows the fundamentals and the new features of the learning layer; and Section 5 describes a scenario used to test the module. Finally, Section 6 is dedicated to conclusions and future works.

2 Data Mining Applied to Smart Environments

Nowadays, the development of data mining techniques to analyze data captured from a set of sensors and actuators in a smart home is gaining importance due to the necessity of satisfying users' whims and anticipating to their habits. There exist a significant number of research projects about AmI systems controlling and automating users' tasks with different degree of success, and most of these works are focused upon different aspects from each other, owing to the complexity of these systems. However, the amount of literature about intelligent systems which make use of ontologies and production rules is not really extensive.

CASAS [2] introduces an adaptive smart home system that uses some interesting techniques to discover frequent and periodic patterns in user's daily activities. Moreover, this system describes an important approach about how to incorporate temporal reasoning to our algorithms, as shown in [3].

The MavHome project [4] develops a home automation model in order to deduce a profile from the inhabitant of the intelligent environment. Then, the system can use it to automate devices and adapt then over time to satisfy the inhabitant needs. In [5] this project presents a method of automatically constructing universal models by taking the output of a sequential data mining algorithm and sequential prediction algorithm, inspired on Active LeZi algorithm [6].

Aztiria et al. [7] propose a process to discover sequences of user actions in a system based on speech recognition. In this approach, patterns are used not only to automate actions or devices, but also to understand users' behavior and act in accordance with it. The speech recognition system allows user to interact with the patterns in order to use his/her acceptance to automate actions.

These works show that there is no a holistic approach yet. In that sense, combining different skills seems a promising strategy. This way, our learning layer uses some of the above techniques, and the concepts established on a previous work [1], but introducing new parameters to create a system where the user can guide the environment to behave in a customized manner.

3 IntelliDomo's System

IntelliDomo (http://www.intellidomo.es) is an intelligent system able to control the devices of a home automation system automatically and in real time using SWRL rules. Its main feature is the ability of reasoning and responding in view of the changes in smart environments.

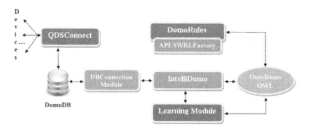

Fig. 1 This figure shows the components of the IntelliDomo architecture and their relations.

Each event captured from sensors and actuators is recorded. To manage all this knowledge, IntelliDomo is built upon an ontology called OntoDomo, whose concepts have been modeled to be in sync with the physic devices that constitute a smart environment, so that it can be stored its outstanding values and properties. Furthermore, the ontology works together with a set of established SWRL rules. The users can modify these rules and create new ones through DomoRules tool.

There also exists a domotic database (DomoDB) where IntelliDomo updates the state values of the devices in real time. The QDSConnect and DBConnection modules are used together to transfer changes instantly from physical (devices) to logical (database) levels and vice versa, maintaining the data integrity.

Finally, the learning module incorporates algorithms to acquire the users' habits and automatically produce rules that satisfy their needs. The parameters governing the execution of these algorithms can be customizable by the users themselves if they consider it necessary. This way, the system can adapt to changes in the discovered patterns based on the user implicit and explicit feedback.

4 Behavior Learning Layer

The learning of patterns is not merely an optional aspect of the system which may bring some advantages to an intelligent environment, rather we consider that an essential contribution to the idea that an environment can be intelligent. It supports environments which adapt itself to its users in an unobtrusive way and one where the users are released from the burden of programming any device. Therefore, the ability to learn patterns of behavior is of paramount importance for the successful implementation of intelligent environments.

Data collected in a smart environment consist of readings by all different types of sensors and inhabitant manual manipulation of devices. Each event from the data set is considered an action; and each one is defined by three fields: date and time of the event, data source and state of the source.

The learning layer includes the following tasks over the data set: identifying frequent patterns on time, finding common sequences of chained actions, and automating both of them with the generation of SWRL rules. In [8], it was introduced an algorithm that learns rules, based on time of day, describing user actions in a smart home in order to ultimately automate those actions. However, it needs some issues and improvements, such as the analysis of the relationships between consecutive actions, that will be explained later. The present module pretends to establish the bases for the development of a global learning approach, using techniques inspired in Apriori's Algorithm from Rekesh Agrawal [8].

The way to carry out this learning involves checking the actions performed by the inhabitant and finding out not only frequent and periodic patterns, but also common sequences of consecutive actions. Next, we describe the main learning phases in order to generate rules from both isolated and sequential behavior patterns.

4.1 First Phase: Identifying Frequent Patterns

The main objective of this phase is to learn and identify human behavior patterns, based on time of day, from the input data. A pattern is described as a frequent, well-defined and isolated event. These features will be explained below.

Let's assume frequency as the number of times a certain action is repeated in the data set. In order to compute it, an *action range* must be defined to establish the time interval to be considered, since user's daily actions would not usually happen at exactly the same time.

An action frequency is believed to be relevant when it reaches a certain threshold referred as *minimum support*. A periodic pattern is considered frequent and well-defined when it exceeds the minimum support.

The *periodicity* represents the regularity of occurrence of actions. IntelliDomo considers several types of periodicity: daily, weekly, monthly... because there are different behaviors depending on these types of temporality.

Due to human beings' erratic nature, different kinds of actions could merge on time and make the detection of a frequent pattern difficult. For this reason, the algorithm establishes a new threshold value to determine whether a set of actions can be identified as a single unit or not. This parameter is referred as *noise percentage*. A frequent, periodic and well-defined pattern is considered isolated when it doesn't exceed the noise percentage.

4.2 Second Phase: Finding Common Sequences of Actions

The second phase aims to discover frequent sequences of consecutive actions. However, the timestamp of these actions is not so important now. The point is to know that whenever an action "A" takes place, a sequence of actions "B", "C", and so on will happen right after. As far as the discovery of actions sequences is concerned, it's no use to exclusively analyze temporal reasoning, because in that case the frequency which determines if an action could be considered as a frequent habit comes from close repetitions over time. In this sense, studying action-action relationships requires the evaluation of other features in order to find out activities from the dataset properly. In other words, if an action isn't clearly happening at a specific time but in a scattered way, it doesn't mean that the system has to ignore the action as it happens in our previous work. In this case, the algorithm looks for events which recur with some periodicity without taken into account if they always take place at around the same time.

From now on, let's assume that a frequency is believed to be relevant when a sequence of two or more actions is repeated in the data set, which is also referred as an activity. A window of size w (initialized to 1) is passed over the data, and every sequence of length that is equal to the window size is processed.

1. First step: window size equal to 1

In this step, the point is to clean the input dataset, leaving the most common actions. An action will be considered usual if its number of repetitions exceeds the minimum support with some periodicity, but now there's no need to look upon a particular time interval. This way, the algorithm won't take as truth some action sequences that appear as the result of a circumstantial situation.

2. Second step: window size >= 2

This step is the main process of the learning algorithm of events sequences. In short, it's a loop that gradually increases in accordance with the window size, until reaching the maximum length of the particular chain of actions.

First of all, the algorithm takes the set of frequent actions from the previous step. Then, for each one of them, analyzes its relationship with the rest, increasing little by little the window size and sequentially linking it with as many actions as possible. The periodicity of the chained actions must be the same. In order to check if two or more actions have a time relationship, the system counts the number of times where the first action is followed by the second one, then goes the third one... and so on. If they take place together in a significant number of times and the time interval between them is approximately the same, then they can be considered as a frequent sequence of actions. In this step, the minimum support determines whether the number of occurrences is relevant or not, while the action range establishes if the time interval among the actions is usually the same.

When the algorithm stop processing an event to verify if it's part of a sequence, it takes the next event within the input dataset and repeats the whole process again until it has analyzed all the frequent actions.

4.3 Third Phase: Validating the Learning Process

Users can fill in the configuration settings of the phases described above, and decide which events from the data collected are going to be evaluated by the module. To solve this question, two additional parameters are needed: the *initial and final dates* that determine which entries from the database must be taken into account. The data mining process is carried out offline at regular time intervals confirmed by the resident. There also exists the option to let the system run the algorithms when it feels like it.

As the learning process stops, the system generates automatically a SWRL rule for each pattern and the user can decide which of them are going to be activated; e.g., if the user is not happy with a certain rule, he/she can turn it down. The transformation of these patterns into rules was described in [1]. In addition, the accepted rules join together with the existent ones in a minimal priority scale.

5   Case Study of the Learning Layer

In this section, a typical scenario that covers the concepts mentioned above has been described in order to better understand the system operation. We have tested the system in an office at the Department of Quercus Software Engineering Group to obtain a real dataset. The users of this environment are four employees.

The devices installed into this office are: two light controls, a dimmable light switch, two motion sensors, an occupancy sensor, a temperature sensor and three Ac power plugs. Each of the AC power plug devices takes control of one of the following appliances: a heater (heating system), a stereo and a Wi-Fi router that provides Internet access in the entire office. Otherwise, one of the light controls is a reading lamp which is used every day to illuminate a desk.

In this context, the hours of opening goes from 9:00 a.m. to 3:00 p.m. However, on Fridays, it's closed at 2:00 p.m. as an exception. The employees usually have a coffee break at about 11:00 a.m. The users tend to be in strict conformity with the timetable but, at the time of tidying up, they leave the office individually at different times, except on Fridays, when they go out together for a meal.

All the actions done by the users over different home automation devices at the office, including data collected from sensors, are registered into the log database. It has been established a period of time of three months between 01/12/2010 and 28/02/2011, in which all the information that is part of the dataset has been registered. Altogether, 621 events have taken place with the format shown in Figure 2.

	Source [PK] text	State [PK] text	Timestamp [PK] timestamp without time zone
1	"Heating1"	"Off"	2010-12-01 13:54:13
2	"Heating1"	"Off"	2010-12-02 14:02:20
3	"Heating1"	"Off"	2010-12-03 14:02:22
4	"Heating1"	"Off"	2010-12-06 14:10:29
5	"Heating1"	"Off"	2010-12-07 14:05:30

Fig. 2 Example of the format used for the trial scenario events.

Once the log database is prepared, it's time to set up the algorithm. In this case, the users have established the following learning settings: minimum support required of 60%, 5 minutes action range, noise of 30%, and both, daily and weekly periodicity. After running this data mining approach over the previous data set, a few behavior rules are suggested (see Table 1) and the users can validate them.

Table 1 This table shows the behavior rules proposed at the trial scenario.

Source	State	Periodicity	Event
Light1 – Heating1	On – On	Daily	09:00:16 – (2 min)
Light1	Off	Daily	11:00:01
Light1	On	Daily	11:25:22
Light1 – Heating1	Off – Off	Friday	13:56:14 – (5 seg)

Taking into account the previous description about the environment, the behavior patterns seem to be consistent with the data collected, and represent a clear example about the periodic actions previously remarked: the users get to work every day at 9:00 a.m., they turn on the reading lamp and 2 minutes later they switch on the heater. This sequence of actions, with a length of 2, could be identified as "arrival to the office". After that, they usually have a coffee break at around 11:00 a.m., turning off the reading lamp before leaving and turning it on once again when they return to the room (11:25 a.m.). Finally, they go home at an uncertain time, so it's not possible to find a behavior rule in this case. However, on Fridays they leave the office together and they are used to do it just before 2:00 p.m., turning off the reading lamp and then (5 seconds later) the heater. So, the system has reached the expected results for the settings selected in this scenario.

In our ongoing work, we plan to test the module in other scenarios to better understand the strengths and weaknesses of the system, because now we're focused on the data collection process.

6 Conclusions and Future Works

The Ambient Intelligence is one of the areas which are rapidly gaining importance in the application of learning models. This work is focused on the development of a global learning approach and that's why we have built new mechanisms and parameters that can be set up by the inhabitants. In that sense, the recognition of sequential behavior patterns helps to improve the learning of frequent user's habits. Thanks to the context analysis, it's possible to identify those frequent and chained actions that usually take place in the same order. This way, combining contextual information and temporal reasoning seems a promising technique to understand human behavior and let the system evolve to a better user-adapted model. So, the IntelliDomo's learning layer has been updated looking for the adaptability and personification in an easy-to-use environment.

As future works we will try to reach a higher level of analysis, including the comprehension of user's motivations by the system, and also adapt this approach for particular scenarios (energy saving, health care,...), because if the environment knows how we behave, it can automate our habits while improving its operation.

Acknowledgments. This work has been developed under support of Telefónica Chair of the University of Extremadura, FEDER, TIN 2008-02985 and Junta de Extremadura.

References

1. Valiente-Rocha, P., Lozano-Tello, A.: Ontology and SWRL-based Learning Model for Home Automation Controlling. In: ISAmI 2010, June 16-18, pp. 79–86 (2010)
2. Rashidi, P., Cook, J.: Keeping the resident in the loop: adapting the smart home to the user. IEEE Transactions on Systems, Man, and Cybernetics, Part A: Systems and Humans 39(5), 949–959 (2009)
3. Singla, G., Cook, D.J., Schmitter-Edgecombe, M.: Incorporating Temporal Reasoning into Activity Recognition for Smart Home Residents. In: AAAI 2008 Workshop, Chicago, USA, July 13-17 (2008)
4. Youngblood, G.M., Cook, D.J., Holder, L.B.: Managing adaptive versatile environments. Pervasive and Mobile Computing 1(4), 373–403 (2005)
5. Youngblood, G.M., Heierman, E.O., Holder, L.B., Cook, D.J.: Automation Intelligence for the Smart Environment. In: IJCAI 2005, Edinburgh, Scotland, pp. 1513–1514 (2005)
6. Gopalratnam, K., Cook, D.J.: Active LeZi: An Incremental Parsing Algorithm for Sequential Prediction. In: FLAIRS 2003, Florida, USA, May 11-15 (2003)
7. Aztiria, A., Izaguirre, A., Basagoiti, R., Augusto, J.C.: Learning about preferences and common behaviours of the user in an intelligent environment. Behaviour Monitoring and Interpretation – BMI – Smart Environments 3, 289–315 (2009)
8. Botón-Fernández, V., Lozano-Tello, A.: Learning Algorithm for Human Activity Detection in Smart Environments. In: HAI 2011 Workshop, Lyon, France, August 22-27 (2011)

Reflective Relational Learning for Ontology Alignment

Andrzej Szwabe, Pawel Misiorek, and Przemyslaw Walkowiak

Abstract. We propose an application of a statistical relational learning method as a means for automatic detection of semantic correspondences between concepts of OWL ontologies. The presented method is based on an algebraic data representation which, in contrast to well-known graphical models, imposes no arbitrary assumption with regard to the data model structure. We use a probabilistic relevance model as the basis for the estimation of the most plausible matches. We experimentally evaluate the proposed method employing datasets developed for the Ontology Alignment Evaluation Initiative (OAEI) Anatomy track, for the task of identifying matches between concepts of Adult Mouse Anatomy ontology and NCI Thesaurus ontology on the basis of expert matches partially provided to the system.

1 Introduction

In the case of 'mainstream' Statistical Relational Learning (SRL) methods that are based on graphical models [2, 6], the probabilistic modeling merely enhances the results of structure learning founded purely on the set of first-order logic propositions explicitly provided to the system [3]. In contrast to such methods, an SRL method that is based on algebraic data representations may impose no arbitrary assumptions with regard to the model's structure, i.e., no arbitrary distinction between the phases of structure learning and parameter learning is necessary. Moreover, a 3rd-order tensor is known as a data object type that allows for very convenient representation of heterogeneous relational data, including the data provided as RDF triples [5, 8].

The Tensor-based Reflective Relational Learning System (TRRLS) presented in this paper is based on one of the first tensor-based approaches to Statistical

Andrzej Szwabe · Pawel Misiorek · Przemyslaw Walkowiak
Institute of Control and Information Engineering, Poznan University of Technology,
M. Sklodowskiej-Curie Square 5, 60-965 Poznan, Poland
e-mail: {Andrzej.Szwabe,Pawel.Misiorek,
         Przemyslaw.Walkowiak}@put.poznan.pl

S. Omatu et al. (Eds.): Distributed Computing and Artificial Intelligence, AISC 151, pp. 519–526.
springerlink.com   © Springer-Verlag Berlin Heidelberg 2012

Relational Learning - referred to as Tensor-based Reflective Relational Learning Framework (TRRLF). One of the key components of TRRLF is a model of probabilities corresponding to all logical propositions that a TRRL system deals with – a model which (in contrast to graphical models) does not require making any conditional independence assumptions at the structure learning phase.

2 Related Work

Tensor-based modeling is a new approach to Statistical Relational Learning; it may be regarded as an alternative to more common solutions based on graphical models. According to [10], this trend has been initiated by research on collective matrix factorization applied to relational data [9]. In [11], the authors employed the tensor factorization approach to the framework. The model is based on Bayesian clustering and is not designed to deal with data provided as RDF triples. Two most recent works which involve using tensors to represent relational data given as RDF triples are [5] and [8]. In [5], the authors used the PARAFAC decomposition. The approach presented in [8] contributes an efficient algorithm to compute the factorization of a 3rd-order tensor, but it is limited to the dyadic relational data model. Neither approach targets the probabilistic interpretation of input data and processing results.

3 Algebraic Model of Relational Data

We have chosen the 3rd-order-tensor as the simplest algebraic structure that is suitable to represent relational data, including the data provided as RDF triples [8]. The proposed model enables probabilistic interpretation of both input data and processing results. In contrast to the existing tensor-based approaches to SRL [5, 8], we model propositional data in a way that enables full flexibility of specifying the roles that any pair of entities plays with regard to any relation. Instead of using the dyadic relational model [8], we represent the 'active' (as the relation's subject) and the 'passive' (as the relation's object) 'views' of a given entity as potentially fully independent entities.

3.1 Tensor-Based Representation of Relational Data

In the paper, we use the concept of tensor which is defined as a multidimensional array. The k-th order tensor is an element of the tensor product of k vector spaces. These vector spaces have their coordinate systems, which may have different lengths. In particular, in this paper we focus on 3rd-order tensors. It is noteworthy (in order to avoid confusion) that the distinction should be drawn between the use of the word 'tensor' in papers on semantic integration [8], and its use in mathematics.

We introduce system T representing subject-predicate-object dependencies as a 3rd-order tensor. System T models the set of relations (predicates) $R = \{r_k\}$ between subjects from set $S = \{s_i\}$ and the objects from set $O = \{o_j\}$. Let us assume that

$|R| = m$ and that $|S| = |O| = n$. The latter assumption is motivated by the fact that we allow each entity, that we would like to represent in the tensor, to play the role of a subject or an object. Let us define the tensor as $T_{i,j,k} = [t_{i,j,k}]_{n \times n \times m}$, where value $t_{i,j,k}$ describes our initial knowledge about relation k from subject i to object j. Let us assume that $t_{i,j,k} \geq 0$ for every i, j, k.

Each element (i.e. subject, object, or predicate) from set $E = S \cup O \cup R$ is represented by a unique slice (2nd-order tensor) of the 3rd-order tensor. In order to incorporate a new relation to the tensor, one needs to add an additional slice corresponding to the relation and ensure that all the subjects and objects, which are involved in this particular relation, are already represented by slices in both the object and subject tensor modes.

Additionally, let us define set F as a set of all known facts (i.e., RDF triples (i, j, k)) which are used to build the input tensor. The number $|F| = f$ determines the number of positive cells in the input tensor.

3.2   Vector-Space for Subjects, Predicates, Objects and Facts

For each element described in T, i.e., for relations, objects, and subjects, as well as for all known facts stored in the input tensor, we provide an additional representation in the d-dimensional vector space called the context vector. Context vectors are stored in matrix $X = [x_{i,j}]_{(2n+m+f) \times d}$. This matrix is used in the training procedure, which is aimed at determining the correspondences between elements from set $E = S \cup O \cup R$, based on the information about connections between them stored in the tensor. We assume that the first n rows of matrix X describe elements from set S, the rows indexed from $n + 1$ to $2n$ describe elements from set O, and the next rows indexed from $2n + 1$ to $m + 2n$ describe elements from set R. Let us denote the top n rows of X by $n \times d$ matrix X^S (of vectors x_i^S for $i = 1..n$), the next n rows (rows indexed from $n + 1$ to $2n$) by $n \times d$ matrix X^O (of vectors x_j^O for $j = 1..n$), and the next m rows by $m \times d$ matrix X^R (of vectors x_k^R for $k = 1..m$). Additionally, we denote the top $2n + m$ rows of X by $(2n + m) \times d$ matrix X^E representing context vectors of all the entities from set $E = S \cup O \cup R$. The bottom f rows of matrix X contain context vectors representing facts from the set F and form $f \times d$ matrix X^F.

The vector space used for modeling context vectors enables their comparable common representation, which may then be used for the purpose of reflective learning on the basis of the connections between entities modeled as a 3-rd order tensor. Such a representation makes it possible to model all the real-world objects described in the system as compatible objects [13]. The TRRL framework allows to configure the length of context vectors by setting parameter d.

3.3   Proposition Probability Space

The tensor reconstruction from the set of context vectors is based on probabilistic relevance modeling used in the area of quantum information retrieval [13].

The probability space $(\Omega, \mathcal{F}, P)$ provides a functional interpretation of any possible state of system T. This space is a conditional probability space representing the system state, under the condition that a given relation between a given subject and a given object is observed. The sample space $\Omega = \{A_{i,j,k}\}$ of the probability space is a set of events such as $A_{i,j,k}$ – an event corresponding to the presence of relation k ($r_k \in R$) from subject i ($s_i \in S$) to object j ($o_j \in O$) is observed. $\mathcal{F}$ is defined as a set of all the possible subsets of Ω. Finally, the probability measure $P: \mathcal{F} \to [0,1]$ is defined according to a distribution which can be presented as the following 3rd order tensor $B^P = [b^P_{i,j,k}]_{n \times n \times m}$, where $b^P_{i,j,k} = P(A_{i,j,k})$ for $i = 1..n$, $j = 1..n$, $k = 1..m$.

A TRRL system builds the input tensor from RDF triples, in such a way that all the known facts (i.e., propositions) are represented by the non-negative tensor cells. Each tensor cell (i,j,k), which corresponds to some RDF triple from set F (i.e., for which we know from input data that relation k for subject i and object j holds), is equal to the same positive value, whereas all the values in the remaining cells are set to be equal 0. Then the tensor is normalized in order to obtain the distribution $b^P_{i,j,k}$ (i.e., in order to ensure that $\sum_{i=1}^{n} \sum_{j=1}^{n} \sum_{k=1}^{m} b^P_{i,j,k} = 1$).

Probabilities of events: A^S_i - the event that some relation from subject i to some object is observed, A^O_j - the event that some relation from some subject to object j is observed, and A^R_k - the event that relation k from some subject to some object is observed, are calculated based on distribution described by B^P, by summarizing the entries in the corresponding slices of this tensor. In particular, we have $P(A^S_i) = \sum_{j=1}^{n} \sum_{k=1}^{m} b^P_{i,j,k}$ for $i = 1..n$, $P(A^O_j) = \sum_{i=1}^{n} \sum_{k=1}^{m} b^P_{i,j,k}$ for $j = 1..n$, and $P(A^R_k) = \sum_{i=1}^{n} \sum_{j=1}^{n} b^P_{i,j,k}$ for $k = 1..m$. These probabilities are used in the procedure of tensor reconstruction after the training.

Following the principle of indifference, for the purposes of learning and matches calculation procedure we use conditional events $A_{i,j,k}|A^S_i$, $A_{i,j,k}|A^O_j$, and $A_{i,j,k}|A^R_k$. We assume that events $A_{i,j,k}|A^S_i$, $A_{i,j,k}|A^O_j$, and $A_{i,j,k}|A^R_k$ are independent. More precisely, we build the tensor $C = [c_{i,j,k}]_{n \times n \times m}$, where

$$c_{i,j,k} = P\left(\left(A_{i,j,k}|A^S_i\right) \cap \left(A_{i,j,k}|A^O_j\right) \cap \left(A_{i,j,k}|A^R_k\right) \right) \tag{1}$$

$$= P\left(A_{i,j,k}|A^S_i\right) P\left(A_{i,j,k}|A^O_j\right) P\left(A_{i,j,k}|A^R_k\right) = \frac{P(A_{i,j,k})}{P(A^S_i)} \frac{P(A_{i,j,k})}{P(A^O_j)} \frac{P(A_{i,j,k})}{P(A^R_k)}.$$

TRRLF does not involve a typical tensor factorization - a cell is not factorized into a set of additive components, but is reconstructed, using the formula derived from quantum probability calculations, which interweaves the influence of the three factors:

$$T^{rec}(i,j,k) = cos^2(x^S_i, x^O_j)cos^2(x^S_i, x^R_k)cos^2(x^O_j, x^R_k). \tag{2}$$

3.4 Context Vector Update Procedure

Both tensor B^P and matrix X are used in the relational learning procedure. Matrix X consists of context vectors which are updated during learning, whereas the tensor is applied as a source of data used in the learning process.

We propose the Tensor-based Reflective Indexing procedure, which allows each vector to be 'learned' by other context vectors according to the 'connections' between entities represented in tensor B^P. Similarly to the method presented in [12], the proposed procedure is based on consecutive reflections conducted in a way typical for the Reflective Random Indexing (RRI) approach [12]. The tensor B^P is used to construct matrix $A = [a_{i,j}]_{(2n+m) \times f}$ which describes correlations between entities and facts represented as tensor B^P. In particular, matrix A is calculated in the following way:

$$a_{i,j} = \begin{cases} (r_i)^{-1/2} & \text{if element } i \text{ is the subject, object or predicate of fact } j, \\ 0 & \text{otherwise,} \end{cases} \quad (3)$$

for $1 \leq i \leq 2n+m$, and $1 \leq j \leq f$, where r_i is the number of facts in F, which concern element i. At the first step of the procedure, for each row of matrix X^F we select s coordinates (uniformly at random from the set of d dimensions) and set them to be equal to 1. Then each row of X^F is normalized using the formula $x_{i,\cdot}^F = x_{i,\cdot}^F / \|x_{i,\cdot}^F\|_2$, for $1 \leq i \leq f$. Each reflection step is based on the following context vector update procedure:

1. $X^E := AX^F$,
2. $\forall_{i=1..2n+m} \forall_{j=1..d} \ x_{i,j}^E := \sqrt{x_{i,j}^E}$,
3. $x_{i,\cdot}^E := \frac{x_{i,\cdot}^E}{\|x_{i,\cdot}^E\|_2}$, for $1 \leq i \leq 2n+m$,
4. $X^F := A^T X^E$,
5. $\forall_{i=1..f} \forall_{j=1..d} \ x_{i,j}^F := \sqrt{x_{i,j}^F}$,
6. $x_{i,\cdot}^F := \frac{x_{i,\cdot}^F}{\|x_{i,\cdot}^F\|_2}$, for $1 \leq i \leq f$.

After each reflection step, for each known fact (i, j, k) from set F we, calculate its probability value using the reconstruction formula (Eq. (2)). The learning procedure stops when all the reconstructed values $T^{rec}(i, j, k)$ are at least as big as the corresponding value $c_{i,j,k}$ (calculated according to Eq. (1)).

4 TRRL-Based Ontology Alignment

To the best of our knowledge, TRRLS is the first tensor-based system that has been applied for any OAEI task [14]. Following the rules of OAEI, TRRLS uses two ontologies as an input (a source and a target) – both in the OWL format [15]. Input OWLs are represented as a set of matrices, where each matrix corresponds to one of the relations.

The tensor for the OAEI Anatomy track scenario is constructed in the following way. Each ontology relation is modeled as a slice consisting of two submatrices describing facts for the source ontology and the target ontology, respectively. The next *hasTerm* and *TermOf* slices represent terms that 'describe' the entities. The use of the *termsOf* slice is optional – this relation is the inverse of the *hasTerm* relation. The *weakIdentity* slice is a diagonal matrix combining the role of an object and the role of a subject for the same entity. The final correspondences between entities are determined using the slice *matchesTo*, which is built using the information about partial matches that is available to the OAEI competitors for the purposes of Subtask #4 of the OAEI Anatomy track [4].

TRRLS matches generation procedure starts after the execution of the context vector update procedure, described in Subsect. 3.4. In order to calculate the tensor values a reconstruction formula (2) is used, where x_i^S corresponds to entities from compared ontologies in the subject mode, x_j^O corresponds to entities from compared ontologies in the object mode, and x_k^R corresponds to *matchesTo* relation. The submatrix obtained this way is used to establish the final matches. The matches selection procedure is based on selecting maximum values over the rows and columns and then applying the thresholding operation.

5 Experimental Evaluation

The experiments presented in this section have been performed using the OAEI 2010 dataset from the Anatomy track [4]. The experiments followed the scenario of identifying matches between the concepts of Adult Mouse Anatomy (MA) and NCI Thesaurus ontology.

We have followed the scenario of OAEI Anatomy Subtask #4, which involves using the data set of the so-called partial matches in the matchmaking process [14], what is equivalent to the ostensive retrieval approach [13]. We have compared our results with those provided by the systems competing in OAEI 2010 Anatomy Subtask #4 [4]: the Agreement Maker (AgrMaker), the Automated Semantic Mapping of Ontologies with Validation system (ASMOV), and the Combinatorial Optimization for Data Integration system (CODI).

For the purposes of tests presented in this paper the TRRL system has built a tensor describing 2737 entities from MA ontology, 3298 entities for NCI ontology, 2193 terms, and the total number of 53427 facts. We evaluate the performance of the system for $d = 1600$. We use $s = 2$ for the random initialization of matrix X^F. Moreover, we set the threshold value for the matches generation procedure in such a way that the system is allowed to provide additional 300 matches more than the available 987 partial matches. In order to obtain the results which are comparable with those obtained by the competitors of OAEI 2010 Anatomy Subtask #4, we strictly follow the methodology described in [4]. We have repeated each experiment 10 times in order to evaluate the impact of the method's randomness on the matching quality [1]. Table 2 presents the Precision, Recall, and F-measure results in terms of the minimum value, the maximum value, the average value and the standard devia-

Table 1 Results of OAEI 2010 Anatomy Subtask #4.

Competitor	Precision	Recall	F-measure
AgreementMaker	0.929	0.851	0.888
ASMOV	0.837	0.808	0.822
CODI	0.968	0.746	0.843

Table 2 The experimentation results.

	Precision	Recall	F-measure
average	0.908	0.770	0.833
standard deviation	0.002	0.002	0.002
minimum	0.904	0.768	0.831
maximum	0.911	0.773	0.836

tion. Although, in contrast to measures based on ROC analysis [12], the Precision, Recall, and F-measure do not provide a probabilistic interpretation of measured results, we have decided to use these measures as they allow for the comparability with the results of the OAEI contest leaders.

Table 1 summarizes the results of Subtask #4 of OAEI 2010 Anatomy track. AgrMaker and ASMOV provide relatively good Recall scores, but these systems use external knowledge sources (they use deeper linguistic analysis based on Word-Net or UMLS) in order to increase their performance. CODI is based on lexical similarity measures combined with the Markov logic approach. The system has relatively good Precision results but suffers from relatively weak Recall [4]. In contrast, our method is neither supported by an external knowledge base nor focuses on the lexical similarity of matched nodes' names. Despite those assumptions, the TRRL system allows to achieve matching quality comparable the leading methods.

6 Conclusions

TRRLS is an application of the reflective relational learning framework that may be regarded as a general-purpose (i.e., domain-unspecific) ontology matching tool. We have demonstrated that despite the fact that the tool does not utilize external knowledge sources, it may be successfully used for the ontology alignment task. We believe that the future system enhancements (e.g., concerning the introduction of domain knowledge) will lead to even better performance of TRRLS.

Being an SRL solution, the TRRL framework is likely to be successfully applicable to schema matching and directory matching systems, as well as to systems from other application areas, especially in scenarios, in which case the probability, rather than the provability of automated inferences, is of key importance [7].

Acknowledgements. This work is supported by the Polish Ministry of Science and Higher Education grant N N516 196737, and by Poznan University of Technology grant 45-085/11 DS-PB.

References

1. Ciesielczyk, M., Szwabe, A.: RI-based Dimensionality Reduction for Recommender Systems. In: Proc. of 3rd International Conference on Machine Learning and Computing, IEEE Press, Singapore (2011)
2. De Raedt, L.: Logical and Relational Learning. Springer (2008)
3. Dietterich, T., Domingos, P., Getoor, L., Muggleton, S., Tadepalli, P.: Structured Machine Learning: the Next Ten Years. Machine Learning 73(1), 3–23 (2008)
4. Euzenat, J., Ferrara, A., Meilicke, C., Nikolov, A., Pane, J., Scharffe, F., Shvaiko, P., Stuckenschmidt, H., Svb-Zamazal, O., Svtek, V., Trojahn dos Santos, C.: Results of the Ontology Alignment Evaluation Initiative 2010. In: Proc. of 5th ISWC Workshop on Ontology Matching (OM), Shanghai, pp. 85–117 (2010)
5. Franz, T., Schultz, A., Sizov, S., Staab, S.: Triplerank: Ranking Semantic Web Data by Tensor Decomposition. In: The Semantic Web-ISWC 2009, pp. 213–228 (2009)
6. Getoor, L., Taskar, B.: Introduction to Statistical Relational Learning. The MIT Press (2007)
7. Haenni, R.: Towards a Unifying Theory of Logical and Probabilistic Reasoning. In: ISIPTA, pp. 193–202 (2005)
8. Nickel, M., Tresp, V., Kriegel, H.-P.: A Three-Way Model for Collective Learning on Multi-Relational Data. In: Proceedings of the 28th International Conference on Machine Learning (2011)
9. Singh, A.P., Gordon, G.J.: Relational Learning via Collective Matrix Factorization. In: Proceeding of the 14th ACM SIGKDD International Conference on Knowledge Discovery and Data Mining, pp. 650–658 (2008)
10. Struyf, J., Blockeel, H.: Relational Learning. In: Sammut, C., Webb, G. (eds.) Encyclopedia of Machine Learning, pp. 851–857. Springer (2010)
11. Sutskever, I., Salakhutdinov, R., Tenenbaum, J.B.: Modelling Relational Data Using Bayesian Clustered Tensor Factorization. In: Advances in Neural Information Processing Systems, vol. 22 (2009)
12. Szwabe, A., Ciesielczyk, M., Misiorek, P.: Long-Tail Recommendation Based on Reflective Indexing. In: Wang, D., Reynolds, M. (eds.) AI 2011. LNCS, vol. 7106, pp. 142–151. Springer, Heidelberg (2011)
13. van Rijsbergen, C.J.: The Geometry of Information Retrieval. Cambridge University Press, New York (2004)
14. Ontology Alignment Evaluation Initiative, Campaign (2011), http://oaei.ontologymatching.org/2011/
15. W3C. Web Ontology Language (OWL), http://www.w3.org/2004/OWL

On the Evolutionary Search for Data Reduction Method

Hanna Lacka and Maciej Grzenda

1 Introduction

One of the key applications of statistical analysis and data mining is the development of the classification and prediction models. In both cases, significant improvements can be attained by limiting the number of model inputs. This can be done at two levels, namely by eliminating unnecessary attributes [3] and reducing the dimensionality of the data [12].Variety of methods have been proposed in both fields.

As far as dimensionality reduction (DR) techniques are concerned, Principal Component Analysis (PCA) [9, 10, 12], Multidimensional Scaling (MDS) [10, 12], Fisher Linear Discriminant (FLD)[10], and Kernel PCA (KPCA) [12] have been used for many years. Other techniques, originally invented not as DR methods, such as Self-Organising Maps (SOMs) [5, 12] or Multilayer Perceptrons [5], can be used to perform DR, too. In a recent version of Matlab Toolbox for Dimensionality Reduction[11], more than 30 different dimensionality reduction techniques are made available. Not surprisingly, this makes the selection of an appropriate technique problematic. Apart from selecting a method reducing a dimension, a partly subjective decision on the reduced dimension has to be made. However, in our recent study it has been shown that a priori selection of a reduced dimension may not yield the optimal quality of the prediction models created with the reduced data [4]. More precisely, the impact of reduced features e.g. principal components on the predicted feature or a class of an object is largely unknown.

Hanna Lacka
Polish Academy of Sciences, Institute of Computer Science, 01-237 Warszawa, ul. J.K. Ordona 21, Poland
e-mail: H.Lacka@phd.ipipan.waw.pl

Maciej Grzenda
Warsaw University of Technology, Faculty of Mathematics and Information Science, 00-661 Warszawa, Pl. Politechniki 1, Poland
e-mail: M.Grzenda@mini.pw.edu.pl

S. Omatu et al. (Eds.): Distributed Computing and Artificial Intelligence, AISC 151, pp. 527–534.
springerlink.com   © Springer-Verlag Berlin Heidelberg 2012

Hence, potential benefits of data reduction methods are largely hindered by the fact that the selection of an appropriate feature selection or dimensionality reduction technique is not an easy task. At the same time, evolutionary methods have been developed to tackle difficult global optimisation problems. Thus, they might be considered as methods used to both select dimensionality reduction settings and eliminate unnecessary attributes. Different approaches can be taken to represent features and feature transformation space within genotype [14]. Assuming, the ultimate objective of data reduction is the improvement of model quality developed with the reduced data, the quality of the model should be the key component of a fitness function. Resulting from such an approach is a hybrid method with a feedback-loop between e.g. a classifier being trained and tuned, and a candidate feature extraction [14]. In our study, a comparison of a group of popular dimensionality reduction techniques and an evolutionary algorithm is made. The role of the algorithm is to eliminate unnecessary attributes and training patterns negatively affecting the training process. To implement classification and prediction models, multilayer perceptrons are used.

The remainder of the work is organised as follows:

- An overview of the proposed evolutionary method is described in Sect. 2,
- The results of the simulations made with the method are summarised in Sect. 3,
- Finally, conclusions and the main directions of future work are outlined in the summary.

2 Evolutionary Selection of Data Set Transformation

2.1 Formal Definition

This section describes the evolutionary selection of data set transformation (ESDT). The purpose of the method is to select a transformation which reduces data set dimensionality, while preserving as much of the information content as possible. Unlike in most works, a proposal is made to reduce the dimensionality of the data in view of the prediction or classification problem the data set is used for. Hence, the data transformation is not performed as a part of data preprocessing. In other words, ESDT is not refraining from the impact of input attributes on the output features. Using a separate "black-box" algorithm for evaluation of a candidate data transformation, it does not rely only on properties of data. ESDT is intended as a wrapper method [13]. The data reduction of a data set $D \subset \mathbb{R}^N$ is defined by a function used to code an element $x \in D$ [12]:

$$\psi : \mathbb{R}^N \longrightarrow \mathbb{R}^R, \ x \longrightarrow \tilde{x} = \psi(x) \tag{1}$$

When the role of ESDT is only to eliminate unnecessary attributes,

$$\tilde{x} = [x_{i_1}, \ldots, x_{i_R}], 1 \leq x_{i_1} < x_{i_2} < \ldots x_{i_R} \leq N \tag{2}$$

Moreover the key criterion is the optimisation of the quality of classification or prediction models M i.e. the search for Ψ_m such that

$$E(M(\Psi_m(D))) = min_i E(M(\Psi_i(D))), \Psi_i : \mathbb{R}^{k \times N} \longrightarrow \mathbb{R}^{k_i \times R_i} \qquad (3)$$

2.2 Main Scheme of the Method

The ESDT method is based on the standard Genetic Algorithm (GA)[7], adapted for the problem. Moreover, some inspiration for the method was taken from a GA-based Instance Selection algorithm [8], that improved generalization skills of a neural network simulated for complex problems with highly correlated data. GA-based IS performs feature and pattern selection, evaluating the candidate solutions with the usage of a nearest neighbour [1] classifier. In comparison to GA-based IS, ESDT method proposes a more complex genotype version and uses a more universal evaluation tool. The evolutionary mechanisms applied include tournament or roulette reproduction. A uniform probability mutation operator is used to change the selection of individual attributes and patterns. A singlepoint crossover operator is applied to swap genotype parts between two parent individuals. Each candidate solution, i.e. the reduced data set, is evaluated using a predefined multilayer perceptron neural network (MLP). The algorithm stops when a certain number of generations or an expected precision of the solution is reached. Finally, the ESDT parameters are as follows:

- recounting: number of MLP training sessions ($fitN$), number of method recounts ($algN$),
- evolution: maximum number of generations ($genN$), population size ($popN$), solution precision($evoE$), fraction of elite individuals in population ($elitFr$), fraction of the offspring deriving from mutation ($mutFr$), mutation probability for single genotype unit ($mProb$), crossover probability for a pair of parents ($cProb$),
- MLP: maximum number of epochs ($epochsN$), precision (nnE), number of hidden layers, the number of neurons in the hidden layer ($L1N$).

2.3 Genotype

Two alternative versions of the genotype of an individual were proposed for the method. The basic version is represented by the concatenation of two binary vectors: an N-element one, coding a selection (1) or omission (0) of the consecutive attribute of the pattern, and an M-element one, coding selection (1) or omission (0) of the consecutive pattern. The advanced version of the genotype consists of two parts:

1. A part representing a transformation method to apply, consisting of:

   a. a subpart representing an alternative selection of a feature extraction method, where 0 - omission of feature extraction, 1 - PCA, 2 - MDS, 3 - FLD,
   b. a subpart representing an alternative (0,1) omission of certain attributes.

2. A part representing the parameters of the chosen methods, consisting of:

   a. a subpart representing the number (L) of left components for the data transformation method or 0 - omission of feature extraction ($L := N$),
   b. a subpart being a binary vector of length N, representing a list of consecutive attributes to select (1) or omit (0), where at least one attribute is selected and a maximum possible position of a selected attribute is L. In the case of $L < N$ this part of the genotype defines the selection of reduced features. Hence, the reduced dimension is no greater than L i.e. $R \leq L$.

2.4   The Fitness Function

The minimalised fitness function is calculated as defined in Alg. 1.

Input: D - a matrix of input patterns, F - vector of output features, Ψ - a function defined by the genotype, performing data set transformation, ϕ - a function defined by the genotype, performing data set transformation without pattern omission, *recounts* - number of MLP training sessions, *params* - user-defined MLP parameters

Data: L - number of attributes left after data set transformation, *inputSize* - input size for MLP, *net* - MLP, T,V,S - matrices representing a training, validation and testing set for MLP, T',V',S' - submatrices of T,V,S representing reduced data sets, *testErrors* - an array of testing errors for each recount, i - an iterator

Result: $median(testErrors)$ - median classification or prediction testing error of MLPs.

begin
  $T,V,S = \text{DivideToTrainValidationTestSets}(D,F)$;
  $T',L = \Psi(T); V' = \phi(V); S' = \phi(S);$
  $inputSize = L$;
  $net = \text{CreateMLP}(inputSize, params)$;
  for $i = 1 \ldots recounts$ **do**
    $net = \text{TrainMLP}(net,T',V')$;
    $testErrors_i = \text{TestMLP}(net,S')$;
  end
  return $median(testErrors)$;
end

Algorithm 1: ESDT. Calculation of the fitness function

3   Experimental Results

3.1   Introduction

In order to check the effectiveness of ESDT method a group of experiments was performed aiming to compare the benefits of ESDT approach and non-evolutionary data set transformation methods. Each experiment tests a different transformation method for 16 data sets from UCI Machine Learning Repository [2]. The following criteria for data sets selection were used:

- Task type. Equal number of data sets representing classification tasks and of those representing regression tasks.
- A ratio defined as the number of attributes to the number of data records of a given data set. At least one representative data set from each of the $G1$, $G2$, $G3$ groups, defined as data sets with the ratio from, respectively, $[0.001, 0.01)$, $[0.01, 0.1]$ and $[0.1, 1]$ intervals.

The prognostic tasks chosen include: Wine Quality (Red) ($G1$), Concrete Compressive Strength ($G1$), Auto MPG ($G2$), Forest Fires ($G2$), Housing ($G2$), Computer Hardware ($G2$), Automobile ($G3$) and Breast Cancer Wisconsin (Prognostic) ($G3$). The chosen classification problems include: Wine Quality (White) ($G1$), Car Evaluation ($G1$), Blood Transfusion Service ($G1$), Image Segmentation ($G1$), Iris ($G1$), Glass ($G2$), Breast Tissue ($G2$) and Connectionist Bench ($G3$).

The experiments of four categories were performed, i.e. the data reduction performed with ESDT and its basic genotype (experiment 1), with ESDT and its advanced genotype (experiment 2), with GA-based IS (experiment 3) and with standard DR techniques (experiments 4-6). Once the final data transformation method was determined, the network training, pruning and assessment process was performed. The pruning process was based on the Optimal Brain Surgeon (OBS) method [6]. The error rates reported for individual experiments are the classification or prediction error values resulting from the evaluation of the final pruned MLP networks. The following evolution and MLP training parameters were set for experiments: $genN$=50, $popN$=50, $evoE$=$1E-2$, $elitFr$=$4E-2$, $mutFr$=$2E-1$, $mProb$=$1E-2$, $epochsN$=$5E+1$, nnE=$1E-2$. Moreover, in the case of PCA, FLD and MDS methods, the reduced dimension was determined individually for every data set. It was set to the number of features explaining 95% of data variance i.e. using the proportion of explained variation (PEV) criterion [9] applied in PCA.

3.2 Detailed Results

Extensive calculations were made to objectively assess the impact of different data reduction techniques on the accuracy of the models created with the reduced data.

The details of experiments 1 and 2 are reported in Tables 1 and 2, respectively. In spite of enabling both the selection of raw input attributes and transformed reduced features, the results reported in Table 2 are not superior to the basic form of the algorithm. The simplicity of the basic genotype results in significantly better results. The results of experiment 3 added to the Table 1 for similar evolutionary methods comparison, indicate the ESDT method using MLP classifier returns better results than GA-based IS.

Table 1 Experiments 1 and 3. Detailed results of data set transformation with ESDT method having first version of genotype and GA-based IS error rates results.

No.	Problem	DD	algN	L1N	fitN	cProb	ESDT v.1 testing errors	ESDT v.1 training errors	GA-based IS testing errors	GA-based IS training errors	Avg. gen. no.	Exec. time
1	Wine (Red)	60-20-20	1	5	10	1.0	**2.943E-01**	3.221E-01	-	-	51	2
2	Concrete CS	60-20-20	10	11	10	1.0	**6.800E-03**	5.000E-03	-	-	2	1
3	Auto MPG	c-v:10	1	3	10	1.0	**7.600E-03**	6.100E-03	-	-	2	0
4	Forest Fires	c-v:10	5	3	10	1.0	**4.800E-03**	6.000E-03	-	-	2	0.25
5	Housing	c-v:10	3	13	10	1.0	**8.268E-03**	5.625E-03	-	-	2	0.25
6	Computer Hardware	c-v:10	10	3	10	1.0	**2.800E-03**	2.700E-03	-	-	2	0.25
7	Automobile	c-v:10	1	5	10	1.0	**1.364E-02**	8.567E-04	-	-	2	0
8	BCW (Progn.)	c-v:10	5	3	10	1.0	**3.940E-02**	2.210E-02	-	-	51	9.5
9	Wine (White)	60-20-20	1	5	10	1.0	**62.880%**	61.930%	**48.880%**	48.980%	51	37.75
10	Car Evaluation	60-20-20	1	13	10	1.0	**10.430%**	10.330%	**29.100%**	16.380%	51	29.25
11	Blood Transf. SC	c-v:10	1	5	10	1.0	**25.472%**	22.652%	**26.920%**	25.960%	51	1.25
12	Image Segment.	60-20-20	1	9	10	1.0	**6.240%**	5.060%	**7.955%**	7.590%	51	33
13	Iris	c-v:10	5	5	10	1.0	**3.330%**	2.840%	**0.000%**	2.564%	17.4	3.5
14	Glass	c-v:10	5	3	10	1.0	**16.230%**	15.840%	**63.333%**	48.228%	51	12
15	Breast Tissue	c-v:10	1	3	10	1.0	**47.730%**	31.580%	**66.667%**	34.783%	51	3.75
16	Connectionist Bench	c-v:10	1	35	3	1.0	**17.020%**	0.000%	**37.500%**	0.000%	51	30

Problem - a data set tested, DD - percentage data division (% of learning sets -% of validation sets -% of testing sets) or k-fold cross-validation (c-v:k), Avg. gen. no. - average no. of generations to stop evolution, Exec. time - execution time in hours, rounded to the nearest quarter of an hour.

In experiments 4 and 5, in all the cases, the number of training sessions $algN$ was set to 50 and the same reduced dimension R_i was applied for each i-th data set. The results of using MDS proved to be quite similar to the results attained when PCA was applied. In fact, it can be shown that PCA and metric MDS minimise the same criterion [12]. Finally, FLD technique was applied for the classification problems. It is worth noting that the FLD method is the only one using the target class data to calculate the transformation simplifying the separation of classes.

3.3 Summary of the Results

The summary of the results attained with all the data transformation methods is reported in Table 3. The comparison points out relatively good performance of the basic version of the evolutionary algorithm. Its advanced version returned unsatisfactory results, also in comparison with non-evolutionary methods. Interestingly, this may suggest that the simplifications of the search space for an evolutionary algorithm caused by the basic genotype, result in substantial improvement of results, in spite of the elimination of both PCA and FLD-based transformation. The advanced form of the genotype makes the crossover operator more problematic as two parent individuals may exhibit two substantially different genotypes, using subsets of the features arising from different data transformations.

Finally, what should be emphasised is the potential of the hybrid approach combining neural networks with dimensionality reduction techniques and evolutionary approach. Moreover, the impact of the hybrid combination can be easily observed, as similar MLP network architectures were used in all the algorithm runs.

Table 2 Experiment 2. Detailed results of data set transformation with ESDT method having second version of genotype.

No.	Problem	DD	$algN$	$L1N$	$fitN$	$cProb$	**Testing errors**	Training errors	Avg. gen. no.	Exec. time
1	Wine Red	60-20-20	1	5	10	5E-3	**4.393E-02**	1.426E-02	51	9
2	Concrete CS	60-20-20	3	11	10	5E-3	**8.141E-03**	3.447E-03	2	1
3	Auto MPG	c-v:10	1	3	10	5E-3	**2.083E-02**	5.250E-03	2	5
4	Forest Fires	c-v:10	1	3	10	5E-3	**4.030E-02**	6.700E-03	2	2.5
5	Housing	c-v:10	1	5	3	5E-3	**3.429E-02**	3.771E-03	2	1.5
6	Computer Hardware	c-v:10	1	3	3	5E-3	**1.127E-02**	3.142E-03	2	0.75
7	Automobile	c-v:10	1	5	10	5E-3	**1.039E-01**	2.085E-03	2	3.25
8	BCW Progn.	c-v:10	2 c-v	3	5	5E-3	**7.644E-02**	3.193E-02	51	11.75
9	Wine (White)	60-20-20	1	5	3	5E-3	**56.692%**	46.066%	51	43.5
10	Car Evaluation	60-20-20	1	13	3	5E-3	**35.362%**	5.207%	51	43
11	Blood Transf. SC	c-v:10	1	5	1	5E-3	**31.778%**	22.717%	51	31.75
12	Image Segment.	60-20-20	1	9	3	5E-3	**84.293%**	6.550%	51	0.5
13	Iris	c-v:10	1	5	1	5E-3	**17.143%**	2.126%	2	0.75
14	Glass	c-v:10	1	3	1	5E-3	**19.048%**	12.865%	2.5	2.75
15	Breast Tissue	c-v:10	1	3	10	5E-3	**71.818%**	33.338%	51	76
16	Connectionist Bench	c-v:10	1 c-v	35	1	5E-3	**40.000%**	32.934%	51	18.25

Table 3 Summary of results. Testing error rates.

No.	Problem	ESDT v.1	GA-based IS	ESDT v.2	PCA	FLD	MDS
1	Wine (Red)	2.943E-01	-	4.393E-02	2.610E-02	-	**2.550E-02**
2	Concrete CS	**6.800E-03**	-	8.141E-03	1.498E-02	-	1.210E-02
3	Auto MPG	**7.600E-03**	-	2.083E-02	1.210E-02	-	1.190E-02
4	Forest Fires	4.800E-03	-	4.030E-02	6.700E-03	-	**3.900E-03**
5	Housing	**8.268E-03**	-	3.429E-02	3.070E-02	-	3.130E-02
6	Computer Hardware	**2.800E-03**	-	1.127E-02	3.000E-03	-	3.000E-03
7	Automobile	1.364E-02	-	1.039E-01	**6.800E-03**	-	7.500E-03
8	BCW Progn.	3.940E-02	-	7.644E-02	**3.680E-02**	-	3.700E-02
9	Wine (White)	62.880%	**48.880%**	56.692%	54.170%	50.630%	53.660%
10	Car Evaluation	10.430%	29.100%	35.362%	6.380%	**5.510%**	6.090%
11	Blood Transf. SC	**25.472%**	26.920%	31.778%	27.570%	28.300%	27.358%
12	Image Segment.	**6.240%**	7.955%	84.293%	21.220%	24.341%	20.144%
13	Iris	3.333%	**0.000%**	17.143%	3.330%	3.333%	3.333%
14	Glass	16.230%	63.333%	19.048%	**16.070%**	57.143%	16.234%
15	Breast Tissue	**47.730%**	66.667%	71.818%	70.000%	81.820%	70.000%
16	Connectionist Bench	17.020%	37.500%	40.000%	14.643%	38.095%	**14.286%**

4 Summary

Dimensionality reduction is often used to reveal underlying dependencies in the data, and by eliminating redundant or noise components, simplify the development of classification and prediction models. In our work, a selection of the most popular methods was compared with the benefits arising from the use of evolutionary approach to eliminate attributes and patterns that do not positively impact the model quality. Experiments performed with well-known UCI data sets, show that the evolutionary approach can often produce results superior or comparable to more sophisticated techniques such as discriminant analysis or principal component analysis. What is interesting, an advanced form of a genotype enabling the evolutionary search for the best dimensionality reduction method and the list of resulting components in a reduced space did not result in better results. In other words, the complexity of the search space might have hindered the potential of the advanced evolutionary method. Future plans include the creation of a multi-stage algorithm based on the initial evolutionary selection of preserved attributes and patterns and the evolutionary search for the optimal dimensionality reduction in the second stage.

References

1. Cover, T.M., Hart, P.E.: Nearest neighbor pattern classification. IEEE Transaction of Information Theory 13, 21–27 (1967)
2. Frank, A., Asuncion, A.: UCI Machine Learning Repository. University of California, School of Information and Computer Scienc, Irvine, CA (2010), http://archive.ics.uci.edu/ml
3. de Haro-García, A., Pérez-Rodríguez, J., García-Pedrajas, N.: Feature Selection for Translation Initiation Site Recognition. In: Mehrotra, K.G., Mohan, C.K., Oh, J.C., Varshney, P.K., Ali, M. (eds.) IEA/AIE 2011, Part II. LNCS, vol. 6704, pp. 357–366. Springer, Heidelberg (2011)
4. Grzenda, M.: Prediction-Oriented Dimensionality Reduction of Industrial Data Sets. In: Mehrotra, K.G., Mohan, C.K., Oh, J.C., Varshney, P.K., Ali, M. (eds.) IEA/AIE 2011, Part I. LNCS, vol. 6703, pp. 232–241. Springer, Heidelberg (2011)
5. Haykin, S.: Neural Networks and Learning Machines. Person Education (2009)
6. Haykin, S.: Neural Networks: A Comprehensive Foundation. Prentice Hall Inc. (1999)
7. Holland, J.H.: Adaptation in natural and artificial systems. University of Michigan Press, Ann Arbor (1975)
8. Ishibuchi, H., Nakashima, T., Nii, M.: Learning of Neural Networks with GA-based Instance Selection. In: Proc. of 9th IFSA World Congress and 20th NAFIPS International Conference, Vancouver, Canada, July 25-28, pp. 2102–2107 (2001)
9. Larose, D.T.: Data Mining Methods and Models. John Wiley & Sons (2006)
10. Lattin, J.M., Carroll, J.D., Green, P.E.: Analyzing Multivariate Data. Thomson (2003)
11. http://homepage.tudelft.nl/19j49/Matlab_Toolbox_for_Dimensionality_Reduction.html
12. Lee, J., Verleysen, M.: Nonlinear Dimensionality Reduction. Springer (2010)
13. Kohavi, R., John, G.H.: Wrappers for feature subset selection. Artificial Intelligence 97(1-2), 273–324 (1997)
14. Raymer, M.L., Punch, W.F., Goodman, E.D., Kuhn, L.A., Jain, A.K.: Dimensionality reduction using genetic algorithms. IEEE Transactions on Evolutionary Computation 4(2), 164–171 (2000)

Handwritten Character Recognition with Artificial Neural Networks

Stephane Kouamo and Claude Tangha

Abstract. Apart from differents techniques studied in an increasing order of difficulty, this work presents a new approach based on the use of a multilayer perceptron with the optimal hidden neurons. Idea is to compute the training stage by using two classes of prototypes, to represent data already known ; hidden layers are then initialized by that two classes of prototypes. One of the advantages of this technique is the use of the second hiden layer which allows the network to filter better the case of nearby data. The results come from the *Yann Le Cun* database [9], and show that the approach based on the use of a multilayer perceptron with two hidden layers is very promising, though improvable.

Keywords: neural network, back propagation, handwritten character recognition.

1 Introduction

Automatic handwritten character recognition has been for several years the object of intense research, justified by the importance of its several potential applications (mail treatment, check, document compression,...) [12]. Automation of mail sorting is a concrete example. In view of the high quantity of mail to be handled every day, a zip code recognition system would permit to reduce advantageously the scale of this unpleasant and boring task. Automatic handwritten character recognition consist to convert an image which are understandable by human, in an interpretable code by computer. Two disciplines can be then loosened : the on-line recognition and the outstanding recognition [12]. In this paper we are only interested in the case of the outstanding recognition of printed characters in a multi-context writer. Automatic handwritten character recognition is done in three stages, namely :

Stephane Kouamo
University of Yaounde I, Department of Computer Science
e-mail: skouamo@gmail.com

Claude Tangha
University of Yaounde I, Polytechnic National High School
e-mail: ctangha@gmail.com

S. Omatu et al. (Eds.): Distributed Computing and Artificial Intelligence, AISC 151, pp. 535–543.
springerlink.com   © Springer-Verlag Berlin Heidelberg 2012

- segmentation,
- pretreatment consisting of grouping together the nearby pixels blocks of the image,
- classification on itself.

Several techniques are used to realize handwritten character recognition. The approaches using Artificial Neural Network (ANN) and particularly the multilayer perceptron are shown very effective [2].

This paper presents with tests to support, an experimental study of some techniques in an increasing order of difficulty, used to realize character recognition. We present then, a method based on the multilayer perceptron with two hidden layers, used to represent the prototypes of every data class already classified, that sometimes produces better results.

The rest of the document is organized in four sections. The first section presents the differents studied methods of data recognition (we shall limit ourselves to non-parametrics and neuronal classifiers). The second section presents the proposed method based on multilayer perceptron with two hidden layers. The third section is reserved for the implementation of studied methods. The fourth section, presents various results obtained from all studied methods. We finish with a conclusion.

2 Studied Methods

There are several handwritten characters recognition techniques. This paper just talks about a few based on non-parametrics and neuronal classifiers. Therefore, we have : the K-nearest neighbours (KNN), the Dynamic Clouds and recognition by Artificial Neural Networks (ANN).

2.1 The K-Nearest Neighbours (KNN)

The KNN is a supervised learning method used for data classification in artificial intelligence, simply having the objective to predict the class of a new pattern based on the examples already known [5]. The shape classification is compared with other forms having been filled, and we assign the most represented class among the k closest to it. The concept of proximity used here is quantified by a similarity measure. The similarity measure used most frequently is the *Euclidean distance*. For the algorithm begining, the use of examples shape whose class is known are required. The new shape will be compared to those who are already classified.

Algorithm Parameters

The various algorithm parameters adjustable by the user are : the number of examples whose class is known (used at the beginning of algorithm), the value of k and the similarity measure.

Algorithm

1. Choose the initial examples whose the class is known.
2. Choose a vector to be classified.
3. Measure the similarity between the vector to be classified and the vectors already classified.
4. Determine the k vectors for which the similarities are high.
5. Determine the class most represented among these k vectors and assign the vector to be classified.

Remark 1

- The computation increases with the number of vectors already classified.
- The classification results depend on the patterns presentation order.
- The choice of k constitutes an arrangement : a little value permits to define complicated borders between classes, but gives a good feeling for noise. A high value of k defines a smooth border and presents an insensitivity to the noise, but allows to treat the case of classes possessing a reduced number.

2.2 The Dynamic Clouds

Here, the main difference with *KNN* is that each vector to be classified is not compared to all already classified examples, but to only one *representative* vector of every class [5] :

- every class is represented by a vector called *representative* or *prototype*.
- this prototype is obtained by the combination of the affected vectors to its class by the algorithm (here it seems to be taken like the mean of those vectors).
- every vector to be classified is compared to all prototypes by applying a similarity measure (euclidian distance). the vector is associated to a class of the prototype which it is most close.
- the prototype is updated at every application of a new vector to a class it is represented by the equation below :

$$m_{n+1} = \frac{n.m_n + x_{n+1}}{n+1},\qquad(1)$$

with m_{n+1} the update prototype, m_n is the old prototype, $x_n + 1$ the vector to be classified and n the number of vectors already classified.

- the initial representant can be chosen in a random way between vectors of one class.

Algorithm

1. Choose the initial representative of every class (or calculate it from the vectors of this class).
2. Select a vector to be classified.
3. Measure the similarity between the vector to be classified and the representatives of the various classes.
4. Determine the representative corresponding to the maximal similarity.
5. Modify this representative to take into account the new vector allocated to the class.

Remark 2

- The computation increase with the data class number, and not with that of vectors already classified.
- The classification results depend on the initial representatives choice.

2.3 Recognition by Artificial Neural Network

An Artificial Neural Network (*ANN*) is a computation model in which the conception was inspired by the functioning of the real neuron (human beings or not), it can be considered as a set of interconnected units which communicates together and with the outside by means of connections called *synapses* [8]. Several techniques based on neural network have shown efficiency in the data classification domain [10]. In this work, we will insist on the case of *LVQ* (Learning Vector Quantization).

LVQ is very near of the dynamics cloud method [7], but rather than to use vectors mean affected to a class to represent it, *prototypes* vectors are adapted by a training rule at every new classification. As with every supervised training algorithm, membership training pattern information is used during the training stage. *Prototypes* are coded into neuron weight vector ; in classic versions we search for example a minimal distance like maximal similarity criteria. In neural versions we search maximal scalar product between vector to classify and neuron weight vector. We can show that these two criteria are the same [12].

The principle consists of two steps :

1. A training stage : in which neural network is trained to adapt itself on data to be treated by connection weight.
2. A recognition stage : that consists of the use of the trained neural network to classify examples not in the training stage.

For an input vector to be classified, it is compared to all prototype of classes already known according to training ratio. The class of the winner neuron weight vector is assigned to the input vector.

Remark 3

- For an input vector to be classified, if the winner neuron weight vector is a *prototype* of the right class, it is modified by equation (2); it is brings closer the input vector. If the winner neuron weight vector is not a prototype of the right class, it is removed from the input vector.
- The training stage allows the network to adapt on the data to be treated in favour of connections weight, this contributes to improve recognition quality.

By using advantages and disadvantages of different studied methods before, we proposed a new direction based on multilayer perceptron with two hidden layers representing the two classes of prototypes, which in case tests done with Yann Le Cun database [9] gives interesting results.

3   Proposed Method : Recognition with Two Prototypes

In this section, we explain a new recognition method based on the multilayer perceptron with optimal hidden neurons. This method is very close to LVQ, but uses a four layer perceptron to train the neural network with : an *input layer*, a *first hidden layer*, a *second hidden layer* and an *output layer*. Input and output layers have the same size N and the hidden layers have the same size K (with $K > N$). The principle is the same like the LVQ, and is based on two steps : a training and a recognition stage.

The four layers of our network are strongly connected among themselves. The input layer is connected to the first hidden layer, the first and second hidden layers are connected together and the second hidden layer is connected to the output layer, by connections weight. We constitute two samples of prototypes of every classes of data already known. The input vector to be classified is applied on the input layer of our neural network. Connections weight of the first hidden layer are initialized by the first sample of prototypes, and the second hidden layer is initialized by the second sample of prototypes.

Input vector is compared to all prototypes of the first hidden layer, and it is assigned to the class of the winner neuron weight. After this, the winner neuron is also compared to all data of the second hidden layer, and it is assigned to the class of the winner neuron weight too. If the winner neuron is not the prototype of right class, commited errors are back propagated.

Remark 4

- Back propagation of errors permits to treat cases of complex classes,
- The use of two hidden layers secure the capacity of network to handle case of neighbour data classes without improving recognition time.
- Only the first prototype is modified when the right class is recognized.

4   Implementation of Studied Methods

4.1   Preparation and Sample Choice

Data used in this work are handwritten digits from 0 to 9, from *Yann LeCun* database
[9]. This database contains 60000 fixed images with size 28×28. We used two
samples methods :

- 50% of the data for the training stage and 50% for the recognition stage and ;
- 90% of the data for the training stage and 10% for the recognition stage.

4.2   Learning Vector Quantization (LVQ)

The structure of our network consists of three layers : an input layer, one hidden
layer and an output layer. Once the network structure defined, we can apply training
recognition algorithm as :

1. Choose initial representative(prototype) for every data class.
2. Initialize every hidden neuron weight with a prototype of each class.
3. Apply an input vector.
4. Apply the similarity measure as :

$$h_j = \sqrt{\sum_{i=1}^{N}(x_i - w_{ij})^2} \quad 1 \leq j \leq N, \tag{2}$$

5. Select the winner neuron.
6. Modify the winner neuron weight vector by training rule :

$$\Delta W_{ij} = \alpha(X_i - W_{ij}), \tag{3}$$

7. Return to (3) till finishing input vector.

Where W_{ij} represent prototype neuron weight, X_i input vector neuron weight, and α
training ratio.

4.3   Proposed Method

Network structure can be represented as : an *input layer*, a *hidden layer* with two
layers (first hidden layer and second hidden layer) and an output layer. Fig. 1 below
shows us the multilayer perceptron neural network structure, where x_i represents
input data of our network, w_{ij} connection weight between input and hidden layer, h_j

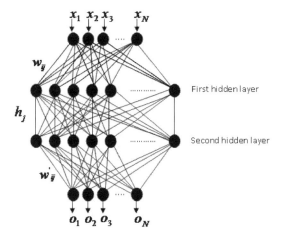

Fig. 1 Multilayer perceptron structure

hidden neurons and o_i output neurons. After the neural network structure is defined, we can apply recognition algorithm like this :

1. Choose initial prototype for every data class.
2. Initialize every first hidden neuron weight with the representative of each first class of data.
3. Initialize every second hidden neuron weight with the representative of each second class of data.
4. Apply an input vector.
5. Apply the similarity measure on the fisrt hidden layer.
6. Select the winner neuron.
7. Apply the similarity measure on the second hidden layer.
8. Select the winner neuron as in (6).
9. Modify the winner weight by :

$$w_{ij} = \delta(x_i - w_{ij})n_j, \tag{4}$$

10. Return to (4) till finishing input vector.

5 Results

After the training stage, we have to use a recognition stage which made successively with a 50% and a 10% data sample coming from *Yann LeCun* database [9], as explained above at section 4.1. Recognition performances will be evaluated by : acceptance or recognition rate, rejection rate, error rate, training time, and recognition time. Before applying one recognition method on a sampling data, treatment consisting of grouping together the blocks of nearly image pixels are made. Obtained results by implementation of the studied methods are given by table 1 below.

Table 1 Studied methods results

	Sample : 50% - 50%					Sample : 90% - 10%				
	KNN	Dynamic Clouds	LVQ	New Method	Parallel method	KNN	Dynamic Clouds	LVQ	New Method	Parallel method
Training time			27m	50m	29m			58m	1h32	50m
Recognition time	27m	28 s	21 s	10 s	6 s	29m	29 s	21 s	10 s	6 s
reconignition rate(in %)	91	89.5	92.1	93	93	95	91.9	96	97	97
Rejection rate(in %)	1	1	1	1	1	1	1	1	1	1
error rate(in %)	7	9.5	6.9	6	6	4	7.1	3	2	2

Table 1 shows that :

(1) Dynamic Clouds give better recognition time than *KNN*. But in return, a decline average of 1.5% is recorded in recognition rate and a rise of 1.5% in that of the error rate. This is due to the fact that in the dynamic clouds, the system takes into account only the representatives of every class and not all the database.

(2) By using a 90% − 10% sampling with regard to that of a 50% − 50%, an important profit of 4% on average can be reached in the quality of recognized images. This can be explain by the much more important number of data to manipulate during the training stage with a 90% − 10% sampling. In return, an increase of almost double is recorded in learning time.

(3) The neuronal classifiers offers better recognition rates than the other classic classifiers. A real profit of 2% can be reached with the neuronal classifiers. This is due to the process of learning which offer neural networks, that allows the network to adapt itself to data to be treated before applying the similarity measure.

(4) The Proposed method based on multilayer perceptron offers better performances in terms of recognition time and recognition rate than the other studied techniques. This can be explain by the fact that proposed method use two prototypes classe's to train network.

(5) The use of parallelism contribute a reduction in the traning and the recognition times as shown in Table 1 above (column 6). A benefit of almost the half can be achieve by computing algorithm in parallel.

6 Conclusion

In this paper, we studied experimentaly some recognition techniques in an increasing order of difficulty based on non-parametrics classifier and neuronal classifier. Obtained results shows that computation time rises with classified data number for all the studied methods. The Increase of computation time allows however a clear improvement of the system recognition quality, a 4% mean profit can be reached on the rate recognition. The neuronal classifiers offer better performances in term of recognition rate and recognition time, than other classification techniques studied. Implementation tests done on the proposed method, based on use of multilayer perceptron with two hidden layers and parallelism, improve a little more recognition results than other methods, and is very promising.

One of the possible ways of improvement, is the use of more prototypes classes than only two, to represent every data class already known. This will allow to treat better the case of the neighbour (or complex) data classes, and will be suited to handling better the case of the online signatures recognition.

References

1. Adhiwiyogo, M., Chong, S., Huang, J.:: Fingerprint recognition, Spring, Final report 18-551 (1999)
2. Belaid, A., Belaid, Y.: Forms recognition, applications and methods. INTEREDITION Paris (1992)
3. Cichocki, A.: Unbehauen: Neural Networks for Optimization and Signal Processing. Wiley (1993)
4. Dollfus, D., Keskes, N., Thierstei, H.: Natural forms recognition by neural network, limestone nannoplancton application. ARIMA 2(6), 270 (1997)
5. Gilloux, M.: Digit handwritten recognition by pseudo-2D Markov model, CNED 1994 Congress N3, Rouen, FRANCE vol. In: CNED 1994 Congress N3, Rouen, FRANCE, vol. 12(6) (156 p.) (10 ref.), pp. 561–566 (1995)
6. Jiang, J.: Image compression with neural networks. A survey. Signal Processing, Image Communication (1998)
7. Jiang, J.: Fast competitive learning algorithm for image compression neural networks. Electronic Lett. 32(15), 1380–1381 (1996)
8. Kouamo, S., Melatagia, P., Ndoundam, R.: Compression d'images avec les reseaux de neurones. 9e CARI, Rabat, Maroc, pp. 127-134 (October 2008)
9. Le-Cun, Y.: A competitive learning procedure for network with asymmetric threshold. In: COGNITIVA 1985, Paris, June 4-7 (1985)
10. LeCun, Y., Baser, B., Denker, J.S., Henderson, D., Howard, R., Hubbard, R., Jackel, L.: Handwritten digit recognition with a back-propagation network. In: Touretzky, D. (ed.) Advances in Neural Information Processing Systems, vol. 2, Morgan Kaufmann (1990)
11. Torres-Moreno, J.M.: Generalisation and training by neural network: Study of new constructive algorithms. PHD thesis, Grenoble National Institut Polytechnic (1997)
12. Yacoubi Abdenaim, E.I., Guy, L.: Handwritten writing Markovian modelling, application to postal adress recognition. INIST-CNRS (307 p.) (bibl. 102 ref.), N 96 REN1 0092 (1996)

Design of a CMAC-Based PID Controller Using Operating Data

Shin Wakitani, Yoshihiro Ohnishi, and Toru Yamamoto

Abstract. In industrial processes, PID control strategy is still applied in a lot of plants. However, real process systems are nonlinear, thus it is difficult to obtain the desired control performance using fixed PID parameters. Cerebellar model articulation controller (CMAC) is attractive as an artificial neural network in designing control systems for nonlinear systems. The learning cost is drastically reduced when compared with other multi-layered neural networks. On the other hand, theories which directly calculate control parameters without system parameters represented by Virtual Reference Feedback Tuning (VRFT) or Fictitious Reference Iterative Tuning (FRIT) have received much attention in the last few years. These methods can calculate control parameters using closed-loop data and are expected to reduce time and economic costs. In this paper, an offline-learning scheme of CMAC is newly proposed. According to the proposed scheme, CMAC is able to learn PID parameters by using a set of closed-loop data. The effectiveness of the proposed method is evaluated by a numerical example.

1 Introduction

In most real processes, PID control [1, 2] has been used because the controller structure is simple. However, the desired control performance is not maintained by using

Shin Wakitani
Graduate School of Engineering, Hiroshima University, 1-4-1 Kagamiyama
Higashi-hiroshima Hiroshima Japan
e-mail: wakitani-shin@hiroshima-u.ac.jp

Yoshihiro Ohnishi
Faculty of Education, Ehime University, 3, Bunkyo-cho, Matsuyama-shi, Ehime Japan
e-mail: ohnishi@ehime-u.ac.jp

Toru Yamamoto
Faculty of Engineering, Hiroshima University, 1-4-1 Kagamiyama Higashi-hiroshima
Hiroshima Japan
e-mail: yama@hiroshima-u.ac.jp

S. Omatu et al. (Eds.): Distributed Computing and Artificial Intelligence, AISC 151, pp. 545–552.
springerlink.com   © Springer-Verlag Berlin Heidelberg 2012

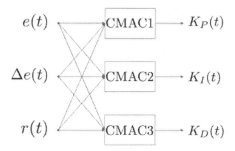

Fig. 1 Conceptual figure of calculating PID gains by CMACs

fixed PID parameters, due to the nonlinearities of systems. Controllers using CMAC (see [3, 4] and references there in), which is a type of neural network [5] and effective schemes for these nonlinear systems, have been proposed. However, the practical realization of this method is impeded because CMAC requires online-learning to acquire optimum control parameters.

VRFT [6] and FRIT [7, 8] have meanwhile attracted much attention in the last few years. According to these methods, control parameters are tuned using one-shot operating data, therefore it is expected that the computational cost can be reduced. Moreover, it is consider that these schemes are useful from a practical application viewpoint because random signals for system identification are not necessary when using these methods.

In this paper, a CMAC offline-learning scheme based on FRIT is newly proposed. According to the proposed method, CMAC learns offline by using a set of operating data; as a result, it is expected that the problem of practical realization will be solved. The effectiveness of the proposed method is evaluated by a numerical example.

2 CMAC-PID Controller Design

2.1 Design of Velocity Type of PID Controller

The velocity type of PID control law is given by (1)

$$u(t) = u(t-1) + K_P(t)\Delta e(t) + K_I(t)e(t) + K_D(t)\Delta^2 e(t) \tag{1}$$

where $u(t)$ is the control input and $e(t)$ is the control error which is defined by (2)

$$e(t) := r(t) - y(t). \tag{2}$$

In (2), $r(t)$ and $y(t)$ are the reference signal and the system output respectively. Moreover, Δ is the differencing operator which is given by $\Delta := 1 - z^{-1}$, where z^{-1} is the back word operator which denotes $z^{-1}e(t) := e(t-1)$. In the CMAC-PID controller, PID gains $K_P(t), K_I(t)$ and $K_D(t)$ are online computed by discrete

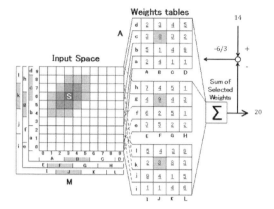

Fig. 2 Learning structure of CMAC.

CMACs. The conceptual figure of calculating PID gains by CMACs is shown in Fig.1.

2.2 CMAC

CMAC is a mathematical model which simulates the information processing mechanism in a human cerebellum. The learning structure of CMAC is explained in more detail using Fig.2. When the vector $(3, 6)$ is inputted, it is replaced with label sets. Next, the CMAC refer to values 8, 9 and 3 from weight tables, and outputs 20 as the sum total of these values. If the desired CMAC output is 14, the number obtained by dividing the difference of an output signal (equal to 20) and a teacher signal (equal to 14) by the number of the weight tables, is added to the loads (equal to 8, 9 and 3).

In the above explanation, a learning process with two-dimensional input space is described in order to explain the method easily; however, the input space actually consists of three-dimensions which comprise $r(t), e(t)$ and $\Delta e(t)$ in the proposed method. If the total size of the labels are set to $size_r, size_e$ and $size_{\Delta_e}$ then the size of input space of CMAC becomes $size_r \times size_e \times size_{\Delta_e}$, and it is discretized to N sheets of tables.

2.3 FRIT

FRIT is a method of calculating control parameters by using one-shot experimental input/output data and a fictitious reference signal. The block diagram of the FRIT is shown in Fig.3. In this figure, $u_0(t)$ is the control input and $y_0(t)$ is the system output in the operating data and $C(z^{-1})$ is the polynomial equation of a controller. Moreover, $\tilde{r}(t)$ is the fictitious reference signal which is given by (3)

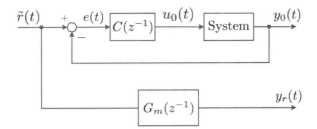

Fig. 3 Block diagram of the FRIT.

$$\tilde{r}(t) = C^{-1}(z^{-1})u_0(t) + y_0(t). \tag{3}$$

In FRIT, the criterion is first defined as (4) and calculates the optimum control parameters to minimize the criterion

$$J = \sum_{\tau=1}^{N} \{y_0(\tau) - y_r(\tau)\}^2. \tag{4}$$

Here, $y_r(t)$ is the output of $G_m(z^{-1})$, which is the desired response model given by a user and it is denoted as (5).

$$G_m(z^{-1}) := \frac{z^{-1}(1+g_1)}{1+g_1 z^{-1}} \tag{5}$$

Where

$$g_1 = -\exp(-T_s/\sigma). \tag{6}$$

In (6), T_s and σ are the sampling interval and rise time which are specified by an operator.

2.4 CMAC-FRIT

In this paper, a new offline-learning method of CMAC using FRIT is proposed. For CMAC to learn offline, the closed loop data is taken by using the stabilized controller in advance. The block diagram of the proposed method is as shown in Fig.4. In Fig.4, $r_0(t)$ and $e_0(t)$ are the expressed reference signal and control error in the closed data respectively. When $r_0(t)$, $e_0(t)$ and $\Delta e_0(t)$ are input to CMACs, they calculate PID gains by load summations according to (7)

$$K_P(t) = \sum_{h=1}^{N} W_{P,h}, \quad K_I(t) = \sum_{h=1}^{N} W_{I,h}, \quad K_D(t) = \sum_{h=1}^{N} W_{D,h}. \tag{7}$$

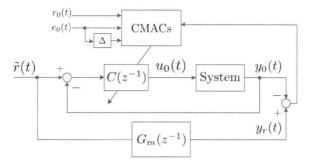

Fig. 4 Block diagram of the proposed method.

Where W is the load of the h-th weight tables and N is the number of the weight tables. The fictitious reference signal $\tilde{r}(t)$ is calculated by (3) and CMACs learn by using the steepest descent method as (8) to minimize the criterion (9).

$$
\left.
\begin{aligned}
K_P^{new}(t) &= K_P^{old}(t) - \eta_P \frac{\partial J(t+1)}{\partial K_P^{old}(t)} \\[2mm]
K_I^{new}(t) &= K_I^{old}(t) - \eta_I \frac{\partial J(t+1)}{\partial K_I^{old}(t)} \\[2mm]
K_D^{new}(t) &= K_D^{old}(t) - \eta_D \frac{\partial J(t+1)}{\partial K_D^{old}(t)}
\end{aligned}
\right\}
\tag{8}
$$

$$
J_c(t+1) := \frac{1}{2} \{ y_0(t) - y_r(t) \}^2
\tag{9}
$$

In (8) η_P, η_I and η_D are learning rates. After that, the weight tables are updated according to (10)

$$
\left.
\begin{aligned}
W_{P,h}^{new} &= \frac{1}{N} K_P^{new}(t) \\[2mm]
W_{I,h}^{new} &= \frac{1}{N} K_I^{new}(t) \\[2mm]
W_{D,h}^{new} &= \frac{1}{N} K_D^{new}(t)
\end{aligned}
\right\} .
\tag{10}
$$
$$
(h = 1, \ldots, K)
$$

3 Numerical Example

To evaluate the effectiveness of the proposed method, Hammerstein model, which is given by (11), was applied:

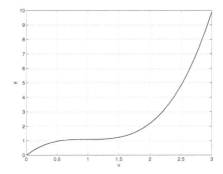

Fig. 5 Static property of the Hammerstein model.

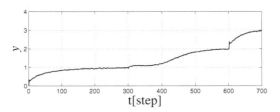

t[step]

Fig. 6 Control result which is used as the operating data.

$$\left. \begin{array}{l} y(t) = 0.6y(t-1) - 0.1y(t-2) + 1.2x(t-1) - 0.1x(t-2) + \xi(t) \\ x(t) = 1.5u(t) - 1.5u^2(t) + 0.5u^3(t) \end{array} \right\}. \quad (11)$$

Where $\xi(t)$ is white Gaussian noise with zero mean and variance 1.0×10^{-4}. The static property of the model is as shown in Fig.5. Moreover, the reference signal values for each instant of time are set as shon in (12):

$$r(t) = \begin{cases} 1.0 \ (0 \le t < 100) \\ 2.0 \ (100 \le t < 200) \\ 3.0 \ (200 \le t < 300) \end{cases}. \quad (12)$$

The fixed PID controller was first employed in order to get the initial operating data, whose PID parameters were determined so that the control system could be stabilized. Where PID gains are set as shown in (13):

$$K_P = 0.1, \quad K_I = 0.01, \quad K_D = 0.1. \quad (13)$$

The control result is as shown in Fig.6. Fig.6 shows that the control result was bad; this is because the controller attached importance to stability, thus the tracking property was poor.

Next, the CMAC-PID controller is constructed and it learns offline using the proposed method. In this simulation, the size of the weight table is set as

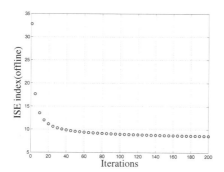

Fig. 7 Trajectory of ISE index of offline-learning.

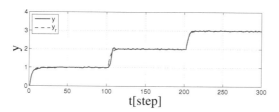

Fig. 8 Control result using the proposed method.

$size_w = size_e = size_{\Delta e} = 9$, the number of the weight tables N is set as 3, and all of learning rates are set as 0.05. In addition, the desired system model is designed as (14)

$$G_m(z^{-1}) = 1 - 0.7165z^{-1}. \tag{14}$$

In this paper, the progress of offline-learning is evaluated by the ISE(Integral of Squared Error) index given by (15), and learning is completed when the ISE index converges on a steady value.

$$\text{Index} = \sum_{i=1}^{M}\{y_0(i) - y_r(i)\}^2 \tag{15}$$

where M is the number of the operating data. The trajectory of the ISE index is as shown in Fig.7. From Fig.7, the index value converges mostly at around 200 iterations. The control result and trajectory of PID gains are shown in Fig.8 and Fig.9. The figures show that the system output tracks the desired model output y_r and PID gains are suitably adjusted at each balance point.

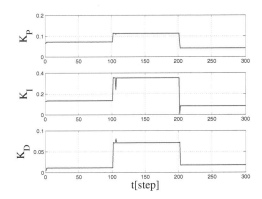

Fig. 9 Trajectory of PID gains corresponding to Fig.8.

4   Conclusions

In this paper a new offline-learning method of CMAC based on FRIT is proposed. According to this method, CMAC is able to learn the loads from weight tables by using a set of closed loop data. As a result, CMAC dose not require online-learning, thus the time and cost to learn is drastically reduced. Moreover, the effectiveness of the proposed method is evaluated by a numerical example. In the simulation, PID gains are adaptively tuned by the learned CMAC and the desired system output can be obtained. In the future work, to validate the practicality of the proposed method, it will be applied to a real system.

References

1. Zieglar, J.G., Nichols, N.B.: Optimum settings for automatic controllers. Transaction of the ASME 64(8), 759–768 (1942)
2. Chien, K.L., Hrones, J.A., Reswick, J.B.: On the automatic control of generalized passive systems. Transaction of the ASME 74, 175–185 (1972)
3. Albus, J.S.: A new approach to manipulator control: The cerebellar model articulation controller. Transaction of the ASME 97(3), 270–277 (1975)
4. Kurozumi, R., Yamamoto, T., Fujisawa, S.: Development of training equipment with an adaptive and learning mechanism using balloon actuator-sensor system. In: Proceedings of SMC 2007, pp. 2624–2629 (2007)
5. Beale, R., Jackson, T.: Neural Computing – An Introduction. Institute of Physics Publishing (1990)
6. Campi, M.C.: Virtual reference feedback tuning (VRFT): A direct method for the design of feedback controllers. Automatica 38(8), 1337–1346 (2002)
7. Kaneko, O., Souma, S., Fujii, T.: Fictitious reference iterative tuning in the two-degree of freedom control scheme and its application to a facile closed loop system identification. Transactions of SICE 42(1), 17–25 (2006)
8. Masuda, S., Kano, M., Yasuda, Y., Li, G.D.: A fictitious reference iterative tuning method with simulations delay parameter tuning of the reference model. IJICIC 6(7), 2927–2939 (2010)

Autonomous Control of Octopus-Like Manipulator Using Reinforcement Learning

So Kuroe and Kazuyuki Ito

Abstract. In this paper, we apply reinforcement learning to an octopus-like manipulator. We employ grasping and calling tasks. We show that by designing the manipulator to utilize properties of the real world, the state-action space can be abstracted, and the real-time learning and lack of generalization ability problems can be solved.

Keyword: Generalization, Abstraction of state-action, Grasping.

1 Introduction

Recently, autonomous robots that operate in unknown complex environments, like rescue robots, agricultural robots, and so on, have attracted considerable attention. Robots with many degrees of freedom to achieve various tasks have been developed. However, with an increase in the degrees of freedom, autonomous control becomes significantly difficult.

In particular, for robots that have learning ability, there are two significant problems: real-time learning and a lack of generalization ability. In general, the required learning time increases exponentially with an increase in the degrees of freedom, which makes it impossible for real robots to complete the learning process within a reasonable time limit (the real-time learning problem). Moreover, even if the learning process is completed, the reduced generality of the obtained policy makes it difficult to apply to other situations (the lack of generalization ability problem). In a real complex environment, the same situations never occur again. Thus, a lack of generality is a very serious problem.

In conventional studies, various approaches to improve the learning efficiency or generalization ability have been proposed. For example, combining reinforcement learning with a neural network [1], fuzzy control [2], genetic algorithm [3], etc. However, a real environment is too complex for real robots to learn through actual trials. Thus, almost all conventional studies utilize simulations for the learning process.

So Kuroe and Kazuyuki Ito
Hosei University 3-7-2, Kajino-chou Koganei, Tokyo 184-8584, Japan
e-mail: so.kuroe.4b@stu.hosei.ac.jp, ito@hosei.ac.jp

S. Omatu et al. (Eds.): Distributed Computing and Artificial Intelligence, AISC 151, pp. 553–556.
springerlink.com   © Springer-Verlag Berlin Heidelberg 2012

On the other hand, real creatures can learn various tasks through a small number of real trials in spite of the fact that they have many degrees of freedom. Among these, it is known that an octopus has a high level of learning ability, and can learn how to grasp various objects. It is even capable of unscrewing the cap on a bottle. The brain of an octopus is very small. Moreover, its body has huge degrees of freedom. What enables the octopus to learn is still an open question.

In our previous works, we formulated a hypothesis that the body, rather than the brain, solves these two problems. We proposed to design a body for abstracting the necessary small state-action space from a huge state-action space by utilizing properties of the real world, like dynamics, mechanical constraints, and so on. This hypothesis was tested by employing snake-like mobile robots [4]. However, more complicated tasks like grasping were not discussed.

In this paper, we consider grasping and calling tasks by an octopus-like manipulator. We show that by designing the manipulator to utilize properties of the real world, the state-action space can be abstracted, and the real-time learning and lack of generalization ability problems can be solved.

2 Task

Fig. 1 shows an outline of the task. The aim of this task is to bring various objects from the start position to the goal position. We consider 4 objects, as shown in Fig. 1. First, the manipulator learns the task using only one object (object 1). Then, the obtained policy is applied to carry other objects, and its generalization ability is examined.

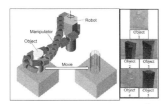 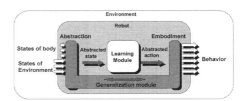

Fig. 1 Task.   **Fig. 2** Proposed framework.

3 Abstraction Using Properties of Environment and Body

Fig. 2 shows our proposed framework [4]. The learning module for reinforcement learning is surrounded by the generalization module. The generalization module is realized by the interaction of the body and environment. By the function of the generalization module, a small state-action space is abstracted from the huge state-action space of the body and environment. Thus, the learning module can learn using the small state-action space. The action of the learning module is embodied by the generalization module, and complex behavior of the body with many degrees of freedom is realized.

By abstracting the small state-action space, the learning time is extremely reduced, and the problem of real-time learning is solved. As trivial differences in the environment are compensated by the generalization module, a similar but different situation can be viewed as the same situation by the learning module. Therefore, the obtained policy is applicable to similar situations, even if the learning algorithm of the learning module has no generalization ability. Thus, the problem of the lack of generalization ability is solved.

The most important point here is that the generalization is not realized by improving the learning algorithm but is realized by properties of the real world.

4   Octopus-Like Manipulator

We next focus on the physical properties of muscles. In general, muscles have elasticity and generate pulling forces. When we relax, the elasticity is very low, and the muscles can be moved passively. When we grasp something, the elasticity and passivity allow the shape of our hand to adapt to the grasped object without precisely controlling the angles of our joints.

In this paper, we assume that the elasticity and passivity can abstract a state-action space, which improves the learning efficiency and generalizes the obtained policy. Fig. 3 and Fig. 4 show our developed octopus-like manipulator. A piece of rubber is embedded between every pair of links. These rubber pieces generate a pulling force to return the manipulator to a straight shape. One long wire is installed on the manipulator through the free pulleys. When the length of the wire is long enough, the manipulator moves passively. By rolling up the wire, a pulling force is generated between the links analogous to muscles. Pieces of a frictional material are embedded on the sides of the links instead of suckers. The manipulator has three motors. One is used for rolling up the wire, and the other is used to rotate the base of the manipulator.

Using this mechanism, the learning machine only has to control the length of the wire and the rotational angle of the base of the manipulator. It does not have to know the precise shapes of the grasped objects. The control of each joint angle to adapt to the shape of the grasped object is realized by the passivity and physical constraints.

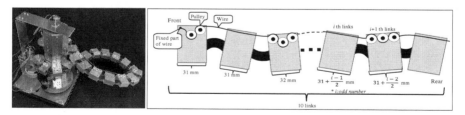

Fig. 3 Developed manipulator.   **Fig. 4** Wire constraint.

5   Experiment

We employ typical Q-learning [5]. The Q-learning states are the wire length (3 divisions) and angle of rotation (3 divisions). The actions are to roll up the wire or loosen it.

Fig. 5 a) shows the learned behavior, while Fig. 5 b) and c) show the behavior realized by applying the acquired policy to other unknown objects without additional learning. For all of the objects shown in Fig. 1, the acquired policy is applicable. These results show that the real-time learning and lack of generalization ability problems have been solved.

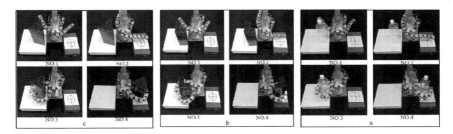

Fig. 5 Realized behavior.

6 Conclusion

In this paper, we addressed the real-time learning and lack of generalization ability problems in applying reinforcement learning to an octopus-like manipulator. We focused on properties of the real world and proposed a framework to solve these problems by utilizing the properties for abstracting a state-action space. We developed an octopus-like manipulator and demonstrated that the proposed framework could be realized by a simple mechanism, which solved these problems mechanically without improving the learning algorithm.

Our future work is to improve our proposed framework to make it applicable for practical uses like rescue operations, agriculture, nursing care, etc.

References

[1] Macek, K., Petrovic, I., Peric, N.: A Reinforcement Learning Approach to Obstacle Avoidance of Mobile Robots. In: 7th Int. Workshop on Advanced Motion Control, pp. 462–466 (2002)
[2] Berenji, H.R.: Fuzzy & Q-Learning for Generalization of Reinforcement Learning. In: Proc. of the 5th IEEE Int. Conf. on Fuzzy Systems Fuzz-IEEE, vol. 3, pp. 2208–2214 (1996)
[3] Goldberg, D.E., Holland, J.H.: Genetic Algorithms and Machine Learning. Machine Learning 3(2-3), 95–99
[4] Ito, K., Fukumori, Y., Takayama, A.: Autonomous control of a real snake-like robot using reinforcement learning - Abstraction of state-action space using properties of real world. Proc. of the 3rd Int. Conf. on Intelligent Sensors, Sensor Networks and Information, 389–394 (2007)
[5] Watkins, C.J.C.H., Dayan, P.: Technical Note: Q-Learning. Machine Learning 8, 279–292 (1992)

Relationship between Quality of Control and Quality of Service in Mobile Robot Navigation

José-Luis Poza-Luján, Juan-Luis Posadas-Yagüe, and José-Enrique Simó-Ten

Abstract. This article presents the experimental work developed to test the viability and to measure the efficiency of an intelligent control distributed architecture. To do this, a simulated navigation scenario of Braitenberg vehicles has been developed. To test the efficiency, the architecture uses the performance as QoS parameter. The measuring of the quality of the navigation is done through the ITAE QoC parameter. Tested scenarios are: an environment without QoS and QoC managing, an environment with a relevant message filtering and an environment with a predictive filtering by the type of control. The results obtained show that some of the processing performed in the control nodes can be moved to the middleware to optimize the robot navigation.

1 Introduction

In mobile robot navigation architectures, different components work at different control nodes that are connected through the communications channels. To measure the efficiency of the communications, and the quality of component's services, system uses the concept of Quality of Service (QoS) [1] through the QoS parameters [2]. The communications management oriented to optimize the QoS parameters is known as QoS policies [3]. Among standards to manage distributed communications systems, the DDS standard [4] implements a large type of QoS policies. DDS is based on publish-subscribe paradigm, extended with some elements that connect the application synchronously (readers and writers) and asynchronous (listeners). A good explanation of the operation can be found at [5]. Therefore, DDS is well suited for implementing distributed intelligent control architectures [6].

José-Luis Poza-Luján · Juan-Luis Posadas-Yagüe · José-Enrique Simó-Ten
University Institute of Control Systems and Industrial Computing (ai2). Universitat Politècnica de València (UPV). Camino de vera, s/n. 46022 Valencia (Spain)
e-mail: {jopolu,jposadas,jsimo}@ai2.upv.es

S. Omatu et al. (Eds.): Distributed Computing and Artificial Intelligence, AISC 151, pp. 557–564.
springerlink.com   © Springer-Verlag Berlin Heidelberg 2012

To measure the control efficiency, currently is used the concept of Quality of Control (QoC) [7]. The QoC measures the quality of the control action through equations, generally using the difference between the input signal and the reference signal. Sometimes the QoC parameters are used as feedback of control action; thus, the QoC measures the control efficiency and it makes easier the control processing.

The control efficiency does not depend exclusively on the algorithms used; the communications efficiency also affects the control action. [8]. To prove the relationship between QoS and QoC, an architecture called FSACtrl [9] and [10] has been developed. FSACtrl allows measuring QoC and QoS parameters in control nodes. Architecture is based on DDS standard, and it uses the DDS QoS policies to manage the communications.

Paper describes tests performed in a simulated mobile robot environment. It shows results obtained by using QoS and QoC to measure the efficiency of control node depending on the communications configuration.

The paper is organized as follows: the following section describes the environment used to perform tests of the architecture: simulation environment and simulated robots. Then, third and four sections describe the QoC and QoS parameters that have been considered in the described environment. The fifth section describes tests performed and results. Finally, the paper ends with conclusions of experiments done and the future work to be developed.

2 Experimental Environment

To test the architecture, the control of first five Braitenberg vehicles [11] has been simulated. The first three vehicles are characterized by the lack of advanced control functions; so that, these vehicles are suitable for evaluating the performance of the communications because messages are processed principally in the middleware.

The interest of Braitenberg vehicles is in the simplicity of control, based on the simple functions that connect sensors and actuators. In addition, the possibility to have different types of sensors that react to different sources provides a lot of messages that are used to test the effect of communications configuration in the control efficiency.

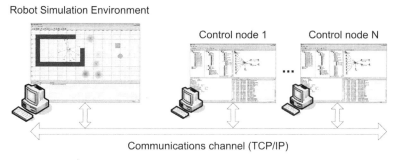

Fig. 1 Experimental environment used to test the FSACtrl architecture.

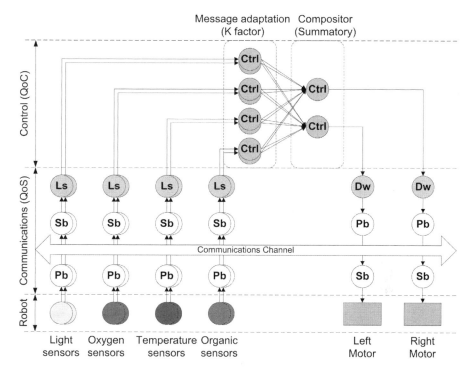

Fig. 2 FSACtrl architecture implementation of the Braitenberg 3.c vehicle.

To test the architecture, a simulation platform has been developed. The platform has a mobile robot simulator and an application to design and implement the control algorithms. Figure 1 shows the topology of the distributed system used to test the architecture.

The simulator allows user to create a 2D environment and insert any number of robots. For each robot, the simulator has twelve different types of sensors. All robots are circular and have two motors. This configuration allows robots to move in any direction in the simulated environment.

The robot simulation environment is composed of a space with different signal sources and rectangular and circular obstacles. The simulator sends via TCP clients the data from the sensors of each robot, and it receives, via a TCP server, speeds assigned to each robot motor.

Control nodes are composed of an FSACtrl elements editor that launches the control processes. The editor allows insert, modify and configure QoS policies and QoS parameters to each FSACtrl architecture element. The system implements the control node over personal computers on a TCP/IP based network. The accuracy of the measurements in the control nodes is nanoseconds; the computation time of the control nodes has been simulated in order to obtain comparable results.

3 Quality of Control in Robot Navigation

Braitenberg vehicle that uses simple control algorithms and works with a large amount of data is the vehicle 3.c. FSACtrl architecture elements are shown in figure 2. The vehicle calculates the direction that should be taken based on information obtained from the four types of sensors available, using equation 1.

$$Output_{compositor} = \sum_{i=1}^{N} K_i \cdot Input_i \qquad (1)$$

The output of each of the composers is calculated from the contribution of each input of the N sensors of the vehicle, weighted by a specific K factor for each sensor. The quality of control in the 3.c Braitenberg vehicle is measured by means of the angle that the vehicle deviates from the planned angle in the theoretical analysis of the vehicle mission (figure 3). The equations to obtain the quality of control parameter can be very different, because of the quality parameter is directly associated to the characteristics of the robot on which it is applied [12]. In the case of vehicle 3.c the quality of control is directly calculated with the parameter ITAE (equation 2)

$$ITAE = \int_0^{t_{END}} t \cdot \left| \varphi_y(t) - \varphi_r(t) \right| dt \qquad (2)$$

In the equation 2, $\varphi_y(t)$ is the value of the desired angle for a time t, while $\varphi_r(t)$ is the real angle obtained in the same instant of time. ITAE parameter considers the navigation error with the same weight during all the navigation time, so that it is very suitable to make global comparisons. The smaller ITAE value, the better quality of navigation of the vehicle is. This is because of the angle obtained from the course is closer than expected angle.

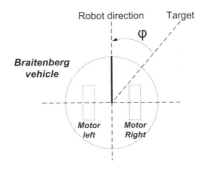

Fig. 3 Path error in the Braitenberg 3.c vehicle used to calculate the ITAE parameter.

4   Quality of Service in Robot Navigation

Middleware manages the QoS. In the case of FSACtrl architecture, QoS is managed by the QoS policies of the DDS standard. In the tests performed the QoS parameters that have been measured are the control component load and the rate of useful messages. The control component load (ρ) is calculated as the rate between the service demand and the service rate of the component. Due to the architecture elements are made by messages queues [13], global load is obtained through the pondered rate of each element load (equation 3). The K factor is used to balance the most important control components.

$$\rho_{global} = \sum_{i=1}^{N}(K_i \cdot \rho_i)\bigg/N \qquad (3)$$

To calculate the load ρ of each component is used the equation 4, where λ is the demand for the services requested from the vehicle control and μ is the rate of service provided by the control component. Both of these parameters are expressed in messages per second so that the load is a dimensionless parameter. Closer load to zero better is the control component load.

$$\rho = \lambda/\mu \qquad (4)$$

The useful messages rate (UM) is obtained by means the equation 5. The concept of utility of a message can be quite large. In the experimental environment a useful message is considered when the message produces a change in vehicle navigation. The variation of navigation is produced when the control action calculated for a measurement is different from the control action calculated for the previous measurement. Closer to one are, better the parameter is. The control action in Braitenberg vehicles is performed on the speed of the motors.

$$UM = N_{output(i)\neq output(i-1)}\big/N_{total} \qquad (5)$$

From the two previous equations, the performance (η) of the control can be obtained (equation 6). Performance is defined as the satisfactory results obtained in relation to the cost in resources used. The control performance is obtained through the parameters from the equations 4 and 5. Through the performance equation, can be verified the effectiveness of the control messages related to the resources consumed from the control service.

$$\eta = UM \cdot \left(1 - \rho_{global}\right) \qquad (6)$$

5 Experimental Tests and Results

Three scenarios on the architecture with the Braitenberg vehicle 3.c have been tested:

1. Control action without filtered messages optimization and without messages selection optimization (not QoS and not QoC management).
2. Control action with filtered messages optimization and without messages selection optimization (QoS managed but not QoC management).
3. Control action with filtered messages optimization and with messages selection optimization (QoS and QoC managed).

Message filtering consists of transmit through the middleware only those messages whose content is different, compared with the preceding message. The message filtering is one of the characteristics specified in DDS standard recommendations for a middleware. The control optimization is performed by inserting control components that predict the change in control action. The prediction is made comparing the messages from different sensors involved in the calculation of control action. The environment is a system without obstacle with the four types of sources associated with the four types of sensors of the vehicle 3.c of Braitenberg (figure 2). The vehicle is configured to be attracted by light and organic matter sources, and to be rejected by heat and oxygen sources. The vehicle follows a path that depends on the location of the sources in the environment (figure 4).

Tests have been performed starting the vehicle in the same position and the sources placed in the same location and changing the middleware according to each scenario described.

Table 1 shows experimental values for each of the scenarios described at the beginning of the paragraph. Columns show the average values of the control load, the usefulness of messages rate, the performance of the control element and the value of ITAE. Each row contains the data for each of the scenarios described above.

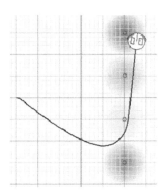

Fig. 4 Example of 3.c Braitenberg vehicle navigation in a multi-source environment.

Table 1 Experimental results based on different scenarios (average values).

Scenarios	ρ	UM	η	ITAE
1. Without QoS and QoC management	0,184	0,212	0,173	0,252
2. With QoS management and without QoC management	0,121	0,323	0,284	0,261
3. With QoS and QoC management	0,119	0,683	0,602	0,284

Due to the response time of control service is the same in all scenarios tested the variation of the control load depends on the message arrival frequency. Because of the scenarios 2 and 3 include a message filtering phase the control load decreases significantly respect the scenario 1.

UM rate changes progressively among the three different scenarios. In the scenario 2, UM value rises respect the scenario 1 because the middleware has filtered some messages that do not generate a control action. However, the most significant improvement of useful message index is produced in the scenario 3. In the scenario 3, the control receives only messages that haven't been filtered in the middleware and in the control prediction. For this reason the message utility rate increases considerably compared with the previous two scenarios.

Figure 5 shows the comparison between the service performance index (η) and the control index (ITAE). The service performance describes the common contributions of the two parameters analysed and it is a good measure of the quality of service that the control component provides. The figure shows how performance is directly related to the optimizations used in each scenario. ITAE parameter is used to check the efficiency of the control service optimizations of the vehicle navigation. In this case, ITAE parameter increases very slightly in relation with the optimized scenario, so that improvements implemented on every scenario do not affect the quality of the robot navigation mission.

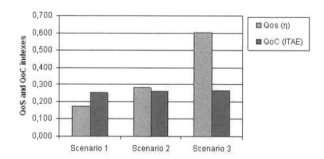

Fig. 5 Comparison chart between the η values (QoS) and the ITAE values (QoC).

6 Conclusions and Future Work

Results of the experimental work carried out are satisfactory. Results show that the FSACtrl architecture is viable as a middleware with support to simple control actions. It is also proves as manager of the communications layer allows to optimize the control layer that affects overall system optimization.

As future work, several studies related with the relation between QoS and QoC can be performed. One of the most interesting questions, to develop, is the dynamic adjustment, through QoS policies, of the robot navigation. The concept of the dynamic variation can be extended to the QoC with the QoC policies. The objective is determine the convenience to adjust the communications and control characteristics, as the sampling frequency, according to certain environmental and design constraints such as energy consumption or the time to complete the mission of the vehicle.

Acknowledgments. The work described in this article is a part of the coordinated project SIDIRELI: (Distributed Systems with Limited Resources) and COBAMI (Mission-Based Control) Education and Science Department, Spanish Government and European FEDER found. MICINN CICYT: SIDIRELI: DPI2008-06737-C02-01/02, COBAMI: DPI2011-28507-C02-02.

References

1. Vogel, A., Kerherve, B., von Bochmann, G., Gecsei, J.: Distributed Multimedia and QoS: A Survey. IEEE Multimedia 2(2), 10–19 (1995)
2. Crawley, E., Nair, R., Rajagopalan, B.: RFC 2386: A Framework for QoS-based Routing in the Internet. IETF Internet Draft, 1–37 (1998)
3. Bradner, S.: RFC 2026: The Internet Standards Process. IETF Internet Draft, sec.10 (1996)
4. Object Management Group (OMG): Data Distribution Service for Real-Time Systems, v1.1. Document formal (April 12, 2005)
5. Poza, J.L., Posadas, J.L., Simó, J.E.: QoS-based middleware architecture for distributed control systems. In: International Symposium on Distributed Computing and Artificial Intelligence. DCAI, Salamanca, Spain (2008)
6. Poza, J.L., Posadas, J.L., Simó, J.E.: A Survey on Quality of Service Support on Middleware-Based Distributed Messaging Systems Used in Multi Agent Systems. In: 9th International Conference on Practical Applications of Agents and Multi-Agent Systems. DCAI, Salamanca, Spain (2011)
7. Dorf, R.C., Bishop, R.H.: Modern Control Systems, 11th edn. Prentice Hall (2008)
8. Soucek, S., Sauter, T.: Quality of Service Concerns in IPBased Control Systems. IEEE Transactions on Industrial Electronics 51(6) (December 2004)
9. Poza, J.L., Posadas, J.L., Simó, J.E.: Multi-Agent Architecture with Support to Quality of Service and Quality of Control. In: 11th International Conference on Intelligent Data Engineering and Automated Learning, Paisley, UK (2010)
10. Braitenberg, V.: Vehicles: Experiments on Synthetic Psychology. MIT Press, Cambridge (1984)
11. Gabel, O., Litz, L.: QoS-adaptive Control in NCS with Variable Delays and Packet Losses – A Heuristic Approach. In: 43rd IEEE Conference on Decision and Control (2004)
12. Poza, J.L., Posadas, J.L., Simó, J.E.: From the Queue to the Quality of Service Policy: A Middleware Implementation. In: Omatu, S., Rocha, M.P., Bravo, J., Fernández, F., Corchado, E., Bustillo, A., Corchado, J.M. (eds.), Part II. IWANN 2009. LNCS, vol. 5518, pp. 432–437. Springer, Heidelberg (2009)

Fusing Facial Features for Face Recognition

Jamal Ahmad Dargham, Ali Chekima, and Ervin Gubin Moung

Abstract. Face recognition is an important biometric method because of its potential applications in many fields, such as access control, surveillance, and human-computer interaction. In this paper, a face recognition system that fuses the outputs of three face recognition systems based on Gabor jets is presented. The first system uses the magnitude of the, the second uses the phase, and the third uses the magnitude with phase of the jets. The jets are generated from facial landmarks selected using three selection methods. It was found out that fusing the facial features gives better recognition rate than either facial feature used individually regardless of the landmark selection method.

Keywords: gabor filter, face recognition, bunch graph, image processing, wavelet.

1 Introduction

Face recognition approaches can be divided into three groups [2]; global, local, and hybrid approaches. The bunch graph method is a local approach that works by first locating a landmark on a face, then convolving a sub-image around each landmark with a group of Gabor filters. This produces a jet from each landmark. These jets will be used for face recognition by computing and comparing similarity scores between jets of two different images. Wiskott et al. introduced a face recognition method called the Elastic Bunch Graph Method [3] and compared the EBGM with several face recognition methods on the FERET and Bochum image databases in different face poses. Their system achieved 98% recognition rate for frontal images. Bolme [4] also used Elastic Bunch Graph Method but he only used one training image per person and the jets were computed from manually selected training images landmarks. These jets were used to find new jet from new image using a displacement estimation method to locate the node on the new image.

Jamal Ahmad Dargham · Ali Chekima · Ervin Gubin Moung
School of Engineering and Information Technology, University Malaysia Sabah, Locked Bag 2073, Teluk Likas, 88999 Kota Kinabalu, Sabah, Malaysia
e-mail: {jamalad,chekima}@ums.edu.my, menirva.com@gmail.com

S. Omatu et al. (Eds.): Distributed Computing and Artificial Intelligence, AISC 151, pp. 565–572.
springerlink.com   © Springer-Verlag Berlin Heidelberg 2012

These new jets are then added to the existing jets database. By using the automatically obtained jets for recognition task an 89.8% recognition rate was reported on the FERET database. Sigari and Fathy [5] proposed a new method for optimizing the EBGM algorithm. Genetic algorithm was used to select the best wavelength of the Gabor wavelet. They had tested the proposed method on the frontal FERET face database and achieved 91% recognition rate. In this paper, a face recognition system that fuse facial features extracted using Gabor wavelet is presented. In section 2 the theory of Gabor wavelet method will be presented while in section 3 the application of bunch graph method to extract facial feature is presented. Section 4 describes the proposed system, section 5 is the experiments discussion and section 6 is the conclusion.

2 Gabor Wavelet Transform

2D Gabor wavelets shown in (1) were used to extract features from landmarks by convolving the wavelet on the landmarks of the faces.

$$g(x,y) = \exp(-\frac{x'^2+\gamma^2 y'^2}{2\sigma^2})\exp(i\left(2\pi\frac{x'}{\lambda}+\psi\right)) \tag{1}$$

where $x' = x\cos\theta + y\sin\theta$, $y' = -x\sin\theta + y\cos\theta$ while λ specifies the wavelength, θ specifies the orientation of the wavelet, ψ specifies the phase of the sinusoid, σ specifies the radius of the Gaussian, and γ specifies the aspect ratio of the Gaussian. The parameter value of the parameter used in this paper are the same as those used by Wiskott in [3], which give 40 Gabor wavelets with different frequencies and orientations. Convolving the same landmark with many Gabor wavelet configurations produces a collection of Gabor coefficients called jets. Each Gabor coefficient has a real and imaginary component. The magnitude and phase of the image's content at a particular wavelet's frequencies can be computed from the complex number. Let J be a complex number Gabor coefficient, the magnitude. $J_{magnitude}$ and the phase angle ϕ of J are given as in (2) and (3) respectively.

$$J_{magnitude} = \sqrt{J_{real}^2 + J_{imaginary}^2} \tag{2}$$

$$\phi = cos^{-1}\left(\frac{J_{real}}{J_{magnitude}}\right) = sin^{-1}\left(\frac{J_{imaginary}}{J_{magnitude}}\right) \tag{3}$$

3 Bunch Graph Method

A face image is represented as a bunch graph. Bunch graph is a collection of jets for an image. Fig. 1(a) shows the landmarks that were selected as point of interest to be convolved with a group of Gabor wavelets. An example of a convolution of a Gabor wavelet at the chin of a person is shown in Fig. 1(b). Face images are zero padded for the convolutions where the wavelet exceeds the image dimensions, which normally occur near the edge of the image.

(a)   (b)

Fig. 1 (a) An image with seven landmarks selected (b) convolution of a Gabor kernel at the chin.

3.1 Jet Extraction and Bunch Graph Creation

The convolution process produces a matrix having the same dimension as the Gabor wavelet dimension. According to [7], when the mask size of the wavelet comes closer to image size, the recognition performance increases. In this paper, the mask size was set 51 x 51 dimensions. Assuming that matrix **A** contains the complex Gabor wavelet coefficients for one landmark. Each matrix A is concatenated into a single vector. A collection of the concatenated version of matrix **A** for one landmark is called a jet. Thus, assuming matrix **B** represent the Jet then, $B = \{A_{J1}, A_{J2}, ..., A_{J40}\}$ contains the entire Gabor coefficient for one landmark. A bunch graph for an image is a collection of jets. Let matrix **C** represent a bunch graph, then Matrix $C = \{B_{N1}, B_{N2}, ..., B_{N7}\}$ will be used for similarity score calculation between images.

4 Proposed System

Fig. 2 shows the block diagram of Gabor based face recognition system. Seven landmarks as shown in Fig.1 (a) selected from face images are convolved with group of Gabor wavelets. Jets from each landmark were then collected together to create a bunch graph as face representation and will be use for matching task. Three systems will be tested.

 i.   System A uses the jets magnitude information in matching task
 ii. System B uses the jets phase information in matching task
 iii. System C uses jets magnitude weighted by similarity of the phase between two different jets

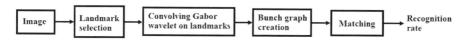

Fig. 2 Bunch graph face recognition system

4.1 Landmark Selection

The landmark selection for training images was done manually. For testing image, three method of landmark selection were conducted. (i) The first method is by manually selecting landmark on the testing image. (ii) The second method is by using the mean coordinate from all training image landmark coordinates,

$$mean\_coord = \frac{1}{M}\sum_{i=1}^{M}\{(x,y)_{i1,}(x,y)_{i2}\dots(x,y)_{iN}\} \qquad (4)$$

(iii) The third method is by using the mode coordinate from all training image landmark coordinate, $mode\_cord = \{mode(x,y)_1, mode(x,y)_2, \dots, mode(x,y)_N\}$ where M = total of training image, N = total landmark.

4.2 Similarity Score

For bunch graph similarity measurement, three similarity measurements are considered [4];

$$S_m(\boldsymbol{B},\boldsymbol{B}') = \frac{\sum_{i=1}^{G}J_iJ'_i}{\sqrt{\sum_{i=1}^{G}J_i^2\sum_{i=1}^{G}J'_i^2}} \qquad (5)$$

$$S_\phi(\boldsymbol{B},\boldsymbol{B}') = \frac{\sum_{i=1}^{G}\phi_i\phi'_i}{\sqrt{\sum_{i=1}^{G}\phi_i^2\sum_{i=1}^{G}\phi'_i^2}} \qquad (6)$$

$$S_p(\boldsymbol{B},\boldsymbol{B}') = \frac{\sum_{i=1}^{G}J_iJ'_i\cos(\phi_i-\phi'_i)}{\sqrt{\sum_{i=1}^{G}J_i^2\sum_{i=1}^{G}J'_i^2}} \qquad (7)$$

Where G is number of wavelet coefficients in a jet, J_i is the magnitude of the jet and ϕ_i is the phase angle. $\boldsymbol{B}$ and $\boldsymbol{B'}$ are the jets for two different images. Equation (5) computes jet similarity score using jet magnitude (System A), (6) computes jet similarity score using jet phase (system B), while (7) use magnitude weighted by similarity of the phase angle to compute jet similarity score (System C). To compute the similarity score between two bunch graphs, (8) was used and N is total number of landmarks.

$$S_{bunch}(\boldsymbol{C},\boldsymbol{C}') = \frac{1}{N}\sum_{i=1}^{N}S(\boldsymbol{B},\boldsymbol{B}') \qquad (8)$$

4.3 Matching

For the matching task, if the score $S_{bunch}(\boldsymbol{C},\boldsymbol{C}')$ produced by (8), between the bunch graphs of a test image y and an image x in the training database is larger than a given threshold t, then images y and x are assumed to be of the same person. The scores produced by equation (7) were normalized so that $0 \leq S_{bunch}(C, C') \leq 1$, and the threshold t value can be tune between 0 and 1. To measure the performance of the system, several performance metrics are used. These are:

i.   For Recall
  • **Correct Classification**. If a test image y_i is correctly matched to an image x_i of the same person in the training database.
  • **False Acceptance.** If test image y_i is incorrectly matched with image x_j, where i and j are not the same person
  • **False Rejection.** If image y_i is of a person i in the training database but rejected by the system.
ii.   For Reject
  • **Correct Classification**. If y_i, from the unknown test database is rejected by the program
  • **False Acceptance**. If image y_i is accepted by the program.
iii. **Equal Error Rate (EER).** Recall correct classification is equal to reject correct classification.

4.4 Data Fusion

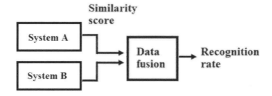

Fig. 3 Block diagram of the fusion system

Fig. 3 shows the block diagram of the fusion of system A, B, and C, mentioned in Sect. 4. The fusion decision stage is a module which is based on an OR operator shown in the Table 1.

Table 1 Fusion Decision Rules

System A	System B	Fusion System output
0	0	0
1	0	1
0	1	1
1	1	1

4.5 Face Database

A total of 500 images with frontal face of a person were selected from the FERET database. They represent 200 different individuals. 100 individuals are used for training & testing, and the other 100 different individuals are used for testing only. All the 500 selected FERET images were cropped to get only the desired face part of a person (from forehead to the chin). All images are adjusted so that both eyes coordinates of an individual are aligned in the same horizontal line and the dimension for each image is set to 60 x 60 pixels. Three images per individual will be used for training. Two testing databases were created. The first database, Known Test Database, has 100 images of the 100 persons in the training database. This database will be used to test the recall capability of the face recognition system. The second database, Unknown Test Database, has also 100 images of 100 different persons. This database will be used to test the rejection capability of the system.

5 Results and Discussion

As stated earlier, the range of the similarity score can be between 0 and 1. The threshold also can be tuned so that the performance of the system can either have high correct recall with high false acceptance rate for application such as boarder monitoring or high correct rejection rate for unknown persons for application such as access control. For this work, the threshold tuning parameter was set so that the system has equal correct recall rate and correct rejection rate. Three landmark selection criteria were tested and three systems were considered.

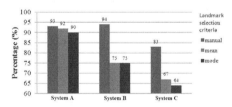

Fig. 4 Recognition rate using magnitude, phase, and magnitude with phase

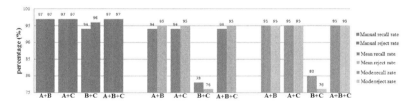

Fig. 5 Recognition rate for data fusion. The '+' sign means two or more systems were OR'ed.

Fig. 4 shows the performance of the system individually. The manual selection outperforms mean and mode selection. Comparing the two automatic selection (mean and mode), the mean outperforms the mode selection criteria for all three system. System A outperforms other systems generally. Fig. 5 shows that system fusion give better performance than single system. The selection method does not significantly affect the performance of the system. It gives better result when system A data were used in the fusion. The performance of our system is also compared with several methods that are based on bunch graph methods and use the same database as shown in Table 2. Our system performs better than both systems reported in [4] and [5] but lower than [3]. This may be due to the fact that [3] uses a precise jets extraction instead of just manually selecting a node on a face, thus creating a very detailed face graph with high precision as well designing the system specifically for in-class recognition task.

Table 2 Comparison of several EBGM-based face recognition methods on FERET database.

Methods	Recognition Rate
Elastic Bunch Graph Method [3]	98%
EBGM (automatic facial feature selection) [4]	89.8%
Gabor wavelength selection based on Genetic Algorithm [5]	91%
Our proposed method	
Mean facial feature coordinate selection	94% (recall), 95% (reject)
Mode facial feature coordinate selection	95% (recall), 95% (reject)

6 Conclusion

In this paper, a system that fuses the outputs of three systems is presented. These systems are based on the bunch graph method but one use magnitude of the jets, one use the phase, and last one used the magnitude with phase. Three methods for selecting the landmarks where the jets are generated are used. It was found that selection method criteria did not significantly affect the performance of the system fusion. In addition, the manual selection gives the highest recognition rate followed by the mean and mode methods. However, the output of the fusion system gives higher recognition rate than all system individually.

References

1. Turk, M.A., Pentland, A.P.: Face recognition using eigenfaces. In: Proc. IEEE Conf. on Computer Vision and Pattern Recognition, pp. 586–591 (1991)
2. Zhao, W., Chellappa, R., Phillips, P.J., Rosenfeld, A.: Face recognition: A literature survey. ACM Computing Surverys (CSUR) 35(4), 399–458 (2003)

3. Wiskott, L., Fellous, J.-M., Kruger, N., Von Der Malsburg, C.: Face Recognition by Elastic Bunch Graph Matching. In: Intelligent Biometric Techniques in Fingerprint and Face Recognition, ch.11, pp. 355–396 (1999)
4. Bolme, D.: Elastic bunch graph matching. Master's thesis, Colorado State University (Summer 2003)
5. Sigari, M.H., Fathy, M.: Best wavelength selection for Gabor wavelet using GA for EBGM algorithm. In: Machine Vision, ICMV 2007, Islamabad, December 28-29, pp. 35–39 (2007)
6. Wiskott, L., Fellous, J.M., Kruger, N., Malsburg, C.V.D.: Face Recognition by Elastic Bunch Graph Matching. IEEE Transaction on Pattern Analysis and Machine Intelligence 19(7), 775–779 (1997)
7. Gökberk, B.: Feature Based Pose Invariant Face Recognition. Master's thesis, Bogazici University (2001)

Hybrid Component-Based Face Recognition System

Jamal Ahmad Dargham, Ali Chekima, and Munira Hamdan

Abstract. Face recognition system is a fast growing research field because of its potential as an eminent tool for security surveillance, human-computer interaction, identification declaration and other applications. Face recognition techniques can be categorized into 3 categories namely holistic approach, feature-based approach, and hybrid approach. In this paper, a hybrid component-based system is proposed. Linear discriminant analysis (LDA) is used to extract the feature from each component. The outputs from the individual components are then combined to give the final recognition output. Two methods are used to obtain the components, namely the facial landmarks and the sub-images. It was found out that the fusion of the components does improve the recognition rate compared to individual results of each component. From the sub-image method, it can be seen that as the size of the components get smaller, the recognition rate tends increase but not always.

Keywords: component-based, face recognition, Linear discriminant analysis, image processing, grid.

1 Introduction

Face recognition system is a fast growing research field because of its potential as an eminent tool for security surveillance, human-computer interaction, identification declaration and other applications. Face recognition techniques can be categorized into 3 categories namely holistic approach, feature-based approach, and hybrid approach. Principal component analysis (PCA) [1, 2] and linear discriminant analysis (LDA) [3, 4] are the common tools for global based techniques while elastic bunch graph matching [5] and local feature analysis (LFA) [6] are the common tool for as feature- based techniques. Some of these techniques has been modified or combined to develop hybrid face recognition technique [7, 8]. Hybrid

Jamal Ahmad Dargham · Ali Chekima · Munira Hamdan
School of Engineering and Information Technology, University Malaysia Sabah, Locked
Bag 2073, Teluk Likas, 88999 Kota Kinabalu, Sabah, Malaysia
e-mail: {jamalad,chekima}@ums.edu.my, ette02@gmail.com

S. Omatu et al. (Eds.): Distributed Computing and Artificial Intelligence, AISC 151, pp. 573–580.
springerlink.com   © Springer-Verlag Berlin Heidelberg 2012

face recognition has gained significant interests in recent years due to its similarity to human's capability to recognize a person [9]. Therefore, this technique is deemed as an ideal system and some recent publications based on this technique have achieved higher performance compared to the holistic and feature-based approaches [10, 11, 12, 13]. A particular topic that sparks an interest for this paper is the component-based method. In [14], an interesting concept of component-based approach and LDA method was developed and showed a promising result. In this paper, a component-based system is proposed. Two methods are used to obtain the components. The first method manually select facial landmarks such as eyes and nose as the components while in the second method, the face image is divided into equal-sized windows. Linear Discriminant Analysis (LDA) is applied to each component and the result is then fused to give the final recognition result. The paper is organized as follows: The next section, section 2, a brief description of LDA is given followed by a description of the proposed system in section 3. In section 4, a description of the database used for the experiments is given. In section 5, the results and discussions are presented and finally section 6 concludes the paper.

2   Linear Discriminant Analysis

Linear discriminant analysis (LDA) is a popular technique for dimensionality reduction and simple classifiers in the reduced feature vector. Let P_1, P_2,..., P_c be the face classes in the database and let each face class P_i, $i = 1,2,...,c$ has n facial images x_j, $j=1,2,...,n$. The mean image μ_i of each class P_i is computed as:

$$\mu_i = \frac{1}{n} \sum_{j=1}^{n} x_j \qquad (1)$$

Where n is the total number of training images for each class.

Then the mean image μ of all the classes in the database is calculated as:

$$\mu = \frac{1}{c} \sum_{i=1}^{c} \mu_i \qquad (2)$$

The within-class scatter matrix is calculated as:

$$S_W = \sum_{i=1}^{c} \sum_{x_j \in X_i} (x_j - \mu_i)(x_j - \mu_i)^T \qquad (3)$$

And the between-class scatter matrix is calculated as:

$$S_B = \sum_{i=1}^{c} N_i (\mu_i - \mu)(\mu_i - \mu)^T \qquad (4)$$

Where N_i is the number of images of the same person. The eigenvectors are calculated as:

$$S_B w_k = \lambda_k S_W w_k \qquad k = 1, 2, ..., m \qquad (5)$$

Where w_k= eigenvector and λ_k= eigenvalue.

Note that there are at most $c - 1$ nonzero generalized eigenvalues, and so an upper bound on m is $c - 1$, where c is the number of classes and m is the number of Fisherface eigenvalue. In face recognition problem, the within-class scatter matrix S_W is always singular. Fisherface method can overcome this problem by using PCA to reduce the dimension of the feature space to $M - c$ where M is total training images, and then applying the standard LDA defined by (5) to reduce the dimension to $c - 1$.

3 Proposed Method

Fig. 1 shows the block diagram of the proposed system. The system has two phases, a training phase and a testing phase. Two types of components are used. The first type is facial landmarks such as nose, eyes, and mouth. The second is facial sub-images obtained by dividing the facial images into sub-images of equal sizes using fixed grid.

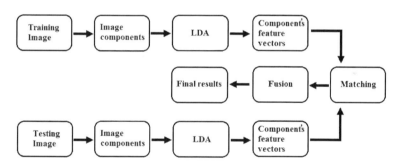

Fig. 1 Block diagram of the system

3.1 Facial Landmarks

All the facial landmarks such as eyes, nose, and mouth are selected manually using a rectangular sized window from the mean face image as it represents general pattern of face images in the database. Fig. 2 shows an example of facial landmarks selection method.

Assuming an image size of M x M along with nose component size of a x b, right eye component size of c x d is the left eye component size of f x g, and mouth component size of k x l. The components are selected based on mean face image.   Important points are selected so that each component contains feature that is required. Nose component is selected where a is the width of nostril and b is the height of nose from nose tip to the brow level. *c* and *f* is determined by the width of each eyebrow. The height *d* and *g* is determined by approximating half the height of the nose. As for the mouth component, *k* is selected from the corners of the mouth and *l* is determined from the line that connects the mouth corners to a point below the lip.

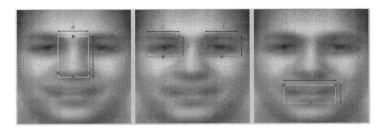

Fig. 2 Facial landmark selection

3.2   *Facial Sub-Images*

Assuming the image size is *M* x *M* pixels and *G* is the number of equal sized sub-images the image will be divided into on each axis. The values of *G*, grid scale, considered in this work are {2, 3, 4, 5, 6, 8, 10, 12, 15, 16, 20, 24}. Since four sizes for images, 30 by 30, 60 by 60, 120 by 120, and 240 by 240 are used not all values of *G* are used for each image size. Only the values of *G* that divide the image into equal sized components are used for the given image size. The total number of components in an image is G^2. The value of *G* is inversely proportional to the size of the components.

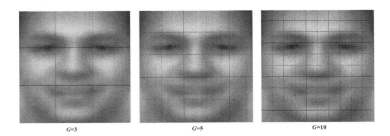

Fig. 3 Examples of the fixed grid when on images.

3.3   *Classification Criteria*

The following classification task will be used to determine recognition rate for this system.

For the image comparison task, Euclidean Distance measures are used,

$$E = \min \|\Omega - \Omega'\|$$

Where $E = Euclidean\ Distance\ Score$ and Ω = Feature Vector.

All the Euclidean score were normalized to be in the range of 0 to 1. A threshold t can be set between 0 and 1 where $0 < t < 1$. If the Euclidean distance between test image y and image x in the training database $d(x, y)$ is smaller than the threshold t then images y and x are assumed to be of the same person. Whereas if $d(x, y) > t$, then the image does not belong to the database. To measure the performance of the system, several performance metrics are used. These are:

a) *For Recall*
- *Correct Classification.* If a test image y_i is correctly matched to an image x_i of the same person in the training database.
- *False Acceptance.* If test image y_i is incorrectly matched with image x_j, where *i* and *j* are not the same person
- *False Rejection.* If image y_i is of a person *i* in the training database but rejected by the system.

b) For Reject
- *Correct Classification.* If y_i, from the unknown test database is rejected by the program.
- *False Acceptance.* If image y_i is accepted by the program.

c) *Equal Correct Rate (ECR).* Recall correct classification is equal to reject correct classification.

3.4 Data Fusion

The recognition result from each feature vector obtained from training and testing database after the LDA process will be used as input for the fusion decision stage. The fusion decision stage is a module which is based on an OR gate operator shown in the Table 1.

Table 1 OR gate fusion system

A	B	Fusion System output
0	0	0
1	0	1
0	1	1
1	1	1

4 Face Database

A total of 500 images with frontal face of persons were selected from the FERET database. They represent 200 different individuals. 100 individuals are used for training and testing, and the other 100 different individuals are used for testing only. All the 500 selected FERET images were cropped to get only the desired face part of a person (from forehead to the chin). All images have been adjusted so that both eyes of an individual are aligned in the same horizontal line. The sizes of the images are not predetermined since the system will load and resize these images according to the experiments. For this paper, image sizes of 30x30, 60x60, 120x120 and 240x240 will be used. Each of the 100 persons in the training and testing database (Known Database) has four images. Three images will be used for training and one image will be used for testing. Fig. 2 shows an example of training and testing database. The four images are of the same person. The three images on the left, Fig. 1(a), will be used for training while the image on the right, Fig. 1(b), will be used for testing. Two testing databases were created. The first database, known test database, has 100 images of the 100 persons in the training database. This database will be used to test the recall capability of the face recognition system. The second database, unknown test database, has also 100 images of 100 different persons. This database will be used to test the rejection capability of the system.

Fig. 4 Examples of known database where (a) are the training images and (b) is the testing image.

5   Results and Discussions

As stated earlier, two set of facial features were used. First, the facial landmarks such as eyes, nose and mouth were used as the features. Fig. 5(a) shows the equal error rate for the whole face as well as each of the facial landmarks. As can be seen from Fig. 5(a), the whole face gives substantially higher recognition rate than the individual facial landmarks. However, when the outputs of the systems using facial landmarks are fused together, the recognition rate improves significantly but still lower than the whole face as can be seen from Fig. 5(b). Thus, using the facial landmark for face recognition is not a viable solution but the concept of fusing the outputs of systems using different features is worth further investigation.

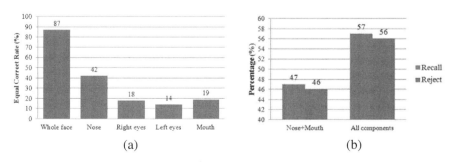

Fig. 5 (a) Equal correct recognition rate for whole face and facial landmarks components. (b) Recall and reject recognition rates for the fusion of components.

Second, the face sub-images were used as the features. As can be seen from Fig. 6, using different values of G does affect the recognition rate. The results suggest that the smaller the size of the facial sub-image the better the recognition rate tends to be.

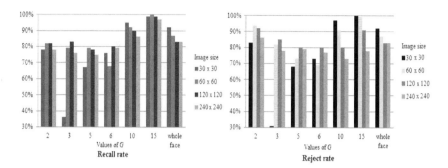

Fig. 6 Recall and reject rates for the fusion of the components for four size images.

6   Conclusion

In this paper, a component based system that fuses the components of the face image is presented. Two methods are used to obtain the components, namely the facial features components and the fixed grid position. This system uses LDA on the component images of the face. The FERET database is used for training and testing. It was found out that the fusion of the components does improve the recognition rate compared to individual results of each component. From the fixed grid method, it can be seen that as the size of the components get smaller, the recognition rate tends increase but not always.

References

1. Moghaddam, B., Wahid, W., Pentland, A.: Beyond Eigenfaces: Probabilistic Matching for Face Recognition. In: Proceedings, pp. 20–35 (1998)
2. Mudrová, M., Procházka, A.: Principal Component Analysis in Image Processing. In: Proceeding of Automatic Face and Gesture Recognition (1998)
3. Lu, J., Plataniotis, K.N., Venetsanopoulos, A.N.: Face Recognition Using LDA Based Algorithms. IEEE Transaction Neural Networkss (2003)
4. Chen, L.-F., Liao, H.-Y.M., Ko, M.-T., Lin, J.-C., Yu, G.-J.: A new LDA-based face recognition system which can solve the small sample size problem. Pattern Recognition 33(10), 1713–1726 (2000)
5. Wiskott, L., Fellous, J.M., Kuiger, N., von der Malsburg, C.: Face recognition by elastic bunch graph matching. IEEE Transactions on Pattern Analysis and Machine Intelligence (1997)

6. Jiao, F., Gao, W., Chen, X., Cui, G., Shan, S.: A Face Recognition Method Based on Local Feature Analysis. In: The 5th Asian Conference on Computer Vision (2002)
7. Lee, T.: Hybrid Approaches for Face Recognition Using Principal Component Analysis, Master's thesis (January 2004)
8. Yu, H., Yang, J.: A Direct LDA Algorithm for High-Dimensional Data with Application to Face Recognition. Pattern Recognition 34, 2067–2070 (2001)
9. Zhao, W., Chellappa, R., Phillips, P.J., Rosenfeld, A.: Face recognition: A literature survey. ACM Computing Surverys (CSUR) 35(4), 399–458 (2003)
10. Gao, Y., Wang, Y., Feng, X., Zhou, X.: Face Recognition Using Most Discriminative Local and Global Features. In: ICPR 18th International Conference on Pattern Recognition (2006)
11. Lee, Y.-J., Lee, K.-H., Pan, S.B.: Local and Global Feature Extraction for Face Recognition. In: Kanade, T., Jain, A., Ratha, N.K. (eds.) AVBPA 2005. LNCS, vol. 3546, pp. 219–228. Springer, Heidelberg (2005)
12. Huang, J., Yuen, P.C., Lai, J.H.: Combining local and global features for face recognition. In: Proc. Asian Conference on Computer Vision (ACCV 2002) (January 2002)
13. Huang, J., Heisele, B., Blanz, V.: Component-based face recognition with 3D morphable models. In: Proceedings, International Conference on Audio- and Video-Based Person Authentication (2003)
14. Huang, J., Yuen, P.C., Chen, W.S., Lai, J.H.: Component-based LDA Method for Face Recognition with One Training Sample. In: Proceedings of the IEEE International Workshop on Analysis and Modelling of Faces and Gestures (2003)

A Mixed Pixels Estimation Method
for Landsat-7/ETM+ Images

Seiji Ito and Yoshinari Oguro

Abstract. In this paper, the estimation method of the mixed pixel for satellite images has been proposed. A mixed pixel consists of several categories and the aim of this study is to estimate the mixture ratios of the categories. The filter of neighborhood pixels had been proposed. In this paper, the optimal filter coefficients have been considered in detail.

Keywords: Mixed pixel estimation, Landsat-7ETM+, Filter of neighborhood pixel.

1 Introduction

Satellite images recognition methods had been proposed[1],[2], but the errors of land cover classification have been generated around the edge of categories especially. A mixed pixel which includes the multiple categories is one of the error factors. All remotely sensed images include the mixed pixels, particularly the middle or low resolution images include the more mixed pixels. The several papers had proposed mixed pixels estimation methods[3]-[6]. The nonlinear model and the filter of the neighborhood pixels had been proposed at our research[7]. The aim of this paper has been considered about the optimal filter coefficients.

2 The Proposed Method

2.1 The Mixed Pixel Model

Mixed pixels include the components of the multiple categories. There are many mixed pixels around the edges of categories. In this paper, two categories have

Seiji Ito · Yoshinari Oguro
Department of Grobal Environment Studies, Hiroshima Institute of Technology, 2-1-1, Miyake, Saeki-ku, Hiroshima, 731-5193, Japan
e-mail: `seiji@cc.it-hiroshima.ac.jp`, `y.oguro.yx@it-hiroshima.ac.jp`

S. Omatu et al. (Eds.): Distributed Computing and Artificial Intelligence, AISC 151, pp. 581–586.
springerlink.com   © Springer-Verlag Berlin Heidelberg 2012

been estimated; "Vegetation" and "Non-vegetation". The mixed pixel of band i ($I(i)$) is given as follows:

$$I(i) = a_v I_v(i) + a_n I_n(i) + a_0 \qquad (1)$$

where, a_v and a_n are the mixture ratio of the "Vegetation" and "Non-vegetation" respectively, a_0 is a constant, $I_v(i)$ and $I_n(i)$ are the representative values of the each category on band i respectively. The aim of this paper is replaced to the estimation of a_v and a_n from Eq.(1). The mixed pixel model has to be considered about the several objects reflection in a pixel. Eq.(1) means that reflection in a pixel are simply added of the two different object reflection, but the overlap of several reflections is complex. The nonlinear model had been proposed as this equation:

$$I(i) = a_v I_v(i) + a_n I_n(i) + a_{vn} I_v(i) I_n(i) + a_0 \qquad (2)$$

where, a_{vn} is the nonlinear term coefficient. $a_{vn} I_v(i) I_n(i)$ is the considered interaction of the two objects reflection. Other nonlinear terms are considered; indirectly reflection by atmosphere or cloud, multiple reflection between two objects or atmosphere. In this paper, we select only the nonlinear term in eq.(2), since it is difficult to describe all of above physics phenomenon by a mathematical model and the mixture ratios cannot be estimated by the several terms mathematically.

In order to estimate a_v, a_n and a_{vn}, "Vegetation" has been regarded as the mixture of "Forest" and "Grass", "Non-vegetation" has been regarded as the mixture of "Building", "Soil" and "Water". The mixture ratios of a_v and a_n have been estimated from $I_v(i)$ and $I_n(i)$ by the method of least squares. If the mixture ratio takes more than 100% or negative, then the mixture ratio has been trimmed into 100% or 0% respectively.

2.2   The Filter of the Neighborhood Pixels

The mixture ratios that have been estimated using the mixed pixel model are modified using the neighborhood pixels filter. In the case of Landsat-7 ETM+[8], the filter size is 7 x 7 pixels[7], since object reflection has an effect until 90m[9]. Modification quantities e_v, e_n of the estimated a_v, a_n by the nonlinear model are given by

$$e_v = \sum_{(p,q)}^{N} \alpha^{(p,q)}\left(a_v - a_v^{(p,q)}\right), \quad e_n = \sum_{(p,q)}^{N} \alpha^{(p,q)}\left(a_n - a_n^{(p,q)}\right) \qquad (3)$$

where, (p,q) is the neighborhood pixel, N is the number of neighborhood pixels, $a_v^{(p,q)}, a_n^{(p,q)}$ are mixture ratios of "Vegetation" and "Non-vegetation" respectively. $\alpha^{(p,q)}$ is the filter coefficients of (p,q), and the filter coefficients correspond to the Euclidian distance from target pixels as shows Fig.1.   The mixture ratios are corrected by the following equations

$$a_v^{new} = a_v^{old} + e_v, \quad a_n^{new} = a_n^{old} + e_n \ .$$ (4)

If the mixture ratio takes more than 100% or negative, then the mixture ratio must be trimmed into 100% or 0% respectively. Reflection intensity distribution is explained for consideration about the filter correlations of neighborhood pixel. An object has properties both of a mirror reflection and a uniform diffuse reflection surfaces (see Fig.2). A target pixel is spherically affected by neighborhood pixels by a property of a uniform diffuse reflection. However, in theory it is difficult to study the quantity of the reflection intensity by the observation conditions (e.g. weather, angles of object surface). The filter coefficients of Fig.1 must be set spherically, since radiant intensity spherically decays with increasing the distance from the target.

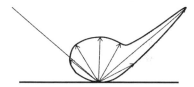

Fig. 1 The filter of neighborhood pixels.   **Fig. 2** A spatial distribution of radiant intensity.

3   Simulations

In order to show the effectiveness of the proposed method, the method has been simulated using around the Hiroshima city in Japan observed by Landsat-7 ETM+ on April 10th in 2003. Fig.3 shows the original image.

Fig. 3 An original image (Hiroshima city, Landsat-7 ETM+ on April 10. 2003).

Hiroshima city has several rivers, and it is an area adjacent to sea and mountains. The size of the image is 1600 x 2000 pixels. Evaluation pixels have been selected around Hiroshima Nishi Airport and Sanyo Express way because these areas do not change much with time.

In the estimation of the nonlinear model, $I_v(i)$ has been calculated by the 5% steps from the pure pixels of Forest and Grass. The pure pixels are considered as averages in the 10 pixels which extracted manually from the image. $I_n(i)$ has been calculated similarly. $I_v(i)$ and $I_n(i)$ have 21 or 231 patterns respectively. All the patterns have been simulated and the optimal mixture ratios at the nonlinear model have been selected the minimum error of the $I(i)$ from the estimated a_v and a_n, and the pixel data.

In the modifying by the filter of the neighborhood pixels, we have simulated the four patterns filters (Fig.4) about the quantities of the filter coefficients; Gauss, Line, Elpsup (ellipse upper) and Elpsdown (ellipse down) types. Filter coefficients of the four patterns are shown in Table1. Where, a suffix number of the type name is sum of filter coefficients. In the case of "Line_100", it means Line type and one to one relation between target pixel and neighborhood pixels. We have considered that the optimal filter coefficients decrease concave down type such as Elpsup type by Fig.2. For the purpose of comparison, we select the three types.

Mixture ratios in the evaluation pixel have been obtained by using high resolution images (QuickBird-2 images) on Google Earth. Quantitative evaluations have been regarded as a correlation coefficient and gradient of regression line on a correlation diagram of evaluation data and simulation results.

Fig.5 shows the correlation diagrams of five simulations (four types filters and without the filter). "Elpsup_100" is the best result in the five simulations, since the correlation coefficients of "Elpsup_100" are the highest and the regression lines' gradients of "Elpsup_100" is the nearest to 1 from Fig.5. Therefore, we have shown that the filter coefficients should be spherically set with the distance from the target in the simulation data.

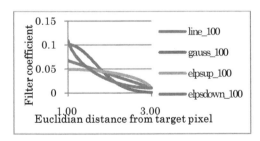

Fig. 4 Four kinds of filter coefficients.

Table 1 The filter coefficients of four types (%).

	A	B	C	D	E	F
Line_100	6.7	5.5	3.9	3.2	1.5	1.0
Gauss_100	10.1	8.2	3.1	1.6	0.2	0.1
Elpsup_100	4.9	4.8	4.3	3.9	2.1	1.0
Elpsdown_100	10.9	5.6	2.6	1.9	1.4	1.0

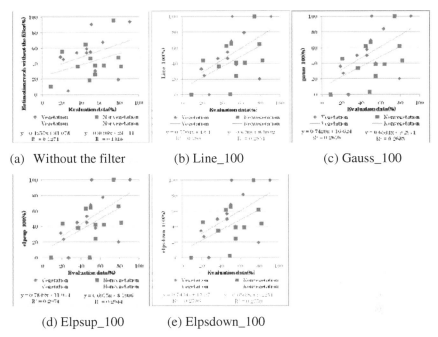

(a) Without the filter   (b) Line_100   (c) Gauss_100

(d) Elpsup_100   (e) Elpsdown_100

Fig. 5 The four results of the mixed pixel estimation. Horizontal axis is a mixture ratio for evaluation data, and vertical axis is an estimation result for each simulation.

4 Conclusion

In this paper, the mixed pixel estimation method with the nonlinear model and the filter of the neighborhood pixel had been proposed. The optimal filter coefficients have been considered. As the future work, we will propose the robust filter coefficients on the other season or the other area.

References

[1] Fujiwara, N., Muramatsu, K., Awa, S., Hasumi, T., Ochiai, F.: Pattern Expand Method for Satellite Data Analysis. Journal of RSSJ 16(3), 17–34 (1996) (in Japanese)
[2] Hayashi, A., Furumi, S., Muramatsu, K., Fujiwara, N.: Analysis of Paddy Fields Using Pattern Expand Method. Journal of RSSJ 17(2), 5–18 (1997) (in Japanese)
[3] Arai, K., Terayama, Y., Matsumoto, M., Fujiku, K., Tsuchiya, K.: Context image classification with propotion estimation of mixed pixels under the assumption of continuously situated boundary pixels with almost same proportion. Journal of RSSJ 11(4), 33–40 (1991) (in Japanese)
[4] Chen, J., Jia, X., Yang, W., Matsushita, B.: Generalization of Subpixel Analysis for Hyperspectral Data With Flexibility in Spectral Similarity Measures. IEEE Transactions on Geoscience and Remote Sensing 47(7), 2165–2171 (2009)

[5] Kawaguchi, S., Yamazaki, K., Nishii, R.: Unsupervised Contextual Classification of Remotely Sensed Imagery by Taking Mixel Information into Account. Journal of RSSJ 26(2), 107–116 (2006) (in Japanese)

[6] Kageyama, Y., Nishida, M.: Lineament Detection due to Land Cover Information in Mixel Using Landsat TM Data. IEEJ Trans. EIS 123(6), 1086–1093 (2003) (in Japanese)

[7] Ito, S., Oguro, Y.: The Mixed Pixels Estimation for the Remotely Sensed Images Using the Nonlinear Model and the Filter of Neighborhood Pixels. IEEJ Electronics, Information and Systems 131(6), 1233–1240 (2011) (in Japanese)

[8] Landsat Science, http://landsat.gsfc.nasa.gov

[9] Kawata, Y., Nishikawa, E.: Atmospheric Correction on Satellite Image Data Over Land. Journal of RSSJ 20(5), 2–12 (2000) (in Japanese)

Phoneme Recognition Using Support Vector Machine and Different Features Representations

Rimah Amami, Dorra Ben Ayed, and Noureddine Ellouze

Abstract. Although Support Vector Machines (SVMs) have been proved to be very powerful classifiers, there performance still depending on the features representations which they used. This paper describes an application of SVM to multiclass phoneme recognition using 7 sub-phoneme units and different features representations as MFCC, PLP and RASTA-PLP. The phoneme recognition system is tested and experimentally evaluated using speech signals of 49 speakers from TIMIT Corpus in order to find the adequate feature coefficient for our phonemes databases. The experimental results show that, MFCC and PLP are significantly superior and get better recognition rates than RASTA-PLP (52% vs. 29%).

Keywords: Phoneme Recognition, SVM, Kernel Tricks, MFCC, PLP, Rasta-PLP.

1 Introduction

Recently Support Vector Machines (SVMs) have been the core of most Automatic Speech Recognition (ASR) systems due to their inherent discriminative learning and generalization capabilities [1]. SVMs are effective discriminant classifiers by maximizing the error margin. They have the advantage of being able to deal with samples of a very higher dimensionality. Besides, instead of only minimizing the empirical risk, they also try to minimize the structural risk in purpose to achieve good generalization [2]. Nevertheless, to use them in a problem of speech recognition, some limitations must be overcome. The successful application of SVMs to the ASR is inherently constrained by features which represent the phonemes in the database.

Rimah Amami · Dorra Ben Ayed · Noureddine Ellouze
Department of Electrical Engineering, ENIT
Ecole Nationale d'Ingénieurs de Tunis, Tunisia

S. Omatu et al. (Eds.): Distributed Computing and Artificial Intelligence, AISC 151, pp. 587–595.
springerlink.com   © Springer-Verlag Berlin Heidelberg 2012

Our goal is to investigate the adequate features representations to SVM. This paper describes a multiclass phoneme recognition task based on SVMs and different features coefficients such as MFCC, PLP and RASTA-PLP. The phoneme database will be split into 7 units which forms our sub-phoneme databases.

This paper is structured as follow: In section 2, the learning algorithm SVM is briefly introduced; the work conditions and experiments are described in section 3 and 4. The results of the phonemes recognition are presented in section 5.

2 Overview of SVM

SVM was introduced by Vladimir Vapnik on 1995 [1] [3]. However the current standard approach (soft margin) was proposed by Vladimir Vapnik and Corinna Cortes in 1995[4].

The SVM approach consists of constructing one or several hyperplanes in order to separate the different classes. Nevertheless, an optimal hyperplane must be found. A hyperplane is defined as follows [5]:

$$Arg \max_{w,w_0} \min_i \left\{ \|x - x_i\| : x \in \mathbb{R}^d, (w^T x + w_0) = 0, i = 1 \dots m \right\} \tag{2.1}$$

Vapnik and Cortes defined an optimal hyperplane as the linear decision function with maximal margin between the vectors of the two classes, see figure 1. The margin is determined by the training examples which are called support vectors (SVs). Those SVs are the training examples close to the separation hyperplane and which have, after solving the quadratic programming problem, a non zero lagrangian (Lagrange multipliers) [6].

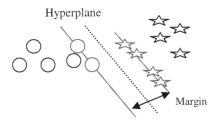

Fig. 1 An example of a linear SVM in a 2 dimensional space

In Figure 1, the SVs of the first class are colored red while the SVs of the second class are in blue.

Whereas, in our case we proceed with a multiclass classification, therefore the training examples are not linearly separable in a finite dimensional space. That's why, a much higher-dimensional space was proposed instead of the original finite-dimensional space, so as to make the separation between the different classes possible.

In the other side, for constructing a nonlinear SVM, Vapnik and al. [7] suggest to apply the kernels tricks in purpose to maximum-margin hyperplanes. In this work, the Platt's Sequential Minimal Optimization (SMO) method [8] was used which allows a fast SVM training and with a fairly high number of samples [9]. Besides, it solves the quadratic programming problem that SVM occurs.

3 Choice of Features Representations

In the speech recognition filed, the feature extraction is considered as the greatest important denominator of all recognition systems. It converts the speech waveform to a set of parametric representation. In fact, those features are used for further analysis and processing in order to improve the system outcomes. Hence, we have used different feature extractor such as MEL frequency cepstral coefficients (MFCC), Perceptual Linear Prediction (PLP) and Relative Spectral   Transform - Perceptual Linear Prediction (RASTA-PLP) in order to compare them and so find the optimal feature representation for our data.

3.1 MFCC

Mel-frequency cepstral coefficients (MFCC) are derived from a type of cepstral representation. They are frequently used as feature in speech and speaker recognition. Davis and Mermelstein [10] were the first who introduced in 8O's the MFCC concept for automatic speech recognition. The main idea of this algorithm consider that the MFCC are the cepstral coefficients calculated from the mel-frequency warped Fourier transform representation of the log magnitude spectrum.

The Delta and the Delta-Delta cepstral coefficients are an estimate of the time derivative of the MFCCs. In order to improve the performance of speech recognition system, an improved representation of speech spectrum can be obtained by extending the analysis to include the temporal cepstral derivative; both first (delta) and second (delta-delta) derivatives are applied [11][12]. Those coefficients have shown a determinant capability to capture the transitional characteristics of the speech signal that can contribute to ameliorate the recognition task.

3.2 PLP

Perceptual Linear Prediction (PLP) is a hybrid of DFT and LP, an acronym for linear predictive, techniques proposed by Hynek Hermansky [13]. In fact, PLP analysis is computationally efficient and yields a low-dimensional representation of speech signal. PLP algorithm is based on the short-term spectrum of speech. It modifies the short-term spectrum of the speech by several psychophysically based transformations.

3.3 RASTA-PLP

Relative Spectral Transform - Perceptual Linear Prediction (RASTA-PLP) is also a popular speech feature representation. PLP was originally proposed by Hynek Hermansky as a way of warping spectra to minimize the difference between speakers while preserving the important speech information [13]. RASTA is a separate technique that applies a band-pass filter to the energy in each frequency subband in order to smooth over short-term noise variations and to remove any constant offset resulting from static spectral coloration in the speech channel [14].

4 Experimental Setup

4.1 Introduction to Phonology

In a spoken language, a phoneme is the smallest phonologically meaningful units of a particular language or dialect. They are a sound pronounced in one or different ways which serves to distinguish between meanings of words. In fact, phonemes are perceived differently. Thus, they are subdivided into different groups. In this study, we split the phonemes into 7 groups shown in table 1.

Table 1 The sub-phoneme units

Sub-Phonemes	Units
Vowels	[/aa/ /ae/ /ah/ /ao/ /aw/ /ax/ /ax-h/ /axr/ /ay/ /eh/ /er/ /ey/ /ih/ /ix/ /iy/ /ow/ /oy/ /uh/ /uw/ /ux/]
Semivowels	[/l/ /r/ /w/ /y/ /hh/ /hv/ /el/]
Stops	[/b/ /d/ /g/ /p/ /t/ /k/ /dx/ /q/ /bcl/ /dcl/ /gcl/ /pcl/ /tcl/ /kcl/]
Other stops	[/pau/ /epi/ /h#/]
Nasals	[/m/ /n/ /ng/ /em/ /en/ /nx/]
Affricates	[/ch/ /jh/]
Fricatives	[/s/ /sh/ /z/ /zh/ /f/ /th/ /v/ /dh/]

4.2 Phoneme Recognition Architecture

The architecture of our recognition system consists of: (1) conversion from speech waveform to spectrogram (2) spectrogram to MFCC/PLP and RastaPLP spectrum using the adequate analysis (3) segmentation of the phoneme database to 7 subphoneme databases (4) initiate the phoneme recognition task using the nonlinear SVM approach, see figure 2.

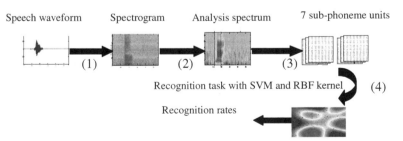

Fig. 2 Phoneme Recognition Process

This work was done using LibSVM toolbox [15]. The goal was to find the suitable feature representation for our data. To evaluate those techniques, we used the dialect region DR1 (New England) from TIMIT corpus [16]. Moreover, for the nonlinear SVM approach, we choose the RBF (Gaussian) Kernel trick, this choice was made after a previous study done on our datasets with different kernel tricks (Linear, Polynomial, Sigmoid). There are several ways to carry out a multiclass SVM classification. In the current work we use the "one-against-one" method [15] and the voting strategy. As the classification performance of SVMs is mainly affected by its model parameters particularly the Gaussian width *Gamma* and the regularization parameter *C*, we set, for all experiments, gamma as a value within 1/K where K is the number of features and C as value within 10, see table 2.

Otherwise, for the recognition we used a training datasets to get the SVMs model and so the support vectors. A test datasets was used for classification. Hence, each phoneme was labeled by the number of class to which it belongs. Furthermore, the input speech signal is segmented into frames of 16 ms with optional overlap of 1/3~1/2 of the frame size. Actually, If the sample rate is 16 kHz and the frame size is 256 sample points, then the frame duration is 256/16000 = 0.016 sec = 16 ms. Additional, if the overlap is 128 points, then the frame rate is 16000/(256-128) = 125 frames per second. Each frame has to be multiplied with a hamming window in order to keep the continuity of the first and the last points in the frame [17]. If the signal in a frame is denoted by s (n), n = 0…N-1, then the signal after Hamming windowing is s (n)*w (n), where w (n) is the Hamming window defined by:

$$w(n, a) = (1 - a) - a\cos(2pn / (N - 1)) \tag{4.2}$$

Moreover, each phoneme has a feature vector which contains 36 coefficients including first delta (Delta) and second delta (Delta-Delta). Indeed, the choice of the feature extractor was made in view of the fact that, those coefficients are the most known and used in pattern recognition researches.

Table 2 Summary of Experiments conditions

Method	SVM
Gamma	0.027
Cost	10
Kernel trick	RBF
Frames	3 Middle Frames
Corpus	TIMIT
Dialect	New England
Frame rate	125
Features number	36

5 Experimental Results

One of the fondamental denominator of all recognition system is the feature representations used since all of the information necessary to distinguish phonemes is preserved during the feature extraction. So, if during the feature extraction there is important information loss, the performance of the following recognition stage will be inherently deteriorated which will affect the system outcome.

To evaluate the performance of each feature representation, the SVM classifier was trained on 7 sub-phoneme databases. A performance comparison is presented through table 3 and 4.

As can be seen from table 3, the MFCC and PLP outperforms the Rasta-PLP coefficients in terms of misclassification (test error) rates and in terms of runtime (training and testing). Additionally, MFCC and PLP for vowels recognition perform similarly (52%). Meanwhile, PLP outreach slightly MFCC with a recognition rates up within 2% to 9% (i.e. 54% Vs 63%)

In fact, Rasta-PLP leads to lowest recognition rates compared with the others representations used. The variation of results with MFCC in Rasta-PLP domain range from +2 to +32% except for the Otherstops datasets where the error rate decrease within 5%.

Meanwhile, the variation with PLP representation range from +12% and +30%, see table 2.

Table 3 Error rates on the test sets

	ER*: Error Rates (%)		ET**: Execution Time (s)			
	MFCC		PLP		Rasta-PLP	
Features	ER*	ET**	ER*	ET**	ER*	ET**
Sub- Database						
Vowels	48	80,65	48	78,25	71	113,50
Semivowels	21	8,02	23	6,75	53	12,16
Stops	56	54,68	51	55,59	63	100,73
Other stops	20	2,74	17	2,67	14	3,61
Nasals	46	6,63	37	6,51	51	11,89
Affricates	28	0,12	31	0,09	30	0,09
Fricatives	29	8,91	26	9,36	56	13,34

Furthermore, as is shown in table 3, the results with MFCC and PLP coefficients are very close and they leads to a significantly lower error rates in comparison with Rasta-PLP representation. Indeed, the accuracy of our phoneme recognition system improved when PLP and MFCC features were used.

Based on tables 3 and 4, the variation of error rates can be explained by the fact that some phonemes are difficult to be recognized. Thus, the accuracy of the system recognition was widely affected which leads to drop its performance i.e. the error rates of 11 vowels phonemes, among 20 vowels phonemes, exceeds 50%.

Table 4 The fifteen most common errors (%)

	MFCC	PLP	Rasta-PLP
/aa/	49	29,95	97
/ux/	93	70	90
/aw/	100	83	100
/uh/	100	100	100
/el/	94	96	100
/pcl/	96	77	100
/em/	100	100	100
/en/	100	100	100
/pau/	100	100	100
/zh/	100	100	100
/tcl/	88	38	71
/th/	84	85	100
/ng/	100	100	100
/nx/	100	100	95
/gcl/	81	64	88

As seen in table 4, the most difficult phonemes to be recognized are /pau/, /zh/, /em/, /en/, /ng/, nx/, /aw/ and /uh/ for all experiments. On the other hand, phonemes /h#/, /n/ produced excellent recognition rates within 90%.

Despite, the similarity of the results based on MFCC and PLP features, the error rates with PLP coefficients are a little inferior. The variation range from 2% to 8%. Even the runtime of both features is very close.

6 Conclusion

The aim of the study was to investigate if the feature representation utilized has an impact on the system robustness? In this paper, a phoneme recognition based on a multiclass SVMs approach is presented. To evaluate our system recognition, we used several representations such as MFCC, PLP and Rasta-PLP. The results produced by our system demonstrate that the performance of MFCC and PLP coefficients outperforms the Rasta-PLP coefficients with less misclassification rates and runtime. Additionally, using MFCC and PLP features, the system recognize different set of phonemes most properly and achieve a better overall performance. Otherwise, in all experiments, the Rasta-PLP feature provides the lowest recognition rates. The comparison of PLP with MFCC shows only minor differences which indicate the possibility of combining these features in order to improve the overall performance.

However, the problem is still open as no approach is both fast and accurate enough to be an efficient recognition system. Besides these, the use of boosting methods to further improve phoneme's recognition rates can be investigated. Currently, some experimental results on pattern recognition filed shows that Boost-SVM has much better performance than SVM on unbalanced problems.

References

[1] Vapnik, V.: The nature of statistical learning theory. Springer-Verlag New York, Inc., New York (1995)
[2] Bebis, G.: Support Vector Machines (SVMs), Mathematical Methods for Computer Vision, Nevada (2003)
[3] Nilsson, N.J.: Robotics Laboratory, Introduction to Machine Learning: An early draft of a proposed textbook. Department of Computer Science Stanford University Stanford, CA (1998)
[4] Cortes, C., Vapnik, V.: Support-Vector Networks. Machine Learning, 20 (1995)
[5] Cornuéjols, A., Miclet, L.: SVM: Support Vector Machines Séparateurs à vaste marge Apprentissage Artificiel: Méthodes et Algorithmes. Eyrolles (2002)
[6] Burges, C.J.C.: A Tutorial on Support Vector Machines for Pattern Recognition. Kluwer Academic Publishers, Boston (1998)
[7] Boser, B.E., Guyon, I.M., Vapnik, V.N.: A training algorithm for optimal margin classifiers. In: 5th Annual ACM Workshop on COLT, Pittsburgh, PA, pp. 144–152 (1992)
[8] Platt, J.C., Scholkopf, B., Burges, C.J.C., Smola, A.J.: Fast training of support vector machines using sequential minimal optimization. In: Advances in Kernel Methods - Support Vector Learning, Cambridge, MA (1998)
[9] Zribi, S., Ben Ayed, D., Ellouze, N.: Support Vector Machines Approaches and its Application to Speaker Identification. In: Digital Ecosystems and Technologies, IEEE, Istanbul (2009)
[10] Davis, S.B., Mermelstein, P.: Comparison of parametric representations for monosyllabic word recognition in continuously spoken sentences. IEEE Trans. Acoust Speech Signal Processing 28(4), 357–366 (1980)
[11] Cilliers, F.D.: Tree-based Gaussian Mixture Models for Speaker Verification. Ph.D. Thesis, Department of Electrical and Electronic Engineering, University of Stellenbosch, South Africa (2005)
[12] Kawtrakul, A., Deemagarn, A.: Thai Connected Digit Speech Recognition Using Hidden Markov Models, Russia (2004)
[13] Hermansky, H.: Perceptual Linear Predictive (PLP) Analysis of Speech. Journal of the Acoustical Society of America 87(4), 1738–1752 (1990)

[14] Hermansky, H., Morgan, N.: RASTA processing of speech. Proc. of IEEE Trans. on Speech and Audio 2(4), 578–589 (1994)

[15] Garofolo, J.S., Lamel, L.F., Fisher, W.M., Fiscus, J.G., Pallett, D.S., Dahlgren, N.L., Zue, V.: TIMIT Acoustic-Phonetic Continuous Speech Corpus. USA (1993)

[16] Chang, C.-C., Lin, C.-J.: LIBSVM: a Library for Support Vector Machines, Department of Computer Science National Taiwan University, Taipei, Taiwan (2011)

[17] Jang, R.: Audio Signal Processing and Recognition, ch. 12. Tsing Hua University, Taiwan (2009)

Localization Systems for Older People in Rural Areas: A Global Vision

L. Martín, I. Plaza, M. Rubio, and R. Igual

Abstract. Location applications can be considered promising tools to improve the quality of life for the elderly. They remove architectural barriers, promote independence and social inclusion and increase personal autonomy. This work presents a review of the status of location systems thought the analysis of several research projects, patents and commercial products. The obtained conclusions are completed with the direct opinions of the users in rural area. Data were collected through several workshops with elderly people and caregivers from Teruel (Spain). The results will provide a basis for researchers, developers and manufactures to discover strengths, weakness and needs that should be met at the future in order to improve the location systems for the elderly in rural areas.

1 Introduction

According to the latest population projections, in 2025, women and men of working age will account for a smaller proportion of population in the EU [1]. In this context of aging, the Information Technology and Communications can be a great help. For instance, mobile devices could improve the quality of life (QoL) components for seniors [2]. Specifically, location applications can be considered promising tools to improve the quality of life for the elderly: They remove architectural barriers, promote independence and social inclusion and increase personal autonomy [3]. Thus this work presents a review of the status of locators to identify strengths, weakness and shortcomings of this technique, as well as, get a perspective on technology trends and a future vision. This work will help developers, researchers and manufacturers to obtain a global vision about opportunities that locators offer to improve the quality of life of the elderly in rural area.

L. Martín · I. Plaza · M. Rubio · R. Igual
EduQTech R&D&i group. University of Zaragoza. C/ Ciudad Escolar s/n 44003 – Teruel
e-mail: {lourdes,iplaza,marubio,rigual}@unizar.es

S. Omatu et al. (Eds.): Distributed Computing and Artificial Intelligence, AISC 151, pp. 597–600.
springerlink.com   © Springer-Verlag Berlin Heidelberg 2012

2   Information Collection

State of the Art on Location Systems

This work presents a review of the status of location systems thought the analysis of several research projects, patents and commercial products:

1. Location applications for dependent people in research projects. It is not difficult to find different research project paying attention to location applications. Without wishing to be exhaustive, several projects have been chosen: a) focusing on locator for elderly or disabled people and b) research projects with access to published papers: *CAALYX: Complete Ambient Assisted Living Experiment* [4], *LifeMap* [5], *Localization of a wheelchair* [6], *Localization of people with mental illness* [7], and *RFID System for the new hospital La Fe* [8]. From a technological point of view, different tecnologies can be identified. Five are particularly worthy of mention: GPS, WiFi, GSM, RFID and ZigBee. A subsequent study of these specific technologies using specialized texts (articles, books…) leads to discover other new technologies that can be used in the location field. Specifically, the following technologies were incorporated in the initial list: WiMAX, Bluetooth, UWB and Ultrasound. Thus, the study leads to eight technologies that can be classified depending on person happens to be: inside or outside.
2. Patents about location systems to dependent people. In order to obtain a global vision, patents should be consulted. Focusing on location, public databases enable to find out a high number of them. However the quantity quickly decreases if the criteria are limited to the topics of this study, and more specifically, to potential users of the application. Nine patents have been selected to be analysed in this study: *Infrared personnel locator system* [9], *Locator device* [10], *IR/IF locator* [11], *Personal locator system* [12], *Anytime/anywhere child locator system* [13], *Apparatus and method for locating missing persons, animals and objects* [14], *Shoes for personal location* [15], *System for locting an alzheimer´s patient* [16], and *Alzheimer´s patient tracking system* [17].
3. Commercial products. To obtain a realistic global vision of the state of locations systems, the commercial products should be also considered. In this way, it will be possible to know the real application of research and development projects. Moreover, this analysis will enable to detect strengths and weaknesses. Eleven devices were found and compared: *GPS- GLOBAL P100* [18], *GPS- GLOBAL P200* [18], *GLOBAL VEGA* [18], *"KERUVE"* [19], *S-911* [20], *SILOVE 4, SPHERA-Mobile* [21], *AEROTEL GEOSPEEKER* [22], *ALARES N-CARD* [23], *BEN* and *PEOPLE TRACKUSA* [24].

User´s Point of View in a Rural Area

To complete the conclusions obtained in the previous section with the direct opinions of the users in a rural area a fieldwork was developed in Teruel (Spain). The fieldwork was developed with fifty-eight persons, aged between 39 and 90 years. Specifically, 23.73% of the participants were under 65 years versus 76.27% higher than 65 years. In this way, it also was possible to incorporate the opinion and information from caregivers (under 65 years old).

3 Conclusions

The review of the state of locator systems for dependent people leads to a high number of technologies that could be used in this kind of applications. But only several technologies are mainly used in patents and commercial products. Thus, it is possible to conclude that real devices do not take advantage of all the technological possibilities. GPS could be considered the best option for outdoor environment. However, systems should work inside and outside without lead to high costs. ZigBee and WiFi are usually used in these two environments, either independently or in mixed technological solutions. The use of IEEE 802.16 standard (WiMAX) in rural villages can be considered as an incipient line of work that should be raised in the following years.

According the date recollected through workshops, devices or systems designed to improve Safety, Security and Health are appreciated. Global systems to integrate location functionalities with other features should be considered. To prevent rejection of technology, the number of mobile devices for the elderly should be minimized and their usability improved. The application of open philosophy (open hardware and software) could facilitate the adaptation and tailoring process.

Despite the substantial body of projects and products, there is still the need to incorporate specific training actions and special support to the elderly as an important part of the technological solutions. This is a new detected opportunity for the future.

Acknowledgments. The authors would like to acknowledge the ARAID foundation and IberCaja for their support for the project "LAD-TERUEL". Thanks to the "Departamento de Ciencia, Tecnología y Univ. del Gob. de Aragón" and to the "Fondo Social Europeo" for their support to the EduQTech group.

References

[1] Eurostat, The life of women and men in Europe. A statistical portrait. European Commission, Office for Official Publications of the European Communities, Luxembourg. Theme: Population and social conditions. Collection: Statistical books (2008)

[2] Plaza, I., et al.: Mobile applications in an aging society: Status and trends. The J. of Syst. & Softw. 0164-1212 84, 1977–1988 (2011)

[3] Benavente-Pece, C., Moracho-Oliva, V.M., Domínguez García, A., Logilde-Rodríguez, M.: Global System for Location and Guidance of Disabled People: indoor and outdoor technologies integration. In: Fifth Internacional Conference on Networking and Servicies, pp. 370–375 (2009) ISBN 978-1-4244-3688-0

[4] Boulos, M.N.K., Rocha, A., Martins, A., et al.: CAALYX: A new generation of location-based services in healthcare. Int. J. of Health Geogr.

[5] Chon, Y., Cha, H.: LifeMap: Smartphone-based Context Provider for Location-based Services. IEEE Pervasive Comput. 15361268 58, 58–67 (2011)

[6] Zhihong, T., Jinsheng, Y., Jianguo, Z.: Location-based Services Applied to an Electric Wheelchair Based on the GPS and GSM Networks. In: International Workshop Intelligent Systems and Applications, pp. 1–4 (2009) ISBN 978-1-4244-3893-8

[7] Liu, H.-H., Chen, Y.-J., Chang, Y.-J., Chen, W.-H.: Mobile guiding and tracking services in public transit system for people with mental illness. In: TENCON IEEE Region 10 Conference, pp. 1–4 (2009)

[8] Rfid (2011) Web Page that contains the digital news "TSB-SPHERAhospital" published on (March 1, 2011), http://www.rfidpoint.com/regiones-y-paises/europa/ (accessed, June 2011)

[9] White, L.L.: Infrared personnel locator system. Patent Number: 4, 275, 385. Country of origin: EEUU (1981)

[10] Ladner, D.L.: Locator device. Patent Number: 5, 594, 425. Country of origin: EEUU (1997)

[11] Crimmins, J.W.: IR/IF locator. Patent Number: 5, 917, 425. Country of origin: EEUU (1999)

[12] Westrick, M.D.: Sistema de localización de personas 2 189 280. Country of origin: Spain (2000)

[13] Elliot, B.D.: Anytime/anywhere child locator system. Patent Number: 6, 917, 425 B1. Country of origin: EEUU (2001)

[14] Toubia, S.: Apparatus and method for locating missing persons, animals, and objects. Patent Number: 6, 317, 049 B1. Country of origin: EEUU (2001)

[15] Pros, R.: Calzado para la localización personal. Patent Number: 1 062 084. Country of origin: Spain (2006)

[16] Garrison, J.: System for locating an Alzheimer's patient. Patent Number: US 2009/0237256 A1. Country of origin: EEUU (2009)

[17] Janetis, E.: Alzheimer's patient tracking system. Patent Number: US 7,825,794 B2. Country of origin: EEUU (2010)

[18] Website of "AVL Global", http://www.globalavl.com/es/localizador-gps-pulsera-brazaletereloj.html (accessed March 2011)

[19] Link with official information about the personal locater Keruve. http://www.keruve.com/2010/ (accessed March 2011)

[20] Website of "Laipac Tech", http://www.laipac.com/bracelet_application.htm (accessed, March 2011)

[21] Website of "Technologies for Health and Welfare/ Tecnologías para la Salud y el Bienestar" (Spain), http://www.tsbtecnologias.es/productos-ebienestar/SPHERAmobile/ (accessed, March 2011)

[22] Website of "Aerotel Medical Systems", http://www.aerotel.com/en/ (accessed, June 2011)

[23] Alares Web Page of the Group "Alares", http://www.alares.es (accessed, June 2011)

[24] Website of "People TrackUSA", http://peopletrackusa.com/ (accessed, June 2011)

A Maritime Piracy Scenario for the n-Core Polaris Real-Time Locating System

Óscar García, Ricardo S. Alonso, Dante I. Tapia, Fabio Guevara,
Fernando de la Prieta, and Raúl A. Bravo

Abstract. There is a wide range of applications where Wireless Sensor Networks
(WSN) and Multi-Agent Systems (MAS) can be used to build context-aware systems. On the one hand, WSNs comprise an ideal technology to develop sensing
systems, as well as Real-Time Locating Systems (RTLS) aimed at indoor environments, where existing global navigation satellite systems, such as GPS, do not
work correctly. On the other hand, agent technology is an essential piece in the
analysis of data from distributed sensors and gives them the ability to work
together and analyze complex situations. In this sense, n-Core Polaris is an indoor
and outdoor RTLS based on ZigBee WSNs and an innovative set of locating and
automation engines. This paper describes the main components of the n-Core
Polaris system, as well as a proposed scenario where n-Core Polaris can be used
to support boarding and rescue operations in maritime piracy environments.

Keywords: Wireless Sensor Networks, Real-Time Locating Systems,
Multi-Agent Systems, Maritime Piracy.

1 Introduction

Wireless Sensor Networks (WSNs) are used for gathering the information useful
to build context-aware environments, whether in home automation, industrial

Óscar García · Ricardo S. Alonso · Dante I. Tapia · Fabio Guevara
R&D Department, Nebusens, S.L., Scientific Park of the University of Salamanca, Edificio
M2, Calle Adaja, s/n, 37185, Villamayor de la Armuña, Salamanca, Spain

Fernando de la Prieta · Raúl A. Bravo
Department of Computer Science and Automation, University of Salamanca. Plaza de la
Merced, s/n, 37008, Salamanca, Spain
e-mail: {oscar.garcia,ricardo.alonso,dante.tapia,
        fabio.guevara}@nebusens.com, {fer,raulabe}@usal.es

S. Omatu et al. (Eds.): Distributed Computing and Artificial Intelligence, AISC 151, pp. 601–608.
springerlink.com   © Springer-Verlag Berlin Heidelberg 2012

applications or smart hospitals [1]. Nevertheless, the information obtained by Wireless Sensor Networks must be managed by intelligent and self-adaptable technologies to provide an adequate interaction between the users and their environment. In this sense, agents and Multi-Agent Systems (MAS) [2] comprise one of the areas that contribute expanding the possibilities of Wireless Sensor Networks.

One of the most interesting applications for WSNs is Real-Time Locating Systems (RTLS). Although outdoor locating is well covered by systems such as the current GPS or the future Galileo, indoor locating needs still more development, especially with respect to accuracy and low-cost and efficient infrastructures [3]. Therefore, it is necessary to develop RTLSs that allow performing efficient indoor locating in terms of precision and optimization of resources. In this sense, the use of optimized locating techniques allows obtaining more accurate locations using even fewer sensors and with less computational requirements [3].

For these reasons, Nebusens and the BISITE Research Group of the University of Salamanca have developed n-Core Polaris, an innovative indoor and outdoor Real-Time Locating System based on ZigBee WSNs and agent technology so that information gathered by WSN nodes is managed by intelligent agents with reasoning mechanisms [4] [5]. The new n-Core Polaris exploits the potential of the n-Core platform [4], a hardware and software platform intended for developing and deploying easily and quickly a wide variety of WSN applications based on the ZigBee standard [6]. n-Core Polaris is an especially useful tool in environments where it is needed to locate people or assets in real-time with a fast deployment, such as natural or nuclear disasters. This paper proposes a maritime piracy scenario where n-Core Polaris can be applied. This scenario consists of a RTLS that can be deployed to support maritime boarding and rescue operations. This way, the system will support special corps when performing rescue operations that involve a hostile boarding. In this regard, the system will provide them with real-time information, facilitating the coordination of the operation and avoiding casualties.

The rest of the paper is structured as follows. The next section explains the problem description, as well as the research areas involved in the development of the n-Core Polaris RTLS. Then, the main characteristics of the innovative n-Core Polaris system are described. After that, a case study where the n-Core Polaris system is proposed to be applied to fight against maritime piracy is depicted. Finally, the conclusions and future lines of work are presented.

2 Problem Description

In recent years, the problem of maritime piracy has become worryingly well-known all over the world due to attacks against fishing ships and oil tankers in Indian Ocean's waters near Somalia coast [7]. These attacks imply substantial human, social and economic costs for the fishing and merchant countries due to military expenses, ransoms, as well as the reduction of the international commerce and fishing. In this regard, the use of technology can help civilian and military personnel both at sea and at ground to face emergency situations, reducing drastically the costs derived from an eventual rescue intervention, as well as the expenses in preventive measures.

Nevertheless, technology should help users to perform surveillance and rescue tasks without distracting them. In addition, technology should increase the knowledge about the environment by users, have a steep learning curve, as well as be non-invasive, context-aware, efficient and inexpensive. Some of the research areas and technologies proposed in this work to fight against maritime piracy are Wireless Sensor Networks, Multi-Agent Systems and Real-Time Locating Systems.

One of the most important technologies used to provide context-awareness for systems and applications is Wireless Sensor Networks (WSN) [1]. Context-aware technologies allow civil and military developments to automatically obtain information from users and their environment in a distributed and ubiquitous way. The context information may consist of many different parameters such as location, the ambient status (e.g., temperature), vital signs (e.g., heart rhythm), etc. Sensor networks need to be fast and easy to install and maintain. In this regard, Wireless Sensor Networks are more flexible and require less infrastructural support than wired sensor networks, existing plenty of technologies for implementing WSNs, such as RFID, UWB, ZigBee, Wi-Fi or Bluetooth [1].

Moreover, the information obtained by WSNs can be managed by intelligent and self-adaptable technologies to provide an adequate interaction between the users and their environment. In this sense, the development of agents is an essential piece in the analysis of data from distributed sensors and gives them the ability to work together and analyze complex situations, thus achieving high levels of interaction with humans [2]. Furthermore, agents can use reasoning mechanisms and methods in order to learn from past experiences and to adapt their behavior according to the context [8].

Tracking the real-time position of people and assets can make the difference in a maritime piracy scenario. One of the most interesting applications for WSNs is Real-Time Locating Systems (RTLS). Real-Time Locating Systems can be categorized by the kind of its wireless sensor infrastructure and by the locating techniques used to calculate the position of the tags (i.e., the locating engine). This way, there is a combination of several wireless technologies, such as RFID, Wi-Fi, UWB and ZigBee, and also a wide range of locating techniques that can be used to determine the position of the tags. Among the most widely used locating techniques we have signpost, fingerprinting, triangulation, trilateration and multilateration [3] [9] [10].

3   The n-Core Polaris Real-Time Locating System

n-Core Polaris is an innovative indoor and outdoor Real-Time Locating System based on the n-Core platform that features a tested accuracy, flexibility and automation integration [4] [5] [11]. The new n-Core Polaris is based on the n-Core platform, also developed by Nebusens and BISITE, taking advantage of the set of features of the n-Core Sirius devices and the n-Core Application Programming Interface [4].

The n-Core platform is based on the IEEE 802.15.4/ZigBee international standard, which operates in the 868/915MHz and 2.4GHz unlicensed bands. Unlike Wi-Fi or Bluetooth, ZigBee is designed to work with low-power nodes and allows up to 65,534 nodes to be connected in a star, tree or mesh topology network [6]. The n-Core platform consists of several modules, fully integrated among them, which provide all the functionalities of the platform.

At the hardware level, the n-Core platform provides a set of radio-frequency devices, called n-Core Sirius A, Sirius B and Sirius D. Each n-Core Sirius device includes an 8-bit RISC (Atmel ATmega 1281v) microcontroller with 8KB RAM, 4KB EEPROM and 128KB Flash memory and an IEEE 802.15.4/ZigBee 2.4GHz (AT86RF230) or 868/915MHz (AT86RF212) transceiver, and several communication ports (GPIO, ADC, I2C, PWM and UART through USB or DB-9 RS-232) to connect to distinct devices, such as computers, sensors and actuators [4].

At the software level, all n-Core Sirius devices are provided with a specific firmware that offers all its functionalities. This way, developers do not have to write embedded code. They can either simply configure the devices functionalities from a specific tool or write high-level code using the n-Core Application Programming Interface (API) from a computer. n-Core also offers through this API different modules/engines to develop specific applications, including an automation engine (for controlling sensors and actuators), a locating engine (includes innovative algorithms to calculate the position of any n-Core device) and a data engine (for transmitting general-purpose data frames among devices).

Therefore, the wireless infrastructure of n-Core Polaris is made up of a set of n-Core Sirius devices. In the n-Core Polaris RTLS, n-Core Sirius B devices are used as tags, while n-Core Sirius D devices are used as readers (i.e., position references). This way, n-Core Sirius B devices are carried by users and objects to be located, whereas n-Core Sirius D devices are placed throughout the environment to detect the tags. Finally, n-Core Sirius A devices are used to connect sensors and actuators through their communication ports.

Figure 1 (left) shows the basic architecture of the n-Core Polaris RTLS. The kernel of the system is a computer that is connected to a ZigBee network formed by n-Core Sirius devices. That is, the computer is connected to an n-Core Sirius D device through its USB port. This device acts as coordinator of the ZigBee network. The computer runs a web server module that offers the innovative locating techniques provided by the n-Core API. On the one hand, the computer gathers the detection information sent by the n-Core Sirius D devices acting as readers to the coordinator node. One the other hand, the computer acts as a web server offering the location info to a wide range of possible client interfaces. In addition, the web server module can access to a remote database to obtain information about the users and register historical data, such as alerts and location tracking.

The operation of the system is as follows. Each user or object to be located in the system carries an n-Core Sirius B acting as tag. Each of these tags broadcasts periodically a data frame including, amongst other information, its unique identifier in the system. The rest of the time these devices are in a sleep mode, so that the power consumption is reduced. This way, battery lifetime can reach even several months, regarding the parameters of the system (broadcast period and transmission power). A set of n-Core Sirius D devices is used as readers throughout the environment. The broadcast frames sent by each tag are received by the readers that are close to them. This way, readers store in their memory a table with an entry per each detected tag. Each entry contains the identifier of the tag, as well as the RSSI (Received Signal Strength Indication) and the LQI (Link Quality Indicator) gathered from the broadcast frame reception. Periodically, each reader sends this table to the coordinator node connected to the computer. The coordinator forwards each table received from each reader to the computer through the USB port. Using these detection information tables, the n-Core API applies a set of locating techniques to estimate the position of each tag in the environment. These locating techniques include signpost, trilateration, as well as an innovative locating technique that takes into account different confidence levels when estimating the distances between tags and readers from the detected RSSI values (due to multipath effects, some detected RSSI intervals are less reliable than others).

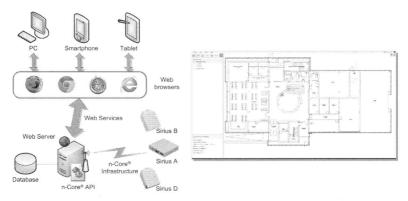

Fig. 1 The Web Services based architecture of the n-Core Polaris RTLS.

Then, the web server module offers the location data to remote client interfaces as web services using SOAP (Simple Object Access Protocol) over HTTP (Hypertext Transfer Protocol). Figure 1 (right) shows a screenshot of the web client interface. This client interface has been designed to be simple, intuitive and easy-to-use. Through the different interfaces, administrator users can watch the position of all users and objects in the system in real-time. Furthermore, administrators can define restricted areas according to the users' permissions. This way, if a user enters in an area that is forbidden to him according to his permissions, the system will generate an alert that is shown to the administrator through the client interfaces. In addition, such alerts are registered into the database, so administrators can check anytime if any user violated his permissions. Likewise, administrators can query the database to obtain the location track of a certain user, obtaining statistical measurements about its mobility or the most frequent areas where it moves.

Furthermore, users can use one of the general-purpose buttons provided by the n-Core Sirius B devices to send an alert to the system. Similarly, administrators can send alerts from the system to a user or a set of users, which can confirm the reception using other of the buttons. The system not only provides locating features, but also scheduling and automation functionalities. The system can be easily integrated with a wide range of sensors and actuators using the variety of communication ports included in the n-Core Sirius A devices. By means of the automation engine provided by the n-Core API, the n-Core Polaris system can schedule automation tasks, as well as monitor all sensors in the environment in real-time.

4 Proposed Scenario

This section describes a case study where the n-Core Polaris RTLS is proposed to be applied to fight against maritime piracy and illegal traffic. The system proposed in this case study consists of a Real-Time Locating System that can be deployed to support maritime boarding and rescue operations. In this scenario, the main objective is to avoid casualties, as well as avoid ransom payments and discourage further hijackings. As n-Core Polaris can be deployed in just few minutes throughout the area of interest and works properly indoors achieving an indoor accuracy with just 1m error [4] [5] [11], these features make it suitable for military applications where is required to monitor the position of people and objects in real-time and with minimal installation and deployment times.

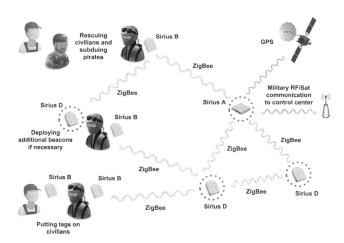

Fig. 2 Schema of the n-Core Polaris RTLS in a boarding and rescue scenario.

Figure 2 shows the basic schema of n-Coe Polaris running in a boarding and rescue scenario. Each member of the rescue military troops carries an n-Core Sirius B device so that the system can locate him in a certain area at all times, both indoors and outdoors. In addition, some soldiers can carry additional n-Core Sirius B devices to be used as tags by civilians. This way, soldiers put an additional n-Core Sirius B device on each civilian, activating it to be tracked by the system. Likewise, some soldiers carry a set of n-Core Sirius D devices to be placed as additional beacons in the environment and acting as distances references in the system.

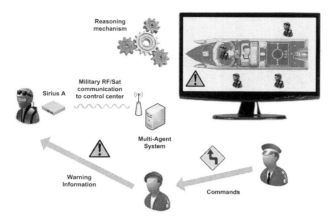

Fig. 3 Multi-Agent System and reasoning mechanisms in the boarding and rescue scenario.

Furthermore, there is an n-Core Sirius A device acting as coordinator node in the system and that can be carried by a soldier or by a boat close to the rescue area. This n-Core Sirius A device is connected to a GPS receiver to obtain its global position as the main reference. In addition, this device is connected to a remote control center through a military radio-frequency or satellite link. The remote control center runs a Multi-Agent System that includes reasoning mechanisms and makes use of the locating techniques provided by the n-Core API, as can be seen in Figure 3. The remote control center gathers the detection data sent by the n-Core Sirius D acting as readers to the coordinator node.

5 Conclusions and Future Work

Piracy and illegal traffic imply human, economic, social and political costs. In this sense, it is necessary to apply non-invasive, context-aware, efficient and inexpensive technology to minimize these costs. Systems based on Multi-Agent Systems, Wireless Sensor Networks and Real-Time Locating Systems can give support to military and civil authorities to deal with these problems. There are different wireless technologies that can be used on RTLS. The ZigBee standard offers interesting features over the rest of technologies, as it allows the use of large mesh networks of low-power devices and the integration with many other applications.

In this regard, the n-Core Polaris RTLS could provide an important competitive advantage to applications where it is necessary to gather sensing data, automate tasks and know the location of people or objects. Amongst its multiple application areas are the healthcare, the industrial or the agricultural sectors, as well as those related to security. Its optimal indoor and outdoor functioning makes n-Core Polaris flexible, powerful and versatile solutions.

Regarding its performance, the n-Core Polaris indoor locating system has been awarded as the winner of the first international competition on indoor localization and tracking, organized by the Ambient-Assisted Living Open Association (AALOA) [11]. These results demonstrate that n-Core Polaris is a robust system suitable to be used in indoor environments and that can locate users and assets with up to 1m accuracy without interfering in the daily-life of people.

Future lines of work include obtaining ideas from specialized military and civilian users to get feedback and improve the proposed case study. Then, it will be performed a detailed analysis and design process to develop and deploy prototypes to test performance and get additional feedback.

Acknowledgments. This project has been supported by the Spanish Ministry of Science and Innovation (Subprograma Torres Quevedo).

References

1. Sarangapani, J.: Wireless Ad hoc and Sensor Networks: Protocols, Performance, and Control. CRC (2007)
2. Wooldridge, M.: An Introduction to MultiAgent Systems. Wiley (2009)
3. Liu, H., Darabi, H., Banerjee, P., Liu, J.: Survey of Wireless Indoor Positioning Techniques and Systems. IEEE Trans. Syst. Man Cybern. Part C-Appl. Rev. 37, 1067–1080 (2007)
4. Nebusens, n-Core®: A Faster and Easier Way to Create Wireless Sensor Networks (2011), http://www.n-core.info (accessed, November 2011)
5. Tapia, D.I., Alonso, R.S., Rodríguez, S., de la Prieta, F., Corchado, J.M., Bajo, J.: Implementing a Real-Time Locating System Based on Wireless Sensor Networks and Artificial Neural Networks to Mitigate the Multipath Effect. In: 2011 Proceedings of the 14th International Conference on Information Fusion (FUSION), pp. 1–8. IEEE/ISIF, Chicago (2011)
6. Baronti, P., Pillai, P., Chook, V.W.C., et al.: Wireless sensor networks: A survey on the state of the art and the 802.15.4 and ZigBee standards. Comput. Commun. 30, 1655–1695 (2007)
7. Maouche, A.: Piracy along the Horn of Africa: An Analysis of the Phenomenon within Somalia. PiraT Arbeitspapier zur Maritimen Sicherheit, 6. Hamburg (2011)
8. Tapia, D.I., de Paz, J.F., Rodríguez, S., Bajo, J., Corchado, J.M.: Multi-Agent System for Security Control on Industrial Environments. International Transactions on System Science and Applications Journal 4(3), 222–226 (2008)
9. Ding, B., Chen, L., Chen, D., Yuan, H.: Application of RTLS in Warehouse Management Based on RFID and Wi-Fi. In: 4th International Conference on Wireless Communications, Networking and Mobile Computing, WiCOM 2008, pp. 1–5 (2008)
10. Stelios, M.A., Nick, A.D., Effie, M.T., et al.: An indoor localization platform for ambient assisted living using UWB. In: ACM (ed.) Proceedings of the 6th International Conference on Advances in Mobile Computing and Multimedia, Linz, Austria, pp. 178–182 (2008)
11. AAL Open Association, Evaluating AAL Systems through Competitive Benchmarking (2011), http://evaal.aaloa.org (accessed October 2011)

A Serious Game for Android Devices to Help Educate Individuals with Autism on Basic First Aid

Zelai Sáenz de Urturi, Amaia Méndez Zorrilla, and Begoña García Zapirain

Abstract. Within the group of individuals with autism, we can find a certain number of people who interact well with mobile devices and other types of technology. To improve their knowledge on first aid, the main aim was to create an application composed of a set of Serious Games oriented towards first aid education: i.e. how to handle specific situations, basic knowledge about healthcare or medical specialties… all employing the use of current technologies such as Smartphones or Tablets, specifically running the Android operating system. Not only technological results have been investigated, but also feedback was taken from opinions and experiences by both users and specialists taking part in the practical validation and testing of the application.

Keywords: autism, First Aid education, Serious Games for health, Android.

1 Introduction

Autism is not a disease; it is a syndrome, a cluster of symptoms that characterize a degenerative disorder of the bio-psycho-social. This disability is a developmental disorder, permanent and profound, and significantly affects and alters the areas of communication, social interaction and behavior. Often these symptoms are accompanied by abnormal behavior, such as activities and interests of repetitive and stereotypical rocking movements, and an unusual obsession towards objects or events [1].

Appears during the first three years of life and it will drastically limit the life of the person who has it besides causing significant stress in the family [2].

Zelai Sáenz de Urturi · Amaia Méndez Zorrilla · Begoña García Zapirain
DeustoTech-Life Unit. Deusto Institute of Technology. University of Deusto. Avda. de las Universidades, 24. 48007 Bilbao, Spain
e-mail: {amaia.mendez,mbgarciazapi}@deusto.es

S. Omatu et al. (Eds.): Distributed Computing and Artificial Intelligence, AISC 151, pp. 609–616.
springerlink.com   © Springer-Verlag Berlin Heidelberg 2012

Studies in North America, Europe and Asia have identified individuals with autism with an approximate prevalence of 0.6% to over 1% [3]. A recent study in South Korea (May 2011) [4] reported a prevalence of 2.64%, i.e. 1 child out of every 38 suffers it. There have also been estimates that between 1 in 88 and 1 in 240 with an average of 1 in 110 children in the United States have autism [5].

The percentage of individuals who have sufficient capacity to live independently (or partially independent) is relatively low, and most of them require a great help throughout life. But it's established, that if they receive appropriate resources and support, are able to co-exist into society properly, improving their quality of life, and acquiring a greater degree of autonomy.

Due to these reasons, this study aims to improve quality of life of individuals with autism and enhance autonomy as much as possible through the use of the latest technologies in mobile devices such as smartphones and tablets, implementing Serious Games [6].

The computer games have been increasingly mediated via mobile devices, such as smartphones, tablets and other touch-based devices. One of the main advantages is that they allow users to use multi-touch and gestures to receive rapid device feedback.

Studies suggest that computer-assisted instruction incorporating multimedia elements, and particularly, visually-rich elements, can be used to engage and effectively teach children with autism [7-9].

The Serious Games have been designed so that, besides their pure entertainment value, they convey relevant ideas or messages about various aspects not related to the gaming industry. A Serious Game is associated with the education and learning of new concepts and skills, but can also works as training and simulation of various activities of real life. In other words, a serious game should have an evident connection between the real and virtual world, and a purpose beyond the scope of just playing a video game.

The project will be aimed at individuals with autism themselves (who study or work) and people who work with them, such as psychologists, instructors, assistants and family members.

These specialists will be provided with objective reporting of activities through a single web application. This kind of information will help them to achieve greater influence on people with this disability.

2 Objectives

The overall objective of this project is to create an application composed of several Serious Games oriented towards first aid education, minor diseases, and knowledge of medical specialties, using Android operating system.

This general approach can be broken down into several specific objectives, as described below:

1. Develop a "Serious Game" aimed at learn how to deal with first aid situations in a fun and dynamic way.
2. Identify and evaluate a range of objective variables such as: the movements made by the user, the level of success or the time needed.
3. Design an interface that suits the needs of the autistic population, making use of ICT, specifically through the use of touch-enabled mobile devices such as smartphones and tablets.
4. Promoting e-inclusion: equal opportunities for access and participation of people with autism in the Information Society.

For the accomplishment of these goals and due to ethical constraints, consent forms were signed by the parents of all children participating in the experiment.

3 The Application

The architecture used to develop the application is a three-layer one. The main goal of the architecture is the segmented design, which separates the logical layer from the business layer. This means that tasks can be divided in order to work comfortably by levels, abstracting some levels from others and coming up with a big scalability and modifiability.

3.1 Architecture Description

The three layers are described below:

- *Presentation Layer.* The various multimedia elements making up the application are situated in this layer. They display and capture the user information in order to deliver it to the business layer.
- *Business Layer.* This layer houses the internal logical needed to provide core functionality of the mini-games. Having an independent business layer provides great adaptability to the application because it allows the incorporation of new mini-games to this Serious Game without compromising other system components. Each mini game is independent, although they can easily communicate between them.

   The resulting information is sent to be displayed in the presentation layer, on the phone screen, or to be used in concurrence with the database, to store all the data generated during the last session.
- *Database Layer.* This layer is made up of the database manager, which will back up the business layer in the storing and data-obtaining processes. In the database some parameters will be stored like the name of the user, the actual date and hour, the number of right and wrong combinations, time needed... etc.

The diagram of Fig. 1 explains in more detail how the application works:

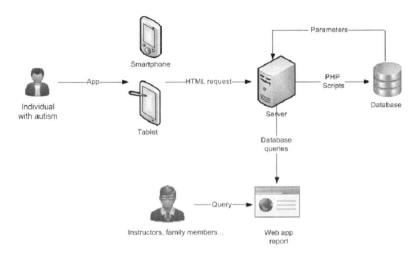

Fig. 1 First Aid Serious Game's General Diagram.

The logic of the application is as follows: First of all, the user has two sets of mini games or activities to choose from, each one with different levels of difficulty.

After finishing the activity, the application (from the phone) will connect to the server using HTTP / HTTPS connection requests. It will request access to the database, store the data generated during gameplay and finally update the record for that user, if necessary. Once the data has been saved, the application will generate a more detailed report about the user, explaining in which activity errors have been made, and what the reason might related to. The instructors can review the reports in a web page to evaluate the improvement grade and also assist them in real life activities for which they failed or had obtained a lesser score.

Encryption

Because private data about the activities of each user is sent through the Internet, it is desirable to encrypt this information.

This module is responsible for signing the data from your phone for later verification by the server, and decrypting the datagrams from the latter. This results, in a security layer for the obtained data to make sure no malicious theft can occur.

It uses an asymmetric encryption system, with a private key that is housed on the server and a public key in the application.

As an independent module, the encryption algorithm can easily be altered without much impact on the other modules of the application. Situations where this might be needed are a security breach or the emergence of more robust or optimized algorithms for mobile devices.

3.2 Description of Mini-Games/Activities

At the beginning of each mini game, the user will be asked to enter the name (like a login method) which will be referenced in the final report. They can also choose a picture from an avatar (an image from a girl, boy, a person who wears glasses, an student or worker) which identifies them during gameplay.

The application consists of two sets of mini-games:

- 1. Order the sequences (see Fig. 2):

It consists of ordering sequences of previously scrambled images being di played at the top of the screen. In order to guess the correct sequence, they can simply touch the images on the screen in the order they desire. The current order selected by the user will be shown in the bottom portion of the screen.

These sequences are related to various aspects of first aid such as: what to do when you get a cut, headache or what to do to call a doctor or to prevent a bur. All the pictures have a little explanation down.

The player will have a limited time to solve as many sequences as possible and get a good score. If the order is right, the user will carry on with more difficult sequences and finally he will obtain his score.

The user playing (or his/her instructors) can choose whether or not to include a time limit. The inclusion of this constraint should depend on the kind of person who is playing in the specific session, their autism grade and his/her level of stress while playing.

Upon completion of this activity and after storing the appropriate parameters in the database, a new report associated with that user will be generated. This report will state the most important data, i.e. everything which is necessary to help the family members or psychologists to know what sequences the user has difficulty understanding or interpreting.

- 2. Images association and medical specialties (Fig. 3)

In this game you are playing:
  - To identify the roles of people in a hospital using the color of uniforms they are working in.
  - To identify a medical specialty, by choosing which area of the body that it deals with.
  - How to move inside a hospital or a health center identifying the indications inside.

The user will have to choose between different pictures, which appear randomly on the screen, which one has the correct association, by tapping the corresponding images. As in the other activity, choosing the right answer will lead to more difficult associations and the user can choose whether or not to include a time-limit.

3.3 Description of the Final Report

This part of the application is aimed at instructors or family members of a specific user. The main objective of this module is to show those people the continuous evolution of the users.

For the design of this module the MVC (Model View Controller) architectural pattern has been proposed. A web page has been created showing a table, with all the usernames that have played, will load. The names displayed here are identical to the ones input by the user at the start of the application. Each name is presented in the form of a link, and clicking it will cause data to be show regarding that specific user: i.e. date, total time, number of the failed sequence, number of errors in each sequence, total errors, time to resolve each sequence...etc. In addition, to prevent unauthorized access, an access control module has been added to the page.

4   Results

This section describes the results obtained in two facets: technical results and the feedback received from the user satisfaction questionnaires completed by the pilot users and experts.

- Test Execution

The users chosen by the experts to participate in the evaluation were 10 children and adults, both male and female.

All people who participated in the pilot demonstrated the ability to perform activities independently. However during testing, monitors were there to assist testers in case of problems.

Testing was performed using a smartphone and two tablets (with 7" and 10" screens).

- Technical Results

Individuals with autism are involved in different situations in their daily life, the game shows most of the typical ones.

 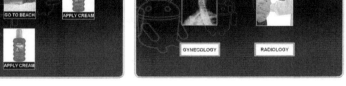

Fig. 2 Playing with the 1st mini-game   **Fig. 3** Image from the 2nd mini-game

Whenever the user starts an activity, he is given instructions on how to play, after which the training exercise starts. The instructions are presented in the form a set of videos with additional written explanations, which the users are able to understand with ease.

- Satisfaction Results

The opinions from the instructors were sought in a simple personal interview, in which they were asked the following questions, answering on a scale from 1 to 5.

Table Instructors questionnaire

QUESTION		SCORE (average)
1-	Did you enjoy using the tool?	3,0
2-	Was the user interface easy to use for individuals with autism?	4,0
3-	The instructor can easily understand the report?	3,5
4-	Is a good idea to use text and images in the video tutorials?	3,2

After the test, participants answered several questions, and following data were obtained:

- 70% of them prefer multi-touch screen to a screen with keyboard.
- 60% like using tablets with big screens instead of smartphones.
- 20% don't like to play with games or use smartphones/tablets.

5 Conclusions

The presented game is a project with real applications aimed at the collective of individuals with autism who has a high degree of independence.

When designing interfaces for learning games, developers should 1) not assume that individuals with autism have the same preferences as most people or 2) that autistics children are a "type," with rigidly defined characteristics. Autism is a spectrum disorder, which means that while affected individuals may show common learning challenges and strengths, they will do so to varying unique degrees.

The application has proved that it is possible to enrich and increase the education/therapy impact through the introduction of information and communication technologies.

Individuals with this disorder have accepted the mobile devices well, but the results are even better with the tablets because of their larger size.

After the experiment and the evaluation of the project, the following benefits can be underlined:

- Social Benefits:
- The concept therapy is changed by game concept, to improve the children motivation [10].
- With educational games, individuals with autism feel more relax doing the activities.
- Medical Benefits:
- The application collects some parameters relevant to the therapy, which permits an objective evaluation of learning/therapy process.

Specialists have recently been demanding that Serious Games of this kind be included in their daily practice because they provide a more recreational vision on activities assessing and measuring the progress of certain therapies.

This paper reflects only the first results. In the nearby future it will increase the number of people with autism who use the application and users will be extended to other groups.

Acknowledgments. This work was supported in part by Telefonica Moviles and the Basque Country Department of Education, Universities and Research. The author also wants to thank the cooperation of APNABI (Association of Parents of individuals with Autism in Bizkaia).

References

1. NINDS, National Institute of Neurological disorders and stroke,
   http://espanol.ninds.nih.gov/trastornos/autismo.htm
2. Ruiz-Lázaro, P.M., Posada de la Paz, M., Hijano Banddera, F.: Trastornos del espectro autista. Detección precoz. herramientas de cribado. Rev. Pediatr Aten Primaria 11(suppl. 17), 381–397 (2009)
3. Summary of Autism/ASD Prevalence Studies,
   http://www.cdc.gov/ncbddd/autism/documents/Autism_Prevalen
   ceSummaryTable_2011.pdf
4. Am J Psychiatry. Prevalence of Autism Spectrum Disorders in a Total Population Sample. American Journal of Psychiatry 9 (May 2011)
5. Catherine Rice. Prevalence of Autism Spectrum Disorders, Autism and Developmental Disabilities Monitoring Network, United States (2006),
   http://www.cdc.gov/mmwr/preview/mmwrhtml/ss5810a1.html
6. Corti, K.: Games-based business & management skills development. White Paper (2001), http://www.pixelearning.com/docs/
   justifying_games_for_learning.pdf
7. Gal, E., Bauminger, N., Goren-Bar, D., Pianesi, F., Stock, O., Zancanaro, M., et al.: Enhancing social communication of children with high-functioning autism through a co-located interface. Artificial Intelligence and Society 24(1), 75–84 (2009)
8. Barakova, E., Gillessen, J., Feijis, L.: Social training of autistic children with interactive intelligent agents. Journal of Integrative Neuroscience 8(1), 23–34 (2009)
9. Aldrich, C.: Learn by doing: a comprehensive guide to simulations, computer games, and pedagogy in e-learning and other educational experiences. John Wiley and Sons, Pfeiffer (2005)
10. Kato, P.M.: Review of General Psychology 14(2), 113–121 (2010)

User-Independent Human Activity Recognition Using a Mobile Phone: Offline Recognition vs. Real-Time on Device Recognition

Pekka Siirtola and Juha Röning

Abstract. Real-time human activity recognition on a mobile phone is presented in this article. Unlike in most other studies, not only the data were collected using the accelerometers of a smartphone, but also models were implemeted to the phone and the whole classification process (preprocessing, feature extraction and classification) was done on the device. The system is trained using phone orientation independent features to recognize five everyday activities: walking, running, cycling, driving a car and sitting/standing while the phone is in the pocket of the subject's trousers. Two classifiers were compared, $k$nn (k nearest neighbours) and QDA (quadratic discriminant analysis). The models for real-time activity recognition were trained offline using a data set collected from eight subjects and these offline results were compared to real-time recognition rates. Real-time recognition on the device was tested by seven subjects, three of which were subjects who had not collected data to train the models. Activity recognition rates on the smartphone were encouraging, in fact, the recognition accuracies obtained are approximately as high as offline recognition rates. The real-time recognition accuracy using QDA was as high as 95.8%, while using $k$nn it was 93.9%.

1 Introduction and Related Work

Human activity recognition using wearable sensors, such as accelerometers, has been widely studied during the recent 20 years. Despite several years of study and promising recognition results, not many commercial products, besides pedometers, exploiting these results are available. There are some exceptions, however, such as Polar Active [14] and ActiGraph [1], which can

Pekka Siirtola · Juha Röning
Department of Computer Science and Engineering
P.O. BOX 4500, FI-90014 University of Oulu, Finland
e-mail: pesiirto@ee.oulu.fi, jjr@ee.oulu.fi

S. Omatu et al. (Eds.): Distributed Computing and Artificial Intelligence, AISC 151, pp. 617–627.
springerlink.com © Springer-Verlag Berlin Heidelberg 2012

be used to detect the intensity of the activity. Nevertheless, in overall, it seems that companies and people have not been willing to spend money on technology enabling activity recognition. Now, things are about to change: Smartphones are sold more and more every year (smartphone shipments: 2009: 169 million, 2010: 295 million [18]). Smartphones include a wide range of sensors, such as accelerometers, magnetometers, gyroscopes, and GPS, all of which are sensors used for activity recognition in the past studies. Therefore, people already have the technology enabling activity recognition and mobile application stores (AppStore, Nokia Store, Android Market, etc.) can be used to distribute activity recognition applications directly to end-users.

Human activity recognition using accelerometers has been carried out in various studies, such as [2, 4, 20, 21]. These studies were done using accelerometers build for research use. Therefore, based on these results, it is not straightforward to build a commercial product. There are also some articles where activity recognition using mobile phones has been studied ([3, 11, 13, 16, 19, 22]). In each of these studies, the data is collected using a mobile phone and the activity recognition is done afterwards on PC, based on collected data. Thus, the activity recognition algorithms are not implemented on the phone, and the classification is not done in real-time on a mobile phone as in our study.

An activity recognition system running purely on a smartphone is presented in [6]. The presented system can be trained on the device and it also does the classification in real-time on the device. The recognition is based on features calculated using geometric template matching and support vector machine (SVM) is used as a classifier. Unfortunately, the article does not include recognition rates: thus, the evaluation of the system is difficult. However, the smartphone application is available from Android Market. The system described in [10] can also be found from Android Market. It seems to recognize activities with high accuracy, but all the features used are not orientation independent.

Activity recognition using mobile phones has some limitations. Because smartphones are expensive products, people do not want to carry a phone while performing activities where there is a danger to break it. Therefore, it is not necessary to recognize most of the sports activities such as playing football or swimming. Thus, this study concentrates on recognizing five everyday activities, *walking, cycling, running, idling (=sitting/standing)* and *driving/riding a car*, using the mobile phone accelerometers. Although the latest smartphones are equipped with processors enabling huge calculation capacity, the activity recognition algorithms must nevertheless be light. The mobile phone can be running several applications simultaneously and the activity recognition algorithms are not allowed to use the whole processing power, nor disturb other applications. Therefore, the recognition must be done using light methods.

The paper is organized as follows: Section 2 describes sensors and data sets. Section 3 introduces the techniques and methods used in this study.

Models trained using offline data and their accuracy are presented in Section 4. Section 5 evaluates the accuracy of the activity recognition based on models trained using offline data, when detection is done in real-time on a mobile phone. Finally, conclusions are discussed in Section 6.

2 Data Set

The data were collected using a Nokia N8 smartphone [12] running Symbian^3 operating system. N8 includes a wide range of sensors: tri-axis accelerometer and magnetometer, two cameras (12 MP and 0.3 MP), GPS, proximity sensor, compass, microphones and ambient light sensor. Still, only the tri-axis accelerometer was used in this study to detect activities. The sampling frequency was set to 40Hz.

The models used in this study were trained based on activity data collected from eight healthy subjects. The trousers' front pocket was fixed as the phone placement, but the subject was allowed to determine whether the phone was placed in the left or right pocket. The participants performed five different activities: walking, running, cycling, driving a car, and idling, that is, sitting/standing. The total amount of the data collected was about four hours. These activities were selected because normal everyday life consists mainly of these five activities. Walking and running are different from the other three because everyone has a personal walking and running style. Other activities are not personal, for instance, while cycling, the movement trajectory is predefined. Therefore, the models to recognize walking and running are most challenging to train.

The real-time classification was tested by seven subjects, three of whom were subjects whose data were not used to train the recognition models. These subjects carried the phone in their trousers' front pocket and performed from one to five activities.

The training data were collected by subjects whose age varied from 25 to 34 years (average 29 years) and height from 1.65 to 1.90 metre (average 1.78 metre) and real-time classification was tested by subjects whose age varied from 27 to 34 years (average 30 years) and height from 1.65 to 1.90 metre (average 1.75 metre). They performaed activies outside the laboratory. Subjects walked inside and outside, mainly on flat surface but also in a staircase. Streets where subjects walked, run, drove a car, and cycled were normal tarmac roads, and the route and speed were determined by subjects themselves. Partly the same roads were employed in offline and real-time tests. The roads used for collecting driving a car data included motorways, as well as roads at the city center. Idling consists mostly of office working but includes also standing.

3 Activity Recognition

In this study, two different activity recognition classifiers were compared: quadratic discriminant analysis [8] (QDA) and k nearest neighbours [5] ($k$nn). In addition, the recognition was performed using two different settings: (1) offline recognition, to compare different features, classifiers and to evaluate models for online recognition, and (2) online recognition on the device, to test the models in realistic real-life conditions. In both cases, the raw data were processed in the same way to obtain comparable results.

Preprocessing and feature extraction: The purpose of this study was to develop a user-independent activity recognition method that runs purely on a smartphone and gives accurate recognition results also when the system is used in non-laboratory conditions. The recognition was supposed to work when the mobile phone is placed in the trousers' front pocket. However, the mobile phone can lay on the pocket in numerous different orientations. There are two ways to eliminate the effect of the orientation: (1) by recognizing the orientation of the phone, or (2) by eliminating the orientation information. On the other hand, the orientation is impossible to recognize using only accelerometers. Therefore, the effect of orientation had to be eliminated. In the preprocessing stage, the three acceleration channels were combined as one using square summing to obtain the magnitude acceleration, which is orientation independent. Moreover, the orientation of the phone has limitations, the screen or the back of the phone is always against the user's leg when the phone is in the pocket. Therefore, it was tested if features extracted from a signal where two out of three acceleration channels were square summed would improve the classification accuracy.

The online activity recognition was done using a sliding window technique. The signals from the sensors were divided into equal-sized smaller sequences, also called windows. From these windows, features were extracted and finally the classification of the sequences was done based on these features. In this study, the windows were of the length of 300 observations, which is 7.5 seconds, because the sampling frequency was 40Hz. In offline recognition, the slide between two sequential window was 75 observations, while in online recognition, the slide was set to 150 observations. To reduce the number of misclassified windows, the final classification was done based on the majority voting of the classification results of three adjacent windows. Therefore, when activity changes, a new activity can be detected when two adjacent windows are classified as a new activity. For instance, if the slide is 150 observations, a new activity can be detected after 450 observations, which is around eleven seconds if the sampling rate is 40Hz.

The total number of 21 features were extracted from magnitude acceleration sequences. These features were standard deviation, mean, minimum, maximum, five different percentiles (10, 25, 50, 75, and 90), and a sum and square sum of observations above/below certain percentile (5, 10, 25, 75, 90, and 95). The same features were also extracted from the signals where two

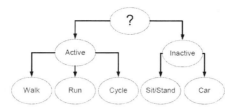

Fig. 1 The decition tree obtained to recognize the type of activity.

Table 1 The results of offline recognition using QDA.

True class / predicted class	idling	walking	cycling	driving	running
idling	**94.3%**	1.2 %	0.3 %	4.2 %	0.0 %
walking	1.0 %	**95.6%**	2.3%	0.0 %	1.3 %
cycling	0.4 %	3.4 %	**94.3%**	1.9 %	0.0 %
driving	3.7 %	0.0 %	2.2 %	**94.2%**	0.0 %
running	0.0 %	0.0 %	0.0 %	0.0 %	**100.0%**

out three acceleration channels were square summed together. It was noted that the combination of x and z axis signal channels improved the classification most. Therefore, from each window, the total number of 42 orientation independent features were extracted, 21 features from the magnitude acceleration signal and 21 features from the signal where x and z were square summed.

The classification result was obtained using the decision tree presented in Figure 1, which classifies activities using a two stage procedure. In the first classification stage, a model is trained to decide if the studied subject is currently active (walking, running or cycling) or inactive (driving a car or idling). In the second stage, the exact activity label is obtained. One model has to be trained to classify an active activity as walking, running or cycling, and the other to classify an inactive activity as idling or driving.

The models were trained offline using the collected data. These models were implemented to a smartphone and also used in online tests. To compare different classifiers, the classification was performed using two different classification methods, $k$nn and QDA. The most descriptive features for each model were selected using a sequential forward selection (SFS) method [7]. QDA classifiers were trained using the whole training data set, similar to $k$nn classifier for the offline recognition. However, because of the limited computational power of the smartphone, the activity recognition on the device was performed using only a limited number of randomly chosen instances from training data.

Table 2 The results of offline recognition using $k$nn.

True class / predicted class	idling	walking	cycling	driving	running
idling	**94.5%**	1.1 %	0.0 %	4.3 %	0.0 %
walking	1.0 %	**90.2%**	8.5%	0.0 %	0.3 %
cycling	0.3 %	1.7 %	**94.6%**	3.4 %	0.0 %
driving	4.2 %	0.0 %	2.1 %	**93.8%**	0.0 %
running	0.0 %	0.4 %	0.0 %	0.0 %	**99.6%**

4 Model Training and Offline Recognition

The purpose of the offline recognition is to build and test accurate models that can later be implemented on a mobile phone to enable user-independent real-time recognition of the activities on the device. Models were trained for $k$nn and QDA classifiers based on the data collected from eight persons.

Results: To obtain reliable user-independent results, the training was performed using the leave-one-out method, so that each person's data in turn was used for testing and the other seven sequences were employed for model training. The results are shown in Tables 1 and 2.

Discussion: The offline recognition results show that both classifiers, QDA and $k$nn, enable accurate results. The average classification accuracy using QDA is 95.4%, while $k$nn enables an accuracy of 94.5%. It should be noted, however that this difference is not statistically significant according to paired t-test. Also, each of the five activities are recognized with high accuracy.

5 Real-Time Experiments on Device

An activity recognition application for Symbian^3 devices was build using Qt [15] programming language. Every Nokia phone running a Symbian^3 operating system has the same kind of accelerometers, and therefore, the results presented in this section can be obtained using any Nokia Symbian^3 phone. The application uses the activity recognition models that were trained using the data presented in Section 2. It should be noted that offline recognition employing $k$nn uses the whole training data set to recognize activities from the test data, making the classification process complex. As mantioned before, because of the limited computational power of the smartphone, the recognition on the device was performed using only a limited number of randomly chosen instances from training data. In this study, eight instances from each activity per subject were chosen as instances of $k$nn-based recognition model. QDA -based real-time classification results were obtained using the very same models used in offline recognition.

The purpose of the application is to make the user conscious of how much and what kind of functional exercise he/she does daily and weekly by showing real-time activity statistics. The application also shows the period of time since the person's most recent active period. It is important to note that the application does not require any data transfer to phone or from phone, everything is done on the device: preprocessing, feature extraction and classification.

Results: The application and the real-time classification were tested by seven persons carrying Nokia N8 smartphone on their trouser's front pocket. Three of these were different from the eight subjects that collected the data for training the recognition models. The recognition results are shown in Tables 3 and 4. Both classifiers were running on the device in parallel; thus, the results are comparable.

Discussion: The real-time experiment showed that the application and models are running smoothly on the device. When activity recognition is done using QDA classifier, the application uses under 5% of the CPU's (680Mhz ARM11 processor) capacity. Therefore, the application can be employed alongside other applications, such as games. In addition, the recognition rates on the device are around as high as offline. The average recognition rate using QDA is 95.8%, while using $k$nn it is slightly lower, 93.9%. According to paired t-test, this difference is not statistically significant, however.

Online recognition was tested by subjects (subjects 1, 2 and 3) whose data was not used for training as well as subjects (subjects 4, 5, 6 and 7) whose data was used for training. In both cases, the average recognition rate is high. However, there are two cases where user-independent classification has not succeed very well. Walking activity of Subject 1 was recognized only with the rate of 65.6% when QDA is used as a classifier and cycling of Subject 3 using $k$nn was recognized correctly only in 76.3% of the cases. In both cases, cycling and walking were mixed together. It seems that inner class variation of these activities is too low causing misclassification. As mentioned above, walking is one of the most difficult activities to recognize user-independently, because every subject has a personal walking style. In addition, not the whole training data were used to train the $k$nn model to keep to recognition process light, which may have caused the weak recognition rates with Subject 3's cycling activity. In overall, the recognition on the device works well, however. It seems that the first phase of the recognition, where observations are classified as active or inactive, appears to work almost perfectly. Therefore, to make recognition more accurate, the second phase of the classification should be improved.

The models used in online recognition are user-independent and it also seems that they are "car-independent". Two different cars were used in the data collecting phase. Although during the online test, subjects 1 and 2 used a car not used to train the models, the recognition rate is still high. On the

Table 3 The results of online recognition on device using QDA.

Subject /Activity	idling	walking	cycling	driving	running	total
Subject 1	98.5%	65.6%	95.7%	99.6%	91.5%	90.2%
Subject 2	99.9%	97.6%	91.2%	88.5%	99.2%	95.3%
Subject 3	99.9 %	96.7 %	91.3%	–	99.8%	96.9%
Subject 4	–	–	–	87.2%	–	87.2%
Subject 5	98.2%	99.3%	97.6%	98.1%	99.9%	98.6%
Subject 6	99.9%	96.0%	93.8%	–	99.9%	97.4%
Subject 7	99.9%	99.8%	98.1%	–	99.9%	99.4%

Table 4 The results of online recognition on device using $k$nn.

Subject /Activity	idling	walking	cycling	driving	running	total
Subject 1	91.5%	99.9%	89.2%	91.4%	87.2%	91.8%
Subject 2	99.9%	99.9%	93.7%	87.8%	92.4%	94.7%
Subject 3	76.3%	99.9%	89.6%	–	92.8%	89.7%
Subject 4	–	–	–	97.6%	–	97.6%
Subject 5	95.6%	99.9%	89.5%	89.4%	97.9%	94.5%
Subject 6	94.1%	99.9%	93.8%	–	99.9%	96.9%
Subject 7	83.3%	99.8%	98.1%	–	99.9%	95.3%

other hand, to make sure that the models are car-independent, more tests should be carried out using different cars and road conditions.

In the real-world scenario, accurate cycling recognition is challenging, because while cycling, subjects tend to just taxi at times so they do not pedal constantly. Therefore, it is a matter of opinion if one concludes these events as cycling or not. This sort of situation was tested and it seems that these events are classified as driving a car. This is logical because both events cause quite similar vibrations due to the roughness of the road.

6 Conclusions

Orientation independent real-time activity recognition of five everyday activities using a mobile phone was introduced in this study. The whole classification process, including preprocessing, feature extraction, and classification, was done on the device. Recognition accuracies were tested using two classifiers ($k$nn and QDA). User-independent models for online recognition were trained offline using a data set collected by eight subjects. Using these models, the recognition rates on the device are around as high as offline. In the offline case, the average classification accuracy based on the data used to train the models using QDA is 95.4%, while $k$nn enables an accuracy of 94.5%. While performing online recognition on the device, the average recognition rate using QDA is 95.8%, while using $k$nn it is slightly lower, 93.9%. However, in some cases, user-independent real-time recognition results on the device are not as high as expected. In order to achieve more accurate online results with every subject, the training data should contain more variation. Now it seems that in some cases the models for online recognition are build using too homogeneous a data set, and therefore, the models are not as good as they could be. Nevertheless, the results are encouraging. Moreover, activity recognition application runs smoothly on the device. It uses under 5% of CPU capacity when QDA is employed as a classifier; thus, other applications can be run alongside.

Real-time recognition on the device was only tested by predefined five activities and not when the subject is doing something else. Null-data recognition is not included in this study, and therefore, such activities cause incorrect classifications. Thus, to improve the accuracy of the application, null-activity recognition should be included. Also building a behavior recognition system based on the activity recognition results could reduce the number of misclassifications. In addition, it should be tested how different trousers affect the results, now every test subject was wearing jeans.

The presented activity recognition application is not body position independent. The system is trained to recognize activities when the phone is placed to the subject's trousers' pocket. Although trousers' pockets are the most common place to carry a phone [9], especially among males, a body position independent approach should be considered. Body position independent recognition is naturally more difficult than position dependent, and therefore, most likely the recognition rates would not be as high as the ones presented in this study.

Although, the recognition accuracy on the device is excellent, there are still some remaining issues. The application uses too much battery and, therefore, even lighter methods should be used. For instance, the sampling frequency could be reduced to half or the number of required classifications could be reduced by using periodic quick-test [17]. However, even now without memory and processing power optimization, the battery of Nokia N8 lasts over 24 hours while the application is running at the background.

In this study, everything except model training is done on the device. Other option would be to send the accelerometer data to the server, perform the classification process there and send the results to a mobile phone. In this case, calculation capacity would not be an issue, but on the other hand, privacy issues should be handled. Moreover, data transfer is not free and can cause exceptionally high costs, especially when the mobile phone and application are used abroad.

Acknowledgment. This work was done as a part of MOPO study (ClinicalTrials.gov Identifier: NCT01376986). Pekka Siirtola would like to thank GETA (The Graduate School in Electronics, Telecommunications and Automation) for financial support. The authors would slso like to thank Infotech Oulu and the Finnish Funding Agency for Technology and Innovation for funding this work. Special thanks to Tero Vallius who did a big part of the coding work.

References

1. ActiGraph, http://www.theactigraph.com/
2. Bao, L., Intille, S.S.: Activity Recognition from User-Annotated Acceleration Data. In: Ferscha, A., Mattern, F. (eds.) PERVASIVE 2004. LNCS, vol. 3001, pp. 1–17. Springer, Heidelberg (2004)
3. Brezmes, T., Gorricho, J.-L., Cotrina, J.: Activity Recognition from Accelerometer Data on a Mobile Phone. In: Omatu, S., Rocha, M.P., Bravo, J., Fernández, F., Corchado, E., Bustillo, A., Corchado, J.M. (eds.) IWANN 2009. LNCS, vol. 5518, pp. 796–799. Springer, Heidelberg (2009)
4. Ermes, M., Pärkkä, J., Mäntyjärvi, J., Korhonen, I.: Detection of daily activities and sports with wearable sensors in controlled and uncontrolled conditions. IEEE Transactions on Information Technology in Biomedicine 12(1), 20–26 (2008)
5. Fix, E., Hodges, J.L.: Discriminatory analysis: Nonparametric discrimination: Consistency properties. Tech. Rep. Project 21-49-004, Report Number 4, USAF School of Aviation Medicine, Randolf Field, Texas (1951)
6. Frank, J., Mannor, S., Precup, D.: Activity Recognition with Mobile Phones. In: Gunopulos, D., Hofmann, T., Malerba, D., Vazirgiannis, M. (eds.) ECML PKDD 2011, Part III. LNCS, vol. 6913, pp. 630–633. Springer, Heidelberg (2011)
7. Haapalainen, E., Laurinen, P., Junno, H., Tuovinen, L., Röning, J.: Feature selection for identification of spot welding processes. In: Proceedings of the 3rd International Conference on Informatics in Control, Automation and Robotics, pp. 40–46 (2006)
8. Hand, D.J., Mannila, H., Smyth, P.: Principles of data mining. MIT Press, Cambridge (2001)
9. Ichikawa, F., Chipchase, J., Grignani, R.: Where's the phone? a study of mobile phone location in public spaces. In: 2005 2nd International Conference on Mobile Technology, Applications and Systems, pp. 1–8 (2005)
10. Kwapisz, J.R., Weiss, G.M., Moore, S.A.: Activity recognition using cell phone accelerometers. SIGKDD Explor. Newsl. 12, 74–82 (2011)
11. Lu, H., Yang, J., Liu, Z., Lane, N.D., Choudhury, T., Campbell, A.T.: The Jigsaw continuous sensing engine for mobile phone applications. In: Proceedings of the 8th ACM Conference on Embedded Networked Sensor Systems, SenSys 2010, pp. 71–84 (2010)
12. Nokia N8, http://europe.nokia.com/find-products/devices/nokia-n8
13. Peebles, D., Lu, H., Lane, N.D., Choudhury, T., Campbell, A.T.: Community-guided learning: Exploiting mobile sensor users to model human behavior. In: Proceedings of the Twenty-Fourth AAAI Conference on Artificial Intelligence, AAAI 2010, Atlanta, Georgia, USA, July 11-15 (2010)

14. Polar Active, http://www.polaroutdoor.com/en/support/product_support?
    product=29451
15. Qt, http://qt.nokia.com/
16. Ryder, J., Longstaff, B., Reddy, S., Estrin, D.: Ambulation: A tool for monitoring mobility patterns over time using mobile phones. In: International Conference on Computational Science and Engineering, CSE 2009, vol. 4, pp. 927–931 (2009)
17. Siirtola, P., Koskimäki, H., Röning, J.: Periodic quick test for classifying long-term activities. In: IEEE Symposium on Computational Intelligence and Data Mining (CIDM 2011), pp. 135–140 (2011)
18. Smartphone shipments, http://www.bgr.com/2011/03/10/
    berg-smartphone-shipments-grew-74-in-2010/
19. Sun, L., Zhang, D., Li, B., Guo, B., Li, S.: Activity Recognition on an Accelerometer Embedded Mobile Phone with Varying Positions and Orientations. In: Yu, Z., Liscano, R., Chen, G., Zhang, D., Zhou, X. (eds.) UIC 2010. LNCS, vol. 6406, pp. 548–562. Springer, Heidelberg (2010)
20. Suutala, J., Pirttikangas, S., Röning, J.: Discriminative Temporal Smoothing for Activity Recognition from Wearable Sensors. In: Ichikawa, H., Cho, W.-D., Satoh, I., Youn, H.Y. (eds.) UCS 2007. LNCS, vol. 4836, pp. 182–195. Springer, Heidelberg (2007)
21. Van Laerhoven, K., Cakmakci, O.: What shall we teach our pants? In: The Fourth International Symposium on Wearable Computers, pp. 77–83 (2000)
22. Wang, S., Chen, C., Ma, J.: Accelerometer based transportation mode recognition on mobile phones. In: Asia-Pacific Conference on Wearable Computing Systems, pp. 44–46 (2010)

Implementing a Spatial Agenda in Android Devices

C.N. Ojeda-Guerra

Abstract. Smartphone are becoming common in everyday's life. They provide new possibilities as the integrate communication, geo-positioning, multi-touch capabilities, and so on. Increasingly, these devices are used as proactive tools assisting people with their day-to-day activities, making everyones life more comfortable. In this paper, we present an Android application which is based on the location technology and Google maps in order to help the user to know what ads or warnings has in a specific location and time.

1 Introduction

With the development of the Information Technology (IT), people can achieve the "all the time everywhere" goal of pervasive computing in order to access to the information.

In this context, *Location-based service* (LBS) [1] provides a user with contents customized by the user's current location, such as the nearest restaurants, hospitals, way points, and so on, which are retrieved from a spatial database stored remotely in the LBS server. Also with the increasing number of mobile devices featuring built-in Global Positioning System (GPS) technology, LBS has experimented a rapid growth in recent years. In an ambient intelligence world, devices work in concert to support people in carrying out their everyday life activities, tasks and rituals in easy, natural way using information and intelligence that is hidden in the network connecting these devices.

In this paper, we present a spatial agenda (location-based service) in which the service and the database are implemented in an Android device with GPS and access to Google maps. The main goal of our application is to help the user to know what ads or warnings has in a specific location and time. To achieve this goal, we have to solve different sub-goals such as:

C.N. Ojeda-Guerra
Departamento de Ingeniera Telemática, Universidad de Las Palmas de Gran Canaria, Spain
e-mail: cnojeda@dit.ulpgc.es

S. Omatu et al. (Eds.): Distributed Computing and Artificial Intelligence, AISC 151, pp. 629–636.
springerlink.com   © Springer-Verlag Berlin Heidelberg 2012

- Implement a service which begins with the terminal's boot, to obtain the user's location using GPS or network provider.
- Implement a data base for the structured storage of all information in the system: warnings, way points and configuration.
- Implement the different views of the application, some of which interact with *Google Maps*.

In the context of this paper, the authors in [2], present a public safety application using Android, which collects speed and location information from a GPS receiver to determine the location of nearby schools, and sounds an alarm if a person drove over the speed limit in a school zone. In [3] the authors present an application which finds the banks, supermarkets, gas stations and other place around users, furthermore provide navigation function. The authors in [4], implement a multimedia M-learning game (m-gymkhanas) that can be performed in a geographic extended area using GPS receiver, Google Maps, a web-based infrastructure and web services. In [5] an Android application for dementia patients (iWander) is presented, which improves the quality of treatment for dementia patients using mobile applications. In [6], the authors present *rider spoke*, where players explore a city on bicycles, equipped with a mobile phone, which invites them to record personal and reflective audio messages at chosen locations, and then search for and listen to other players' messages which are located nearby.

The remainder of this paper is organized as follows: Section 2 introduces the Android system; Section 3 describes the technical details of our implementation; Section 4 outlines the system functionalities; and the final section offers the conclusions of this work.

2 Android Architecture

Android software frame is shown in Fig. 1 and the software structure can be divided into the following several levels from top to down: Application, Application Framework, Libraries and Android RunTime and Operating System.

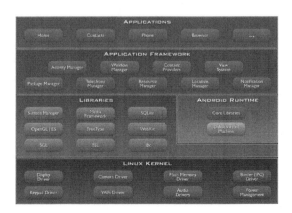

Fig. 1 The major components of the Android operating system

Android application usually involves interaction between user interface and user. Currently the Android application is developed by Java language. Android will ship with a set of core applications including an email client, SMS program, calendar, maps, browser, contacts, and others. At the same time, the application developers can use the API of application framework to achieve their programs; it is also the source of the enormous potential of Android [7].

Application components are the essential building blocks of an Android application. These are:

- *Activity*: is an application component that provides a screen with which users can interact in order to do something, such as dial the phone, take a photo, send an email, or view a map. Each activity is given a window in which to draw its user interface.
- *Service*: is an application component that can perform long-running operations in the background and does not provide a user interface. Another application component can start a service and it will continue to run in the background even if the user switches to another application.
- *Broadcast receiver*: responds to system-wide broadcast announcements, such as: boot completed, battery is low, picture is captured and so on.
- *Content provider*: providers store and retrieve data and make it accessible to all applications. They're the only way to share data across applications.

3 Technical Details

The application presented in this paper, has four packages: *dataBase*, *dialog*, *location* and *spatialAgenda* (Fig. 2 show the class diagram of the application). The package *dataBase* (which contains the classes: *DBManagement*, *ActionBase* and *WaypointBase*) is responsable of the management and control of the storage of all information in the system. It uses a SQLite data base with two different tables: *waypoints* (where the waypoint are saved) and *actions* (where the warning are stored) and its classes implement several methods in order to insert, delete, update and search information.

The package *dialog* (which contains the classes: *CheckDialog*, *DateDialog*, *TimeDialog* and *TextDialog*) manages the different dialog of the application. In Fig. 8(a) we can see one of these dialogs. The dialogs and the rest of interactive views are simple and easy to understand.

The package *location* (which contains the classes: *LocationManagement* and *LocationService*) implements a service which begins with the terminal's boot, to obtain the user's location using GPS receiver or network provider (chosen by the user). The *LocationManagement* class updates the alarms based on the location and put the application in foreground if it was in background. The user can stop the alarm touching the volume button.

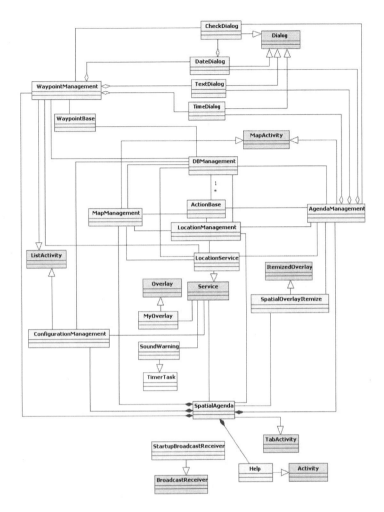

Fig. 2 UML diagram of the spatial agenda (the classes in blue colour belong to the Android's APIs)

The package *spatialAgenda* (which contains the rest of classes in Fig. 2) is responsable of the different views of the application, some of which interact with *Google Maps*. When an instance of the *StartupBroadcastReceiver* class detects that the boot of the terminal has finished, it launches the service (represents by the *LocationService* class) which shows a notification, in the notification bar. When the location of a warning (which is stored in the data base) matches with the user location, the alarm is activated and the map view is put in foreground.

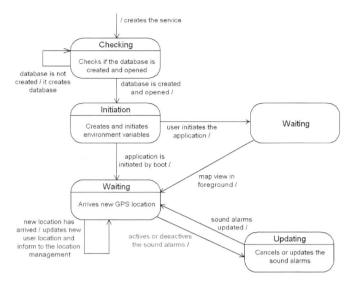

Fig. 3 State diagram of the service object

When a new GPS user location arrives (Fig. 3) the service takes the control of the application and if at least, one warning in the data base matches with the new user location (Fig. 4), the agenda view is shown in the terminal and the sound and visual alarms are activated (Fig. 5). If there is not any warning which matches with the user location, the user is not aware of this fact.

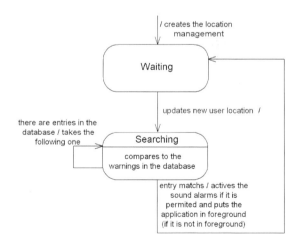

Fig. 4 State diagram of the location management object

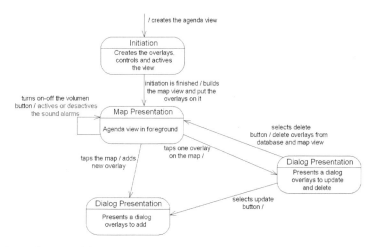

Fig. 5 State diagram of the agenda view object

4 System Functionalities

In this section, we show the inner working of the spatial agenda. This application helps the user to know what ads or warnings has in a specific location and time. The application has four tabs: *map*, *agenda*, *waypoint* and *configuration* (Fig. 6).

Fig. 6 Tabs of the application

The **Map tab** allows the user to view the ads and the user's current position. Selecting each warning, the user can see the description date and time of the warning. In Fig. 7(a) we can see some warnings which are in red color because or the user location is not known or the warnings are not close to it. In Fig. 7(b) we select the central warning and we show the information attach to this ad (it is a meeting to celebrate on Monday, Wednesday and Friday at 17:00).

The **Agenda tab** allows the user to insert and show the ads and the user's current position. Selecting each warning, the user can update and delete the specific warning. Also, in this window the user can add new way points. In Fig. 8(a) we insert a new ad or way point (with a single touch) adding its description, date, time, area of searching and if the audible alarm is on for this specific ad. In Fig. 8(b) we select the previous ad and show the information attach to it and the possibility of deleting or updating it.

The **Waypoint tab** allows the user to edit and delete the way points (by moving the finger on the item and selecting the proper button such as is shown in Fig. 9).

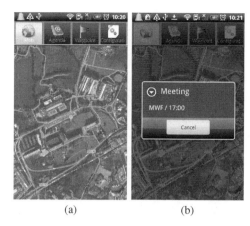

Fig. 7 Map tab view   (a)   (b)

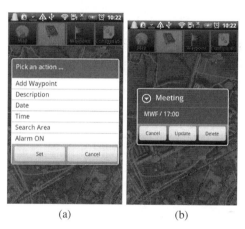

Fig. 8 Agenda tab view   (a)   (b)

Fig. 9 Waypoint tab view

Also, the user can insert new ads associated with a particular way point (with a single touch).

Finally, the **Configuration tab** allows the user to configure some parameters such as: The distance warning, the alarm sound interval, the range of GPS readings, if the map is centered on the user, if the application represents the map or satellite image and the location provider (GPS, network or default).

5 Conclusion

In this paper, we have presented an application for Android terminals, which implements a spatial agenda. The main goal of our application is to help the user to know what ads or warnings has in a specific location and time. Some features of this application are the following:

- Starts at device boot (showing a bell in the notification bar).
- Locates to the user on the map using the GPS receiver of the mobile device or network provider (use a yellow mark to represent the user location).
- Notifies the user if the GPS provider is selected and a location can not be found (through an audible alarm).
- Places way points and ads (warning) on the map with a touch. The ads can associate with specific dates, ranges days or any date, as well as specific times or anytime.
- Plays an audible and visual when the user is near the warning.
- Uses the volume button to stop the alarm sound (volume down) or active it (volume up).
- Shows a help button on the map window.

References

1. Shu, W., Jungwon, M., Byung, K.: Location Based Services for Mobiles: Technologies and Standards. In: Proceedings of the IEEE International Conference on Communication (ICC) 2008, Beijing, China (2008)
2. Whipple, J., Aresnman, W., Starr Boler, M.: A Public Safety Application of GPS-Enabled Smartphones and the Android Operating System. In: Proceedings of the 2009 IEEE International Conference on Systems, Man, and Cybernetics, San Antonio, TX, USA (2009)
3. Dong, L., Gaoshi, Y., Hai, T.: Android Based Wireless Location and Surrounding Search System Design. In: 2010 Ninth International Symposium on Distributed Computing and Applications to Business, Engineering and Science, Hong Kong, China (2010)
4. Robles, G., González-Barahona, J., Fernández-González, J.: Implementing Gymkhanas with Android Smartphones. In: 2011 IEEE Global Engineering Education Conference (EDUCON), Amman, Jordan (2011)
5. Sposaro, F., Danielson, J., Tyson, G.: iWander: An Android Application for Dementia Patients. In: 32nd Annual International Conference of the IEEE EMBS, Buenos Aires, Argentina (2010)
6. Isaacman, S., Becker, R., Cáceres, R., Kobourov, S., Martonosi, M., Rowland, J., Varshavsky, A.: Identifying Important Places in People's Lives from Cellular Network Data. In: Lyons, K., Hightower, J., Huang, E.M. (eds.) Pervasive 2011. LNCS, vol. 6696, pp. 133–151. Springer, Heidelberg (2011)
7. What is Android?, http://developer.android.com/guide/basics/what-is-android.html

M-Learning for Elderlies: A Case Study

Fernando de la Prieta, Antonia Macarro, Amparo Jiménez, Amparo Casado,
Kasper Hallenborg, Juan F. De Paz, Sara Rodríguez, and Javier Bajo

Abstract. In this article a case study about m-learning for elderlies using mobile devices is presented. The study focuses on a practical study about language learning for elderly students that attend classes in the so-called Inter-university program for Elderly People. The recent surge of university programs for elderly people requires novel solutions related to learning methods, directed by the special needs of this sector of the society.

Keywords: elderlies, m-learning, mobile devices, disabled people.

1 Introduction

There is an ever growing need to supply constant care and support to the disabled and elderly and the drive to find more effective ways to provide such care has become a major challenge for the scientific community [3]. During the last three decades the number of Europeans over 60 years old has risen by about 50%. Today they represent more than 25% of the population and it is estimated that in 20 years this percentage will rise to one third of the population, meaning 100 millions of citizens [3]. In the USA, people over 65 years old are the fastest growing segment of the population [1]and it is expected that in 2020 they will represent about 1 of 6 citizens totaling 69 million by 2030. Furthermore, over 20%

Fernando de la Prieta · Juan F. De Paz · Sara Rodríguez
Computer and Automation Department, University of Salamanca, Salamanca, Spain
e-mail: {fer,fcofds,sgr}@usal.es

Antonia Macarro · Amparo Jiménez · Amparo Casado · Javier Bajo
Universidad Pontificia de Salamanca, Spain
e-mail: {mamacarroal,ajimenezvi,acasadome,jbajope}@upsa.es

Kasper Hallenborg
University of Shouthern Denmark, Denmark
e-mail: hallenborg@mmmi.sdu.dk

S. Omatu et al. (Eds.): Distributed Computing and Artificial Intelligence, AISC 151, pp. 637–645.
springerlink.com   © Springer-Verlag Berlin Heidelberg 2012

of people over 85 years old have a limited capacity for independent living, requiring continuous monitoring and daily care [2]. Some estimations of the World Health Organization show that in 2025 there will be more than 1000 million people aged over 60 in the world, so if this trend continues, by 2050 will be double, with about the 80% concentrated in developed countries [6].

Education is the cornerstone of any society and it is the base of most of the values and characteristics of that society. The new knowledge society offers significant opportunities for AmI applications, especially in the fields of education and learning [5]. The new communication technologies propose a new paradigm focused on integrating learning techniques based on active learning (learning by doing things, exchange of information with other users and the sharing of resources), with techniques based on passive learning (learning by seeing and hearing, Montessori, etc.) [4]. While the traditional paradigm, based on a model focused on face to face education, sets as fundamental teaching method the role of the teachers and their knowledge, the paradigm based on a learning model highlights the role of the students. In this second paradigm the students play an active role, and build, according to a personalized action plan, their own knowledge. Moreover, they can establish their own work rhythm and style. The active methodology proposes learning with all senses (sight, hearing, touch, smell and taste), learn through all possible methods (school, networking, etc.), and have access to knowledge without space or time restrictions (anywhere and at any time).

There are different studies that have used the Ambient Intelligence to facilitate learning. In [3], Bomsdorf shows the need to adapt intelligent environments to changes depending on the educational context and the characteristics of users. Morken et al. [6] analyze the characteristics of intelligent environments for learning. They focus on the role of mobility in educational environments and the role that acquire the mobile devices. Naismith et al. [7] conducted a detailed study describing the role of mobile devices in education, analyzing the characteristics of the devices and their capacity for learning in educational environments. All these approaches are focused on the role of learning in Ambient Intelligence environments, but none of them is oriented on learning for dependents or elderly people. The following section presents a multiagent architecture that facilitates learning methodology using an active through mobile devices.

This work presents a practical study about language learning for elderly students that attend classes in the so-called Inter-university program for Elderly People at the Pontifical University of Salamanca, Spain. The recent surge of university programs for elderly people requires novel solutions related to learning methods, directed by the special needs of this sector of the society. With the aim of obtaining an improvement in the French language learning, we have tested different methodologies. The one presented in this paper is an empirical one, based on the M-learning paradigm. This paper focuses in the combination of the new information technologies along with the traditional teaching. In this way it will be possible to combine the advantages of the face to face teaching with the advantages of distance learning. It will be necessary to upgrade the systems of evaluation/accreditation to assess the knowledge or skills acquired during the learning process. To achieve this objective, we propose the use of mobile devices, intelligent systems and wireless communications. The aim is to provide complementary methods to the traditional learning strategies. The proposed mechanism was tested in a case study, trying to evaluate the impact of the new approach on the elderly students.

The rest of the paper is structured as follows: Next section introduces the problem that motivates most of this research. Section 3 describes a case study to test the proposal and, finally, Section 4 shows the results and the conclusions obtained.

2 Background

This section presents the problem that motivates this research. More specifically, we will focus in two main concepts: Learning techniques oriented to elderly people and mobile learning techniques. In the following paragraphs we revise the related work about these two concepts.

2.1 *Learning Techniques Oriented to Elderly People*

Elderly student are acquiring a relevant role in Spanish universities. Different factors as the improvement in the quality of life and the educational interests of this sector of the population contribute to these new educational needs. In 1973, professor Pierre Vellas created the first university program for elderlies in Toulouse. Since then, the growth of these university programs has been unstoppable, and now they exist in practically all the continents. Focusing on Spain, the University of Alcalá de Henares and the Pontifical University of Salamanca were the first university in including a university program for elderlies in 1993. These educational programs presented a new challenge regarding learning methods adapted to the special needs of the new students. These students present a series of special characteristics: They usually compose a heterogeneous group, with different academic background (we can find students with basic education and students with university degrees), different ages (it can vary from 55 to 80 years), and different level of language domain. In Spain, most of these people studied a foreign language, French in most of the cases, but most of them don't have memories about this language. In general they are students that attend the classes and are interested in continuing with the students in the subsequent years.

Apart from the special characteristics of this new student profile, there is another obstacle: there not exists a didactic method oriented to elderly students.

All the existing methods are oriented to children, teenagers or young people. It has only been possible to find an English method oriented to elderlies [10]. The existing learning methods are not appropriated for the special characteristics of these new students. Thus, it is necessary to investigate in new learning techniques and methods oriented to satisfy the special requirements of this social sector.

2.2 Mobile Learning Techniques

The experience acquired after teaching French during a decade to elderly students, it is possible to conclude that the new information and communication technologies (ICT) are still a challenge for these students. The impact and growing use of the ICTs in our society it is a reality and it is not possible to live without them. The use of the ICTs can help to notably improve the quality of life of elderlies and facilitate their integration in the information society. The advances in ICTs in this century are very important, and in parallel, the traditional educational model has been substituted for new paradigms that incorporate e-learning methods. One of the advantages of these methods is the elimination of temporal and location barriers. There are different types of e-learning methods. It is possible to distinguish:

- E-learning is electronic learning, and makes use of communication networks as Internet or Intranets and platforms as Moodle, specialized in tele-learning.
- B-learning or blended learning is a combination of traditional teaching and electronic learning (E-learning). It combines attendance modules and non-attendance modules.
- M-learning or mobile learning represents a step ahead in educational models. It adapts the learning methods to mobile devices such as mobile phones, PDAs, tablets, pocket pcs, i-pads, etc. An specialization of m-learning is MALL (Mobile assisted language learning), focused on language learning from mobile devices [11].

In this paper we focus on M-learning, and more specifically in MALL. Klopfer 10 indicates that the mobile devices should incorporate five characteristics to be appropriated for mobile learning: portability, social interactivity, context-awareness, connectivity and individuality. M-learning can be combined with traditional learning methods. It is necessary to remark that M-learning is a ubiquitous method and avoids temporal and location restrictions. Moreover, M-learning becomes more important as the mobile devices become more important. According to a study of ITAD consulting, for the year 2015, the mobile Internet penetration worldwide will reach 37%. According Soichi Nakajima [12] the mobile Internet has reached a final take-off stage in Western Europe and North America, so far only been seen in Japan for almost a decade and to a lesser extent, South Korea. This growth in the number of smartphones has a big impact in the information society, and most of the users are now replacing the typical mobile phone. Besides applications for mobile phones makes smartphones much more attractive to users and also be the way to make more mobile connections [16]. The World Summit Award Mobile Content (WSA-mobile), is a global initiative of the European Academy of Digital Media – EADiM to select and promote contents and creativity in websites [14]. The project MOBIlearnof the European Comission join together universities and telecommunications companies from Europe, Israel, USA and Australia, with the aim of defining theoretical models for learning process carried out by means of mobile technologies. The Learning and Skills Network (LSN) organization promotes MoLeNET (Mobile Learning Network) to design educational products based on mobile devices [13]. In Spain, the University RoviraiVirgilicreated apodcasting Project to improve language learning [8]. Lingling Yang remarks the importance of language learning through mobile devices [8]. In [15], Jan Herrington describes a case study in australian universities to learn using smartphones and iPods.

There are different studies that have used the Ambient Intelligence to facilitate learning. In [4], Bomsdorf shows the need to adapt intelligent environments to changes depending on the educational context and the characteristics of users. Naismith *et al.* [18] conducted a detailed study describing the role of mobile devices in education, analyzing the characteristics of the devices and their capacity for learning in educational environments. All these approaches are focused on the role of learning in Ambient Intelligence environments, but none of them is oriented on learning for dependents or elderly people. The following section presents a case study to evaluate a learning methodology using an active through mobile devices.

3 Case Study: University Program Oriented to Elderlies

This study has two objectives: i) To improve language learning in university programs oriented to elderlies, and ii) to contribute to integrate the elderly people in the information society, making use of the information technologies. At present there are various tools to facilitate the active learning, such as forums, wikis, email, chat, virtual campuses, and so on. However, none of them is focused to language learning for elderly people. This paper presents an interactive system specifically designed for language learning for elderly people and people with visual disabilities. The proposed approach includes new interaction techniques adapted to be applied in mobile devices. Interaction techniques have been used to design Ambient Intelligence interaction mechanisms suitable for use in teaching languages to people involved in courses at the university program for elderlies. Therefore, interfaces have been made simple and straightforward and have facilitated interaction through touch devices. The application developed for mobile devices contain a series of tests that individuals can complete to carry out language learning. Thus, the student uses a type of interface that allows him to interact with the system.

After revising the related work, it has been observed that the existing approaches are oriented to language learning for young students. Young students use mobile devices in their daily life, and it has not been possible to find any project oriented to people up to 55 years. In this sense, the approach presented in this paper proposes an innovative perspective where the students are elderlies.

The application is based on a navigation menu using the accelerometer available in many mobile devices, especially in the iPhone platform. This application may solve the problems that elderlies have while using a menu with different options. Cutting-edge devices have touch screens and it's impossible for them to identify where they are pressing. A movement recognition algorithm has been created in order to collect all data necessary for the successful operation of the system [19].

4   Results

To evaluate the proposed approach, the system was tested with two different groups of students at the university program for elderlies at the Pontifical University of Salamanca, Spain. The tests consisted of 2 tests for two different groups, one performed in Group 1 and another made to Group 2. The first test involved 12 individuals, while the second test involved 15 individuals. The sex of individuals was not taken into consideration to perform this test, because this parameter has not been considered as significant for this study. Each of the individuals completed one of the tests proposed in the mobile application, and noted by 10 questions. After the test, each individual completed a form on which he was asked about the evaluation of the test. The test assessed 5 items:

- Usefulness of the test. This item is valued feedback from users about the usefulness of the test as a learning tool.
- Easy to use mobile phone. This item assesses the usability of the proposed application.
- Utility as a tool for teaching languages. This item is valued feedback from users about the usefulness of the tool for language learning.
- Using mobile phone before for similar activities. This item is valued the ease and frequency of use of mobile devices.
- Overall assessment of the test. This item assesses the overall opinion about the test users.

The results obtained in the experiment are show in the following table:

Table 5 Results for the items.

	G1	G2	G1+G2
Utility of the test			
Very Useful	75 %	60 %	55,55 %
Quite Useful	0 %	40 %	33,33 %
Useful	25 %	0 %	11,11 %
Not very useful	0 %	0 %	0 %
Not useful	0 %	0 %	0 %
Easy to use mobile phone			
Very Easy	50 %	20 %	11,11 %
Easy	0 %	60 %	55,55 %
Normal	50 %	0 %	22,22 %
Difficult	0 %	0 %	0 %
Very difficult	0 %	25 %	11,11 %
Utility of the approach as a tool for language learning			
Very appropriated	25 %	20 %	33,33 %
Quite appropriated	0 %	20 %	11,11 %
Appropriated	75 %	0 %	0 %
Not Very appropriated	0 %	20 %	11,11 %
Inappropriate	0 %	40 %	22,22 %
Previous use of the mobile phone for similar activities			
Yes	100 %	80 %	88,88 %
No	0 %	0 %	0 %
Not answer	0 %	20 %	12,12 %
Global evaluation of the test			
Very satisfied	25 %	40 %	33,33 %
Quite satisfied	25 %	60 %	44,44 %
Satisfied	50 %	0 %	22,22 %
Not Very satisfied	0 %	0 %	0 %
Unsatisfied	0 %	0 %	0 %

Taking into account these results, it is possible to conclude that the approach is promising and can be of interest for this sector of the population. The participants in this experiment provided some feedback about the proposed approach: they indicated that the approach is very interesting for certain aspects, as verbs learning, and it is innovative, since it is not useful for them the use of mobile devices in the classes. Some of them also indicated that it would be of interest to improve the proposed approach with accessibility facilities, specifically to augment the size of the fonts.

Acknowledgements. This work has been supported by the Junta de Castilla y León project PON147A11-2.

References

1. Anastasopoulos, M., Niebuhr, D., Bartelt, C., Koch, J., Rausch, A.: Towards a Reference Middleware Architecture for Ambient Intelligence Systems. In: ACM Conference on Object-Oriented Programming, Systems, Languages, and Applications (2005)
2. Angulo, C., Tellez, R.: Distributed Intelligence for smart home appliances. Tendencias de la minería de datos en España. Red Española de Minería de Datos. Barcelona, España (2004)
3. Bajo, J., Corchado, J.M., de Paz, Y., de Paz, J.F., Rodríguez, S., Martín, A., Abraham, A.: SHOMAS: Intelligent Guidance and Suggestions in Shopping Centres. Applied Soft Computing 9(2), 851–862 (2009)
4. Bomsdorf, B.: Adaptation of Learning Spaces: Supporting Ubiquitous Learning in Higher Distance Education. In: Davies, N., Kirste, T., Schumann, H. (eds.) Dagstuhl Seminar Proceedings. Mobile Computing and Ambient Intelligence: The Challenge of Multimedia, Dagstuhl, Germany (2005)
5. Brown, T.H.: Beyond constructivism: Exploring future learning paradigms. In: Education Today, issue 2 of 2005. Aries Publishing Company, Thames (2005)
6. Friedewald, M., Da Costa, O.: Science and Technology Roadmapping: Ambient Intelligence in Everyday Life (AmI@Life). Working Paper. Seville: Institute for Prospective Technology Studies IPTS (2003)
7. Kurniawan, S.H., King, A., Evans, D.G., Blenkhorn, P.L.: Personalising web page presentation for older people. Interacting with Computers 18, 457–477 (2006)
8. Naismith, L., Lonsdale, P., Vavoula, G., Sharples, M.: Futurelab Literature Review in Mobile Technologies and Learning, Technical Report for Futurelab (2004), http://www.futurelab.org.uk/research/reviews/reviews_11_and12/11_01.htm
9. Martí, C., Mar, González, E., Vanessa, Cervera, G., Mercè: Speak new words, speak new languages: a podcasting experience in multicultural settings
10. Revista MASTER NEW MEDIA; 17 de enero de (2006), http://www.masternewmedia.org/es/2006/01/17/aprendizaje_movil_mlearning_es_lo.htm
11. Klopfer, E., Squire, K., Jenkins, H.: Environmental Detectives: PDAs as a window into a virtual simulated world. In: Proceedings of IEEE International. Workshop on Wireless and Mobile Technologies in Education. IEEE Computer Society, Vaxjo (2002)
12. Chinnery, G.M.: EMERGING TECHNOLOGIES Going to the MALL: Mobile Assisted Language Learning, vol. 10(1), pp. 9–16. University of Maryland, Baltimore County (2006)
13. Soichi Nakajima 2015. En 2015 el 37% de la población utilizará internet móvil (2011), http://congresoweb.es/2011/01/en-2015-el-37-de-la-poblacion-utilizara-internet-movil/
14. Douch, R., Savill-Smith, C., Parker, G., Attewell, J.: Work-based and vocational mobile learning - Making IT work. Learning and Skills Network (2010), http://www.molenet.org.uk/pubs/
15. EADIM, http://www.eadim.com/

16. Herrington, J., Herrington, A., Mantei, J., Olney, I., Ferry, B. (eds): New technologies, new pedagogies: Mobile learning in higher education, Faculty of Education, University of Wollongong, p. 138 (2009)
17. Frohberg, D.: Mobile Learning is Coming of Age: What we have and what we still miss. In: Proceedings for DeLFI 2006, September 11-14 (2007)
18. Naismith, L., Lonsdale, P., Vavoula, G., Sharples, M.: Futurelab Literature Review in Mobile Technologies and Learning, Technical Report for Futurelab (2004), http://www.futurelab.org.uk/research/reviews/reviews_11_and12/11_01.htm
19. Sánchez, A., Villarrubia, G., Macarro, A., Jiménez, A., Zato, C., Bajo, J., Rodríguez, S., Hallenborg, K., Corchado, J.M.: Menu navigation in mobile devices using the accelerometer. In: Proceedings of ISAMI 2011 (in press)

Mobile Device to Measure Ubiquitous Danger in a Great City Based on Cultural Algorithms

Alberto Ochoa[*], Erick Trejo, Daniel Azpeitia, Néstor Esquinca,
Rubén Jaramillo, Jöns Sánchez, Saúl González, and Arturo Hernández

Abstract. Beginning in 2007, Juarez City has been a victim of the consequences brought about organized crime, such as killing of innocent people as the main fact assaults, kidnappings, multi homicides, burglary, among other consequences. What has pushed the population of the city to take action of different kinds to minimize the violence in a society with 15700 violent homicides during this period of time; the reason which we present the following project, which aims to provide people moving in this city a technological tool that provides an indicator to the user information based statistics compiled by the Centre for Social Research at Juarez City University and public sources so uncertain is the place where you are. This research try to combine a Mobile Device based on Cultural Algorithms and Data Mining to determine the danger of stay in a part of the city in a specific time.

1 Introduction

According to tests show that it is feasible to use an application such as a level of insecurity of the most frequent areas that provides the user aware that activities performed or not, that time is more convenient to visit a certain place, is it better to take some alternative route to the destination, among other

Alberto Ochoa · Erick Trejo · Daniel Azpeitia · Néstor Esquinca · Saúl González · Arturo Hernández
Juarez City University, México

Rubén Jaramillo · Jöns Sánchez
CIATEC Research Center, México
and
CIMAT Research Center, México.

[*] Corresponding author. alberto.ochoa@uacj.mx

S. Omatu et al. (Eds.): Distributed Computing and Artificial Intelligence, AISC 151, pp. 647–655.
springerlink.com © Springer-Verlag Berlin Heidelberg 2012

advantages. The thought of insecurity is never rest, and feasible to have a number of tools at hand to achieve this problem, which in this case technology, information and some mathematics. All of them joining with the aim of this project consist in a mobile geographic information system (SIGMA) on levels of insecurity in areas of Ciudad Juarez, which is presented in detail below, this research include the analysis of Cultural Algorithms to determine the danger of occurs a contingence during a specific time. We realize an exhaustive analysis of other similar research, the only similar context is explain in [6], where is calculated the insecurity of a vehicular group which requires delivery products in different places with randomly scheduling, but this research don't considering real statistics on time and the perspective to suggest a different route to travel or stay during many time.

2 Project Development

To do this research project was developed by dividing into three sections which are modules of application development, implementation of the server and the intelligent module associated with Cultural Algorithms and Data Mining. Android is the operating system that is growing in the U.S. rivaling iPhone and Motorola along with some other manufacturers are propelling the Latin American landing on Android with inexpensive equipment, and on the other hand, some complain about the fragmentation of the platform due to the different versions. Android is free software, so any developer can download the SDK (development kit) that contains your API [2]. This research tries to improve the security of the people lives on Juarez City where the deaths by 100,000 people are 327 during 2010.

3 Components of the Application

The first step is to get the coordinates where the user is located, and later sent to the server when it receives and calculates the number of incidents within the radio closest specified in the configuration of the mobile device, the results are processed to determine a numerical index which will then be represented by a color for better visualization of the user, all this information is obtained from a criminal databases and analyzed with Data Mining. After the values are sent to the mobile device and interpreted to construct a URL to be sent to Google Maps API with which to get a map with the indicators in the area. The figure 1 shown below represents generally the operation of the Sigma, which is divided into two parts, the mobile application and Web services hosted on the server.

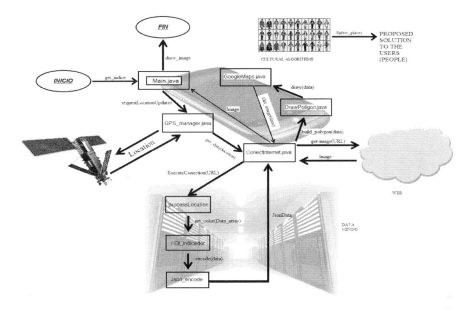

Fig. 1 Functional diagram of the SIGM including hybrid methodology.

The following figure 2 shows the diagram relational database in which data is stored in the incidences occurred in Ciudad Juarez, which has 3 tables:

Type_crime: In which are characterized the types of crimes that were defined.

Neighborhood: Here is the georeference of the neighrborhood.

Incidents: It lists the incidents raised in the different regions of Juarez City, which are also expressed in decimal degrees georeferenced to facilitate further calculations.

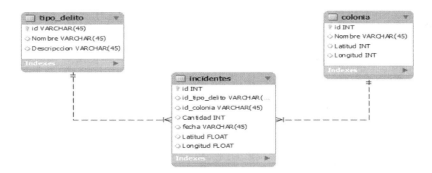

Fig. 2 Diagram SIGM Databases.

The structure of the database has a table called events, where each record contains the geographic coordinates expressed in decimal degrees, which determine the position of the incident [1]. To locate nearby points within a radius requires four parameters: latitude, longitude, and distance equatorial radius, the latter determines the maximum distance the search radius, these parameters are part of a method based on Havesine formula, which is used to calculate great circle distances between two points on a sphere. Haversine function that is different from Haversine formula is given by the function of semiverseno where:

$$semiversin(\theta) = haversin(\theta) = \frac{versin(\theta)}{2} = sin^2\left(\frac{\theta}{2}\right)$$

Haversine formula for any two points on a sphere:

d is the distance between two points (along a great circle of the sphere),
R is the radius of the sphere,
φ 1 is the latitude of point 1,
φ 2 is the latitude of point 2, and
Δ λ is the length difference

$$haversine\left(\frac{d}{R}\right) = haversin(\varphi 1 - \varphi 2) + cos(\varphi 1)\,cos(\varphi 2)haversine(\Delta \lambda)$$

Solving Haversine formula can calculate the distance, either by applying the function inverse Haversine or by using the arcsine where $h = haversine\left(\frac{h}{R}\right)$

$$d - R\,haversine^{-1}(h) - 2R\,arcsin(\sqrt{h})$$

With this formula one can construct the following instructions, taking into account the role of semiverseno we have: $h = sin^2\left(\frac{\varphi 1 - \varphi 2}{2}\right) + cos(\varphi 1).cos(\varphi 2).sin^2\left(\frac{\Delta \lambda}{2}\right)$

$$d = 2R\,arcsin(\sqrt{h})$$

For performance reasons when Haversine implement the function using SQL statements used a similar formula but in terms of spherical cosine law called cosine.

$$cos(c) = cos(a)\,cos(b) + sin(a)\,sin(b)\,cos(C)$$

This is only an approximation when applied to land, and that this is not a perfect sphere radius of the earth varies as we approach the poles. The SQL statement based on the formula Haversine find all locations that are within the range of the variable $ radius and groups them by type of crime.

Calculate of the uncertainty is modeled on the human development index (HDI) prepared by the United Nations Program for Development (PNDU)

The insecurity index is composed of two components Number of Events (Q) and time (H) each component represents half the total value of the index:

Each expressed with a value from 0 to 1 for which we use the following general formula:

Component index = (Value – minimum) / (Maximum – minimum)

The component Q is calculated by the number of events in the area, determined by the position to calculate the maximum and minimum events are grouped by neighborhood, according to the CIS-ICSA at Juarez City University is the colony with more incidents is "El Granjero" with 53 homicides, which is the maximum, then a colony could exceed this number and then would be the new high which means that the maximum is dynamic depending on the data entering the database. To calculate the H component data are grouped in ranges of time, for example suppose that areas of 300 m radius 13 homicides are grouped in the following form and consult the index at 4:38 pm, as is shown in Table 1.

Table 1 Example of Neighborhood number of homicides range by hour.

Amount	Time Range
5	**5:00 a 6:00 pm.**
3	**4:00 a 5:00pm.**
3	**11:00 a 12:00pm.**
2	**9:00 a 10:00am.**

The time when we consulted put us in the range of 4:00 to 5:00 pm. And the calculation of the H component would be: $H = \dfrac{3-2}{5-2} = .333$

Using data from the previous example would be the component Q:

$$Q = \frac{13-2}{53-2} = .215$$

The index would be as follows $indice = \dfrac{1}{2}(215) + \dfrac{1}{2}(333) = .274$

Having calculated the numerical index is assigned a color depending on the range in which is positioned according to the table 2 divided into eight classes.

Table 2 Ranges of colors according to the numeric index.

0 - .125		.56 - .625	
.126 - .25		.626 - .75	
.26 - .375		.751 - .875	
.376 - .5		.876 - 1	

4 Functionalty of the Application

When designing an interface for mobile devices has to take into account that the space is very small screen, plus there are many resolutions and screen sizes so it is necessary to design an interface that suits most devices. This module explains how to work with different layout provided by the Android API. The programming interface is through XML. Obtaining the geographical position of a device can be made by different suppliers, the most commonly used in this project through GPS and using access points (Wi-Fi) nearby, and perform the same action but differ in some as accuracy, speed and resource consumption. Data Server Communication is the module most important because it allows communication with the server, allowing you to send the GPS position obtained by receiving the processed image and map of our location, thus showing the outcome of your application that is the indicator of insecurity. To communicate to a server requires a HTTP client which can send parameters and to establish a connection using TCP / IP, client for HTTP, can access to any server or service as this is able to get response from the server and interpreted by a stream of data. The Android SDK has two classes with which we can achieve this, HttpClient and HttpPost. With the class HttpClient is done to connect to a remote server, it needs HttpPost class will have the URI or URL of the remote server. This method receives a URL as a parameter and using classes HttpPost HttpClient and the result is obtained and received from the server, in this specific case is only text, which can be JSON or XML format. Here, the server responds with a JSON object which would give the indicator is then used to create the map. For the construction of the polygon that indicates the rate of incidents in a certain radius of the current position is not possible to create it using the GPS coordinates that yields, as these are specified in "degrees" and requires the unit to convert to meters. For this you need to know how an arc equals the terrestrial sphere, which depends on the place on earth where it is located and the address where you are, the simplest case is to measure an arc in Equator, considering that the earth is 3670 km radius, the perimeter of serious Equator radio 2, which would be equal to 40.024 miles. With this you can get a relationship that would be as follows. If 360 degrees is 40.024 miles then a degree is 111,000.18 miles, this relationship can add and subtract yards to the position, as shown in the figure 3.

Fig. 3 Acquisition of polygon map with the position of the Android application.

For the preparation of graphics, we propose use a class supported with Cultural Algorithm proposed which facilitates the manipulation of data to express visually using different types of graphs, as in the figure 4.

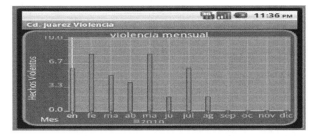

Fig. 4 Statistics Graphics related with the month violence and supported with the Cultural Algorithm.

5 Implementation of the Intelligent Application

To implement the application is installed in operating system devices with Android 2.2 or higher, which tests the system in different areas of the city based on the previously research related with Cultural Algorithms on Urban Transport [5], by answering a questionnaire of seven questions to all users after a scholar semester have elapsed since installing the application, the questions are to raise awareness of the performance, functionality and usability of the system, the demonstrate use of this application is shown in figure 5. To understand in an adequate way the functionality of this Intelligent Tool, we proposed evaluate our hybrid approach and compare with only data mining analysis and random select activities to protect in the city, we analyze this information based on the unit named "époques", which is a variable time to determine if exist a change in the proposed solution according at different situation of violence.

Fig. 5 Implementation and use of Hybrid Inteliigent Application based on Cultural Algorithms and Data Mining.

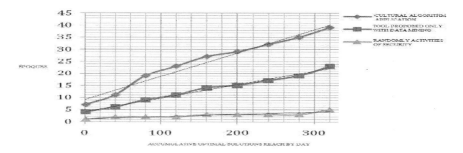

Fig. 6 Solutions proposed to the problem of insecurity: (blue) our hybrid approach; (red) only using data mining analysis and (green) using randomly actions to improve the safety of the users.

We consider different scenarios to analyze during different time, as is possible see in the Figure 6, and apply a questionnaire to a sample of users to decide search a safety place in the City, when the users receive information of different trends of danger (Data Mining Analysis) try to improve their space of solution but when we send solutions amalgam Cultural Algorithms and data Mining was possible determine solutions of security by the users and describe the real situation of danger, the use of our proposal solution improve in 78% against a randomly action and 35% against only use Data Mining analysis the possibilities of recommend leave a danger place, these messages permits in the future decrease the possibility of suffer a violent attack.

6   Conclusions

With the use of this innovative application combine Cultural Algorithms and Data Mining based on a mobile dispositive is possible determine the level of insecurity in areas of Juarez City by an alert sent to a mobile device with GPS, providing statistical information through a Web server that returns the level of insecurity in the area consulted [4]. The future research will be to improve the visual representation of insecurity to a family or a social networking to this we proposed an Intelligent Dyoram with real on time information of each one of their integrants. The most important contribution is prevent more deaths in the city because stay on an incorrect place on a wrong time, our future research is adequate the information to actualize from the central server of security of the Police Department of the city, if the people considering that 75000 people died during this last six years in Mexico, this innovative application is possible to use in another great cities in Latin America as Caracas in Venezuela, Medellin in Colombia or Río de Janeiro in Brazil with similar values of homicides.

References

1. Ignacio, A.F.: Las coordenadas geográficas y la proyección UTM, Universidad de Valladolid, Ingeniería cartográfica, geodésica y fotogrametría (Febrero 2001)
2. Alejandro, A.R.: Estudio del desarrollo de aplicaciones RA para Android. Trabajo de fin de Carrera. Catalunya, España (2011)
3. Orlando, B.J.: Sistema de Información Geográfica Móvil Basado en Comunicaciones Inalámbricas y Visualización de Mapas en Internet. M.C. tesis, Ensenada, Baja California, México (2011)
4. Cáceres, A.: Sistemas de Información Geográfica. Profesorado en Geografía. Instituto Formación Docente P.A.G (2007)
5. Cruz, L., Ochoa, A., et al.: A Cultural Algorithm for the Urban Public Transportation. In: HAIS 2010, pp. 135–142 (2010)
6. Glass, S., Vallipuram, M., Portmann, M.: The Insecurity of Time-of-Arrival Distance-Ranging in IEEE 802.11 Wireless Networks. In: ICDS Workshops 2010, pp. 227–233 (2010)

Simulated Annealing for Constructing Mixed Covering Arrays

Himer Avila-George, Jose Torres-Jimenez, Vicente Hernández,
and Loreto Gonzalez-Hernandez

Abstract. Combinatorial testing is a method that can reduce costs and increase the effectiveness of software testing for many applications. It is based on constructing test-suites of economical size, which provide coverage of the most prevalent configurations of parameters. Mixed Covering Arrays (MCAs) are combinatorial structures which can be used to represent these test-suites. This paper presents a new Simulated Annealing (SA) algorithm for Constructing MCAs. This algorithm incorporates several distinguishing features including an efficient heuristic to generate good quality initial solutions, a compound neighborhood function which carefully combines two designed neighborhoods and a fine-tuned cooling schedule. The experimental evidence showed that our SA algorithm improves the obtained results by other approaches reported in the literature, finding the optimal solution in some of the solved cases.

Keywords: Mixed Covering Array, Combinatorial Testing, Simulated Annealing.

1 Introduction

A good strategy to test a software component involves the generation of the whole set of cases that participate in its operation. While testing only individual values may not be enough, testing all possible combinations (exhaustive approach) is not always feasible. An alternative technique to accomplish this goal is called

Himer Avila-George · Vicente Hernández
Instituto de Instrumentación para Imagen Molecular (I3M). Centro mixto CSIC Universitat Politècnica de València - CIEMAT, camino de Vera s/n, 46022 Valencia, Spain
e-mail: hiavgeo@posgrado.upv.es, vhernandez@dsic.upv.es

Jose Torres-Jimenez · Loreto Hernandez-Gonzalez
Information Technology Laboratory, CINVESTAV-TAMAULIPAS, Ciudad Victoria, Tamaulipas, Mexico
e-mail: jtj@cinvestav.mx, agonzalez@tamps.cinvestav.mx

S. Omatu et al. (Eds.): Distributed Computing and Artificial Intelligence, AISC 151, pp. 657–664.
springerlink.com   © Springer-Verlag Berlin Heidelberg 2012

combinatorial testing. Combinatorial testing is a method that can reduce costs and increase the effectiveness of software testing for many applications. It is based on constructing test-suites of economical size, which provide coverage of the most prevalent configurations of parameters. Mixed Covering Arrays (MCAs) are combinatorial structures which can be used to represent these test-suites.

A mixed covering array [3], denoted by $MCA(N;t,k,v_1v_2\dots v_k)$, is an $N\times k$ array where $v_1v_2\dots v_k$ is a cardinality vector that indicates the values for every column. An MCA has the following two properties: (1) Each column $i(1\le i\le k)$ contains only elements from a set S_i with $|S_i|=v_i$. (2) The rows of each $N\times t$ subarray cover all t-tuples of values from the t columns at least once. The minimum N for which there exists an MCA is called *mixed covering array number* and is formally defined as $MCAN(t,k,v_1v_2\dots v_k)=\min\{N\mid \exists\ MCA(N;t,k,v_1v_2\dots v_k)\}$. A short notation for the MCA can be given using the exponential notation $MCA(N;t,k,v_1^{q_1}v_2^{q_2}\dots v_g^{q_w})$; the notation describes, that there are q_r parameters from the set $\{v_1,v_2,\dots,v_k\}$ that takes v_s values.

Because of the importance of the construction of (near) optimal MCAs, much research has been carried out in developing effective methods for construct them. There are several reported methods for constructing these combinatorial models, among them are: algebraic methods, recursive methods, greedy methods, and meta-heuristics methods. Mathematicians use some algebraic and combinatorial methods to construct MCAs. Colbourn et al. [5] studied a construction method for strength 2 MCAs. Williams and Probert [16] proposed a method for constructing MCAs based on algebraic methods and combinatorial theory. Sherwood [12] described some algebraic constructions for strength-2 MCAs developed from index-1 orthogonal arrays, ordered designs and covering arrays. Williams [17] presented a tool called TConfig to construct MCAs. TConfig constructs MCAs using recursive functions that concatenate small MCAs to create MCAs with a larger number of columns. Moura et al. [11] introduced a set of recursive algorithms for constructing MCAs based on MCAs of small sizes. Colbourn and Torres-Jimenez [4] presented a recursive method to construct CAs using *perfect hash families* for CAs construction. The advantage of the recursive algorithms is that they construct almost minimal arrays for particular cases in a reasonable time. Their basic disadvantage is a narrow application domain and impossibility of specifying constraints.The majority of commercial and open source test data generating tools use greedy algorithms for MCAs construction (AETG [2], TCG [15], ACTS [9], DDA [1] and All-Pairs [10]); the greedy algorithms provide the fastest solving method. Some stochastic algorithms in artificial intelligence, such as TS (Tabu Search) [6], GA (Genetic Algorithms), ACO (Ant Colony Optimization Algorithm) [13] and SA (Simulated Annealing) [3, 14] provide an effective way to find approximate solutions.

A SA metaheuristic has been applied by Cohen et al. [3] for constructing MCAs. Their SA implementation starts with an initial solution created at random. A series of iterations is then carried out to visit the search space according to a neighborhood. At each iteration, a neighboring solution M' is generated by changing the value of the element $a_{i,j}$ by a different legal member of the alphabet in the current solution M. The cost of this iteration is evaluated as $\Delta_c=c(M')-c(M)$. In their implementation, Cohen et al. use a simple linear function $T_n = 0.9998T_{n-1}$ with an initial temperature fixed at $T_i = 0.20$. They conclude that their SA implementation is able to produce smaller MCAs than other computational methods, sometimes improving upon algebraic constructions. However, they also indicate that their SA algorithm fails to match the algebraic constructions for larger problems, especially when $t = 3$.

This paper aims at developing a powerful Simulated Annealing (SA) algorithm for finding near-optimal MCAs. In contrast to other existing SA implementations developed for constructing MCAs, our algorithm has the merit of improving three key features that have a great impact on its performance: (1) a method designed to generate initial solutions with maximum Hamming distance; (2) instead of using a single neighborhood function, the algorithm chooses between a set of predefined neighborhood functions according to a assigned probability to each function; and (3) a fine-tuned cooling schedule. The performance of the proposed SA algorithm is assessed with a benchmark, composed by 19 mixed covering arrays of strengths two and three taken from the literature. The computational results are reported and compared with previously published ones, showing that our algorithm was able to find 4 new upper bounds and to equal 15 previous best-known solutions on the selected benchmark instances.

The next section presents the SA approach proposed in this paper to construct MCAs, also the components of our new SA algorithm are discussed in depth.

2 Proposed Approach

In this section we present a Simulated Annealing algorithm for constructing MCAs. Simulated Annealing (SA) is a general-purpose stochastic optimization method that has proven to be an effective tool for approximating globally optimal solutions to many types of NP-hard combinatorial optimization problems (see [8] for details of the SA heuristic). The following paragraphs will describe each of the components of the implementation of our SA.

The **initial solution** M is constructed by generating M as a matrix with maximum Hamming distance. The Hamming distance $d(x,y)$ between two rows $x, y \in M$ is the number of elements in which they differ. Let r_i be a row of the matrix M. To generate a random matrix M of maximum Hamming distance the following steps are performed: (1) generate the first row r_1 at random; (2) generate two rows c_1, c_2 at random, which will be candidate rows; (3) select the candidate row c_i that maximizes the Hamming distance and added it to the ith row of the matrix M; (4) repeat from step 2 until M is completed.

The **evaluation function** $C(M)$ of a solution M is defined as the number of combination of symbols missing in the matrix M. Then, the expected solution will be zero missing.

Two **neighborhood functions** were implemented to guide the local search of our SA algorithm. The neighborhood function $\mathcal{N}_1(s)$ makes a random search of a missing t-tuple, then tries by setting the jth combination of symbols in every row of M. The neighborhood function $\mathcal{N}_2(s)$ randomly chooses a position (i,j) of the matrix M and makes all possible changes of symbols. During the search process a combination of both $\mathcal{N}_1(s)$ and $\mathcal{N}_2(s)$ neighborhood functions is employed by our SA algorithm. The former is applied with probability P, while the latter is employed at a $(1-P)$ rate. This combined neighborhood function $\mathcal{N}_3(s,x)$ is defined in the Ecuation 1, where x is a random number in the interval $[0, 1]$.

$$\mathcal{N}_3(s,x) = \begin{cases} \mathcal{N}_1(s) & \text{if } x \leq p \\ \mathcal{N}_2(s) & \text{if } x > p \end{cases} \tag{1}$$

The **cooling schedule** determines the degree of uphill movement permitted during the search and is thus critical to the SA algorithm's performance. The parameters that define a cooling schedule are: an initial temperature, a final temperature or a stopping criterion, the maximum number of neighboring solutions that can be generated at each temperature, and a rule for decrementing the temperature. In our SA implementation we used a geometrical cooling scheme mainly for its simplicity. It starts at an initial temperature T_i which is decremented at each round by a factor α using the relation $T_k = \alpha T_{k-1}$. For each temperature, the maximum number of visited neighboring solutions is L. It depends directly on the parameters (N, k and $\mathcal{V}$, where $\mathcal{V}$ is the maximum cardinality of M) of the studied mixed covering array. This is because more moves are required for MCAs with alphabets of greater cardinality.

The **stop criterion** for our SA is either when the current temperature reaches T_f, when it ceases to make progress, or when a valid MCA is found. In the proposed implementation a lack of progress exists if after ϕ (frozen factor) consecutive temperature decrements the best-so-far solution is not improved.

3   Experimental Design

The SA algorithm was coded in C and compiled with *gcc* using the optimization *flag* -O3. It was run sequentially into a CPU Intel(R) Xeon(TM) a 2.8 GHz, 2 GB of RAM with Linux operating system. In all the experiments the following parameters were used for our SA implementation: (1) Initial temperature $T_i = 4.0$; (2) Final temperature $T_f = 1.0E - 10$; (3) Cooling factor $\alpha = 0.99$; (4) Maximum neighboring solutions per temperature $L = Nk\mathcal{V}^2$; (5) Frozen factor $\phi = 11$; (6) According to the results shown in section 3.1, the neighborhood function $\mathcal{N}_3(s,x)$ is applied using a probability $P = 0.3$.

3.1   Fine Tuning of the Neighborhood Functions

It is well-known that the performance of a SA algorithm is sensitive to parameter tuning. In this sense, we follow a methodology for a fine tuning of the two neighborhood functions used in our SA algorithm. The fine tuning was based on the next linear Diophantine Equation, $P_1 x_1 + P_2 x_2 = q$. Where x_i represents a neighborhood function and its value set to 1, P_i is a value in $\{0.0, 0.1, .., 1.0\}$ that represents the probability of executing x_i, and q is set to 1.0 which is the maximum probability of executing any x_i. A solution to the given linear Diophantine Equation must satisfy $\sum_{i=1}^{2} P_i x_i = 1.0$. This Equation has 11 solutions, each solution is an experiment that tests the grade of participation of each neighborhood function in our SA implementation to accomplish the construction of an MCA. Every combination of the probabilities was applied by our SA to construct the set of MCAs shown in Table 1(a) and each experiment was run 31 times, with the obtained data for each experiment we calculate the median. A summary of the performance of our SA with the probabilities that solved the 100% of the runs is shown in Table 1(b).

Table 1 (a) A set of 7 MCAs configurations; (b) Performance of our SA with the 11 combinations of probabilities which solved the 100% of the runs to construct the MCAs listed in (a).

(a)

Id	MCA description
mca_1	$MCA(81; 2, 16, 9^2 8^2 7^2 6^2 5^2 4^2 3^2 2^2)$
mca_2	$MCA(42; 2, 19, 7^1 6^1 5^1 4^5 3^8 2^3)$
mca_3	$MCA(36; 2, 20, 6^2 4^9 2^9)$
mca_4	$MCA(30; 2, 19, 6^1 5^1 4^6 3^8 2^3)$
mca_5	$MCA(29; 2, 61, 4^{15} 3^{17} 2^{29})$
mca_6	$MCA(360; 3, 7, 10^1 6^2 4^3 3^1)$
mca_7	$MCA(49; 2, 10, 7^2 6^2 4^2 3^2 2^2)$

(b)

p_1	p_2	mca_1	mca_2	mca_3	mca_4	mca_5	mca_6	mca_7
0	1	4789.763	3.072	46.989	12.544	3700.038	167.901	0.102
0.1	0.9	1024.635	0.098	0.299	0.236	344.341	3.583	0.008
0.2	0.8	182.479	0.254	0.184	0.241	173.752	1.904	0.016
0.3	0.7	224.786	0.137	0.119	0.222	42.950	1.713	0.020
0.4	0.6	563.857	0.177	0.123	0.186	92.616	3.351	0.020
0.5	0.5	378.399	0.115	0.233	0.260	40.443	1.258	0.035
0.6	0.4	272.056	0.153	0.136	0.178	69.311	2.524	0.033
0.7	0.3	651.585	0.124	0.188	0.238	94.553	2.127	0.033
0.8	0.2	103.399	0.156	0.267	0.314	81.611	5.469	0.042
0.9	0.1	131.483	0.274	0.353	0.549	76.379	4.967	0.110
1	0	7623.546	15.905	18.285	23.927	1507.369	289.104	2.297

Finally, given the results shown in Figure 1, the best configuration of probabilities was $P_1 = 0.3$ and $P_2 = 0.7$ because it found the MCAs in smaller time (median value). The values $P_1 = 0.3$ and $P_2 = 0.7$ were kept fixed in the second experiment.

Fig. 1 Performance of our SA algorithm. We used a Diophantine equation with 11 solutions, every combination of the probabilities was applied by our SA to construct the set of MCAs shown in Table 1(a). Each experiment was run 31 times and we used the median.

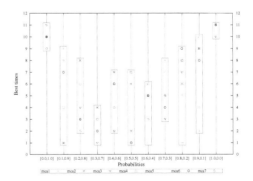

3.2 Comparing Our SA with the State-of-the-Art Algorithms

The purpose of this experiment is to carry out a performance comparison of the best bounds achieved by our SA with respect to those produced by the following state-of-the-art procedures: AETG [2], TCG [15], SA [3], GA [13], ACO [13], DDA [1], Tconfig [17], ACTS [9], AllPairs [10], Jenny [7] and TS [6]. Table 2 displays the detailed computational results produced by this experiment. The benchmark is shown in the column two; from column 3 to 13 the results reported by some of the state-of-the-art approaches are presented. The previous best-known (β) solution is shown in column 14. The results of constructing the MCAs for the benchmark using our SA

Table 2 For each instance shown in column 2, the best solution, in terms of the size N, found by AETG, TCG, SA, GA, ACO, DDA, Tconfig, ACTS, AllPairs, Jenny, TS and our SA are listed. The * means that the solution is optimal. The difference between the best result produced by our SA and the previous best-known solution ($\Delta = \Theta - \beta$) is depicted in the last column.

		N											Best	Our SA	Improvements
ID	MCA description	AETG [2]	TCG [15]	SA [3]	GA [13]	ACO [13]	DDA [1]	Tconfig [17]	ACTS [9]	AllPairs [10]	Jenny [7]	TS [6]	β	Θ	Δ
1	$t2k11v5^1 3^8 2^2$	20	20	15	15	16	21	21	19	20	23	15	15*	15	0
2	$t2k9v4^5 3^4$	-	-	-	-	-	25	28	24	22	26	19	19	19	0
3	$t2k75v4^1 3^{39} 2^{35}$	27	-	21	27	27	27	30	28	26	31	22	21	21	0
4	$t2k21v5^1 4^4 3^{11} 2^5$	28	30	21	26	25	27	32	26	27	32	22	22	22	0
5	$t2k61v4^{15} 3^{17} 2^{29}$	37	33	30	37	37	35	40	33	35	39	30	30	29	-1
6	$t2k19v6^1 5^1 4^6 3^8 2^3$	35	-	30	33	32	34	50	36	34	40	30	30*	30	0
7	$t2k20v6^2 4^9 2^9$	-	-	-	-	-	-	90	39	38	44	36	36*	36	0
8	$t2k16v6^4 4^5 2^7$	-	-	-	-	-	-	90	44	45	53	38	38	38	0
9	$t2k19v7^1 6^1 5^1 4^5 3^8 2^3$	44	45	42	42	42	43	91	43	43	50	42	42*	42	0
10	$t2k14v6^5 5^5 3^4$	-	-	-	-	-	58	90	56	53	56	50	50	46	-4
11	$t2k18v6^7 4^8 2^3$	-	-	-	-	-	-	90	54	55	63	47	47	47	0
12	$t2k19v6^9 4^3 2^7$	-	-	-	-	-	-	90	61	59	64	51	51	51	0
13	$t2k8v8^2 7^2 6^2 5^2$	-	-	-	-	-	74	64	72	64	76	64	64*	64	0
14	$t3k9v4^5 3^4$	-	-	-	-	-	-	103	138	-	115	85	85	80	-5
15	$t3k6v5^2 4^2 3^2$	114	-	100	108	106	-	106	111	-	131	100	100*	100	0
16	$t3k7v10^1 6^2 4^3 3^1$	377	-	360	360	361	-	372	383	-	399	360	360*	360	0
17	$t3k12v10^2 4^1 3^2 2^7$	-	-	-	-	-	-	472	400	-	413	400	400*	400	0
18	$t3k14v6^5 5^5 3^4$	-	-	-	-	-	-	400	420	-	414	370	370	370	0
19	$t3k8v8^2 7^2 6^2 5^2$	-	-	-	-	-	-	594	614	-	645	540	540	535	-5

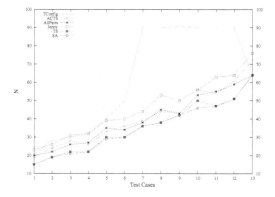

Fig. 2 Graphical comparison of the best bounds achieved by our SA with respect to those produced by the state-of-the-art procedures (TConfig [17], ACTS (IPOG) [9], AllPair [10], Jenny [7] and TS [6]), when the strength $t = 2$. Note that the performance of our SA improves or equals the best-known solutions.

are shown in column 15. The difference between the best result produced by our SA and the previous best-known solution ($\Delta = \Theta - \beta$) is depicted in the last column. Next, Figure 2 compares the results shown in Table 2.

From the data presented, we can make the following main observations. The quality solution attained by the proposed SA is very competitive with respect to that produced by the state-of-the-art procedures. In fact, it is able to improve the previous best-known solutions on 4 benchmark instances. For the remaining of the instances in the test-suite, our SA equals the previous best-known solutions.

4 Conclusions

This paper focused on constructing MCAs with a new approach of SA, which integrates three key features that importantly determines its performance. First, an efficient method to generate initial solutions with maximum Hamming distance. Second, a carefully designed composed neighborhood function that allows the search to quickly reduce the total cost of candidate solutions, while avoiding to get stuck on some local minimal. Third, an effective cooling schedule allowing our SA algorithm to converge faster, producing at the same time good quality solutions.

The empirical evidence presented in this paper showed that our SA improved the size of some MCAs in comparison with the tools that are among the best found in the state-of-the-art of the construction of MCAs. The performance of the proposed SA algorithm was assessed with a benchmark, composed by 19 MCAs of strengths two and three taken from the literature. The computational results are reported and compared with previously published ones, showing that our algorithm was able to find 4 new upper bounds and to equal 15 previous best-known solutions on the selected benchmark instances.

Finally, the new MCAs are available in CINVESTAV Covering Array Repository (CAR), which is available under request at http://www.tamps.cinvestav.mx/~jtj/CA.php.

Acknowledgements. The authors thankfully acknowledge the computer resources and assistance provided by Spanish Supercomputing Network (TIRANT-UV). This research work was partially funded by the following projects: CONACyT 58554, Cálculo de Covering Arrays; 51623 Fondo Mixto CONACyT y Gobierno del Estado de Tamaulipas.

References

[1] Bryce, R.C., Colbourn, C.J.: The density algorithm for pairwise interaction testing. Software Testing, Verification and Reliability 17(3), 159–182 (2007)

[2] Cohen, D.M., Dalal, S.R., Parelius, J., Patton, G.C.: The combinatorial design approach to automatic test generation. IEEE Software 13(5), 83–88 (1996)

[3] Cohen, M.B., Colbourn, C.J., Ling, A.C.H.: Augmenting simulated annealing to build interaction test suites. In: Proceedings of the 14th International Symposium on Software Reliability Engineering, ISSRE 2003, pp. 394–405. IEEE Computer Society (2003)

 [4] Colbourn, C.J., Torres-Jimenez, J.: Heterogeneous Hash Families and Covering Arrays. In: Error-Correcting Codes, Finite Geometries and Cryptography, vol. 523, pp. 3–15. Contemporary Mathematics (2010)

 [5] Colbourn, C.J., Martirosyan, S.S., Mullen, G.L., Shasha, D., Sherwood, G.B., Yucas, J.L.: Products of mixed covering arrays of strength two. Journal of Combinatorial Designs 14(2), 124–138 (2006)

 [6] Gonzalez-Hernandez, L., Rangel-Valdez, N., Torres-Jimenez, J.: Construction of Mixed Covering Arrays of Variable Strength Using a Tabu Search Approach. In: Wu, W., Daescu, O. (eds.) COCOA 2010, Part I. LNCS, vol. 6508, pp. 51–64. Springer, Heidelberg (2010)

 [7] Jenkins, B.: Jenny: a pairwise testing tool (2011), `http://burtleburtle.net/bob/math/jenny.html` (last accessed on June 22, 2011)

 [8] Kirkpatrick, S., Gelatt, C.D., Vecchi, M.P.: Optimization by simulated annealing. Science 4598(13), 671–680 (1983)

 [9] Lei, Y., Kacker, R., Kuhn, D.R., Okun, V., Lawrence, J.: IPOG: A general strategy for t-way software testing. In: Proceedings of the 14th Annual IEEE International Conference and Workshops on the Engineering of Computer-Based Systems, ECBS 2007, pp. 549–556. IEEE Computer Society (2007)

[10] McDowell, A.G.: All-pairs testing (2011), `http://www.mcdowella.demon.co.uk/allPairs.html` (last accessed on June 21, 2011)

[11] Moura, L., Stardom, J., Stevens, B., Williams, A.: Covering arrays with mixed alphabet sizes. Journal of Combinatorial Designs 11(6), 413–432 (2003)

[12] Sherwood, G.B.: Optimal and near-optimal mixed covering arrays by column expansion. Discrete Mathematics 308(24), 6022–6035 (2008)

[13] Shiba, T., Tsuchiya, T., Kikuno, T.: Using artificial life techniques to generate test cases for combinatorial testing. In: Proceedings of the 28th Annual International Computer Software and Applications Conference, COMPSAC 2004, pp. 72–77. IEEE Computer Society (2004)

[14] Torres-Jimenez, J., Rodriguez-Tello, E.: New bounds for binary covering arrays using simulated annealing. Inf. Sci. 185, 137–152 (2012)

[15] Yu-Wen Tung, W.S., Aldiwan: Automating test case generation for the new generation mission software system. In: IEEE Aerospace Conference Proceedings, vol. 1, pp. 431–437. IEEE Computer Society (2000)

[16] Williams, A.W., Probert, R.L.: A practical strategy for testing pair-wise coverage of network interfaces. In: Proceedings of the Seventh International Symposium on Software Reliability Engineering, ISSRE 1996, pp. 246–256. IEEE Computer Society (1996)

[17] Williams, A.W.: Determination of test configurations for pair-wise interaction coverage. In: Proceedings of the 13th International Conference on Testing Communicating Systems: Tools and Techniques, TestCom 2000, pp. 59–74. Kluwer, B.V (2000)

Improve the Adaptive Capability of TMA-OR

Jungan Chen, Qiaowen Zhang, and Zhaoxi Fang

Abstract. In T-detector Maturation Algorithm with Overlap Rate (TMA-OR), the parameters Omin and self radius r_s are required to be set by experience. To solve the problem, negative selection operator and self radius learning mechanism are proposed. The results of experiment show that the proposed algorithm can achieve the same effect when KDD and iris are as data set.

Keywords: Artificial immune system, negative selection, variable size detector.

1 Introduction

Nowadays, Artificial Immune System (AIS) has been applied to many areas such as computer security, classification, learning and optimization [1]. Negative Selection Algorithm, Clonal Selection Algorithm, Immune Network Algorithm and Danger Theory Algorithm are the main algorithms in AIS [2][3].

A real-valued negative selection algorithm with variable-sized detectors (V-detector Algorithm) applied in abnormal detection is proposed to generate detectors with variable r. A statistical method (naïve estimate) is used to estimate detect coverage [4]. But as reported in Stiboret later work, the performance of V-detector on the KDD Cup 1999 data is unacceptably poor[5]. So a new statistical approach (hypothesis testing) is used to analyze the detector coverage [6]. But hypothesis testing requires np>5, n(1-p)>5 and n>10. When p is set to 90%, n must be set to at least 50. Because the number of detectors affect the detect performance, the hypothesis has its shortage. Actually in naïve estimate method, V-detector algorithm tries to maximize the distance among valid detectors. So with the number of valid detectors increasing, it is difficult to find valid detector.

Jungan Chen · Qiaowen Zhang · Zhaoxi Fang
Electronic Information Department
Zhejiang Wanli University
No.8 South Qian Hu Road
Ningbo, Zhejiang, 315100, China
e-mail: `friendcen21@hotmail.com`, `cn_hnzqw@yahoo.com.cn`,
`zhaoxifang@gmail.com`

S. Omatu et al. (Eds.): Distributed Computing and Artificial Intelligence, AISC 151, pp. 665–671.
springerlink.com   © Springer-Verlag Berlin Heidelberg 2012

To choose the appropriate distance among valid detectors and achieve less number of detectors generated, a parameter overlap rate (Omin) in T-detector Maturation Algorithm (TMA) is proposed to control the distance among detectors [7].But the optimized Omin is required to be set by experience. To solve this problem, a suppression operator called Negative Selection operator (NS operator) is used in TMA. NS operator is first proposed to eliminate those network cells which are recognized by others in optaiNet [8]. So there is no parameter Omin in TMA with NS operator. But there is another problem in both V-detector algorithm and TMA. As reference[9] metioned "self radius is an important control parameter of V-detector to balance between high detection rate and low false alarm rate in more general cases." , the self radius is difficult to set. In this paper, a self radius learning mechanism is proposed to achieve the adaptive self radius. By combining NS operator and self radius learning mechanism, an augmented TMA called TMA with Adaptive Capability (TMA-AC) is proposed.

2   Algorithm

2.1   Match Range Model

$U=\{0,1\}^n$,n is the number of dimensions. The normal set is defined as selves and abnormal set is defined as nonselves. selves∪nonselves=U. selves∩nonselves=Φ. There are two points $x=x_1x_2...x_n$, $y=y_1y_2...y_n$. The Euclidean distance between x and y is:

$$d(x, y) = \sum_{i=1}^{n} (x_i - y_i)^2 \tag{1}$$

The detector is defined as dct = {<center, selfmin, selfmax > | center ∈ U, selfmin, selfmax∈N}. center is one point in U. selfmax is the maximized distance between dct.center and selves. selfmin is the minimized distance. The detector set is defined as DCTS. Selfmax and selfmin are calculated by setMatchRange(dct, selves), dct.center∈U, i∈[1, |selves|], $self_i$∈selves :

$$setMatchRa\,nge = \begin{cases} selfmin = min(\{d(sel\,f_i, dct.center\,)\}) \\ selfmax = max(\{d(sel\,f_i, dct.center\,)\}) \end{cases} \tag{2}$$

[selfmin,selfmax] is defined as self area. Others is as nonself area. Suppose there is one point x∈U and one detector dct ∈ DCTS. When d(x,dct) ∉ [dct.selfmin, dct.selfmin], x is detected as abnormal. So one rule called Range Match Rule (RMR), RMRMatch(x,dct) shown in equation 3, is proposed. In equation 3, value 1 means that x is abnormal.

$$RMRMatch = \begin{cases} 0, d(x, dct.center) \in [dct.selfmin, dct.selfmax] \\ 1, d(x, dct.center) \notin [dct.selfmin, dct.selfmax] \end{cases} \tag{3}$$

Based on RMR, the detect procedure detect(x,DCTS) is defined as equation 4, true means that x is abnormal.

$$Detect = \begin{cases} true, RMRMatch(x, dct_k) = 1, \exists dct_k \in DCTS \\ false, \qquad\qquad\qquad others \end{cases} \tag{4}$$

2.2 Self Radius Learning Mechanism

To learning the appropriate self radius, a property minselfList is added to the properties of detector. The property minselfList has three element {self0,self1,self2} $\in$ selves, which have the minimized distance with a given detector dct.center.

Self radius r_s can be achieved by the equation 5.

$$\begin{aligned} d01 &= d(self0, self1) \\ d12 &= d(self1, self2) \\ d02 &= d(self0, self2) \\ r_s &= (d01 + d12 + d02)/3 \end{aligned} \tag{5}$$

2.3 NS Operator

NS operator is first proposed to eliminate those network cells which are recognized by others in optaiNet [8]. In this work, it is defined as following. Parameter UnValid is computed through equation 7.

$$IsValidAnd = \begin{cases} false, & UnValid >= 3 \\ true, & others \end{cases} \tag{6}$$

$$\begin{aligned} &\exists dct_k \in DCTS \\ &if(NSMatchAnd(dctx, dct_k)) \qquad UnValid = UnValid + 1 \end{aligned} \tag{7}$$

$$NSMatchAnd = \begin{cases} true, d < dctx.selfm\ in \wedge d < dctk.selfm\ in \\ false, \qquad\qquad\qquad others \end{cases} \tag{8}$$

$$d = d(dctx.center, dctk.center) \tag{9}$$

2.4 The Model of Algorithm

The algorithm, called TMA-AC (TMA with Adaptive Capability), is shown in Fig.1. Step 2~5 is used to generate candidate detector which does not covered by self with r_s. Step 10 is used to estimate the detect coverage. Step 6 is used to decide whether candidate detector is a valid detector according equation 6.

```
1.   Set the desired coverage pc

2.   Generate one candidate detector dctx randomly

3.   setMatchRange(dctx,selves)

4.   compute the self radius r_s according equation 5

5.   if dctx.selfmin< r_s then Go to 2;

6.   if isvalidAnd(dctx,DCTS) then // equation 6

7.        dctx is added to detector set DCTS

8.            covered=0

9.   Else

10.           covered ++

11.  If covered <1/(1- pc) then goto 2
```

Fig. 1 TMA-AC algorithm model

3   Experiments

For the purpose of comparison, experiments are carried out using 2-dimensional synthetic data in table 1,which is described in Zhou's paper[10]. Over the unit square $[0,1]^2$,various shapes are used as the self region. In every shape, there are training data (self data) of 1000 points and test data of 1000 points including both self points and nonself points. In the famous benchmark Fisher's Iris Data, one of the three types of iris is considered as normal data, while the other two are considered abnormal [4]. As for KDD data, 20 subsets were extracted from the enormous KDD data using a process described in [5]. Self radius and Omin used in TMA-OR are given in table.1. All the results shown in these figures are average of 100 or 20 (see table 1) repeated experiment with coverage rate 99%.

When KDD and iris are as data set, TMA-AC can get the higher detect rate (Fig.2)with lower false alarm rate(Fig.3) and smaller number of detectors(Fig.4). As reference[7], TMA-OR(Omin=0.7) can achieve the best effect. By comparing with TMA-OR(Omin=0.7),TMA-AC can achieve the same effect when KDD and iris are as data set. But when 2-dimensional synthetic data is used, TMA-AC shows less effective than TMA-OR.

Table 1 data set and parameters used in experiments

Data set		Parameters		
		r_s	*Omin*	Repeated times
2-dimensional synthetic data	Comb	0.03	0, 0.7	100
	Cross			
	Ring			
	Triangle			
	Stripe			
	Intersection			
	Pentagram			
Iris data	Setosa as self data			
	Versicolor as self data			
	Virginica as self data			
KDD data		0.05		20

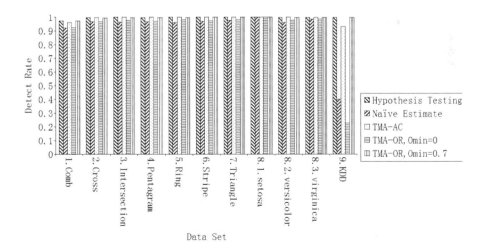

Fig. 2 Detect Rate.

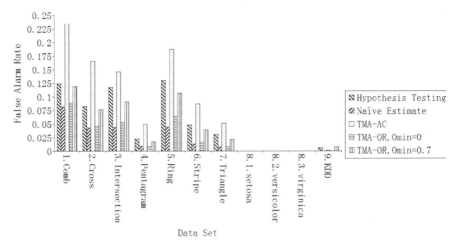

Fig. 3 False Alarm Rate.

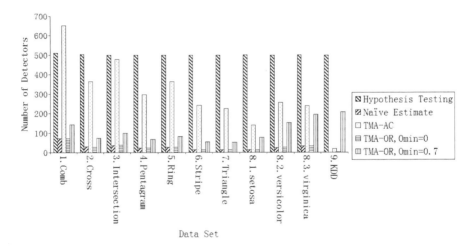

Fig. 4 Number of Detectors.

4 Conclusion

As the parameters Omin and self radius r_s in TMA-OR are required to be set by experience. To solve the adaptive problem, NS operator and self radius learning mechanism are proposed, and then an augmented TMA called TMA-AC is proposed. There is no Omin and r_s in TMA-AC. The results of experiment show that the proposed algorithm can achieve the same effect when KDD and iris are as data set. But TMA-AC shows less effective than TMA-OR when 2-dimensional synthetic data is as the data set. One reason for the result of 2-dimensional synthetic data is that the data set is generated by manually and not reflected the real world problem. Of course, more data sets are required to be used in experiments for evaluating correctly.

Acknowledgments. This work is supported by This work is supported by National Natural Science Foundation of China 71071145, Zhejiang Provincial Nature Science Foundation Y1110200, Y6090027, Thanks for the assistance received by using KDD Cup 1999 data set [http://kdd.ics.uci.edu/databases / kddcup99/ kddcup99.html], the 2-dimensional synthetic data set [https:// umdrive.memphis.edu/ zhouji/ www/ vdetector.html].

References

1. Hart, E., Timmis, J.: Application areas of AIS: The past, the present and the future. Journal of Applied Soft Computing 8(1), 191–201 (2008)
2. Timmis, J., et al.: An interdisciplinary perspective on artificial immune systems. Evolutionary Intelligence 1(1), 5–26 (2008)
3. Greensmith, J., Aickelin, U., et al.: Information Fusion for Anomaly Detection with the Dendritic Cell Algorithm. Information Fusion 11(1), 21–34 (2010)
4. Ji, Z., Dasgupta, D.: Real-Valued Negative Selection Algorithm with Variable-Sized Detectors. In: Deb, K., et al. (eds.) GECCO 2004. LNCS, vol. 3102, pp. 287–298. Springer, Heidelberg (2004)
5. Stibor, T., Timmis, J.I., Eckert, C.: A Comparative Study of Real-Valued Negative Selection to Statistical Anomaly Detection Techniques. In: Jacob, C., Pilat, M.L., Bentley, P.J., Timmis, J.I. (eds.) ICARIS 2005. LNCS, vol. 3627, pp. 262–275. Springer, Heidelberg (2005)
6. Ji, Z., Dasgupta, D.: Estimating the Detector Coverage in a Negative Selection Algorithm. In: Genetic and Evolutionary Computation Conference (2005)
7. Chen, J.: T-detector Maturation Algorithm with Overlap Rate. WSEAS Transactions on Computers 7(8), 1300–1308 (2008)
8. Chen, J.: A novel suppression operator used in optaiNet. BSBT 57, 17–23 (2009)
9. Ji, Z., Dasgupta, D.: Augmented Negative Selection Algorithm with Variable-Coverage Detectors. In: Congress on Evolutionary Computation, vol. 1, pp. 1081–1088 (2004)
10. Ji, Z.: Negative Selection Algorithms: from the Thymus to V-detector. PhD Dissertation, University of Memphis (2006)

A New Hybrid Firefly Algorithm for Complex and Nonlinear Problem

Afnizanfaizal Abdullah, Safaai Deris, Mohd Saberi Mohamad,
and Siti Zaiton Mohd Hashim

Abstract. Global optimization methods play an important role to solve many real-world problems. However, the implementation of single methods is excessively preventive for high dimensionality and nonlinear problems, especially in term of the accuracy of finding best solutions and convergence speed performance. In recent years, hybrid optimization methods have shown potential achievements to overcome such challenges. In this paper, a new hybrid optimization method called Hybrid Evolutionary Firefly Algorithm (HEFA) is proposed. The method combines the standard Firefly Algorithm (FA) with the evolutionary operations of Differential Evolution (DE) method to improve the searching accuracy and information sharing among the fireflies. The HEFA method is used to estimate the parameters in a complex and nonlinear biological model to address its effectiveness in high dimensional and nonlinear problem. Experimental results showed that the accuracy of finding the best solution and convergence speed performance of the proposed method is significantly better compared to those achieved by the existing methods.

Keywords: Firefly Algorithm, Differential Evolution, hybrid optimization, parameter estimation, biological model.

1 Introduction

Global optimization is an important task in most scientific and engineering problems. These problems include finding the minimal vehicle routing [1-2] and

Afnizanfaizal Abdullah · Safaai Deris · Mohd Saberi Mohamad
Artificial Intelligence and Bioinformatics Group (AIBIG), Faculty of Computer Science and Information Systems, Universiti Teknologi Malaysia, 81310 UTM, Johor, Malaysia

Siti Zaiton Mohd Hashim
Soft Computing Research Group (SCRG), Faculty of Computer Science and Information Systems, Universiti Teknologi Malaysia, 81310 UTM, Johor, Malaysia

S. Omatu et al. (Eds.): Distributed Computing and Artificial Intelligence, AISC 151, pp. 673–680.
springerlink.com   © Springer-Verlag Berlin Heidelberg 2012

the optimal design in electronic systems [3]. For the past few years, many global optimization methods have been proposed to solve these problems. Most of these methods are metaheuristics methods such as Genetic Algorithm (GA) [4], Particle Swarm Optimization (PSO) [5] and Evolutionary Programming (EP) [6]. These methods have received remarkable attentions as they are known to be derivative free, robust and often involve a small number of parameter tunings. However, applying such single methods is sometimes too restrictive, especially for high dimensional and nonlinear problems [8]. This is because these methods usually require a substantially huge amount of computational times and are frequently trapped in one of the local optima. Recently, different methods have been combined to overcome these disadvantages. The hybrid optimization methods have proved their effectiveness in several high dimensional and nonlinear problems including in bioinformatics [7] and electrical engineering [8].

In this paper, a new hybrid optimization method is introduced. The proposed method, called Hybrid Evolutionary Firefly Algorithm (HEFA), combines the recently introduced Firefly Algorithm (FA) [9] with the evolutionary operations adopted from the Differential Evolution (DE) [10]. In this method, the population is firstly ranked according to the fitness value. Then, the sorted population is divided into two sub-populations. The first sub-population; which contains the solutions with potential fitness values, is subjected to undergo neighborhood-based optimization, whereas the other sub-population is subjected to perform the evolutionary operations. The proposed method is used to estimate parameters in a complex and nonlinear biological model. The experimental results showed that the accuracy and speed performance of the HEFA method had outperformed the other existing methods. This paper is organized as follows: in Section 2, the proposed HEFA method is introduced and the details of the method are presented. In Section 3, the experimental results of evaluating the effectiveness of the proposed method to the parameter estimation of nonlinear biological model are described. Lastly, in Section 4, the conclusion of the contribution and the future works are discussed.

2 Hybrid Evolutionary Firefly Algorithm (HEFA) Method

The proposed HEFA method is basically a combination of the FA [9] and DE [10] methods. In this method, each solution in a population represents a solution which is located randomly within a specified searching space. The ith solution, X_i, is represented as follows:

$$X_{i(t)} = \left\{ x_{i1(t)}, x_{i2(t)}, \dots, x_{id(t)} \right\} \tag{1}$$

where x_{ik} is the vector with $k = 1, 2, 3, \dots, d$, and t is the time step. Initially, the fitness value of each solution was evaluated. The solution that produced the best fitness value would be chosen as the current best solution in the population. Then, a sorting operation was performed. In this operation, the newly evaluated solutions were ranked based on the fitness values and divided into two sub-populations.

The first sub-population contained solutions that produced potential fitness values. The fitness value of each ith solution in this sub-population was then compared with its jth neighboring solution. If the fitness value of the neighboring solution was better, the distance between every solution would then be calculated using the standard Euclidean distance measure. The distance was used to compute the attractiveness, β:

$$\beta = \beta_0 e^{-\gamma r_{ij}^2} \tag{2}$$

where β_0, γ and , r_{ij} are the predefined attractiveness, light absorption coefficient, and distance between ith solution and its jth neighboring solution, respectively [9]. Later, this new attractiveness value was used to update the position of the solution, as follows:

$$x_{id} = x_{id} + \beta\left(x_{jd} - x_{id}\right) + \alpha(\delta - \tfrac{1}{2}) \tag{3}$$

where α and δ are uniformly distributed random values between 0 to 1. Thus, the updated attractiveness values assisted the population to move towards the solution that produced the current best fitness value [9, 11].

On the other hand, the second sub-population contained solutions that produced less significant fitness values. The solutions in this population were subjected to undergo the evolutionary operations of DE method. Firstly, the trivial solutions were produced by the mutation operation performed on the original counterparts. The ith trivial solution, V_i, was generated based on the following equation:

$$V_{i(t)} = \left\{v_{i1(t)}, v_{i2(t)}, \dots, v_{id(t)}\right\} \tag{4}$$

$$v_{i(t)} = x_{best(t)} + F \cdot (x_{r1(t)} - x_{r2(t)}) \tag{5}$$

where $x_{best(t)}$ is the vector of current best solution, F is the mutation factor, x_{r1} and x_{r2} are randomly chosen vectors from the neighboring solutions [10]. Next, the offspring solution was produced by the crossover operation that involved the parent and the trivial solution. The vectors of the ith offspring solution, Y_i, were created as follows:

$$Y_{i(t)} = \left\{y_{i1(t)}, y_{i2(t)}, \dots, y_{id(t)}\right\} \tag{6}$$

$$y_{i(t)} = \begin{cases} v_{i(t)} \ if \ R < CR \\ x_{i(t)} \ Otherwise \end{cases} \tag{7}$$

where R is a uniformly distributed random value between 0 to 1 and CR is the predefined crossover constant [10]. As the population of the offspring solution was produced, a selection operation was required to keep the population size constant. The operation was performed as follows:

$$X_{i(t+1)} = \begin{cases} Y_{i(t)} \ if \ f(Y_{i(t)}) \le f(X_{i(t)}) \\ X_{i(t)} \ if \ f(Y_{i(t)}) > f(X_{i(t)}) \end{cases} \tag{8}$$

This indicates that the original solution would be replaced by the offspring solution if the fitness value of the offspring solution was better than the original solution. Otherwise, the original solution would remain in the population for the next iteration. The whole procedure was repeated until the stopping criterion was met. Figure 1 shows the outline of the proposed HEFA method.

Hybrid Evolutionary Firefly Algorithm (HEFA)

Input: Randomly initialized position of d dimension problem: X_i
Output: Position of the approximate global optima: X_G
Begin
    Initialize population; Evaluate fitness value;
    X_G ←Select current best solution;
    For $t \leftarrow 1$ to *max*
        Sort population based on the fitness value;
        $X_{good} \leftarrow first\_half(X)$; $X_{worst} \leftarrow second\_half(X)$;
        For $i \leftarrow 0$ to number of X_{good} solutions
            For $j \leftarrow 0$ to number of X_{good} solutions
                If $(f(X_i) > f(X_j))$ **then**
                    Calculate distance and attractiveness;
                    Update position;
                End If
            End For
        End For
        For $i \leftarrow 0$ to number of X_{worst} solutions
            Create trivial solution, $V_{i(t)}$;
            Perform crossover, $Y_{i(t)}$;
            Perform selection, $X_{i(t)}$;
        End For
        $X \leftarrow combine(X_{good}, X_{worst})$;
        X_G ←Select current best solution;
        $t \leftarrow t + 1$;
    End For
End Begin

Fig. 1 The outline of proposed HEFA method

3 Results

To address its effectiveness, the proposed method was used to estimate the parameters in a complex and nonlinear biological model. A general kinetic model of the Maillard reaction occurring in heated monosaccharide casein systems [12] was used in this experiment. The model is downloaded from the BioModels repository database [14]. The model observed the dynamic of the metabolites concentrations involved in the systems through different reaction environments. Sugars, including glucose and fructose, were utilized to analyze the effect of the reaction kinetics. The model describes the reactions of the glucose, Glu, and fructose, Fru, concentrations as [12]:

$$\dot{Glu} = -k_1(Glu) + k_2(Fru) - k_3(Glu) - k_4(Glu)(Lys) \qquad (9)$$

$$\dot{Fru} = k_1(Glu) - k_2(Fru) - k_5(Fru) - k_6(Fru) - k_7(Fru)(Lys) \qquad (10)$$

where Glu, Fru, and Lys are the concentrations of glucose, fructose and lysine, respectively. The parameter values are $k_1 = 0.01$ min^{-1}, $k_2 = 0.00509$ min^{-1}, $k_3 = 0.00047$ min^{-1}, $k_4 = 0.00018$ L mmol^{-1} min^{-1}, $k_5 = 0.0011$ min^{-1}, $k_6 = 0.00712$ min^{-1} and $k_7 = 0.00015$ L mmol^{-1} min^{-1} [12]. Figure 2 shows the dynamics of glucose and fructose concentration as depicted in Equation 9 and 10.

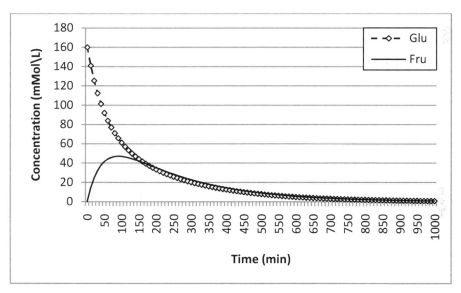

Fig. 2 The dynamics of glucose (Glu) and fructose (Fru) concentration in the system

Parameter estimation was implemented to find the optimal parameter values of the system so that the error difference between the experimental and simulated data would be minimized, as follows:

$$\min f(X_i) = \sum_{n=1}^{N} \sum_{m=1}^{M} (x_{mn}^{exp} - x_{mn}^{model})^2 \qquad (11)$$

where N is the number of parameter values, M is the number of observable state variables, x_{mn}^{exp} and x_{mn}^{exp} are the experimental and simulated data points for the mth parameter value in the nth state time, respectively.

The parameter estimation results of four existing optimization methods, GA, PSO, FA, and EP, were compared with the proposed HEFA. In this comparison, the population size and the number of iterations of all methods were 50 and 1000, respectively. All methods were executed 50 times independently. For the proposed method, the initial attractiveness value, β_0, was set to 0.5 and the light absorption coefficient, γ, value was 0.01 [11]. Furthermore, for GA, EP, and HEFA methods, the mutation factor, F, and crossover constant, CR, were both set to 0.9 [10]. Table 1 presents the overall performance of these methods. For 50 runs, the average fitness value of the HEFA method was better compared to the other methods. The error percentage of the method for both glucose and fructose concentration are calculated as

$$Error = \sum_{m=1}^{M} \frac{x_m^{exp} - x_m^{model}}{x_m^{exp}} \times 100\% \qquad (12)$$

From the table, it shows that the error percentage produced by the HEFA method was substantially small compared to other methods. In term of the computational time, the result showed by GA method was better than the proposed HEFA method. However, the small number of evaluated functions by GA indicated that only a small number of possible solutions were considered through the whole iterations. Thus, even though HEFA method required more computational time, the method tended to evaluate more functions than GA. Figure 3 shows the convergence performance of all tested methods. This proved that the HEFA method managed to escape the local optima more effectively compared to other methods. Overall, the advantages of the HEFA method were majorly due to two main factors. The first factor was the utilization of the solutions that produced least significant fitness values. The results showed that the use of these solutions had increased the exploration capability which allowed the method to escape the local optima effectively. The second factor was the neighborhood information sharing scheme by the evolutionary operations. It had been proven that the used of evolutionary operations could enhance the exploitation of each solution, thus improving the accuracy of finding the optimum solutions.

Table 1 Performance of different methods

	Method				
	GA	PSO	FA	EP	HEFA
Accuracy performance					
Error (Glucose)	1.07%	1.05%	0.42%	1.05%	0.42%
Error (Fructose)	0.28%	0.28%	0.17%	0.28%	0.01%
Average Fitness Value	6.43×10^{8}	6.38×10^{8}	9.91×10^{8}	6.43×10^{8}	7.79×10^{-18}
Standard Deviation	1.47×10^{5}	1.47×10^{5}	2.17×10^{8}	1.47×10^{5}	1.66×10^{-25}
Speed performance					
No. of Evaluated Functions	109	25059	29510	25009	35100
Time (second)	0.046	5.413	4.131	5.444	3.521

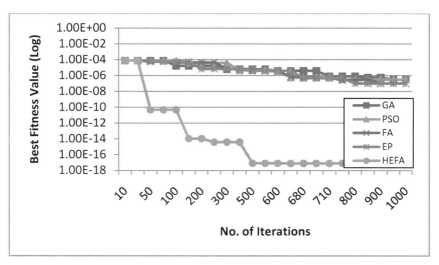

Fig. 3 Convergence behavior of GA, PSO, FA, EP and HEFA methods

4 Conclusion and Future Work

In this paper, the new hybrid optimization called HEFA is introduced. The proposed method combined the FA method with the evolutionary operations adopted from DE. The proposed method is used to estimate the parameters in a biological model. The experimental results showed that the accuracy and speed performance of HEFA had significantly outperformed the results produced by GA, PSO, EP, and the standard FA methods. Moreover, the convergence analysis showed that the proposed method was capable to escape from the local optima more effectively. For the future research, several improvements are suggested to further enhance the performance of the proposed method. Firstly, the adaptive control parameter can be introduced to enhance the function evaluation scheme by the evolutionary operations [13]. This is important to ensure that the speed performance will not be affected by the problem complexity. Secondly, the direction of the fireflies can be added to the method so that the firefly movements can be improved substantially [1]. Lastly, the proposed method should be tested to estimate the parameters in more complex problems such as noise and identifiability.

References

1. Jati, G.K., Suyanto: Evolutionary Discrete Firefly Algorithm for Travelling Salesman Problem. In: Bouchachia, A. (ed.) ICAIS 2011. LNCS, vol. 6943, pp. 393–403. Springer, Heidelberg (2011)
2. Shao, Z., Gao, S., Wang, S.: A Hybrid Particle Swarm Optimization Algorithm for Vehicle Routing Problem with Stochastic Travel Time. In: Fuzzy Info. and Engineering, ASC, vol. 54, pp. 566–574 (2009)

3. dos Santos Coelho, L., Mariani, V.: Combining of Differential Evolution and Implicit Filtering Algorithm Applied to Electromagnetic Design Optimization. In: Soft Computing in Industrial Applications, ACS, vol. 39, pp. 233–240 (2007)
4. Goldberg, D.E., Holland, J.H.: Genetic algorithms and machine learning. Machine Learning 3(2), 95–99 (1988)
5. Kennedy, J., Eberhart, R.: Particle swarm optimization. In: Proc. IEEE International Conference on Neural Networks, pp. 1942–1948 (1995)
6. Fogel, L.J., Owens, A.J., Walsh, M.J.: Artificial intelligence through simulated evolution. Wiley (1966)
7. Lillacci, G., Khammash, M.: Parameter estimation and model selection in computational biology. PLoS Computational Biology 6(3), e1000696 (2010)
8. Das, S., Abraham, A., Konar, A.: Particle swarm optimization and differential evolution algorithms: technical analysis, applications and hybridization perspective. SCI, vol. 116, pp. 1–38 (2008)
9. Yang, X.-S.: Firefly Algorithms for Multimodal Optimization. In: Watanabe, O., Zeugmann, T. (eds.) SAGA 2009. LNCS, vol. 5792, pp. 169–178. Springer, Heidelberg (2009)
10. Storn, R., Price, K.: Differential evolution—a simple and efficient heuristic for global optimization over continuous spaces. Journal of Global Optimization 11, 341–359 (1997)
11. Łukasik, S., Żak, S.: Firefly Algorithm for Continuous Constrained Optimization Tasks. In: Nguyen, N.T., Kowalczyk, R., Chen, S.-M. (eds.) ICCCI 2009. LNCS, vol. 5796, pp. 97–106. Springer, Heidelberg (2009)
12. Brands, C.M.J., van Boekel, M.A.J.S.: Kinetic modeling of reactions in heated monosaccharide-casein systems. Journal of agricultural and food chemistry 50(23), 6725–6739 (2002)
13. Noman, N., Iba, H.: Accelerating differential evolution using an adaptive local search. IEEE Transactions on Evolutionary Computation 12(1), 107–125 (2008)
14. Li, C., Donizelli, M., Rodriguez, N., Dharuri, H., Endler, L., Chelliah, V., Li, L., He, E., Henry, A., Stefan, M., et al.: BioModels Database: An enhanced, curated and annotated resource for published quantitative kinetic models. BMC Systems Biology 4(1), 92 (2010)

A General Framework for Naming Qualitative Models Based on Intervals

Ester Martínez-Martín, M. Teresa Escrig, and Angel P. del Pobil

Abstract. In qualitative spatial and temporal reasoning we can distinguish between *comparing* and *naming* magnitudes. In particular, *naming* qualitative models allow humans to express spatio-temporal concepts such as "That horse is really *fast*". In colloquial terms, *naming* concepts are called relative. In this paper we present a general framework to solve the representation magnitude and a general algorithm to solve the basic step of inference process of qualitative models based on intervals. The general method is based on the definition of two algorithms: the qualitative sum and the qualitative difference.

1 Introduction

A widely used way to model commonsense reasoning in the spatial domain is by using qualitative models. In fact, qualitative reasoning can help to express poorly defined problem situations, support the solution process and lead to a better interpretation of the final results [20]. In nature, most of the knowledge about time and space is qualitative such that it is not necessary to know the exact amount of a spatio-temporal aspect to perform context-dependent comparisons. Humans are a clear example since they are not good at determining accurate lengths, volumes, etc., while they can easily perform context-dependent comparisons and make correct decisions from those comparisons [11].

Ester Martínez-Martín
Universitat Jaume-I, Castellón, Spain
e-mail: emartine@icc.uji.es

M. Teresa Escrig
Universitat Jaume-I, Castellón, Spain
e-mail: escrigm@icc.uji.es

Angel P. del Pobil
Universitat Jaume-I, Castellón, Spain
e-mail: pobil@icc.uji.es

S. Omatu et al. (Eds.): Distributed Computing and Artificial Intelligence, AISC 151, pp. 681–688.
springerlink.com   © Springer-Verlag Berlin Heidelberg 2012

Therefore, a qualitative representation can be defined as that representation that *makes only as many distinctions as necessary to identify objects, events, situations, etc., in a given context* [11]. Note that the way to define those distinctions depends on two different aspects. The first one is the level of *granularity*. In this context, *granularity* refers to a matter of precision in the sense of the amount of information which is included in the representation. Thus, a fine level of *granularity* will provide a more detailed information than a coarse level.

The second aspect corresponds to the distinction between *comparing* magnitudes and *naming* magnitudes [4]. This distinction refers to the usual comparison between *absolute* and *relative*. From a spatial point of view, that controversy corresponds to the way the relationships among objects in the world are represented. Levinson [16] pointed out that *absolute* defines an object's location in terms of arbitrary bearings such as, for instance, cardinal directions (e.g. North, South, East, West), by resulting in binary relationships. Instead, *relative* leads to ternary relationships. Consequently, for *comparing* magnitudes, an object b is any compared relationship to another object a from the same Point of View (PV). The comparison depends on the orientation of both objects *with respect to* (wrt) the PV, since objects a and b can be at any orientation wrt the PV. On the other hand, *naming* magnitudes divide the magnitude of any concept into intervals (sharply or overlapped separated, depending on the context such that a qualitative label is assigned to each interval. Note that the result of reasoning with this kind of regions can provide imprecision. That imprecision is solved by providing disjunction in the result. That is, if an object can be found in several qualitative regions, q_i or q_{i+1} or ... or q_n, then all possibilities are listed as follows $\{q_i, q_{i+1}, \ldots, q_n\}$ by indicating this situation.

Note that, although qualitative reasoning is an established field of pursued by investigators from many disciplines including geography [19], psychology [15], ecology [2, 18], biology [14, 10], robotics [17, 12] and Artificial Intelligence [5], the number of practical applications that make use of it is comparatively small. One reason for this can be seen in the difficulty for people from outside the field to incorporate the required reasoning techniques into their methods. So, the aim of this paper is to design a general, easy-to-use framework that overcomes that problem. So, the aim of this paper is to present a general systemic algorithm that integrates and solves the reasoning process of all qualitative models based on intervals. From the starting point that the development of any qualitative model consists of a representation of the magnitude at hand and the reasoning process, the structure of this paper is as follows: Section 2 describes the representation of a magnitude. The Reasoning Process is introduced in Section 3, and discussed in Section 4.

2   Magnitude Representation

In qualitative spatial reasoning, it is common to consider a particular aspect of the physical world, i.e. a magnitude like distance, and to develop a system of qualitative relationships between entities which cover that aspect of the world to some degree.

Thus, the first issue to be solved refers to how to properly represent the magnitude to be modelled.

Focusing on qualitative naming models based on intervals, any magnitude is represented by the following three elements:

1. The **number of objects** implied in each relation. A relationship is *binary* when there are only two objects implied (object b *wrt* object a, i.e. b *wrt* a). So, an object acts as reference (a) and the other one is referred (b). For instance, how far an object is *wrt* another object is a binary relationship as defined in [13]. On the contrary, a relationship is *ternary* when three objects are implied (c *wrt* ab). In this case, two objects form the reference system (ab) and the other object (c) is referenced *wrt* such reference system (e.g. qualitative orientation [9])

2. The **set of relations between objects**. It depends on the considered level of *granularity*. In a formal way, this set of relations between objects is expressed by means of a *Reference System (RS)*. A RS will contain, at least, a couple of components:

   - A *set of qualitative symbols*, in increasing order, represented by $Q = \{q_0, q_1, ..., q_n\}$, where q_0 is the qualitative symbol closest to the *Reference Object (RO)* and q_n is the one furthest away, going to infinity. Here, by cognitive considerations, the acceptance areas have been chosen in increasing size. Note that this set defines the different areas in which the workspace is divided and the number of areas depends on the *granularity* of the task at hand, as above-mentioned
   - The *structure relations*, $\Delta r = \{\delta_0, \delta_1, ..., \delta_n\}$, describe the acceptance areas for each qualitative symbol q_i. So, δ_0 corresponds to the acceptance area of qualitative symbol q_0; δ_1 to the acceptance area of symbol q_1 and so on. These acceptance areas are quantitatively defined by means of a set of closed or open intervals delimited by two extreme points: the initial point of the interval j, δ_j^i, and the ending point of the interval j, δ_j^e. So, the acceptance area of a particular entity of a magnitude, $AcAr(entity)$, is δ_j if the entity value is between the initial and ending points of δ_j, that is, $\delta_j^i \leq value\,(entity) \leq \delta_j^e$

3. The **operations**. The number of operations associated to a representation corresponds to the possible change in the PV. For example, if the relationship is binary (b *wrt* a), only one operation can be defined: inverse (a *wrt* b). Nevertheless, it is possible to define five different operations when the relationship between objects is ternary (c *wrt* ab) [9]: inverse (c *wrt* ba), homing (a *wrt* bc), homing-inverse (a *wrt* cb), shortcut (b *wrt* ac) and shortcut-inverse (b *wrt* ca).

3   The Reasoning Process

The *reasoning process* is divided into two steps:

- **The Basic Step of the Inference Process (BSIP)**. It can be defined as: "given two relationships, (1) the object b *wrt* a reference system, *RS1*, and (2) the object *c wrt another reference system, RS2*, such that the object *b* is included into the *RS2*, the BSIP obtains the relationship *c wrt RS1*". In Spatial Reasoning, the BSIP is usually represented by *composition tables*. These tables can be obtained either by hand or automatically by using algorithms, if they exist.

- **The Complete Inference Process (CIP)**. It is necessary when more than two objects (in binary relationships) or three objects (in ternary relations) are involved in the reasoning mechanism. Mainly, it consists of repeating the BSIP as many times as possible with the initial information and the information provided by some BSIP until no more information can be inferred

3.1   The Basic Step of the Inference Process

Basically, the BSIP is defined as the process of inferring the relationship between two (or three) entities of a magnitude from the knowledge of two other relationships such that there is an object in common in both relationships. The way to infer the new relationship depends on the considered magnitude. However, all the qualitative models based on intervals define the magnitude in the same way, as above-mentioned. For that reason, an abstraction can be done by resulting in a general algorithm. Here, we propose a general algorithm based on qualitative sums and differences that solves the inference process for all models based on intervals.

3.1.1 The General Algorithm

As previously introduced, magnitudes are represented by three different elements: (1) the number of objects implied in each relationship, (2) the set of relationships between entities and (3) the operations that can be defined. Note that there is a difference between concepts of commonsense knowledge in terms of dimensionality. For example, *time* is a scalar magnitude, whereas *space* is much more complex because it is inherent multi-dimensional. This inherent feature leads to a higher degree of freedom and an increased possibility of describing entities and relationships between entities. Because of the richness of space, its multi-dimensionality and its multiple aspects, most work in qualitative reasoning has focused on single aspects of space such as, for instance, topology, orientation or distance. Nevertheless, as pointed out in [8], relationships between entities can be seen as movements in the space or spatial deformations in physical space. As a result, when the relationships between entities are considered as directed vectors and using a reference orientation *ab*, three different situations can take place: (1) relationships between entities are in the same orientation; (2) relationships between entities are in the opposite orientation; and (3) relationships between entities are at any orientation. As a result, the inferred relationship will be composed of all possible relationships between the entities by considering the three possible orientations. According to a deeper analysis of the possible orientations, it is clear that the extreme cases are obtained when the implied objects are in the same orientation and when they are in the opposite one. Consequently, if both extreme cases are solved, the result will be built as a disjunction of qualitative symbols from the inferred area closest to the RO to the furthest one. With the aim of automatically solving these extreme cases, the qualitative sum and difference of intervals are defined.

3.1.2 The Qualitative Sum

Let q_i be the qualitative symbol which represents a relationship b *wrt a reference system RS1*, and let q_j be the qualitative symbol referred to the relationship c *wrt another reference system RS2*, such that b is included into the RS2. Supposing that those relationships are binary, we would have a situation similar to the one illustrated in Figure 1. In this example, from the knowledge b *wrt* $a = q_3$ and c *wrt* $b = q_2$, c *wrt a* should be inferred.

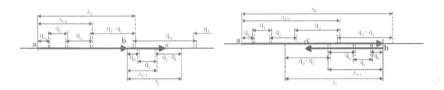

Fig. 1 Example of qualitative sum when structure relations with overlapped acceptance areas are used (left); example of qualitative difference when structure relations with overlapped acceptance areas are used (right)

However, it is desirable to achieve the same solution from a mathematical point of view. The development of such a method (the *qualitative sum*) has several advantages. Firstly, it does not require to represent the relationships for any composition. This is especially important when the dimensionality of the magnitude is high. Moreover, it can be applied to all the models based on intervals since the reasoning mechanism is the same in all of them. Therefore, the *qualitative sum* of the two intervals δ_i and δ_j, results in a range of qualitative symbols given by:

$$AcAr\left(\Delta_{i-1}+\Delta_{j-1}\right)\ldots AcAr\left(\Delta_i+\Delta_j\right) \tag{1}$$

where Δ_k represents the distance from the origin to δ_k, i.e. the sum of consecutive intervals from the origin to δ_k.

Therefore, the developed method proposed to solve the *qualitative sum* of intervals is divided into three steps:

1. **Obtaining the Upper Bound (UB) of the Result.** With the aim to obtain the UB of the disjunctino of qualitative symbols for the relationship c *wrt a*, the case in which entities b and c are set to the ending points of their respective acceptance areas is studied. Under this hypothesis, three different cases can occur:

   - The distance from the origin of qualitative areas to δ_j, Δ_j, is much lower than the sum of acceptance areas to δ_i, i.e. Δ_i. In this case, the absorption rule is applied. This rule, stated in [3], means that if an interval δ_i is k times greater than other (δ_j), then it can be assumed, without loss of information, that the sum or difference of them is δ_i, that is, $(\Delta_i \gg \Delta_j) \Leftrightarrow \Delta_i \geq k * \Delta_j \Rightarrow \Delta_i \pm \Delta_j \cong \Delta_i$ where k is a constant which depends on the context

- δ_i corresponds to the last defined qualitative area. Therefore, δ_i will be the UB
- Otherwise, an iterative procedure has been defined to recursively seek for the minimum qualitative area δ_k that satisfies $\Delta_j \leq \delta_{i+1} + \delta_{i+2} + \ldots + \delta_k \Leftrightarrow \Delta_j \leq \Delta_{(i+1)..k}$ Note that it stops when it comes to the last qualitative defined region or when the sum of acceptance areas from the origin to δ_j, i.e. Δ_j, is less than or equal to the sum of acceptance areas starting from δ_{i+1} to δ_k with $k > i$.

2. **Obtaining the Lower Bound (LB) of the Result.** Unlike the previous case, the values of the entities b and c are supposed to be set to the initial points of their respective acceptance areas. Therefore, the expression to be satisfied in this case is $\Delta_{j-1} \leq \delta_i + \delta_{i+1} + \ldots + \delta_k$ It will stop when it comes to the last qualitative region of the structure relations or when the distance from the origin to the qualitative area previous to δ_j, that is, Δ_{j-1}, is less or equal than the sum of acceptance areas starting from δ_j to δ_k with $k \leq i$.

3. **Building the Result.** It builds the list of qualitative areas from the LB to the UB

3.1.3 The Qualitative Difference

When the given relationships are opposite directed, as in the example shown in Figure 1, the qualitative difference of intervals must be solved. With this aim, a new method has been designed. With a similar reasoning mechanism to the qualitative sum, the qualitative difference of two intervals δ_i and δ_j is given by:

$$\begin{cases} AcAr(\Delta_i - \Delta_j) \ldots AcAr(\Delta_{i-1} - \Delta_{j-1}) & \text{when } \Delta_i \geq \Delta_j \\ AcAr(\Delta_j - \Delta_i) \ldots AcAr(\Delta_{j-1} - \Delta_{i-1}) & \text{otherwise} \end{cases} \quad (2)$$

From that definition, the process to obtain the qualitative difference consists of the following three steps:

1. **Obtaining the UB**. The UB of the range of acceptance areas is computed by considering that the entity values are set to the initial points of their acceptance areas. Under this hypothesis, a recursively function that seeks for the minimum acceptance area δ_k that satisfies the comparison $\Delta_{j-1} \leq \delta_i + \delta_{i-1} + \ldots + \delta_k$, has been implemented. From its definition, it will stop when it comes to consider the first region of the relation structure or when the sum of acceptance areas from the origin to δ_j (without including δ_j), i.e. Δ_{j-1}, is less than or equal to the sum of acceptance areas starting from δ_i to δ_j with $k \leq i$.

2. **Obtaining the LB**. The LB is obtained by supposing than the entity values are set to the ending points of their acceptance areas. Thus, as in the case of the qualitative sum UB, three cases can occur:

- If the absorption rule is satisfied, the LB will be δ_i or δ_j by depending on $\Delta_i \geq \Delta_j$ or $\Delta_j > \Delta_i$ respectively

- If δ_i, or δ_j when $\Delta_j > \Delta_i$, is the first defined acceptance area, then δ_i, or δ_j respectively, is the LB
- Otherwise, a recursive backward search among the defined qualitative areas is applied. It is aimed at finding the qualitative area δ_k that satisfies the relationship $\Delta_j \leq \delta_{i-1} + \delta_{i-2} + \ldots + \delta_k$ ($\Delta_i \leq \delta_{j-1} + \delta_{j-2} + \ldots + \delta_k$).

3. **Building the result**. It builds the list of qualitative areas from the LB to the UB

The remaining issue is to know which operation (i.e. qualitative sum or qualitative difference) solves the reasoning process of any magnitude. That choice mainly depends on two different aspects: (1) the sign of the magnitude values (i.e. positive and/or negative) and (2) the *physical* definition of the magnitude.Hence, the resulting disjunction of qualitative areas for the inferred relationship will be built from different ways to obtain the LB and UB.

Regarding the CIP, it can be formalized as a Constraint Satisfaction Problem (CSP) since knowledge about relationships between entities is often given in the form of constraints. Note that CSP is *consistent* if it has a *solution*. Consequently, a path consistency algorithm can be used as a heuristic test for if the defined constraint network is *consistent* [1], and, therefore, if the CSP has a *solution*.

4   Conclusions

Physical space and its properties play essential roles in all sorts of actions and decisions. As a result, the ability to reason in and about space is crucial for any system involved in theses actions and decisions. So, given that qualitative representations are one source of flexibility in commonsense reasoning, they are more stable than quantitative representations, and provide a level of description that can be more easily matched and reasoned with. Although positional information has been the starting point, the spatial reasoning can imply different physical properties by requiring higher dimensional descriptions. For that reason, both magnitude representation and reasoning process have been analysed by leading to a general algorithm which solves the representation and the inference process of any qualitative model based on intervals in n dimensions. Actually, the general framework presented in this paper allows investigators from other disciplines to easily incorporate the required reasoning techniques into their methods. In addition, focused on assessing the approach's performance, it has been instanced to two different qualitative models: (1) naming distance, and (2) qualitative velocity by obtaining the same results to the presented ones in [6] [7], respectively.

Acknowledgements. The authors want to thank the support received from Generalitat Valenciana (PROMETEO/2009/052).

References

1. Allen, J.: Maintaining knowledge about temporal intervals. Communications of the ACM 26, 832–843 (1983)
2. Cioaca, E., Linnebank, F., Bredeweg, B., Salles, P.: A qualitative reasoning model of algal bloom in the danube delta biosphere reserve (ddbr). Ecological Informatics 4(5-6), 282–298 (2009)
3. Clementini, E., Felice, P.D., Hernández, D.: Qualitative representation of positional information. Tech. rep., Technische Universität München (1995)
4. Clementini, E., Felice, P.D., Hernández, D.: Qualitative representation of positional information. Artificial Intelligence 95(2), 317–356 (1997)
5. Cohn, A., Hazarika, S.: Qualitative spatial representation and reasoning: An overview. Fundamenta Informaticae 46(1-2), 1–29 (2001)
6. Escrig, M., Toledo, F.: Qualitative Spatial Reasoning: Theory and Practice. In: Application to Robot Navigation. Frontiers in Artificial Intelligence and Applications, vol. 47, IOS Press (1998)
7. Escrig, M., Toledo, F.: Qualitative Velocity. In: Escrig, M.T., Toledo, F.J., Golobardes, E. (eds.) CCIA 2002. LNCS (LNAI), vol. 2504, Springer, Heidelberg (2002)
8. Freksa, C.: Using Orientation Information for Qualitative Spatial Reasoning. In: Frank, A.U., Formentini, U., Campari, I. (eds.) GIS 1992. LNCS, vol. 639, Springer, Heidelberg (1992)
9. Freksa, C., Zimmermann, K.: On the utilization of spatial structures for cognitively plausible and efficient reasoning. In: SMC, Chicago, USA, pp. 261–266 (1992)
10. Guerrin, F., Dumas, J.: Knowledge representation and qualitative simulation of salmon redd functioning. part i: qualitative modeling and simulation. Biosystems 59(2), 75–84 (2001)
11. Hernández, D.: Qualitative Representation of Spatial Knowledge. LNCS, vol. 804. Springer, Heidelberg (1994)
12. Holzmann, C.: Rule-based reasoning about qualitative spatiotemporal relations. In: 5th MPAC, Newport Beach, California, pp. 49–54 (2007)
13. Jong, J.: Qualitative reasoning about distances and directions in geographic space. Ph.D. thesis, Dept. of Surveying Engineering, University of Maine (1994)
14. King, R., Garrett, S., Coghill, G.: On the use of qualitative reasoning to simulate and identify metabolic pathways. Bioinformatics 21(9), 2017–2026 (2005)
15. Knauff, M., Strube, G., Jola, C., Rauh, R., Schlieder, C.: The psychological validity of qualitative spatial reasoning in one dimension. Spatial Cognition & Computation 4(2), 167–188 (2004)
16. Levinson, S.: Space in Language and Cognition. In: Explorations in Cognitive Diversity. Cambridge University Press, United Kingdom (2003)
17. Liu, H.: A fuzzy qualitative framework for connecting robot qualitative and quantitative representations. IEEE Trans. on Fuzzy Systems 16(6), 1522–1530 (2008)
18. Salles, P., Bredeweg, B.: Modelling population and community dynamics with qualitative reasoning. Ecological Modelling 195(1-2), 114–128 (2006)
19. van de Weghe, N., Cohn, A., de Tré, G., de Maeyer, P.: A qualitative trajectory calculus as a basis for representing moving objects in geographical information systems. Control and Cybernetics 35(1), 97-119 (2006)
20. Werthner, H.: Qualitative Reasoning. Modeling and the generation of behaviour. Springer (1994)

Approach of Genetic Algorithms with Grouping into Species Optimized with Predator-Prey Method for Solving Multimodal Problems

Pablo Seoane, Marcos Gestal, Julián Dorado, J. Ramón Rabuñal,
and Daniel Rivero

Abstract. Over recent years, Genetic Algorithms have proven to be an appropriate tool for solving certain problems. However, it does not matter if the search space has several valid solutions, as their classic approach is insufficient. To this end, the idea of dividing the individuals into species has been successfully raised. However, this solution is not free of drawbacks, such as the emergence of redundant species, overlapping or performance degradation by significantly increasing the number of individuals to be evaluated. This paper presents the implementation of a method based on the predator-prey technique, with the aim of providing a solution to the problem, as well as a number of examples to prove its effectiveness.

Keywords: Genetic Algorithms, Multimodal Problems, Species Evaluation, Predator-Prey Approach.

1 Introduction

In the Genetic Algorithms [1] there is a simulation of a population of individuals that evolves until reaching a solution within a given search space. With the aim of achieving various solutions in a multimodal environment, that is, with several optimal valid values, a division of the population into species is carried out, so that each can specialize in a solution of the problem. Using this technique, the obtained results were satisfactory [2]. However, it has some drawbacks. Due to the fact that new species are created with each generation, and therefore new individuals, the population grows exponentially. The immediate consequences are an increased

Pablo Seoane · Marcos Gestal · Julián Dorado · J. Ramón Rabuñal · Daniel Rivero
Univ. A Coruña, Fac. Informática, Campus Elviña, 15071, A Coruña, Spain
e-mail: `pablo.amor@udc.es, mgestal@udc.es, julian@udc.es,`
`juanra@udc.es, drivero@udc.es`

S. Omatu et al. (Eds.): Distributed Computing and Artificial Intelligence, AISC 151, pp. 689–699.
springerlink.com   © Springer-Verlag Berlin Heidelberg 2012

consumption of computational resources, as well as a slowing down of the whole system.

To avoid the created drawbacks, the concept of predator-prey [3] is introduced in the system applying it to species in the area. The technique is firstly designed to distribute individuals behaving as prey randomly in an area and then do the same with some individuals called predators, which, according to some rules, deal with deleting some of the prey in the neighborhood. The deleted individual is replaced by another, obtained as a result of a mutation of a randomly chosen nearby prey and predators move around, looking for a new victim.

This technique could be used in a similar way when dealing with grouping into species, so that species could compete against each other, as preys or predators, according to some previously defined rules. The role is assigned dynamically, after the meeting of the two species. In the same way, the species which enhances the values of the chosen rule is helped to continue its evolution, whereas the individuals of the other species disappear. The species and individuals that are considered dispensable for obtaining solutions will be removed from the system.

The predator-prey method is aimed at overcoming the limitations that arise when applying the technique of the Genetic Algorithms grouped into species to a multimodal problem, obtaining the best results provided by the grouping of species, but using the fewest elements possible, so that the species could maintain their numbers or even suffer losses during the development of the method and as a result the total amount of individuals decreases. Hence, the method implementation is optimized.

A first approximation is performed to test the system using a multimodal Rastrigin function. This is a preliminary study whose aim will be to apply the solution to complex problems.

2 Genetic Algorithms

The Genetic Algorithms are adaptive methods, generally used in search problems and parameter optimization, based on the principle of sexual reproduction and survival of the fittest. The development of Genetic Algorithms is largely due to John Holland, a researcher at the University of Michigan.

To obtain the solution to a problem, we start from an initial set of individuals, called population, generated randomly. Each of these individuals, coded similarly to chromosomes, represents a possible solution to the problem. These individuals will evolve according to the mechanism of natural selection proposed by Darwin [4]. Each of these individuals will be associated with their fitness value, which quantifies its validity as a solution to the problem, obtained by means of an objective function. Depending on this value, each solution will have more or fewer possibilities of reproduction. Once evolved, not only the best individuals are maintained in the population, but also those which are not so good and even the worst ones, as diversity should be maintained. In addition, mutations of these chromosomes may be carried out with the aim of maintaining the diversity of the population. The reason of maintaining the diversity is that, when facing a changing situation, individuals which at a given moment are good, at another may cease to be like this and, if so, other individuals would be needed, individuals which would be able to respond well to the situation. Consequently, diversity is necessary to obtain a valid solution at any time.

3 Techniques of Grouping into Species with Genetic Algorithms

The Genetic Algorithm-based approach is able to obtain a good approximation to the solution of the problem to be solved within a few generations. However, the tendency of finding a single solution becomes a disadvantage when dealing with multimodal problems, since in such cases it is preferred to find several solutions.

One of the options to try to solve this drawback involves using the technique of grouping into species which, broadly speaking consists of grouping the initial population of individuals into classes with similar characteristics [2]. Hence, the aim is that each group will be specialized in a particular area of search space. Thus, each species will tend to find an existing solution in its area, other than those provided by other species. This is an attempt of modeling the species distribution of individuals in the natural environment, and their evolution separately. For example, there are individuals adapted to live in cold areas, others in dry or hot ones, etc. Each group manages life in a given environment, adapting to this end specific characteristics that differentiate it from the other groups.

However, this technique has its drawbacks. Thus, certain conditions are required for proper operation, conditions which are not usually obtained in the initial distribution of the problem. For example, it would be recommended that the population should be evenly distributed throughout the search space, and moreover, that groups should be well distributed and in accordance with the number of solutions to the problem. If there are no such characteristics, there may be unexplored areas in contrast to others that are highly explored and in which, depending on how the groups are formed, several species can coexist.

To overcome these drawbacks, we use the crossing of individuals from different species through several generations. In doing so, the offspring resulting from these crosses mix knowledge of their predecessors' species and there emerges the possibility of creating a new species in an area different from their parents'. In this way, stagnation of species is avoided, new areas are explored and new knowledge emerges, that is, diversity is achieved in the environment. Once again, individuals' behavior is being modeled in their natural environment, whereas migration or expulsion of individuals takes place and if they find compatible individuals of other groups, they end up creating new species.

For the implementation of these techniques an initial population is created. The overall process - starting from creating the initial population - is to carry out successive iterations in which the following steps are required:

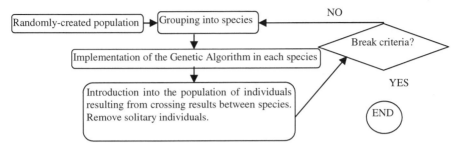

Fig. 1 Operation diagram of grouping into species

One of the main criteria is to check whether the number of iterations, also called evolutions, reaches the maximum number allowed; another is to check whether the population reaches its upper bound of individuals. If either of these conditions is met, the implementation of the algorithm is concluded. Another possible break criterion is that the error value of individuals had fallen below a threshold set in advance.

For the second step of the algorithm, there is no standard way to divide the population of the Genetic Algorithms into species. To solve this problem, techniques of unsupervised grouping of individuals are employed, as there is no a priori knowledge but simply an input data set. The classification is carried out according to some specific parameters of each grouping algorithm type. Two of these techniques that could be used are the Adaptive Method [5], which is a simple and efficient incremental heuristic method using only two parameters and Batchelor and Wilkins' algorithm or the Maximum Distance algorithm [6]. In this case, we also deal with an incremental heuristic method, but it uses a single parameter.

4 Predator-Prey Interaction Method

Biologically, predation occurs when one of the animals (the predator) devours another living animal (the prey) to use the energy and nutrients in the body of the prey for growth, maintenance or reproduction. Using the predator-prey idea, a model was proposed, adapting the predator-prey concept to Genetic Algorithms [7]. There are software projects dedicated to designing and analyzing predator-prey models [8].

The original operation consists of the fact that each individual representing a solution in the genetic population plays the role of prey and of the fact that other individuals in the system play the role of predators, choosing their prey according to the objective function and the fitness of each prey. The method imitates the natural phenomenon in which a predator eliminates the weaker prey, which means that a predator eliminates the most unfit individual in the environment, which corresponds to the worst value obtained in such individuals with the objective function.

To implement this idea a network was proposed, in which the prey are randomly distributed at each node and wherein one or more predators are also placed randomly at some of the nodes.

From that moment, each predator evaluates all prey in its area and deletes the prey corresponding to the worst objective value. Then, a nearby prey is chosen and mutated. The mutated individual replaces the deleted prey and the predator moves to one of the neighboring nodes. This procedure is followed for all predators.

The use of this method is widely associated with the multi-objective Genetic Algorithms [9]. These are Genetic Algorithms which are aimed at finding solutions that optimize several objective functions simultaneously. In such a case, one or more predators are created for each of the objective functions involved; or even a predator that takes into account several objective functions involved in the system is created. The initial diagram regarding the operation of the classical algorithm of the predator-prey approach had already taken this aspect into account. Although modifications and optimizations have been made since its inception, the diagram we have broadly followed is classical [3], as detailed below.

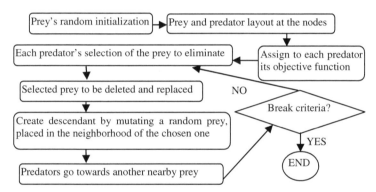

Fig. 2 Operation diagram of the predator-prey approach

The break criterion is either a maximum number of generations or the fact that the objective functions to be optimized have an error below a certain threshold.

Several set-up changes can be performed, as using more predators for every objective function, mutate the best neighbor after predation or move the predator to the box of the best neighbor, instead of moving it randomly. In some cases [9], the outcome of the overall population is improved.

5 Predator-Prey Application in Genetic Algorithms with Grouping into Species

In spite of the fact that the Genetic Algorithms provide optimal solutions to many problems, they have some drawbacks when used to find several solutions in scenarios with multiple optimal points. With the aim of trying to overcome these drawbacks, other solutions were searched for. Among these solutions we mention the grouping the population of individuals into species.

However, it is verifiable that, in the implementations of grouping into species carried out, there are some drawbacks which arise due to the fact that both the number of species and number of individuals tend to continue to grow indefinitely throughout the different evolutions [2]. Such an increased number of individuals and species leads to a continuous increase of the necessary computational resources.

In order to optimize the number of elements used in computing, we suggest applying the benefits of the predator-prey approach to the system made up of Genetic Algorithms grouped into species. This new system is aimed at reducing the number of elements involved in computing, allowing the predation of individuals and, if the choice of which ones should be deleted is made correctly, maintaining similar results to those obtained without the predator-prey approach.

In order to apply the predator-prey approach to the developed system, some changes are necessary, using the main idea of the method as base. Thus, the elements involved are as follows:

- Space for action. This refers to the search space itself, where individuals are distributed.
- Prey. Any species in the system can become prey, being devoured by a predator, which would mean the removal of the species and individuals within it.
- Predator. Any species in the system can become predatory. A prey species devours other prey species. As a benefit, individuals of the species that devours, as well as the species itself, will be able to continue to evolve.
- Objective Function. In this approach, the system will use only the objective function that the Genetic Algorithm employs in each case to calculate the fitness of individuals if necessary.
- Interaction criterion. While in the classical predator-prey algorithm the predator devoured the worst prey in the neighborhood, in this case we need to know the criterion involved so that a species can try to devour another. An example in this sense would be when the species is close enough to the area of another species.
- Role determination criterion. Besides the classical players of the predator-prey approach, it is necessary to define a new concept, the winning rules. In the classical algorithm, some individuals behaved as prey and others as predators. In this case, the same entity – a species – can behave as predator on some occasions and as prey on others. It is necessary, therefore, to define a rule specifying, when appropriate, which of the species will behave as predator and which as prey, and therefore to know which species will survive (predator) and which will be deleted (prey). To this end, the concept of role determination criterion is defined. The role determination criterion refers to a series of algorithms by which it is decided which species would behave as predator and which as prey. The direct consequence is that the predator species will devour the prey species, the latter disappearing from the system.

An example in this regard would be that the predator species (and therefore the survivor) is the one whose individual has the best possible fitness.

The general operation is described below. Once the grouping of individuals into species is performed, and before applying the Genetic Algorithm to each of them, the predator-prey algorithm is applied as follows. The flow chart of the method is detailed in Figure 3.

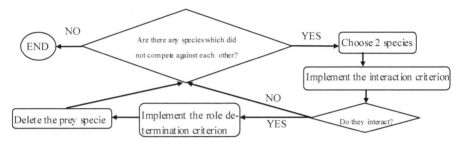

Fig. 3 Flow chart of the predator-prey approach and grouping into species

As shown, two species are chosen from the set created by grouping into species. Then, the interaction criterion is applied. If the species meet the criterion so that the predator-prey approach could be applied, we continue with the next step; otherwise, we should check again whether there are still species that do not meet the interaction criterion. If the species interact, the next step is to apply the role determination criterion to see the role each species assumes, predator or prey. Finally, the predator species is maintained and the prey species is deleted. Individuals belonging to the prey species are marked for removal. The process is repeated until there are no more species that could compete with each other.

To be able to work with this approach, a type of rules should be defined. There will be a more in depth discussion on interaction criterion, providing some examples to decide when two species should behave as predator and prey, and on role determination criterion to indicate how the decision is made regarding which species should be prey and which predator.

1) Examples Interaction criteria. By grouping the individuals of the population into species, it is not known whether they should interact so that one becomes a prey and the other predator. Thus, the interaction criteria are used to decide when two species in the environment should establish a predator-prey relationship.

If the interaction criterion implemented for two species in the environment decides which species behave as predator and prey, we continue with the next step where it is decided which will behave as prey and which as predator. If the criterion is not met, the search is continued among the total number of species, until finding a pair to which the interaction criterion has not been applied yet. If there is no such pair of species, the proceedings will be completed.

An example of interaction criterion is making two species face each other if the distance between them is below a certain threshold.

In order to observe an example of interaction criterion of distance between species, a hypothetical scenario is shown in Figure 4, where a decision is made regarding which species will interact with the "E1" specie. With a threshold value of 5, only the "E3" specie is below the threshold, so this would be the only species that E1 interacts with.

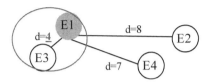

Fig. 4 Example of interaction criterion of distance between centroids

2) Examples Role determination criteria. Unlike the original algorithm of the predator-prey approach in which the roles of individuals are static, as there are prey and predators from the beginning in the system and there will be as long as they are used, in this case we are dealing with a dynamic allocation, since any species can behave as prey or predator.

A role determination criterion is an algorithm that decides which of the two interacting species survives and behaves as a predator and which one behaves as prey and is therefore deleted. These rules can be usually generalized to any number of species.

A possible Role Determination Criterion is that of the Best Individual. Using this criterion, the winner is the species that has the best individual out of the two species. This rule could be generalized to choose a predator from any number of species. In Figure 5 it is shown an example where the species with the best individual is the predator and the one with the worst individual is the prey, being thus deleted. In this example, the greater is the fitness of the individual, the better it is. E3 is the predator specie.

Fig. 5 Example of role determination criterion - the Best Individual

6 Tests

To check the performance of the predator-prey approach, we use a Genetic Algorithm with Grouping into Species and the corresponding algorithms of the method are applied throughout each evolution.

Before starting the application, we need on the one hand to choose the problem to be solved and on the other hand, considering the problem in question, to select the parameters to be used both to group the individuals into species and for the Genetic Algorithm to be applied to each species in each evolution step.

Hence for a first approximation, the Rastrigin function [10] is chosen, which is widely used to show the effectiveness and study of the multimodal problem solving methods. This is a function that has many local minimum and maximum values. In this test are sought maximums.

To work with the predator-prey approach, the values previously used in the Genetic Algorithm as well as the Grouping one are set. These are values that obtain satisfactory results in the system of Genetic Algorithms with Grouping into Species. Once set, we compare the results obtained when using or not using the predator-prey approach.

To show the results of applying the predator-prey approach, we implement the distance between species as interaction criterion and the best individual as role determination criterion. The results are compared to those obtained after running the system without using this method.

Once implemented the predator-prey system, it is run. Are only required 10 evolutions to show good results. As follows we present the solutions in the contour plot of the Rastrigin function. The red dots represent the best individuals of each species (the individuals with the least error). Red and green contours are maximum.

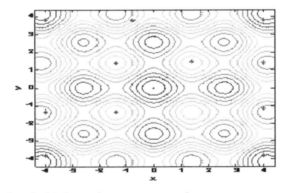

Fig. 6 Solutions found with the predator-prey approach

To obtaining such results, practically identical to those obtained without applying the predator-prey approach, it was necessary to employ a number of species and individuals at all times lower than those employed in the execution without the predator-prey approach, as shown in Figure 7. It is noted that in this case the trend in the number of individuals is increasing, but to a much lesser extent than when the method is not used, since in this case there are deletions of entire species. In Figure 7, we can see the decrease in the number of species and individuals throughout the various evolutions after using the predator-prey approach and in Figure 8 the error evolution.

Evolution	Without predator-prey		With predator-prey	
	Species	Individuals	Species	Individuals
1	15	70	12	49
5	22	129	16	86
10	17	210	11	123

Fig. 7 Increased number of individuals and species

Evolution	Without predator-prey		With predator-prey	
	Average	Individual	Average	Individual
1	2,5	1,9	2	1,4
5	2,6	1,3	1,7	1,4
10	2	1,3	1,7	1,4

Fig. 8 Evolution of the error and of the best individual's error

Note that the error persists, and the number of individuals and species is much lower, decreasing approximately 40%. The number of solutions reached in this case is slightly lower than in the case of not using the predator-prey approach.

Therefore, the objectives of reducing the number of species and individuals under study are fulfilled, maintaining the error and optimizing the number of necessary resources. Similarly, we found a large number of solutions in any of the two executed cases, approaching to the one found without applying the predator-prey approach.

7 Conclusions

The tests conducted showed that efficiency Genetic Algorithm with Grouping into Species for multimodal problems was improved. Only a few evolutions were necessary to verify the benefits of the application of predator-prey.

In general, the results were maintained or even improved since species in which individuals did not provide good solutions and increased the error were deleted of the environment.

After this preliminary study the next step is to perform extensive testing in variable selection problems. Specifically, a chemometric problem with previous results in different approaches related with the work performed [2].

Bibliography

1. Goldberg, D.E.: Genetic Algorithms in Search, Optimization and Machine Learning. Addison-Wesley, Reading (1989)
2. Gestal Pose, M.: Computación evolutiva para el proceso de selección de variables en espacios de búsqueda multimodales. PhD Thesis (2010)
3. Laumanns, M., Rudolph, G., Schwefel, H.P.: A spatial predator-prey approach to multi-objective optimization: A preliminary study. In: Proceedings of the Parallel Problem Solving from Nature (1998)

4. Darwin, C.: On the Origin of Species by Means of Natural Selection (1859)
5. Cortijo Bon, F.J.: Técnicas no Supervisadas: Métodos de Agrupamiento (2001)
6. Batchelor, B.G., Wilkins, B.R.: Method for location of clusters of patterns to initialise a learning machine. Electronic Letters, 481–483 (1969)
7. Chen, H., Li, M., Chen, X.: A Predator-Prey Cellular Genetic Algorithm for Dynamic Optimization Problems. In: Information Engineering and Computer Science, ICIECS (2010)
8. Blom, H., Küch, C., Losemann, K.: PEPPA: a project for evolutionary predator prey algorithms. In: GECCO 2009 (2009)
9. Kalyanmoy, D., Bhaskara, U.: Investigating predator-Prey Algorithms for Multi-Objective Optimization. Department of Mechanical Engineering Indian Institute of Technology Kanpur (2005)
10. Torn, A., Zilinskas, A.: Global Optimizacion. Springer (1989)

BTW: A New Distance Metric for Classification

Julio Revilla Ocejo and Evaristo Kahoraho Bukubiye

Abstract. In this paper, BTW, a new method for similar case search, is presented. The main objective is to optimize the metrics employed in classical approaches in order to obtain an intense compression in the data and a deterministic real-time behavior; and without compromising the performance of the classification task. BTW tries to conjugate the best of three well-known techniques: Nearest Neighbor, Fisher discriminant and optimization.

Keywords: Data fusion, metrics, K-NN, Fisher discriminant, classification.

1 Introduction

Distance based classification tasks rely heavily on the metrics employed in them; data coming from very different sources, ranges, etc. must be seamlessly merged into a final decision. BTW is an algorithm based on a new ∞-NN distance measure that replaces the euclidean metric by a Fisher-oriented one, and optimizes it selecting the best neighborhood radius. Real-time behavior, data compression and the ability to cope with huge databases are its innovative contributions.

The rest of this paper is organized as follows: in section 2 it is described the traditional approaches of classification solving. Sections 3 and 4 provide details of the proposed BTW algorithm, and how to employ it in real-time tasks. Finally, in section 5, its results are compared with their close related algorithms.

2 Metrics, the Nucleus of Classification

A new case is characterized by a set of numerical, ordinal and/or nominal values formed by its P attributes, which is supposed to belong to one of the J possible classes $\{\ y_j\ \}$ previously defined; the target of the classification algorithm is

Julio Revilla Ocejo · Evaristo Kahoraho Bukubiye
Dept. of Industrial Technologies, Area of Automation
Faculty of Engineering, University of Deusto
48007 Bilbao, Spain
e-mail: jrevilla@deusto.es, kahoraho@deusto.es

S. Omatu et al. (Eds.): Distributed Computing and Artificial Intelligence, AISC 151, pp. 701–708.
springerlink.com　　　　　　　　　　　　　　　　© Springer-Verlag Berlin Heidelberg 2012

assigning it to the correct one. The classification algorithms based on empirical data have at their disposal N test cases, consisting of the values of the attributes and their classes $\{\mathbf{x}_n, y_n\}$, $n = 1,..., N$. The number of items per class is termed N_j.

2.1 Nearest Neighbor Methods

The search for the nearest neighbors (NN) was the base of the first non-parametric methods used in Case Based Reasoning (CBR). Despite the significant advances that have been done in this area, they continue providing good performance [1].

1-NN algorithm [2] [3] assigns the new cases the same class as its nearest neighbor, according to a metric defined in the attributes space. This technique partitions the attribute space by creating non-linear hypersurfaces.

k-NN method is similar to 1-NN, it differs in that it chooses the class most often repeated among the k nearest cases. The optimal value of k depends on the problem and it should be chosen, by means of a thorough search, for each scenario. The main advantage of k-NN is that the boundaries among classes are less irregular that in the 1-NN method.

2.2 Metrics in the Nearest Neighbor World

To calculate what the nearest neighbor is, some kind of "measure of dissimilarity" must be used. The most habitual is the euclidean metric:

$$d^2(\mathbf{x}_i, \mathbf{x}_m) = (\mathbf{x}_i - \mathbf{x}_m)^T \, \Sigma \, (\mathbf{x}_i - \mathbf{x}_m) \tag{1}$$

where $d^2(\mathbf{x}_i, \mathbf{x}_m)$ is the distance between cases i and m, $\mathbf{x}_i$ is the vector of attributes for the ith case and Σ is a symmetric, positive, semidefinite matrix.

This metric possesses several drawbacks:

- If an identity matrix is chosen for Σ, attributes with large values dominate the metric; but they do not need to be the most relevant in classifying a case. Normalizing all attributes could solve the problem (though the ideal solutions would be those that weighted attributes in proportion to their predictive power).
- Prediction ability degrades exponentially as the dimensionality of the attribute space grows. This phenomenon is termed as "the curse of dimensionality".
- The number of parameters to be estimated (or optimize) in the dense Σ matrix is very large; it grows according to $O(P^2)$.

Multiple researchers have tackled these difficulties. Well-known solutions are the reduction in dimensionality by Principal Component Analysis (PCA) [4] or the optimization of weights applied to each attribute proposed by Lowe [5] (using a diagonal matrix with adjustable coefficients for Σ). Although Lowe's method represents an important step trying to solve these three problems, it has an important shortcoming: it only generates isometric curves parallel to the coordinated axes (Fig. 1a). However, in those cases where a strong correlation between attributes exists, the isometric curves should be ellipses in inclined axes (Fig.1b).

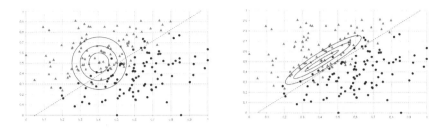

Fig. 1 a) Lowe's isometric curves,   b) Isometric curves not parallel to the coordinate axes

The study of more elaborate metrics has received much attention in recent years. Hastie and Tibshirani proposed the DANN metric [6] [7] with the following Σ:

$$\Sigma = \mathbf{W}^{-1/2}\left(\mathbf{W}^{-1/2}\mathbf{B}\ \mathbf{W}^{-1/2} + \varepsilon\ \mathbf{I}\right)\mathbf{W}^{-1/2} \qquad (2)$$

where $\mathbf{W}$ is the weighted covariance matrix (P_xP) of the attributes of each class (locally estimated); and $\mathbf{B}$ is the covariance matrix between the centers of the classes.

More recently, J.Goldberger et al.[8] introduced an algorithm, named NCA, which optimizes a Mahalanobis matrix, trying to improve the k-NN classification.

Bar-Hillel et al.[9] have proposed to create a Mahalanobis-like metric based on a semi-supervised classification (it only distinguishes between similar and dissimilar cases) which they called RCA.

Weinberger et al.[10] introduced the LMNN algorithm, which increases the margin of separation of cases from different classes by means of optimizing a metric based on a complete Mahalanobis matrix.

3   The BTW Algorithm

In reference [1] it is stated that for classification methods of type Nearest Neighbor: "An area that requires further study is in fast data-based methods for choosing distance appropriate measures, appropriate variable selection and the number of neighbors".

In this line of research, we have developed a new algorithm named BTW. A revision of the DANN algorithm is proposed so that the metric presented here, instead of local, becomes global, transforming the attributes according to the LDA approach. BTW algorithm proceeds in three stages:

- First, a coordinate transform that projects the values of attributes to the new set of axes is introduced (to improve the separation between classes).
- Then, a new function to move from the concept of "distance" to "similarity" is provided. The idea is to extend the strategy of the algorithm k-NN to ∞-NN by inversely weighting the influence of each case to the distance to the new one.
- Finally, there exists an optimization phase that calculates the optimal weights for the metric in the new axes. This metric is based on a diagonal matrix.

3.1 An Oriented Metric

The first step is to orient the main axis of the metric in the direction connecting the center of both classes, according to the LDA approach. Following the DANN metric, now applied to the whole space of attributes, the proposed Σ would be:

$$\Sigma = \mathbf{W}^{-1/2} \left(\mathbf{W}^{-1/2} \, \mathbf{B} \, \mathbf{W}^{-1/2} \right) \mathbf{W}^{-1/2} \tag{3}$$

$$\mathbf{W} = \frac{\sum\limits_{j=1}^{J} \sum\limits_{n=1}^{N_j} \pi_n \left(\mathbf{x}_n - \mathbf{m}_j \right) \left(\mathbf{x}_n - \mathbf{m}_j \right)^T}{\sum\limits_{n=1}^{N} \pi_n} \qquad \mathbf{B} = \frac{\sum\limits_{j=1}^{J} \left(\mathbf{m}_j - \mathbf{m} \right) \left(\mathbf{m}_j - \mathbf{m} \right)^T \left(\sum\limits_{n=1}^{N_j} \pi_n \right)}{\sum\limits_{j=1}^{J} \sum\limits_{n=1}^{N_j} \pi_n} \tag{4}$$

where $\mathbf{W}$ is the weighted sum of the covariance matrices of the attributes; $\mathbf{B}$ is the covariance matrix between the means of two classes; π_n is the a priori probability of a certain case; $\mathbf{m}_j$ is the vector of attribute means of the class j (globally estimated); and $\mathbf{m}$ is the average of attribute means vectors for the two classes.

Our main interest is in the Σ's direction that provides the maximum discrimination between classes; thus, we proceed with a PCA on the matrix Σ, looking for the vector with the largest singular value. As Σ is real, square, symmetrical and has a large null subspace (of dimension P-1 for two classes), a Singular Value Decomposition (SVD) algorithm will be used to carry out the calculation: $\Sigma = \mathbf{U} \, \sigma \, \mathbf{V}^T$.

The next step is to project the original attribute values onto the axes defined by $\mathbf{U}$:

$$\mathbf{z} = \mathbf{U}^{-1} . \mathbf{x} = \mathbf{U}^T . \mathbf{x} \tag{5}$$

where $\mathbf{z}$ is the attributes vector in the new coordinates after this transformation.

3.2 The Similarity Function

Once data are represented in the new reference, it is proposed to classify a new case by means of a new algorithm of type ∞-NN; that is, taking into account all the cases, but weighting them inversely to their distance to the case to be classified.

Quantitatively, a multidimensional Gaussian probability density function, centered at zero and with standard deviation adjusted by the weights σ_p is chosen.

Thus, the "similarity" $s(z_i, z_m)$ between the cases i and m will be defined by:

$$s\left(\mathbf{z}_i, \mathbf{z}_m \right) = e^{-d^2 \left(\mathbf{z}_i, \mathbf{z}_m \right)} = e^{-\left(\mathbf{z}_i - \mathbf{z}_m \right)^T \Sigma_{diag} \left(\mathbf{z}_i - \mathbf{z}_m \right)} \tag{6}$$

where Σ_{diag} is the same diagonal matrix as in Lowe metric. It contains the weights σ_p, $1 \leq p \leq P$ associated to each new dimension.

Nearby cases will present a similarity close to unity, while for those located far away the similarity will be approximately zero (they will not contribute to decide the resulting class for the case). The final decision to classify the case i in one or another class is made based on the ratio expressed in Eq. 7; thus, for class 1:

$$\mathrm{pr}\left(y_1 \mid \mathbf{z}_i\right) = \frac{\sum\limits_{n=1}^{N_1} s\left(\mathbf{z}_i, \mathbf{z}_n\right)}{\sum\limits_{n=1}^{N_1} s\left(\mathbf{z}_i, \mathbf{z}_n\right) + \sum\limits_{n=1}^{N_2} s\left(\mathbf{z}_i, \mathbf{z}_n\right)} \tag{7}$$

where all the similarities within the N_1 cases of class 1 are summed up and divided by the sum of the similarities within both classes.

3.3 Optimizing the Weights

Based on the probabilities defined in Eq. 9, an error function is proposed:

$$E = \sum_{n=1}^{N} \sum_{j=1}^{J} \left(c_{n,j} - \mathrm{pr}\left(y_j \mid \mathbf{z}_n\right)\right)^2, \quad J = 2 \tag{8}$$

where: $c_{n,j}$ is 1 if the nth case belongs to class j, 0 otherwise; and $\mathrm{pr}\,(y_j \mid \mathbf{z}_n)$ is the calculated probability the nth case belongs to class j.

The error function implicitly depends on the parameters σ_p (see Eqs. 6 to 8). Minimizing the error function involves finding the optimal weights σ_p that maximize the probability of correctly classifying all the cases.

Our first optimization attempt employed the Levenberg-Marquardt algorithm; the results were not as satisfactory as expected, after a thorough analysis, it was shown that in all scenarios the weights for attributes of the null subspace only contributed to worsen the quality of class prediction. Eliminating all the dimensions but the first, improved the classification performance. In a second attempt, and now reducing the problem to a single attribute, it was decided to implement a thorough examination of its entire range of variation. To decrease the number of function evaluations the golden section exhaustive search was employed, preceded by a routine that narrows the intervals producing the minimums.

To prevent over-learning in the estimation of the parameters, we proceeded to employ a "bootstrap" technique[11]. Two hundred minimizations were performed, with randomly chosen cases, to estimate the best weight for this unique dimension.

4 The Use of BTW Algorithm

The use of a single attribute can significantly accelerate the search of neighbors. It requires an initial ("offline") data preconditioning:

1. Transform the original attributes of the training data to the new directions according to the proposed transformation (Eq. 5). Retain only the first attribute for each case.

2. Calculate/provide the optimal weight (radius) to improve classification.
3. Sort numerically the cases according to the value of their first attribute and store them in an array.

At runtime, it will be necessary:

1. Take/measure the attributes of the new case and apply the transform to them (by Eq. 5). Retain only the first attribute.
2. Using a bisection search technique in the sorted array, search for the two elements that are closer to the current case. The index of those items will be saved as a "start" and "end" index of an integer range.
3. Read out the elements in array at the start and end indexes. Calculate which of them is closest to the current case (this can be easily evaluated by comparing, in absolute value, the difference between the attribute of the current case and the attribute of both selected cases). Choose the closest, and adjust either the "start" index (one position back) or "end" index (one position forward) according.
4. By Eq. 6 calculate the similarity between the new case and the chosen element. Recalculate the probability of belonging to both classes.
5. Tentatively, consider that all the remaining cases in the array belong to the actual minority class and that they possess the same degree of similarity that the last found (this is clearly an upper bound). In this hypothetical scenario: would the class that is currently the most probable be changed by this last decision?

   • The response is 'no': the search is over; although all the remaining cases belonged to the other class, it would be insufficient to change the decision.
   • The response is 'yes': the algorithm iterates by moving again to step 3.

Although the case database had a few dozens or hundreds of thousands of cases, in a very few iterations the algorithm finds the desired classification (without the need to calculate the distance to every case in the case database, as it is usual in the Nearest Neighbor algorithms).

The only aspect that depends on the size of the database is to search in the sorted array and it has a complexity of type $O(log\ N)$. The remaining operations have a similar cost to the calculation of the distance to another case.

5 Results and Comparative Analysis

To test the feasibility of the proposed algorithm, experiments with 11 traditional problems in the world of artificial intelligence have been conducted. Most of the problems have been obtained from the database of the UCI [12], other are common in classification research. The performance of the BTW classification algorithm, estimated by LOO techniques, has been compared to its related algorithms: LDA, QDA, 1-NN and k-NN.

Table 1 Classification performance for various algorithms

	J	P	Classification accuracy in %					
			LDA	QDA	1-NN	k-NN		BTW
Iris flowers	3	4	98,00	96,67	96,00	(19)	98,00	97,33
Australian Credit	2	14	86,09	80,72	79,86	(24)	85,94	86,52
Breast Cancer	2	9	95,99	95,85	95,14	(8)	97,00	96,85
German Credit	2	24	78,90	79,40	66.10	(16)	71,60	75,50
Glass	6	9	66,36	63,55	73,36	(2)	73,36	74,30
Heart	2	13	84,81	87,78	57,41	(25)	68,52	84,81
Image Segmentation	7	19	91,64	48,01	96,71	(1)	96,71	96,10
Satellite Image	6	36	82,45	84,55	89,35	(8)	90,65	87,85
Shuttle Control	7	9	94,53	93,65	99,88	(2)	99,88	97,84
Vehicle Silhouettes	4	18	79,78	91,37	70,45	(6)	72,81	82,39
WaveForms	3	21	88,33	93,67	79,00	(24)	83,00	90,33
	Average:		**86,08**	**83,20**	**83,01**		**85,22**	**88,17**

In Table 1, it can be seen that the BTW algorithm, with only one parameter, usually offers an intermediate performance between the LDA and k-NN, tending to approach to the best one (and in many cases surpassing both of them). Averaging the results in the 11 scenarios, the BTW algorithm provides the best mean.

This result is very promising since we have designed a Nearest Neighbor metric that not only performs well, but that fusions all the information of a case in a single attribute without severe loss of information.

6 Conclusions

The use of the BTW method in expert systems based on CBR provides not only a reliable classification of the new problem, but also lists the cases that have been used to make this decision. Compared to other techniques that do not supply this info (such as linear regression, LDA, neural nets ...), with BTW a human expert can judge the goodness of classification but, in addition, he/she can also get information or reuse the strategies/solutions that were employed to solve old problems.

The most important innovation offered by the BTW method (over other Nearest Neighbor algorithms) is its applicability to tasks that need deterministic execution times; furthermore, on average, the quality of forecasts improves both k-NN and LDA algorithms; and, due to the fusion of all the attributes in a unique value, the RAM storage requirements decrease drastically.

Other novelty of our approach is that ∞-NN do not need to establish an optimal number of neighbors. It only considers the cases of the database that are "near" to the new one.

Some aspects of this algorithm will require further study. Due to the singular directions in the attribute's space is different for each pair of classes; if the classification problem includes multiple classes, it is necessary to adopt an algorithm that allows working with them in pairs and then combine the results of all these predictions in a single result. J. Friedman [13] discusses several solutions for this aspect. The variant implemented in this work would be the equivalent of an election where the voting is over two rounds. For each pair of classes: find the probability that the new case belongs to each class; sum up these probabilities to a totalizing variable for each class; and finally, choose the two most likely classes, and between them decide the winner.

In addition, it will be necessary to explore in depth the behavior of the BTW algorithm when the original number of attributes of the problem is very large. To reduce a large number of values to only one could fail in some scenarios; in those cases, it will be necessary to study whether more dimensions could improve the classification performance.

References

[1] Michie, D., Spiegelhalter, D.J., Taylor, C.C. (eds.): Machine Learning, Neural and Statistical Classification, Hertfordshire. Ellis Horwood Series in Artificial Intelligence, p. 216
[2] Duda, R., Hart, P., Stork, D.: Pattern Classification, 2nd edn. John Wiley, New York (2000)
[3] Cover, T.M.: Rates of convergence for nearest neighbor procedures. In: Proc. of the Hawaii Inter. Conference on System Sciences, pp. 413–415. Western Periodicals, Honolulu (1968)
[4] Jolliffe, I.T.: Principal Component Analysis. Springer, New York (1986)
[5] Lowe, D.G.: Similarity metric learning for a variable-kernel classifier. Neural Computation 7, 72–85 (1995)
[6] Hastie, T., Tibshirani, R.: Discriminant adaptive nearest neighbor classification. IEEE Transactions on Pattern Analysis and Machine Intelligence 18, 607–616 (1996)
[7] Hastie, T., Tibshirani, R., Friedman, J.: The Elements of Statistical Learning: Data Mining, Inference, and Prediction, 2nd edn., New York. Springer Series in Statistics (2009)
[8] Goldberger, J., Roweis, S., Hinton, G., Salakhutdinov, R.: Neighbourhood Component Analysis. In: Advances in Neural Information Processing Systems, vol. 17, pp. 513–520 (2005)
[9] Bar-Hillel, A., Hertz, T., Shental, N., Weinshall, D.: Learning a Mahalanobis metric from equivalence constraints. Journal of Machine Learning Research 6(1), 937–965 (2006)
[10] Weinberger, K.Q., Saul, L.K.: Distance Metric Learning for Large Margin Nearest Neighbor Classification. Journal of Machine Learning Research 10 (2009)
[11] Efron, B., Tibshirani, R.: An Introduction to the Bootstrap. Chapman and Hall, N.York (1996)
[12] Blake, C.L., Merz, C.J.: UCI repository of machine learning databases (2010), http://www.ics.uci.edu/~mlearn/MLRepository.html
[13] Friedman, J.: Another Approach to Polychotomous Classification, Internal report (1996)

Using an Improved Differential Evolution Algorithm for Parameter Estimation to Simulate Glycolysis Pathway

Chuii Khim Chong, Mohd Saberi Mohamad, Safaai Deris, Shahir Shamsir, Afnizanfaizal Abdullah, Yee Wen Choon, Lian En Chai, and Sigeru Omatu

Abstract. This paper presents an improved Differential Evolution algorithm (IDE). It is aimed at improving its performance in estimating the relevant parameters for metabolic pathway data to simulate glycolysis pathway for yeast. Metabolic pathway data are expected to be of significant help in the development of efficient tools in kinetic modeling and parameter estimation platforms. Nonetheless, due to the noisy data and difficulty of the system in estimating myriad of parameters, many computation algorithms face obstacles and require longer computational time to estimate the relevant parameters. The IDE proposed in this paper is a hybrid of a Differential Evolution algorithm (DE) and a Kalman Filter (KF). The outcome of IDE is proven to be superior than a Genetic Algorithm (GA) and DE. The results of IDE from this experiment show estimated optimal kinetic parameters values, shorter computation time and better accuracy of simulated results compared to the other estimation algorithms.

Keywords: Parameter Estimation, Differential Evolution Algorithm, Kalman Filter, Simulation.

Chuii Khim Chong · Mohd Saberi Mohamad · Safaai Deris · Afnizanfaizal Abdullah · Yee Wen Choon · Lian En Chai
Artificial Intelligence and Bioinformatics Research Group, Faculty of Computer Science and Information Systems, Universiti Teknologi Malaysia, 81310 Skudai, Johor, Malaysia
e-mail: ckchong2@live.utm.my, saberi@utm.my, safaai@utm.my,
    afnizanfaizal@utm.my, ywchoon2@live.utm.my,
    lechai2@live.utm.my

Shahir Shamsir
Department of Biological Sciences, Faculty of Biosciences and Bioengineering, Universiti Teknologi Malaysia, 81310 UTM Skudai, Johor, Malaysia
e-mail: shahir@fbb.utm.my

Sigeru Omatu
Department of Electronics, Information and Communication Engineering, Osaka Institute of Technology, Osaka 535-8585, Japan
e-mail: omatu@rsh.oit.ac.jp

S. Omatu et al. (Eds.): Distributed Computing and Artificial Intelligence, AISC 151, pp. 709–716.
springerlink.com   © Springer-Verlag Berlin Heidelberg 2012

1 Introduction

Metabolic Engineering is an approach which alters host cells to enable them to generate a novel or enhance the production of compounds for industrial and medical use. Recent studies have focused on the mean of analysis by modifying the computer readable data from the biological process. Hence, scientists are able to simulate the process inside the cell by using mathematical modeling and studying the metabolic pathway. Glycolysis is the metabolic pathway which produces pyruvate from glucose and is studied in this paper.

Parameter estimation is one of the critical steps in constructing a mathematical model. Unfortunately, it possesses several problems; on one hand the existence of noisy data leads to low accuracy [1], and on the other hand the increasing number of unidentified parameters and equations in the model makes it a complex model [2]. Thus, we proposed IDE which is a hybrid of DE and KF, to solve the problems regarding the increasing number of unidentified parameters which consequently adds to the difficulty of the model in estimating the kinetic parameters, and the existence of noisy data that leads to low accuracy for estimated result.

The benefits of using DE are efficiency, high speed, straightforwardness, and ease of use as it contains only few control parameters [3]. The use of KF can improve DE's performance as it updates the population with Kalman gain value which handles noisy data [1]. DE has been implemented as a parameter estimation approach by Christophe Chassagnole et al. [4] and Moonchai Sompop et al. [5] to enhance the production of bacteriocin, aspartate, beer, and the simulation of the real process in cell by estimating the kinetic parameters and control parameters. Parameter estimation with DE is done without noisy data handling process. Noisy data occurs when the obtained results differ from each other and this is caused by the apparatus limitation or human error. IDE takes advantage of KF which includes getting feedback from the noisy measurement to improve the performance of each result that was generated by DE.

2 An Improved Differential Evolution Algorithm

This paper proposes an improved differential evolution algorithm (IDE), which is a hybrid combination of DE [6] and KF [7]. In parameter estimation, existing algorithms [4, 5] solely use DE whereas IDE uses a hybrid of DE and KF. Fig. 2.1 shows the details of the IDE. Kinetic parameters existed in the glycolysis pathway model for yeast [8] go through IDE to estimate their optimal values. Fixed control parameter values used in this study are population size, $NP=10$, mutation factor, $F=0.5$, and crossover constant, $CR=0.9$.

In IDE, we added the process of updating the population as a new step that improved the conventional DE. This is a self-adapting approach. In conventional DE, the original population which is an m x n population matrix, is generated from the first generation (Gen_1) and continues until it reaches the maximum generation (Gen_i) in the initialization process. m represents the number of generations and n represents the number of identifiable parameters. In evaluation process, the fitness function, J represented as

$$J = \sum_{i=1}^{N} |f(X, X0, \varnothing 0) - f(Y, X0, \varnothing)|^2 \qquad (1)$$

is applied to evaluate the fitness of each individual. X represents the state vector for measurement system, Y represents the state vector for simulated system, $\varnothing 0$ represents a set of original parameters, $\varnothing$ represents a set of estimated parameters, $X0$ represents the initial state, N=the ending index, and i=the index variable.

In mutation process, three individuals (*Ind1*, *Ind2* and *Ind3*) first being selected then treated with the formula showed in Fig 2.1. In the mutation section, *temp_population* represents the mutated population matrix, F represents the mutation factor, and *Pop* represents the original population matrix. The subsequent crossover process is mainly performed based on *CR*, which indicates crossover constant value, and *Randb(i)* which indicates i-th random evaluation of a uniform random number generator [0,1]. If the *randb(i)* value of the individual in mutated population is lower than the *CR* value then that individual becomes the individual for the resultant population of the crossover process and vice versa. This is followed by the updating process that is performed according to the Equation 2.2. This step updates the population, which is generated by the crossover process and it is based on the Kalman gain value K, retrieved from the Equation 2.3. The Kalman gain value from the Equation 2.3 takes the process noise covariance and measurement noise covariance into account. These noisy data values were obtained from the experiment and in this study the noisy data values used are 0.1. After handling the noisy data, the updated population once again undergoes the evaluation process and the whole process is repeated till the stopping criterion is met. The stopping criteria are set via predefined maximum loop values or when the fitness functions have converged. The updating population process is highlighted with the dotted box in Fig. 2.1 and is carried out according to the following formula.

$$temp\_population = (temp\_population' + K)' \qquad (2.2)$$

$$K = P * H' * inv(H * P * H' + R) \qquad (2.3)$$

Where K=Kalman gain value, A=state transition matrix, B=input matrix, H=observation matrix, Q=process noise covariance, R=measurement noise covariance, P=covariance of the state vector estimate, and H'=inverse of matrix H.

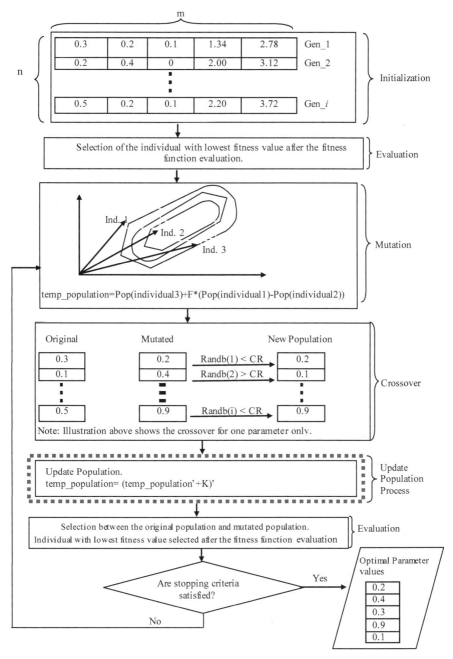

Note: Updating population process is added after the crossover process to improve DE performance and it is highlighted with the dotted box.

Fig. 2.1 Schematic Overview of IDE.

3 Experimental Results

In this study, three estimation algorithms (GA, DE, and IDE) are compared. Kinetic parameter values in Table 3.1 are retrieved from literature review [8] and generated by the estimation algorithms. To evaluate the accuracy of each estimation algorithm, time series data for concentration of adenosine monophosphate (AMP) and pyruvic acid (PYR) were generated. AMP and PYR are significant metabolites. AMP acts as an energy regulator and sensor while PYR acts as an energy supplier in presence of oxygen through citric acid cycle. From the time series data, we calculate the average error rate. The details of the accuracy measurement are discussed in this section.

Table 3.1 Kinetic parameter values of IDE compared with GA and DE.

Kinetic parameters	Measurement kinetic parameter values[8]	Simulated kinetic parameter values		
		GA	DE	IDE
V1	0.5	1.1492	0.1934	0.1524
K1GLC	0.1	0.0695	0.3064	0.4450
K1ATP	0.063	0.0568	0.0451	0.0432
V2	1.5	7.8890	2.0267	1.4561
K2	0.0016	0.0010	0.0024	0.0037
k2	0.0017	0.0492	0.0019	0.0493
K2ATP	0.01	0.0185	0.0229	0.0099
K3f	1	0.3510	0.1984	5.5314
K3b	50	48.4765	72.5643	12.1150
V4	20	9.8508	9.1572	4.0841
K4GAP	1	0.5554	0.7145	0.4708
K4NAD	1	0.9907	1.2682	7.6088
K5f	1	0.4973	1.7580	0.7912
K5b	0.5	0.1361	0.6089	0.1241
V6	10	23.9385	16.5141	47.7182
K6PEP	0.2	0.1357	0.8816	0.8593
K6ADP	0.3	0.0618	0.4417	0.1116
V7	2	10.5984	0.4764	0.6771
K7PYR	0.3	0.7196	0.2546	0.2508
k8f	1	0.3443	0.3972	0.2114
k8b	0.000143	0.0002	0.0005	0.0003
k9f	10	26.5712	1.1240	2.2127
k9b	10	6.1839	54.3684	10.1453
k10	0.05	0.1264	0.0047	0.2720

Note: Shaded rows are the kinetic parameter values contributed to the calculation of average error rate for metabolite AMP in Table 3.2 and PYR in Table 3.3.

The measurement kinetic parameter values and simulated kinetic parameter values were substituted into the ordinary differential equations (ODEs) (Equation 3.1 and Equation 3.2) of AMP and PYR respectively.

$$\frac{dAMP}{dt} = -AMPflow - reaction_9 \tag{3.1}$$

$$\frac{dPYR}{dt} = -PYRflow + reaction\_6 - reaction\_7 \tag{3.2}$$

Where *reaction_6=compartment* * (V6 * *adp* * *pep* / ((K6PEP + *pep*) * (K6ADP + *adp*), *reaction_7=compartment* * (V7 * *pyr* / (K7PYR + *pyr*)), *reaction_9= compartment* * (k9f * *amp* * *atp* - k9b * power(*adp*,2)), *AMPflow=compartment* * *amp* * *flow*, *PYRflow= compartment* * *pyr* * *flow*, *compartment*=constant value of 1, *flow*=value fixed to 0.011, *amp*=concentration of AMP, *pyr*=concentration for PYR, *adp*=concentration for adenosine diphosphate, *atp*=concentration of adenosine triphosphate, and *pep*=concentration for phosphoenolpyruvic acid.

Time series data for concentration of AMP and PYR were subsequently generated from Equation 3.1 and Equation 3.2. The time series data contain measurement results, *y,* and simulated results *yi* for IDE, DE, and GA respectively. Error rate (*e*) and Average error rate (*A*) are calculated based on Equation 3.3, and Equation 3.4 respectively.

$$e = \sum_{i=1}^{N} (y - yi)^2 \tag{3.3}$$

$$A = \frac{e}{N} \tag{3.4}$$

Table 3.2 and Table 3.3 show the average error rate for AMP and PYR respectively.

Table 3.2 Average error rate for AMP.

Evaluation criteria	GA	DE	IDE
Average error rate, *A*	0.000247729	0.059147889	0.000099818

Note: Shaded column represents the best results.

Table 3.3 Average error rate for PYR.

Evaluation criteria	GA	DE	IDE
Average error rate, *A*	0.000038704	0.000109841	0.000004786

Note: Shaded column represents the best results.

For AMP (Table 3.2), IDE presented the lowest average error rate with 0.0000998178. DE showed the worst performance with the average error rate of 0.059147889. GA showed more moderate performance with average error rate of 0.000247729. On the other hand, for PYR (Table 3.3), IDE once again performed better than other estimation algorithms where its average error rate is 0.000004786. The average error rate for GA and DE are 0.000038704 and 0.0000109841 respectively

Table 3.4 shows execution time of each estimation algorithm on a Core i5 PC with 4GB main memory. The result shows that DE required the longest time (9 minutes and 30 seconds) to find the optimal value for all kinetic parameters compared to IDE which took the shortest time (6 minutes 55 seconds). It is shown that IDE tends to use less computation time than DE and GA.

Table 3.4 Execution time of IDE compared with GA and DE.

Computation usage	GA	DE	IDE
Execution time (hh:mm:ss)	00:07:12	00:09:30	00:06:55

Note: Shaded column represents the best results.

IDE exhibits lesser computation time and possesses a higher accuracy when compared to both GA and DE. The implementation of DE that aims to estimate the relevant kinetic parameters and the additional of Kalman gain value which targets to handle the noisy data has improved the computational time and accuracy. Hence, the IDE which is a hybrid of DE and KF minimizes the computational time and also increases the accuracy between the simulated results and measurement results.

4 Conclusion and Future Work

In this paper, the experiment to compare the performances of three different estimation algorithms using glycolysis pathway data in yeast [8] showed that an improved algorithm, IDE, which is a hybrid algorithm of DE and KF with the shortest execution time and the lowest average error rate performed better than the rest of algorithms. It successfully reduces the high difficulty of the system in estimating the relevant kinetic parameters resulting in shorter computation time. The ability to handle noisy data has contributed to an improved accuracy of the estimated results. In conclusion, IDE is shown to be superior compared to both GA and DE in terms of computational time and accuracy. IDE can be generalized where it can be implemented in the areas which its data consists of noisy for example electrical and electronic engineering field [9].

DE shows to be very delicate to control parameters: population size (*NP*), crossover constant (*CR*), and mutation factor (*F*) [10]. Thus, for future work, self-adapting approach to these control parameters can be implemented to enhance the performance of the IDE. Moreover, additional steps can be added to the process of generating new populations with the aim of improving the performance of IDE.

Acknowledgments. This work is financed by Institutional Scholarship MyPhd provided by the Ministry of Higher Education of Malaysia. We also would like to thank Universiti Teknologi Malaysia for supporting this research by UTM GUP research grants (vot number: QJ130000.7123.00H67 and QJ130000.7107.01H29).

References

[1] Lillacci, G., Khammash, M.: Parameter Estimation and Model Selection in Computational Biology. PLoS Computational Biology 6(3), 1–17 (2010), doi:10.1371/journal.pcbi.1000696

[2] Chou, I.C., Voit, E.O.: Recent developments in parameter estimation and structure identification of biochemical and genomic systems. Mathematical Biosciences 219, 57–83 (2009), doi:10.1016/j.mbs.2009.03.002

[3] Wang, F.S., Chiou, J.P.: Differential evolution for dynamic optimization of differential-algebraic systems. In: IEEE International Conference on Evolutionary Computation, April 13-16, pp. 531–536 (1997), doi:10.1109/ICEC.1997.592367

[4] Chassagnole, C., Doncescu, A., Manyri, L., Yang, L.T.: Parameters Estimation by Differential Evolutionary Algorithms for Simulation of metabolic pathways in Escherichia coli. In: International Conference on AINA 2006, April 18-20, pp. 593–598. IEEE Computer Science (2006), doi:10.1109/AINA.2006.258

[5] Moonchai, S., Madlhoo, W., Jariyachavalit, K., Shimizu, H., Shioya, S., Chauvatcharin, S.: Application of a mathematical model and Differential Evolution algorithm approach to optimization of bacteriocin production by Lactococcus lactis C7. Bioprocess Biosyst. Eng. 28, 1–17 (2005), doi:10.1007/s00449-005-0004-5

[6] Storn, R., Price, K.: Differential Evolution – A Simple and Efficient Heuristic for Global Optimization over Continuous Spaces. Journal of Global Optimization 11(4), 341–359 (1997), doi:10.1023/A:1008202821328

[7] Cetto, J.A.: The Kalman Filter. Institut de Robotica i Informatica Industrial, UPC-CSIC. Llorens i Artigas 4-6, Edifici U, 2a pl. Barcelona 08028, Spain (2002), http://digital.csic.es/bitstream/10261/30069/1/doc1.pdf (accessed January 22, 2011)

[8] Nielson, K., Sorensen, P.G., Hynne, F., Busse, H.G.: Sustained oscillations in glycolysis: an experimental and theoretical study of chaotic and complex periodic behavior and of quenching of simple oscillations. Biophysical Chemistry 72, 49–62 (1998), doi:10.1016/S0301-4622(98)00122-7

[9] Ijaz, U.Z., Khambampati, A.K., Lee, J.S., Kim, S., Kim, K.Y.: Nonstationary phase boundary estimation in electrical impedance tomography using unscented Kalman filter. Journal of Computational Physics 227(15), 7089–7112 (2008), doi:10.1016/j.jcp.2007.12.025

[10] Feng, L., Yang, Y.F., Wang, Y.X.: A New Approach to Adapting Control Parameters in Differential Evolution Algorithm. In: Li, X., Kirley, M., Zhang, M., Green, D., Ciesielski, V., Abbass, H.A., Michalewicz, Z., Hendtlass, T., Deb, K., Tan, K.C., Branke, J., Shi, Y. (eds.) SEAL 2008. LNCS, vol. 5361, pp. 21–30. Springer, Heidelberg (2008), doi:10.1007/978-3-540-89694-4_3

Structural Graph Extraction from Images

Antonio-Javier Gallego-Sánchez, Jorge Calera-Rubio, and Damián López

Abstract. We present three new algorithms to model images with graph primitives. Our main goal is to propose algorithms that could lead to a broader use of graphs, especially in pattern recognition tasks. The first method considers the q-tree representation and the neighbourhood of regions. We also propose a method which, given any region of a q-tree, finds its neighbour regions. The second algorithm reduces the image to a structural grid. This grid is postprocessed in order to obtain a directed acyclic graph. The last method takes into account the skeleton of an image to build the graph. It is a natural generalization of similar works on trees [8, 12]. Experiments show encouraging results and prove the usefulness of the proposed models in more advanced tasks, such as syntactic pattern recognition tasks.

1 Introduction

Many fields take advantage of graph representations, these fields include but are not limited to: biology, sociology, design of computer chips, and travel related problems. This is due to the fact that graph are suitable to represent any kind of relationships among data or their components. The expressive power of graph primitives has been specially considered in pattern recognition tasks, in order to represent objects [7] or bioinformatics [13]. Therefore, the development of algorithms to handle graphs is of major interest in computer science.

In this paper we present three new methods to model images using graph primitives. The main goal we aim to achieve is to propose algorithms that could lead

Antonio-Javier Gallego-Sánchez · Jorge Calera-Rubio
Departamento de Lenguajes y Sistemas Informáticos, University of Alicante, Spain
e-mail: {jgallego,calera}@dlsi.ua.es

Damián López
Departamento de Sistemas Informáticos y Computación, Technical University of Valencia, Spain
e-mail: dlopez@dsic.upv.es

S. Omatu et al. (Eds.): Distributed Computing and Artificial Intelligence, AISC 151, pp. 717–724.
springerlink.com   © Springer-Verlag Berlin Heidelberg 2012

to a broader use of graphs. In that way, the work of [4] shows that this kind of contribution is suitable and interesting.

The first method we propose (Section 2) takes into account the neighbour paths within a q-tree [2]. We present algorithms that extend other works for neighbourhood calculations in spatial partition trees [5, 11]. These works generally calculate only the 4-neighbours of each region, or use two steps to calculate the corner neighbours. The proposed method completes these works with an estimation of equal complexity for the 8-neighbours at any level of resolution (equal, higher or lower level). Moreover, neighbour calculation can help to improve other applications, speeding up the process to locate the next element, such as in: image representation, spatial indexing, or efficient collision detection.

The second method (Section 3) considers a structural grid of the image which is then postprocessed to obtain the graph model. This method maintains those desiderable features of q-tree representation, that is: the resolution is parametrizable and allows to reconstruct the image.

The third method (Section 4) extends a similar method used on trees [12, 8]. The initial image is preprocessed to obtain the skeleton. This skeleton is then traversed to build a graph that summarizes the core structure of the image.

2   Neighbourhood Graphs

Using the tree defined by a q-tree (see Fig. 1) and the proposed algorithm for neighbourhood calculation (see section 2.1), a directed acyclic graph (dag) can be extracted to represent the sample and to provide information such as: neighbourhood, structural traversal, resolution or thickness, in addition to allow its reconstruction.

The graph is created walking top-down the tree, and from left to right. For each nonempty node, its 8-neighbours are calculated. It is important that: 1) Do not add arcs to processed nodes. 2) If it has to create an arc from a non-pointed node to one that is already pointed: then the direction of the arc is changed. These criteria ensures that the graph is acyclic.

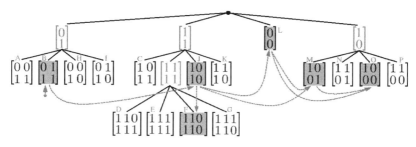

Fig. 1 Tree obtained from the q-tree of the Fig. 2(c), and its corresponding neighbour graph.

2.1 Calculating Neighbours

Binary encoding is used to encode the position of partitions whithin the image and
their corresponding traversal path down the q-tree [5]. Each dimension is encoded
with 1 bit coordinate, describing the direction as positive (1) or negative (0). A
location array is defined using this encoding (Fig. 2(a)). In q-trees, this array has
two rows: horizontal and vertical coordinates. Positive directions (up and right) are
represented by a 1, and the negative directions (down and left) by 0 (Fig. 2(b)). In
general, the location array associated with a given node is defined as the location
array of his father, but adding on the right a new column with its own encoding.
For each new partition, the origin of the reference system is placed in the center
of the area where the subdivision is done. In Fig. 2(c), the second level node K is
represented by the $\begin{bmatrix} 1 \\ 1 \end{bmatrix}$ of his father, and then adding the $\begin{bmatrix} 1 \\ 0 \end{bmatrix}$ for its own level.

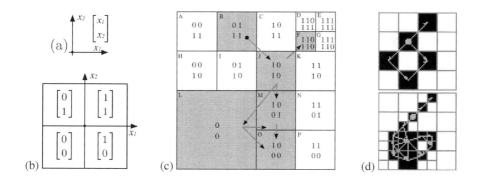

Fig. 2 (a) Reference system and location array, (b,c) Q-trees with the corresponding location
arrays, (d) Graphs extracted using different truncate levels.

2.1.1 Algorithm

To calculate the neighbours of a given node, the location array described above is
used. The proposed algorithm distinguishes two cases: face neighbours (two regions
share more than one point, such as A with B in Fig. 2(c)), and corner neighbours
(they share one point, A with I in Fig. 2(c)).

2.1.2 Face Neighbours

To calculate the face neighbour of a given node, the algorithm needs to know its
location array (M), and the dimension dim and the direction dir in which the neighbour has to be calculated. The algorithm (see Sec. 2.1) visits each bit of the location
array (M) in the given dimension from right to left, and negates the bits until the
result is equal to the explored direction (1: positive, 0: negative).

As an example, the horizontal neighbour of *I* (Fig. 2(c)) in the positive direction (1) is calculated. The algorithm begins with the location array and negates each bit in the top row (dim 1) from right to left (indicated with a bar) until reaching the halt condition: the result is equal to the explored direction.

$$I = \begin{bmatrix} 0 & \bar{1} \\ 1 & 0 \end{bmatrix} \Rightarrow \begin{bmatrix} \bar{0} & 0 \\ 1 & 0 \end{bmatrix} \Rightarrow \begin{bmatrix} 1 & 0 \\ 1 & 0 \end{bmatrix} = J$$

2.1.3 Corner Neighbours

In the case of corner neighbours, the algorithm has to modify the two dimensions of the location array. As for the face neighbours, it begins with the location array *M*, but in this case it receives as input an array *dir* with both horizontal and vertical directions. Below is the complete algorithm to compute face and corner neighbours:

```
function GetNeighbour( M, dim, dir )
  level := |M[0]|    // Level is equal to the number of columns
  if( |dir| == 1 ) {    // Face neighbours
    do {
      M[dim][level] := ! M[dim][level]
      level := level − 1
    } while( level ≥ 0 ∧ M[dim][level] != dir )
  } else {    // |dir| == 2 → Corner neighbours
    flag1 := false ; flag2 := false
    do {
      if( !flag1 )
        M[0][level] := ! M[0][level]
        if( M[0][level] == dir[0] )  flag1 := true
      if( !flag2 )
        M[1][level] := ! M[1][level]
        if( M[1][level] == dir[1] )  flag2 := true
      level := level − 1
    } while( level ≥ 0 ∧ ( !flag1 ∨ !flag2))
  }
  return M
```

As example, the corner neighbours of *N* (Fig. 2(c)) are calculated:

$$N = \begin{bmatrix} 1 & \bar{1} \\ 0 & \bar{1} \end{bmatrix} \Rightarrow \begin{bmatrix} 1 & 0 \\ \bar{0} & 0 \end{bmatrix} \Rightarrow \begin{bmatrix} 1 & 0 \\ 1 & 0 \end{bmatrix} = J \quad \| \quad N = \begin{bmatrix} 1 & \bar{1} \\ 0 & \bar{1} \end{bmatrix} \Rightarrow \begin{bmatrix} 1 & 0 \\ 0 & 0 \end{bmatrix} = O$$

2.1.4 Type of Nodes

It is important to differentiate the type of a node within the structure (Fig. 1). The function *TypeOfNode(M)* returns this type, it can be: *leaf-node* (the node exists and has no children), *inner-node* (the node exists and has children) or *no-node* (the node does not exist). If the location array is a *leaf-node* the algorithm ends. If it is an *inner-node*, then the algorithm has to calculate the higher-level neighbours

(Sec. 2.1). If the node does not exist, the algorithm has to calculate the lower-level neighbours (Sec. 2.1).

2.1.5 Lower Resolution Neighbours

In this case, the algorithm first calculates the same resolution neighbour using $M :=$ *GetNeighbour(M, dim, dir)*; and then eliminates the final column/s (on the right side) until it finds a *leaf-node*:

---
function *GetLowerLevel(M)*
  do {
    $M_{2 \times n} \leftarrow M_{2 \times n-1}$
    } **while**(*TypeOfNode(M)* == *no-node*)
  return *M*

Example: region *D* in Fig. 2(c).

$$D = \begin{bmatrix} 1 & 1 & \bar{0} \\ 1 & 1 & 1 \end{bmatrix} \Rightarrow \begin{bmatrix} 1 & \bar{1} & 1 \\ 1 & 1 & 1 \end{bmatrix} \Rightarrow \begin{bmatrix} 1 & 0 & 1 \\ 1 & 1 & 1 \end{bmatrix} \Rightarrow no\text{-}node \Rightarrow \begin{bmatrix} 1 & 0 & 1 \\ 1 & 1 & 1 \end{bmatrix} \Rightarrow \begin{bmatrix} 1 & 0 \\ 1 & 1 \end{bmatrix} = C$$

2.1.6 Higher Resolution Neighbours

The algorithm first calculates the neighbour at the same level of resolution using $M := GetNeighbour(M, dim, dir)$. If M corresponds to an *inner-node*, then the higher-level neighbour/s is/are calculated. It is worth to note that, face neighbours may have two or more neighbours (e.g. *L* with *H* and *I* in Fig. 2(c)), whereas corner neighbours only have one neighbour. In this process, columns are added on the right side of the matrix *M* until reaching a *leaf-node*. These columns are created in the opposite direction of the dimension/s of interest. See the pseudocode below:

---
function *GetHigherLevel(M, dim, dir)*
  $\mathscr{L} := \langle \varnothing \rangle$ *// Result set of higher-level neighbours*
  if($|dir| == 1$) {   *// Face neighbours*
    for($i = 1; i \leq 2, ++i$) {
      $N_i[dim][0] :=!dir$; $N_i[\,!dim][0] := i - 1$
      $M_i := M \oplus N_i$
      if(*TypeOfNode(M_i)* == *leaf-node*) $\mathscr{L} \leftarrow M_i$
      else $\mathscr{L} \leftarrow GetHigherLevel(M_i, dim, dir)$
    }
  } **else** {   *// $|dir| == 2 \to$ Corner neighbours*
    do {
      $N[0][0] :=!dir[0]$; $N[1][0] :=!dir[1]$
      $M := M \oplus N$
      } **while**(*TypeOfNode(M)* != *leaf-node*)
    $\mathscr{L} \leftarrow M$
  }
  return $\mathscr{L}$

The symbol $\oplus$ denotes the concatenation of a matrix *M* with a vector *N*:

$$M \oplus N = \begin{bmatrix} x_0^1 & x_1^1 \\ x_0^2 & x_1^2 \end{bmatrix} \oplus \begin{bmatrix} y^1 \\ y^2 \end{bmatrix} = \begin{bmatrix} x_0^1 & x_1^1 & y^1 \\ x_0^2 & x_1^2 & y^2 \end{bmatrix}$$

Below is an example of higher-level neighbours calculation (see Fig. 2(c)):

$$C = \begin{bmatrix} 1 & \bar{0} \\ 1 & 1 \end{bmatrix} \Rightarrow \begin{bmatrix} 1 & 1 \\ 1 & 1 \end{bmatrix} \Rightarrow \textit{inner-node} \Rightarrow \left\langle \begin{bmatrix} 1 & 1 & 0 \\ 1 & 1 & 1 \end{bmatrix} = D; \quad \begin{bmatrix} 1 & 1 & 0 \\ 1 & 1 & 0 \end{bmatrix} = F \right\rangle$$

2.1.7 Neighbourhood of Border Regions

When a region is on the border of the structure, do not present neighbours on that border side. In this case, the algorithm ends without reaching the halt condition. E.g. node A (horizontal-left):

$$A = \begin{bmatrix} 0 & \bar{0} \\ 1 & 1 \end{bmatrix} \Rightarrow \begin{bmatrix} \bar{0} & 1 \\ 1 & 1 \end{bmatrix} \Rightarrow \begin{bmatrix} 1 & 1 \\ 1 & 1 \end{bmatrix} \Rightarrow \textit{border}$$

3 Structural Grid Graph

We here propose an algorithm to model images using graph primitives. The main goal of this new proposal is to maintain the better properties of q-trees but introducing higher variability in the incoming and outgoing degrees. Thus, the properties to fulfill are: 1) The model should consider the resolution as a parameter, allowing higher and lower resolution representations. 2) The representation should contain enough information to reconstruct the original image.

Without loss of generality, we will consider directed acyclic graphs and two colour images: *background* and *foreground* colours. Starting from the top-left corner, the algorithm divides the image into squared regions of a given size k. Those regions with foreground colour will be considered the nodes of the graph. The edges are defined from a foreground region (node) to those adjacent-foreground regions to the east, south and southeast. Figure 3 shows an example.

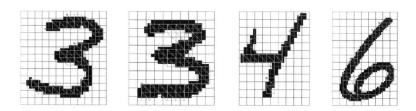

Fig. 3 Some examples of the grid graph modeling are shown. Those regions with at least 20% black pixels are considered foreground. Root nodes are marked in gray.

4 Skeleton Graphs

The last method models the graph using the skeleton that defines the shape of the figure. It is a natural extension of the algorithms proposed for trees in [8, 12]. First, a thinning process is applied to the image [1], and then the graph is generated as:

1. The top left pixel of the skeleton is chosen as the root of the graph.

2. Every node in the graph will have as many children as unmarked neighbour pixels has the current pixel. For each child, an arc is created following that direction until one of the following criteria is true:

   a. The arc has the maximum fixed parameter size (window parameter).
   b. The pixel has no unmarked neighbours (terminal node).
   c. The pixel has more than one unmarked neighbour (intersection).
   d. The pixel is marked (existing node).

3. A new node is assigned to every end of arc pixel obtained by the previous step (a,b,c), or it is joined with the corresponding node (d). If the node has unmarked neighbours, then go to step 2, otherwise it terminates.

Fig. 4 Skeleton and graph results.

5 Experimentation

This section aims to demonstrate the usefulness of the proposed models. For this reason, it uses simple classification methods, without stochastic layer and using only the structure of graphs (without labels). Results prove the correct separation of classes, in addition to be a baseline for future research.

To perform the experiments, 5 thousand images of digits (10 classes), with resolution of 64x64 pixels, from the NIST dataset are used. Graphs have been extracted using the three proposed methods, but varying the parameters (truncate level, size of cells, and window size) (see Fig. 5). Six methods are used to perform the classification: First, a feature vector is calculated using the graph degree distribution [10], and then classified using Naive Bayes, Random Forest, SVM and k-NN [6]. The 75% of the dataset is used for training and the rest for test. A preliminar error correcting distance is also defined to test the proposed methods. It is a simple algorithm with linear complexity. For this distance, the classification is performed using leaving-one-out. A semi-structural method of Flow Complexity is also used [3]. It analyzes the structure of the graph based on feature selection via spectral analysis and complexity.

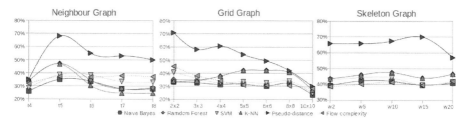

Fig. 5 Results using different classification methods.

6 Conclusions

This paper presents three methods for extracting graphs from images. The obtained graphs introduce a variability that should create new characteristic features in the modelling of the classes of a pattern recognition task, and thus, it should ease a syntactic approach. Experiments show encouraging results which prove the usefulness of the proposed models, the correct separation of classes, in addition to be a baseline for future work.

The first method creates neighbourhood paths or segmentations within the q-tree, providing more data and creating new possibilities of operations and applications. The second algorithm maintains the features of q-tree representation, that is: the resolution is parametrizable and allows to reconstruct the original image. The third method takes into account the skeleton of the shape to build the graph, summarizing the core structure of the image.

Acknowledgments. This work is partially supported by Spanish MICINN (contract TIN2011-28260-C03-01, contract TIN2009-14205-C04-C1) and CONSOLIDER-INGENIO 2010 (contract CSD2007-00018).

References

1. Cychosz, J.M.: Thinning algorithm from the article: Efficient binary image thinning using neighbourhood maps. In: Graphics Gems IV, pp. 465–473. Academic Press (1994)
2. de Berg, M., van Kreveld, M., Overmars, M., Schwarzkopf, O.: Computational geometry, pp. 291–306. Springer (2000)
3. Escolano, F., Giorgi, D., Hancock, E.R., Lozano, M.A., Falcidieno, B.: Flow Complexity: Fast Polytopal Graph Complexity and 3D Object Clustering. In: Torsello, A., Escolano, F., Brun, L. (eds.) GbRPR 2009. LNCS, vol. 5534, pp. 253–262. Springer, Heidelberg (2009)
4. Flasinski, M., Myslinski, S.: On the use of graph parsing for recognition of isolated hand postures of Polish Sign Language. Pattern Recognition 43, 2249–2264 (2010)
5. Goodchild, M.: Quadtree algorithms and spatial indexes. Technical Issues in GIS, NC-GIA, Core Curriculum 37, 5–6 (1990)
6. Hall, M., Frank, E., Holmes, G., Pfahringer, B., Reutemann, P., Witten, I.H.: The WEKA Data Mining Software: An Update. SIGKDD Explorations 11(1), 10–18 (2009)
7. Liu, J., Li, M., Liu, Q., Lu, H., Ma, S.: Image annotation via graph learning. Pattern Recognition 42, 218–228 (2009)
8. López, D., Piñaga, I.: Syntactic Pattern Recognition by Error Correcting Analysis on Tree Automata. In: Amin, A., Pudil, P., Ferri, F., Iñesta, J.M. (eds.) SPR 2000 and SSPR 2000. LNCS, vol. 1876, pp. 133–142. Springer, Heidelberg (2000)
9. Luque, R.G., Comba, J.L.D., Freitas, C.: Broad-phase collision detection using semi-adjusting bsp-trees. In: ACM i3D, pp. 179–186 (2005)
10. Newman, M.E.J.: The structure and function of complex networks. SIAM 45 (2003)
11. Poveda, J., Gould, M.: Multidimensional binary indexing for neighbourhood calculations in spatial partition trees. Comput. Geosci. 31(1), 87–97 (2005)
12. Rico-Juan, J.R., Micó, L.: Comparison of AESA and LAESA search algorithms using string and tree edit distances. Pattern Recognition Letters 24, 1427–1436 (2003)
13. Shin, H., Tsuda, K., Schölkopf, B.: Protein functional class prediction with a combined graph. Expert Systems with Applications 36, 3284–3292 (2009)

Mutagenesis as a Diversity Enhancer and Preserver in Evolution Strategies

José L. Guerrero, Alfonso Gómez-Jordana, Antonio Berlanga, and José M. Molina

Abstract. Mutagenesis is a process which forces the coverage of certain zones of the search space during the generations of an evolution strategy, by keeping track of the covered ranges for the different variables in the so called gene matrix. Originally introduced as an artifact to control the automated stopping criterion in a memetic algorithm, ESLAT, it also improved the exploration capabilities of the algorithm, even though this was considered a secondary matter and not properly analyzed or tested. This work focuses on this diversity enhancement, redefining mutagenesis to increase this characteristic, measuring this improvement over a set of twenty-seven unconstrained optimization functions to provide statistically significant results.

1 Introduction

Evolutionary computation [2] arose as a powerful technique to deal with optimization and search problems, particularly evolution strategies [12], focused on real value representations, which have been successfully tested on a wide variety of problems, both theoretical and practical in nature. The increase in the computational resources of computers and the increasing number of parallel implementations [4] have lead this growth, making them more appealing for practitioners focused on solving particular problems, rather than theoretical research of the algorithms themselves. There are, however, a number of issues for these applications.

Local optima constitute a drawback for evolutionary algorithms, since they do not provide (as most metaheuristics [13]) a measurement of the proximity of the solutions found to global optima, performing a best-effort approach. Early convergence

José L. Guerrero · Alfonso Gómez-Jordana · Antonio Berlanga · José M. Molina
University Carlos III of Madrid, Computer Science Department, Group of Applied Artificial Intelligence. Avda. Universidad Carlos III, 22, Colmenarejo, Spain
e-mail: jguerrer@inf.uc3m.es,
　　　alfonso.gomez-jordana@alumnos.uc3m.es,
　　　aberlan@ia.uc3m.es, molina@ia.uc3m.es

S. Omatu et al. (Eds.): Distributed Computing and Artificial Intelligence, AISC 151, pp. 725–732.
springerlink.com　　　　　　　　　　　　　© Springer-Verlag Berlin Heidelberg 2012

arises as a concern regarding this topic, being closely related to the diversity preservation in the population as the algorithm progresses. Many approaches have been proposed to deal with this issue, from the restriction of certain operator applications (such as the incest prevention proposed in [6]) or multi-objective approaches [5] where the diversity of the population is treated as an additional objective function [14]. General stopping criteria are also a concern for practitioners using evolutionary techniques, which is in fact shared by many different iterative processes [1], but the stochastic nature of evolutionary computation makes it probably more important and, at the same time, harder to solve. Finally, evolutionary algorithms tend to favor the exploratory nature of the search process, leading to a slow convergence towards the minimum. This handicap has been faced with memetic algorithms [11], which combine evolutionary techniques with local search in an attempt to obtain a good exploration of the search space with a fast information exploitation once the most interesting zones have been determined.

ESLAT (Evolution Strategies Learned with Automated Termination criteria) technique was introduced to deal with some of those issues. ESLAT is a memetic algorithm which combined the evolutionary cycles of an evolution strategy with two different local search procedures which were applied sequentially: the *Broyden-Fletcher-Goldfarb-Shanno* Quasi-Newton method (BFGS) with a cubic line search procedure [3] and Nelder Mead's algorithm, based on Kelley's modification [10]. Along with the local search techniques, it introduced an artifact used to control the stopping generation: the *gene matrix*, which tracked the zones of the search space covered and forced the exploration of those which had not been reached by the main search cycle, also determining the generation at which the algorithm should be stopped according to that search space coverage.

ESLAT presented a series of difficulties in its results, which were faced in R-ESLAT (Robust ESLAT) [8]. These faced difficulties included control over the search space, the configuration of the local search techniques used and the stopping criterion used which led to a comparison with CMA-ES algorithm [9] including promising results in terms of solution quality. However, the basis of the proposal was the increased exploratory capabilities introduced by the gene matrix, which were not individually tested, only as a part of the overall algorithm performance.

This work analyzes the gene matrix as a diversity preservation technique in evolution strategies, modifying its original behavior according to this new role and testing its performance against a canonical evolution strategy. This test is based on a set of twenty-seven unconstrained optimization functions with different characteristics, in order to highlight the statistical significance of the obtained results.

The paper is structured according to the following sections: the second section will introduce mutagenesis and gene matrix concepts, along with the paper proposal for their definition and use. The third section will present the experimental setup used and the obtained results, along with their analysis. Finally the fourth section will present the conclusions obtained from the available results and the future lines of the work.

2   Gene Matrix and Diversity Preservation

The *gene matrix* (GM) is responsible of tracking the exploration process and keeping the diversity in the population. It is composed of n by m elements, where n is the number of genes in the chromosome and m is the number of sub-ranges in which the search space of that chromosome is divided. This matrix is initialized with zeros, and those zeros are updated to ones as elements with genes covering the different sub-ranges are found in the different populations as the evolution progresses. Figure 1 shows an example of a GM with two variables. Variable x_1 (horizontal) has individuals in subrages 1, 2 and 4 (represented with 1's in the first row of the GM) and variable x_2 (vertical) in subranges 1, 3 and 4 (represented in the second row of the GM).

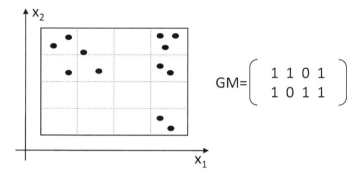

Fig. 1 Gene Matrix example

The *GM* is used, therefore, as a measurement of the depth in the exploration process. In order to use it to keep the diversity in the population as well, the *mutagenesis* operator is introduced. At the end of every generation, the mutagenesis operator chooses the N_w worst individuals which have survived to the next generation and changes the values of one of their genes in order to cover new zones of the search space (according to the information in the GM). Specifically, for each of the N_w worst individuals in the population, one of the sub-ranges containing a zero value in the *GM* is selected randomly, that GM position updated to a one, and the value in the correspondent gene of the individual is updated according to a random value within the sub-rage boundaries. Figure 2 presents an example of this process. The individual altered by the mutagenesis process fills an empty subrange of variable x_1, being this change reflected in the GM (the fourth element of the first row is changed to a 1).

According to its original definition, once the gene matrix had been completely filled with ones, the mutagenesis procedure was stopped, and the algorithm stopped after a certain number generations (the problem dimensionality). This stopped the diversity increase once the search space had been covered, and focused the gene matrix use on a simple mechanism to control the automated stopping criterion. As a

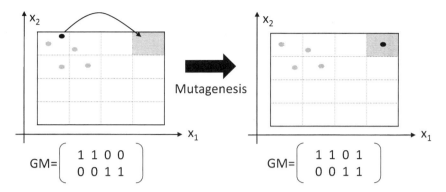

Fig. 2 Mutagenesis example

diversity enhancer, however, this behavior is not acceptable. Another issue concerning the gene matrix was the number of subranges required, fixed to 30. However, preliminary tests showed that this choice may not be optimal.

The novel gene matrix proposal is focused on diversity enhancement, rather than its application as a termination criterion. To do so, an initial number of subranges is set a-priori. Once the gene matrix is filled with ones at a certain generation, it is *restarted*, reinitializing it with zeros and updating it with the individuals in the population which caused this reinitialization. Every time the gene matrix is reinitialized, its number of contained subranges is doubled. This mechanism achieves a constant diversity enhancement and also a more thorough coverage of the search space as the algorithm progresses, depending on the dimensionality of the problem faced.

The mutagenesis procedure has also been reviewed. As previously explained, it originally introduced a certain number of modifications on the worst individuals of the population, changing concrete values from the chromosome to unexplored subranges of the chosen gene. This behavior may not introduce enough diversity in a population heavily dominated by the best individual, so an additional probability is added to the algorithm configuration: p_{rm}, random mutagenesis probability. According to this probability, mutagenesis may generate a random individual covering the chosen subrange instead of modifying just one gene from one of the worst individuals in the population.

Additional controls have also been added to mutagenesis. If an individual has covered a new subrange in current generation, it is never changed any further by the mutagenesis procedure, regardless of its rank. This allows the new information introduced during the evolutionary cycle to survive at least one generation, in order to give the new individual the chance to procreate and mutate before any directed change is applied to it. This also implies a change in the mutagenesis configuration. Instead of N_w changed individuals, the user configures a more versatile N_c parameter, establishing the number of new subranges covered each generation. If the evolutionary cycle covers the required number of changes, no mutagenesis is applied. In other case, the worst individuals (as many as required in order to cope with the desired N_c changes) are picked to go through the mutagenesis procedure.

Finally, the stopping criterion used in R-ESLAT implied that the best fitness was repeated over a certain window of generations. This exact repetition may be too strict for a stopping criterion, since very small changes in fitness values (which might even be affected by the representation precision) would lead to a continuation in the evolutionary algorithm once the search process had stagnated regarding all practical purposes. For these reasons this exact comparison was changed to the comparison quotient presented in equation 1, which provides a more flexible mechanism to control the relevance of the changes.

$$\frac{previous_{best} - current_{best}}{previous_{best}} \leq Improvement_{factor} \tag{1}$$

3   Experimental Validation

For the experimental validation of the proposed technique, a complete dataset of twenty-seven different functions for unconstrained optimization has been used. These functions test different aspects, such as the presence of multiple local optima (Griewank), different sizes of the search spaces (from small search spaces, such as in Hartmann, up to much larger ones, as the one exhibited by Schwefel), different ranges for the different variables of the search space (Branin), search spaces showing ranges where the fitness is unable to provide any information to the evolutionary algorithm (Easom) and the performance of the different techniques regarding problem dimensionality, with values ranging from two (Beale) up to thirty (Ackley). Table 3 shows a brief description of the complete dataset.

Several parameters (according to their description included in the previous section) have to be established for the proposed technique, which are presented in

Table 1 Experimental configuration

Parameter	Description	Value
μ	Population size	5, 10, 15, 30
$init_{sr}$	Initial subranges	10
min_{sr}	Minimum subranges covered per generation	$\mu/5$
p_{rm}	random mutagenesis probability	0.5
I_f	Improvement factor	1E-05

Table 2 Results comparison for the different considered population sizes

Population size	Statistical Best	Statistical Worst	Best
5	7	3	19
10	7	8	14
15	10	6	15
30	3	10	13

Table 3 Test-set functions overview and results for population size five

id	Name	n	min. bound.	max. bound.	Gene Matrix results mean	std	Canonical results mean	std	normal ttest	wilcoxon	statistical best	best
f1	Ackley	30	-15	30	1,41E+00	1,45E+00	1,54E+00	1,95E+00	1	0	0	-	Gene Matrix
f2	Beale	2	-4,5	4,5	3,05E-02	1,51E-01	3,04E-01	4,13E-01	1	1	1	Gene Matrix	Gene Matrix
f3	Bohachecsky	2	-100	100	6,12E-02	1,78E-01	8,31E-03	5,85E-02	1	1	1	Canonical	Canonical
f4	Booth	2	-10	10	8,19E-08	2,62E-07	1,53E-07	5,86E-07	1	0	0	-	Gene Matrix
f5	Branin	2	-5	15	4,58E-07	4,44E-07	4,79E-02	3,26E-01	1	0	0	-	Gene Matrix
f6	Colville	4	-10	10	6,04E-01	1,39E+00	1,02E+00	2,87E+00	1	0	0	-	Gene Matrix
f7	Dixon-Price	30	-10	10	2,45E+00	1,98E+00	1,80E+00	1,46E+00	1	1	0	-	Canonical
f8	Easom	2	-100	100	8,83E-01	3,21E-01	9,50E-01	1,78E-01	1	0	0	-	Gene Matrix
f9	Goldstein-Price	2	-2	2	2,16E+00	7,40E+00	9,92E+00	2,66E+01	1	0	0	-	Gene Matrix
f10	Griewank	30	-600	600	1,21E-01	2,15E-01	1,29E-01	1,97E-01	1	0	0	-	Gene Matrix
f11	Hartmann	6	0	1	3,82E-02	5,62E-02	3,36E-02	5,39E-02	1	0	0	-	Canonical
f12	Hump	2	-5	5	1,63E-02	1,15E-01	6,03E-06	2,96E-05	1	0	0	-	Canonical
f13	Levy	30	-10	10	1,14E-01	3,50E-01	4,76E-01	1,31E+00	1	1	1	Gene Matrix	Gene Matrix
f14	Matyas	2	-10	10	5,17E-09	1,23E-08	5,86E-06	4,13E-05	1	0	0	-	Gene Matrix
f15	Michalewicz	10	0	pi	5,81E-01	3,35E-01	1,01E+00	5,01E-01	0	1	1	Gene Matrix	Gene Matrix
f16	Perm	30	-30	30	2,59E+85	4,57E+85	1,35E+86	1,56E+86	1	1	1	Gene Matrix	Gene Matrix
f17	Powell	28	-4	5	1,08E-02	1,97E-02	6,51E-03	1,11E-02	1	1	1	Canonical	Canonical
f18	Power Sum	4	0	4	4,05E-02	3,24E-02	1,24E-01	4,16E-01	1	0	0	-	Gene Matrix
f19	Rastrigin	30	-5,12	5,12	2,35E-01	1,20E-01	4,25E+01	1,86E+01	0	1	1	Gene Matrix	Gene Matrix
f20	Rosenbrock	30	-5	10	9,90E-01	1,66E+02	7,72E+01	1,21E+02	1	0	0	-	Canonical
f21	Schwefel	30	-500	500	7,69E+02	4,72E+02	2,35E+03	4,07E+02	0	1	1	Gene Matrix	Gene Matrix
f22	Shekel	4	0	10	5,54E+00	3,27E+00	5,28E+00	3,41E+00	0	0	0	-	Canonical
f23	Shubert	2	-10	10	2,72E-02	1,84E-01	2,65E+00	1,87E+01	1	0	0	-	Gene Matrix
f24	Sphere	30	-5,12	5,12	5,14E-05	2,01E-04	1,46E-05	5,58E-05	1	0	1	Canonical	Canonical
f25	Sum Squares	30	-10	10	2,84E-04	1,03E-03	1,53E-03	9,70E-03	1	0	0	-	Gene Matrix
f26	Trid	10	-100	100	6,00E-02	1,36E-01	1,07E-01	4,56E-01	1	0	1	Gene Matrix	Gene Matrix
f27	Zakharov	30	-5	10	1,97E+01	4,37E+01	2,25E+01	4,39E+01	1	0	0	-	Gene Matrix

table 1. As included in that table, four different population sizes are used to cover the comparison of the two different techniques. The complete results for population size five are presented in table 3. Following [7], the individual comparison for the different test functions is performed according to parametric and non-parametric tests. The normality test used is the Shapiro-Wilk test, the parametric test is Student's t-test and the non-parametric test is Wilcoxon signed-rank test. The statistical best results are provided according to the t-test if the data follows a normal distribution and according to the non-parametric test otherwise. Fifty iterations have been run in order to establish the statistical significance of the results.

To test the final performance comparison, a Wilcoxon rank-sum test is carried out over the mean results for the twenty-seven functions and the four considered population sizes. The p-value obtained is 0.0275, which implies that with a significance level as low as 3% (lower than the usual 5% considered for these tests) the proposed gene matrix diversity enhancer allows evolution strategies to perform better.

Analyzing the individual results, the effectiveness of the diversity enhancement is, in general, more representative at lower population sizes (where the risk of falling into local optima is higher and the exploration capabilities are reduced) but, at the same time, since the number of required changes per generation are configured as a certain percentage of the population, the use of the gene matrix is more accused on higher population sizes. The balance between these two factors determines the effectiveness of the mutagenesis changes. This is reflected in the variable number of significant best and worst results obtained for the different population sizes.

Finally, regarding the individual analysis of the results for the different test functions, it must be noted that the non-parametric tests do not seem to be able to properly measure some behavior differences (due to their zero median null hypotheses). This can be seen, for instance, in table 3, function f14, where, even though the mean value obtained by the evolution strategy using mutagenesis is several orders of magnitude better, the Wilcoxon test does not determine it to be the best. This points to the requirement of mean based statistical tests not requiring normality distribution over their measures to perform quality comparisons between algorithms.

4   Conclusions

Gene matrix and mutagenesis were originally presented as part of the ESLAT memetic algorithm. They were used as guidance for the automated stopping criterion, even though they also increased the exploration capabilities of the evolution strategy included. This work isolates these artifacts and focuses on their diversity enhancement, redefining the processes in order to maximize these characteristics, and tests the results comparing them to the performance of canonical evolution strategies. The obtained results show that the exploration improvements lead the algorithm to an overall better performance, with a different impact regarding the population size and the percentage of the population which goes through mutagenesis processing. For a set of twenty-seven unconstrained optimization functions, the algorithm is statistically better considering four different population sizes and fifty iterations, providing a fair statistical significance. The testing process also highlights the requirement for mean centered statistical tests, since non-parametric alternatives may not be able to measure performance differences under certain specific circumstances due to their median analysis. Future lines include the redefinition of the original memetic algorithm, the inclusion of these techniques in different algorithms and the study of novel performance comparison measures to cover the possible lacks of non-parametric statistical tests.

Acknowledgements. This work was supported in part by Projects CICYT TIN2011-28620-C02-01, CICYT TEC2011-28626-C02-02, CAM CONTEXTS (S2009/TIC-1485) and DPS2008-07029-C02-02.

References

1. Arioli, M., Duff, I., Ruiz, D.: Stopping criteria for iterative solvers. SIAM Journal on Matrix Analysis and Applications 13, 138 (1992)
2. Back, T., Fogel, D., Michalewicz, Z.: Handbook of evolutionary computation. IOP Publishing Ltd. (1997)
3. Broyden, C.: The convergence of a class of double-rank minimization algorithms 1. general considerations. IMA Journal of Applied Mathematics 6(1), 76 (1970)
4. Cantu-Paz, E.: Efficient and accurate parallel genetic algorithms. Springer, Netherlands (2000)
5. Coello, C., Lamont, G., Van Veldhuizen, D.: Evolutionary algorithms for solving multi-objective problems. Springer-Verlag New York Inc. (2007)
6. Eshelman, L., Schaffer, J.: Preventing premature convergence in genetic algorithms by preventing incest. In: Proceedings of the Fourth International Conference on Genetic Algorithms, pp. 115–122. Morgan Kaufmann Publishers (1991)
7. García, S., Molina, D., Lozano, M., Herrera, F.: A study on the use of non-parametric tests for analyzing the evolutionary algorithms behaviour: a case study on the cec2005 special session on real parameter optimization. Journal of Heuristics 15(6), 617–644 (2009)
8. Guerrero, J., Berlanga, A., Molina, J.: A robust memetic algorithm with self-stopping capabilities. In: Proceedings of the 13th Annual Conference Companion on Genetic and Evolutionary Computation, pp. 113–114. ACM (2011)
9. Hansen, N., Müller, S., Koumoutsakos, P.: Reducing the time complexity of the derandomized evolution strategy with covariance matrix adaptation (CMA-ES). Evolutionary Computation 11(1), 1–18 (2003)
10. Kelley, C.: Detection and remediation of stagnation in the Nelder-Mead algorithm using a sufficient decrease condition. SIAM Journal on Optimization 10(1), 43–55 (2000)
11. Krasnogor, N., Smith, J.: A tutorial for competent memetic algorithms: model, taxonomy, and design issues. IEEE Transactions on Evolutionary Computation 9(5), 474–488 (2005)
12. Michalewicz, Z., Michalewicz, Z.: Genetic Algorithms + Data Structures = Evolution Programs, vol. 19. Springer, Berlin (1992)
13. Talbi, E.: Metaheuristics: From design to implementation. Wiley Online Library (2009)
14. Toffolo, A., Benini, E.: Genetic diversity as an objective in multi-objective evolutionary algorithms. Evolutionary Computation 11(2), 151–167 (2003)

Nonlinear Approaches to Automatic Elicitation of Distributed Oscillatory Clusters in Adaptive Self-organized System

Elena N. Benderskaya and Sofya V. Zhukova

Abstract. Chaotic neural networks find more and more applications in pattern recognition systems. However hybrid multidisciplinary solutions that combine advances from physics and artificial intelligence fields tend to enrich the complexity of designed systems and add more discussion points. This paper questions the applicability of well known chaotic time-series metrics (Shannon entropy, Kolmogorov entropy, Fractal dimension, Gumenyuk metric, complete and lag synchronization estimations) to simplify elicitation of distributed oscillatory clusters that store clustering results of a problem. Computer modeling results gives evidence that in case of clustering simple datasets the metrics are rather effective; however the concept of averaging out agent's dynamics fails when the clusters in the input dataset are linearly non-separable.

Keywords: chaotic neural network, self-organization, distributed clusters, chaotic synchronization, primitive agents, entropy, fractal dimension.

1 Introduction

Neural networks with chaotic activity of primitive bio-inspired agents attract scientific attention due to different reasons [1-3]. The one under the focus of this paper is neurophysiology evidence about chaotic synchronization effects that take

Elena N. Benderskaya
St. Petersburg State Polytechnical University, Faculty of Computer Science, Russia,
194021, St. Petersburg, Politechnicheskaya 21
e-mail: helen.bend@gmail.com

Sofya V. Zhukova
St. Petersburg State University, Graduate School of Management, Russia, 199004,
St. Petersburg Volkhovsky Per. 3
e-mail: sophya.zhukova@gmail.com

S. Omatu et al. (Eds.): Distributed Computing and Artificial Intelligence, AISC 151, pp. 733–741.
springerlink.com   © Springer-Verlag Berlin Heidelberg 2012

place in human brain when it solves clustering problems. Investigations of self-organization as the product of chaotic multi-agent system functioning add knowledge on perception and recognition processes. The desired outcome is understanding of brain structures formation principles in order to create highly intelligent systems.

Chaotic neural network is one of the formal models that comprise self-organization phenomena applied to solve clustering problems (2D, 3D, N-dimension). Chaotic dynamics of neurons is independent from initial conditions in the stationary regime and represents self-organization effects in terms of fragmentary synchronization within oscillatory clusters. Moreover, structure of chaotic neural network, comprised by primitive chaotic agents, adapts to the input dataset. Thus it makes possible to apply this model to cluster datasets with minimal or without any a priori information. However complex dynamics of agents is very hard for automatic interpretation. Full search of pairs of neurons that belong to the same cluster is too resource consuming to call clustering by means of chaotic neural network a real-time technique. Considered chaotic multi-agent system gives clustering answer in the form of oscillatory clusters to be interpreted further on. This paper aims to test applicability of different nonlinear metrics to elucidate data clusters from oscillatory clusters.

2 Adaptive Multi-agent System

Chaotic neural network (CNN) is a recurrent neural network with one layer of n neurons (agents). Each neuron corresponds to one point in the input dataset which in general case consists of n objects, each described by p features (p-dimensional dataset). CNN is a dynamic neural network, where each processing unit changes its state depending on the dynamics of all other neurons:

$$\begin{cases} y_1(t+1) = \dfrac{1}{C_1} \sum_{i=1}^{N} w_{1i} f(y_i(t)), t = \overline{1, T}_n \\[2mm] y_2(t+1) = \dfrac{1}{C_2} \sum_{i=1}^{N} w_{2i} f(y_i(t)), t = \overline{1, T}_n \\[2mm] \dots \\[2mm] y_N(t+1) = \dfrac{1}{C_N} \sum_{i=1}^{N} w_{Ni} f(y_i(t)), t = \overline{1, T}_n \end{cases} \tag{1}$$

$$C_i = \sum_{\substack{i \ne j}}^{N} w_{ij}, \quad i, j = \overline{1, N}, \tag{2}$$

$$W = \{w_{ij}\} = \exp\!\left(-d_{ij}^2 / 2a\right), \quad i, j = \overline{1, N}, \tag{3}$$

$$f(y(t)) = 1 - 2y^2(t), \tag{4}$$

where N – number of neurons, w_{ij} - strength of linkage between elements i and j, d_{ij} - euclidean distance between neurons i and j (each neuron represents a point from input dataset described by p coordinates), a – local scale calculated by means of triangulation metric [4], T_n – evolution period. The initial state of neural network is described by random values in the range [-1, 1]. Information about input dataset is given to CNN by means of linkage coefficients calculated via (3) - this is the first learning stage, the second coincides with iterative oscillation. Number of neurons N corresponds to n objects in the input dataset. To demonstrate paper results two input datasets are considered (Fig. 1.c, Fig. 1.f): simple one with linearly separable groups of points and complex one – with linearly inseparable groups. Transfer function of each neuron is calculated via (4) (as it was shown in [5] the role of transfer function can accomplish any logistic map that generates chaotic oscillations).

The key point of CNN functioning is the emergence of cooperative dynamics between neuron's outputs via time when all agents evolve due to (1). After some transition period they start to change states synchronously. Thus in respond to input datasets (Fig. 1.c, Fig. 1.f) CNN generates time sequences that comprise oscillatory clusters (Fig. 1.b, Fig. 1.e).

3   Self-organization via Internal Synchronization of Agents

To analyze synchronous activity chaotic neural network is considered as an ensemble of nonlinear agents. Space-time self-organization via synchronization of ensemble elements is one of the main research directions in modern nonlinear dynamics. Advances on chaotic synchronization processes that take place in distributed chaotic oscillatory systems demonstrate rich properties of such systems [6-8]. Some of them, comprised by identical elements (like CNN), are able to form synchronous clusters without any external influence. In opposite to external synchronization such phenomenon was called internal synchronization.

It was shown in [6, 9-10] that in ensembles of identical elements with weak linkage may take place complete synchronization, in-phase synchronization, phase synchronization, generalized synchronization and lag-synchronization. In chaotic neural network output dynamics is mainly chaotic. So there arise subsidiary task of chaotic synchronization elucidation to translate oscillatory clusters into clustering results. Dynamic regimes that occur in system (1) under condition of homogeneous mean field influence are described in [6].

In CNN mean field is inhomogeneous thus complete synchronization within oscillatory clusters occur only when simple dataset (with bunches of linear separable objects) is under process (Fig 1.c). To explore the chaotic neural dynamics (oscillatory clusters) visualization of CNN output evolution is provided (Fig. 1.b). The gray-scale gradient points (white color corresponds to 1.0 value, black color corresponds to -1.0 values) represent instant values of neurons outputs that change during observation period of 500 iterations (thus a representative sample is formed).   One can see vividly that in opposite to complete synchronization (Fig. 1.b) in case of clustering complex input dataset (Fig. 1.f) instant output values in one cluster may coincide neither by amplitude nor by phase and there can be no fixed synchronization lag (Fig. 1.e). In spite of everything joint mutual synchronization exists within each cluster. This synchronization is characterized by individual oscillation cluster melodies, by some unique "music fragments" corresponding to each cluster (Fig. 1.b, Fig. 1.e). From this follows the name we give to this synchronization type - fragmentary synchronization [11].

Investigation of different techniques to reveal clusters melodies lead us to the following conclusions. Fragmentary synchronization detection is to be based on the comparison of instant absolute values of outputs but not approximated in this or that way values. Asynchronous oscillations within one cluster in case of fragmentary synchronization can be nevertheless classified as similar. On Fig. 1.*a* and Fig. 1.*d* is represented the detailed dynamics of 5 neurons that correspond to 5 points in test datasets. These 5 points belong to 3 different clusters: 1,10,30 point – cluster 1, 80 point – cluster 2, 120 point – cluster 3 (this is relevant for both simple and complex image as their size and structure are chosen to be commensurable). On Fig. 1.*a* and Fig. 1.*d* first 50 counts of 500 counts statistics are considered. One can see that in case of complex image neurons 1, 10, 30 (Fig. 1.*d*) evolve asynchronously though they belong to the same cluster. However by means of human eyes they can be classified to the same group because clustering wave is spread between cluster agents.

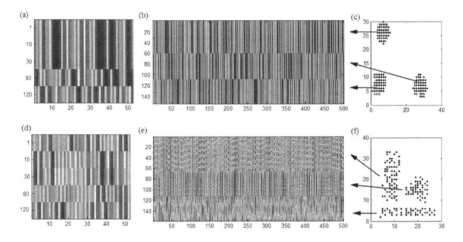

Fig. 1 Fragmentary synchronization of CNN outputs: : (a) – detailed first 50 counts of CNN dynamics statistics for 1, 10, 30, 80, 120 neurons (in respond to simple dataset); (b) – CNN dynamics statistics comprised by completely synchronized neurons within each of clusters in respond to simple input image; (c) – simple input dataset comprised by 61, 46, 50 points in three clusters (clusters are compact); (d) – detailed first 50 counts of CNN dynamics statistics for 1, 10, 30, 80, 120 neurons (in respond to complex dataset); (e) - CNN dynamics statistics comprised by fragmentary synchronized neurons in respond to complex input image; (f) – complex input image comprised by 65, 47, 46 points in three clusters (clusters are linear inseparable).

Method to detect automatically fragmentary synchronization [11] is based on comparison of neurons instant values in pairs and between them. The method is so resource consuming that prevents CNN application to solve real clustering problems, because full search of synchronized pairs take ten times more than formation of oscillatory clusters (solution time expands exponentially with increase of points in input dataset). To simplify decoding of self-organized clusters, approaches to calculation nonlinear metrics are considered.

4 Nonlinear Approaches for Synchronization Assessment

To detect synchronized pairs of neurons to be further interpreted as points of some cluster one should consider advances in nonlinear dynamics, synchronization theory, chaos theory. To prevent misunderstanding in highly multidisciplinary environment of this paper we consider as synonyms "dynamics of neurons outputs", "time-series", "output sequences", "symbolic chains", "neuron's trajectories", "agent dynamics", "output signals". There exist two main approaches to diversify chaotic time-series. First is based on substitution of each sequence (vector) with a scalar (fractal dimension, Gumenyuk integrated characteristic of elements order). Second is based on calculation of aggregated characteristic for a pair of sequences (Shannon information, Kolmogorov entropy, complete synchronization metric, lag synchronization metric). As the outputs of chaotic neural network have continuous values (varying in the interval of [-1;1]) the first step in calculation of six mentioned characteristics is discretization of these values.

Shannon information (I) based on Shannon entropy [10] is calculated by coarsening of positive CNN outputs with ones and negative outputs with zeros. Thus we calculate number of two different states in each sequence by means of non-conditional and conditional probabilities:

$$I_{ij} = H_i + H_j - H_{ij},\qquad(5)$$

where H_i, H_j – Shannon entropy of i and j sequence correspondingly, H_{ij} – mutual entropy for sequences i and j.

Gumenyuk integrated characteristics of elements order (V) [12] allows to take into consideration not only the number of repeating states in a sequence but also their order within each sequence:

$$V_j = \prod_{i=1}^{n_j} \Delta_{ij},\qquad(6)$$

where j – number of sequence, Δ_{ij} – interval between i and $i+1$ occurrences of value j in the symbolic chain (sequence), n_j – number of j-value entries in sequence.

Kolmogorov entropy (K) speaks for the speed of information decrease [8]. This characteristic allows taking into account the order of transitions between different states within a sequence and between two of them:

$$K = \lim_{\tau \to 0} \lim_{l \to 0} \lim_{N \to \infty} \frac{1}{N\tau} \sum_{n=0}^{N-1} (K_{n+1} - K_n),\qquad(7)$$

where τ is the interval of trajectory measurements, l - precision of measurement, N – sequence length, where K_n, K_{n+1} is the information needed to predict which hypercube the trajectory will go up to [8].

Fractal dimension (D) is applied to measure spatial arrangement of trajectories [8] which is very attractive because in CNN each trajectory contributes to the formation of separate cluster macroscopic attractor:

$$D = -\lim_{\varepsilon \to 0} \frac{\log N(\varepsilon)}{\log \varepsilon}, \qquad (8)$$

where $N(\varepsilon)$ the number of self-similar structures of linear size ε needed to cover the whole structure of attractor.

Complete synchronization metric (v) helps to reveal completely synchronized chaotic signals without any averaging out of outputs [10]:

$$v = \frac{\sigma_{x_1 + x_2}^2}{2\left(\sigma_{x_1}^2 + \sigma_{x_2}^2\right)}, \qquad (9)$$

where x_1, x_2 time sequences, $\sigma_{x_1}^2$ и $\sigma_{x_2}^2$ – squared standard deviations of sequences, $\sigma_{x_1 + x_2}^2$ - standard deviation of sequences sum.

Lag synchronization metric (k) reveals not only complete but also synchronization with delays [10]. As it is impossible to determine phase concept for a CNN system of difference equations, other metrics to detect phase synchronization are not considered in the paper:

$$\kappa = \min_\tau G(\tau), \; G(\tau) = \frac{\left\langle (x_2(t+\tau) - x_1(t))^2 \right\rangle}{\sqrt{\left\langle x_1^2(t) \right\rangle \left\langle x_2^2(t) \right\rangle}}, \qquad (10)$$

where $G(\tau)$ – similarity function that measures synchronization of sequences $x_1(t)$, $x_2(t)$ with τ delay, $\langle . \rangle$ - value averaged within ensemble.

5 Results

Computer Modeling Experiment
To find differences in between chaotic time-series (dynamic outputs of agents) the statistics (visualized on Fig.1.c for simple image and on Fig. 1.f for complex image) is considered. As it follows from metrics descriptions we consider two types of characteristics: pair-trajectory characteristics (that measure similarity/difference between time-series, e.g. Shannon information, Kolmogorov entropy, complete-synchronization metric, lag-synchronization metric), individual trajectory characteristics (integral numerical characteristic of elements order, fractal dimension). Considered characteristics were calculated for all points (pairs of points) for datasets from Fig. 1.c, Fig. 1.f and many other test clustering problems. In this paper we generalize our experimental base on the example of a representative, most common situation analyzing mutual fluctuations of neurons, that belong to the same cluster (neurons 1, 10, 30) and to different cluster (neurons 80,120). So we extract 4 rows from Fig. 1.b with numbers 1, 10, 30, 80, 120 and 4 rows from Fig. 1.e with numbers 1, 10, 30, 80, 120. Calculation of characteristics should allow us to diversify time-series in two clusters without component-wise analysis of instant values. The calculation results for both simple (Fig. 1.c) and complex (Fig.1.f)

images are presented on Fig. 2. The ideal situation is represented on Fig. 2.*g* for desired ideal metric IM. For the points 1, 10, 30 of the same cluster IM has large value of similarity (0.8, 0.9) and small value of similarity (0.1, 0.2) for points that belong to different clusters. In other words the set of metrics values calculated to substitute time-series with scalars should provide enough information for appropriate automatic differentiation of quasi-synchronized chaotic time-series.

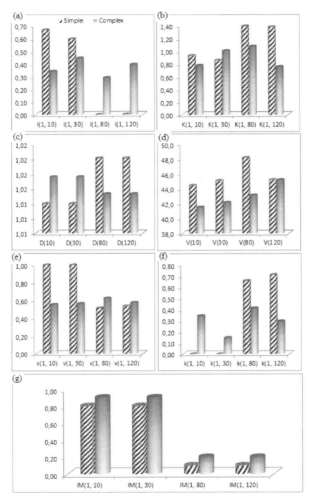

Fig. 2 Applicability of different nonlinear metrics to automatic joining of sequences (1 and 10 to one cluster) and automatic diversification of sequences (80, 120 to different clusters): (a) – Shannon information (I) calculated for pairs (1,10) (1,30), (1,80), (1,120); (b) – Kolmogorov entropy (K) calculated for pairs (1,10) (1,30), (1,80), (1,120); (c) – fractal dimension (D) calculated for points 10, 30, 80, 120; (d) – Gumenyuk integral characteristic of element order in sequences 10, 30, 80, 120 ; (e) – complete synchronization metric (v) calculated for pairs (1,10) (1,30), (1,80), (1,120); (f) – lag synchronization metric (k) calculated for pairs (1,10) (1,30), (1,80), (1,120); (g) – desired ideal metric IM values.

Analysis of Nonlinear Metrics Application

Analysis of characteristic brings us to following outcomes.

(a) Shannon information metric is applicable in case of clustering simple datasets when complete synchronization (Fig. 1.*a*, Fig. 1.*b*) within clusters can be produced: I-values differ greatly for neurons that correspond for points from different clusters (Fig. 2.*a*). For linear non-separable dataset from (Fig. 1.*f*) it is hard two find automatically the threshold for I that would allow to differ neurons 80, 120 from neurons 1, 10. That means that if Shannon metric would be used as part of clustering method points 1,10, 30, 60, 80 would be related to the same cluster and clusterization mistake would increase. This happens because of very similar values of entropy, discussed in [13].

(b) Kolmogorov entropy seems to complicate diversification of time-series, because of insignificant changes between K-values of points pairs that belong to the same cluster and K-values of points pairs that belong to different clusters (Fig. 2.*b*). It is too hard to find a proper threshold automatically.

(c) Fractal dimension happens to be inapplicable. Values of fractal dimension for different sequences are very similar to each other (Fig. 2.*c*) and thus all considered neurons are marked as belonging to the same cluster. This result supports the idea that required characteristic should measure not individual but relative dynamics of elements.

(d) Complete and lag synchronization metric application (Fig. 2.*e*, Fig. 2.*f*) results are somewhat similar to Shannon entropy's. The concept of averaging out that take place in case of these three metrics destroys or smooths out information about intricate structure of mutual fluctuations of neurons values in time and space. Though it is necessary to conclude that these metrics are applicable to cluster linear separable groups of objects in input dataset.

(e) Gumenyuk integrated characteristic of elements order have rather similar values for chains that belong even to different clusters (Fig. 2.*d*). This can be explained that in spite of belonging to different clusters the nature of chains is chaotic thus sequences have very similar structure in terms of alphabet. The question of application homogeneous chains within inhomogeneous one remains for open discussion.

6   Conclusions

This paper finds out that Shannon information, Kolmogorov entropy, complete and lag synchronization metrics can be successfully applied to cluster compact, standing far from each groups of objects in input dataset. Application of fractal dimension and integrated characteristic of elements order leads to erroneous clustering results due to losses information about unique structure of dynamic cluster (mutual interactions between neurons outputs are not considered). For all six considered characteristics in case of clustering complex datasets the following statement is true: the more complex is the data, the more intricate oscillation clusters are formed, and the greater clustering mistake is produced when the considered metrics are applied. Discover of proper characteristic to decode self-organized clusters of chaotic nature is the issue of further investigation.

References

1. Kazanovich, Y.B.: Nonlinear dynamics modeling and information processing in the brain. Optical Memory & Neural Networks 16(3), 111–124 (2007)
2. Borisyuk, R., Hoppensteadt, F.: Oscillatory models of the hippocampus: A study of spatio-temporal patterns of neural activity. Biological Cybernetics 81, 359–371 (1999)
3. Kuzmina, M.G., Manykin, E.A., Surina, I.I.: Oscillatory network with Self-organized dynamical connections for synchronization-based image segmentation. Bio. Systems 76, 43–53 (2004)
4. Benderskaya, E.N., Zhukova, S.V.: Clustering by Chaotic Neural Networks with Mean Field Calculated Via Delaunay Triangulation. In: Corchado, E., Abraham, A., Pedrycz, W. (eds.) HAIS 2008. LNCS (LNAI), vol. 5271, pp. 408–416. Springer, Heidelberg (2008)
5. Benderskaya, E.N., Zhukova, S.V.: Oscillatory Chaotic Neural Network as a Hybrid System for Pattern Recognition. In: IEEE Workshop on Hybrid Intelligent Models and Applications (HIMA), Paris, pp. 39–45 (2011)
6. Junji, I., Kaneko, K.: Spontaneous structure formation in a network of dynamic elements. Phys. Rev. E 67(14), 119–129 (2003)
7. Ivanova, A.S., Kuznetsov, S.P., Osbaldestin, A.H.: Universality and Scaling in Networks of Period-Doubling Maps with a Pacemaker. Discrete Dynamics in Nature and Society 2006, Article ID 74723, 1–19 (2006)
8. Mikhailov, A.S., Loskutov, A.: Foundation of Synergetics II. Complex Patterns, p. 210. Springer, Berlin (1995)
9. Pikovsky, A., Rosenblum, M., Kurths, J.: Synchronization: A Universal Concept in Nonlinear Sciences. Cambridge University Press, CNSS (2003)
10. Anishchenko, V.S., Astakhov, V.V., Neiman, A.B., Vadivasova, T.E., Schimansky-Geier, L.: Nonlinear Dynamics of Chaotic and Stochastic Systems. In: Tutorial and Modern Development, 2nd edn. Tutorial and Modern Development, p. 460. Springer, Heidelberg (2007)
11. Benderskaya, E.N., Zhukova, S.V.: Large-dimension image clustering by means of fragmentary synchronization in chaotic systems. Pattern Recognition and Image Analysis 19(2), 306–314 (2009)
12. Gumenyuk, A., Kostyshin, A., Simonova, S.: An approach to the research of the structure of linguistic and musical texts. Glottometrics (3), 61–89 (2002)
13. Dunin-Barkowski, W.L.: On Neuroscience-Based Informational Technologies for Authentic Models of Individual Mind. Optical Memory and Neural Networks 16(1), 47–50 (2007)

Classifier Ensemble Framework Based on Clustering

Hamid Parvin, Sajad Parvin, Zahra Rezaei, and Moslem Mohamadi

Abstract. This paper proposes an innovative combinational method how to select the number of clusters in the Classifier Selection by Clustering (CSC) to improve the performance of classifier ensembles both in stabilities of their results and in their accuracies as much as possible. The CSC uses bagging as the generator of base classifiers. Base classifiers are kept fixed as either decision trees or multilayer perceptron during the creation of the ensemble. Then it partitions the classifiers using a clustering algorithm. After that by selecting one classifier per each cluster, it produces the final ensemble. The weighted majority vote is taken as consensus function of the ensemble. Here it is probed how the cluster number affects the performance of the CSC method and how we can switch to a well approximation option for a dataset adaptively. We expand our studies on a large number of real datasets of UCI repository to reach a well conclusion.

Keywords: Decision Tree, Classifier Ensembles, Bagging, AdaBoosting, Clustering.

1 Introduction

Two of the most common problems which have always been the hot spot for the researches in pattern recognition are Classification and Regression. In classification the aim is to determine a discriminative function, so that the function maps the n-dimensional feature space to decision regions. In regression the aim is to find a model of an unknown function that can map the n-dimensional feature space to an approximation of the target function.

In other words, classification is the process of assigning data samples to predefined classes. Classification is also considered as a kind of supervised

Hamid Parvin · Sajad Parvin · Zahra Rezaei · Moslem Mohamadi
Nourabad Mamasani Branch, Islamic Azad University Nourabad Mamasani, Iran
e-mail: hamidparvin@mamasaniiau.ac.ir, s.parvin@iust.ac.ir,
    b_minaei@iust.ac.ir, mohamadi@iust.ac.ir

S. Omatu et al. (Eds.): Distributed Computing and Artificial Intelligence, AISC 151, pp. 743–750.
springerlink.com   © Springer-Verlag Berlin Heidelberg 2012

schemes [11], [12]. One of the most important goals of artificial intelligence is to design models with high recognition rates. In pattern recognition, the input space is mapped into the high dimensional feature space, and in the feature space it is tried to determine the optimal hyperplane(s), so that the mapped function better approximates the main function for each unseen data.

Although the more accurate classifier leads to a better performance, there is another approach to use many inaccurate classifiers specialized for a few data in the different problem-spaces and using their consensus vote as the classifier. This can lead to a better performance due to the reinforcement of the consensus classifier in the error-prone feature spaces. In General, it is ever-true sentence that combining diverse classifiers usually results in a better classification [6]-[7].

Giacinto and Roli propose a clustering and selection method to deal with the diversity generation problem [5]. They first produce a large number of classifiers with different initializations, after that they select a subset of them according to their distances in their output space. They don't take into consideration how the base classifiers are created. In this paper it is explored that usage of Bagging and Boosting as the sources of generators of diversity how can affect on their methods. Besides it is explored that selection of a classifier from each of clusters how can be done better.

Generally, there are two important challenging approaches to combine a number of classifiers that use different train sets. They are Bagging and Boosting. Both of them are considered as two methods that are sources of diversity generation.

The term Bagging is first used by [2] abbreviating for Bootstrap AGGregatING. The idea of Bagging is simple and interesting: the ensemble is made of classifiers built on bootstrap copies of the train set. Using different train sets, the needed diversity for ensemble is obtained. Breiman [3] proposes a variant of Bagging which it is called Random Forest. Random Forest is a general class of ensemble building methods using a decision tree as the base classifier. Random Forest algorithm which is one of the well known versions of Bagging classifier [7] is implemented and compared with the proposed method. Boosting is inspired by an online learning algorithm called Hedge(β) [4]. Another version of these algorithms is arc-x4 which performs as another version of recently ADAboost [7].

Giacinto and Roli propose a clustering and selection method [5]. They first produce a large number of MLP classifiers with different initializations. After that they partition them. They select one classifier from each cluster of the partition. Finally they consider them as an ensemble and aggregate them to produce final decisions.

Parvin et al. propose a framework for development of combinational classifiers. In their framework, a number of train data-bags are first bootstrapped from train data-set. Then a pool of weak base classifiers is created; each classifier is trained on one distinct data-bag. After that to get rid of similar base classifiers of the ensemble, the classifiers are partitioned using a clustering algorithm. The partitioning is done considering the outputs of classifiers on train dataset as feature space. In each partition, one random classifier is selected to participate in final ensemble. Then, to produce consensus vote, different votes (or outputs) are gathered out of ensemble. After that the weighted majority voting algorithm is applied over them. The weights are determined using the accuracies of the base classifiers on train dataset [9].

2 Classifier Selection by Clustering

The main idea behind the classifier selection by clustering method is to use the most diverse set of classifiers obtained by Bagging or Boosting mechanism. Indeed a number of classifiers are first trained by the two well-known mechanisms: Bagging or Boosting. After that the produced classifiers partitioned according their outputs. Then the nearest classifier to the head of each produced cluster is selected. Since each cluster is produced according to classifiers' outputs, it is highly likely that selecting one classifier from each cluster, and using them as an ensemble can produce a diverse ensemble that outperforms the traditional Bagging and Boosting, i.e. usage of all classifiers as an ensemble.

In training phase of the classifier selection by clustering method by Bagging, n subsets of dataset are bootstrapped with b percent of the train dataset. Then a classifier is trained on each of those subsets. Each classifier is also then tested over the whole of train dataset and its accuracy is calculated after that. The output of ith classifier over train dataset is denoted by O_i and its accuracy is denoted by P_i.

In training phase of the classifier selection by clustering method by Boosting, a subset of dataset containing b percent of train dataset is again selected. Then the first classifier is trained on this subset. After that the first classifier is tested on the whole train dataset which this results in producing the O_1 and P_1. Using O_1, the next subset of b percent of train dataset is obtained. This mechanism is continued in such a way that obtaining ith subset of b percent of train dataset is produced considering the $O_1, O_2, \ldots, O_{i-1}$. For more information about the mechanism of Boosting, the reader can refer to Kuncheva [7].

In final part of the method a dataset whose ith dataitem is O_i is first produced. Features of this dataset are real dataitems of under-leaning dataset. A new dataset having n classifiers and N features is available, where n is a predefined value showing the number of classifiers produced by Bagging or Boosting and N is the cardinality of under-leaning datasets. After producing the mentioned dataset, the dataset is partitioned by use of a clustering algorithm that this results in some clusters of classifiers. Each of the classifiers of a cluster has similar outputs on the train dataset; it means these classifiers have low diversities, so it is better to use one of them in the final ensemble rather than all of them. For escaping from outlier classifiers, the clusters which contain number of classifiers smaller than a threshold are ignored.

Table 1 Details of used dataset.

Dataset Name	# of dataitems	# of features	# of classes	Data distribution per classes
Breast Cancer*	404	9	2	444-239
Balance Scale*	625	4	3	288-49-288
Bupa*	345	6	2	145-200
Glass*	214	9	6	70-76-17-13-9-29
Galaxy*	323	4	7	51-28-46-38-80-45-35
Half-Ring*	400	2	2	300-100
SAHeart*	462	9	2	160-302
Ionosphere*	351	34	2	126-225
Iris*	150	4	3	50-50-50
test Monk1	412	6	2	216-216
test Monk2	412	6	2	216-216
test Monk3	412	6	2	216-216
train Monk1	124	6	2	62-62
train Monk2	169	6	2	105-64
train Monk3	122	6	2	62-60
Wine*	178	13	3	59-71-48
Yeast*	1484	8	10	463-5-35-44-51-163-244-429-20-30

3 Experimental Study and Discussion

The accuracy is taken as the evaluation metric throughout all the paper. All the experiments are done using 4-fold cross validation. The results obtained by 4-fold cross validation are repeated as many as 10 independent runs. The averaged accuracies over the 10 independent runs are reported.

The proposed method is examined over 13 different standard datasets and one artificial dataset. It is tried for datasets to be diverse in their number of true classes, features and samples. A large variety in used datasets can more validate the obtained results. Brief information about the used datasets is available in Table 1. These real datasets are available at UCI repository [1]. The details of half-ring dataset can be available in [8].

Note that some of datasets which are marked with star (*) in Table 1 are normalized. All experiments are done over the normalized features in the stared dataset. It means each feature is normalized with mean of 0 and variance of 1, $N(0, 1)$.

The measure of decision for each employed decision tree is taken as Gini measure. The threshold of pruning is set to 2. Also the classifiers' parameters are fixed in all of their usages.

All multilayer perceptrons which are used in the experiments have two hidden layers including 10 and 5 neurons respectively in the first and second hidden layers, as well as they are iterated 100 epochs.

In all experiments n, b and threshold of accepting a cluster are set to 151, 30 and 2 (i.e. only the clusters with one classifier is dropped down) respectively. All the experiments are done using 4-fold cross validation. Clustering is done by k-means clustering algorithm with different k parameters.

Table 2. Comparison of the results by considering Decision Tree as base classifier. * shows the dataset is normalized, and 4 fold cross validation is taken for performance evaluation. ** shows that the train and test sets are predefined and averaged over 10 independent runs.

	Arc-X4	Random Forest	Classifier Selection By RF	Classifier Selection By Arc-X4	Cluster and Selection
Breast Cancer*	95.74	96.32	96.47	95.05	93.68
Balance Scale*	94.44	93.60	94.72	94.24	94.44
Bupa*	70.64	72.09	72.97	66.28	64.53
Glass*	65.04	70.28	70.28	62.26	60.85
Galaxy*	70.59	73.07	72.45	70.28	70.94
Half-Ring*	97.25	95.75	97.25	95.75	95.75
SAHeart*	70.00	71.30	72.61	69.70	68.04
Ionosphere*	90.31	92.31	91.45	89.74	87.64
Iris*	96.62	95.27	96.62	95.95	94.59
Monk problem1**	98.11	97.49	98.76	97.37	98.34
Monk problem2**	97.01	86.64	97.62	86.73	97.14
Monk problem3**	87.29	96.92	96.97	96.34	87.31
Wine*	96.07	97.19	97.19	95.51	92.61
Yeast*	53.17	53.98	53.98	52.09	54.51
Average	84.45	85.16	86.38	83.38	82.88

Table 2 shows the accuracies of different methods by considering a DT as each of the base classifiers. Table 3 shows the accuracies of different methods by considering a MLP as each of the base classifiers. The parameter r is set to 33 to reach the results of the Table 2 and Table 3.

While we choose only at most 22 percent of the base classifiers of Bagging, the accuracy of their ensemble outperforms the full ensemble of them, i.e. Bagging Method. Also it outperforms Boosting method and proposed method based on Boosting method.

Because the classifiers selected in this manner (by Bagging along with clustering), have different outputs, i.e. they are as diverse as possible, they are more suitable than ensemble of all them. It is worthy to mention that the Boosting is inherently diverse enough to be an ensemble totally; and the reduction of ensemble size by clustering destructs their Boosting effect. Take it in the consideration that in Boosting ensemble, each member covers the drawbacks of the previous ones.

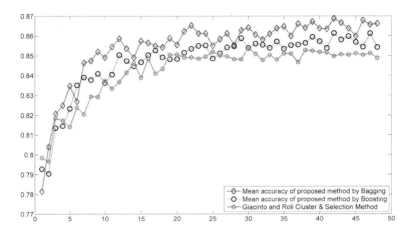

Fig. 1 The performance of CSC methods averaged over 14 datasets of Table 1 with *n=151*
and different *r* and MLP as base classifier

To see how the parameter *r* affects over the performance of classification over
CSC methods (by bagging, boosting and Gianito [17]) with two base classifiers
(MLP, DT), take a look at Fig. 1 and Fig. 2.

These figures depict the accuracies of different method by 4-fold cross
validation on some benchmarks. As it is inferred from these figures, increasing the
cluster number parameter *r* is not always resulted in the improvement in the
performance.

Indeed an *r*=15 is a well choice for all of the datasets. It means that if the
classifier number parameter, *n*, is 151 then *r*=15 is a well value for the cluster

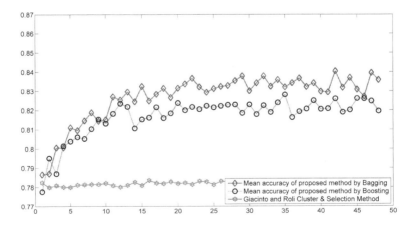

Fig. 2 The performance of CSC methods averaged over 14 datasets of Table 1 with *n=151*
and different *r* and DT as base classifier

number parameter. In other words, using 10 percent of the base classifiers in the final ensemble can be a good option.

Indeed in this option, a classifier is selected from a cluster of classifiers that contains about 10 classifiers, so it gives the method the ability to select each classifier from a good coverage of classifiers.

The Fig. 1 and Fig. 2 depict the averaged accuracies over all 14 different datasets. Fig. 1 reports the performances of the framework by using the MLP as base classifier.

Fig. 2 reports the performances of the framework by using the DT as base classifier.

As it is illustrated in the Fig. 1 and Fig. 2, usage of bagging as generator of classifiers for CSC method is better than Boosting and Giacinto and Roli's methods. Also it is concluded that using $r=33$ instead of $r=15$ is a better choice for all of the 14 datasets. In other words, using 22 percent of the base classifiers in the final ensemble can be a better option.

Table 3. Comparison of the results by considering MLP as base classifier. * shows the dataset is normalized, and 4 fold cross validation is taken for performance evaluation. ** shows that the train and test sets are predefined and averaged over 10 independent runs.

	Arc-X4	Bagging	Classifier Selection By Bagging	Classifier Selection By Arc-X4	Cluster and Selection
Breast Cancer*	97.06	96.91	96.91	96.47	96.19
Balance Scale*	93.27	91.99	91.35	92.95	95.75
Bupa*	70.06	71.22	72.09	68.02	71.98
Glass*	66.04	66.98	67.45	66.04	67.05
Galaxy*	87.00	85.62	85.62	84.52	87.00
Half-Ring*	97.25	95.75	97.25	97.25	97.25
SAHeart*	73.04	72.39	71.52	71.09	70.18
Ionosphere*	90.03	88.51	90.31	87.64	88.51
Iris*	96.62	96.62	97.97	97.33	93.33
Monk problem1**	98.06	92.23	98.43	97.87	98.34
Monk problem2**	87.35	85.68	87.41	87.23	87.21
Monk problem3**	97.09	95.87	97.33	96.99	96.77
Wine*	96.59	96.06	97.19	95.51	95.23
Yeast*	60.85	61.19	61.19	60.85	60.56
Average	86.45	85.50	86.57	85.70	86.10

Comparing the Fig. 1 and Fig. 2 one can find out that the using the decision tree as base classifier increases the gap between the three approaches to generate the base classifiers. It is due to special feature of the decision tree. Because it is very sensitive to its train set, the use of decision tree as base classifier is very consistent with the Bagging mechanism.

4 Conclusion and Future Works

In this paper, we have proposed a new method to improve the performance of classification. The proposed method uses Bagging as generator of the base classifiers. Then using k-means we partition the classifiers. After that we select one classifier per a validated cluster.

Also it is concluded that using 22 percent of the base classifiers in the final ensemble can be a well option generally.

Using the decision tree as base classifier increases the gap between the three approaches to generate the base classifiers. It is due to special feature of the decision tree. Because it is very sensitive to its train set, the use of decision tree as base classifier is very consistent with the Bagging mechanism. While we choose only at most 22 percent of the base classifiers of Bagging, the accuracy of their ensemble outperforms the full ensemble of them. Also it outperforms Boosting.

As a future work, one can turn to research on the variance of the method. Since it is said about Bagging can reduce variance and Boosting can simultaneously reduce variance and error rate.

References

[1] Blake, C.L., Merz, C.J.: UCI Repository of machine learning databases (1998), http://www.ics.uci.edu/~mlearn/MLRepository.html

[2] Breiman, L.: Bagging Predictors. Journal of Machine Learning 24(2), 123–140 (1996)

[3] Breiman, L.: Random Forests. Machine Learning 45(1), 5–32 (2001)

[4] Freund, Y., Schapire, R.E.: A Decision-Theoretic Generalization of On-Line Learning and an Application to Boosting. J. Comput. Syst. Sci. 55(1), 119–139 (1997)

[5] Giacinto, G., Roli, F.: An approach to the automatic design of multiple classifier systems. Pattern Recognition Letters 22, 25–33 (2001)

[6] Günter, S., Bunke, H.: Creation of Classifier Ensembles for Handwritten Word Recognition Using Feature Selection Algorithms. In: Proceedings of the Eighth International Workshop on Frontiers in Handwriting Recognition (IWFHR 2002), August 06-08, p. 183 (2002)

[7] Kuncheva, L.I.: Combining Pattern Classifiers, Methods and Algorithms. Wiley, New York (2005)

[8] Minaei-Bidgoli, B., Topchy, A.P., Punch, W.F.: Ensembles of Partitions via Data Resampling. In: ITCC, pp. 188–192 (2004)

[9] Parvin, H., Minaei-Bidgoli, B., Beigi, A.: A New Classifier Ensembles Framework. Knowledge-Based and Intelligent Information and Engineering Systems, 110–119 (2011)

[10] Yang, T.: Computational Verb Decision Trees. International Journal of Computational Cognition 4(4), 34–46 (2006)

[11] Duda, R.O., Hart, P.E., Stork, D.G.: Pattern Classification, 2nd edn. John Wiley & Sons, NY (2001)

[12] Jain, A.K., Duin, R.P.W., Mao, J.: Satanical pattern recognition: a review. IEEE Transaction on Pattern Analysis and Machine Intelligence PAMI-22(1), 4–37 (2000)

Resource Allocation Strategies to Maximize Network Survivability Considering of Average DOD

Frank Yeong-Sung Lin, Pei-Yu Chen[*], and Quen-Ting Chen

Abstract. In this paper, an innovative metric called Average Degree of Disconnectivity (Average DOD) is proposed. The Average DOD combining the concept of the probability calculated by contest success function with the DOD metric would be used to evaluate the damage degree of network. The larger value of the Average DOD, the more damage degree of the network would be. An attack-defense scenario as a mathematical model would be used to support network operators to predict that all the likelihood strategies both cyber attacker and network defender would take. The attacker could use the attack resources to launch attack on the nodes of network. On the other hand, the network defender allocates existed resources of defender to protect survival nodes of network. In the process of problem solving, the "gradient method" and "game theory" would be adopted to find the optimal resource allocation strategies for both cyber attacker and network defender.

Keywords: Average Degree of Disconnectivity, Average DOD, Contest Success Function, Gradient Method, Network Survivability, Optimization, Game Theory, Resource Allocation, Network Attack and Defense.

Frank Yeong-Sung Lin · Pei-Yu Chen · Quen-Ting Chen
Department of Information Management, National Taiwan University,
Taipei, Taiwan, R.O.C.
e-mail: yslin@im.ntu.edu.tw, d96006@im.ntu.edu.tw,
    r98043@im.ntu.edu.tw

Pei-Yu Chen
CyberTrust Technology Institute, Institute for Information Industry2 Taipei,
Taiwan, R.O.C.

[*] Correspondence should be sent to d96006@im.ntu.edu.tw.

S. Omatu et al. (Eds.): Distributed Computing and Artificial Intelligence, AISC 151, pp. 751–758.
springerlink.com   © Springer-Verlag Berlin Heidelberg 2012

1   Introduction

In the past, the security state of systems or infrastructures was classified in terms of two states: safe or compromised [1]. However, the network often faces many situations such as natural disasters, malicious attacks, and random error conditions which could result in different outcomes. Network security professionals must ensure the available and continuous services. Therefore, the binary concept is insufficient to describe a system's state. As a result, more and more researchers studied the issue of network survivability.

Traditionally, when measuring the survivability of a network, some researchers use the concept of Greatest Residual Region (GRR, or the largest component in remaining network). After partial of the network are disabled by attacks or failures, GRR is then used to assess the impact in terms of the proportion of nodes in the largest region of a broken network. However, this metric is insufficient to describe the sensibility of partial dysfunctional network yet.

In [2], this paper proposed another metric of the network survivability which is called the Degree of Disconnectivity (DOD). The DOD metric could be used to evaluate the partial damage degree of network. The calculated value seems to be effectively represented the damage degree of network. However, there is still a disadvantage to DOD, which has assumed that the attacker would launch the attack either successfully or unsuccessfully. This assumption is limited, since the attack might not be 100% successful or unsuccessful. As a result, in this paper, an evolved DOD metric is proposed, which is narrated the average probability of each attack and defense case, the more details is explained in the next sections.

Moreover, the interaction between cyber attackers and network defenders is similar to information warfare. The cyber attackers expect to minimize the degree of network survivability. On the other hand, the network defenders always expect to maximize the degree of network survivability. This attack-defense situation would become a min-max or max-min problem. The researchers solve this kind of attack-defense problem of network security by mathematical programming approach, such as game theory [3], Lagrangean Relaxation Method. [4] is considering one or several important nodes or systems in the topology. However, this assumption restricts to evaluate the generalize situation in the network. To enhance or reduce network survivability, both network defender and cyber attacker usually need to invest the fixed number of resources in the network. In the end, it would be a significant issue about how to efficiently allocate resources to the network both cyber attacker and network defender. The more details are described in the following.

2   Problem Formulation and Notations

The DOD metric, proposed in [3], assumed that the cyber attacker launches the attack either successfully or unsuccessfully, but this assumption is limited to describe the attack result that might not be completely compromised, which is a disadvantage to this metric. To improve this problem, we propose a new metric of the network survivability which is called Average DOD. Average DOD combined the concept of probability calculated by the contest success function [5] with the DOD metric. The definition of contest success function is showed in Table 1 and equation (1).

According to the definition of contest success function, if the attacker allocated more resource on a node, the more probability of the attacker could compromise the node. Similarly, if the defender allocated more resource on a node, the more probability of the defender could protect the node. Besides, m is a parameter which describes the intensity of the contest [6]. Here, we demonstrate an example to describe Average DOD. In a network, each network configuration would have a probability determined by the attack success or failure probability of each node; the method to calculate the probability of each kind of network configuration would be to multiply the attack success or failure probability of each node. If all the nodes of network are compromised by the attacker, the probability of this network configuration would be $\prod\limits_{i=1}^{9} S_i$ (Where S_i means the attack success probability of the node i). However, if all the nodes of network are still functional, the probability of this network configuration would be $\prod\limits_{i=1}^{9}(1-S_i)$. Furthermore, each kind of network configuration would lead to different damage degree of network. The Degree of Disconnectivity (DOD) could be adopted to measure the damage degree of network. If all the nodes of network are still functional, the DOD value would be 0.

The concept of expectation value the predicted mean value of the result of an experiment of statistics would be adopted to evaluate average damage degree of whole network. The calculated expectation value is defined as the Average DOD here. The larger number of the Average DOD value is, the more damage degree of the network would be. Meanwhile, the Average DOD value is influenced by the attack success probability calculated by the resource allocation of both cyber attacker and network defender. Therefore, the Average DOD value could be adopted to measure the damage degree of the network and find the optimal resource allocation in each node for both cyber attacker and network defender.

Table 1 Given Parameters and Decision Variables

Given parameter	
Notation	Description
$S_i(T_i, t_i)$	the attack successful probability on node i
T_i	the attack resource allocated on node i
t_i	the defensive resource allocated on node i
m	contest intensity

$$S_i(T_i,t_i) = \frac{T_i^m}{T_i^m + t_i^m}, where \frac{\partial S}{\partial T} \geq 0, \frac{\partial S}{\partial t} \geq 0, m \geq 0 \qquad (1)$$

2.1 Problem Description

Cyber attacker and network defender are always limited by the invested resources. How to make the decision to efficiently allocate resources to each node is an extremely significant issue for both cyber attacker and network defender. We proposed a new mathematical model to support both cyber attack and network defender to make the optimal decision. In this model, the damage degree of network is evaluated by the Average DOD value.

In this attack-defense problem, cyber attacker and network defender are through some strategies to attend their goals. From the perspective of network defender, the defender is usually looking forward to minimizing the damage degree of network. On the other hand, the cyber attacker expects to maximize the damage degree of network. Both of them could take some strategies to attend their goal. It is usually constrained by the allocated resources. The cyber attacker needs to determine how to allocate resources to attack targeted network. Besides, the attacker could accumulate some experience that could help the attacker having higher probability to compromise network in next time.

2.2 Problem Formulation

The above problem is formulated as a maximization mathematical model as follows. Note that the network discussed here is at the AS level. Both the attacker and the defender have complete information about the targeted network topology and the budget allocation is assumed. For simplicity, since the targeted network is at the AS level, the attacker cannot simply attack any node directly. The notations used in this paper and problem formulation is defined in Table 2.

Table 2 Given Parameters and Decision Variables

Given parameter	
Notation	Description
V	Index set of nodes
$\hat{A}$	Total budget of attacker
$\hat{B}$	Total budget of defender
θ_i	Existing defense resource allocated on node i, where $i \in V$
d_{ri}	The discount rate of defender reallocate resources on node i in round r, where $i \in V$ and $r \in R$

Decision variable	
Notation	Description
$\vec{a}$	Attacker's budget allocation, which is a vector of attack cost a_1, a_2 to a_i, where $i \in V$
$\vec{b}$	Defender's budget allocation, which is a vector of defense cost a_1, a_2 to a_i, where $i \in V$
a_i	Attacker's budget allocation on node i, where $i \in V$
b_i	Defender's budget allocation on node i, where $i \in V$
$\bar{D}(\vec{a},\vec{b})$	The Average DOD, which is considering under attacker's and defender's budget allocation are $\vec{a}$ and $\vec{b}$

The problem is then formulated as the following problem:

Objective function:

$$\min_{\vec{b}} \max_{\vec{a}} \bar{D}(\vec{a},\vec{b}),$$ (IP 1)

Subject to:

$$\sum_{i \in V} b_i \leq \hat{B} + \sum_{i \in V} \theta_i d_i$$ (IP 1.1)

$$\sum_{i \in V} a_i \leq \hat{A}.$$ (IP 1.2)

The objective function is to minimize the maximum the Average DOD. IP 1.1 describes the sum of the allocated defense budgets in each node should not exceed the sum of the new allocated and reallocated budgets in that round. IP 1.2 calculates the sum of the allocated attack budgets in each node should not exceed the attack budgets in that round.

3   Solution Approach

Here, how to optimize resource allocation in each node for both cyber attacker and network defender and to evaluate damage degree of network by the Average DOD value would be introduced. The gradient method [7] is used to calculate the Average DOD value and to find the optimal resource allocation strategy in each node for both cyber attacker and network defender. The detailed solution procedure would be discussed in first section. In addition, the concept of gradient method and the detailed method to calculate the Average DOD value would be introduced in second part. Finally, the time complexity of the solution approach would be analyzed.

3.1   Gradient Method

The gradient method is a general framework used to solve the optimization problems what is to maximize or minimize functions of continuous parameters. This problem is a min-max formulation and both cyber attacker and network defender are assumed that they could allocate continuous resources in each node. Here, the gradient method is adopted to solve this problem.

The gradient method usually could be categorized into two types, one is gradient descent and the other one is gradient ascent [9]. The gradient descent method could be used to solve the optimal minimization problem. On the other hand, the optimal maximization problem could be solved by the gradient ascent method. The concept of gradient descent and gradient ascent is extremely similar, so both of them could adopt the following algorithm: The detailed process flow of the gradient method is also described in Table 3.

Table 3 The Algorithm of the Gradient Method

Step1.Get a start point
Step2.Determine a positive or negative direction
Step3.Determine a step size
Step4.Repeat
a.   Find the most impact of all dimensions
b.   Move a step of the most of all dimensions
c.   Update the start point
Until a stop criterion is satisfied

3.2 Using Game Theory to Find the Optimal Solution

In this paper both cyber attacker and network defender need to determine how to efficiently allocate resources simultaneously in each node. This problem could be viewed as a simultaneous or imperfect information game. In addition, both cyber attacker and network defender have complete information about the strategies. Hence, this problem also could be regarded as a complete information game. Here, two players (cyber attacker and network defender), zero-sum, complete and imperfect information game would be used to solve this problem.

The representation of game theory normally has two types, one is the extensive form and the other one is the normal form. The normal form would be introduced to solve this problem in this model, which is represented by a matrix which shows the players, strategies, and payoff values. For example, two players, one is on the first column and the other one is on the first row of the matrix, own lots of different strategies, respectively. For example, both two players have five different strategies (S_{11} to S_{15} and S_{21} to S_{25}). The combination of two players with different strategies would produce 25 (U_{11} to U_{55}) different results (the Average DOD value).

Both cyber attacker and network defender have different strategies about the percentage resource allocation in each stage. In addition, the results of each kind of percentage resource allocation of each player would be calculated by the Average DOD. The solution approach of the complete and imperfect information game would be introduced in the following. Generally, the solution procedure of the complete and imperfect information game is shown as following.

Step1. Dominant strategy eliminating. The dominant strategy means that no matter what kind of strategy that the opponent to take is better than other strategies.

Step2. If only one strategy is left of each player, it would be the optimal strategy. Otherwise, go to step 3.

Step3. Using the minmax strategy to find the optimal strategy of each player. If minmax strategy still could not find the optimal strategy, go to step4.

Step4. Using the mixed strategy (Linear programming) to find the optimal strategy of each player.

3.3   Time Complexity Analysis

The time complexity of the algorithm quantifies the amount of time taken by an algorithm to run as a function of the size of the input to the problem. It would influence the efficiency of the proposed algorithm. To calculate the Average DOD value, the gradient method would be used to find the optimal resource allocation in each node. In addition, the DOD value would be used to measure the damage degree of each configuration. The time complexity of the gradient method would be $O(mV)$, since the impact degree of each node would be checked in each round. (Where m is the maximum number of the checked round and V is the node number in the network) In addition, the time complexity of the DOD value of each configuration would be $O(WV^2)$ (Where W is the number of the O-D pair), because the broken nodes of each O-D pair are needed to be checked. The checked method would adopt the Dijkstra's shortest path algorithm, so the time complexity would be $O(V^2)$. However, to compute the Average DOD value, the 2^V different kinds of network configuration would need to be considered. The time complexity to compute the Average DOD value would be $O(2^V WV^2)$. As a result, once the number of node is too huge, it must take much time to compute the Average DOD value.

4   Computational Experiments

Because of the complexity of this problem, the number of network node considered in the experiments is 9. In addition, three kinds of network topology are considered, the grid network (GD), the random network (RD), and the scale-free network (SF). The feature of the GD is really regular network. Besides, the SF is a kind of network whose degree distribution follows a power law. Finally, the RD is randomly connected with other nodes. Three kinds of network topology adopted to take the experiments in this paper are demonstrated in.

4.1   Experiment Environment

To find the optimal resource allocation strategy for both cyber attacker and network defender, there is something needing to be determined firstly. The resource reallocation policy of the defender would be that the defense resources of each stage would not be accumulated (RR1) and node recovery policy of the defender would be that all the compromised nodes would not be recovered (NR2). In addition, the accumulated experience of the attacker would not be considered in this experiment (NAE). The experiment results would be demonstrated in Table 4.

Table 4 The Experiment under Different Topology

Network Topology	Grid	Random	Scale-free
Average DOD	1.49871	1.49876	1.49886
Strategy of Attacker	20	20	20
Strategy of Defender	20	20	20

5   Conclusion

In this paper, we first evaluate the network survivability with a new proposed survivability metric called Average DOD, which is more sensible to respond the largest connected component of the network. The metric combined the concept of the probability calculated by the contest success function with the DOD metric would be as a new tool to evaluate network survivability.

In addition, an efficient attack-defense scenario is formulated. Considering the scenario of both cyber attacker and network defender utilize their resource and in each node are solved. The model is demonstrated under 3 different topologies and discussed. It seems that the grid network is fully connected is more robust among these topologies. However, considering of the interaction of attacker and defender, the model shall improve to solve multi-stage attack-defense issue. Moreover, some advanced technology, such as parallel processing system, could be considered to be adopted to improve efficiency in this model in the future.

Acknowledgments. This research was supported by the National Science Council of Taiwan, Republic of China, under grant NSC-100-2221-E-002-174.

References

1. Ellison, R.J., Fisher, D.A., Linger, R.C., Lipson, H.F., Longstaff, T., Mead, N.R.: Survivable Network Systems: An Emerging Discipline. Technical Report CMU/SEI-97-TR-013 (November 1997)
2. Lin, F.Y.S., Yen, H.H., Chen, P.Y., Wen, Y.F.: Evaluation of Network Survivability Considering Degree of Disconnectivity. In: Corchado, E., Kurzyński, M., Woźniak, M. (eds.) HAIS 2011, Part I. LNCS (LNAI), vol. 6678, pp. 51–58. Springer, Heidelberg (2011)
3. Jiang, W., Fang, B.X., Zhang, H.l., Tian, Z.H.: Optimal Network Security Strengthening Using Attack-Defense Game Model. In: Sixth International Conference on Information Technology (2009)
4. Lin, Y.S., Tsang, P.H., Chen, C.H., Tseng, C.L., Lin, Y.L.: Evaluation of Network Robustness for Given Defense Resource Allocation Strategies. In: Proceedings of the First International Conference on Availability, Reliability and Security (2006)
5. Skaperdas, S.: Contest Success Functions. Economic Theory (1996)
6. Peng, R., Levitin, G., Xie, M., Ng, S.H.: Defending simple series and parallel systems with imperfect false targets. Reliability Engineering & System Safety 95(6), 679–688 (2010)
7. Hassoun, H.: Fundamentals of Artificial Neural Networks. MIT Press (1995)

Comparative Analysis of Two Distribution Building Optimization Algorithms

Pavel Galushin, Olga Semenkina, and Andrey Shabalov

Abstract. This paper proposes the modification of genetic algorithm, which uses genetic operators, effecting not on particular solutions, but on the probabilities distribution of solution vector's components. This paper also compares reliability and efficiency of basic algorithm and proposed modification using the set of benchmark functions and real-world problem of dynamic scheduling of truck painting.

1 Introduction

Modern simulation models become more and more complex. As the rule researcher has no knowledge about properties of the model of interest, and model is not mathematical formula, but computer program. Moreover simulation model can contains stochastic elements. This facts heavily complicates model's optimization l, since traditional optimization methods use values of objective's gradients, which exist not always. In this situation, it is necessary to use direct search optimization methods. Important class of such optimizers is the class of evolutionary and distribution building algorithms. This paper proposes new model building algorithm and compares it with Population Based Incremental Learning technique using set of benchmark problems and practical problem of dynamic scheduling.

Population-based incremental learning (PBIL) is an attempt to create an algorithm similar to the genetic algorithm (GA), which preserves the basic properties of the genetic operators, but defined in terms of the theory of incremental learning [2]. The PBIL explicitly computes the components of the vector of probabilities and updates it using incremental learning. The PBIL scheme is as follows:

Pavel Galushin · Olga Semenkina · Andrey Shabalov
Institute of Computer Science and Telecommunication, Siberian State Aerospace University, Krasnoyarsk, Russia
e-mail: galushin@gmail.com, semenkina.olga@mail.ru,
　　　　shabalov-andrey@mail.ru

S. Omatu et al. (Eds.): Distributed Computing and Artificial Intelligence, AISC 151, pp. 759–766.
springerlink.com　　　　　　　　　　　　　　© Springer-Verlag Berlin Heidelberg 2012

1. Set all components of probability vector to 0.5 (uniform distribution).
2. A given number of solutions are generated from the probability vector.
3. The fitness of each solution is evaluated.
4. The best solution is selected.
5. Update probabilities vector: increase probability of solution's components which is equal to 1 and decrease others.
6. Mutate probability vector.
7. Repeat steps 1-5 until termination condition was met.

Thus PBIL explicitly uses some statistical information about objective. Baluja also mentioned [2] that this is possible to develop optimization method with mutation and selection operators, effecting not on the individuals, but on genes' values distribution, but preserve statistical properties of standard genetic operators.

2   Selection

Let us consider selection in the genetic algorithm. On this step, intermediate population is generated using given selection and crossover scheme (probability to be selected into the intermediate population is higher for individuals with better objective values).

Let population contain individuals $x_1,...,x_n$, probabilities to be selected (in one experiment) are $g_1,...,g_n$. Expected value of probability that i-th gene will be equal to 1, is

$$p^{(i)} = \sum_{k=1}^{n} x_k^{(i)} g_k , \qquad (1)$$

Where $x_k^{(i)}$ is the value of i-th variable of k-th individual. Using this formula, it is possible to calculate distribution of genes in the intermediate population and generate solution according to this distribution without explicit selection procedure. Such approach can be called "asymptotic", since it gets genes distribution which is the limit for distributions, generated by traditional approach, if population size tends to infinity. Proportional and ranking selection methods calculate selection probabilities explicitly, and therefore asymptotic approach can be applied to these methods directly.

During tournament selection explicit selection probabilities are not calculated, therefore asymptotic approach can't be applied here without modifications. However, tournament selection is known to be more efficient and reliable than other selection methods in many cases, and therefore an expanding of asymptotic approach on tournament selection is an important task.

To construct asymptotic selection method, which is equivalent to tournament selection; it is necessary to find out dependency between selection probabilities and fitness ranks.

Let population size be equal to N, tournament group size be equal to t. $y_1,...,y_N$ are objective values of current population solutions. We need to calculate probabilities $g_1,...,g_N$ of reproduction. Let $Y_1,...,Y_K$ be a set of unique objective values sorted in ascendant order, where K is the number of unique objective values ($K \leq N$). Let us also suppose that n_i solutions have objective value Y_i. It is clear that following equality holds:

$$\sum_{i=1}^{K} n_i = N .$$ (2)

As b_i we will denote the reproduction probability of value Y_i, cumulative probabilities of reproduction will be denoted as B_i:

$$B_i = \sum_{j=1}^{i} b_j .$$ (3)

By its definition, B_i is the probability of reproduction for solution which objective value is less or equal than Y_i. Since tournament groups are created regardless of objective values, then this probability is equal to number of tournament groups, containing solutions with objective values not greater than Y_i, divided by total number of tournament groups of given size.

Let us denote number of tournament group with size t for population of size m as $s(m,t)$, and partial sums of numbers of values as

$$\sum_{j=1}^{i} n_j = N_i .$$ (4)

The following equality will take place:

$$B_i = \frac{s(N_i,t)}{s(N,t)} .$$ (5)

If cumulative probabilities are known, we can obtain probabilities of reproduction of objective values as adjacent differences. Since reproduction probabilities of objective values are calculated, we can return to calculation of reproduction probabilities for solutions. Number of solutions with objective value Y_i is equal to n_i, and all solutions with given objective values must have same reproduction probability. Therefore selection probability of solution may be expressed by the formula

$$g_i = \frac{b_j}{n_j}, j : Y_j = y_i .$$ (6)

Now we must obtain expression for function $s(k,t)$ defining number of tournament groups. If random sampling for tournament group admits repetitions, then this function has simple expression:

$$s_1(k,t) = k^t .$$ (7)

If tournament group is generated without repetitions, then formula becomes

$$s_2(k,t) = \begin{cases} 0, & k \leq t \\ k^{\underline{t}} = \dfrac{k!}{(k-t)!}, & k \geq t \end{cases}.$$ (8)

Let us estimate difference between probabilities in these two schemes:

$$D_{t,N} = \sqrt{N} \max_{0 \leq k \leq N} \left| \frac{s_1(N_k,t)}{s_1(N,t)} - \frac{s_2(N_k,t)}{s_2(N,t)} \right|.$$ (9)

This difference is analogous to statistic of Kolmogorov-Smirnov test. If $D_{t,N}$ is sufficiently small, we can say, that two sampling schemes difference is not statistically significant.

Since expansion of function domain cannot result in decreasing of its exact higher bound on this domain, then

$$D_{t,N} \leq \sqrt{N} \max_{0 \leq x \leq N} d_{t,N}(x),$$ (10)

$$d_{t,N}(x) = \left| \frac{s_1(x,t)}{s_1(N,t)} - \frac{s_2(x,t)}{s_2(N,t)} \right|.$$ (11)

In this formula, variable x can be real. If $x < t$, then $s(x,t)=0$ and therefore

$$d_{t,N}(x) = \left(\frac{x}{N} \right)^t \leq \left(\frac{t-1}{N} \right)^t, x < t.$$ (12)

For the case $t=2$ upper bound can be obtained easily:

$$d_{2,N}(x) \leq \frac{1}{(N-1)} \frac{x}{N} \left(1 - \frac{x}{N} \right) \leq \frac{1}{4(N-1)}.$$ (13)

Therefore we can conclude, that

$$D_{2,N}(x) \leq \frac{\sqrt{N}}{4(N-1)} \sim O\left(\frac{1}{\sqrt{N}} \right).$$ (14)

Using methods of calculus and discrete mathematics we can prove that

$$d_{t,N}(x) \leq \sum_{k=1}^{t-1} \frac{1}{N-k}.$$ (15)

Sum in this formula can be expressed using harmonic numbers:

$$\sum_{k=1}^{t-1}\frac{1}{N-k} = \sum_{k=1}^{N-1}\frac{1}{k} - \sum_{k=1}^{N-t}\frac{1}{k} = H_{N-1} - H_{N-t}. \tag{16}$$

Since holds following equality

$$H_n = \ln(n) + \gamma + \frac{1}{2n} - \frac{1}{12n^2} + \frac{\varepsilon_n}{120n^4}, \tag{17}$$

where $\gamma \approx 0.577$ is Euler constant [3], therefore we can conclude that

$$H_{N-1} - H_{N-t} \leq \ln\left(\frac{N-1}{N-t}\right) + \frac{1}{2(N-1)}. \tag{18}$$

Now we can write following inequality for $d_{t,N}$:

$$d_{t,N} \leq \ln\left(\frac{N-1}{N-t}\right) + \frac{1}{2(N-1)}. \tag{19}$$

Thus, difference between reproduction probabilities in two sampling schemes of tournament selection goes to zero if population size grows and tournament group size is sufficiently small (relatively to population size).

This result allows us to use less computationally complex formula to calculate selection probabilities for tournament selection method in the case of relatively small tournament group.

3 Mutation

Now we can consider mutation in the genetic algorithm from statistical point of view as we did it for selection. Since genes mutate independently, we can study only one gene. Let p denotes a probability, that gene was equal to 1 before mutation. Now we will determine the probability p' to be equal to 1 for same gene after mutation. Mutation probability will be denoted as p_m.

Gene can be equal to 1 after mutation in two cases: it was equal to 1 before mutation and was not mutated or it was equal to 0 before mutation and was mutated. Therefore following equality holds

$$p' = p_m + p(1 - 2p_m). \tag{20}$$

Such transformation can be called *asymptotic mutation* operator. Time consumption of this step does not depend on population size and does not need random numbers generation. Probabilities of genes' values are computed independently. This approach does not contain conditional logic and is well-suited for modern processors [1].

Using introduced asymptotic mutation and selection operators, we can define pseudo-Boolean optimization algorithm with following scheme:

1. Set all components of probability vector to 0.5 (uniform distribution).
2. A population is generated from the probability vector.
3. The fitness of each population member is evaluated.
4. Create new probabilities vector using asymptotic selection (1).
5. Apply asymptotic mutation to the new probabilities vector (20).
6. Replace old probabilities vector with new one.
7. Repeat steps 1-5 until termination condition was met.

This optimizer can be called Asymptotic Genetic Algorithm (AGA), since its mutation and selection operators behave as corresponding operators of standard genetic algorithm if population size tends to infinity.

4 Algorithm Comparison Using Benchmark Problems

Since proposed algorithm is a stochastic optimization algorithm, it is necessary to perform many experiments and average their results to estimate its equality. Primary quality measure is the reliability (number of experiments in which algorithm found global optimum divided by total number of experiment)

We use following settings: population size is equal to 100, maximum iterations number – 50, number of experiments for quality characteristics estimation – 100.

First used benchmark problem is pseudo-Boolean optimization problem:

$$f_1(x) = \sum_{k=2}^{n} |x_k - x_{k-1}|, \ x_k \in \{0,1\}. \tag{21}$$

Second benchmark function has one real argument and strongly oscillates:

$$f_2(x) = \exp(-x^2) + 0.01\cos(200x), \ x \in [-2;2]. \tag{22}$$

Third function is the function with two real variables:

$$f_3(x, y) = \frac{1 - \sin\left(\sqrt{x^2 + y^2}\right)^2}{1 + 0.001(x^2 + y^2)}, \ x, y \in [-5;5]. \tag{23}$$

Following function of two real variables have four local minima with equal function values; three of them have irrational coordinates.

$$f_4(x, y) = (x^2 + y - 11)^2 + (x + y^2 - 7)^2,$$
$$x, y \in [-4;4]. \tag{24}$$

Fifth function has one global maximum and many local one:

$$f_5(x, y) = x^2 - 10\cos(2\pi x) + y^2 - 10\cos(2\pi y),$$
$$x, y \in [-5;5].$$

(25)

Reliabilities of algorithms are summarized in the Table 1.

Table 1 Maximal reliability of optimizers

Problem	PBIL	AGA
f_1 (max)	0.29	**0.45**
f_2 (max)	**1.00**	**1.00**
f_3 (min)	0.25	**0.59**
f_4 (min)	**1.00**	**1.00**
f_5 (min)	0.97	**1.00**

It is easy to see, that AGA is better than PBIL on some problems and has same efficiency on others. We can conclude that proposed asymptotic genetic algorithm may be more reliable that PBIL for some optimization problems. Therefore it may be promising to use some hybridization of these two algorithms to get optimizer that combines their advantages.

5 Algorithm Comparison Using Dynamic Scheduling Problem

Let us consider following problem [4]: assigning trucks to paint botches in a truck facility: trucks get out of the assembly one per minute. The color that truck should have is determined by customer order. It takes 3 minute to paint a truck, but it takes more time if the color of paint in the booth must be changed: it takes 3 minute more. Booths can randomly go down; fixing breakdown takes random amount of time (1 to 20 minutes). One of the heuristic algorithms of dynamic scheduling is market based algorithm [4].

In the numerical experiments we use following settings: booths'number is 6, trucks number is 420, breakdown probability – 0.01. Objective function is time needed to complete painting of trucks: we run model 10 times and use average value of completion time. Population size was equal to 50, generations' number – 20. AGA uses tournament selection with the group size 2. Each algorithm was run 30 times. Average completion time for PBIL was equal to 3.8, and for AGA – 3.5. Student's t-test shows that difference between these two values is significant on the level 0.05 (standard deviation of mean is equal to 0,14).

6 Conclusion

We can conclude that proposed Asymptotic Genetic Algorithm can be used to dynamic scheduling algorithms' parameter tweaking. It outperforms Population

Based Incremental Learning technique on particular real-world problem and performs comparable with PBIL on set of benchmark optimization problems.

References

1. Aho, A.V., Lam, M.S., Sethi, R., Ullman, J.D.: Compilers: Principles, techniques, and tools, 2nd edn. Addison-Wesley (2007)
2. Baluja, S.: Population-Based Incremental Learning: A method for integrating Genetic Search Based Function Optimization and Competitive Learning, Technical Report. Carnegie Mellon University, Pittsburgh (1994)
3. Knuth, D.: The Art of Computer Programming, 3rd edn. Seminumerical algorithms, vol. 2. Addison-Wesley (1997)
4. Kurose, J.F., Simha, R.: A microeconomic approach to optimal resource allocation in distributed computer systems. IEEE Trans. Computers. 38, 707–717 (1989)

Development of a Computational Recommender Algorithm for Digital Resources for Education Using Case-Based Reasoning and Collaborative Filtering

Guadalupe Gutiérrez[*], Lourdes Margain, Alberto Ochoa, and Jesús Rojas

Abstract. This paper describes the development proposal of a Computational Recommender Algorithm (ReCom), to support user to find the correct Digital Resource of Education (DRE), in this case the learning objects (LO) that meet the needs and study preferences of the user. The search is performed on a database that contains a collection of metadata of learning objects of different topics related to Software Architect. The algorithm ReCom proposed uses the technique Collaborative Filtering (CF) and artificial intelligence technique known as Case-Based Reasoning (CBR) using for it the framework jCOLIBRI2. The preliminary test plan is presented to evaluate the effectiveness of the recommendations for the user, considering the user profile and the value of variables of influence. It also presents the proposal of a mathematical equation to measure the degree of satisfaction of user recommendations.

1 Introduction

At present there is a wide variety of LO in warehouses or databases called repositories. Search engines have been the pillar for the process of locating this type of

Guadalupe Gutiérrez · Lourdes Margain
Universidad Politécnica de Aguascalientes, México
e-mail: {guadalupe.gutierrez,lourdes.margain}@upa.edu.mx

Alberto Ochoa
Universidad Autónoma de Ciudad Juárez, México
e-mail: alberto.ochoa@uacj.mx

Jesús Rojas
Cloud Technologies Consulting
e-mail: jrojashdz@gmail.com

[*] Corresponding author.

S. Omatu et al. (Eds.): Distributed Computing and Artificial Intelligence, AISC 151, pp. 767–774.
springerlink.com   © Springer-Verlag Berlin Heidelberg 2012

DRE, but the results presented are regularly inadequate because they do not consider information for each individual, therefore arise recommender systems which considering information of a user profile provide the major resource suited to needs and preferences user [1].

The CBR technical proposes the reuse of past experiences, or the previous cases to solve new problems of the user.

To determine the reliability of ReCom is required to undertake a process of experimentation with a database that contains at least one hundred metadata of LO and user data. The organization of this paper is as follows: firstly describes the literature of the recommender systems, CBR and LO. Then presents the description of the architecture diagram of the algorithm ReCom, as well as a preliminary test plan and the equation proposed to evaluate the effectiveness of the recommendations for the user and finally the results and conclusions made.

2 Recommender System

The use of recommender systems began in the area of electronic commerce to supporting the user to choose a service or resource best suited to your preferences. Wang [2] indicates that a recommender system considers the needs of the user to select the resources that are stored and characterized based on their attributes.

The recommender systems are based on differents recommendation techniques to work, among which are: content-based, collaborative filtering and based on knowledge. The collaborative filtering is one of the techniques applied in commercial recommender systems (e.g. Amazon[1] y Netflix[2]), however, also has been used in recommender systems of LO along with the content-based technique.

In the algorithm ReCom described in this paper is applied the recommendation technique known as knowledge-based using CBR and the CF to generate more recommendations to users and avoid falling into the problem of over-specialization. In the next section presents the basis for CBR theory.

3 Case-Based Reasoning (CBR)

The CBR technique is based on a collection of cases decided considering a problem from the past that has one or more successful solutions. The CBR cycle proposed by Aamondt [3] shown in the Fig. 1 is the reasoning process of this technique. The purpose of the CBR technique is adapt solutions to new problems considering previous similar cases stored in a Library of Case (The "Library of Case" means like as the set of cases that have a successful solution). Ochoa [4] says that solutions similar to the current problems are the basis for the solution of a new problem, and can make changes or repairs to the cases to improve the similarity with the proposed solution.

---

[1] www.amazon.com
[2] www.netflix.com

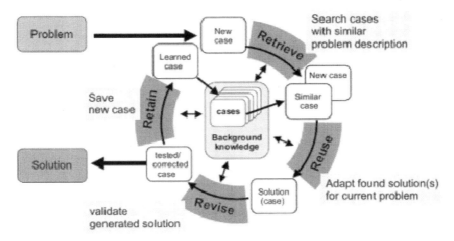

Fig. 1 CBR cycle (Aamondt, 1994)

The CBR cycle is a continuous model of reasoning composed of four main processes.

- Retrieve process: find relevant cases according to the query executed by the user.
- Reuse Process: the cases found are used for the construction of a first solution.
- Revise Process: adapt the solution to the current situation.
- Retain process: the solution is stored in the library of cases for future reuse.

3.1 jCOLIBRI2 as CBR Framework

There are several tools to facilitate the development of an algorithm or CBR system. jCOLIBRI (short for "Cases and Ontology Libraries Integration for Building Reasoning Infrastructures") is an academic framework developed in the GAIA group (Group for Artificial Intelligence Applications) of the Universidad Complutense de Madrid, which provides a domain-independent architecture. There are currently two versions jCOLIBRI and jCOLIBRI2. The first provides a graphical user interface to guide in the designing of a CBR system. The second architecture is divided into two parts: one aimed at designers and other aimed at developers, also has four methods of implementing a system or algoritm CBR.

jCOLIBRI2 is the framework used in ReCom due to its independent architecture and its application schema.

3.2 Other Technologies Used in ReCom

- ICEfaces is the main framework in Java for web applications that are used in the algorithm recommend, for setting the whole structure of the client side and server side.

- Hibernate is a tool for object-relational mapping in Java applications. This technology is used to recommend for all transactions in the database by providing a structure for storing and accessing information.
- Spring is a framework used to recommend facilitating the manipulation of objects and xml files eliminating the need for different and varied types of configuration files.
- Web Services technology is applied to recommend for processing queries and recommendations on different servers.

In the following section briefly describes the theory of LO, which are the resources that the algorithm uses to suggest to the user.

4 Learning Objetcs (LO)

The LO have emerged in order to support users to learn about a specific topic. Wiley [5] states that the LO are digital entities that can be used or reused by multiple users for different learning contexts. It is important that each of these digital resources for education is accompanied by its respective metadata, which as discussed Margain [6] are used to describe the resource and facilitate their search, management and location. The following section shows the architecture of the algorithm ReCom.

5 Computational Recommender Algorithm (ReCom)

The algorithm ReCom consists of five phases: Phase of Retrieve, Phase of Reuse, Phase of Revise, Phase of Retain and Phase of Collaborative Filtering (see Fig. 2). The first four phases are part of the CBR technique.

5.1 ReCom with CBR

Phase of Retrieve: after configure and load the Library of Cases in memory by the methods: set() and precyle(), it proceeds to the execution of the method of cycle(), starting with the retrieval phase in which is considered a query of user to obtain the resource most similar to your preference or need. In this process is used the algorithm k-NN (k Nearest Neighbor) to compare the influence of user (user's profile) with the attributes of the LO showing only those that meet the user's profile.

- Phase of Reuse: this phase also known as adaptation process, seeks to adjust the solution of the retrieved cases with the user's query using a direct numerical ratio between the values of consultation and case.
- Phase of Review: in this phase, the proposed solution is evaluated, so that is necessary the user evaluates the recommended resources in order to know how accurate are the recommendations established and if it finds one of them as a solution to consultation, if so the case will be kept in the Case Library.

- Phase of Retain: This phase is executed only if the user answer is yes the question "Does the RDE satisfy your need to study?" This question is a restriction to the user to continue. If the user response is positive, the case is stored in the Case Library for future reuse. Otherwise it will display a feedback screen which the user can express why they do not feel comfortable with the recommended resource.

There is a fifth phase called Collaborative filtering, which also generate recommendations to the user considering the resources that other users with similar preferences (close neighbors) have evaluated positively.

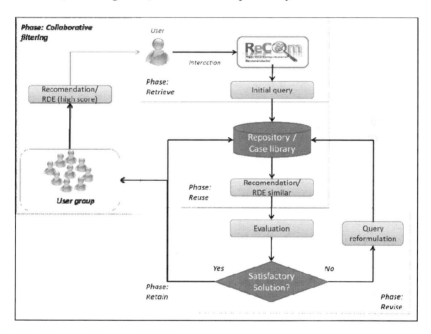

Fig. 2 Architecture Diagram of ReCom

5.2   ReCom with Collaborative Filtering (CF)

To implement the collaborative filtering is necessary establish classifications based on the User Profile similar, as well as objects that contain similar attributes. Once established a classification of users and objects, a prediction technique is used to estimate the value of the user on certain resources. The following describes the methods used for CF.

5.2.1   Similarity Measure

To establish the similarity between objects and users must define a measure that allows us to assess the degree of similarity between them, for this is used the cosine coefficient. This method assumes that two resources are represented by vectors x and y in space, so the similarity between them is given by the cosine of the angle. The equation (2) for the calculation is as follows:

$$S(x,y) = \frac{\sum_{i=1}^{n} x_i y_i}{\sqrt{\sum_{i=1}^{n}(x_i)^2 \sum_{i=1}^{n}(y_i)^2}} \tag{1}$$

5.2.2 Prediction

Once made the similarity process between the RDE is necessary to make the prediction process of the user for RDE, for this is used the technique known as weighted sum, which calculates the prediction of a "resource i" by "user u" as the sum of the evaluations of "users ua" about items similar to i. Each of these assessments is weighted by the corresponding similarity s (i, j) between i and j. The equation of weighted (3) sum is:

$$p(u_a, i_a) = \frac{\sum_{h=1}^{k} s(i_a i_h)(ru_a, i_h)}{\sum_{h=1}^{k} |s(i_a i_h)|} \tag{2}$$

Where k is the k most similar to user resources ia. Through this technique attempts to capture how the active user evaluates similar resources that want to predict. The refining process on the recommendations of RDE will be better when the users evaluate new resources and increasing their User Profile with more attributes.

5.3 Equation Proposed to Measure the Satisfaction of the Acceptance of the Recommendation

To measure user satisfaction factor with respect to the recommendations made is proposed a mathematical formula by which statistical evaluation will be made. This formula is proposed considering Ochoa [1]. Following is the mathematical equation (4) proposed for measuring user satisfaction factor with respect to the recommendation.

$$Sj = \sum_k \beta_k t_{ik} / \propto S_i \tag{3}$$

Where:
S = factor of satisfaction of the recommendation
i = expected satisfaction
j = satisfaction received
α = attributes associated with satisfaction
k = recommendation
β = weight of the attributes (class)
t = current satisfaction

To measure the satisfaction factor of recommendation is necessary to evaluate the expected satisfaction versus the satisfaction received.

6 Preliminary Test Plan to Evaluate the Effectiveness of the Recommendations

To test the technique of RBC have a situation in which a user named "Albert" is looking for an OA about cryptology, which interacts with the algorithm and capture your User Profile, and some of the attributes required to have the object. The requirement's user is:

KEYWORDS:	SEGURIDAD, CRIPTOGRAFÍA
LEARNING CHANNEL:	VISUAL
LANGUAGE:	ESPAÑOL
SPECIALIZATION:	SISTEMAS
USER TYPE:	ESTUDIANTE
RDE TYPE:	OA
LEVEL OF KNOWLEDGE:	BAJO
AREA OF KNOWLEDGE:	SISTEMAS
RDE LANGUAGE:	ESPAÑOL

7 Results and Conclusions

The results presented with CBR on the recommendation after applying the solution are:

Object returns = criptografía
Priority / return value objetc (max) = 4 / .866666
Priority / return value object (min) = 1 / .533333

As you can see "ReCom" returns the user object with a higher priority Cryptography (4) and a value of .86666, and it recommends another resource the lowest priority (1) with a value of .533333. Below are the values obtained by the equation (4) to measure the factor of acceptance of recommendation.

- When value = .86666 the acceptance factor of the recommendation = 87%
- When value = .53333 the acceptance factor of the recommendation = 32%

The objetc of highest priority (when $k = 1$) has a value of 8666 in terms of similarity with the user, obtaining 87% in terms of acceptance of the recommendation. However, when $k = 2$ we obtain 32% in the effectiveness of the recommendation because the recommended order is not accurate enough for the user. Applying now the sum of the recommendations is a factor obtained acceptance of the recommendation equal to 74% under the RBC technique, which is considered feasible in this case.

References

1. Casali, A., Gerling, V., Deco, C.Y., Bender, C.: Sistema inteligente para la recomendación de objetos de aprendizaje. Revista Generación Digital 9(1). Edición 16. Enero de 88 2011 (2010) ISSN 1909-9223
2. Wang, P.: Why recommendation is special? In: Papers from the, Workshop on Recommender Systems, part of the 15 th National Conference on Artificial Intelligence (1998)
3. Aamondt, A.U.P., Enric: Case-Based Reasoning: Foundational Issues, Methodological Variations, and System Approaches. AI Communications 7(1), 39–59 (1994)
4. Ochoa, A., Hernández, J., Álvarez, F., Burlak, G., Muñoz, J.: Más allá del Razonamiento Basado en Casos y una Apróximación al Modelado de Sociedades utilizando Minería de Datos. In: ADC 2006, pp. 127–136 (2006) ISBN: 968-878-273-4
5. Wiley, D.: Connecting Learning Objects to Instructional Design Theory: A definition, a metaphor, and a taxonomy. In: Wiley, D.A. (ed.) Instructional Use of Learning Objects, E.A.f.I.T (2002)
6. Margain, L., Álvarez, J.M.F., Vanderdonkt, J., Orey, M.: Mirros: Intermediary Model to Recovery Learning Objects. Computación y Sistemas 13(4), 373–384 (2010)

Author Index